T0181489

Lecture Notes in Artificial Intelligence   13653

Subseries of Lecture Notes in Computer Science

Series Editors

Randy Goebel
University of Alberta, Edmonton, Canada

Wolfgang Wahlster
DFKI, Berlin, Germany

Zhi-Hua Zhou
Nanjing University, Nanjing, China

Founding Editor

Jörg Siekmann
DFKI and Saarland University, Saarbrücken, Germany

More information about this subseries at https://link.springer.com/bookseries/1244

João Carlos Xavier-Junior ·
Ricardo Araújo Rios (Eds.)

Intelligent Systems

11th Brazilian Conference, BRACIS 2022
Campinas, Brazil, November 28 – December 1, 2022
Proceedings, Part I

Editors
João Carlos Xavier-Junior ⓘ
Federal University of Rio Grande do Norte
Natal, Brazil

Ricardo Araújo Rios ⓘ
Federal University of Bahia
Salvador, Brazil

ISSN 0302-9743   ISSN 1611-3349 (electronic)
Lecture Notes in Artificial Intelligence
ISBN 978-3-031-21685-5   ISBN 978-3-031-21686-2 (eBook)
https://doi.org/10.1007/978-3-031-21686-2

LNCS Sublibrary: SL7 – Artificial Intelligence

This Springer imprint is published by the registered company Springer Nature Switzerland AG
The registered company address is: Gewerbestrasse 11, 6330 Cham, Switzerland

Preface

The 11th Brazilian Conference on Intelligent Systems (BRACIS 2022) was one of the most important events in Brazil for researchers interested in publishing significant and novel results related to Artificial and Computational Intelligence. The Brazilian Conference on Intelligent Systems (BRACIS) originated from the combination of the two most important scientific events in Brazil in Artificial Intelligence (AI) and Computational Intelligence (CI): the Brazilian Symposium on Artificial Intelligence (SBIA, 21 editions), and the Brazilian Symposium on Neural Networks (SBRN, 12 editions). The event is supported by the Brazilian Computer Society, the Special Committee of Artificial Intelligence (CEIA), and the Special Committee of Computational Intelligence (CEIC).

The conference aims to promote theoretical aspects and applications of Artificial and Computational Intelligence, as well as the exchange of scientific ideas among researchers, practitioners, scientists, and engineers.

This year, BRACIS was held in Campinas, Brazil, from November 28 to December 1, 2022, in conjunction with five other events: the National Meeting on Artificial and Computational Intelligence (ENIAC), the Symposium on Knowledge Discovery, Mining and Learning (KDMiLe), the Concurso de Teses e Dissertações em Inteligência Artificial e Computacional (CTDIAC), the Brazilian competition on Knowledge Discovery in Databases (KDD-BR) and the Workshop of the Brazilian Institute of Data Science (WBIOS).

BRACIS 2022 received 225 submissions. All papers were rigorously double-blind peer-reviewed by an international Program Committee (with an average of three reviews per submission), which was followed by a discussion phase for conflicting reports. After the review process, 89 papers were selected for publication in two volumes of the Lecture Notes in Artificial Intelligence series (an acceptance rate of 39.5%).

The topics of interest included, but were not limited to, the following:

- Agent-based and Multi-Agent Systems
- Cognitive Modeling and Human Interaction
- Constraints and Search
- Foundations of AI
- Distributed AI
- Information Retrieval, Integration, and Extraction
- Knowledge Representation and Reasoning
- Knowledge Representation and Reasoning in Ontologies and the Semantic Web
- Logic-based Knowledge Representation and Reasoning
- Natural Language Processing
- Planning and Scheduling
- Evolutionary Computation and Metaheuristics
- Fuzzy Systems
- Neural Networks

- Deep Learning
- Machine Learning and Data Mining
- Meta-learning
- Reinforcement Learning
- Molecular and Quantum Computing
- Pattern Recognition and Cluster Analysis
- Hybrid Systems
- Bioinformatics and Biomedical Engineering
- Combinatorial and Numerical Optimization
- Computer Vision
- Education
- Forecasting
- Game Playing and Intelligent Interactive Entertainment
- Intelligent Robotics
- Multidisciplinary AI and CI
- Foundation Models
- Human-centric AI
- Ethics

We would like to thank everyone involved in BRACIS 2022 for helping to make it a success.

November 2022

João C. Xavier Júnior
Ricardo A. Rios

Organization

General Chairs

João M. T. Romano   UNICAMP, Brazil
Leonardo T. Duarte   UNICAMP, Brazil

Program Committee Chairs

João C. Xavier Júnior   UFRN, Brazil
Ricardo A. Rios   UFBA, Brazil

Organizing Committee

João M. T. Romano   UNICAMP, Brazil
Leonardo T. Duarte   UNICAMP, Brazil
Gisele Baccaglini   UNICAMP, Brazil
Cristiano Torezzan   UNICAMP, Brazil
Denis Gustavo Fantinato   UFABC, Brazil
Henrique N. de Sá Earp   UNICAMP, Brazil
Ricardo Suyama   UFABC, Brazil
Rodolfo de C. Pacagnella   UNICAMP, Brazil
João B. Florindo   UNICAMP, Brazil
Jurandir Z. Junior   UNICAMP, Brazil
Leandro Tessler   UNICAMP, Brazil
Luiz Henrique A. Rodrigues   UNICAMP, Brazil
Priscila P. Coltri   UNICAMP, Brazil
Renato Machado   ITA, Brazil
Renato da R. Lopes   UNICAMP, Brazil
Rosângela B.   UNICAMP, Brazil

Program Committee

Adenilton da Silva   UFPE, Brazil
Adrião D. Dória Neto   UFRN, Brazil
Alexandre Falcao   UNICAMP, Brazil
Alexandre Ferreira   UNICAMP, Brazil
Aline Neves   UFABC, Brazil
Aline Paes   UFF, Brazil

Alneu Lopes	USP, S.Carlos, Brazil
Aluizio Araújo	UFPE, Brazil
Ana Bazzan	UFRGS, Brazil
Ana Carolina Lorena	ITA, Brazil
Ana C. B. Kochem Vendramin	UTFPR, Brazil
Anderson Soares	UFG, Brazil
André Britto	UFS, Brazil
André Ponce de Leon F. de Carvalho	USP, S.Carlos, Brazil
André Rossi	UNESP, Brazil
André Takahata	UFABC, Brazil
Andrés Eduardo Coca Salazar	UTFPR, Brazil
Anna Helena Reali Costa	USP, Brazil
Anne Canuto	UFRN, Brazil
Araken Santos	UFERSA, Brazil
Ariane Machado-Lima	USP, Brazil
Aurora Pozo	UFPR, Brazil
Bruno Masiero	UNICAMP, Brazil
Bruno Nogueira	UFMS, Brazil
Bruno Pimentel	UFAL, Brazil
Carlos Ribeiro	ITA, Brazil
Carlos Silla	PUCPR, Brazil
Carlos Thomaz	FEI, Brazil
Carolina Paula de Almeida	UNICENTRO, Brazil
Celia Ralha	UnB, Brazil
Cephas A. S. Barreto	UFRN, Brazil
Claudio Bordin Jr.	UFABC, Brazil
Claudio Toledo	USP, Brazil
Cleber Zanchettin	UFPE, Brazil
Cristiano Torezzan	UNICAMP, Brazil
Daniel Dantas	UFS, Brazil
Danilo Sanches	UTFPR, Brazil
Debora Medeiros	UFABC, Brazil
Denis Fantinato	UFABC, Brazil
Denis Mauá	USP, Brazil
Diana Adamatti	FURG, Brazil
Diego Furtado Silva	USP, Brazil
Edson Gomi	USP, Brazil
Edson Matsubara	UFMS, Brazil
Eduardo Borges	FURG, Brazil
Eduardo Costa	Corteva, Brazil
Eduardo Palmeira	UESC, Brazil

Contents – Part I

Contents – Part II

Mortality Risk Evaluation: A Proposal for Intensive Care Units Patients Exploring Machine Learning Methods

Alexandre Renato Rodrigues de Souza[1,6]($\boxtimes$) , Fabrício Neitzke Ferreira[2] ,
Rodrigo Blanke Lambrecht[3,4] , Leonardo Costa Reichow[4] ,
Helida Salles Santos[5] , Renata Hax Sander Reiser[6] ,
and Adenauer Correa Yamin[3,6]

[1] Instituto Federal de Educação, Ciência e Tecnologia do Rio Grande do Sul (IFRS),
Rua Engenheiro Alfredo Huch, 475, Rio Grande/RS, Brazil
`alexandre.souza@riogrande.ifrs.edu.br`
[2] Instituto Federal de Educação, Ciência e Tecnologia Sul-Rio-Grandense (IFSUL),
Pelotas, Brazil
[3] Universidade Católica de Pelotas (UCPEL), Pelotas, Brazil
[4] Lifemed Ind. de Equip. e Artigos Médicos e Hospitalares S.A. (LIFEMED),
Sao Paulo, Brazil
[5] Universidade Federal do Rio Grande (FURG), Rio Grande, Brazil
[6] Universidade Federal de Pelotas (UFPEL), Pelotas, Brazil
`https://ifrs.edu.br/riogrande/`

Abstract. The high availability of clinical data and a heterogeneous and complex patient population makes Intensive Care Units (ICUs) environments opportune for developing a system that analyzes large amounts of raw data, which human specialists can neglect. Quantifying a patient's condition to support the definition and adjustment of clinical treatment and predict future outcomes is a significant research problem in intensive care. This work's main objective is to conceive an approach to predicting ICUs mortality risk. Therefore, the designed approach is a binary classification task that aims to predict whether patients will die or survive during their ICU stay. A cohort of 17,734 patients was used from the MIMIC-III database, considering 10 input predictor variables and 8 Machine Learning methods. Sensitivity, specificity, F1 score, AUC, and ROC curve are used to compare different models of mortality risk prediction in a 48-h window of data acquisition. The best performance was achieved by the Gradient Boosting Machine (GBM) method, which obtained 0,843 ($\pm$0,015) of AUC and 0,503 ($\pm$0,048) for the F1 score. The approach conceived enables the generation of robust models capable of detecting hidden patterns and having greater power of discrimination in classifications. The results are promising and, in some cases, superior to those obtained by other proposals identified in the literature review.

Keywords: Mortality · Machine learning · ICU · Deterioration · Early warning · Vital signs

Supported by CAPES, CNPq and IF-RS.

J. C. Xavier-Junior and R. A. Rios (Eds.): BRACIS 2022, LNAI 13653, pp. 1–14, 2022.
https://doi.org/10.1007/978-3-031-21686-2_1

1   Introduction

Hospital Intensive Care Units (ICUs) have attracted relevant research efforts, as patients require continuous monitoring of their physiological parameters due to the severity of their health condition and a high risk of rapid clinical deterioration [19,20].

A primary outcome of interest in intensive care is mortality, as its rates in ICUs are the highest among hospital units, around 10 to 29%, depending on age and disease. Thus, identifying patients at most significant risk is critical to improving treatment outcomes [12].

It is worth bearing in mind that traditional clinical warning scores, such as Mortality Probability Models (MPM), Early Warning Score (EWS), National Early Warning Score (NEWS), Modified Early Warning Score (MEWS), Sequential Organ Failure Assessment (SOFA), Quick Sequential Organ Failure Assessment (qSOFA), Simplified Acute Physiology Score (SAPS), and Acute Physiology and Chronic Health disease Classification System (APACHE), already have the premise of classifying patients according to the risk of death. These scores use simple mathematical models to predict the clinical outcome. Although these scores are still widely used in the hospital setting, several studies show that custom models for mortality risk assessment using Machine Learning outperform these traditional scoring systems [19].

In 2012, the Laboratory for Computational Physiology (LCP) of the Massachusetts Institute of Technology (MIT) proposed a challenge to encourage the development of new Machine Learning techniques to identify the risk of hospital mortality in patients admitted to ICUs. This challenge promoted increased interest in this topic among the international scientific community and open health datasets. In this sense, databases such as the Medical Information Mart for Intensive Care (MIMIC) are gradually becoming available, contributing to research like the one developed in this work [13].

Considering this scenario, the present research aims to investigate the use of 8 classification methods exploring Machine Learning in developing a mortality risk prediction approach in ICUs to assist physicians in decision making. The Machine Learning methods evaluated are Random Forest (RF), Gradient Boosting Machine (GBM), Gaussian Naive Bayes (GNB), Multilayer Perceptron (MLP), Adaptive Boosting (AdaBoost), eXtreme Gradient Boosting (XGBoost), Support Vector Machines (SVM) and Linear Discriminant Analysis (LDA). The prediction is made by analyzing different clinical data from patients collected in the first 48 h after admission.

In turn, the specific objectives of this study are: (i) to consider the main challenges in using Machine Learning techniques to predict the risk of hospital mortality of patients admitted to ICUs; (ii) to propose an approach to predicting the risk of mortality for these patients; and, (iii) evaluate and compare the performance of different Machine Learning methods, using a database built from real information from ICUs (MIMIC-III).

The following Section discusses related works to the study area of this article, which was identified from a Systematic Literature Review carried out during the research, which is the focus of another publication.

2   Related Works Exploring Machine Learning to Predict Mortality Risk in ICUs

As part of the efforts associated with the research presented in this article, a Systematic Literature Review (SLR) was carried out, identifying several studies on the theme related to the prediction of mortality risk in ICUs.

Priority was given to related works that contemplated the following aspects: (i) use of MIMIC as a database in order to consider cohorts of patients admitted to ICUs from a single hospital; (ii) use of a maximum of 20 clinical variables originating from the database; and (iii) use of the AUC metric (Area Under ROC Curve) as a way of evaluating the performance of the proposed models, thus making possible a comparison between the different works, which are briefly described in Table 1.

It is worth noting that the literature has pointed out the AUC as recommended "single number" metric to evaluate the performance of Machine Learning methods because it has several desirable properties compared to other alternatives. [8,17].

Table 2 presents a comparison between the related works, contemplating the central aspects considered in the conception of the approach proposed by this study: (i) identification of the work; (ii) number of clinical variables; (iii) description of clinical variables; (iv) version of the MIMIC database; (v) time window for measuring clinical variables; (vi) prediction methods evaluated (when a study evaluated more than one method, the one with the best performance is highlighted in blue); (vii) metrics used to measure model performance; (viii) better performance obtained by the work considering the AUC metric; and, (ix) cohort of patients or admissions.

3   Prediction of Mortality Risk in ICUs: Approach Design

This Section describes the proposed approach to deal with the prediction of mortality risk in ICU patients, considering the design decisions that guided its construction. An overview of this approach is shown in Fig. 1.

3.1   Discussion of the Research Problem

Traditional clinical warning scores (like EWS, SOFA, APACHE, and SAPS) have been used to identify the deterioration of the patient's condition. These scoring

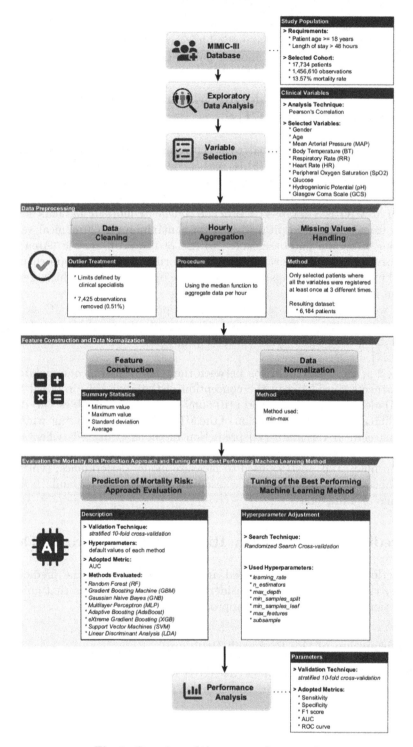

Fig. 1. Overview of the proposed approach

Table 1. Description of related works

#	Description
1	[3] proposed a hybrid model that combines CNN and BiLSTM to predict mortality risk from statistics that describe variation in heart rate, blood pressure, respiratory rate, blood oxygen levels, and body temperature. The best-performing model obtained 0.88 of AUC. The work concluded that using a CNN-BiLSTM hybrid network effectively determines mortality risk for the 3, 7, and 14-day windows of vital signs. The results show that it is possible to implement an accurate system to continuously and automatically predict the risk of mortality, reducing the burden on health professionals and improving the clinical outcomes of patients.
2	[1] developed models for predicting ICU length of stay and mortality risk based on the MIMIC-III database. Six commonly used Machine Learning methods were applied to predict mortality risk, using 11 input variables (demographic data and vital signs) in each model. The best AUC achieved in the mortality model was 0.78 using the Random Forest algorithm. The novelty in this approach was the construction of models to predict ICU length of stay and mortality risk with reasonable accuracy based on a combination of Machine Learning and the quantile approach that uses only the vital signs available in the patient's profile. The technique used is based on the feature engineering of vital signs, including their modified means, standard deviations, and quantiles of the original variables, which provided a more appropriate dataset to obtain a better predictive power of the models.
3	[19] presented the performance results for several clinical prediction tasks, such as mortality risk, length of stay, and ICD-9 code using models from Deep Learning, ensembles of Machine Learning models (Super Learner algorithms), SAPS II, and SOFA scores. ICD-9 is the official code system for diagnoses and procedures in hospitals in the United States. The MIMIC-III set was used as a data source. The results showed that Deep Learning models consistently outperform all other approaches, mainly when raw clinical time series data are used as input variables for the models. The MMDL deep learning method (Multimodal Deep Learning Model) achieved an AUC of 0.87 using 17 predictor variables and 48 hours of data.
4	[6] proposed a new algorithm for predicting ICU mortality risk to solve the class imbalance problem. The method is based on the transformation of predictor variables to reduce the existing correlation. The effectiveness of the algorithm was demonstrated in simulated datasets and MIMIC-II. An advantage of the proposal is using only 6 patient clinical data (mean blood pressure, heart rate, body temperature, sodium level, potassium level, and magnesium level). In comparison, other scoring methods or systems use measures that may not be available to all patients and may therefore require manual intervention or review of clinical scores. The model developed, called CHISQ-NEW by the authors, obtained an AUC of 0.87.
5	[18] developed a Super Learner algorithm for predicting mortality risk for ICU patients, comparing its performance with traditional scores. The calibration, discrimination, and risk classification of predicted hospital mortality based on Super Learner compared to SAPS-II, APACHE-II, and SOFA were evaluated. The clinical data of 24,508 patients from the MIMIC-II database was used as a primary data source. Two sets of predictions were produced based on Super Learner; the first based on the 17 variables as shown in the SAPS-II score (SL1), and the second based on the original variables without transformations (SL2). Super Learner achieved an AUC of 0.85 when using SL1 and 0.88 with SL2. Compared with traditional scores, the proposed model better predicted the risk of hospital mortality in ICU patients
6	[12] presented four clinical prediction tasks using data derived from MIMIC-III. These tasks cover a range of clinical problems, including mortality risk modeling, length of stay prediction, physiological deterioration detection, and phenotype classification. Linear and neural network models were proposed for all four tasks. The effect of deep supervision, multitask training, and modifications of specific data architectures on the performance of neural models were evaluated. The work identified that LSTM-based models significantly outperformed linear models and presented the advantages of using channel-wise LSTMs to predict multiple tasks using a single neural model. The highest AUC achieved by the work for predicting mortality risk was 0.87.
7	[2] investigated how the risk of hospital mortality can be predicted for ICU patients. The results showed that the discriminating power of Machine Learning classification methods after 6 hours of admission outperformed the traditional scoring systems used in intensive care medicine (APACHE, SAPS, and SOFA) after 48h of admission. The best performing classifier was RF (AUC of 0.90), followed by NB and PART in different experimental settings. The authors concluded that: (i) there is a marked improvement in performance at the 6th hour of ICU admission; (ii) the percentage of missing values in the dataset drastically reduces at the 6th hour of ICU admission and continues to decrease gradually until the 48th hour. The work alerts to the problem of the missing values of the variables collected to emphasize the importance of collecting specific measures from the beginning of the hospitalization, as this will influence the predictive performance of mortality risk prediction models.
8	[This work], whose technical-scientific contributions are discussed in this article, also appears in Table 2 to promote a summary of its characteristics in the literature. A discussion of this comparison is given in Section 4.2.

Table 2. Comparison of related works

#	n	Clinical Variables Description	MIMIC	Time	Prediction Methods	Performance Metrics	AUC	Cohort
1	9	BT, DBP, HR, age, MAP, RR, SBP, gender, SpO2	III	24h	**hybrid neural network (CNN-BiLSTM)**	AUC, AUPRC, accuracy, specificity, sensitivity, ROC curve	0,88	51.279 patients
2	11	height, BT, DBP, glucose, HR, age, weight, RR, SBP, gender, SpO2	III	-	RF, LR, LDA, kNN, SVM, XGBoost	accuracy, sensitivity, specificity, NPV, PPV, AUC, ROC curve	0,78	44.626 admissions
3	17	AIDS, bicarbonate level, bilirubin level, BT, urinary output, GCS, HR, age, hematologic malignancy, metastatic cancer, PaO2/FiO2, potassium level, SBP, sodium level, admission type, urea level, WBC count	III	24/48h	super learner, **MMDL**	AUC, AUPRC	0,87	35.627 admissions
4	6	BT, HR, magnesium level, MAP, potassium level, sodium level	II	-	**CHISQ-NEW**, SVM, RF, LR, LDA, QDA, Adaboost	AUC	0,87	4.000 patients
5	17	AIDS, bicarbonate level, bilirubin level, BT, urinary output, GCS, HR, age, hematologic malignancy, metastatic cancer, PaO2/FiO2, potassium level, SBP, sodium level, admission type, urea level, WBC count	II	-	**super learner**	AUC, ROC curve	0,88	24.508 patients
6	17	height, BT, DBP, capillary refill rate, FiO2, GCS total, GCS motor response, GCS eye opening, GCS verbal response, glucose, HR, MAP, weight, pH, RR, SBP, SpO2	III	-	LR, channel-wise LSTM, multitask standard LSTM, standart LSTM, deep supervision, **multitask channel-wise LSTM**	AUC, AUPRC, ROC curve	0,87	21.139 admissions
7	8	BT, creatinine, GCS, HR, age, PaO2, RR, SBP	II	48h	RF, NB, PART, DT, SVM, JRip	AUC, ROC curve	0,90	11.722 patients
8	10	BT, GCS, glucose, HR, age, MAP, pH, RR, gender, SpO2	III	48h	RF, **GBM**, GNB, MLP, Adaboost, XGBoost, SVM, LDA	sensitivity, specificity, F1 score, AUC, ROC curve	0,84	17.734 patients

Description of variables:
HR - Heart Rate; **SBP** - Systolic Blood Pressure; **DBP** - Diastolic Blood Pressure; **MAP** - Mean Arterial Pressure; **RR** - Respiratory Rate; **SpO2** - Peripheral Oxygen Saturation; **BT** - Body Temperature; **GCS** - Glasgow Coma Scale; **FiO2** - Fraction of Inspired Oxygen; **PaO2** - Partial Pressure of Oxygen; **WBC** - White Blood Cell Count; **AIDS** - Acquired Immunodeficiency Syndrome; **pH** - Hydrogenionic Potential.

Prediction Methods:
CNN - Convolutional Neural Networks; **LSTM** - Long Short-Term Memory; **BiLSTM** - bidirectional Long Short-Term Memory; **RF** - Random Forest; **LR** - Logistic Regression; **kNN** - k-Nearest Neighbors; **SVM** - Support Vector Machines; **XGBoost** - eXtreme Gradient Boosting; **MMDL** - Multimodal Deep Learning Model; **LDA** - Linear Discriminant Analysis; **QDA** - Quadratic Discriminant Analysis; **AdaBoost** - Adaptive Boosting; **NB** - Naive Bayes; **DT** - Decision Tree; **PART** - partial Decision Tree; **JRip** - Repeated Incremental Pruning to Produce Error Reduction (RIPPER); **RNN** - Recurrent Neural Network; **GNB** - Gaussian Naive Bayes; **MLP** - Multilayer Perceptron; **GBM** - Gradient Boosting Machine; **LDA** - Linear Discriminant Analysis.

Performance Metrics:
ROC - Receiver Operating Characteristic; **AUC** - Area Under the Curve; **AUPRC** - Area Under Precision-Recall Curve; **NPV** - Negative Predictive Value; **PPV** - Positive Predictive Value.

systems, however, only take into account health data from a given moment in time without considering their tendency to vary during ICU stay [10].

In turn, the increasing implementation of Electronic Health Records (EHR) in hospitals has made it possible to record patients' historical data, whose vital signs and laboratory test results collected over time can be interpreted as time series. This scenario has promoted the application of computational techniques that process these data and make it possible to produce predictions of the evolution of the clinical status of patients.

The approach whose design is discussed in this article aims to predict whether patients will die or survive during their stay in ICUs. The MIMIC-III will be used as the database for this binary classification task. The performances of various machine learning methods were compared using data collected in the first 48 h after the patient was admitted to an ICU. This time window was established from the Systematic Literature Review, which showed that a sufficiently accurate mortality risk estimate is already possible with this time interval [2].

3.2 Database and Study Population

The relevant patient cohort to the approach design was extracted from the Medical Information Mart for Intensive Care (MIMIC-III). MIMIC-III [14] is a public access clinical data repository created based on patients from the Beth Israel Deaconess Medical Center (BIDMC), a teaching hospital of Harvard Medical School. For the development of the model used in the approach, only records of patients aged 18 years or older who remained hospitalized in ICUs for a minimum of 48 h were included. These requirements resulted in a cohort of 17,734 patients and 1,456,610 observations. Of these patients, 15,328 survived, and 2,406 died, resulting in a mortality rate of 13.57%.

3.3 Exploratory Data Analysis and Variable Selection

Clinical variables selected and extracted from MIMIC-III to develop the proposed approach contains patient demographic information, laboratory test results, vital signs, and the Glasgow Coma Scale, as listed in Table 3.

Table 3. Selected clinical variables

Variable	Abbreviation	Type	Unit	Observations	Incidence
Gender	-	demographic data	-	1.456.610	100%
Age	-	demographic data	years	1.456.610	100%
Mean Arterial Pressure	MAP	vital sign	mmHg	1.006.905	69%
Body Temperature	BT	vital sign	°C	320.929	22%
Respiratory Rate	RR	vital sign	RPM	1.064.918	73%
Heart Rate	HR	vital sign	BPM	1.055.868	72%
Peripheral Oxygen Saturation	SpO2	vital sign	%	1.063.682	73%
Glucose	-	lab test	mg/dL	246.892	17%
Hydrogenionic Potential	pH	lab test	-	122.296	8%
Glasgow Coma Scale	GCS	score	-	158.842	11%

Figure 2 presents the matrix with Pearson's correlation coefficients (r) of the clinical predictor variables, whose function is quantifying the linear relationships between their pairs.

Correlation coefficients are in the range of –1 to +1. Thus, in Fig. 2, two variables have a perfect positive correlation if r = +1, no correlation if r = 0 and a perfect negative correlation if r = –1.

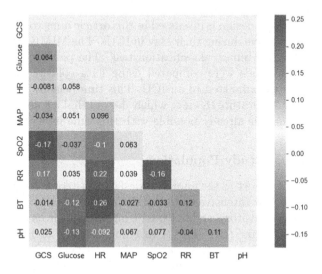

Fig. 2. Correlation between predictor clinical variables

A timely procedure should select variables that contribute the best gain of information to predict the result, minimizing the redundancy of information [15]. As shown in the matrix, there is no strong correlation between any pairs of variables since the highest coefficient is 0.26 between BT and HR.

3.4 Data Preprocessing

The quality and quantity of helpful information are essential factors that will influence the efficiency of a Machine Learning method. Based on this premise, it should be taken into account that the existing data in Electronic Health Records have different formats, dimensions, and characteristics, so they are generally not ready to be entered directly into these methods. Therefore, data preparation is essential before feeding the models [4]. This task is performed by preprocessing, which is composed of the following steps:

– **Data Cleaning**. Data extracted from MIMIC-III have erroneous values due to noise, incorrect records, typographical errors, and inconsistent information imputation or units [19]. The specifications in work [12] were used to address these outliers present in the database, which was defined by clinical specialists based on their knowledge of valid measurement intervals. In the proposed model, each numerical variable is associated with upper and lower limits defined by specialists to detect unusable values (outliers). The observed value will be excluded if it is outside these limits. When applying these rules to generate the model cohort, 7,425 observations (0.51%) classified as extreme outliers were removed.

- **Hourly Aggregation**. MIMIC-III data have a date/time register of capture for each measurement performed. However, most measurements are sampled on a non-periodic time basis, so the time series for each raw variable is considerably sparse. Thus, to obtain a denser representation of the physiological data to provide a better inference by the algorithms, the observations of each time series were aggregated in hourly intervals by calculating the median [21].
- **Missing Values Handling**. To minimize the impact on prediction performance due to lack of data, only patients who had all variables recorded at least once at three different times within the 48-h measurement window were included. This criterion resulted in a dataset of 6,184 patients.

3.5 Feature Construction and Data Normalization

Feature construction addresses the problem of finding the transformation of variables that have the most helpful information. For the treatment of missing information in the MIMIC-III time series, the proposed model calculates each variable's statistical data (minimum value, maximum value, standard deviation, and mean) within the 48-h time window. This strategy reduces the complexity of the Machine Learning model generated, as it uses more relevant information as input for the prediction task.

Determining the minimum and maximum values of each series aims to show the extreme events during ICU stay. The standard deviation was used to quantify the variability of events. The mean was calculated to provide a representation of the mean event for each variable.

Many Machine Learning methods require the selected variables to be on the same scale for better performance [11]. This requirement is usually achieved by turning the features into the interval [0, 1] or a standard normal distribution with zero mean and unity variance. This technique prevents the model from being influenced by variables with higher values than the others. The normalization method called âĂIJmin-maxâĂİ was chosen for this work, through which the values were transformed to a minimum-zero and maximum-one scale.

4 Prediction of Mortality Risk in ICUs: Approach Evaluation

The training dataset was used to evaluate the mortality risk prediction performance of 8 Machine Learning methods. Table 4 presents the results of mean AUC and standard deviation, which were quantified using the default adjustment of the hyperparameters of each algorithm. The best performance was achieved by the GBM method, which obtained 0.841 (±0.024) of AUC. Performance values were obtained by cross-validation to avoid biased evaluations. The technique Stratified 10-fold Cross-Validation (StratifiedKFold) was used, a variation of k-fold that returns stratified data subsets, where each data subset contains the same percentage of surviving and dead patients as the complete set.

Figure 3 presents the ROC curves of the evaluated prediction methods. Despite the different AUC performances, it can be seen from the curves that the RF, GBM, MLP, Adaboost, XGBoost, SVM, and LDA methods have very similar behavior considering Specificity and Sensitivity. The lower-left corner of the ROC curves shows that the GNB algorithm showed lower sensitivity at low false-positive rates than the other methods, which indicates a lower ability to predict correctly patients who died.

4.1 Tuning of the Best Performing Machine Learning Method

After identifying that the GBM Machine Learning method achieved the best prediction performance, its hyperparameters were adjusted to increase the model's performance. This adjustment was accomplished using the training dataset, and the performance was evaluated using the AUC metric. The method tuning was performed using Randomized Search Cross-Validation [5]. This technique aims to find the combination of hyperparameter values that result in the highest performance of the model. For this, a search is implemented where each configuration is sampled from a list of specified values. The hyperparameters considered were: *learning_rate, n_estimators, max_depth, min_samples_split, min_samples_leaf, max_features,* and *subsample.*

Table 4. Classification performances

Method	Average AUC	Standard Deviation
RF	0.830	0.025
GBM	**0.841**	0.024
GNB	0.786	0.029
MLP	0.838	0.025
Adaboost	0.815	0.031
XGBoost	0.823	0.019
SVM	0.821	0.025
LDA	0.830	0.025

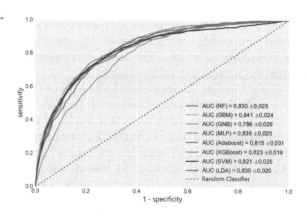

Fig. 3. ROC curves of prediction methods

4.2 Performance Analysis of the Best Performance Method

After finding the optimized hyperparameter values for the GBM method, the test dataset was used to perform the final performance evaluation. Cross-validation was employed to investigate variability in model performance across 10 different data partitions. It is time to reiterate that the test dataset had not yet been applied along with the prediction model.

Sensitivity, Specificity, F1 score, and AUC were used to evaluate the model for predicting mortality risk. Therefore, the designed approach is a binary classification task that aims to predict whether patients will die or survive during their ICU stay. The Sensitivity indicates the ability of the classification method to predict patients who died (positive class) correctly. The Specificity indicates the ability of the classification method to predict the patients who survived (negative class) correctly. F1 score is a harmonic average calculated based on Precision and Sensitivity [11]. The AUC is a scalar quantity between 0 and 1 representing the area under the ROC curve and measuring the quality of model predictions regardless of the classifier operating point [17]. Accordingly, AUC measures the inherent ability of the model to discriminate between patients who died and those who survived.

An AUC performance of 0.843 (±0.015) was achieved after tuning the GBM method by adjusting its hyperparameters, as shown in Table 5. The measure was obtained through cross-validation (10-folds). These results of AUC, Sensitivity, Specificity and F1 Score show the robustness of the proposed approach. They indicate that the model can identify patients at high risk of dying and avoid false mortality classifications of surviving patients.

Figure 4 shows the average ROC curve, also obtained through cross-validation. The gray region shows the variance of the ROC curve when the data is split into different subsets for training and testing, showing how the classifier's output is affected by changes in the data [9].

Table 5. GBM model performance

Method	Average AUC	Standard Deviation
AUC	0,843	0,015
Sensitivity	0,754	0,031
Specificity	0,761	0,035
F1 Score	0,503	0,048

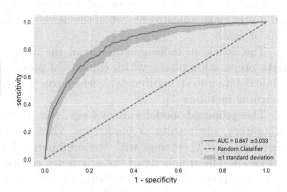

Fig. 4. Average ROC curve obtained through cross-validation (10-folds)

The performance comparison of the proposed approach with other works in the literature will be made based on Table 2. The AUC metric was used for this comparison because the literature points it out as the most relevant for measuring performance in predicting the risk of ICUs mortality [17].

The AUC performance of 0.84 achieved by the proposed approach is superior to the work of [1], which obtained an AUC of 0.78. In turn, the works [19] (AUC of 0.87), [18] (AUC of 0.88), and [12] (AUC of 0.87) present better performance. However, they require a total of 17 predictor variables, depending on up to 6 laboratory tests or requiring the individual values of the 4 criteria of the Glasgow scale, of which often only the total value is recorded. These aspects can make their implementation difficult, as health professionals would have to update them during the patient's hospitalization, and, as a rule, these values are not measured regularly.

Although the proposed approach does not use the least amount of predictor variables, it still presents a performance similar to works [2,3,6] that include other strategies for the prediction effort, which will be the object of study in the continuity of the search. Finally, it is worth emphasizing that the performance obtained by the approach discussed in this article is superior to the traditional SAPS-II (AUC of 0.78) and SOFA (AUC of 0.71) [18] scores, which is an excellent indicator significant for research continuity.

5   Conclusions

This work contributed to designing an approach to predicting the risk of mortality, exploring Machine Learning, and using data collected during the first 48 h of hospitalization in ICUs. MIMIC-III was used as a database of clinical information collected from the real patients.

We understand that a mortality prediction system implemented in an ICU hospital environment could provide valuable information about the patient's response to clinical treatment as a whole, allowing medical professionals to promptly make adjustments when the model predicts an increase in the patient's risk of death.

The article includes a synthesis of the model development stages from the data preparation phase, which consists of processes necessary to transform raw clinical data into structured data, which can then be used as input to the Machine Learning methods [7].

The proposed model employs age, sex, GCS score, vital signs, and some laboratory test results as predictors. These necessary attributes are recorded regularly in ICU settings with the minimum effort required by healthcare professionals. The simplicity of the attributes chosen also makes it possible for the mortality prediction to be recalculated continuously and automatically during the patient's stay in the ICU. The small number of variables used also improves the interpretability of the model, thus increasing the likelihood that healthcare professionals will trust their predictions.

Several Machine Learning methods were tested: Random Forest (RF), Gradient Boosting Machine (GBM), Gaussian Naive Bayes (GNB), Multilayer Perceptron (MLP), Adaptive Boosting (AdaBoost), eXtreme Gradient Boosting (XGBoost), Support Vector Machines (SVM) e Linear Discriminant Analysis (LDA). Their performances were quantified according to the metrics indicated by the literature in evaluating models for mortality prediction. The highest performance was achieved by the GBM algorithm, which obtained 0,843 (±0,015) of AUC and 0,503 (±0,048) for F1 score, using data collected in the first 48 h of ICU stay. As shown in Table 2, the model results are promising, use a small number of variables, and performance is equivalent to other architectures proposed in the literature.

Compared to traditional clinical scores, the proposed approach uses data mining and Machine Learning techniques that generate more sophisticated, robust models capable of detecting hidden patterns and having greater power of discrimination in classifying mortality risk in ICUs.

As a prospect of continuing the research, the expectation for this approach will be explored in the hospital ICU environment. To this end, this work will be integrated into the efforts of the Laboratory of Ubiquitous and Parallel Systems (LUPS/UFPel). This research group employs the middleware EXEHDA [16] and Lifemed's multiparametric monitors in studies cases involving inpatients. Lifemed is an industrial partner that, with its different certifications, enables the provision of solutions applied in real clinical environments.

Acknowledgments. The present work was supported by *Conselho Nacional de Desenvolvimento Científico e Tecnológico* (CNPq), *Coordenação de Aperfeiçoamento de Pessoal de Nível Superior* - Brazil (CAPES) - Financing Code 001 and *Instituto Federal do Rio Grande do Sul* (IF-RS).

References

1. Alghatani, K., Ammar, N., Rezgui, A., Shaban-Nejad, A.: Predicting intensive care unit length of stay and mortality using patient vital signs: machine learning model development and validation (2021). https://doi.org/10.2196/21347
2. Awad, A., Bader-El-Den, M., McNicholas, J., Briggs, J., El-Sonbaty, Y.: Predicting hospital mortality for intensive care unit patients: time-series analysis. Health Inf. J. **26**(2), 1043–1059 (2020). https://doi.org/10.1177/1460458219850323, http://journals.sagepub.com/doi/10.1177/1460458219850323
3. Baker, S., Xiang, W., Atkinson, I.: Continuous and automatic mortality risk prediction using vital signs in the intensive care unit: a hybrid neural network approach (2020). https://doi.org/10.1038/s41598-020-78184-7
4. Barella, V.H., Garcia, L.P.F., de Carvalho, A.C.P.L.F.: Simulating complexity measures on imbalanced datasets. In: 9th Brazilian Conference on Intelligent Systems (BRACIS) (2020)
5. Bergstra, J., Bengio, Y.: Random search for hyper-parameter optimization. J. Mach. Learn. Res. **13**, 281–305 (2012)
6. Bhattacharya, S., Rajan, V., Shrivastava, H.: ICU mortality prediction: a classification algorithm for imbalanced datasets (2017)
7. Bosco, A.D., Vieira, R., Zanotto, B., da Silva Etges, A.P.B.: Ontology based classification of electronic health records to support value-based health care. In: 10th Brazilian Conference on Intelligent Systems (BRACIS) (2021)
8. Bradley, A.P.: The use of the area under the ROC curve in the evaluation of machine learning algorithms. Patt. Recogn. **30**(7), 1145–1159 (1997). https://doi.org/10.1016/S0031-3203(96)00142-2

9. Cagnini, H.E.L., Freitas, A.A., Barros, R.C.: An evolutionary algorithm for learning interpretable ensembles of classifiers. In: 9th Brazilian Conference on Intelligent Systems (BRACIS) (2020)

10. Churpek, M.M., Yuen, T.C., Winslow, C., Meltzer, D.O., Kattan, M.W., Edelson, D.P.: Multicenter comparison of machine learning methods and conventional regression for predicting clinical deterioration on the wards. Critical Care Med. **44**(2), 368–374 (2016). https://doi.org/10.1097/CCM.0000000000001571, http://journals.lww.com/00003246-201602000-00016

11. Faceli, K., Lorena, A.C., Gama, J., de Carvalho, A.C.P.d.L.F., de Almeida, T.A.: Inteligência Artificial - Uma Abordagem de Aprendizado de Máquina. 2a edição edn (2021)

12. Harutyunyan, H., Khachatrian, H., Kale, D.C., Ver Steeg, G., Galstyan, A.: Multitask learning and benchmarking with clinical time series data. Sci. Data **6**(1) (2019). https://doi.org/10.1038/s41597-019-0103-9

13. Johnson, A.E.W., Dunkley, N., Mayaud, L., Tsanas, A., Kramer, A.A., Clifford, D.: Patient Specific Predictions in the Intensive Care Unit Using a Bayesian Ensemble (Mimic), pp. 249–252 (2012)

14. Johnson, A.E., et al.: MIMIC-III, a freely accessible critical care database. Sci. Data **3**(1), 160035 (2016). https://doi.org/10.1038/sdata.2016.35, http://www.nature.com/articles/sdata201635

15. Liu, J., et al.: Mortality prediction based on imbalanced high-dimensional ICU big data. Comput. Indust. **98**, 218–225 (2018). https://doi.org/10.1016/j.compind.2018.01.017, https://linkinghub.elsevier.com/retrieve/pii/S0166361517304505

16. Machado, R.d.S., Almeida, R.B., da Rosa, D.Y.L., Lopes, J.L.B., Pernas, A.M., Yamin, A.C.: EXEHDA-HM: a compositional approach to explore contextual information on hybrid models. Future Gener. Comput. Syst. **73**, 1–12 (2017). https://doi.org/10.1016/j.future.2017.03.005, https://doi.org/10.1016/j.future.2017.03.005

17. Muralitharan, S., et al.: Machine learning-based early warning systems for clinical deterioration: systematic scoping review. J. Med. Internet Res. **23**(2) (2021). https://doi.org/10.2196/25187

18. Pirracchio, R., Petersen, M.L., Carone, M., Rigon, M.R., Chevret, S., van der Laan, M.J.: Mortality prediction in intensive care units with the Super ICU Learner Algorithm (SICULA): a population-based study. Lancet Respiratory Med. **3**(1), 42–52 (2015). https://doi.org/10.1016/S2213-2600(14)70239-5, https://linkinghub.elsevier.com/retrieve/pii/S2213260014702395

19. Purushotham, S., Meng, C., Che, Z., Liu, Y.: Benchmarking deep learning models on large healthcare datasets (2018). https://doi.org/10.1016/j.jbi.2018.04.007, https://www.sciencedirect.com/science/article/pii/S1532046418300716

20. Segato, T.H.F., Serafim, R.M.d.S., Fernandes, S.E.S., Ralha, C.G.: Mas4gc: multiagent system for glycemic control of intensive care unit patients. In: 10th Brazilian Conference on Intelligent Systems (BRACIS) (2021)

21. Zhu, Y., Fan, X., Wu, J., Liu, X., Shi, J., Wang, C.: Predicting ICU mortality by supervised bidirectional LSTM networks (2018). http://ceur-ws.org/Vol-2142/paper4.pdf

Requirements Elicitation Techniques and Tools in the Context of Artificial Intelligence

André Filipe de Sousa Silva , Geovana Ramos Sousa Silva⬡,
and Edna Dias Canedo⁽⊠⁾⬡

Department of Computer Science, University of Brasília (UnB), P.O. Box 4466,
Brasília–DF, Brazil
`ramos.geovana@aluno.unb.br`, `ednacanedo@unb.br`

Abstract. Context: During software development in the context of Artificial Intelligence (AI), just like any other software, there is the requirements elicitation phase. In this phase, developers alongside stakeholders make use of various techniques, methodologies, and tools available to elicit software requirements. Problem: The need of understanding which techniques, methodologies, and tools are suitable in requirements elicitation in the context of AI, taking into consideration ethical issues. Solution: Investigation of the ICT practitioners' perception about their approaches regarding the requirements elicitation process for AI systems. Method: We have conducted a survey with ICT practitioners and reviewed the literature to identify requirements elicitation practices in the context of AI. Summary of Results: Most ICT practitioners work with the techniques and methodologies found in the literature. Regarding tools, our findings were inconclusive, as most practitioners do not use the tools identified in the literature, or even do not use any tools. As for ethical requirements, some were well consolidated with practitioners but others were not, such as the principle of equity and inclusion. Contributions and Impact in the IS area: An overview of how AI systems are being developed across organizations and the treatment that is given to ethical requirements. Our findings reveal that there is a need for organizations to consolidate ethical and legal notions with developers so that they can be applied during the requirements elicitation phase.

Keywords: Requirements elicitation · Artificial intelligence · Ethical issues techniques · Tools

1 Introduction

Artificial Intelligence (AI) is used in everyday life by many people, helping to solve various problems in many different areas and streamlining services. With the increase in the use of AI in recent years, the problems involving it also grow and show the need for care regarding several ethical issues associated with the

J. C. Xavier-Junior and R. A. Rios (Eds.): BRACIS 2022, LNAI 13653, pp. 15–29, 2022.
https://doi.org/10.1007/978-3-031-21686-2_2

creation of AI-based systems. Some of these issues have made global headlines posing questions to users about the security of their data and their privacy. In this context, AI systems are still far from trouble-free, even though there are many tools available that help to improve security in these systems [38]. Moreover, ethical issues persist with the advancement of technology, since the virtualization process requires numerous data from users who usually do not know how their data is being treated [18,32,37].

This shows the need for greater coordination in the development of AI systems, which leads to the use of tools that help to follow ethical principles during development. A more ethical approach to developing, deploying, and using AI can be a competitive advantage for any company [31]. However, many of these tools are relatively immature making it difficult to encourage adoption by developers. AI faces other issues in addition to ethical barriers, which are equally important and may be related to development, deployment, and use. A research conducted by IBM shows that 78% of the leaders of large organizations which use AI or intend to use it, consider it "very" or "critically important" knowing that their AI models are being built fairly, securely, and reliably [22]. The survey also shows that 26% see development tools as the biggest obstacle to adopting AI in their systems [22].

The reliability, integrity, and privacy concerns of AI systems need to be addressed at the earliest steps of development when system requirements are being defined. The elicitation of requirements is an initial process that continues even after development has begun. It aims to determine the scope and specification of the product [36], being the initial door to bar possible ethical and technical problems that may occur during and after creating an application in the AI context [11,12,16].

Knowing the best requirements elicitation techniques for developing an application in the context of ethics-centric AI is an important factor to better understand what stakeholders and even users expect from the software and define a better path to a final system without ethical and technical problems. Bearing in mind the difficulties inherent in the development of an AI system, it is necessary to have a strategy to face them. Therefore, the general objective of this work is to investigate the techniques, methodologies, and tools of Requirements Engineering to carry out the elicitation of functional and non-functional software requirements in the context of AI.

2   Background

The term Artificial Intelligence (AI) was created by a study proposal by John MacCarthy and Marvin Minsky when an invitation was made for a study to be carried out in 1956 at Dartmouth College [17]. It consisted of creating an Artificial Intelligence in which a machine could simulate every aspect of learning or any characteristic of intelligence, given that these aspects and characteristics are precisely described [30].

Artificial Intelligence is considered a multidisciplinary field of study, coming from computing, psychology, engineering, cybernetics, and mathematics, with the main objective of building intelligent behavior systems that perform tasks with competence equivalent or superior to what a human specialist would perform [34]. Most current AI systems perform only a fraction of main activities such as pattern recognition (recognition of plant or animal images or human faces), language processing (language understanding, translating or answering questions), practical suggestions (recommend purchases, provide information, logistical planning or optimize industrial processes), and so on [9]. There are even systems that can combine many of these capabilities, such as autonomous vehicles or assistance robots [33].

The High-Level AI Expert Group (HLEG) [33] characterized the scope of AI research as a scientific discipline, which includes various approaches and techniques such as machine learning (deep learning and reinforcement learning for example), machine reasoning (planning, programming, knowledge representation and reasoning, research and optimization) and robotics (which includes control, perception, sensors, and actuators). To this definition, we could also add communication as well, and particularly the understanding and generation of language, as well as the domains of perception and vision.

2.1 Software Requirements

Software requirements can be classified as descriptions and limitations of what a system and its services must provide. Such limitations meet the needs of stakeholders during software development. Requirements engineering is critical since misinterpreting what services are required of the system can lead to problems during the implementation phase. Requirements Engineering has four main activities [36]: Feasibility study; Requirements elicitation and analysis; Requirements specification; and Requirements validation.

Requirements can be statements in natural language of the services the system provides and its operational constraints or a structured document with detailed descriptions of the system, the first being a user requirement and the later a system requirement [29]. Moreover, they can also be divided into functional requirements or non-functional requirements. The first are statements about the functioning, services, and behavior that the system should perform and it is expected that the description of a functional requirement is as complete and consistent as possible, as it determines much of the scope and specification of the software. Non-functional requirements are the characteristics and restrictions applied to the software's functionalities, which are generally more critical than the functional requirements [40].

Several methods can be used to elicit software requirements. The requirements elicitation process for a project is not restricted to the use of just one technique, but a combination of them [24]. In this way, from a given scenario, it is possible to define the characteristics associated with the context and obtain the most appropriate techniques for a given configuration, observing which techniques are used together.

2.2   Ethical Requirements for AI

The High-Level AI Expert Group (HLEG) [33] characterized the scope of AI research as a scientific discipline, which includes various approaches and techniques such as machine learning (deep learning and reinforcement learning for example), machine reasoning (planning, programming, knowledge representation and reasoning, research and optimization) and robotics (which includes control, perception, sensors, and actuators). To this definition, we could also add communication as well, and particularly the understanding and generation of language, as well as the domains of perception and vision.

From the beginning of the construction of an AI application, it is necessary to keep in mind the ethical issues applicable to it. There are several issues that permeate this theme and one of them is data privacy and security. Siau and Wang [35] showed that in 7 out of 8 frameworks on ethics applied to AI, privacy is one of the most relevant factors. The development of an AI system relies heavily on a huge amount of data, including personal data and private data. With more data generated in societies and companies, there is a greater chance of misuse of that data. Thus, data must be properly managed to prevent misuse [20].

To keep the data safe, every action taken with the data must be detailed and recorded. Both the data itself and the transaction record can cause risks related to privacy. It is also important to consider what should be recorded, who should be in charge registration action and who can have access to the data and records. In the elicitation process, it must be considered how the data processed by the AI will be saved and made available to its users [20].

Another important point raised by Bibal et al. [6] was the ability to explain why the AI got that result after analyzing the data, which is a practice imposed by privacy laws, such as the GDPR. It is extremely important that organizations that build AI systems are able to explain and understand the result of it and thus provide a reliable service to people. Therefore, in the requirements elicitation process, it is necessary to guarantee the explainability of AI models, leaving aside black-box models that hinder the understanding of the model. Once the explainability is guaranteed, it makes it easier to ensure that justice, equity and transparency are being applied, which are other practices required by privacy laws [10]. Moreover, these aspects make the application more reliable because users will be fully aware of its operation and all the resources behind its design.

With responsibility and accountability we can also prevent malfeasance caused by an AI [19], as those who develop or sponsor it will not want their names related to something that harms human beings. Even though it is not possible to prevent maleficence, such responsibilities can lead to the application's creators being arrested or paying fines, setting an example for other developers and sponsors not to do the same. Charity and freedom are also frequently cited ethical principles and must clearly be respected from the stage of eliciting requirements [28]. With them, it is possible can ensure the promotion of well-being, peace and happiness, the creation of socioeconomic opportunities and economic prosperity with an alignment of AI with human values. It also guarantees the freedom of expression or self-determination.

Freedom and beneficence can be guaranteed through transparency and predictable (explicable) AI, as it does not reduce citizens' choices and knowledge, increasing people's knowledge about AI, giving notice and consent, or, conversely, avoiding actively collecting and disclosing data in the absence of consent information [15]. Regarding dignity and sustainability, we know that many AI systems can end up replacing humans doing their job equally or even better [20], therefore it is important during the elicitation phase to have a discussion about the possible impacts of the application on jobs and people in general in view of what AI will provide for its users. It is also important to emphasize a possible impact on the environment, as issues related to the environment are currently widely discussed and an AI that can cause damage to it must be avoided in order to preserve the quality of life.

2.3   Related Works

Aiming to broaden the discussion on ethics in AI, Cerqueira et al. [12] conducted a study of the literature alongside mining GitHub to find projects related to ethics in AI, exploring practical implementations and relating them with the discoveries within the literature. They found total of 182 abstracts in SCOPUS related to ethics in AI and 21 tools from GitHub for implementing AI ethics. The study also says that it is critical that the main focus of regulations should be on the work of the software developer and that implementing ethics in AI is not an easy task, it takes constant dedication.

In order to identify the methods and tools available to help ML developers, engineers and designers reflect and apply "ethics", Morle et al. [31], designed a typology for the ML community focused on the practice that "combined" the tools and methods identified according to the ethical principles (beneficence, non-maleficence, autonomy, justice and explainability). Based on the fact that companies do not provide their practitioners with tools focused on the ethical development of AI, the study moves on to the study of the literature on the subject discussed. Research has shown that there is an uneven distribution of efforts across the entire "AI Ethics" typology, and many of the tools included are relatively immature.

Vakkuri et al. [38] discussed the ethics of AI and concluded that it is currently an area with a large gap between research and practice. Much of the research done is theoretical and conceptual, with a focus on defining the principles for AI ethics and how to address them. Many studies have tried to fill this gap to bring these principles to developers, but it didn't seem to have much success and the author tries to bring another approach, and proposes a method to implement AI Ethics, ECCOLA. ECCOLA is a guide that is designed to provide developers with an actionable tool to implement AI ethics and to assist in using the various AI ethics guidelines in practice. As for the results obtained, there is currently no way to do an ethical benchmarking in the context of AI ethics, so it becomes indisputably a limitation for any method. It is equally difficult for a method to have a quantitatively proven effect because ethical issues are often context specific and require contextual reflection. ECCOLA raises awareness of ethics

AI by making its users aware of various ethical issues and facilitates ethical discussion within the team and also produces system development transparency [38].

Requirements for AI systems are derived from ethical principles or ethical codes (standards) and are similar to legal requirements [13]. Guizzardi et al. [16] is interested in defining the sources of ethical requirements, ethical principles and ethical codes as well as outlining a systematic process to differentiate requirements from their sources. The authors' main thesis is that techniques developed in requirements engineering that have been practiced for decades can also be used to make AI systems compatible with ethical principles and codes found in the literature. As for the implementation of functional and quality requirements derived from ethical requirements, the authors emphasize that the system must be able to perform as well as well-trained humans performing the same task. AI systems must explain their reasoning, rather than just providing results and making decisions.

Aiming to help software managers, mainly in the prioritization of features and improvements for a large amount of information, Mangabeira [29] aimed to create a classifier of user requests, expressed in natural language, automating this solution with processing natural language and machine learning algorithms. The author highlighted the difficulty of software managers with regard to communication with end users of developed software products, and also based on authors in the area, it is possible to say that traditional elicitation methods miss the opportunity to involve a large number of users. Even when performed for a target audience, such as questionnaires, elicitations need reasonable human and financial resources to be applied and evaluated considering a large group of people. On the other hand, today there are many aspects that can be directly elicited by the users of the products themselves, through the investigation of social networks and comments from these users about the use of those products. These aspects can be very useful in the requirements elicitation process.

The work of Cerqueira et al. [14] made use of Design Science Research to identify the guidelines and ethical principles for systems based on Artificial Intelligence existing in the literature. Vogelsang and Borg [39] conducted an interview with four data scientists define characteristics and challenges unique to Requirements Engineering (RE) for ML-based systems. The interview approached elicitation, specification, and assurance of requirements and expectations and showed the need of adaptation for requirements engineering process according to the development paradigm. To the best of our knowledge, we are not aware of any work that investigates the use of existing techniques in the literature to elicit ethical requirements in the context of AI in industry.

3   Research Methodology

This work have investigated the requirements elicitation techniques, methodologies and tools used in the industry in the context of AI through a survey with Information and Communications Technology (ICT) practitioners. The survey has as main objective to collect the perception of ICT practitioners in relation to the findings in the literature review. Survey questions are show in Table 1. Questions Q1 to Q3 are to understand the profile of the survey ICT practitioner. Questions Q4 to Q9 investigate the techniques, methodologies and tools used in requirements elicitation and also seek to know which of the techniques, methodologies and tools the ICT practitioner does not know, giving a notion of the respondent's knowledge in relation to the techniques identified in the literature. Questions Q10 to Q17, on the other hand, have questions about ethics applied to the AI context, to get an idea of how much the subject is being talked about within organizations and also to get an idea of the knowledge of ICT practitioners about the projects they work with.

Table 1. Survey questions

ID	Question
Q1	What is the nature of your organization's operations?
Q2	What is your position in the organization?
Q3	How many years have you worked in your position?
Q4	What requirements elicitation techniques have your organization used? (You can choose more than one if your organization uses multiple techniques)
Q5	Which of these requirements elicitation techniques are you not familiar with? (You can choose more than one if you don't know some of these techniques)
Q6	What requirements elicitation methodologies have your organization used? (You can choose more than one if your organization uses multiple methodologies)
Q7	Which of these requirements elicitation methodologies are you unfamiliar with? (You can choose more than one if you don't know some of these methodologies)
Q8	What requirements elicitation tools have your organization used? (You can choose more than one if your organization uses multiple tools)
Q8	Which of these requirements elicitation tools are you not familiar with? (You can choose more than one if you don't know some of these tools)
Q10	During and after the requirements elicitation phase, is the data needed for AI modeling/construction transparently obtained and its origins well known?
Q11	During and after the requirements elicitation phase, are the tools used to obtain the data needed for AI modeling/construction secure and is your organization fully aware of the tools' data warehousing policy?
Q12	During and after the requirements elicitation phase, is it well defined for what purpose the AI will be used?
Q13	During and after the requirements elicitation phase, is the privacy of the user who will use it already taken into account (AI Security, LGPD Application, etc)?
Q14	During and after the requirements elicitation phase, is your organization concerned about trust in the tools and methods that will be used in development?
Q15	During and after the requirements elicitation phase, is it taking into account the positions/jobs that the AI can take away from other people by doing their work?
Q16	During and after the requirements elicitation phase, is it taking into account diversity, non-discrimination and equity, so that the AI does not make decisions based on something that could be framed as prejudice?
Q17	During and after the requirements elicitation phase, is accountability taken into account (definition of AI providers' responsibility and its results)?

4 Survey Results and Discussion

The survey was prepared by two researchers, one of them with more than 20 years of experience, and the guidelines proposed by Kitchenham et al. [26,27]. We carried out a pilot with 5 professionals who work in the requirements area and, after the pilot, we made some adjustments to the questions. The pilot's responses were not considered in the analysis of the results. The survey consisted

of closed questions with multiple choices. In some questions we used a Likert scale
[5]. The survey was available online from November 9 until February 7, 2021.
We contacted professionals in our relationship network and these professionals
suggested other professionals in their relationship network. In total, we contacted
98 professionals, but only 50 agreed to respond to the survey.

The survey with 17 questions, as shown in Table 1. 56% of people were from
private organizations and 44% from public organizations. According to results
of Q2 which questions ICT practitioners' roles, 16% are project managers, 24%
are application developers, 12% are students, 34% are from areas related to data
analysis, 4% are Researchers, 6% are from other ICT areas, 2% are proofreaders
and 2% are financial analysts. Thus, a total of 80% of the ICT practitioners
work with ICT. As for the length of time that ICT practitioners have held their
current positions in organizations, 32% have been less than 1 year, 22% 1 to 3
years, 12% 3 to 5 years, 14% of 6 to 9 years and finally 20% over 10 years. It
is important to notice that only 46% of the ICT practitioners have more than 3
years of experience in their current position.

Regarding the elicitation techniques used by the ICT practitioners, Fig. 1(a)
shows the most used techniques according to results of Q4. It is possible to notice
results very similar to those found in the literature [3,4,21,23] considering that
interviews, meetings and document analysis are most used ones. However, techniques such as questionnaires and usability tests are commonly cited in the literature as being widely used, although in the survey this result cannot be observed.
Other techniques were selected by less than 30% of the ICT practitioners.

Figure 1(b) shows the results of question Q5, which seeks to know the ICT
practitioner's knowledge about the techniques presented, and also to have a
relationship between unused techniques and unknown techniques. It is easy to
notice that the Laddering, JAD, and Cognitive Mapping are the most unknown
techniques by the ICT practitioners which supports the results of the previous
question, in which these techniques are not part of the majority of answers.

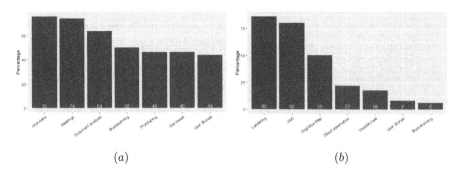

(a)   (b)

Fig. 1. Figure (a) shows requirements elicitation techniques used by the ICT practitioners, while (b) shows the requirements elicitation techniques that ICT practitioners
do not know.

As for the methodologies, Fig. 2(a) shows the results of question Q6, and provides a view of the methodologies used. 64% have used design thinking, 42% have used user-centered design and 14% have used participatory design. These results are similar to the results of Ainhoa Aldave et al. [3] in which design thinking is also the most used followed by user-centered design, except that they appear with lower percentages.

Figure 2(b) shows the results of question Q7 regarding participants' knowledge about techniques. 58% of the ICT practitioners do not know participatory design, 36% do not know user-centered design and only 12% do not know design thinking. This indicates that the low percentage of usage of these techniques is due to participants' lack of knowledge about them.

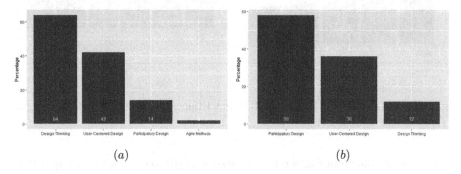

(a)   (b)

Fig. 2. Figure (a) shows methodologies used by ICT practitioners, while (b) shows the methodologies that the ICT practitioners do not know about.

Figure 3(a) shows the results of question Q8 and aims to know the tools used by the ICT practitioners in the requirements elicitation. Of the ICT practitioners, 36% use WebRatio (UML), 18% use Analyst Pro, 14% use no tools, 14% use Objectiver, and 12% use RequisitePro. As for the tools, both the literature review and the survey could not identify the most used tools, since the most chosen tool has only 36% of the total responses, and the other tools are very old and unknown by users.

As presented by Aguilar et al. [1] and Jazlyn Hellman et al. [8] WebRatio is a well-established tool in the market even though it is not recommended for large-scale projects. In general, WebML or UML tools are the most used for their ease of access and also their simplicity in understanding and building diagrams, so this would be the most suitable tool. Analyst Pro, which was right behind WebRatio, is still a used tool that has great scalability and version control [7], [2], and even though it is an old tool, it still has support from its developers. Other tools will not be considered as many ICT practitioners did not choose them and are very old and no longer supported by their developers.

The results for question Q9 are shown in Fig. 3(b). It is possible to observe that many of the presented tools are not known by the ICT practitioners. Our findings reveal that 80% of ICT practitioners do not know the DOORS tool,

60% do not know the Objectiver tool, 60% do not know the Optimal Trace tool, 58% Analyst Pro, 50% RequisitePro and finally 34% do not know the WebRatio (UML) tool. Thus, it is evident that the knowledge of developers about possible tools is very small or the tools found in our literature review are very outdated or unknown in Brazil, for example. Therefore, the results about the tools are weaker when compared to the techniques and methodologies.

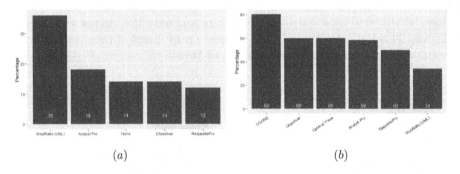

(a)   (b)

Fig. 3. Figure (a) shows tools used by ICT practitioners, while (b) shows the tools the ICT practitioners do not know about.

The results shown in Fig. 4 refer to questions Q10 to Q17, which are questions related to ethics in several areas applied to AI and also to requirements elicitation. According to the results of Q10 in Fig. 4, 56% of respondents strongly agree or agree that the data used during and after requirements elicitation is from known sources or is obtained transparently, 26% neither agree or disagree and 18% strongly disagree or disagree. With this it is easy to see that the ICT practitioners do not have full knowledge about the origin of the data, even using such data in the construction of software.

Similarly, according to the results of Q11 in Fig. 4, only 52% of respondents agree or fully agree on their knowledge of the data storage policy of the tools used in requirements elicitation, 34% neither agree or disagree, and 14% strongly disagree or disagree. We know from previous discussions [20,35] that data is a very important part in the development of an AI, so its correct storage and acquisition need to be taken into account from the requirements elicitation phase, and in the elicitation phase itself as discussions about the software begin in it and this generates data that may be sensitive.

About 74% of ICT practitioners agree or fully agree that the purpose for which the AI will be used is well defined, from the requirements elicitation phase, which is shown in the results of Q12 in Fig. 4. By looking at this result, it is evident that the ICT practitioners' knowledge about the use of the AI even after its completion is already determined from the elicitation of the requirements. Alongside a good definition of usage, AI systems need to guarantee privacy laws are being followed. According to the results of Q13 in Fig. 4, 72% of ICT practitioners agree or fully agree that the privacy of the user who will use the

AI is already taken into account from the requirements elicitation phase, 22% neither agree or disagree and only 6% strongly disagree or disagree. As stated by Jobin et al. [25], explainability is very important and has been increasingly demanded.

Analyzing the responses of the ICT practitioners regarding the organization's concern about the tools and methodologies being adequate for development in Q16, we see that 74% of the ICT practitioners agree or totally agree that the organization is concerned with trust in the tools and methods that will be used in the development, 18% neither agree or disagree and 8% disagree or totally disagree. In the same line of reasoning in Q17, only 36% of respondents agree or fully agree that they care about the issue of developed AI taking out positions/jobs from other people doing their work, during or after the requirements elicitation phase. As for diversity, inclusion and equity approached in Q18, 58% of the ICT practitioners agree or totally agree that, since the elicitation phase, the AI is designed so that it does not make decisions based on something that can be classified as prejudice.

Along with these results, 64% of the ICT practitioners agree or totally agree that since the AI elicitation phase there is a definition of the responsibility of the AI account providers and their results according to results of Q19. These results are in agreement with Adrien Bibal et al. [6] and Tobias Krafft et al. [28] regarding equity and accountability which can be guaranteed by explainability, since knowing how to explain the results can ensure that choices based on prejudice and racism do not happen and consequently the accountability of those responsible for the AI becomes more clear.

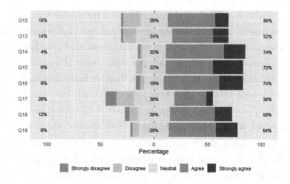

Fig. 4. ICT practitioners' perception of ethical issues

4.1 Threats to Validity

The conclusions obtained from the survey were derived from a limited number of ICT practitioners working with software development in some organizations from Brazil. Moreover, we cannot guarantee that all of the respondents have worked with AI although we did mentioned it was targeted this audience, which impacts interpretations about practices in AI based on these responses. Another threat to validity is the response of practitioners with less than a year of experience, since their limited experience may hinder working with varied tools and methodologies. Regarding the tools and methodologies listed in the survey that ICT practitioners could select from, we mitigated the possible threat to validity of not having found all possible tools and methodologies by adding open fields to the questionnaire. That way, if the tool or methodology used by the ICT practitioner was not listed, he could add this option himself.

5   Conclusions

In this paper, we carry out a survey to understand the perception of ICT practitioners in relation to techniques, methodologies and tools for eliciting requirements and ethical requirements for AI, seeking to know how the organizations they work to deal with the need to develop systems compatible with ethical requirements. Most of the ICT practitioners who responded to the survey have more than three years of experience, 80% work in the ICT field and 34% work in data analysis related fields. Survey results show that the techniques and methodologies found in the literature are widely used by practitioners, but the tools found are not well known and thus became the most inconclusive part of the survey. As for ethical requirements, most practitioners are aware as well as the organization of the importance of the requirements and most of them already use them, and of all the least applied requirement was the issue of jobs that an AI can affect.

The results of this work also demonstrate the need for organizations and developers to experiment with different development techniques and methodologies since many are unaware of distinct techniques and methodologies and that it could perhaps improve the work. As for the ethical requirements, some are already well consolidated with practitioners but others are not, such as in relation to equity and inclusion and also the jobs that can be taken away by AI. The answers were much more varied when compared to questions such as defining the purpose of the AI and accountability. This shows that it is still necessary to consolidate such notions with developers so that they can be applied in the construction of the AI from the requirements elicitation phase.

For future work, it is possible to carry out a survey with a broader population and control the area of work of practitioners to have a large majority working directly with AI. It would also be interesting to include more methodologies and especially tools, as those found in the literature are very old and unknown by survey ICT practitioners.

References

1. Aguilar, J.A., Zaldívar-Colado, A., Tripp-Barba, C., Misra, S., Bernal, R., Ocegueda, A.: An analysis of techniques and tools for requirements elicitation in model-driven web engineering methods. In: Gervasi, O., et al. (eds.) ICCSA 2015. LNCS, vol. 9158, pp. 518–527. Springer, Cham (2015). https://doi.org/10.1007/978-3-319-21410-8_40

2. Ahmed Abbasi, M., Jabeen, J., Hafeez, Y., e Benish Batool, D., Fareen, N.: Assessment of requirement elicitation tools and techniques by various parameters. Softw. Eng. **3**, 7–11 (2015). https://doi.org/10.11648/j.se.20150302.11

3. Aldave, A., Vara, J.M., Granada, D., Marcos, E.: Leveraging creativity in requirements elicitation within agile software development: a systematic literature review. J. Syst. Softw. **157**, 1–31 (2019)

4. Alflen, N.C., Prado, E.P.V.: Técnicas de elicitação de requisitos no desenvolvimento de software: uma revisão sistemática da literatura. AtoZ: novas práticas em informação e conhecimento **10**(1), 39–49 (2021). https://doi.org/10.5380/atoz.v10i1.77393

5. Alismail, S., Zhang, H.: Exploring and understanding participants' perceptions of facial emoji likert scales in online surveys: a qualitative study. ACM Trans. Soc. Comput. **3**(2), 12:1–12:12 (2020)

6. Bibal, A., Lognoul, M., de Streel, A., Frénay, B.: Legal requirements on explainability in machine learning. Artific. Intell. Law **29**(2), 149–169 (2020). https://doi.org/10.1007/s10506-020-09270-4

7. Bokhari, M.U., Siddiqui, S.T.: A comparative study of software requirements tools for secure software development. BVICAM's Int. J. Inf. Technol. **2**, 1–12 (2010)

8. Brambilla, M., Fraternali, P.: Large-scale model-driven engineering of web user interaction: the webml and webratio experience. Sci. Comput. Program. **89**, 71–87 (2014). https://doi.org/10.1016/j.scico.2013.03.010, https://www.sciencedirect.com/science/article/pii/S0167642313000701. (special issue on Success Stories in Model Driven Engineering)

9. Buxmann, P., Hess, T., Thatcher, J.B.: Ai-based information systems. Bus. Inf. Syst. Eng. **63**(1), 1–4 (2021)

10. Canedo, E.D., et al.: Proposal of an implementation process for the brazilian general data protection law (LGPD). In: Filipe, J., Smialek, M., Brodsky, A., Hammoudi, S. (eds.) Proceedings of the 23rd International Conference on Enterprise Information Systems, ICEIS 2021, Online Streaming, 26–28 April 2021, vol. 1. pp. 19–30. SCITEPRESS (2021). https://doi.org/10.5220/0010398200190030

11. Cemiloglu, D., Arden-Close, E., Hodge, S., Kostoulas, T., Ali, R., Catania, M.: Towards ethical requirements for addictive technology: the case of online gambling. In: REthics@RE, pp. 1–10. IEEE (2020). https://doi.org/10.1109/REthics51204.2020.00007

12. de Cerqueira, J.A.S., Althoff, L.D.S., Almeida, P.S.D., Canedo, E.D.: Ethical perspectives in AI: a two-folded exploratory study from literature and active development projects. In: 54th Hawaii International Conference on System Sciences, HICSS 2021, 5 Jan 2021. pp. 1–10. Kauai, Hawaii, USA. ScholarSpace (2021). http://hdl.handle.net/10125/71257

13. de Cerqueira, J.A.S., de Azevedo, A.P., Tives, H.A., Canedo, E.D.: Guide for artificial intelligence ethical requirements elicitation - re4ai ethical guide. In: 55th Hawaii International Conference on System Sciences, HICSS 2022, 3 Jan 2022. pp. 1–10. Kauai, Hawaii, USA. ScholarSpace (2022). http://hdl.handle.net/10125/80015

14. de Cerqueira, J.A.S., Leão, H.A.T., Canedo, E.D.: Ethical guidelines and principles in the context of artificial intelligence. In: Brazilian Symposium on Information Systems (SBSI). pp. 36:1–36:8. ACM (2021). https://doi.org/10.1145/3466933.3466969
15. Floridi, L., et al.: AI4People-an ethical framework for a good AI society: opportunities, risks, principles, and recommendations. Minds Mach. **28**(4), 689–707 (2018). https://doi.org/10.1007/s11023-018-9482-5, http://link.springer.com/10.1007/s11023-018-9482-5
16. Guizzardi, R.S.S., Amaral, G.C.M., Guizzardi, G., Mylopoulos, J.: Ethical requirements for AI systems. In: Goutte, C., Zhu, X. (eds.) Canadian Conference on AI. LNCS, vol. 12109, pp. 251–256. Springer, Cham (2020). https://doi.org/10.1007/978-3-030-47358-7_24(2020)
17. Haenlein, M., Kaplan, A.: A brief history of artificial intelligence: on the past, present, and future of artificial intelligence. California Manage. Rev. **61**(4), 5–14 (2019)
18. Hagendorff, T.: The ethics of AI ethics: an evaluation of guidelines. Minds Mach. **30**(1), 99–120 (2020). https://doi.org/10.1007/s11023-020-09517-8
19. Hagendorff, T.: The ethics of AI ethics: an evaluation of guidelines. Minds Mach. **30**(1), 99–120 (2020). https://doi.org/10.1007/s11023-020-09517-8, http://link.springer.com/10.1007/s11023-020-09517-8
20. HLEG, A.: Ethics guidelines for trustworthy AI (2019). https://digital-strategy.ec.europa.eu/en/library/ethics-guidelines-trustworthy-ai
21. OI Huchenko, I., Gobov, D.: Requirement elicitation techniques for software projects in ukrainian IT: an exploratory study. In: Proceedings of the 2020 Federated Conference on Computer Science and Information Systems. Annals of Computer Science and Information Systems, vol. 21, pp. 673–681. FedCSIS (2020). https://doi.org/10.15439/2020F16
22. IBM: From Roadblock to Scale: The Global Sprint Towards AI (2020). https://filecache.mediaroom.com/mr5mr_ibmnews/183710/Roadblock-to-Scale-exec-summary.pdf
23. Ignácio, R.C., Benitti, F.B.V.: Improving the selection of requirements elicitation techniques: a faceted guide. In: Hadad, G.D.S., Pimentel, J.H., Brito, I.S.S. (eds.) Anais do WER20 - Workshop em Engenharia de Requisitos, 24–28 Aug 2020. pp. 1–14. São José dos Campos, SP, Brasil. Editora PUC-Rio (2020). http://wer.inf.puc-rio.br/WERpapers/artigos/artigos_WER20/01_WER_2020_paper_1.pdf
24. Ignácio, R.C.: Guia facetado de técnicas de elicitação de requisitos. Bachelor's thesis, Federal University of Santa Catarina (2018). https://repositorio.ufsc.br/handle/123456789/192155
25. Jobin, A., Ienca, M., Vayena, E.: The global landscape of AI ethics guidelines. Nat. Mach. Intell. **1**(9), 389–399 (2019). https://doi.org/10.1038/s42256-019-0088-2, http://www.nature.com/articles/s42256-019-0088-2
26. Kitchenham, B.A., Pfleeger, S.L.: Principles of survey research part 2: designing a survey. ACM SIGSOFT Softw. Eng. Notes **27**(1), 18–20 (2002)
27. Kitchenham, B.A., Pfleeger, S.L.: Personal opinion surveys. Shull, F., Singer, J., Sjoberg, D.I.K. (eds.) In: Guide to Advanced Empirical Software Engineering, pp. 63–92. Springer, London (2008). https://doi.org/10.1007/978-1-84800-044-5_3
28. Krafft, T., et al.: From principles to practice - an interdisciplinary framework to operationalise AI ethics (2020)
29. Mangabeira, G.S.: Elicitação de requisitos baseada em análise de multidão : uma abordagem orientada a aprendizado de máquina. Bachelor's thesis, University of Brasília (2018). https://bdm.unb.br/handle/10483/23046

30. McCarthy, J., Minsky, M.L., Rochester, N., Shannon, C.E.: A proposal for the dartmouth summer research project on artificial intelligence, august 31, 1955. AI Magazine **27**(4), 12 (2006). https://doi.org/10.1609/aimag.v27i4.1904, https://ojs.aaai.org/index.php/aimagazine/article/view/1904

31. Morley, J., Floridi, L., Kinsey, L., Elhalal, A.: From what to how. an overview of AI ethics tools, methods and research to translate principles into practices. CoRR abs/1905.06876, pp. 1–28 (2019). arxiv.org/abs/1905.06876

32. Ryan, M., Stahl, B.C.: Artificial intelligence ethics guidelines for developers and users: clarifying their content and normative implications. J. Inf. Commun. Ethics Soc. **19**(1), 61–86 (2021). https://doi.org/10.1108/JICES-12-2019-0138, https://doi.org/10.1108/JICES-12-2019-0138

33. Sartor, G., European Parliament, European Parliamentary Research Service, Scientific Foresight Unit: The impact of the General Data Protection Regulation (GDPR) on artificial intelligence: study. European Parliament (2020). http://www.europarl.europa.eu/RegData/etudes/STUD/2020/641530/EPRS_STU(2020)641530_EN.pdf, oCLC: 1160193938

34. Sellitto, M.A.: Inteligência Artificial: uma aplicação em uma indústria de processo contínuo. Gestão & Produção **9**(3), 363–376 (2002). https://doi.org/10.1590/S0104-530X2002000300010, http://www.scielo.br/scielo.php?script=sci_arttext&pid=S0104-530X2002000300010&lng=pt&tlng=pt

35. Siau, K., Wang, W.: Artificial Intelligence (AI) ethics: ethics of AI and ethical AI. J. Database Manage. **31**(2), 74–87 (Apr 2020). https://doi.org/10.4018/JDM.2020040105, http://services.igi-global.com/resolvedoi/resolve.aspx?doi=10.4018/JDM.2020040105

36. Sommerville, I.: Software Engineering. Pearson Education, Boston (2011)

37. Stix, C.: Actionable principles for artificial intelligence policy: three pathways. Sci. Eng. Ethics **27**(1), 1–17 (2021). https://doi.org/10.1007/s11948-020-00277-3

38. Vakkuri, V., Kemell, K., Jantunen, M., Halme, E., Abrahamsson, P.: ECCOLA - a method for implementing ethically aligned AI systems. J. Syst. Softw. **182** (2021)

39. Vogelsang, A., Borg, M.: Requirements engineering for machine learning: Perspectives from data scientists. In: RE Workshops. pp. 245–251. IEEE (2019). https://doi.org/10.1109/REW.2019.00050

40. Zhou, Z., Zhi, Q., Morisaki, S., Yamamoto, S.: An evaluation of quantitative non-functional requirements assurance using archimate. IEEE Access **8**, 72395–72410 (2020)

An Efficient Drift Detection Module for Semi-supervised Data Classification in Non-stationary Environments

Arthur C. Gorgônio[1]([envelope]) [iD], Cephas A. da S. Barreto[1] [iD],
Song Jong Márcio Simioni da Costa[2] [iD], Anne Magály de P. Canuto[1] [iD],
Karliane M. O. Vale[3] [iD], and Flavius L. Gorgônio[3] [iD]

[1] Department of Informatics and Applicated Mathematics (DIMAp), Federal University of Rio Grande do Norte (UFRN), Natal, Brazil
arthurgorgonio@ppgsc.ufrn.br, anne.canuto@ufrn.br
[2] Digital Metropolis Institute (IMD), Federal University of Rio Grande do Norte (UFRN), Natal, Brazil
[3] Deparment of Computing and Technology (DCT), Federal University of Rio Grande do Norte (UFRN), Caicó, Brazil
{karliane.vale,flavius.gorgonio}@ufrn.br

Abstract. In the data stream (DS) context, data is received at high speed, and it must be processed as soon as possible. Furthermore, it is not possible to guarantee that all data is labelled. Consequently, semi-supervised learning (SSL) becomes an efficient attempt to build an effective model in this context. Dynamic Data Stream Learning (DyDaSL) is a framework that uses a SSL algorithm to build a model able to classify instances in a data stream context. In this paper, an extension of the DyDaSL drift detection module is proposed. Its main aim is to make drift detection more flexible and, in turn, to improve the whole data stream process. An empirical analysis is conducted using real and synthetic datasets. The proposed approach achieved better results than the original one and some state-of-art drift detection methods.

Keywords: Semi-supervised · Data stream classification · Concept drift

1 Introduction

The Machine Learning (ML) area provides a set of tools and techniques to develop the learning capacity of machines. These techniques can be divided into several different taxonomies and the most frequent is using the degree of supervision: supervised, semi-supervised, and unsupervised. The main difference between these types of learning is the dataset: an instance is labelled when a tuple (instance) has a target (desired output); otherwise, it is an unlabelled instance. So, when all labelled instances are available, the supervised machine learning technique is selected to carry out a task. However, the unsupervised

J. C. Xavier-Junior and R. A. Rios (Eds.): BRACIS 2022, LNAI 13653, pp. 30–44, 2022.
https://doi.org/10.1007/978-3-031-21686-2_3

machine learning technique is recommended in cases where all dataset instances are unlabelled. Finally, when both types of instances are present in a dataset and the unlabelled instances are more frequent than labelled, the semi-supervised technique builds models to carry out the tasks [1,2].

Each learning type has particular tasks; both supervised and semi-supervised are used to solve the same problems, classification and regression. In these tasks, the objective is to predict the label of an instance, but the difference is that the regression is used when the output is continuous. The output must be a previously known category or class in a classification task. On the other hand, unsupervised learning has the clustering task to segregate instances into groups with similar characteristics based on their attributes (features) [2].

There is a model training phase in all these tasks, in which the ML algorithm is applied to the data to build a model. In traditional ML, once the model is trained, it is expected that it does not need to be adjusted unless the databases have a concept drift. *Concept drift* is a phenom in which a change is identified in the original data distribution. In non-stationary environments, the original data distribution may change during the training and testing phases, and this model must be adapted when this occurs.

In these environments, data stream classification is a widespread application. Hence, these applications have some common characteristics, such as i) following non-stationary data distributions, ii) having limited availability of hardware resources, and iii) having to process data when it is available [3]. In addition, the data stream can have few labelled instances, and the classification task in data streams leads to the semi-supervised context.

The framework DyDaSL for semi-supervised classification in the DS context was proposed in [4]. However, that proposal uses a fixed threshold in the drift detection module, which is very time-consuming, mainly in the DS context. So, in order to solve the drawbacks of the drift detection module of DyDaSL, this paper proposes an extension in the drift detection module of DyDaSL in order to make a more effective drift detection on the streaming data. Therefore, the main contributions of this paper can be summarised as follows:

a) This study performs a more robust investigation in the drift detection module, investigating different methods and assessing them in several datasets;
b) This study proposes the use of a flexible threshold based on the effectiveness of the ensemble in the drift detection module; and
c) This study extends [4], exploring more datasets and batches sizes.

This paper has seven sections, Sect. 2 describes the main concepts related to the context of this research, while Sect. 3 presents some related studies. Section 4 presents the proposed approach. The experimental methodology is presented in Sect. 5, while the results are described in Sect. 6. Finally, Sect. 7 presents the main conclusions and future works.

2   Background

2.1   Flexible Confidence of a Classifier Semi-supervised Technique

In a semi-supervised context, a dataset (D) has two types of data, labelled (L) when an instance has an associated target and unlabelled (U) when an instance does not have a target. Also, a dataset represents the union of both sets, $D = L \cup U$. Hence, when the labelled set can be expressed as $L = \{(\mathbf{x}_i, y_i)\}_{i=1}^{l}$ and unlabelled set $U = \{(\mathbf{x}_j)\}_{j=l+1}^{l+u}$, the dataset can be expressed as $D = \{(\mathbf{x}_1, y_1), (\mathbf{x}_2, y_2), \ldots, (\mathbf{x}_l, y_l), \mathbf{x}_{l+1}, \ldots, \mathbf{x}_{l+u}\}$.

In [5], it was proposed a new approach to select instances in a semi-supervised context. This approach, named Flexible Confidence of a Classifier (FlexCon-C), builds a model with selected instances to evaluate the effectiveness of the classification process during the training phase. FlexCon-C has a dynamic threshold that is calculated in each iteration of the classification process. This threshold refers to the premise that if the classifier has acceptable effectiveness in initially labelled instances, the threshold can be decreased. On the other hand, if the classification does not achieve the minimum acceptable value, more restrictive is the selection step to avoid the noise.

2.2   Classifier Ensemble

An ensemble of classifiers is a model built using two or more classifiers (base classifiers) to solve a classification task, reducing the variance of a single model [6,7]. One characteristic of an ensemble is that the type of all base classifiers determines if the structure of the ensemble is homogeneous or heterogeneous. The first one occurs when all base classifiers of an ensemble are the same type, and the last one occurs when at least one of the base classifiers is a different type from the others [7].

The idea of creating an ensemble of classifiers to combine their different outputs and improve the performance of a classification task is a promising possibility. However, it is necessary to select only one label for each instance, combining the outputs of base classifiers through majority, weighted vote, or an oracle [8]. The majority vote selects the most voted label among those presented by the base classifiers, while the weighted vote computes the vote using weight criteria to weigh the classifiers with the most effectiveness. On the other hand, the oracle classifier is a model with all knowledge and always correctly predicts an instance [8]. In the DS context, this classifier has the best evaluation compared to the ensemble or base classifiers since the oracle is trained with the most recent instances while the others are not. Also, this classifier helps the labelling process when the base classifiers do not agree with a label of an instance.

2.3   Data Stream Classification

As mentioned earlier, the dataset has as many instances as possible in a data stream context. These instances must be processed in sequential order one-to-one (online) or in batches (offline). In addition, it is impossible to store all instances into a dataset to create a model in one training phase since there is not enough memory to store the amount of incoming data in the data stream. In this context, the traditional approaches to the classification task are ineffective, and they can not be adapted to the drifts that occur in a data stream [3].

In the traditional approaches, there are two well-defined classification phases: training and testing. The training phase focuses on building a model using all instances in the dataset, while the test phase evaluates the built model in the previous phase. However, when the number of instances grows to infinite, it is impossible to finish the training to start the testing phase. Hence, the data stream approaches change both phases to satisfy these criteria.

The training phase is continuous in the data stream applications since the built model must be constantly adapted to incoming data. So, a data stream algorithm must be able to identify a drift and adapt itself to maintain the same effect during the whole process. The drift detection strategy can be of two types, active and passive. The active strategy tries to detect the drift before it occurs, making some processes with the new data available. On the other hand, the passive strategy waits for the drift and slowly evolves the model to adapt to the new data [3]. In addition, the active strategy detects abrupt drifts easier since the data distribution changes very quickly.

In contrast, the passive approaches have more efficient in detecting the incremental and gradual types because it has a transition window where the data distribution slowly evolves inner this window. Usually, the passive approach has more computational cost than the active, because the model is updated regardless of whether a drift occurs. However, the passive approach is more simple than the active one since the last one needs to evaluate some metric to determine the occurrence of the drift in the data. Sometimes, it is impossible to guarantee that enough instances were labelled to train a supervised model. However, SSL can be used instead of that learning type. Also, the data must be processed sequentially to generate models capable of classifying the new instances in the data stream processing. Therefore, SSL can be applied in these applications to generate a model with few labelled instances, and the results were auspicious [4,9–11].

3   Related Work

This section describes research that uses semi-supervised approaches and/or data stream classification tasks once the main topic of interest of this paper is a semi-supervised ensemble approach in the context of data streams [9,11–13].

In [11] proposed a semi-supervised drift detector for the data stream context. Their proposal was based on a weighted detector ensemble, where each ensemble member consists of a set of weights to detect the drift. The detection step consists of how each base member of the ensemble predicts a drift occurrence, and the ensemble weighs the outputs and determines whether a drift occurred.

Another classification system in the same context was proposed in [9]. That approach uses cluster algorithms in Self-training to build the learning model. The ensemble update strategy was based on two approaches: i) when the ensemble determines; or ii) when drift occurs, a newly built cluster model was added to the ensemble by *Kullback-Leibler* divergence. In [12] the authors propose a drift detector, called CPSSDS, that uses the *Kolmogorov-Smirnov* statistical test to determine the occurrence or not of the drift. In this approach, a drift is detected when the *p-value* of this test is below 0.05, comparing the current batch and the previous one. This detector was used and assessed using supervised learning.

Another drift detector (Hinkley) was proposed in [13], their proposal an approach based on an instance-by-instance drift detector using the *Page-Hinkley* statistical test [14]. This detector performs a cumulative sum of the classification effectiveness for each instance in the batch. A drift is detected when the cumulative sum is below a threshold representing the magnitude of changes allowed after drift. This detector was also used and assessed using supervised learning.

All mentioned studies proposed a drift detector using another approach or apparatus. Unlike these studies, this paper presents a different drift detector that works based on the performance of its main component, a classifier ensemble. Furthermore, the drift detection threshold is dynamically updated to better adjust to the dataset characteristics, leading to better performance for problems in the non-stationary context.

4   The Proposed Approach

As previously mentioned, this paper extends [4], in which an improvement in the detection module of the DyDaSL framework is proposed. In the original DyDaSL framework, the drift detection module, or simply the detection module, uses a fixed threshold to determine a drift occurrence. However, it was noted that this fixed threshold approach could make drift detection a complex task. Several mistakes can be triggered using a high threshold; on the other hand, with a low threshold, a proper drift can be detected too late for the reaction or never seen. Therefore, in this paper, two new ways to make the threshold of the detection module flexible during the execution (running time) are proposed.

In this paper, we propose to use classifier ensembles to determine a flexible threshold for drift detection. This flexible threshold strategy was developed using the effectiveness of the current ensemble in the labelled instances of the subsequent batch in the analysed stream. Based on this flexible threshold, two versions of this approach are developed: one using the simple vote (DyDaSL - N) and one with a weighted vote (DyDaSL - W) using a classification effectiveness metric to calculate the confidence (weight) of each base classifier.

The DyDaSL workflow is presented in Fig. 1 and Algorithm 1. Initially, each dataset is divided into batches - to simulate a data stream environment - that will be used as input data. In addition, each batch contains two groups of instances labelled and unlabelled. In the first batch, an ensemble is trained using the FlexCon-C method. In the immediately subsequent batch, the ensemble calculates the classification effectiveness and determines a new threshold to detect a drift in the next batches. This new threshold is determined by ensemble effectiveness (classification metric such as accuracy score, f-measure, and others) in the subsequent batch. A drift is detected when the ensemble classification effectiveness is below the current threshold. The reaction module starts training a new classifier (an oracle) to substitute the worst base classifier in the ensemble. This process is repeated until the last batch is processed, and the ensemble dynamically changes its structure in terms of base classifiers.

Algorithm 1: The DyDaSL algorithm

---

while *data_stream has instances* **do**

    create a *batch* with next n instances;

    if *first batch* **then**

        train an ensemble using FlexCon-C with labelled instances of the *batch*;

    else

        if *detection module* **then**

            train oracle with labelled instances;

            exchange the worst ensemble base classifier by oracle based on oracle prediction;

        end

        classify instances of the current batch;

    end

end

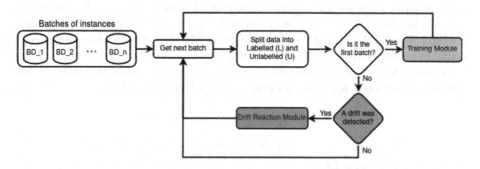

Fig. 1. Workflow of the DyDaSL method

Algorithm 2 shows the steps of the detection module, which define a drift occurrence. In this module, the ensemble predicts the labelled instances in the current batch and measures a classification metric using a simple vote or weighted vote as an ensemble combination module. It is important to emphasise that the ensemble is updated in the current batch when a drift is detected, and the new threshold is only defined for the next batch using the labelled instances.

Algorithm 2: DyDaSL detection module

---

ensemble_pred ← classify_batch(ensemble, D_L);
metric ← calculate(ensemble_pred, D_L);
if *metric < threshold* **then**
   | threshold ← new_threshold(ensemble, current_batch);
   | **return** *Drift detected*;
else
   | **return** *No drift detected*;
end

---

5   Experimental Methodology

This section describes the main aspects of the empirical study conducted in this paper. The main aim is to evaluate the accuracy, F-Score, and kappa metrics of the proposed drift detection approach in the DyDaSL framework. For a comparative analysis, some existing state-of-art drift detectors will also be evaluated, which are: i) Conformal prediction for semi-supervised classification on data streams (CPSSDS) [12]; and ii) Hinkley [13]. The CPSSDS is a semi-supervised drift detector based on statistical evaluation of the chunks, while the Hinkley performs an instance-by-instance detection. These approaches were selected in other to compare the effectiveness of the proposals versus these different drift detectors.

The source code is developed in R using the "RMOA" package, an API for Massive Online Analysis (MOA) framework [15] to simulate the data stream environment and to process all used datasets. Four classification algorithms are used as base classifiers of the FlexCon-C method, which are: Naïve Bayes (NB) [16], Decision Tree (DT) [17], RIPPER [18] and k-NN [19]. These algorithms were available in the "RWeka" package, an API for WEKA [20].

Table 1 shows the characteristics of the analysed datasets in terms of the number of instances, features and classes for each dataset [15,21].

Table 1. Description of the used datasets

Datasets	Number of instances	Features	Classes
Real datasets			
Adult	32 561	14	2
Airlines	539 383	7	2
Connect-4	67 557	42	3
Fars	100 968	29	8
Forest Cover Type	581 012	54	7
Poker	829 201	10	10
Shuttle	58 000	9	7
Synthetic datasets			
Gears2C2D	200 000	2	2
UG2C3D	200 000	3	2

In order to simulate a data stream environment, a conversion of the static datasets into dynamic ones is required. Therefore, subsets with n instances are grouped to generate data conglomerates (batches), and the same batch is used to train all analysed approaches. This study assesses seven batch sizes, with 100, 250, 500, 750, 1000, 2500, and 5000 instances.

Only 10% of data maintains its label for each batch, and the remaining data have their label removed. All proposed drift detectors are evaluated using the DyDaSL framework changing the drift detector to evaluate their performance in the semi-supervised context versus this paper proposal. It is essential to highlight that the parameters of the FlexCon-C are set as proposed in [5].

6   Experimental Results

The results obtained from the empirical analysis performed in this paper are presented in the subsections below, using the following strategies: 1) Analysing the performance by batch size; 2) Comparing the proposed versions of the DyDaSL versus the original DyDaSL, selecting two batch sizes; 3) Comparing the proposed methods versus the state-of-art, also selecting two batch sizes.

Additionally, the Critical Difference (CD) diagrams are used to present the ranking of the evaluated approaches, since these diagrams are easier to analyse and clearly show which algorithm is better than the others. It uses a post-hoc Friedman test, the Nemenyi test. The leftmost method obtained the lowest ranks (better results), and the rightmost method obtained the highest (worst results) when the results were ranked. Moreover, methods not covered by a horizontal line (critical difference) are statistically different. Otherwise, the null hypothesis of the Friedman test cannot be refuted.

6.1   Batch Size Analysis

Table 2 shows the results of all analysed methods. For simplicity reasons, values in this table are averaged over all datasets since the main aim is to evaluate the overall behaviour of all analysed methods when changing the batch size from 100 to 5000. This table is subdivided into three parts: one for each performance metric, Accuracy, F-Score, and Kappa. Then, the rows represent a combination of batch size and a metric, while the columns represent the different concept drift methods. In this table, columns 2 to 6 present, respectively, the results of the following methods: CPSSDS; Hinkley; DyDaSL - FT; DyDaSL - N and DyDaSL - W. Finally, the highest result for each row is highlighted in bold and coloured in purple, while the yellow cells are used to highlight the DyDaSL results that outperformed both values of the state-of-art methods (CPSSDS and Hinkley).

The results in Table 2 point out that the proposed approaches outperformed the state-of-art results in all analysed scenarios for all three metrics. On the other hand, only the accuracy of DyDaSL - N using 100 instances batch is not numerically superior to both state-of-art methods. This approach achieved better results than Hinkley and worse than CPSSDS. Additionally, it can be observed

Table 2. Results of the analysed methods by batch size

Batch size	State-of-art		Standard	Proposals		
	CPSSDS	Hinkley	DyDaSL - FT	DyDaSL - N	DyDaSL - W	
Accuracy						
100	60.79	54.15	73.15	60.12	**73.68**	
250	68.15	63.57	75.13	71.79	**76.08**	
500	68.62	64.65	75.13	71.27	**75.61**	
750	69.25	68.04	74.85	72.10	**77.07**	
1000	68.21	68.81	**76.17**	72.87	75.33	
2500	71.46	68.22	75.87	72.23	**77.52**	
5000	70.24	66.95	**76.19**	71.80	76.12	
Average	68.10	64.91	75.21	70.31	**75.92**	
F-Score						
100	49.06	44.01	60.06	50.48	**60.28**	
250	54.87	51.18	60.14	58.61	**61.82**	
500	53.64	51.01	59.33	57.10	**62.57**	
750	55.67	54.60	59.90	57.71	**64.04**	
1000	52.86	55.11	61.36	58.12	**61.62**	
2500	55.66	52.68	60.05	57.00	**63.10**	
5000	52.25	52.25	59.76	56.51	**61.30**	
Average	53.43	51.55	60.09	56.50	**62.10**	
Kappa						
100	8.69	5.76	**32.44**	25.38	25.91	
250	28.61	22.82	36.52	35.01	**37.18**	
500	28.12	23.19	38.76	35.29	**40.48**	
750	29.00	27.05	38.91	35.11	**43.97**	
1000	26.94	29.78	**44.13**	37.88	40.70	
2500	37.65	31.64	44.40	40.85	**49.25**	
5000	31.53	33.29	46.00	41.96	**48.51**	
Average	27.22	24.79	40.17	35.93	**40.86**	

that the worst results occur in the "Forest Cover Type" dataset, particularly for semi-supervised drift detection since this dataset contains a high number of features and classes. Furthermore, this dataset proved to be quite challenging in the semi-supervised context with 10% of labelled instances. Despite that, the DyDaSL - N method achieved higher average results when compared to CPSSDS.

Among all analysed methods, the highest average results for each metric are obtained by DyDaSL - W, showing that the use of a flexible threshold to detect drift can lead to a performance improvement in a data stream task. In general, the average results of all versions of the DyDaSL outperformed the state-of-art methods in all analysed metrics, with the DyDaSL - W the DyDaSL - FT approaches delivering the best results.

The CD Diagrams presented in Fig. 2 demonstrate the superiority of DyDaSL approaches when compared to the state-of-art methods. The Kappa and F-Score diagrams show that DyDaSL - W is statistically better than all other methods and that the DyDaSL - N is statistically similar to DyDaSL - FT for both metrics. Besides that, the accuracy diagram (Fig. 2a) shows that the two best results, obtained by DyDaSL - W and DyDaSL - FT, are statistically superior to all other methods. These statistical results are expected since the numerical results indicate the superiority of the DyDaSL - W and DyDaSL - FT. Furthermore, the statistical similarity between DyDaSL - N and CPSSDS is also expected, whereas the numerical results of these approaches are very similar.

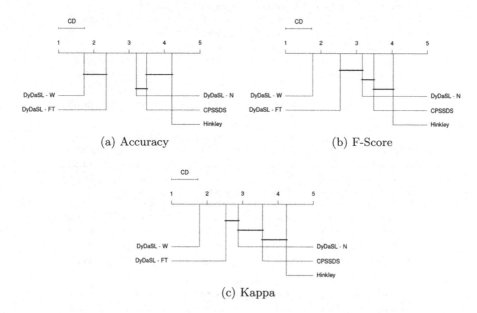

(a) Accuracy   (b) F-Score

(c) Kappa

Fig. 2. CD diagram for all analysed methods

6.2   The Proposed Methods versus DyDaSL - FT

In this section, two different scenarios are analysed, one considering 750 instances and the other considering 5000 instances per batch. These scenarios were selected to analyse the behaviour of all analysed methods in terms of classification effectiveness when a few labelled (75 of 750) instances and a large number of labelled (500 of 5000) instances are available to build the learning model. Table 3 presents the results for all datasets when 750 instances are processed for each batch. Considering the last three columns, DyDaSL - W achieves the best results in 17 cases, out of 27 (62.96%). Then, when analysing both proposed methods, they outperformed DyDaSL - FT results in 21 out of 27 cases (77.78%).

Figure 3 presents the critical difference diagrams for all evaluated metrics when 750 are processed for each batch. In all evaluated cases, DyDaSL - W is statistically superior to the original DyDaSL.

Table 4 shows the results for each dataset when 5000 instances are processed. Once again, DyDaSL - W achieves the best results in 21 cases out of 27 (77.78%).

Table 3. Results of the analysed methods with a 750 instances batch

Dataset	State-of-art		Standard	Proposals	
	CPSSDS	Hinkley	DyDaSL - FT	DyDaSL - N	DyDaSL - W
Accuracy					
Adult	76.23	76.23	**76.47**	76.29	76.33
Airlines	55.44	55.46	58.60	58.09	**61.02**
Connect-4	**65.23**	**65.23**	**65.23**	**65.23**	64.38
Fars	74.22	68.81	73.02	73.46	**75.22**
ForestCover	56.19	50.60	**72.43**	48.75	67.52
GEARS2C2D	95.59	95.58	95.58	**95.67**	95.64
Poker	49.88	49.63	51.99	49.50	**63.09**
Shuttle	91.53	91.51	92.31	91.61	**99.46**
UG2C3D	58.91	59.34	87.99	90.32	**91.00**
Average	69.25	68.04	74.85	72.10	**77.07**
F-Score					
Adult	63.02	63.02	62.14	**63.05**	62.68
Airlines	51.35	51.74	54.07	54.00	**56.28**
Connect-4	43.16	43.16	43.16	43.16	**43.55**
Fars	52.10	48.27	47.71	49.38	**58.58**
ForestCover	41.69	37.94	**54.86**	35.58	50.84
GEARS2C2D	95.63	95.63	95.63	**95.70**	95.68
Poker	29.20	26.55	29.68	25.10	**37.91**
Shuttle	62.84	62.80	63.51	62.85	**79.58**
UG2C3D	62.01	62.28	88.33	90.55	**91.24**
Average	55.67	54.60	59.90	57.71	**64.04**
Kappa					
Adult	1.81	1.81	**3.70**	2.18	2.48
Airlines	0.19	0.03	1.10	3.79	**7.28**
Connect-4	0.00	0.00	0.00	0.00	**9.22**
Fars	64.73	57.44	62.58	63.36	**66.11**
ForestCover	9.94	2.06	**36.45**	0.04	33.02
GEARS2C2D	91.17	91.16	91.16	**91.33**	91.27
Poker	3.09	0.17	2.49	0.61	**5.94**
Shuttle	72.25	72.12	76.78	74.06	**98.44**
UG2C3D	17.84	18.70	75.95	80.64	**81.98**
Average	29.00	27.05	38.91	35.11	**43.97**

In contrast to the previous results, DyDaSL - W wins in three cases (all in the Gears dataset) when analysing both proposed methods; once again, their superiority against the DyDaSL - FT was presented in 24 cases, out of 27 (88.89%).

Figure 4 presents the results of the statistical test using batches with 5000 instances. The accuracy CD (Fig. 4a) shows statistically similarity between DyDaSL - FT and DyDaSL - W. In the other evaluated metrics (Figs. 4b, 4c), DyDaSL - W is statistically superior among other evaluated approaches.

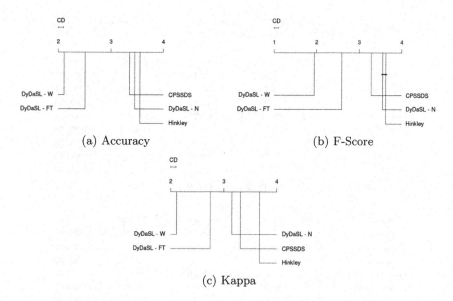

(a) Accuracy   (b) F-Score

(c) Kappa

Fig. 3. CD diagram for 750 instances batch

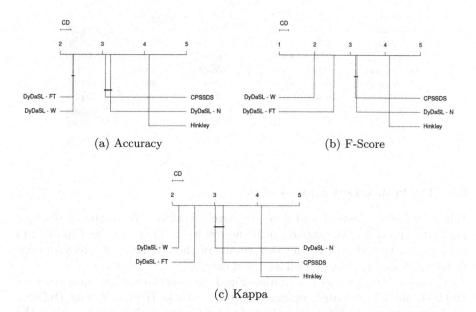

(a) Accuracy   (b) F-Score

(c) Kappa

Fig. 4. CD diagram for 5000 instances batch

Table 4. Results of the analysed methods with a 5000 instances batch

Dataset	State-of-art		Standard	Proposals	
	CPSSDS	Hinkley	DyDaSL - FT	DyDaSL - N	DyDaSL - W
Accuracy					
Adult	81.30	81.30	81.30	**82.19**	81.06
Airlines	59.18	55.97	62.15	59.65	**62.37**
Connect-4	65.76	64.70	65.66	64.04	**65.84**
Fars	71.76	72.97	74.65	73.59	**76.15**
ForestCover	58.04	30.06	**67.79**	49.64	60.41
GEARS2C2D	95.31	95.31	95.31	95.47	**95.76**
Poker	59.09	50.94	58.95	53.83	**62.58**
Shuttle	83.24	92.79	99.40	99.40	**99.52**
UG2C3D	58.52	58.52	80.50	68.40	**81.39**
Average	70.24	66.95	**76.19**	71.80	76.12
F-Score					
Adult	71.60	71.60	71.60	**73.03**	70.93
Airlines	56.60	55.68	58.56	58.21	**59.40**
Connect-4	40.63	39.92	42.26	39.30	**47.66**
Fars	45.98	48.60	48.29	49.17	**53.30**
ForestCover	38.89	23.04	**49.16**	34.73	43.64
GEARS2C2D	95.31	95.31	95.31	95.47	**95.77**
Poker	29.28	25.63	29.18	27.51	**32.46**
Shuttle	35.23	53.72	61.48	61.48	**66.24**
UG2C3D	56.75	56.75	82.01	69.65	**82.29**
Average	52.25	52.25	59.76	56.51	**61.30**
Kappa					
Adult	33.77	33.77	33.77	**38.94**	33.29
Airlines	5.86	2.36	10.15	7.31	**16.57**
Connect-4	4.77	7.42	2.44	5.48	**19.50**
Fars	60.25	62.34	64.81	63.23	**66.90**
ForestCover	19.62	5.21	**32.74**	20.30	26.22
GEARS2C2D	90.61	90.61	90.61	90.92	**91.53**
Poker	20.86	2.39	20.13	16.33	**21.11**
Shuttle	30.97	78.47	98.32	98.32	**98.65**
UG2C3D	17.06	17.06	60.99	36.80	**62.80**
Average	31.53	33.29	46.00	41.96	**48.51**

6.3   DyDaSL versus State-of-Art

When analysing Tables 3 and 4 to compare DyDaSL - W to the state-of-art methods, DyDaSL - W outperformed the results in 49 of 54 cases (90.74%) in both tables. In addition, the average results of the DyDaSL - W were superior to CPSSDS and Hinkley, for all analysed cases.

For the 750 instances scenario, Fig. 3 demonstrates the superiority of DyDaSL, for all evaluated metrics, in which both DyDaSL - W and DyDaSL - FT are superior in 100% of evaluated cases (6 wins/0 draws/0 losses) to the state-of-art methods, while the DyDaSL - N achieves superiority in 50% of the analysed cases (3 wins/1 draw/2 losses). When grouping proposals DyDaSL versions, the overall performance improved in 75% of the analysed cases (9 wins/1 draw/2 losses), from a statistical point of view.

For the 5000 instances scenario, Fig. 4 shows that proposals DyDaSL approaches are statistically superior to Hinkley. When compared to the CPSSDS approach, DyDaSL - N is similar, and DyDaSL - W is superior to this approach. Considering all CD diagrams of this figure, DyDaSL improved the drift detection performance in 75% of the analysed cases (9 wins/3 draws/0 losses).

7   Final Remarks

This paper presented an extension of the DyDaSL framework in which an improvement in the drift detection method is proposed, using a flexible threshold in the detection module. This improvement led to two DyDaSL new versions, named DyDaSL - N and DyDaSL - W.

An empirical analysis was conducted to assess the feasibility of the new DyDaSL versions in a non-stationary DS environment. Thus, a comparative analysis was designed to compare the proposed approaches to two state-of-art methods, CPSSDS and Hinkley. Additionally, this analysis was designed considering: nine datasets, seven batch sizes scenarios (100, 250, 500, 750, 1000, 2500, and 5000), and three performance metrics: Accuracy, F-Score and Kappa statistics.

When comparing only the proposed methods the DyDaSL - W provided the best results in 46 cases, out of 54 (85.18%). When comparing the best proposed version to the original one, in general, DyDaSL - W is superior to DyDaSL - FT in 42 out of 54 (77.78%) cases. Finally, when comparing the proposed method, to the other methods, in a scenario with a large batch (5000 instances), for instance, the proposed version outperformed both the original DyDaSL version (77.78% of the analysed cases) and the state-of-art methods (88.89%) of evaluated cases.

In the statistical analysis, when comparing only the best proposed DyDaSL version (DyDaSL - W) and the original DyDaSL version, DyDaSL - W delivered the best results in 92% of the analysed cases, from a statistical point of view. Based on these results, it can state that the use of a flexible threshold in the drift detection method led to an improvement in the DyDaSL performance.

In future work, we can investigate other percentages of initially labelled data or evaluate the performance of this framework using other ensemble settings. Additionally, it is possible to evaluate other updates to make the ensemble more dynamic regarding the number of base classifiers. Finally, a deeper analysis of these approaches could consider other datasets and more specific drift detection metrics to assess the effectiveness and robustness of our proposal.

Acknowledgment. This study was financed in part by the Coordenação de Aperfeiçoamento de Pessoal de Nível Superior - Brasil (CAPES) - Finance Code 001.

References

1. Chapelle, O., Schölkopf, B., Zien, A.: Semi-Supervised Learning. The MIT Press, Cambridge (2006)

2. Gollapudi, S.: Pratical Machine Learning. Packt Publishing Ltd., Livery Place (2016)
3. Gama, J., Žliobaitė, I., Bifet, A., Pechenizkiy, M., Bouchachia, A.: A survey on concept drift adaptation. ACM Comput. Surv. **46**(4), 44:1–44:37 (2014)
4. Gorgônio, A.C., de P. Canuto, A.M., Vale, K.M.O., Gorgônio, F.L.: A semi-supervised based framework for data stream classification in non-stationary environments. In: International Joint Conference on Neural Networks (2020)
5. Vale, K.M.O., et al.: Automatic adjustment of confidence values in self-training semi-supervised method. In: International Joint Conference on Neural Networks (2018)
6. Zhou, Z.-H.: Ensemble Methods, 1st edn. Chapman & Hall/CRC, New York (2012)
7. Gharroudi, O.: Ensemble Multi-label Learning in Supervised and Semi-supervised Settings. Université de Lyon, Theses (2017)
8. Kuncheva, L.I., Rodriguez, J.J.: Classifier ensembles with a random linear oracle. IEEE Trans. Knowl. Data Eng. **19**(4), 500–508 (2007)
9. Khezri, S., Tanha, J., Ahmadi, A., Sharifi, A.: STDS: self-training data streams for mining limited labeled data in non-stationary environment. Appl. Intell. **50**, 1448–1467 (2020)
10. Bi, X., Zhang, C., Zhao, X., Li, D., Sun, Y., Ma, Y.: Codes: efficient incremental semi-supervised classification over drifting and evolving social streams. IEEE Access **8**, 14024–14035 (2020)
11. Zhang, S., Jung Huang, D.T., Dobbie, G., Koh, Y.S.: Sled: semi-supervised locally-weighted ensemble detector. In 2020 IEEE 36th International Conference on Data Engineering (ICDE), pp. 1838–1841. IEEE, Dallas, Texas (2020)
12. Tanha, J., Samadi, N., Abdi, Y., Razzaghi-Asl, N.: Cpssds: conformal prediction for semi-supervised classification on data streams. Inf. Sci. **584**, 212–234 (2022)
13. Sebastião, R., Fernandes, J.M.: Supporting the page-hinkley test with empirical mode decomposition for change detection. In: Kryszkiewicz, M., Appice, A., Ślęzak, D., Rybinski, H., Skowron, A., Raś, Z.W. (eds.) Foundations of Intelligent Systems, LNAI, vol. 10352, pp. 492–498. Springer, Cham (2017). https://doi.org/10.1007/978-3-319-60438-1_48
14. Page, E.S.: Continuous inspection schemes. Biometrika **41**(1/2), 100–115 (1954)
15. Bifet, A., Holmes, G., Kirkby, R., Pfahringer, B.: MOA: massive online analysis. J. Mach. Learn. Res. **11**, 1601–1604 (2010)
16. Dimitoglou, G., Adams, J.A., Jim, C.M.: Comparison of the c4.5 and a naive bayes classifier for the prediction of lung cancer survivability. J. Comput. **4**(8) (2012)
17. Breiman, L., Friedman, J., Olshen, R.A., Stone, C.J.: Classification and Regression Trees. CHAPMAN & HALL/CRC, New York (1984)
18. Cohen, W.W.: Fast effective rule induction. In: Machine Learning, ML95, pp. 115–123. Morgan Kaufmann Publishers, Tahoe City, California, USA, (1995)
19. Atkeson, C.G., Moore, A.W., Schaau, S.: Locally weighted learning. Artific. Intell. Rev. **11**(1), 11–73 (1997)
20. Hall, M., Frank, E., Holmes, G., Pfahringer, B., Reutemann, P., Witten, I.H.: The weka data mining software: an update. ACM SIGKDD Explor. Newsletter **11**(1):10–18 (2009)
21. Olson, R.S., La Cava, W., Orzechowski, P., Urbanowicz, R.J., Moore, J.H.: Pmlb: a large benchmark suite for machine learning evaluation and comparison. BioData Mining **10**(1), 36 (2017)

The Impact of State Representation on Approximate Q-Learning for a Selection Hyper-heuristic

Augusto Dantas[(✉)] and Aurora Pozo

Department of Computer Science, Federal University of Paraná (UFPR),
Curitiba, Brazil
{aldantas,aurora}@inf.ufpr.br

Abstract. The choice of heuristic operators is strongly related to the performance of a (meta-)heuristic algorithm. Hence, applying an automated selection approach can increase the robustness of an optimization system. In this work, we investigate the use of a reinforcement learning technique as the selection mechanism of a hyper-heuristic algorithm. Specifically, we use the approximate Q-learning using an Artificial Neural Network as function approximation. Moreover, we evaluate different sets of metrics for representing the state of the environment, which in this scenario, must indicate the search stage of the optimization algorithm. The experiments conducted on six combinatorial problem domains indicate that, with simple state measures (combining the last action vector and fitness improvement rate), our approach yields better results compared to a state-of-the-art Multi-Armed Bandit approach, which does not have state representation.

Keywords: Hyper-heuristic · Reinforcement learning · Combinatorial optimization

1 Introduction

Heuristic approaches are useful techniques for many complex real-world optimization problems, where exact methods are often unfeasible [2]. However, their performance are highly dependent on the configuration setting, which must be tuned for the problem-domain at hand [2]. Because of that, there are several adaptive search methodologies that aim to tackle this issue, which are normally termed in the literature as Hyper-Heuristics (HH) [3] or Adaptive Operator Selection (AOS) [5].

Reinforcement Learning (RL) [15] techniques have been widely investigated for HH and AOS applications. However, most of them are traditionally simple additive reinforcement strategies, such as Probability Matching (PM) and Adaptive Pursuit (AP) [5], that use the received feedback to update a probability vector. Others are based on selection rules, that takes into account the feedback

J. C. Xavier-Junior and R. A. Rios (Eds.): BRACIS 2022, LNAI 13653, pp. 45–60, 2022.
https://doi.org/10.1007/978-3-031-21686-2_4

and the frequency of appliances to deal with the exploration versus exploitation dilemma (e.g., Choice Function and Multi-Armed Bandit based strategies [5]).

Although those approaches presented good overall results, they lack a state representation according to the formal RL definition [15], in which an agent learns a policy (directly or not) by interacting with an environment based on the observed state and the received feedback (reward or penalty). Moreover, approaches that make use of a state representation for this selection task has been shown to be advantageous over stateless strategies [16].

There is a large literature on stateless Reinforcement Learning techniques for HH and AOS tasks. For instance, the winning algorithm from the CHeSC 2011 competition, AdapHH [9], dynamically updates the selection probabilities of the heuristics based on the number of best improvements found with respect to the execution time taken. Moreover, it also uses RL for controlling the parameter values (intensity of mutation and depth of search) of each low-level heuristics.

However, the literature is much more scarce when it comes to modeling the selection task as a RL environment with a state representation. Therefore, we must investigate which metrics can be used to correctly define a search state and how to properly reward and penalize a selected operator under this setting. Then, we can explore the existing RL techniques in order to learn a policy that is able to attain high quality solutions on different problem domains, preferably with minimal additional configuration.

There are a few works in the literature that have successfully defined a state representation for HH and AOS. [6] applies the Q-learning algorithm to update the state-action values, which are used to select the crossover operator of an evolutionary algorithm applied on the Quadratic Assignment Problem. Their state definition contains three information: a binary state indicating if a restart has been triggered; a binary state indicating if a new best solution has been found; and a discretized diversity level of the population (low, medium, or high). The experimental results demonstrated that the approach is competitive with classical credit assignment mechanisms, while being less sensitive to the number of operators.

Similarly, [4] applied Q-Learning to select crossover and mutation operators for the Traveling Salesman Problem. The state is defined by a 2-tuple containing the current generation and the fitness improvement of the current best individual over the initial best individual. These values are in the range [0, 1] and are then discretized into 4 intervals of equal size (bins of 0.25). Their approach outperformed a random selection, indicating that the agent was able to learn a working policy while solving the instances.

Meanwhile, [10] proposed a Simulated Annealing (SA) based HH that uses Q-Learning to select the moving operators. Each action is a triplet of three operators, and the state is the number of times that the previous actions succeed (times that an operator generated an accepted solution under the SA conditions). The approach was significantly superior to other versions of SA and two software packages, with respect to both the quality of the solution and the computation time.

One limitation of these works is the use of a discrete state space, which may limit the representation of the search stage [16]. However, when defining a continuous state representation, the classical Q-Learning becomes infeasible due to the high dimensional Q-Table. Therefore, a function approximation model is necessary to estimate the state-action values [15]. The work from [16] defined a continuous state space that includes landscape measures about the current population and some parent-oriented features. The landscape measures are some properties of the population, such as diversity and the proportions of improving, equal or worsening offsprings. The remaining features are measures correlating the offsprings to their respective parent solutions. Then, a Self-Organizing Neural Network is trained offline to select the crossover operator. The performance of this approach was competitive with other selection mechanisms (including a tabular Q-Learning) and even better on some instances, thus highlighting the advantages of using a continuous MDP-based selection strategy.

In [13], the authors use a Double Deep Q-Network to select mutation operators of a Differential Evolution algorithm applied on several CEC2005 benchmark functions. The target network, which is trained offline during the training phase, receives as input 99 continuous features, where 19 are related to the current population, such as diversity of fitness and chromosomes, and the remaining 80 characterize the performance of each operator, such as the improvement of offspring over the parent, best solution, and the median population fitness. This approach outperformed other non-adaptive algorithms and was competitive with state-of-the-art adaptive approaches.

In this work, we apply a standard online selection Hyper-Heuristic, where the operators are chosen based on an approximate Q-learning algorithm. More specifically, we evaluate a few set of measures for defining the state representation and use an Artificial Neural Network as function approximator. Moreover, we also evaluate a discrete state representation with both the approximate and the tabular versions of Q-learning. In contrast with the later related works, that also apply some form of approximate Q-Learning, our approach has an online configuration. In other words, we do not train the approximation model beforehand, instead, it learns while it is solving one instance. Additionally, our work evaluates the Hyper-Heuristic in a cross-domain setup, in which we select low-level heuristics for six combinatorial optimization problems.

In summary, our goal with this work is to answer the following research questions:

- **R1**: Among the features we evaluate, which is the best combination of features for representing the search state?
- **R2**: Can the approximate Q-learning using a discrete state beat the tabular Q-learning?
- **R3**: Is defining a state better than using a stateless reinforcement learning approach (e.g., the Fitness Rate-Rank Multi-Armed Bandit)?

The remainder of this paper is organized as follows: Sect. 2 introduces the main concepts of Reinforcement Learning, also giving a brief explanation of the Q-learning algorithm. In Sect. 3 we give the outline of the Hyper-Heuristic algorithm and framework that we use throughout the experiments, and in Sect. 4 we present our proposed approach. The experimental setup and results are given in Sect. 5 and Sect. 6, respectively. Finally, we draw some conclusions and indicate future works in Sect. 7.

2   Reinforcement Learning

Reinforcement Learning is a computational approach that learns a mapping from situations to actions by interacting with an environment [15]. It differs from other machine learning paradigms, such as supervised and unsupervised learning, in the sense that there is no pre-available dataset. Instead, the learning agent must be able to sense the state of its environment to some extent, and with that, it must decides which action to take based on its observation, with the goal to maximize a numerical reward signal [15].

The task of learning from interaction to achieve a goal can be framed as a Markov Decision Process (MDP). MDPs are classical formalization of sequential decision making, in which actions influence not only the immediate rewards, but also the subsequent situations [15]. Figure 1 illustrates this interaction, where at each time step t the agent receives some representation of the environment's state S_t and, based on that, selects an action A_t. Then, after acting, the agent moves to a new state S_{t+1} and receives a numerical reward R_{t+1}.

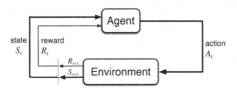

Fig. 1. Agent-environment interaction of a Markov Decision Process [15]

The definition of the state is a fundamental component of a Reinforcement Learning system. In general, the state can be any information that is available to the agent about its environment.

Then, for learning the mapping of states to actions, we use some reinforcement learning algorithm. Among them, Q-learning [17] is an off-policy Temporal-Difference control algorithm [15] that makes estimates on the Q-values, i.e., the estimate of state-action values. Hence, Q-learning gives quality estimates for choosing an action $a \in A$ from state $s \in S$ at time step t, and updates the estimates by

$$Q_{t+1}(s_t, a_t) \leftarrow Q(s_t, a_t) + \alpha \left[R_{t+1} + \gamma \max_a Q(s_{t+1}, a) - Q(s_t, a_t) \right] \tag{1}$$

This is called the tabular Q-learning, since all state-action pairs are stored in a table, which is only possible for small discrete problems. For large and continuous problems, we must rely on a funcion approximation model (called the Q-model). Here, we use as Q-model an Artificial Neural Network (ANN), in which the inputs are the current observed state representation, and the output layer yields the predicted Q-values for the current state-action pairs.

After performing an action, receiving the reward and observing the next state, the Q-model is updated by running one iteration of gradient descent on the ANN, with the target value T_{t+1} defined as

$$T_{t+1} = R_{t+1} + \gamma \max_a Q(s_{t+1}, a) \tag{2}$$

where s_{t+1} is the next state after performing the action, and $\max_a Q(s_{t+1}, a)$ is the highest Q-value of all possible actions from state s_{t+1}. The discount factor γ ($[0, 1]$) controls the influence of the future estimate rewards.

Finally, when dealing with trial-and-error methods, such as the Q-learning, we face the exploitation vs exploration dilemma. While learning, it is desirable to have a proper balance between choosing the actions with highest values known so far (exploitation) and trying out different actions that can hopefully lead to higher rewards (exploration) [15]. A common policy to handle this is the ϵ-greedy policy, that selects a random action with probability ϵ, and selects the action with the highest value with probability $1 - \epsilon$. Thus, ϵ is a parameter that controls the degree of exploration of the agent and is usually set to a small value [15].

3 Selection Hyper-heuristic

As a search methodology, selection HHs explore the search space of low-level heuristics (e.g., evolutionary operators) [3]. To avoid getting stuck into local optima solutions, good HHs must know which is the appropriate low-level heuristic to explore a different area of the search space at the time [3]. Algorithm 1 shows a standard selection Hyper-Heuristic algorithm. Iteratively, it selects and applies a low-level heuristic on the current solution and computes the reward. Then, the acceptance criteria decides if the new solution is accepted and, at last, the HH calls the update method of the corresponding selection model.

Algorithm 1: Selection Hyper-Heuristic

Input: A initial solution ϕ with size n
Output: The best found solution
repeat
    heuristic ← `SelectHeuristic()`
    ϕ' ← `ApplyHeuristic(`ϕ`, heuristic)`
    reward ← `GetReward(f(`ϕ`), f(`ϕ'`))`
    if `AcceptSolution(`ϕ'`)` **then**
       | $\phi \leftarrow \phi'$
    end
    `UpdateSelectionModel(reward)`
until *stopping criteria is not met*

The reward must be a metric that reflects the recent performance of the selected operator. Normally, fitness and diversity measures are used. Moreover, the acceptance criteria must also be defined.

In this work, we use the Hylex Framework, which is a tool that has become the standard benchmark for comparing cross-domain search methods [11]. It implements all of the problem-specific components, such as the representation, initialization, objective function and low-level heuristics.

The HyFlex provides 6 combinatorial optimization problem domains: One Dimensional Bin Packing (BP), Flow Shop (FS), Personal Scheduling (PS), Boolean Satisfiability (MAX-SAT), Traveling Salesman Problem (TSP), and Vehicle Routing Problem (VRP). For each domain, there is an available set of low-level heuritics that are classified into 4 types: mutational, ruin-and-recreate, local search, and crossover. The HyFlex was the benchmark of a competition, the Cross-Domain Hueritic Search Challenge (CHeSC 2011)[1], which attracted significant international attention.

One common approach for the selection mechanism in the HH, is to use any variation of a Multi-Armed Bandit framework, which is composed of N arms (e.g., operators) and a selection rule for selecting an arm at each step. The goal is to maximize the cumulative reward gathered over time [14]. Among several algorithms to solve the MAB, the Upper Confidence Bound (UCB) [1] is one of the most known in the literature, as it provides asymptotic optimality guarantees. The UCB chooses an action based on the following rule.

$$p_{i,t} + C\sqrt{\frac{2log(\sum_{j=1}^{N} n_{j,t})}{n_{i,t}}} \qquad (3)$$

where $n_{i,t}$ is the number of times the ith arm has been chosen, and $p_{i,t}$ is the average reward it has received up to time t. The scaling factor C gives a balance between selecting the best arm so far ($p_{i,t}$, i.e., exploitation) and those that have not been selected for a while (second term in the Eq. 3, i.e., exploration).

The FRRMAB [8] variation proposes the use of Fitness Improvement Rate (FIR) to measure the impact of the application of an operator i at time t (see Eq. 6). Moreover, the FFRMAB uses a sliding window of size W to store the indexes of past operators, and their respective FIRs. This sliding window is organized as a First-in First-out structure and reflects the state of the search process. Then, the empirical reward $Reward_i$ is computed as the sum of all FIR values for each operator i in the sliding window.

In order to give an appropriate credit value for an operator, the FRRMAB ranks all the computed $Reward_i$ in descending order. Then, it assigns a decay value to them based on their rank value $Rank_i$ and on a decaying factor $D \in [0, 1]$

$$Decay_i = D^{Rank_i} \times Reward_i \qquad (4)$$

The D factor controls the influence for the best operator (the smaller the value, the larger influence). Finally, the Fitness-Rate-Rank (FRR) of an operator i is given by

---

[1] http://www.asap.cs.nott.ac.uk/external/chesc2011/.

$$FRR_{i,t} = \frac{Decay_i}{\sum_{j=1}^{N} Decay_j} \qquad (5)$$

The $FRR_{i,t}$ value is set as the value estimate $p_{i,t}$ in the UCB equation (3). Also, the $n_{i,t}$ value considers only the amount of time that the operator appears in the current sliding window. This differs from the traditional MAB, where the value estimate $p_{i,t}$ is computed as the average of all rewards received so far.

4 Proposed Approach

The goal of our approach is to treat the task of selecting low-level heuristics as a formal Reinforcement Learning problem. Figure 2 gives an overview of the proposed system, in which the RL components interacts with the problem domain by selecting the operator and receiving the fitness of the resulting solution.

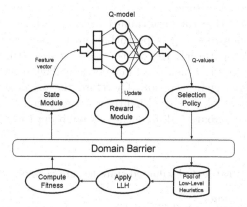

Fig. 2. Diagram of the proposed system

Hence, we must define three main aspects of the system: the state module, the reward module, and the agent module (Q-model and Selection Policy).

4.1 State Module

In this work, we define a simple set of features for representing the search state, and then we evaluate the RL-based Hyper-Heuristics strategy using different combinations of features.

The first state feature is the Fitness Improvement Rate (FIR), which measures the change of fitness when applying an operator i at time t as

$$FIR_{i,t} = \frac{pf_{i,t} - cf_{i,t}}{pf_{i,t}} \qquad (6)$$

where $pf_{i,t}$ is the fitness value of the original solution, and $cf_{i,t}$ is the fitness value of the offspring. Additionally, we also tested a discretized version of this feature, the Discrete Fitness Improvement (DFI), which is defined as

$$DFI_{i,t} = \begin{cases} -1, & \text{if } FIR_{i,t} < 0 \\ 0, & \text{if } FIR_{i,t} = 0 \\ 1, & \text{if } FIR_{i,t} > 0 \end{cases} \tag{7}$$

Another measure that can indicate useful information to the agent is the elapsed time of the search. In fact, there are several classic and new algorithms that uses this notion to control the search in some degree. The Simulated Annealing, for example, has a more explorative behavior at first, and gradually increases its exploitation based on the elapsed time [7]. Here, the elapsed time is measured as

$$\text{Elapsed Time} = \frac{\text{current time}}{\text{max time}} \tag{8}$$

where time can be either evaluation functions, iterations or CPU time, depending on the defined stopping criteria.

Finally, it is known that the performance of an operator is often related to the appliance of past operators (e.g. a perturbation operator can be advantageous after a local search operator). Hence, the third feature that we evaluated is the Last Operator Vector, which is a binary vector flagging which operator was applied in the previous agent iteration. Therefore, the size of this feature depends on the number of available low-level heuristic for a given domain.

In summary, we compare 5 different sets combining these three features, as displayed in Table 1.

Table 1. Evaluated state feature vectors

State Name	Features
S1	Last Operator Vector
S2	Fitness Improvement Rate, Last Operator Vector
S3	Elapsed Time, Last Operator Vector
S4	Fitness Improvement Rate, Elapsed Time, Last Operator Vector
S5	Discrete Fitness Improvement, Last Operator Vector

4.2   Reward Module

The reward must be a measure that gives the agent a notion of goodness or badness of its decisions. In an optimization context, a straightforward reward is the fitness improvement, which indicates how much the last operator could improve over the current or the best solution. Hence, we set as reward the Fitness Improvement Rate (6) between the current and the previous solutions. Although using only the fitness improvement as the measure of reward may penalize long-term strategies, the Q-learning algorithm introduces a long-term reward mechanism, while the ϵ-greedy policy introduces exploration [15].

4.3   Agent Module

This module defines the agent that will act as the selection rule within the Hyper-Heuristic. As discussed in Sect. 2, our approach uses an Artificial Network model estimating the quality of the state-action pairs (the Q-values). The weights of the network are updated using (2) according to the Q-Learning algorithm. Moreover, we also evaluated a standard tabular Q-learning version, using the discrete state representation (S5 in Table 1). Finally, as the selection policy, we use the ϵ-greedy policy.

5   Experimental Setup

Our first set of experiments aimed to answer the research question **R1** (see Sect. 1). Therefore, we evaluated the approximate Q-learning approach using each set of features from Table 1 for representing the state. Here, we assessed their overall performance in terms of final achieved solution across all domains.

Then, after identifying the best set of features for this task, we moved on to answer the research questions R2 and R3. Hence, we compared the approximate Q-learning approach against a tabular Q-learning and also against a state-of-the-art selection rule for Hyper-Heuristics (the FRRMAB).

For each approach, we executed the Hyper-Heuristic 31 times on every instance with different random seeds. We set the stopping criteria as 300 s of CPU running time on a Intel(R) Core(TM) i7-5930K CPU @ 3.50 GHz. These configurations were set following the CHeSC 2011 competition rules.

Moreover, Table 2 displays the parameters we used throughout the experiments. For the FRRMAB parameters, we followed the recommendations from [5]. Then, we kept the same window size (W) for DQN and set γ and ϵ win the intent to give high importance to long-term rewards and a low exploration degree, respectively. For the ANN (Q-model), we set a low learning rate with a state-of-the-art optimizer, and a network architecture with an appropriate size for its input. We used the Multi-Layer Perceptron Regressor implementation from the scikit-learn library [12].

Table 2. Parameters setting

	Parameter	Value
FRRMAB	C	8
	W	100
	D	1
Approximate Q-Learning	γ	0.9
	W	100
	ϵ	0.05
	ANN hidden layers	(30, 20)
	Learning rate	0.001
	Solver	Adam

6   Results and Discussion

First we evaluated the approximate Q-learning approach using each of the feature sets from Table 1. For this, we compared the mean performance obtained by each approach on all domains using the Friedman hypothesis test and a pairwise post-hoc test with the Bergmann correction. Figure 3 displays this comparison, in which the approaches are displayed according to their rank (the smaller the better), and the connected bold lines indicate the approaches that are statistically equivalent ($p < 0.05$).

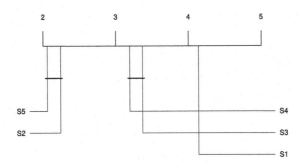

Fig. 3. Friedman ranking of the five compared states on all instances with post-hoc tests

As we can observe, the top 2 best states (S5 and S2) are the ones using the Fitness Improvement information, either continuous or discrete, and the Last Operator Vector. Nevertheless, S5 ranked better, even tough it is a discrete state, which could benefit from the possibility of using the tabular Q-learning instead (we investigate this next).

Meanwhile, states S3 and S4, that includes the elapsed time, were equivalent to each other, but worse than the best ones. This indicates that, in this scenario, the progress of the search in terms of time is not a very informative measure for the Reinforcement Learning agent.

Finally, the state set with the worst performance was S1, which uses only the Last Operator Vector. Since this feature was present in the best sets, we can conclude that it is more useful when coupled with additional information regarding the impact of the last operator. In summary, knowing which was the last action and if it impacted positively or negatively the search worked better on guiding the agent into learning a proper selection rule.

After assessing the best set for representing the state, we compared the approximate Q-learning against its tabular counterpart and also against the FRRMAB selection mechanism, aiming at answering research questions R2 and R3. Here, we compared the results of the three approaches on each instance individually, using the Kruskal-Wallis hypothesis test, followed by a pairwise Dunn's test, with the best ranked approach set as the control variable. Next we present these results for each problem domain.

6.1   Bin Packing

Table 3 reports the average and standard deviation of the best solution found by each selection strategy for the 31 runs. Bold values indicate that the corresponding approach achieved a better performance with statistical difference ($p < 0.05$), and gray background highlights all approaches that were statistically equivalent to the approach with the best rank on that instance.

As we can see, in the Bin Packing domain, our approximate approach outperformed the other two on 9 out of 10 instances.

Table 3. Performance comparison on Bin Packing

ID	Approximate	FRRMAB	Tabular
0	0.1096 (0.0083)	0.1866 (0.0093)	0.1132 (0.0157)
1	**0.0219 (0.0029)**	0.0535 (0.0033)	0.0505 (0.0029)
2	**0.1331 (0.009)**	0.181 (0.0085)	0.146 (0.0122)
3	**0.0619 (0.0032)**	0.1577 (0.0054)	0.1517 (0.0055)
4	**0.0186 (0.0024)**	0.0875 (0.0082)	0.0431 (0.014)
5	**0.0203 (0.0025)**	0.0894 (0.0071)	0.042 (0.0117)
6	**0.0242 (0.0006)**	0.0882 (0.0187)	0.0262 (0.0014)
7	**0.0255 (0.0007)**	0.0782 (0.0162)	0.0277 (0.0013)
8	**0.0087 (0.0019)**	0.0419 (0.0039)	0.0153 (0.005)
9	**0.0071 (0.0018)**	0.0425 (0.0029)	0.0127 (0.005)

6.2   Flow Shop

Similarly, on the Flow Shop we could achieve better results on 9 instances using the approximate Q-learning, as shown in Table 4

Table 4. Performance comparison on Flow Shop

ID	Approximate	FRRMAB	Tabular
0	**26750.0323 (25.0554)**	26894.5161 (44.7083)	26774.9355 (24.4658)
1	26923.5484 (32.4622)	27077.2581 (58.2187)	26947.2581 (36.9742)
2	**26418.0968 (24.4017)**	26629.6774 (52.228)	26456.0 (25.164)
3	**10962.8387 (5.7311)**	11046.0645 (20.8434)	10978.4516 (12.5154)
4	**10530.0968 (7.1091)**	10599.0968 (25.9744)	10548.6452 (8.2874)
5	**6372.2258 (8.1822)**	6481.5161 (25.7768)	6387.3871 (11.1693)
6	**6392.5806 (8.1509)**	6511.8387 (25.4395)	6408.1935 (9.3098)
7	**6315.7419 (7.0709)**	6427.5484 (22.4325)	6330.7097 (10.2774)
8	**6453.4839 (9.1541)**	6575.3871 (28.1318)	6473.6452 (12.9504)
9	**6365.9677 (9.2718)**	6476.7742 (23.5957)	6385.2581 (11.3364)

6.3   MAX-SAT

In the MAX-SAT domain, besides getting statically better results, there was a high difference in the scale of the obtained solution objective functions on several instances, as we can see in Table 5.

Table 5. Performance comparison on MAX-SAT

ID	Approximate	FRRMAB	Tabular
0	5.5484 (0.5587)	27.2903 (19.9874)	6.0 (0.8032)
1	**5.4194 (0.4935)**	30.0645 (21.8513)	6.3871 (1.0057)
2	**12.4516 (3.3873)**	104.3226 (99.0058)	25.9032 (12.6296)
3	**9.4839 (1.5213)**	49.1613 (44.7632)	13.4839 (2.4346)
4	**6.3548 (1.7879)**	53.9677 (47.0055)	11.0 (5.0289)
5	7.1935 (1.4902)	34.0645 (26.4928)	8.0968 (1.146)
6	**8.6129 (4.0614)**	71.4839 (110.7412)	22.3226 (3.7793)
7	**21.0968 (2.2195)**	119.0 (105.6186)	24.0 (2.6274)
8	210.4194 (1.0403)	263.9355 (63.5594)	211.7097 (1.887)
9	**26.2581 (2.7704)**	142.8387 (119.6026)	28.7742 (2.1659)

6.4   Personnel Scheduling

The Personnel Scheduling domain was the only one in which all three approaches were overall equivalent, as displayed in Table 6, thus indicating that there are still room for improvement regarding the state representation.

Table 6. Performance comparison on Personnel Scheduling

ID	Approximate	FRRMAB	Tabular
0	27.5161 (4.7374)	30.3548 (4.8894)	28.1613 (4.9324)
1	1390.2581 (516.5635)	1267.5161 (83.1038)	1789.3226 (2503.471)
2	28.2581 (19.5216)	27.3871 (5.1159)	25.9677 (4.028)
3	3340.7419 (27.8799)	3353.1935 (26.5396)	13227.7419 (18073.267)
4	818.871 (713.1831)	526.9677 (296.2138)	513.7419 (316.738)
5	2583.5806 (562.6071)	2467.2903 (222.0301)	2618.0645 (647.2554)
6	2813.7097 (2242.7199)	2379.1613 (102.3385)	2398.3871 (159.6132)
7	14077.4194 (13894.9304)	**9836.4194 (120.1678)**	62573.8387 (46163.5512)
8	3502.7742 (435.8312)	3419.8065 (119.6747)	5217.7742 (10068.5096)
9	**26.5806 (4.0781)**	30.2903 (4.7055)	39.7742 (38.447)

6.5   Traveling Salesman Problem

Again, the approximate Q-learning presented better performance in the Traveling Salesman Problem (Table 7). However, this time it was only statically better than the tabular counterpart on 5 out of 10 instances.

Table 7. Performance comparison on Traveling Salesman Problem

ID	Approximate	FRRMAB	Tabular
0	**9079.3939 (24.0413)**	9340.6525 (37.203)	9154.316 (25.4277)
1	23304981.3829 (1294778.9725)	25088764.1283 (144515.4733)	24126892.4793 (921926.7273)
2	**70693.6559 (1039.2317)**	79291.076 (1208.747)	73228.5042 (1476.128)
3	48646.3748 (113.3912)	53221.526 (649.4599)	48653.3713 (290.5162)
4	**6948.8743 (14.781)**	7129.7775 (22.2897)	6992.6067 (36.2744)
5	**58976.9946 (258.2069)**	62975.4419 (470.6619)	60130.984 (535.9722)
6	110809.0592 (994.5147)	135768.9563 (3802.5602)	110545.4157 (910.252)
6	61341.4821 (1949.9552)	61675.8326 (1474.6458)	61040.8453 (1715.0968)
8	714626.5632 (41446.5249)	792607.6786 (3129.0284)	719052.5763 (39618.5471)
9	**42834.1968 (124.6904)**	44936.0668 (344.2229)	43264.6181 (235.2842)

6.6   Vehicle Routing Problem

Finally, in the Vehicle Routing domain, our approach achieved the highest performance among the three methods, yielding statistically better results on 6 instances and being at least equivalent on other 3, as we can see in Table 8.

Table 8. Performance comparison on Vehicle Routing Problem

ID	Approximate	FRRMAB	Tabular
0	15191.3914 (387.2835)	19885.8136 (2575.4348)	15363.451 (326.6109)
1	6215.9084 (288.108)	8364.8595 (1773.5206)	6282.1684 (212.7539)
2	5284.6472 (37.7519)	7053.5233 (1344.2861)	**5244.3875 (35.8729)**
3	14452.5897 (188.9685)	17494.6077 (1987.017)	14468.0252 (350.7753)
4	**21625.9059 (247.9696)**	28615.1606 (3593.0337)	22723.4958 (1782.7663)
5	**184561.7378 (1791.5329)**	263334.8528 (54411.5145)	215440.5099 (17378.5754)
6	**72418.9902 (2014.5027)**	122517.6802 (15800.6161)	87703.4683 (10489.226)
7	**170200.6952 (2333.6021)**	215798.7689 (30716.4777)	194088.2141 (14213.8397)
8	**226102.5325 (8135.8159)**	339904.5134 (108719.3493)	305037.2796 (52841.9473)
9	**206931.991 (5423.4539)**	304608.2208 (73962.4556)	264701.1088 (48061.9512)

6.7   Overall Comparison

We have shown that the approximate Q-learning presented better results on 5 out of 6 problems, which highlights the robustness of the approach on a cross-domain environment. Figure 4 shows the Friedman statistical comparison when considering the results in all domains, where we can see that our approached outperformed the tabular Q-learning, which, in turn, also outperformed the FRRMAB approach.

Moreover, Fig. 5 displays the amount of instances that each approach was better than the others (black bars) and the amount in which it was at least among the best equivalent ones (gray bars).

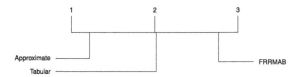

Fig. 4. Friedman ranking with post-hoc tests off the approaches on all instances

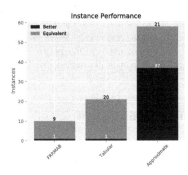

Fig. 5. Instance-wise performance comparison on all problem domains

7   Conclusion

This work investigated the use of straightforward measures for representing the state of an optimization search, in order to learn a policy to guide a selection Hyper-Heuristic algorithm. For this end, we employed an approximate Q-learning algorithm with an Artifical Neural Network for function approximation. Our goal was mainly to answer three questions: which is the best subset of measures?; is approximate Q-learning better than the traditional tabular version?; is this approach better than a state-of-the-art Multi-Armed Bandit? (stateless approach).

We conducted experiments on a cross-domain framework, considering six different combinatorial optimization problems. With the results we could conclude that informing the agent which was the last action and how it impacted the fitness is a feasible state representation for selecting low-level heuristics.

Moreover, we also demonstrated that, even with a discrete state, using the approximate Q-learning is more robust in this scenario than the standard Q-learning, where all state-action pairs are kept in a table. Finally, both Q-learning approaches achieved statistically better results than a stateless agent, namely, the Fitness-Rate-Rank Multi-Armed Bandit.

Future works include extracting other measures that can possibly represent the state of the search, such as metrics derived from Fitness Landscape Analysis. Besides, the selection Hyper-Heristic that we used was a standard single-solution based algorithm. We can instead define complex and populational high level strategies, which would allow us to get some additional measures such as population statistics. Moreover, with a properly defined Reinforcement Learning environment, we can evaluate and compare different and novel RL algorithms.

References

1. Auer, P., Cesa-Bianchi, N., Fischer, P.: Finite-time analysis of the multiarmed bandit problem. Mach. Learn. **47**(2), 235–256 (2002)
2. Blum, C., Puchinger, J., Raidl, G.R., Roli, A.: Hybrid metaheuristics in combinatorial optimization: a survey. Appl. Soft Comput. **11**(6), 4135–4151 (2011)
3. Burke, E.K., Hyde, M., Kendall, G., Ochoa, G., Özcan, E., Woodward, J.R.: A classification of hyper-heuristic approaches. In: Gendreau, M., Potvin, JY. (eds.) Handbook of Metaheuristics. ISOR, vol. 146, pp. 449–468. Springer, Boston (2010). https://doi.org/10.1007/978-1-4419-1665-5_15
4. Buzdalova, A., Kononov, V., Buzdalov, M.: Selecting evolutionary operators using reinforcement learning: initial explorations. In: Proceedings of the Companion Publication of the 2014 Annual Conference on Genetic and Evolutionary Computation, pp. 1033–1036 (2014)
5. Fialho, Á.: Adaptive Operator Selection for Optimization. Université Paris Sud - Paris XI (Dec, Theses (2010)
6. Handoko, S.D., Nguyen, D.T., Yuan, Z., Lau, H.C.: Reinforcement learning for adaptive operator selection in memetic search applied to quadratic assignment problem. In: Proceedings of the Companion Publication of the 2014 Annual Conference on Genetic and Evolutionary Computation, pp. 193–194. GECCO Comp 2014, Association for Computing Machinery, New York, NY, USA (2014)
7. Kirkpatrick, S., Gelatt, C.D., Jr., Vecchi, M.P.: Optimization by simulated annealing. Science **220**(4598), 671–680 (1983)
8. Li, K., Fialho, Á., Kwong, S., Zhang, Q.: Adaptive operator selection with bandits for a multiobjective evolutionary algorithm based on decomposition. IEEE Trans. Evolution. Comput. **18**(1), 114–130 (2014)
9. Mısır, M., Verbeeck, K., De Causmaecker, P., Berghe, G.V.: An intelligent hyper-heuristic framework for CHeSC 2011. In: Hamadi, Y., Schoenauer, M. (eds.) LION 2012. LNCS, pp. 461–466. Springer, Heidelberg (2012). https://doi.org/10.1007/978-3-642-34413-8_45
10. Mosadegh, H., Ghomi, S.F., Süer, G.A.: Stochastic mixed-model assembly line sequencing problem: mathematical modeling and q-learning based simulated annealing hyper-heuristics. Eur. J. Oper. Res. **282**(2), 530–544 (2020)
11. Ochoa, G., et al.: HyFlex: a benchmark framework for cross-domain heuristic search. In: Hao, J.K., Middendorf, M. (eds.) European Conference on Evolutionary Computation in Combinatorial Optimisation(EvoCOP 2012), LNCS, vol. 7245, pp. 136–147. Springer, Heidelberg (2012). https://doi.org/10.1007/978-3-642-29124-1_12
12. Pedregosa, F., et al.: Scikit-learn: machine learning in Python. J. Mach. Learn. Res. **12**, 2825–2830 (2011)
13. Sharma, M., Komninos, A., López-Ibáñez, M., Kazakov, D.: Deep reinforcement learning based parameter control in differential evolution. In: Proceedings of the Genetic and Evolutionary Computation Conference, pp. 709–717 (2019)
14. Soria-Alcaraz, J.A., Ochoa, G., Sotelo-Figeroa, M.A., Burke, E.K.: A methodology for determining an effective subset of heuristics in selection hyper-heuristics. Eur. J. Oper. Res. **260**(3), 972–983 (2017)

15. Sutton, R.S., Barto, A.G.: Reinforcement Learning, Second Edition: An Introduction. MIT Press (2018)
16. Teng, T.H., Handoko, S.D., Lau, H.C.: Self-organizing neural network for adaptive operator selection in evolutionary search. In: Festa, P., Sellmann, M., Vanschoren, J. (eds.) Learning and Intelligent Optimization. LNTCS, vol 10079, pp. 187–202. LNCS, Springer, Cham (2016). https://doi.org/10.1007/978-3-319-50349-3_13
17. Watkins, C.J.C.H., Dayan, P.: Q-learning. Mach. Learn. 8(3), 279–292 (1992)

A Network-Based Visual Analytics Approach for Performance Evaluation of Swarms of Robots in the Surveillance Task

Claudio D. G. Linhares[1]([✉]), Claudiney R. Tinoco[2], Jean R. Ponciano[3], Gina M. B. Oliveira[2], and Bruno A. N. Travençolo[2]

[1] Institute of Mathematics and Computer Sciences, University of São Paulo, São Carlos, Brazil
claudiodgl@usp.br

[2] School of Computer Science, Federal University of Uberlândia, Uberlândia, Brazil
{claudineyrt,gina}@ufu.br, travencolo@gmail.com

[3] School of Applied Mathematics, Getulio Vargas Foundation, Rio de Janeiro, Brazil
jean.ponciano@fgv.br

Abstract. Effectiveness in swarm robotics relies on aspects such as coordination and collective knowledge about the environment. By considering the evolution of intra-swarm communications over time as a temporal network, different strategies can be used in the data analysis. Information visualisation techniques are useful in this context because they can enhance the analysis of individual and global performances by including the user in the data exploration. This work proposes a visual analytics approach that considers a new matrix-based layout and other well-established ones to assess the swarm's efficiency. To analyse this approach, we also propose a temporal network dataset that models the evolution of the communications of a swarm of robots in the surveillance task, including eventual failures. We performed visual analyses in this network and demonstrated that the proposed approach allows easy identification of patterns, trends, and anomalies related to communication and task evolution. As a consequence, the decision-making process and eventual adjustments become faster and more reliable.

Keywords: Swarm robotics · Information visualisation · Complex networks · Data analysis · Surveillance task · Evolutionary models

1 Introduction

The employment of swarm robotics in tasks such as surveillance, exploration and foraging, allows environments to be covered more efficiently without *a-priori* spatial knowledge, as the combination of efforts of each robot results in a complex global behaviour [6]. Although robots can perform their individual tasks without collective knowledge, effective communication provides, on a

J. C. Xavier-Junior and R. A. Rios (Eds.): BRACIS 2022, LNAI 13653, pp. 61–76, 2022.
https://doi.org/10.1007/978-3-031-21686-2_5

global basis, updated information on current conditions, improving the overall performance [4]. Failures related to communication (e.g., the ones caused by a robot's hardware failure or wireless signal loss [12]) can affect swarm efficiency and impair the progress of the investigated task.

Understanding the communication patterns of robots throughout time can play a key role in the evaluation of swarm robotics. These patterns can be modelled as a complex network, commonly used to model parts of a system through instances (nodes) and their connections (edges) [1]. A complex network that considers the information of *when* each connection occurs is denominated temporal network [11]. In our context, a network is composed of robots (nodes) that interact with each other according to their communication (edges) in specific time steps. The modelling of swarm robotics communication using temporal networks provides means to identify temporal communication patterns, anomalies and other behaviours (e.g., whether a robot stops communicating during a particular time interval or a high flow of information between two robots).

Statistical analyses of temporal networks represent a useful resource for identifying specific trends and patterns. Numeric outputs, however, may represent a "black-box" that impairs pattern comprehension [17,19]. In this sense, the use of Information Visualisation techniques includes the user in the data analysis process through graphical and interactive computational tools [31]. The visual analysis of swarm communication networks opens new ways to perform comparisons between different system configurations, identify (un)desired behaviours, assess robots' performance and, finally, support fast and reliable decision-making.

Existing visualisation layouts, such as node-link diagrams [2], *Massive Sequence View* [18], and *Temporal Activity Map* [18], can be used to analyse aspects related to swarm communication, for example, to observe whether a robot is communicating more frequently than the others and to identify who communicates with whom and when. Although such aspects, especially those involving communication failures, highly affect the efficiency of the task, none of these layouts allows us to fully assess the task execution progress of the robots.

In this paper, we employ visual analytics to evaluate swarm of robots performing the surveillance task. We propose a new layout designed for this task, but also show the usefulness of some of the aforementioned layouts in this context. Our main contributions can be summarised as: (i) a matrix-based layout designed to analyse the surveillance performance of swarm of robots; (ii) a new dataset (temporal network) that models the evolution of intra-swarm communications during a surveillance task execution. It is composed of a task evaluation metric and contains several simulated communication failures over time; and, (iii) a visual analytics approach that combines our proposed layout and well-established ones to allow easy identification of patterns, trends, and anomalies related to communication and task evolution.

2   Related Work

This section introduces relevant concepts and a review of the related literature. It presents strategies for visual analysis of temporal networks and shows how they can be used in our context. It also describes the swarm coordination model used as a basis for the temporal network generation.

2.1 Visualisation in Robotics

Different visualisation techniques are used in the swarm robotics literature [14]. Each technique has its purposes, advantages and disadvantages. Among the most common ones, are the statistical charts [8,30] (e.g., lines, bars, regression, and box-plots charts), which, in most cases, show final outcomes. These charts summarise comparative results (e.g., mean and standard deviation) of mass tests with different configurations of the model being analysed. Network modelling and visualisation are also used in swarm robotics [13,22]. Networks provide a mathematical formalism and abstraction for the problem under study.

In other studies, snapshots are used as a visualisation technique [5,28,32]. Snapshots illustrate the configuration and disposition of robots in an environment at a specific time step. Informally, they can be defined as a photo of the environment. For instance, Masar et al. [20] used snapshots to visualise swarms of robots performing the surveillance and exploration tasks. Although the use of snapshots in modelling is remarkable, they have limited information: they show only one time-step and the notion of temporality is very limited.

Another important technique is the heatmap [24,29]. It graphically describes an environment through heat signatures, and can be used to represent the concentration of chemical substances and spatial coverage [25]. Besides, due to their natural compatibility, heatmaps have a strong relationship with some bio-inspired techniques. Although heatmaps are important for parameter calibration, they do not allow the visualisation of the evolution of the task.

Trail charts illustrate the paths taken by the robots during the execution of a task [15,21,23]. To produce these charts, the coordinates of the robots are stored between pre-specified time intervals, or the robots are equipped with some mechanism that allows them to mark the floor of the environment. However, with the evolution of the task, these charts become difficult to interpret. This is because the trails can overlap, and in tasks such as surveillance, which in turn must be performed cyclically, the information becomes obscured.

In the work of Calvo et al. [7], a visualization technique was applied to identify each robot's positioning throughout its evolution. The objective was to verify, in the surveillance task, the frequency and efficiency that the rooms in an environment were visited. In a two-dimensional chart, the x-axis represented the time evolution and the y-axis the combination of each robot with each room that composes the environment. The authors used three robots and a 6-room environment in the experiments, resulting in eighteen possibilities on the vertical axis. Thus, due to its combinatorial search-space, this technique becomes unfeasible to represent swarms of robots.

2.2   Visualisation of Temporal Networks

Visual analysis of temporal networks comprehends an effective resource for fast identification of patterns, trends, anomalies, and other properties existent in real-world temporal data [17]. Although the tabular description (Fig. 1a) is convenient for statistical analysis and data storage, qualitative and visual analysis can be performed by employing different layouts [18]. Two of them, namely *structural* and *Temporal Activity Map (TAM)* layouts (Fig. 1b–c), are widely used in the analysis of network evolution [17, 18].

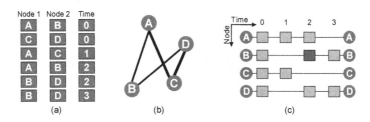

(a)   (b)   (c)

Fig. 1. Temporal network visualisation: (a) Tabular data; (b) Structural Layout (also known as node-link diagram); and (c) Temporal Activity Map (TAM).

In the *structural layout* (also called node-link diagram), nodes are spatially placed with edges (straight lines) linking them (Fig. 1b). It allows analyses that vary from global perspectives (identification of patterns considering the entire/aggregated network) to local ones (analysis of particular time steps—possible when adopting animation to represent time information [9]). For better layout exploration, different node positioning methods are commonly used (e.g., force-based, hierarchical, and circular algorithms [2]).

The *Temporal Activity Map (TAM)* is a timeline-based layout that uses the vertical and horizontal axes to represent nodes and time steps (Fig. 1c). TAM's main objective is to highlight node activity. For this purpose, it omits all edges (fewer visual elements decrease visual clutter) and adopts squares instead of circles to represent nodes, which leads to a better sense of continuity [31]. The node activity over time is highlighted by a colour scale that represents the level of connectivity of each node [18], which can be measured by a network centrality, for example the node degree adopted in the example of Fig. 1(c). Node ordering and edge sampling strategies, such as [17, 18], can be applied to TAM as well.

2.3   *PheroCom* Model

The PheroCom [27] is a model to coordinate a swarm of robots, mainly in tasks like surveillance, foraging, and exploration. Its movement mechanism is based on three bio-inspired strategies: Inverted Ant System (IAS) [7], Cellular Automata (CA) [16] and the Vibroacoustic Based Indirect Transmission (ViBIT) [27].

Briefly, the CA are applied in the discretization of the environment in two grids of identical squared cells (Fig. 2), that are maintained in the internal memory of each robot of the swarm. The first grid (Fig. 2a) is used to represent physical objects that are inserted into the environment and the second one (Fig. 2b) to simulate the dynamics of the pheromone deposited by the robots (feature inherited from the IAS). Whereas the physical grid possesses discrete states {*Robot, Obstacle, Free*}, the pheromone grid has continuous states, which are defined by values between $[0.0, 1.0]$.

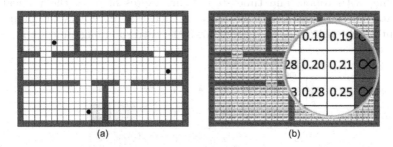

(a)   (b)

Fig. 2. Cellular Automata grids in an environment with 6 rooms and size (20×30) cells in: (a) Physical grid; (b) Pheromone grid. Black dots in (a) refer to robots, white squares are free cells, and grey squares are obstacles [Extracted from [27]]. (Color figure online)

Considering the information provided by the CA, the robots use it to coordinate themselves throughout the environment. Since the IAS is applied in the decision-making process, the pheromone information represents the probability of the robot moving to a cell in a determined time step. This pheromone is repulsive, so the robots tend to spread rather than stay close to each other [26]. In turn, applying the ViBIT protocol, the robots can share information regarding the pheromone, characterising the symbiosis of the model.

Tinoco and Oliveira [27] used an evaluation methodology based on task points to validate the PheroCom model. Task points can be formally defined as follows:

Definition 1 (Task-Point). *Let E be an environment composed of 'm' rooms and S a swarm composed of 'n' robots. A room i belonging to the environment E is described as $\{r_i \mid (i \leq m)$ and $(i \in \mathbb{N}^*)\}$. Similarly, a robot i belonging to the swarm S is described as $\{s_i \mid (i \leq n)$ and $(i \in \mathbb{N}^*)\}$. Therefore, a task point is reached iff every room $r_i \in E$ receives a visit from at least one robot $s_i \in S$.*

It is noteworthy that, when a task-point is reached, the count of visited rooms is restarted to start the counting of a new task-point. Besides, in the time step subsequent to the reset of the counting, all rooms that have the presence of robots are considered visited in the current task-point. Accordingly, it is an optimisation problem where the objective is to visit all areas of a given environment, at the same time that the intervals between task points are minimised.

3   Visualisation Proposal

Network visualisation is commonly used to understand and explore complex systems behaviours, such as existing patterns and anomalies. In our context, it has already been used to analyse tasks in swarm robotics [13]. In this sense, we propose a matrix-based visualisation divided into three parts: (i) *Visited Rooms*; (ii) *Unique Rooms*; and (iii) *Repeated Rooms*. Figure 3 illustrates the proposed layout considering one task point (visiting cycle) and a hypothetical environment with 22 rooms and four robots. Fig. 3(a) shows the *Visited Rooms* for each robot over time, with robots and time steps being represented by the vertical and horizontal axes, respectively. The value of each cell represents the number of different rooms visited by such robot until the corresponding time step.

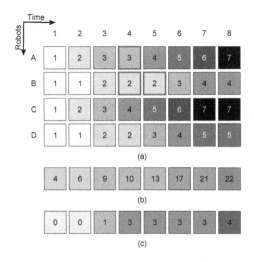

Fig. 3. Proposed layout of visited rooms evolution. (a) Visited Rooms for each robot over time, with robots and time steps being represented by the vertical and horizontal axes, respectively; (b) Unique Rooms, i.e., the number of different rooms visited so far; and (c) Repeated Rooms, i.e., the number of rooms visited more than once so far.

A cell with red borders indicates that the robot is visiting a room already visited by itself, i.e., the robot left a room at a previous time step and entered it again. Thus, it shows robots that are not progressing in the task. The red border only disappears when the robot visits a room not previously visited by itself (it can be hidden via user interaction). In Fig. 3(b–c), two global values are considered: how many different rooms were visited so far (*Unique Rooms* – Fig. 3b); and how many rooms were visited more than once so far (*Repeated Rooms* – Fig. 3c). When a robot visits a room not previously visited by itself but that was already visited by other robots, we increment the number of visited rooms for this robot (Fig. 3a) and the overall number of repeated rooms (Fig. 3c). Stronger and darker colours indicate higher numbers of visited rooms.

In the example of Fig. 3, each robot is positioned in a different room at time step 1, and so the number of visited rooms is one for each one of them (e.g., $VisitedRooms(A, 1) = 1$). Therefore, there are four unique rooms visited and no repeated rooms at time step 1. At time step 2, robots A and C go to unvisited rooms while robots B and D do not leave their rooms $(VisitedRooms(B, 2) = VisitedRooms(B, 1)$ and $VisitedRooms(D, 2) = VisitedRooms(D, 1))$. Although all robots reach new rooms at time step 3, one of them has entered a room already visited by another robot (as indicated by the change in the number of Repeated Rooms at this time step $(RepeatedRooms(time\ step\ 3) = 1))$. Robots A and B leave their rooms and enter rooms already visited by themselves at time step 4 (cells with red borders in the layout). This pattern tends to be more frequent near the end of a visiting cycle, since the robots are searching for the remaining unvisited rooms but end up visiting repeated ones. At time step 8, robot A visits the remaining unvisited room and then the cycle (task point) of 22 visited rooms is finished. In this example, all 22 rooms were visited after 8 time steps (cycle/task point duration) and with 4 repeated visitations.

Since a robot may visit a room that is new to it but that was already visited by others, the number of individual visitations does not necessarily match the number of unique visited rooms. For instance, at time step 8 of Fig. 3, the four robots visited 23 different rooms $(7 + 4 + 7 + 5)$, but there is only 22 unique rooms (Fig. 3b). We show individual (Fig. 3a) and global performances (Fig. 3b–c) to allow analyses of the task-points under both perspectives.

At the end of the cycle, the number of unique rooms is equal to the number of rooms in the environment. The robot with the highest number of visited rooms is responsible for visiting more unique rooms in relation to the others. When a new cycle begins, all counters and colours reset. Ideally, cycles should present short duration, few repeated rooms, and uniform activity among the robots.

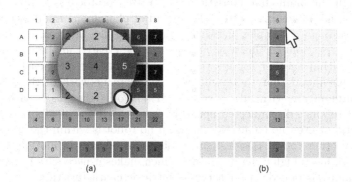

Fig. 4. Interactively, the user can freely explore the layout: (a) Zoom; (b) Selection.

Initially, all visiting cycles are exhibited in the layout. Interactive tools were included so the user can freely explore the layout, as established in the visualisation mantra *"overview first, zoom and filter, then details-on-demand"* [18]. By using pan and zoom, the user may analyse the performance of the robots in a global view or in a more detailed perspective (Fig. 4a). Moreover, the user is able to (i) follow a specific group of robots during the task (by selecting specific robots in the layout) and (ii) consider a particular time interval, even particular cycles (by selecting specific time steps, as illustrated in Fig. 4b). By comparing different cycles in the layout, one may identify similarities and discrepancies among them, which may lead to further investigations and help the user in decision-making and eventual adjustments.

4   Case Study

This section presents a visual analytics approach that considers well-established layouts and our proposed one to evaluate swarms of robots performing the surveillance task from global to local perspectives. We also describe our surveillance network dataset that is used in the experiments.

4.1   Surveillance Network

Taking into account that the PheroCom model [27] (see Sect. 2.3) is based on swarm intelligence, a complex global behaviour must arise through simple local interactions during the execution of the tasks. In this case, its global behaviour is a direct consequence of the information propagation from the pheromone grids of each robot with the whole swarm. Since the robots use the information present in local pheromone grids to decide their movements, when the pheromone information from other robots' grids is aggregated, decision-making becomes more efficient. In fact, in the aggregation, the local grids start to represent the swarm as a whole and not just the information of a single robot.

Although the main characteristic of the ViBIT protocol is indirect communication (based on gossiping [10]), the spread of pheromone information can be described through a complex network representing the information dissemination and aggregation. In this way, an edge can be created in the network when a robot is within another robot's transmission area. Thus, for each time step, there is a likelihood that there will be a complete network in the swarm (all robots communicate with all), a connected network (there is a path between all pairs of robots), disconnected networks forming groups of robots (connected components) or even no connection at all, if no robot is within the transmission area of any other robot. Besides, since robots are always in motion, this complex network has a high dynamism rate, which makes it necessary to perform its temporal analysis in order to observe its intrinsic characteristics.

Here, we have injected failures in the intra-swarm communications to substantiate our proposed visualisation approach. The failures were described as *robot failure*, *cluster failure* and *swarm failure*. Robot failure represents local failures of the robots' communication system, i.e., robots continue to perform the task, however, without propagating local information. In turn, cluster failure describes a subgroup of robots that are in a failure state, considering that these robots are in the same region of the environment. Finally, in swarm failure, all robots of the swarm are unable to communicate.

By applying the PheroCom model to the surveillance task, a network[1] was generated containing 12 nodes (robots) and 63, 743 edges distributed in 7 visiting cycles (task points). There is a total of 20 failures: 14 robot failures, 04 cluster failures and 02 swarm failures. The environment is composed of 40 rooms and dimensions equal to (80×120) cells. The robots were arranged in different rooms at the beginning of the simulation. Usually, a task-end is not defined in the surveillance, since it must be performed cyclically. However, to allow the analysis of the proposed layout, it was applied a limit of $T = 10,000$ time steps. Considering the configurations described, a transmission radius equal to thirteen cells $(r = 13)$ allows satisfactory outcomes [27]. As mentioned, the transmission radius represents the possibility of communication, i.e., if two or more robots are within this transmission radius of each other, data transmissions may occur (i.e., an edge might be created) depending on the occurrence of failures; otherwise, there is no data propagation (no edge).

4.2   Experiments

Different layouts provide different perspectives of analysis, so we used DyNetVis [18], a free interactive software for visualising temporal networks, to complement our analyses with the structural layout and TAM (see Sect. 2.2).

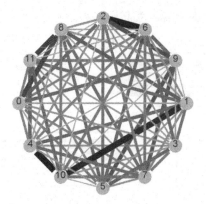

Fig. 5. Structural layout with circular node positioning. The more interactions between two nodes, the darker and thicker is the edge.

Figure 5 shows the structural layout with circular node positioning. The communications appear to involve all nodes (complete network) when considering the aggregated network, i.e., not taking into account the times in which each

---

[1] Freely available at www.github.com/claudiodgl/PheroComNetwork.

of them occurs. Such layout provides an overview of the network, facilitating the identification of global patterns involving the robots. In the figure, the more interactions between two nodes, the darker and thicker is the edge linking them. We see that the pairs of nodes (2,6), (4,10) and (1,10) are those with more connections between themselves over time. On the other hand, several other pairs of nodes have few connections between themselves, for example, the pairs (5,10) and (1,9). Since it is expected a uniform distribution involving robots' communication for better task execution, the perception of too many or too few interactions may require further investigations to optimise the system.

Figure 6(a) presents an overview of all cycles of the network using TAM. At least four failure events are visible (Fig. 6b–e). Figure 6(b) shows a *cluster failure*, i.e., a particular group of robots that lost communication during a time interval (perceived by white spaces over time). Figures 6(c–d) show *swarm failures*, i.e., time intervals in which there are no communications at all (recalling that the network contains exactly two swarm failures). Not least, Fig. 6(e) presents some blank horizontal lines, indicating that the corresponding robots are unable to communicate in the respective interval. Robots without communication for several consecutive time steps (such as the one indicated in Fig. 6e) are probably in a *robot failure* state. Without communicating with others, a robot with failure may impair the visiting task. The analysis of both communication behaviour and presence of failure events is important to support decision-making related to system optimisation. TAM thus represents a useful tool for this purpose.

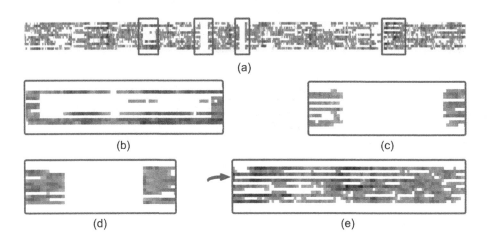

Fig. 6. Network cycles visualised with Temporal Activity Map (TAM). (a) Overview; (b) *cluster failure*; (c–d) *swarm failure*; (e) *robot failure*.

Figure 7 shows the duration of the cycles (number of time steps) when decomposing our proposed layout according to the beginning of each of them. Cycles 1, 3, 4, and 7 had a duration shorter than average. Since the goal is to perform the visiting task as fast as possible, the robots' performances in cycles 1 and 7 were the best ones. In turn, cycles 2 and 6 presented the worst performances and thus represent good candidates for further investigation.

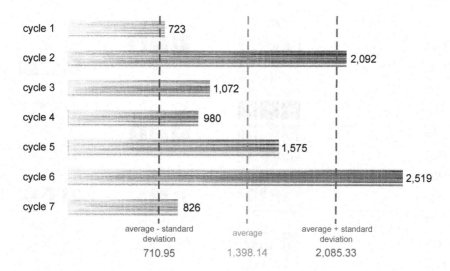

Fig. 7. Duration of the cycles exhibited when decomposing our proposed layout (with hidden red borders) according to the beginning of each of them.

To analyse the performance of the robots, Fig. 8 shows the proposed layout for the last three time steps of cycles 1, 2, 5, and 6. Recall that darker and stronger colours in the layout refer to more visited rooms, which is expected to occur more frequently in long cycles. During cycle 2 (the second longest duration), each robot visited 12.83 rooms on average and 343 repeated rooms were necessary. In contrast, during cycle 1 (the shortest one), each robot visited 5.83 rooms on average and 73 repeated rooms were needed to complete the task. Considering the number of robots in the network (12) and the number of rooms (40), a best-case scenario would require 3.33 visited rooms per robot on average, which supports cycle 1 good efficiency. During cycle 5, robots R01 and R05 visited 17 of the 40 rooms whilst the other robots visited 7.5 rooms on average each. Further investigation considering communication-based aspects (amount and evolution, up/out-dated information, failure events) could be used to analyse and fix such discrepancies in the individual performances.

To exemplify how the aforementioned communication-based aspects affect the swarm performance and, consequently, the task efficiency, Fig. 9 shows our proposed layout (Fig. 9a) along with the corresponding TAM layout (Fig. 9b) for the cycle 6 execution. A few time steps after the cycle begins, the 12 robots had reached 35 out of the 40 rooms (Fig. 9a). However, they got stuck between unique rooms 36 and 38 for approximately 75% of the time used to complete the cycle. During this time interval, at least one cluster failure, one swarm failure, and a few robot failures occurred in the network (Fig. 9b). Each failure event impairs robot communication and may lead to delays, outdated information and, consequently, lack of efficiency.

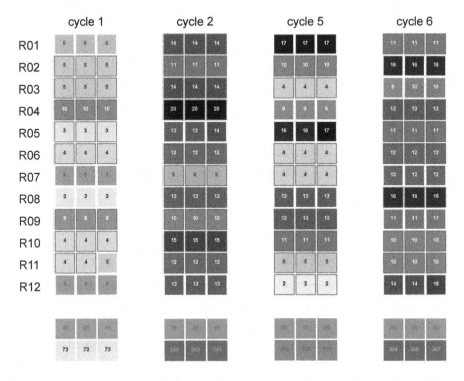

Fig. 8. Proposed layout showing the last three time steps of cycles 1, 2, 5, and 6. Darker and stronger colours refer to more visited rooms, which occur more in long cycles.

5   Limitations

Depending on the number of robots and rooms in the environment, there may be a lot of visual information in the layout. Matrix-based layouts represent a useful approach as they improve readability even when applied to large networks (high number of nodes and edges) when compared with other layouts [3]. In our case, besides the adoption of such representation, we also provide interactive tools, such as zoom in/out and selections, for better visual analysis.

Furthermore, methods for node reordering and edge sampling, among others, are commonly used in the structural and TAM layouts. Some of these methods are even visually scalable (e.g., *Community-based Node Ordering* [17]) and would allow joint analysis using these layouts and ours in large temporal networks. Besides, each time step in the layout refers to a pre-defined duration time interval. The adoption of different temporal resolution scales may affect the perception of patterns [18]. Changes in the resolution scale and the employment of

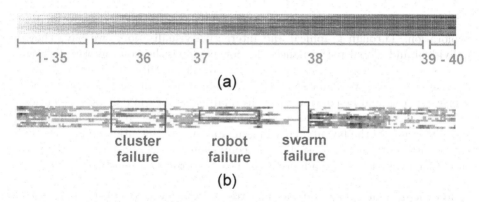

Fig. 9. Analysis of the longest cycle of the network (cycle 6). (a) Proposed visualisation with the number of *Unique Rooms* highlighted in green; (b) Corresponding TAM layout highlighting three failure events.

node reordering and edge sampling methods affect the layouts' quality, but a detailed analysis is outside the scope of this paper.

The parameters applied to the *PheroCom* model were based on the outcomes obtained in [27]. Among these parameters, it is worth mentioning the number of robots in the composition of the swarm, the communication radius, the pheromone evaporation rate, the size of the environment, and the number of rooms. Each configuration highly affects global efficiency, considering, for instance, a good spreading of the robots, avoiding agglomerations in specific areas. Finding the best configuration is not trivial and requests an initial exploratory analysis. The proposed visual analysis facilitates system performance assessment and leads to faster and more reliable adjustments.

Not least, *PheroCom* considers the IAS in its composition. The inverted characteristic of the pheromone, i.e., instead of being attractive, the pheromone generates a repulsive behaviour, causing robots to avoid areas with high pheromone concentrations. Thus, the robots can create blockages in some areas due to pheromone deposits, which, consequently, would compromise the efficiency in reaching task points. Nevertheless, this is not a limitation of our proposal but an intrinsic characteristic of the IAS.

6   Conclusion and Future Work

The analysis of different layouts allows understanding the network data from different perspectives. This paper presented a visual analytics approach that combines our proposed matrix-based layout with TAM and structural, which are well-established in network visualisation. With this approach, one can assess and further investigate (un)desired aspects that affect surveillance performance. We also proposed a temporal network dataset that models the evolution of swarm robotics communication and considers three types of eventual failures.

A major advantage of visual analytics is the insertion of the user in the data exploration. Through graphical and interactive visualisations, we have shown individual and global performances, as, for example, the non-uniform distribution of (i) robots' communication; (ii) cycle duration; (iii) individual and overall visitation rate. As demonstrated, communication failures represent relevant aspects that should be considered in the analysis as they impair task execution.

In future works, we intend to perform user evaluation to better understand the advantages of the layouts regarding mental map preservation and perceptual complexity. Future plans also include complementing the visual analysis with statistical evaluations and adapting the layout to run in a real-time fashion.

Acknowledgments. This work was supported by the following Brazilian support agencies: FAPESP [grants #2020/10049-0, #2020/07200-9 and #2016/17078-0], FAPEMIG [grant #11308], CNPq [grant #456855/2014-9], and CAPES [Finance Code 001].

References

1. Albert, R., Barabási, A.L.: Statistical mechanics of complex networks. Reviews of modern physics (2002)
2. Battista, G.D., Eades, P., Tamassia, R., Tollis, I.G.: Algorithms for drawing graphs: an annotated bibliography. Comput. Geom. **4**(5), 235–282 (1994)
3. Beck, F., Burch, M., Diehl, S., Weiskopf, D.: A taxonomy and survey of dynamic graph visualization. Comput. Graph. Forum **36**(1), 133–159 (2017)
4. Beni, G.: From swarm intelligence to swarm robotics. In: Şahin, E., Spears, W.M. (eds.) SR 2004. LNCS, vol. 3342, pp. 1–9. Springer, Heidelberg (2005). https://doi.org/10.1007/978-3-540-30552-1_1
5. Berman, S., Halász, Á., Hsieh, M.A., Kumar, V.: Optimized stochastic policies for task allocation in swarms of robots. IEEE Trans. Robot. **25**(4), 927–937 (2009)
6. Brambilla, M., Ferrante, E., Birattari, M., Dorigo, M.: Swarm robotics: a review from the swarm engineering perspective. Swarm Intell. **7**(1), 1–41 (2013)
7. Calvo, R., de Oliveira, J.R., Figueiredo, M., Romero, R.A.F.: Bio-inspired coordination of multiple robots systems and stigmergy mechanism to cooperative exploration and surveillance tasks. In: 5th International Conference on Cybernetics and Intelligent Systems (CIS). pp. 223–228. IEEE, Qingdao, China (2011)
8. Couceiro, M.S., Vargas, P.A., Rocha, R.P., Ferreira, N.M.: Benchmark of swarm robotics distributed techniques in a search task. Robot. Autonom. Syst. **62**(2), 200–213 (2014)
9. Crnovrsanin, T., Chu, J., Ma, K.L.: An incremental layout method for visualizing online dynamic graphs. In: Revised Selected Papers of the 23rd International Symposium on Graph Drawing and Network Visualization, vol. 9411. pp. 16–29. GD 2015, Springer, New York (2015)
10. Haas, Z.J., Halpern, J.Y., Li, L.: Gossip-based ad hoc routing. In: IEEE Information Communications Society, vol. 3, pp. 1707–1716. IEEE (2002)
11. Holme, P., Saramäki, J.: Temporal networks. Phys. Rep. **519**(3), 97–125 (2012)
12. Khadidos, A., Crowder, R.M., Chappell, P.H.: Exogenous fault detection and recovery for swarm robotics. IFAC-PapersOnLine **48**(3), 2405–2410 (2015)
13. Kolling, A., Nunnally, S., Lewis, M.: Towards human control of robot swarms. In: Proceedings of the Seventh Annual ACM/IEEE International Conference on Human-robot Interaction, pp. 89–96. ACM (2012)

14. Kolling, A., Walker, P., Chakraborty, N., Sycara, K., Lewis, M.: Human interaction with robot swarms: a survey. IEEE Trans. Hum. Mach. Syst. **46**(1), 9–26 (2015)
15. Kuyucu, T., Tanev, I., Shimohara, K.: Superadditive effect of multi-robot coordination in the exploration of unknown environments via stigmergy. Neurocomputing **148**, 83–90 (2015)
16. Lima, DA., Tinoco, C.R., Oliveira, G.M.B.: A cellular automata model with repulsive pheromone for swarm robotics in surveillance. In: El Yacoubi, S., Was, J., Bandini, S. (eds.) ACRI 2016. LNCS, vol. 9863, pp. 312–322. Springer, Cham (2016). https://doi.org/10.1007/978-3-319-44365-2_31
17. Linhares, C.D.G., Ponciano, J.R., Pereira, F.S.F., Travençolo, B.A.N., Paiva, J.G.S., Rocha, L.E.C.: A scalable node ordering strategy based on community structure for enhanced temporal network visualization. Comput. Graph. **84**, 185–198 (2019)
18. Linhares, C.D.G., Travençolo, B.A.N., Paiva, J.G.S., Rocha, L.E.C.: DyNetVis: a system for visualization of dynamic networks. In: Proceedings of the Symposium on Applied Computing. pp. 187–194. SAC 2017. ACM, Marrakech, Morocco (2017)
19. Linhares, C.D.G., Ponciano, J.R., Pereira, F.S.F., Rocha, L.E.C., Paiva, J.G.S., Travençolo, B.A.N.: Visual analysis for evaluation of community detection algorithms. Multimed. Tools. Appl. **79**, 17645–17667 (2020)
20. Masár, M.: A biologically inspired swarm robot coordination algorithm for exploration and surveillance. In: 2013 IEEE 17th International Conference on Intelligent Engineering Systems (INES), pp. 271–275. IEEE (2013)
21. Mintchev, S., Zappetti, D., Willemin, J., Floreano, D.: A soft robot for random exploration of terrestrial environments. In: International Conference on Robotics and Automation, pp. 7492–7497. IEEE (2018)
22. Muhammad, A., Jadbabaie, A.: Dynamic coverage verification in mobile sensor networks via switched higher order laplacians. In: Robotics: Science and Systems, p. 72 (2007)
23. Prorok, A., Hsieh, M.A., Kumar, V.: The impact of diversity on optimal control policies for heterogeneous robot swarms. IEEE Trans. Robot. **33**(2), 346–358 (2017)
24. Sousselier, T., Dreo, J., Sevaux, M.: Line formation algorithm in a swarm of reactive robots constrained by underwater environment. Expert Syst. Appl. **42**(12), 5117–5127 (2015)
25. Tinoco, C.R., Lima, D.A., Oliveira, G.M.B.: An improved model for swarm robotics in surveillance based on cellular automata and repulsive pheromone with discrete diffusion. Int. J. Parallel Emergent Distrib. Syst. 1–25 (2017)
26. Tinoco, C.R., Oliveira, G.M.B.: Pheromone interactions in a cellular automata-based model for surveillance robots. In: Mauri, G., El Yacoubi, S., Dennunzio, A., Nishinari, K., Manzoni, L. (eds.) ACRI 2018. LNCS, vol. 11115, pp. 154–165. Springer, Cham (2018). https://doi.org/10.1007/978-3-319-99813-8_14
27. Tinoco, C.R., Oliveira, G.M.B.: Pherocom: decentralised and asynchronous swarm robotics coordination based on virtual pheromone and vibroacoustic communication. arXiv preprint arXiv:2202.13456 (2022)
28. Valentini, G., Hamann, H., Dorigo, M.: Efficient decision-making in a self-organizing robot swarm: on the speed versus accuracy trade-off. In: Proceedings of the 2015 International Conference on Autonomous Agents and Multiagent Systems, pp. 1305–1314. International Foundation for Autonomous Agents and Multi-Agent Systems (2015)

29. Varley, J., Weisz, J., Weiss, J., Allen, P.: Generating multi-fingered robotic grasps via deep learning. In: 2015 IEEE/RSJ International Conference on Intelligent Robots and Systems (IROS), pp. 4415–4420. IEEE (2015)
30. Wang, Y., Liang, A., Guan, H.: Frontier-based multi-robot map exploration using particle swarm optimization. In: 2011 IEEE Symposium on Swarm Intelligence (SIS), pp. 1–6. IEEE (2011)
31. Ware, C.: Information Visualization, 3rd edn., p. 514. Morgan Kaufmann, Boston, Interactive Technologies (2013)
32. Zhang, G., Fricke, G.K., Garg, D.P.: Spill detection and perimeter surveillance via distributed swarming agents. IEEE/ASME Trans. Mechatron. **18**(1), 121–129 (2013)

Ulysses-RFSQ: A Novel Method to Improve Legal Information Retrieval Based on Relevance Feedback

Douglas Vitório[1]([✉])(iD), Ellen Souza[2,3](iD), Lucas Martins[3](iD),
Nádia F. F. da Silva[3,4](iD), André Carlos Ponce
de Leon Ferreira de Carvalho[3](iD), and Adriano L. I. Oliveira[1](iD)

[1] Centro de Informática, Universidade Federal de Pernambuco, Recife, PE, Brazil
{damsv,alio}@cin.ufpe.br
[2] Unidade Acadêmica de Serra Talhada, Universidade Federal Rural de Pernambuco,
Serra Talhada, PE, Brazil
ellen.polliana@ufrpe.br
[3] Instituto de Ciências Matemáticas e de Computação, Universidade de São Paulo,
São Carlos, SP, Brazil
lucasfmartins16@usp.br, nadia.felix@ufg.br, andre@icmc.usp.br
[4] Instituto de Informática, Universidade Federal de Goiás, Goiânia, GO, Brazil

Abstract. Obtaining relevant legal documents fast, from very large datasets, is essential for the proper functioning of justice and legislative institutions. Nevertheless, legacy systems currently used by these institutions in Brazil are usually outdated, requiring a large deal of manual work. Legal Information Retrieval focuses on building new methods to deal with the large amount of legal texts, allowing the retrieval of relevant information from them. Relevance Feedback, an important aspect of information retrieval systems, uses the information given by the user to improve the document retrieval for a specific request. However, expanding its use to other queries is a difficult task. A possible approach is to use Relevance Feedback information from past, similar queries. In this paper, we propose Ulysses-RFSQ, a method based on this approach which gives a bonus for the documents marked as *relevant* for similar queries, and, through this bonus, updates the ranking created by a relevance score based Information Retrieval algorithm, which measures the similarity between the query text and the documents to be retrieved. Due to the lack of available datasets containing relevance information for similar queries, we used a corpus of legislative requests from the Brazilian Chamber of Deputies, which are in most cases redundant, allowing the assessment of the proposed method. According to the experimental results, adding the Relevance Feedback bonus to the documents score improved the Recall@20 of a BM25 algorithm by almost 3% in the legal dataset used.

Keywords: Relevance feedback · Similar queries · Legal Information Retrieval · Legislative data

J. C. Xavier-Junior and R. A. Rios (Eds.): BRACIS 2022, LNAI 13653, pp. 77–91, 2022.
https://doi.org/10.1007/978-3-031-21686-2_6

1   Introduction

As consequence of the expanding use of Information Retrieval (IR) to fetch textual documents, there is a strong interest in its use in the legal domain. This is largely due to the high, and fast growing, number of legal texts being produced, resulting in a large workload for the professionals working with legal documents. This created a new IR subarea: Legal Information Retrieval (LIR), which includes tasks such as jurisprudence analysis, as well as to support the law-making process [18].

Within the scope of legislative activity, the need to adopt automated IR techniques is consequence of the increasing growth in the number of documents created by parliamentarians. This growth, together with the non-structured nature of these documents, makes their organization, access, and retrieval a challenging task [7]. As an example, in the Brazilian legislative process, before a Parliament member proposal becomes a bill that can be voted by the Chamber of Deputies, one of the departments of the House, called Legislative Consulting (Conle), must retrieve and analyze similar, previously submitted proposals. This is a very time-consuming process whose automation will enable Conle to deal with the large number of proposals submitted every year: since its founding, the Chamber has processed more than 144,000 bills [6], most of them redundant. Despite the high demand for this task, we found only one other study applying IR to legislative text in Portuguese [3].

One way to improve the document retrieval process is through the use of Relevance Feedback (RF), in which the relevance of the retrieved documents is evaluated by users [12]. RF uses this feedback iteratively to improve its results, usually by expanding the query or using the relevant and non-relevant documents information as a training set for a supervised Machine Learning algorithm. For such, it usually aims to improve only the retrieval for the current query, i.e., the feedback provided by the user will only be used in that session [33].

An alternative for RF to be used in order to make the retrieval model better in a way that impacts other searches is the storage and utilization of this feedback to improve IR for similar queries. Few studies, though, have been performed aiming to use past queries information for this purpose, namely improving Information Retrieval for new queries. This is due to the lack of available datasets containing relevance information for similar queries [12].

In this sense, this paper presents *Ulysses-RFSQ*, a novel IR method that considers the past queries RF information aiming to improve the retrieval for new queries. It is based on the BM25 algorithm [23], but can be easily used together with any IR algorithm that computes a relevance score for the documents. Experiments using legislative documents and requests from the Brazilian Chamber of Deputies were performed to compare Ulysses-RFSQ with the use of BM25 variants without the RF information. Legislative documents are used in the experiments because: 1) the importance of IR for the legal task; and 2) the aforementioned presence of redundancy in the parliamentarian requests, which makes the assessment of the proposed method possible.

This research was conducted in the context of the Ulysses project, an institutional set of artificial intelligence initiatives with the purpose of increasing transparency, improving the Chamber's relationship with citizens, and supporting the legislative activity with complex analysis [2].

The rest of this paper is organized as follows: Sect. 2 presents a literature review for Legal Information Retrieval and for the use of Relevance Feedback for similar queries; Sects. 3 and 4 detail the proposed method and the experimental setup used, respectively; the results are presented and discussed in Sect. 5; finally, Sect. 6 points out the main conclusions from this work.

2   Literature Review

2.1   Legal Information Retrieval

LIR is an important topic in the application of Artificial Intelligence in Law. The fast retrieval of relevant legal documents from a very large dataset is a strong requirement for the proper functioning of the juridical and legislative institutions. This requirement has become stronger with the information revolution and the Open Data movement, which increased the availability of legal data, particularly on the Internet. However, data accessibility did not keep up with this growth [32].

In the juridical scope, a court decision for a legal case should be based on jurisprudence: previous decisions for cases similar to the current case. For such, similar cases should be retrieved and made available for judges and lawyers. Nevertheless, the concept of similarity between the documents is not well defined, needing specialized opinions [4]. Courts commonly employ computational systems to retrieve similar cases. Nevertheless, most of these systems are usually inefficient legacy systems, based on Boolean logic [11], which uses keywords and operators to formulate the query, being complex and depending on the user's knowledge of the problem in order to choose the right keywords [24].

To analyze the retrieval of similar juridical documents, [11] used jurisprudential data from the Brazilian Superior Court of Justice (STJ). Their goal was to compare STJ's legacy system, which uses Boolean queries, with IR approaches based on document similarity, such as TF-IDF, BM25, and word embeddings language models. According to their experimental results, the IR techniques were able to overcome the legacy system both in performance and usability. Meanwhile, [21] used data from the Court of Justice of the State of Sergipe (TJSE) to evaluate the efficiency and impact of Stemming algorithms for IR from juridical texts. For such, the authors compared four radicalization algorithms (Porter, RSLP, RSLP-S, and UniNE) to evaluate: 1) their gain in dimensionality reduction; 2) their predictive performance regarding legal document retrieval; and pointed out that the use of radicalization deteriorated the BM25 performance.

For the legislative scenario, the situation is more complex. Legislative information produced in the course of the law-making process can largely impact and promote changes in the citizens' lives, providing very relevant information for the retrieval of similar legislation. For such, this information must also be properly stored, organized and made available, making its access easier for citizens and parliamentarians [5]. To efficiently access information regarding the legal text and the law-making process, and keep up with the growing need for information, new and more efficient legal document retrieval methods must be developed.

For the analysis of the retrieval of legislative texts, [3] proposed a new method to compute the similarity between documents in a non-supervised way, based on a synset, i.e, a set of synonyms. They used it to rank the legislative documents according to their relevance to a query, regardless of the language used, performing multilingual IR with data written in four languages, from the *JRC-Acquis* dataset[1]. [7], in their turn, investigated document retrieval from the Spanish Congress of Deputies, such as debate transcripts and law proposals, which are part of the *Parlamento2030* dataset. For such, they added a semantic relation measure to the Vector Space Model (VSM) [25], combining it with an ontology-based document representation model.

Finally, [9] investigated regulatory compliance in European Union and United Kingdom legislation using IR. They proposed a new approach: the retrieval of relevant documents according to similarity and their re-ranking using BERT (Bidirectional Encoder Representations from Transformers) [10] neural networks. IR algorithms, such as BM25, and different versions of BERT were evaluated.

2.2 Relevance Feedback and Its Use for Similar Queries

Relevance Feedback (RF) consists of using a user's annotation about the relevance of a document as a way to improve IR for a specific query. Usually, this information is used to select terms and expressions from the relevant documents in order to expand and create a new query [26]. The IR process becomes iterative, repeating the processing a few times aiming to achieve better results for a query. However, as aforementioned, this improvement is only for that specific query, not being used for future ones.

Another way to use the RF information takes place through Supervised Machine Learning, in which IR is understood as a two classes classification problem: *relevant* and *irrelevant*. A classifier is trained using the user's feedback as a training set, and then the classification algorithm is used to label new documents as *relevant* or *irrelevant* for that query. [20,22] used Relevance Feedback to interactively train a Support Vector Machine (SVM) classifier with the goal of improving the document retrieval performance. The authors used a simple IR technique based on VSM to select the first set of documents, which were manually labeled according to their relevance, and an SVM trained with this data was used to provide the final list of documents.

Nevertheless, due to the fact that RF is commonly used only to improve the retrieval for the current query, it is necessary to investigate new ways to use this information also for future queries. As the retrieval process for each query is

---

[1] https://joint-research-centre.ec.europa.eu/language-technology-resources/jrc-acquis_en.

unique, since the documents that are relevant to one query may not be relevant to any other, the alternative is to consider past queries that are similar to the current one. If there are very similar queries in the model usage history, the documents labeled as *relevant* to those queries probably also are relevant to the one currently being processed [14].

Although it is possible to note the use of this historical feedback for image retrieval for at least two decades [33], there are only a few studies in the literature that deal with this kind of use for textual document retrieval. According to [12], this is due to the fact that there are no available benchmark datasets with relevance information for similar queries, as popular IR evaluation collections, such as the ones from TREC[2] and CLEF[3], only provide sets of dissimilar queries. So, most of the techniques that perform IR considering historical data are in the Personalized Information Retrieval area, in which information about a user is used to improve the retrieved information for that user, within a session. This process is commonly used in search engines on the Internet [12].

Considering the use of old queries and their lists of relevant documents in order to improve document retrieval regardless of session, [16] proposed techniques for query expansion. [8], in their turn, computed the similarity between the current query and past ones for resource selection in a Distributed Information Retrieval system. They could estimate the usefulness of available information sources for a new query by combining the results of past queries. Aiming to re-rank search engines results, [19] also used past queries information, but without considering the Relevance Feedback. They built a similarity graph to obtain a set of features and train a Decision Tree classifier. The authors stated that the main challenge was to find the similar queries set.

[29] focused on using historical information aiming to build a new term-weighting method for document retrieval. They assumed that the role of a specific term in previous queries is important to the IR process, and used the ranking and similarity of relevant and irrelevant documents to compute the weight of a term. According to their results, the proposed method outperformed traditional IR techniques, such as TF-IDF and BM25, besides language-based methods. They reported an increase from 0.22% to 1.5% in Mean Average Precision (MAP) for the BM25-based methods in the TREC collections used.

Finally, due to the lack of benchmark datasets containing similar queries, the authors of [12–15] had to simulate an RF dataset to perform their experiments. Their objective was to store and use the list of documents considered relevant for old queries to improve the IR precision for new queries, evaluating four randomized algorithms to select and reuse documents from the most similar past query. When considering this set of relevant documents, the authors pointed out better results in comparison with traditional IR methods.

These last two studies were the only ones found in the literature using RF for similar queries to improve the IR method itself, without modifying the query or training a classifier. Thereby, they are the most similar works to the one

---

[2] https://trec.nist.gov.

[3] http://www.clef-initiative.eu.

presented in this paper. However, they differ from our method as they proposed new IR algorithms and/or a new term weighting scheme, while we presented a method to be used together with existing IR algorithms, such as BM25.

3   Ulysses-RFSQ: Improving LIR With Relevance Feedback

In this section, the proposed method, which is called *Ulysses-RFSQ* (**RFSQ**: **R**elevance **F**eedback for **S**imilar **Q**ueries), is described. It consists of rewarding the documents that were labeled as *relevant* for old queries similar to the current one. It is composed by four steps, which are detailed in the following subsections: 1) the preliminary ranking of documents by an IR algorithm; 2) the similar queries selection; 3) the ranking update; and 4) the Relevance Feedback acquisition.

Figure 1 presents the method's pipeline, in which the new parts added by Ulysses-RFSQ are in yellow, while the blue elements represent the standard IR process, and the RF process is represented by the pink ones. The numbers point out the four mentioned steps.

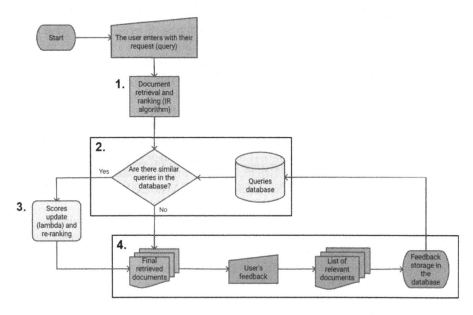

Fig. 1. Ulysses-RFSQ pipeline, pointing out the stages added by it (in yellow) to the standard IR process (in blue), as well as the Relevance Feedback stages (in pink). (Color figure online)

3.1   Step 1: Ranking The Documents

The first step consists in the simple scoring and ranking of documents through an IR algorithm: in this paper, we used BM25 as the base algorithm. However, as aforementioned, any algorithm that results in a score for the documents can be used. The choice for BM25 lies in the simplicity of this technique and in its usage for retrieving legal documents [9, 11, 21].

In the standard BM25 process, the algorithm computes a score for each document and, according to it, the documents are ranked from the highest relevant to the lowest. Then, a list containing the top n documents usually is presented to the user. Okapi BM25 [23] is the original BM25 algorithm, whose scoring function estimates the relevance of a document d to a query q based on the query terms appearing in d, regardless of their proximity within d.

As the scores computed by BM25 don't have an upper limit value, they need to be normalized, in order to be in the range of $[0, 1]$. Without this normalization, the bonus given for the documents might not be enough to have any effect in the posterior document re-ranking. For this, we used the Min-Max Normalization (Eq. 1).

$$normalized\_score(d, q) = \frac{score(d, q) - min(all\_scores)}{max(all\_scores) - min(all\_scores)}, \qquad (1)$$

in which d is a document, q is a query, and all_scores is the set of scores for all documents.

3.2   Step 2: Selecting Similar Queries

As Ulysses-RFSQ focuses on using the feedback given for past queries to improve the documents retrieval, it is necessary to maintain and store the old queries in a database. This database contains the query text and its RF information, with data about the documents marked as *relevant* for that query.

In the second step, the similarity between the current query and each query stored in the database is computed, then those queries that have a similarity greater than a cut-off threshold are selected. This threshold, which we called *cut*, is a parameter of the method that needs to be set and can vary between 0 and 1, according to the chosen similarity measure. If there are no queries that have a similarity greater than the threshold, the third step is skipped and the base IR algorithm list of ranked documents is presented to the user, without any modification.

For this paper, the cosine measure was chosen to compute the similarity between the queries, being this choice based on the work of [12], which also used this measure.

3.3   Step 3: Updating The Ranking

After selecting the similar queries, each document has its score updated by the addition of a bonus. We call this bonus *lambda* (λ) and it is computed by Eq. 2:

$$\lambda(d, q) = \ln\left(\sum_{q_j \in Q} (sim(q, q_j) \cdot normalized\_score(d, q_j)) + 1\right), \qquad (2)$$

in which q is the current query, Q is the set of similar old queries, and $sim(q, q_j)$ is the similarity between q and q_j. We decided to use the natural logarithm to keep the bonus value in a small range, and we added 1 to the sum to prevent the bonus from being negative.

As can be noticed by Eq. 2, the *lambda* bonus is computed based on two values: the similarity between the past query and the current one, and the IR algorithm (e.g. BM25) normalized score computed to that document according to the past query, which is also stored in the database.

Another characteristic of this equation is that if a document is present in more than one similar query, its bonus increases, as *lambda* considers the sum of all occurrences of that document in the similar queries set. On the other hand, if the document is not present in any similar query, its bonus is 0, i.e., its score will remain the same.

Thus, using this method, the final score for each document is computed by Eq. 3:

$$final\_score(d, q) = normalized\_score(d, q) + \lambda(d, q). \qquad (3)$$

With these final scores, the documents are re-ranked and this new ordered list is the result of the IR process.

3.4   Step 4: Acquiring The Relevance Feedback Information

Finally, the n documents with the highest final score are presented to the user. The user, then, provides a feedback, pointing out which documents they consider relevant for their request. This list of *relevant* documents is stored with their respective scores in the database, as well as the query. Table 1 presents a fictional example of how the RF data can be stored in the database: the *Relevance Feedback* column contains, for each document, its *id*, the BM25 score for that document, and the score after normalization.

Table 1. Fictional examples of data stored in the database.

ID	Timestamp	Query text	Relevance feedback
1	2022-01-01 10:01:35	"query 1"	(DOC025, 124.75, 1.00); (DOC011, 115.02, 0.85)
2	2022-01-01 13:19:21	"test 2"	(DOC112, 201.04, 0.97); (DOC114, 196.98, 0.93)
3	2022-01-01 21:56:02	"sample 3"	(DOC066, 110.42, 0.89)

These data will be used for future requests, so the IR system is always being improved by the feedback provided by users. It is worth mentioning that this method can be used in two ways: 1) without any previous stored queries, so, for the first use, the queries database is empty and Step 3 is skipped until this IR system is sufficiently used; or 2) using a previous feedback database, with which the *lambda* bonus might impact the performance from the start.

4 Experimental Setup

4.1 Corpora

To perform the experiments and evaluate our method, two different corpora from the Brazilian Chamber of Deputies were used: one containing bills, and other containing job requests (user's queries) and their RF information. The former is publicly available[4], however the latter contains private information and it is confidential.

Bills Corpus. The Bills corpus was used to perform the IR main process and has a total of 147,008 documents. These documents corresponds to the different types of Brazilian bills. The three most common types were selected for the experiments: Law Project (Projeto de Lei - PL), Complementary Law Project (Projeto de Lei Complementar - PLC), and Constitutional Amendment Proposal (Proposta de Emenda Constituicional - PEC), resulting in a final corpus with 48,555 bills. The corpus is composed by seven attributes describing the bills: *codProposicao, txtNome, txtEmenta, txtExplicacaoEmenta, txtIndexacao, imgArquivoTeorPDF, idTema*, which are, respectively, (1) the unique code of the proposition; (2) the name of the bill (e.g., PL 4395/1998); (3) the bill summary; (4) an explanation of the bill summary; (5) the keywords; (6) the bill itself; and (7) the bill theme. For the experiments, we used the attribute *imgArquivoTeorPDF*.

Job Requests + Relevance Feedback Corpus. Job requests (legislative consultations) are demands from parliamentarians to the Legislative Consulting (Conle) department of the Brazilian Chamber of Deputies. In order to create a new bill, or to verify if there is a similar bill being discussed in the House, a parliamentarian requests, through a query in the SisConle[5] system, a list of relevant documents: active or inactive bills and other job requests. Then, the Conle team searches for similar documents using the two legacy systems available: SisConle and SiLeg[6], within a task called *Preliminary Search*. However, a large amount of this work is manual, as the search engine uses Boolean logic and the consultants build the final list adding the documents manually. So, there is a strong need for a computational tool able to automate this process, reducing manual labor, cost, subjectivity, and human error.

The corpus used in this paper contains the result of preliminary searches, i.e., a set of job requests (queries) and the list of bills found for that query by the

---

[4] https://drive.camara.leg.br/s/c3p2nLgLRcMz6eX.
[5] Legislative Consulting Job Request and Monitoring System - SisConle.
[6] Legislative Information System - SiLeg.

Conle team. As we have a dataset of queries and their list of relevant documents, we can use it as the RF information needed to evaluate Ulysses-RFSQ, and, as most of requests received by Conle are redundant, this corpus is well-suited for this evaluation.

It has a total of 2,420 job requests, which were used for two different steps: we split the corpus into a *testing set*, used to evaluate the method, and a *Relevance Feedback set*, working as the queries database presented in Fig. 1. Thus, the job requests from the *testing set* were used as queries to perform the IR task, returning a list of documents from the Bills corpus, while their lists of relevant documents were used to be compared with the retrieved documents, evaluating the method's performance. Meanwhile, the job requests from the *Relevance Feedback set* worked as the past queries, had their similarities computed, and their lists of documents were used to update, through the *lambda* bonus, the document ranking.

To perform a fair evaluation, we used the 10-fold cross-validation technique to split the corpus into the two aforementioned sets, in order to ensure that all data were used for both steps. We also had to process this corpus using the standard BM25 algorithm to compute the documents score and store this information in a way similar to that presented in Table 1. Finally, as we used this corpus containing queries already stored together with their RF information, the Step 4 of our method was not performed for this evaluation.

4.2   Pre-processing

In a previous work, we presented an IR pipeline for the Brazilian legislative domain [31] using the Bills corpus described in Sect. 4.1. We evaluated many different pre-processing techniques, from which the combination of punctuation, accentuation, and stopwords removal + Stemming (with the Savoy algorithm [27], which was the best Stemming technique in another previous work [30]) + unigram and bigram achieved the best results. So, we opted to use this combination to pre-process both queries and documents during the IR task in this paper.

4.3   BM25 Algorithms

We opted to perform the evaluation using two variants of BM25 as the base algorithm, so we evaluated Ulysses-RFSQ-Okapi, which uses Okapi BM25 [23] and Ulysses-RFSQ-BM25L, using BM25L [17], which fixes the Okapi's preference for shorter documents. The former is the original version of BM25, while the latter achieved the best results in our pipeline evaluation [31].

For the evaluation presented here, we followed the recommendations of the original paper's authors to set the BM25 parameters.

4.4   Cut-off Parameter

As mentioned in Sect. 3.2, to perform the Step 2 of our method, a cut-off threshold needs to be set. This threshold defines whether an old query is considered similar to the current query or not: if its similarity is greater or equal than the threshold, it is a similar query and its *relevant* documents will receive the *lambda* bonus. Thus, for this paper, we set *cut* = 0.3, in order to select a greater number of queries, although disregarding queries not too similar.

4.5   IR Evaluation Measure

Finally, we used the Recall measure, which is the fraction of relevant documents that are retrieved, to obtain the results of our experiments: we analyzed the results from R@1 (Recall at 1 document) to R@20 (Recall at 20 documents).

5   Results and Discussion

To evaluate the main contribution of this study: the proposal of a novel Relevance Feedback based method to improve the IR system also for future queries, we compared the results achieved by Ulysses-RFSQ and the standard BM25 algorithms without the *lambda* bonus.

Figures 2 and 3 present the comparison, in terms of Recall and considering the retrieval from 1 to 20 documents, using Ulysses-RFSQ-Okapi and Ulysses-RFSQ-BM25L, respectively. The results point out that Ulysses-RFSQ improved the IR performance for any amount of retrieved documents and for both algorithms.

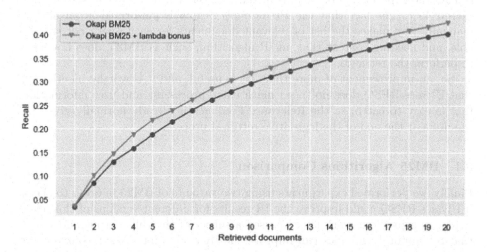

Fig. 2. Comparison between Ulysses-RFSQ-Okapi (with *lambda* bonus) and standard Okapi BM25 without RF.

Using Okapi BM25 as the base algorithm, the average improvement in Recall was 0.0210, i.e., 2.1% more relevant documents were successfully retrieved, on average, with a maximum improvement of 0.0310 (3.1%) when retrieving 5

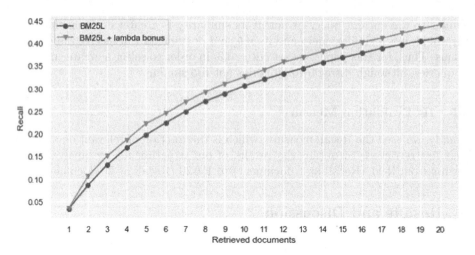

Fig. 3. Comparison between Ulysses-RFSQ-BM25L (with *lambda* bonus) and standard BM25L without RF.

documents. Using BM25L, on the other hand, Ulysses-RFSQ improved the Recall by 2.18%, with a maximum improvement of 0.0294 (2.94%), for the retrieval of 20 documents.

This is a considerable improvement, mainly in view of the fact that we performed the experiments on a difficult dataset, for which the Recall@20 did not exceed 45%. The fact that the *Preliminary Search* is performed in an almost manually way makes the list of relevant documents for each query not so reliable, and needing context that an IR algorithm, such as BM25, does not have to perform the retrieval.

So, an improvement of more than 2% is useful, and it is worthy mentioning that Ulysses-RFSQ does not need much extra processing and uses information that is easy to catch, as the Relevance Feedback is already normally given by the users at the end of the legislative retrieval process.

5.1   BM25 Algorithms Comparison

Finally, we performed experiments using two variants of BM25 in order to check if Ulysses-RFSQ could improve the IR results for different versions of this algorithm. In a previous work [30], we used a subset of the Bills corpus and found that Stemming techniques can improve the IR results for BM25L, but not for Okapi BM25. So, we needed to confirm if the Relevance Feedback *lambda* bonus works well for both variants.

Considering all the results presented before and comparing the results for the two BM25 algorithms, we can confirm the finding of [31] that BM25L performs better than Okapi BM25 for the Bills corpus. Meanwhile, considering only the impact of the *lambda* bonus on the BM25 variants, we could notice that the average improvement is practically the same for both of them.

6   Conclusion

In this paper, we presented Ulysses-RFSQ, a novel method based on the Relevance Feedback given to past queries to improve the Information Retrieval performance, and evaluated it using legislative data from the Brazilian Chamber of Deputies. The method adds a bonus, called *lambda*, to documents that were considered relevant to old queries similar to the current one, wherewith the ranking created by a scoring based IR algorithm, such as BM25, is updated.

The results showed that Ulysses-RFSQ improves both Okapi BM25 and BM25L performances considering different amounts of retrieved documents. This is a considerable improvement, in view of the simplicity of this method, the need of progress in this domain, and the ease of obtaining the RF information, as this method feeds itself back.

As future work we intend to evaluate BERT [10] and other language models, such as the one from [28], which was built from Brazilian legislative data, to select the similar queries from the database. As pointed out by [19], one of the great challenges is to find the similar queries set, and using techniques that can find semantic similarities between the queries may improve our method; as well as techniques for Named Entity Recognition (NER), using a dataset created by us for the Brazilian legislative domain and called *UlyssesNER-Br* [1].

We also plan to modify the *lambda* equation in order to consider different relevance levels that may be presented by the user, such as *a little relevant* or *very relevant*. Finally, an evaluation of the cut-off threshold, as well as investigations using others datasets, have to be performed.

References

1. Albuquerque, H.O., et al.: UlyssesNER-Br: a corpus of Brazilian legislative documents for named entity recognition. In: Pinheiro, V., et al. (eds.) PROPOR 2022. LNCS (LNAI), vol. 13208, pp. 3–14. Springer, Cham (2022). https://doi.org/10.1007/978-3-030-98305-5_1
2. Almeida, P.G.R.: Uma jornada para um Parlamento inteligente: Câmara dos Deputados do Brasil. Red Información **24** (2021). https://www.redinnovacion.org/revista/red-informaci'on-edici'on-n-24-marzo-2021
3. Badenes-Olmedo, C., García, J.L.R., Corcho, Ó.: Legal document retrieval across languages: topic hierarchies based on synsets. CoRR abs/1911.12637 (2019)
4. Bhattacharya, P., Ghosh, K., Pal, A., Ghosh, S.: Methods for computing legal document similarity: a comparative study. ArXiv abs/2004.12307 (2020)
5. Brandt, M.B.: Ethical aspects in the organization of legislative information. KO Knowl. Organiz. **45**(1), 3–12 (2018). https://doi.org/10.5771/0943-7444-2018-1-3
6. Brandt, M.B.: Modelagem da informação legislativa: arquitetura da informação para o processo legislativo brasileiro. Ph.D. thesis, Faculdade de Filosofia e Ciências da Universidade Estadual Paulista (UNESP) (2020)
7. Cantador, I., Sánchez, L.Q.: Semantic annotation and retrieval of parliamentary content: a case study on the Spanish congress of deputies. In: Proceedings of the First Joint Conference of the Information Retrieval Communities in Europe (CIRCLE 2020). CEUR Workshop Proceedings, vol. 2621 (2020)

8. Cetintas, S., Si, L., Yuan, H.: Using past queries for resource selection in distributed information retrieval. Technical report, Department of Computer Science, Purdue University (2011)

9. Chalkidis, I., Fergadiotis, M., Manginas, N., Katakalou, E., Malakasiotis, P.: Regulatory compliance through Doc2Doc information retrieval: a case study in EU/UK legislation where text similarity has limitations. In: Proceedings of the 16th Conference of the European Chapter of the Association for Computational Linguistics: Main Volume, pp. 3498–3511 (2021). https://doi.org/10.18653/v1/2021.eacl-main. 305

10. Devlin, J., Chang, M.W., Lee, K., Toutanova, K.: BERT: pre-training of deep bidirectional transformers for language understanding. In: Proceedings of the 2019 Conference of the North American Chapter of the Association for Computational Linguistics: Human Language Technologies, Volume 1 (Long and Short Papers), pp. 4171–4186 (2019). https://doi.org/10.18653/v1/N19-1423

11. Gomes, T., Ladeira, M.: A new conceptual framework for enhancing legal information retrieval at the Brazilian superior court of justice. In: Proceedings of the 12th International Conference on Management of Digital EcoSystems, MEDES 2020, pp. 26–29 (2020). https://doi.org/10.1145/3415958.3433087

12. Gutiérrez Soto, C.: Exploring the reuse of past search results in information retrieval. Ph.D. thesis, Université de Toulouse, Université Toulouse III-Paul Sabatier (2016)

13. Gutiérrez-Soto, C., Hubert, G.: Probabilistic reuse of past search results. In: International Conference on Database and Expert Systems Applications - DEXA 2014, vol. 1, pp. 265–274 (2014)

14. Gutiérrez-Soto, C., Hubert, G.: Randomized algorithm for information retrieval using past search results. In: 2014 IEEE Eighth International Conference on Research Challenges in Information Science (RCIS), pp. 1–9 (2014)

15. Gutiérrez-Soto, C., Hubert, G.: On the reuse of past searches in information retrieval: study of two probabilistic algorithms. Int. J. Inf. Syst. Model. Des. (IJISMD) **6**(2), 72–92 (2015)

16. Hust, A.: Introducing query expansion methods for collaborative information retrieval. In: Reading and Learning, pp. 252–280 (2004)

17. Lv, Y., Zhai, C.: When documents are very long, BM25 fails! In: Proceedings of the 34th International ACM SIGIR Conference on Research and Development in Information Retrieval, SIGIR 2011, pp. 1103–1104 (2011). https://doi.org/10.1145/2009916.2010070

18. Maxwell, K.T., Schafer, B.: Concept and context in legal information retrieval. In: Proceedings of the 2008 Conference on Legal Knowledge and Information Systems: JURIX 2008: The Twenty-First Annual Conference, pp. 63–72 (2008)

19. Moshfeghi, Y., Velinov, K., Triantafillou, P.: Improving search results with prior similar queries. In: Proceedings of the 25th ACM International on Conference on Information and Knowledge Management, CIKM 2016, pp. 1985–1988 (2016). https://doi.org/10.1145/2983323.2983890

20. Murata, H., Onoda, T., Yamada, S.: Comparative analysis of relevance for SVM-based interactive document retrieval. J. Adv. Comput. Intell. Intell. Inform. **17**(2), 149–156 (2013). https://doi.org/10.20965/jaciii.2013.p0149

21. de Oliveira, R.A.N., Junior, M.C.: Experimental analysis of stemming on jurisprudential documents retrieval. Information **9**(2), 28 (2018). https://doi.org/10.3390/info9020028

22. Onoda, T., Murata, H., Yamada, S.: SVM-based interactive document retrieval with active learning. New Gener. Comput. **26**(1), 49–61 (2007)

23. Robertson, S.E., Walker, S.: Some simple effective approximations to the 2-Poisson model for probabilistic weighted retrieval. In: Proceedings of the 17th Annual International ACM SIGIR Conference on Research and Development in Information Retrieval, SIGIR 1994, pp. 232–241 (1994)

24. Russell-Rose, T., Chamberlain, J., Azzopardi, L.: Information retrieval in the workplace: a comparison of professional search practices. Inf. Process. Manag. **54**(6), 1042–1057 (2018). https://doi.org/10.1016/j.ipm.2018.07.003

25. Salton, G., Wong, A., Yang, C.S.: A vector space model for automatic indexing. Commun. ACM **18**(11), 613–620 (1975). https://doi.org/10.1145/361219.361220

26. Salton, G., Buckley, C.: Improving retrieval performance by relevance feedback. J. Am. Soc. Inf. Sci. **41**(4), 288–297 (1990)

27. Savoy, J.: Light stemming approaches for the French, Portuguese, German and Hungarian languages. In: Proceedings of the 2006 ACM Symposium on Applied Computing, SAC 2006, pp. 1031–1035 (2006). https://doi.org/10.1145/1141277.1141523

28. da Silva, N.F.F., et al.: Evaluating topic models in Portuguese political comments about bills from Brazil's chamber of deputies. In: Britto, A., Valdivia Delgado, K. (eds.) BRACIS 2021. LNCS (LNAI), vol. 13074, pp. 104–120. Springer, Cham (2021). https://doi.org/10.1007/978-3-030-91699-2_8

29. Song, S.K., Myaeng, S.H.: A novel term weighting scheme based on discrimination power obtained from past retrieval results. Inf. Process. Manag. **48**(5), 919–930 (2012). https://doi.org/10.1016/j.ipm.2012.03.004

30. Souza, E., et al.: Assessing the impact of stemming algorithms applied to Brazilian legislative documents retrieval. In: Proceedings of the 13th Brazilian Symposium in Information and Human Language Technology, SBC, pp. 227–236 (2021). https://doi.org/10.5753/stil.2021.17802

31. Souza, E., et al.: An information retrieval pipeline for legislative documents from the Brazilian chamber of deputies. In: Legal Knowledge and Information Systems, pp. 119–126. IOS Press (2021). https://doi.org/10.3233/FAIA210326

32. van Opijnen, M., Santos, C.: On the concept of relevance in legal information retrieval. Artif. Intell. Law **25**(1), 65–87 (2017). https://doi.org/10.1007/s10506-017-9195-8

33. Yin, P.Y., Bhanu, B., Chang, K.C., Dong, A.: Improving retrieval performance by long-term relevance information. In: International Conference on Pattern Recognition, vol. 3, pp. 533–536 (2002)

Adaptive Fast XGBoost for Regression

Fernanda Maria de Souza, Julia Grando, and Fabiano Baldo[✉]

Universidade do Estado de Santa Catarina (UDESC), Joinville, SC, Brazil
`fabiano.baldo@udesc.br`

Abstract. The increasing generation of data by devices, people and systems arises the need for processing non-stationary data streams, which continuously change over time. It was noticed that when compared to data stream classification, there is a lack of data stream regression studies. This work proposes AFXGBReg-D, an Adaptive Fast regression algorithm using XGBoost and active concept drift detectors. AFXG-BReg uses an alternate model training strategy to achieve lean models adapted to concept drift, combined with a set of drift detector algorithms: ADWIN, KSWIN and DMM. We compared two AFXGBReg variants with other regressors and data stream regressors, simulating using synthetic datasets with different kinds of concept drifts. We show that AFXGBReg models have similar MSE to ARFReg, with these models achieving the best performance than others as proven statistically. Also AFXGBReg is 33 times faster than ARFReg, meaning that it is able to keep the same MSE level while being much faster. Another improvement is its ability of doing a faster recovery from concept drifts, having a smaller MSE peak.

Keywords: Data stream regression · Concept drift · XGBoost

1 Introduction

Currently, with the evolution of cloud storage technologies, the improvement of communications infrastructure and the popularization of mobile devices, data is continuously produced on an increasing scale. With this abundance of data, the concept of *Big Data* emerged, whose data science tries to extract potential benefits from its analysis for society [13]. The name *Big Data* aims to describe the exponential growth of data based on five principles: i) the volume of data, ii) the variability of data types, iii) the speed at which data is generated, captured and processed , iv) its veracity, in order to obtain true data, according to reality and v) the value, the costs involved in this operation and the added value of all this work [12].

Considering the potential benefits of extracting patterns and relevant information from large volumes of data, the concept of data mining was consolidated within data science. In it, using machine learning algorithms, it is possible to explore a set of data in order to establish relationships that are difficult to visualize manually.

J. C. Xavier-Junior and R. A. Rios (Eds.): BRACIS 2022, LNAI 13653, pp. 92–106, 2022.
https://doi.org/10.1007/978-3-031-21686-2_7

In the context of machine learning, the vast majority of research focuses on building static models, which are trained once on a training dataset and then applied to the analysis of new data [9]. However, the increasing generation of data by sensors, IoT (Internet of Things) devices, computer network traffic, telephone conversations and financial transactions, arises the demand for processing non-stationary data streams, which continuously change over time. Data streams can be defined as sequences of non-stationary data generated constantly and indefinitely [26]. These characteristics mean that mining algorithms cannot store the data and that they have to deal with changes in concepts inherent to the non-stationary characteristic of streams [11].

Thus, data stream mining deals with challenges on the following dimensions: i) speed: data is processed in limited time; ii) memory efficiency: previously processed data needs to be discarded; iii) non-iterative: data is processed only once; and iv) adaptive: the learning model must be adjusted to concept drifts [16].

As on classification problems, in the context of data stream regression the predictive model needs to adapt to changes in concept. Therefore, for dynamically changing data, traditional supervised learning methods are inappropriate and can cause loss of model performance. To try to maintain the best possible precision, the trained model needs to adapt using new input data. Therefore, predictive models need to be trained incrementally, either by continuous updating or by retraining using recent batches of data [8].

Although there are some proposed approaches to data stream regression, when compared to data stream classification studies, it was not yet much explored by the scientific community, as it is considered a more complex problem and therefore requires more effort to solve [26].

In machine learning, a concept drift means that the statistical properties of the target variable, which the model is trying to predict, change over time in an unforeseen way [14]. Therefore, this results in problems in the predictions as the model becomes less accurate over time. Concept drift detection methods for adaptive machine learning can be classified into two main categories: passive and active. In passive approaches, learning models are updated without prior detection of the concept drift [6], therefore, they tend to be slower in detecting the change. In active mode, approaches are focused on first detecting the deviation/change of concept and, later, accelerating the process of updating the model so that the change is adsorbed by the model as soon as possible, making the detection process faster than in the passive one [7].

To deal with concept drifts in non-stationary data, passive approaches are able to adapt with continuous updates, but require more computational efforts because they are slower [19]. On the other hand, even though active approaches tend to be faster, they introduce to the learning process one more task to be performed, which is the execution of the drift detection algorithm. In this work, the use of active concept drift detection will be investigated in order to explore its potential benefits and limitations in relation to passive concept drift detection in data stream regression.

Among the studies on classification and regression of data streams, the ensemble of models stands out as the most robust and accurate. Looking specifically at classifier ensembles, boosting algorithms are a promising approach, as they are able to produce an accurate general model from the combination of several moderately imprecise models [23]. The basic idea of these algorithms is to iteratively apply classifiers and combine their solutions to get a better predictive result [18]. Two of its most famous algorithms are *AdaBoost* and *Gradient Boosting*, which work by sequentially adding models that correct the residual prediction errors of the previous model, usually made up of decision trees.

However, building models based on the boosting approach can be slow, especially for large data sets. Therefore, to mitigate this problem, the algorithm eXtreme Gradient Boosting (XGBoost) was proposed, which creates individual trees using several processors and the data is organized to minimize the search time, which decreases the training time and improves the performance [22]. The XGBoost algorithm is effective for a wide range of classification and regression problems [4], however, its application in data streams is still not much explored. The main studies are dedicated to the classification of data streams, such as *Adaptive Extreme Gradient Boosting* (AXGB). AXGB uses an incremental strategy to create the set, instead of using the same data to select each function, it uses subsamples of data taken from non-overlapping windows [20]. An adaptation of AXGB was proposed in [3], called AFXGB, which made use of an alternating model training strategy instead of an ensemble of classifiers. It achieved great improvement on running time while keeping the same level of accuracy of AXGB.

Therefore, considering that AFXGB technique has shown good results in classification problems, but has not yet been applied in data stream regression, this work explores AFXGB for regression (AFXGBReg) combined with active concept drift detection techniques. We compare AFXGBReg runtime and mean squared error with other data streams regressors synthetic datasets with different kinds of concept drifts.

The paper is organized as follows: Sect. 2 presents an overview of concept drift detector techniques and data stream regressors, Sect. 3 describes the proposed algorithm and in Sect. 4 we present the testing methodology and discuss the results. Lastly, Sect. 5 features the conclusion and future work.

2   Related Works

This section presents the summary of the literature review that was carried out on themes involving decision trees, data stream regression and strategies to deal with concept drifts.

On the matter of dealing with concept drifts, the detection of drifts can be done in two ways [11]: i) Active: designed to detect concept drifts using different types of detectors, so if there is a concept deviation, the model is updated and ii) Passive: the model is continually updated whenever new data becomes available, regardless of whether the change is taking place or not [17].

One of the most popular active concept drift detector algorithms is the Adaptive Sliding Window (ADWIN). It monitors a sequence of real value inputs using sliding windows [2]. It uses two windows: a reference window and a test window. ADWIN notices the occurrence of concept drift by identifying a large difference between the averages of the two subwindows. Once a drift point is detected, all old data samples before that drift time point are discarded [25]. A more recent method for concept drift detection is the KSWIN, which is based on the Kolmogorov-Smirnov (KS) [15] statistical test. Similar to ADWIN, KSWIN also implements two functions: one for adding the new one-dimensional data in the local sliding window and the second to inform if the deviation was detected or not. Another algorithm called Drift Detection Method (DDM) monitors the error rate, assuming it will decrease as the number of instances increases and the data distribution is stationary in the incremental learning process [24]. DDM was the first algorithm to define an warning level and change level for concept change detection [1]. If the confidence level of the observed error rate reaches the warning level, DDM starts building a new learner by using the old learner for predictions. If the change has reached the concept drift level, the old learner will be replaced by the new learner for other prediction tasks [16]. DDM works best with sudden change data where concepts can gradually pass without triggering the change level [8].

In [14] it was developed a work involving regression in data streams with passive adaptation to concept drifts, called *Rival Learner*. The algorithm uses an ensemble of regressors and has two submodels, one based on historical samples, called the global model, and the other based on samples from current sliding windows, called the local model. These two models compete with each other to see who has the best performance and the winner takes over as the global model. In the tests performed by the authors, when a concept deviation occurs, the proposed algorithm keeps the model effectively updated without the need for active detection.

An algorithm called Adaptive eXtreme Gradient Boosting (AXGB) was presented in [20], using XGBoost for classification of data streams with concept drifts. The central logic of AXGB is the incremental creation/updating of an ensemble of classifiers, with a fixed size, and when this size is reached the oldest classifiers are discarded to make way for the new models. AXGB uses ADWIN to detect concept drifts, triggering a routine to update the ensemble based on the change detection signal. In [3] it was proposed an adaptation of AXGB [20] called AFXGB. In order to reduce its training and testing time, it eliminated the set of classifiers from the original proposal and made use of an alternating model training strategy. This means that the classifier has a lifetime and before this lifetime is reached, an alternate model is trained in background to replace the current model. Using this strategy the model becomes incremental without the need to use a set of models and it also avoids over-specialization.

On the ensemble of classifiers and active concept drift detectors, in [25] it was developed a framework called *Performance Weighted Probability Averaging Ensemble* (PWPAE). The proposed framework is based on learning a set of models that uses combinations of two detection methods, ADWIN and DDM, and two drift adaptation methods, Adaptive Random Forest (ARF) and Streaming Random Patches (SRP), to build basic learners. The results presented by the authors show that the algorithm reached high levels of accuracy compared to recent works.

Regarding data stream regression, in [10] it was proposed an adaptation to Adaptive Random Forest (ARF) for data stream regression task, called ARF-Reg. The algorithm works in a voting system responsible for averaging the individual predictions to obtain the final prediction, with its base learner being a regression tree, with the FIMT-DD algorithm. The update dynamics in ARF-Reg are both passive and active using ADWIN as concept drift detectors. As a result, ARF-Reg achieved considerably small error rates, especially in real-world scenarios, thus showing its effectiveness.

It is noticed that the regression task, despite being one of the most common topics in the area of machine learning, is not a topic widely discussed and studied in the analysis of data streams. Therefore, this work aims to adapt the AXGB algorithm to support data stream regression with active concept change detection.

3   Proposal: Adaptive Fast XGBoost Regressor

The original proposal of AXGB uses an ensemble of XGBoost classifiers, however this considerably harmed the performance of the model. Therefore, the AFXGB algorithm proposed by [3] will be used as the basis of this work, an adaptation to the AXGB that obtained a shorter training time while maintaining the same accuracy as the original AXGB. In the proposal of AFXGB, only one XGBoost model is trained and incrementally updated. However, after this classifier exceeds a limit of trained data, another classifier is trained as the replacement, thus avoiding the overfitting of the first one.

Although the incremental AFXGB model update strategy can handle passive detection of concept change as the model is updated based on more recent data, adjustment for abrupt concept drifts is not adequately supported, due to its delay in the perception of change. Therefore, this work proposes the inclusion of active concept drift detection techniques to AFXGB, while porting it to regression. We call it Adaptive Fast XGBoost Regressor (AFXGBReg).

Therefore, this work proposes to explore the following active concept change detection techniques: ADWIN, KSWIN and DDM. These detectors are intended to identify concept drifts and trigger the model update mechanism. This mechanism consists of resetting the size of the data window coming from the stream used to update the model, causing the model to be updated faster, resulting in a quicker response to the change in concept.

3.1   Implementation

A pseudo-code for the proposed algorithm is presented in Algorithm 1. The modifications implemented AFXGB started with the inclusion of an active detection step of concept drift performed by the set of detectors composed by: ADWIN, KSWIN and DDM. Afterwards, the classification model was ported to regression and a new flag which allows the execution of the algorithm with a window-size *reset* mechanism was implemented, happening when the regressor is replaced by the new one.

Algorithm 1: Adaptive Fast XGBoost Regressor algorithm

Input: (x, y) ϵ Data Stream
Data: w = maximum number of samples on the window, M = XGBoost
     classifier model, W = window, SW = sliding window of data, *life_time* =
     lifetime of each classifier, *training_time* = training time of each regressor,
     count = how many times the window has run

1 Adds (x, y) to window (W)
2 **if** $|W| > w$ **then**
3   **if** $M <> NULL$ **then**
4   **if** *training_time* $\geq$ *(life_time - cont)* **then**
5   # Checks if it is inside the training time of next regressor
6   nextM = trains next XGBoost regressor;
7   **else if** *count* $\geq$ *life_time* **then**
8   # resets counter and starts to use the new classifier
9   count = 0;
10   M = nextM;
11   **if** *active_reset* $==$ 'Y' **then**
12   w = 0;
13   M = loads the previous model;
14   M' = trains regressor with W using M and adds to M;
15   M = M';
16   Saves new M model recently trained;
17   **else**
18   M = Trains regressor with W;
19   Saves new M model recently trained;
20   **if** *detect_drift* $==$ *True* **then**
21   # Calculate the absolute error
22   error = abs(self.predict(X) - y);
23   **foreach** *active_detector of D* **do**
24   **if** *drift_detection(active_detector, error)* **then**
25   # If any active concept drift detector detects a drift based on
                 the absolute error, the window size is reset to zero.
26   w = 0;
27   # In the first detection of concept drift by some detector the
                 loop is broken, advancing to next instances
28   break;
29   W = W - W';
30   count = count + 1; # counts how many times the window has run
31 returns M;

As input, the set x and y belonging to the data stream are present: with x being the input data while y is the value that will be predicted. In the algorithm, when the sliding window of data coming from the stream is full, the algorithm checks if there is any already trained XGBoost regression model (Line 3). If it does not exist, an XGBoost regressor is trained (Line 17). If it already exists, other conditions are tested with the proposal of verifying the life time of the regressor (Lines 4 to 12) and then the regressor is loaded (Line 13) and updated (Line 15).

The adaptation to port XGBoost for regression is carried out in order to change the objective function of the algorithm in the configurable parameters of boosting of the XGB to *reg:squarederror*, having as an internal metric for data validation by the algorithm the Root Mean Square Error (RMSE). In addition, the prediction function was also transformed, in order to return floating values from X instances instead of labels as in classification. With the model loaded, an array of active detectors is iterated to detect a concept drift. Each detector must be complementary to each other, in order to aggregate in accuracy/MSE or R^2, not considerably increasing the execution time or memory weight of the model (Lines 20 to 21). For this implementation, ADWIN, KSWIN and DDM detectors were chosen.

Since AXGB was developed for classification, the instances passed to each detector were the ones incorrectly classified. Now, to make the portability of the algorithm for regression, a function was created (Line 22) to calculate the Mean Absolute Error (MAE), whose value is treated in order to identify changes in concept by the active detection algorithms. MAE is calculated by the absolute difference between observed (actual) and predicted (hypothesis) values. Each time a change in concept is detected by some detector, the size of the sliding window is reset and the iteration loop over all other detectors for that instant analyzed is broken by means of a *break* (Line 28).

With the implementation of the break after a detection, the final execution time of the algorithm is reduced without harming the calculated MSE, because if all the detectors used detected a change in concept at a certain moment, they would all need to be called, making it more costly in time and memory the process, as well as causing an unnecessary *reset* of the same sliding window times the number of detectors used.

3.1.1   Drift Detectors Scheduling

In addition to choosing which detectors would integrate the algorithm, an analysis was also necessary considering their order of execution and scheduling, as well as hyperparameter values for each detector. Analyzing in depth the characteristics of the chosen detectors, the sensitivity for detecting changes in concept by the KSWIN detector was identified through experiments and empirical analysis. Considering aspects related to the amount of concept changes by the KSWIN detector and detection quality by the ADWIN detector, the detectors were staggered in an orderly manner:

1. **ADWIN**: It was chosen as the first detection option due to its assertiveness for recurrent and gradual concept changes, while maintaining a good detection rate for abrupt changes;
2. **DDM**: Effective and neutral performance for abrupt concept changes;
3. **KSWIN**: As a last option, if a change in concept is not detected by any of the other detectors mentioned above, KSWIN has a high quantitative percentage of detections for changes in concept, especially abrupt ones.

3.1.2 Parameters

- **Learning Rate (eta)**: a value between 0 and 1. The learning rate applies a weighting factor to new trees added to the XGBoost model. When next to 0 the algorithm will make less corrections, this resulting in more trees and slower processing time.
- **Maximum depth (*max_depth*)**: the maximum depth which the tree can reach. Increasing this number created a more complex model and increases the memory consumption.
- **Maximum window size (*max_window_size*)**: maximum size of the window which stores the data from the data stream. The algorithm updated the models only when maximum window size is reached.
- **Classifier life time (*life_time*)**: Lifetime of each alternate classifier. This value is based on the number of times which the sliding window was reset.
- **Training time (*training_time*)**: training time of each alternate classifier. This value is based on the number of times which the sliding window was reset.
- **Reset window size when regressor is exchanged (*active_reset*)**: Boolean used to identify whether the window size reset strategy in the regressor exchange will be used or not;
- **Concept drift (*detect_drift*)**: This boolean determines whether the concept change with the array of detectors will be applied during learning.

4 Results Assessment

This section presents the results of tests performed with the AFXGBReg algorithm. The Sect. 4.1 outlines the methodology that was used for the tests and the Sect. 4.2 details the results obtained.

4.1 Testing Methodology

To evaluate of the proposed algorithm, six implementations were considered. Of these six algorithms, two were proposed by the author and four external for comparison, are described below:

- AFXGBReg-Dr: algorithm proposed in this work, using alternating regressors, ensemble of active detectors of concept drift and window reset in concept drift detections as well as in the step where the regressor is replaced by a new one previously trained.

- AFXGBReg-D: algorithm also proposed in this work, using alternate regressors, ensemble of active detectors of concept drift and window reset **ONLY** in detections of concept drift.
- HTR (Hoeffding Tree Regressor): is a regression adaptation of the incremental tree algorithm of the same name for classification. HTR uses Hoeffding's inequality to control its node division decisions.
- HTRA (Hoeffding Adaptive Tree Regressor): similar to the previous implementation, but with the addition of the tree using the ADWIN concept change detector and PERCEPTRON (single layer neural network) to make predictions.
- KNN (k-Nearest Neighbors regressor): nonparametric regression method where predictions are obtained by aggregating the values of stored samples from n_nearest neighbors against a query sample.
- ARFReg (Adaptive Random Forest Regressor): adaptation for regression of the existing algorithm for classification, using the ADWIN concept drift detector.

Evaluated aspects include runtime, Mean Squared Error (MSE), and the ability to adapt to different types of concept drift. The evaluation method selected is *Prequential Evaluation*, where the data are analyzed sequentially when they arrive at the model.

The databases used for the tests were taken from the *scikit-multiflow* library provided by [21], being synthetic databases generated through the *Hyperplane Generator* and *ConceptDriftStream* generation methods. Table 1 presents the bases selected for the evaluation.

Table 1. Datasets used on evaluation.

Dataset	Drift type	# Instances	# Drifts
CDS_A	Abrupt	500.000	10
HYP	Incremental	500.000	1
CDS_G	Gradual	500.000	3

The Prequential Evaluation simulation was executed 5 times for each algorithm and dataset to obtain the average MSE, training time, testing time and total time.

The AFXGBReg hyperparameters were selected empirically based on numerous simulations and were defined as the following: No. Estimators = 30, Learning Rate = 0.3, Max Window Size = 1000, Max Tree Depth = 6. The active drift detectors hyperparameters were defined as: *adwin_delta* = 0.0000001, *kswin_alpha* = 0.0000001, *kswin_window_size* = 100, *kswin_stat_size* = 30, *ddm_min_num_instances* = 30, *ddm_warning_level* = 2, *ddm_out_control_level* = 3.

4.2   Analysis of the Results

We present on Table 2 the average MSE of the 5 executions of the considered models for each dataset. The best MSE is highlighted, and the corresponding ranking of the model for that dataset is next to the MSE. The model rank is defined by ordering all model's MSE for that dataset from lowest to highest, this meaning that the smallest ranking is the better one.

Table 2. Average MSE and rankings for the six algorithms considered in the study over 3 datasets.

Dataset	AFXGBReg-Dr	AFXGBReg-D	HTR	HTRA	KNN	ARFReg
CDS_A	**3459.59** [(1)]	3833.69 [(2)]	9806.61 [(6)]	7971.14 [(5)]	7190.04 [(4)]	5064.42 [(3)]
HYP	0.14 [(5.2)]	0.15[(5.6)]	0.109 [(2.2)]	0.105 [(2.4)]	0.13 [(4.2)]	**0.10** [(1.4)]
CDS_G	**4037.43**[(1)]	4399.73 [(2)]	11587.09 [(6)]	9268.18 [(5)]	7663.68 [(4)]	5273.96 [(3)]
Avg. MSE	2499.06	2744.52	7131.27	5746.48	4951.28	3446.16
Avg. rank	2.4	3.2	4.733	4.133	4.067	2.467

In the following sections we analyze the MSE, runtime and concept drift recovery results.

4.2.1   MSE Analysis

It is noticed that the average MSE of all algorithms varies a lot. To confirm the performance difference between the algorithms, we apply a pair-wise *post-hoc* test for multiple comparisons [5]. The *post-hoc* test applied was a Nemenyi test using the algorithms average ranking. This test determines a Critical Difference (CD), meaning that two algorithms whose ranking difference is greater that the critical difference are considered significantly different. The Nemenyi resulted on a CD of 1.38 and pair-wise comparisons are shown on Fig. 1.

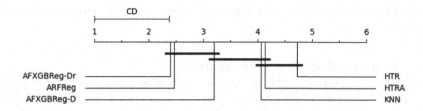

Fig. 1. Nemenyi test of model's average rankings.

It is noticeable that AFXGBReg-Dr and ARFReg have similar performances, followed by AFXGBReg-D. Since the difference of average ranks between these three models is smaller than the value of CD, this test confirms that the difference in regression capability is not significant. This also means that the models AFXGBReg-Dr, ARFReg and AFXGBReg-D are superior than the other models in terms of MSE.

4.2.2   Runtime Analysis

We now focus on comparing the simulated models runtimes, especially of AFXGBReg-Dr, AFXGBReg-D and ARFReg, which were the ones who presented best performance of MSE. Table 3 presents the average training time, average testing time and average total time on minutes of all simulations on all datasets.

Table 3. Average times of simulations for the algorithms considered in the study.

Model	Time (minutes)		
	Avg. training	Avg. testing	Avg. total
AFXGBReg-Dr	3.53	1.88	5.41
AFXGBReg-D	3.53	1.87	5.40
HTR	1.52	0.24	1.75
HTRA	5.89	0.23	6.12
KNN	0.22	3.79	4.01
ARFReg	179.30	3.27	182.57

It is apparent that HTR and KNN have the smaller total times. This is expected because of the algorithms simplicity, since they were not developed for data streams. On the other hand, they also had worse MSE performance than other models. Regarding the superior MSE group, AFXGBReg-Dr and AFXGBReg-D have similar times, as expected since the algorithms are related, but ARFReg shows an outstanding training time when compared to them. The average total time of ARFReg model is 33 times larger than AFXGBReg models.

It is worth mentioning that AFXGBReg's runtime could be reduced even further by using a smaller lifetime, thus decreasing the model's build-up by resetting it more frequently, but this needs to be balanced with possible MSE increase.

4.2.3   Concept Drift Recovery Analysis

To evaluate the ability of the models on recovering from a concept drift we analyzed the accuracy of each model over time after each drift. We selected one simulation of CDS_A to show the model's behavior on abrupt concept drift.

On Fig. 2 we observe the models MSE for the simulation of CDS_A dataset, where both AFXGBReg models have a smaller peak of MSE than the others. This can be perceived in more detail on Fig. 3 and 4, where the drifts of points 50k and 450k are highlighted. It is seen that HTR and HTRA have higher MSE peaks, and also that HTR and KNN do not recover completely from the drift. This happens because HTR and KNN do not have features to reset the model, accumulating obsolete concepts.

On the 50k CDS_A drift depicted by Fig. 3, we can see that AFXGBReg models recovery is faster than other models, showing the effectiveness of the array of active concept drift detectors and model reset. This can be further observed on Table 4, where the MSE of all models is displayed for five data points related to the 50k concept drift of CDS_A.

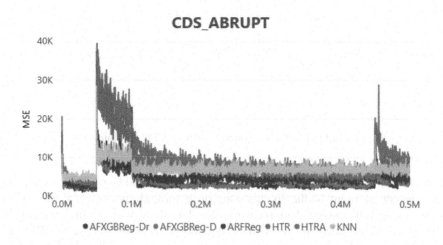

Fig. 2. MSE over time for evaluated models on CDS_A dataset.

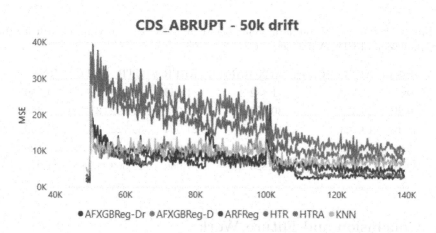

Fig. 3. Highlight of 50k concept drift of CDS_A simulation.

On data 50k the drift has not happened yet, so all models are at their lowest MSE, then on the following data register after the drift (50.2k) all models suffer a great increase in MSE, prominent on ARFReg, HTR and HTRA models. On the next observed data point of 50.4k the models start to adjust to the new

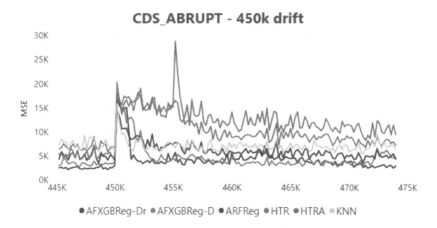

Fig. 4. Highlight of 450k concept drift of CDS_A simulation.

concept, decreasing MSE and are able to arrive on stable predictions for the new concept by point 55.6k. In this scenario it took about 5.6k of data volume for achieving full concept drift recovery. The last data point highlighted on the table is the lowest MSE obtained on this concept, achieved by AFXGBReg-Dr on 82.4k. We can see that none of the models were able to achieve MSE as low as they previously had, but in comparison to AFXGBReg variants and ARFReg, the other models had much larger error.

Table 4. MSE of models on data points before and after the 50k concept drift for the 1st simulation of CDS_A dataset.

# data	AFXGBReg-Dr	AFXGBReg-D	ARFReg	HTR	HTRA	KNN
50k	1,326	1,432	1,963	2,918	2,991	4,750
50.2k	26,358	30,496	35,018	37,292	37,220	28,408
50.4k	18,703	16,580	21,904	32,722	32,722	22,855
55.6k	8,520	8,726	11,613	28,335	24,888	11,196
82.4k	5,646	9,823	7,950	22,663	15,872	11,912

5   Conclusion and Future Work

In this work we proposed a Fast Adaptive XGboost Regression model with active concept drift detetction (AFXGBReg-D) to handle data streams. AFXGBReg uses an alternate model training strategy in order to reduce the model piling up complexity and adapt faster to concept drifts. It was also proposed a variation of AFXGBReg-D with a fixed window size reset each time the regressor is replaced by the new model, called AFXGBReg-Dr.

Comparing their MSE, speed and ability to adapt to concept drift with other models on synthetic datasets with different kinds of concept drifts, it was seen that AFXGBReg variations are able to achieve the same level of MSE of ARFReg, obtaining superior performance than the other models, as proven statistically. Meanwhile, AFXGBReg models are also 33 times faster than ARFReg, meaning that it is able to keep the same MSE performance while being much faster.

Regarding to concept drifts adaptation, AFXGBReg-Dr model presented a smaller peak in MSE and faster adaptation than the other models, including AFXGBReg-D. It was seen that it needs less volume of data to comeback from the MSE increase and it is able to keep a low long-term average MSE, since old concepts are forgotten when the model is resetted and substituted.

For future works we intend to explore AFXGBReg models with semi-supervised learning. Still, we intend to study more about other detection strategies besides the active one, as well as explore the AFXGBReg algorithm in order to make it increasingly performant for different types of concept drift.

Acknowledgments. We would like to specially thanks FAPESC – Fundação de Amparo à Pesquisa e Inovação do Estado de Santa Catarina – to partially funded this research work.

References

1. Abbaszadeh, O., Amiri, A., Khanteymoori, A.R.: An ensemble method for data stream classification in the presence of concept drift. Front. Inf. Technol. Electron. Eng. **16**(12), 1059–1068 (2015). https://doi.org/10.1631/FITEE.1400398
2. Barddal, J.P.: Vertical and horizontal partitioning in data stream regression ensembles. In: 2019 International Joint Conference on Neural Networks (IJCNN), pp. 1–8. IEEE, Curitiba (2019)
3. Bonassa, G.: Adaptação de classificador utilizando a biblioteca XGBoost para classificação rápida de fluxos de dados parcialmente classificados com mudança de conceito (2021)
4. Chen, T., He, T., Benesty, M., Khotilovich, V., Tang, Y., Cho, H., et al.: Xgboost: extreme gradient boosting. R Package Version 0.4-2 **1**(4), 1–4 (2015)
5. Demšar, J.: Statistical comparisons of classifiers over multiple data sets. J. Mach. Learn. Res. **7**, 1–30 (2006)
6. Ditzler, G., Roveri, M., Alippi, C., Polikar, R.: Learning in nonstationary environments: a survey. Comput. Intell. Mag. **10**(4), 12–25 (2015).https://doi.org/10.1109/MCI.2015.2471196
7. Elwell, R., Polikar, R.: Incremental learning of concept drift in nonstationary environments. IEEE Trans. Neural Netw. **22**(10), 1517–1531 (2011)
8. Gama, J., Žliobaitundefined, I., Bifet, A., Pechenizkiy, M., Bouchachia, A.: A survey on concept drift adaptation. ACM Comput. Surv. **46**(4) (2014). https://doi.org/10.1145/2523813
9. Gamage, S., Premaratne, U.: Detecting and adapting to concept drift in continually evolving stochastic processes. In: Proceedings of the International Conference on Big Data and Internet of Thing, BDIOT 2017, pp. 109–114. Association for Computing Machinery, New York (2017). https://doi.org/10.1145/3175684.3175723

10. Gomes, H.M., Barddal, J.P., Ferreira, L.E.B., Bifet, A.: Adaptive random forests for data stream regression. In: ESANN. IEEE, Curitiba (2018)
11. Krawczyk, B., Minku, L.L., Gama, J., Stefanowski, J., Woźniak, M.: Ensemble learning for data stream analysis: a survey. Inf. Fus. **37**, 132–156 (2017)
12. Laney, D.: 3D data management: controlling data volume, velocity, and variety. Technical report, META Group, EUA (2001). http://blogs.gartner.com/doug-laney/files/2012/01/ad949-3D-Data-Management-Controlling-Data-Volume-Velocity-and-Variety.pdf
13. Larson, D., Chang, V.: A review and future direction of agile, business intelligence, analytics and data science. Int. J. Inf. Manag. **36**(5), 700–710 (2016)
14. Liao, Z., Wang, Y.: Rival learner algorithm with drift adaptation for online data stream regression. In: Proceedings of the 2018 International Conference on Algorithms, Computing and Artificial Intelligence, ACAI 2018, Association for Computing Machinery, New York (2018). https://doi.org/10.1145/3302425.3302475
15. Lopes, R.H., Reid, I., Hobson, P.R.: The two-dimensional kolmogorov-smirnov test (2007)
16. Lu, J., Liu, A., Dong, F., Gu, F., Gama, J., Zhang, G.: Learning under concept drift: a review. IEEE Trans. Knowl. Data Eng. **31**(12), 2346–2363 (2018)
17. Mahdi, O.A., Pardede, E., Ali, N., Cao, J.: Fast reaction to sudden concept drift in the absence of class labels. Appl. Sci. **10**(2), 606 (2020)
18. Mayr, A., Binder, H., Gefeller, O., Schmid, M.: The evolution of boosting algorithms. Methods Inf. Med. **53**(06), 419–427 (2014)
19. Mehmood, H., Kostakos, P., Cortes, M., Anagnostopoulos, T., Pirttikangas, S., Gilman, E.: Concept drift adaptation techniques in distributed environment for real-world data streams. Smart Cities **4**(1), 349–371 (2021)
20. Montiel, J., Mitchell, R., Frank, E., Pfahringer, B., Abdessalem, T., Bifet, A.: Adaptive XGBoost for evolving data streams. In: 2020 International Joint Conference on Neural Networks (IJCNN), pp. 1–8. IEEE, Hamilton (2020)
21. Montiel, J., Read, J., Bifet, A., Abdessalem, T.: Scikit-multiflow: a multi-output streaming framework. J. Mach. Learn. Res. **19**(72), 1–5 (2018). http://jmlr.org/papers/v19/18-251.html
22. Ramraj, S., Uzir, N., Sunil, R., Banerjee, S.: Experimenting XGBoost algorithm for prediction and classification of different datasets. Int. J. Control Theory Appl. **9**, 651–662 (2016)
23. Schapire, R.E.: The boosting approach to machine learning: an overview. In: Nonlinear Estimation and Classification, pp. 149–171 (2003)
24. Yan, M.M.W.: Accurate detecting concept drift in evolving data streams. ICT Express **6**(4), 332–338 (2020)
25. Yang, L., Manias, D.M., Shami, A.: Pwpae: an ensemble framework for concept drift adaptation in iot data streams. arXiv preprint arXiv:2109.05013 (2021)
26. Yu, H., Lu, J., Zhang, G.: Morstreaming: a multioutput regression system for streaming data. IEEE Trans. Syst. Man Cybern. Syst., 1–13 (2021). https://doi.org/10.1109/TSMC.2021.3102978

The Effects of Under and Over Sampling in Exoplanet Transit Identification with Low Signal-to-Noise Ratio Data

Fernando Correia Braga[✉][ID], Norton Trevisan Roman[ID], and Diego Falceta-Gonçalves[ID]

Universidade de São Paulo, São Paulo, Brazil
{fernando.braga,norton,dfalceta}@usp.br

Abstract. This paper presents the results of experiments with under-sampling and oversampling applied to machine learning classifiers used in the identification of exoplanet transits with low signal-to-noise ratio (*SNR*) data. We start by giving an overview of the most popular method for exoplanet detection, followed by an analysis of the Kepler Object of Interest (*KOI*) data set, along with an overview of the state of the art machine learning models applied to this problem, and how complex it is to correctly identify exoplanets on low SNR data. We then briefly discuss our signal-to noise ratio reduction procedure, used to generate the low *SNR* data for our experiments. Finally we use our low *SNR* data set to train and evaluate some models in scenarios with no sampling strategy and with oversampling and undersampling, using repeated holdout validation. Results show that current classifiers can identify transits in low *SNR* data sets, with accuracy varying between 69% and 81%, and that sampling strategies can affect simpler classifiers, making them less conservative, but do not show significant effects on more complex classifiers.

Keywords: Exoplanet transit identification · Low signal-to-noise exoplanet detection · Oversampling · Undersampling

1   Introduction

Exoplanets, *i.e.* planets found outside the boundaries of our solar system [8,10], and the search for such celestial bodies have recently gained large attention by the scientific community and by the media in general, as a way of expanding the limits of human knowledge. Finding these planets not only allows us to better understand how solar systems are formed and how they evolve, but also represents an important step in the search for habitable planets and even for extraterrestrial life [9].

There are currently various ways of finding exoplanets [35], but among them, the detection through the transit method is the most successful one, with about 78% of all confirmed exoplanets being first detected through this method [21]. The idea behind this approach is to detect periodic small changes in the star's

© The Author(s), under exclusive license to Springer Nature Switzerland AG 2022
J. C. Xavier-Junior and R. A. Rios (Eds.): BRACIS 2022, LNAI 13653, pp. 107–121, 2022.
https://doi.org/10.1007/978-3-031-21686-2_8

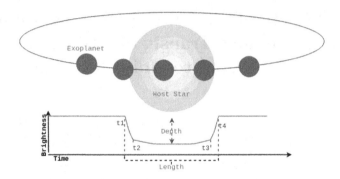

Fig. 1. Simplified diagram of a transit. Adapted from [21].

brightness, which might be due to a planet transiting between the star and the observer, as shown in Fig. 1.

To detect such events, the star's brightness is measured at regular intervals. These time ordered measures then build up a time series known as a *light curve* – (*LC*) (bottom part of the figure). The expected behavior of a planetary transit consists of a "U" shaped event in the *LC* (as shown in the figure), with a reduction in the star's brightness (t1 to t2), followed by a period of lower brightness (t2 to t3), and then a return to the star's previous brightness (t3 to t4) [20]. Finding this kind of behavior at the *LC* is what is called *transit detection*.

After the detection of a transit-like event in an *LC*, it is necessary to identify its origin. Aside from an exoplanet orbiting that star, other causes can produce a similar behavior in the *LC*. These are normally grouped into two classes of False Positives [15,25,32]: the Non-transit Phenomenon, when a transit-like event is detected but no transit exists (*i.e.* a "real" false positive); and the Astronomical False Positives, when a transit really occurred in the *LC*, but which was not caused by an exoplanet orbiting that star, but instead by, for example, the interaction of the stars in a binary system, the "centroid offset" (when the detected transit corresponds to that of some nearby star), and contamination from other nearby celestial bodies, among others [15].

At the beginning, the identification step of the transit method, *i.e.* the definition of the origin of some transit-like event (*cf.* [22]), was manually carried out by a group of specialists which must reach a consensus on the cause of the transit, based on the *LC* and other information regarding the host star [25,32]. This manual identification became however a bottle neck, and Machine Learning (*ML*) came into the picture to allow the whole process to scale up to the large amounts of data being captured. Many different *ML* techniques were successfully applied in this scenario, such as Random Forests – *RF* (*e.g.* [5,6,11,12,19,21,23,25,26,30]), Convolutional Neural Networks – *CNN* (*e.g.* [2,4,13,14,16,21,27,28,30,32,36,37]), Multilayer Perceptrons – *MLP* (*e.g.* [5,14,21,28]), and Support Vector Machines – *SVM* (*e.g.* [11,19,21,28,30]), among others.

Existing efforts, however, either focus on high *signal-to-noise ratios (SNR)*, that is the relation between the strength of the signal and the noise present in the data (we will return to this concept in Sect. 3), removing data below a specific threshold *(e.g.* [4,5,16,25,26,32,36]), or do not tackle this issue at all (*e.g.* [3,11,19,23,30]). This procedure can nevertheless miss the identification of some exoplanets which, despite still remaining unknown to science, may be already present in the available data [32], although being kept away from current detection methods, given its low SNR.

However, in recovering these low *SNR* exoplanets, we could not only find smaller exoplanets (and, consequently, with smaller SNRs), which would be more similar to Earth, but also discover exoplanets with longer transit periods, which currently require long observations times, so as to increase their SNR (see Sect. 3 for details). The ability to detect low SNR transits could even allow us to scan a larger portion of the galaxy, by detecting exoplanets which, despite being large enough to be captured through current methods, present a low SNR given their distance to Earth.

In this article, we aim at moving one step further in this direction, by reporting on the results regarding the performance of current state of art *ML* classifiers for transit identification in low *SNR* real data, along with the effect some sampling strategies, such as undersampling and oversampling, can have in their performance. To do so, we also introduce our "data impoverishment" procedure, whereby one can reduce the SNR of current data so as to train and test ML techniques in low SNR scenarios. As it turned out, the performance of current classifiers with low SNR data lies between 71% and 81% accuracy, with sampling techniques being effective with some, but not all, of them.

The rest of this article is organized as follows. Section 2 gives a brief overview of current uses of ML to exoplanet detection. In Sect. 3, we describe the process researchers follow to build the light curves (known as folding), whereas in Sect. 4 we give details of the experiments we carried out, such as data sets, our procedure to reduce SNR and tested methods. Results are shown in Sect. 5 and discussed in Sect. 6. Finally, in Sect. 7 we present our final remarks and directions for future work.

2   Related Work

Dating back to 2012, Random Forests (*RF*) were one of the first classifiers applied, in the context of the transit method, to identify transits previously detected by Kepler's pipeline [23], achieving a True Positive Rate (*TPR*) of 90% with a 1% False Positive Rate (*FPR*). They were also applied later on, using only four features directly derived from the *LC*, and achieving a 97.0% accuracy over Kepler's data and 96.8% over simulated data [26].

Self-Organizing Maps (*SOM*) were also used in this task, to map the transits' characteristics into its topology, and then extract a statistic representing the probability of that transit being an exoplanet [7]. The classifier was tested with Kepler's first mission data, with 87% accuracy, 86% precision and 87% recall, and with K2's data, with 93% accuracy, 95% precision and 97% recall. This work was later expanded by using SOM's results as input to an RF classifier, along with other LC attributes. Tests of this system over $NGTS$' data achieved 90% precision, 91% recall and an AUC of 0.98.

It did not take longer for Convolutional Neural Networks (CNN) to be tested. Initial experiments, with simulated data, achieved 0.96 AUC, 1.0 FPR and 94.00 TPR [37]. Using Kepler's data, CNNs were found to deliver 96% accuracy and 0.988 AUC [32]. Although reporting the discovery of new exoplanets, this work also points out that low SNR transits could have been previously misclassified or discarded by the pipeline. By the same time, comparative experiments with simulated data, between a CNN, an MLP and other techniques, reported on a 99.6% accuracy, 88.5% recall and 0.96 AUC for one of the tested CNNs, with the MLP delivering 99.8% accuracy, 91.5% recall and a 0.96 AUC [28].

Other initiatives evolved over Shallue and Vanderburg's model [32], being tested in Kepler's data (*e.g.* [4]), K2's data (*e.g.* [16]), $WASP$'s data (*e.g.* [30]) and $TESS$' data (*e.g.* [27,36]), with different values for accuracy, precision and AUC. In a large comparative study [21], where different ML techniques were tested, along with a Discrete Wavelet Transform (DWT) implementation, a CNN model outscored its counterparts over simulated data, with 99.1% accuracy, precision and recall. When running with artificially altered Kepler data, however, best results were delivered by an RF model, with 98.5% accuracy, 98.6% precision and 98.4% recall.

Although sometimes not addressed by current research (*e.g.* [23,30]), the effect SNR has on ML results has been studied mostly with simulated or artificially altered data (*e.g.* [6,27,28,37]), since simulated scenarios make it possible to control for the desired SNR values. Within this set-up, it was possible to determine that the lower the SNR, the worse a classifier's performance, although by how much depended on the classifier, the simulation model and tampering strategy used. Performance losses range from a 21% decrease [27] to an almost 50% decrease in accuracy [28].

Another reason for the focus in artificial data, when studying how different SNRs affect classifiers' performance, may lie in the fact that research based on real data is bound to comply with the limitations of its data source. As an example, initiatives using Kepler's data (*e.g.* [4,16,25,26,32] are subject to its original data processing pipeline's significance threshold, which only considered for further evaluation events with a measure of Multi Event Statistic[1] $MES \geq 7.1\sigma$, with some of existing efforts adopting even higher thresholds (*e.g.* [36]).

In this research, we not only present a new methodology to derive low SNR data from high SNRs, so that researchers can take advantage of already mapped exoplanets to train and test their models, but also move one step further, by presenting our results with some common sampling strategies, designed to reduce data imbalance.

---

[1] MES is a significance metric derived, among other things, from the transit SNR, so that the greater a transit's SNR, the greater its MES [15].

3   Folding the Light Curve

The transit method's greatest challenge is to correctly detect and identify transit events within the noise present in the *LCs*. Consider, for example, the variation the transit of a Jupiter-sized planet causes to its hosting star's brightness. It is estimated that this variation lies around 1%, whereas for an Earth-sized planet it would be less than 0.01%. Such a small fluctuation in the signal's amplitude can be completely hidden or absorbed by the noise present in the *LC* [17]. As a result, the *signal-to-noise ratio – (SNR)*, used to indicate the strength of the real signal relative to the noise present in the data, is one of the most limiting factors to the discovery of new exoplanets.

To address this problem, a folding procedure is usually applied [22], whereby the *LC* for a long period of observations is folded, as illustrated in Fig. 2, around a set of approximate transit periods. The resulting *LC* is then compared to a predefined function which approximates the expected transit behavior, in the search for the best matching configuration. If this best matching configuration crosses a predefined threshold, the transit event is considered valid for identification [25]. This process not only helps to estimate the transit period, *i.e.* the time taken by the exoplanet to orbit its host star, but also reduces noise by diluting the local fluctuations in the *LC*.

As it will be made clearer in Sect. 4, this folding process lies at the heart of our "impoverishment" procedure since, as the number of transits considered in the folding increases, more noise can be softened, thereby increasing the transit's Signal do Noise Ratio (*SNR*). Hence, all one has to do to have a lower *SNR* from a high *SNR* transit is to "unfold" it to the desired *SNR*.

4   Materials and Methods

Kepler's data are amongst the most well studied exoplanet's data sets, so we've decided to use it as our source. It comes separated in two different groups: The first group comprises transit events data, such as the *Kepler Object of Interest (KOI)* catalog, which has information regarding some attributes of transit events, such as its class, if the transit is a "PC" or some false positive according to the result of the identification step and the post processing validation, the transit's *SNR*, its duration, period, the timestamp of the first transit in the sequence, and the id of the observed star in which the transit was detected, among others [15]; The second comprehends *LC* data, which contains the star's raw brightness time series, captured in the first step of the pipeline, along with its preprocessed equivalent, generated in the preprocessing step of the pipeline [22].

Due to the limitations noise can impose on the discovery of exoplanets, as discussed in Sect. 1, even well studied data sets, like the one from Kepler, lack a significant amount of labeled low *SNR* transits, which would allow the training of a *ML* classifier. By looking at Fig. 3, which shows an analysis of the *KOI* catalog's transit events grouped by ranges of *SNR*, we can see that there are

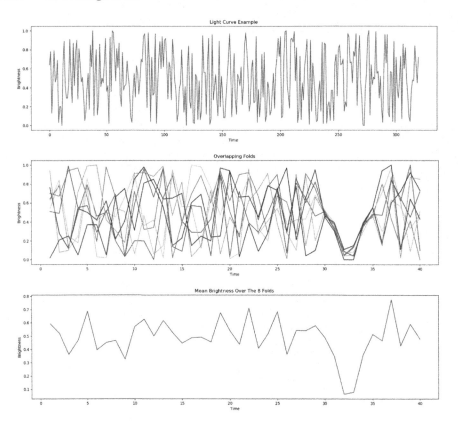

Fig. 2. Example of a folding process. Top: the *LC* across 320 measurements of the star's brightness. Middle: the overlapped curve, folded at every 40 measurements, generating 8 overlapping splits. Bottom: the mean brightness of the 8 overlapping folds.

considerably fewer analyzed transit events (as represented by the green and red bars) with low *SNR*, independently of their cause[2].

In face of this matter, and taking into account our desire to analyze the effects of sampling strategies in low *SNR* scenarios, we devised an "impoverishment" procedure, loosely based on the folding method shown in Fig. 2, so as to lower the *SNR* of transit events to a desired approximated value without including any artificial modification to the *LCs*. To do so, we began with an estimate for a transit's *SNR* [34]:

---

[2] Kepler's *KOI* catalog has three possible classes for transit events. "CONFIRMED" is a transit which was identified as a *PC* and lately confirmed by another method; "FALSE POSITIVES" are transits confirmed to fall into one of the false positive categories from Sect. 1; and "CANDIDATE" are transits identified as *PC* by the pipeline, but which were not yet confirmed by another method.

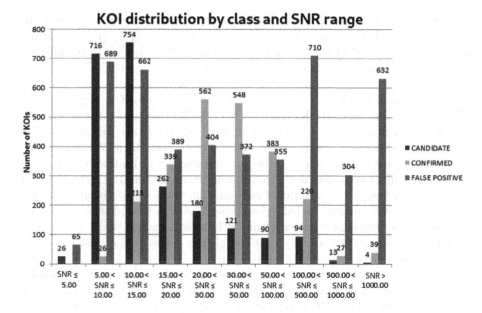

Fig. 3. *KOI* distribution by class and *SNR* range.

$$SNR = \frac{(R_P/R_*)^2 \sqrt{n_{Obs}}}{CDPP_{6h} \sqrt{6h/Dur}}$$

where a transit's *SNR* is a function of the planet radius (R_P), the host star's radius (R_*), the number of observed transits (n_{Obs}), and the Combined Differential Photometric Precision taken on a interval (*e.g.* 6 h), weighted by the duration of that transit in regard of that interval ($CDPP_{6h} \sqrt{6h/Dur}$). Although all these values are actually available in the *KOI* catalog, we can take them to be constant (except for the number of observed transits), since a light curve reflects different transits of the same planet around the same star and these values would not change, and work the equation out to obtain the expected number of observed transits needed to reach some target *SNR*:

$$N_{Target\_Obs} = \left(\frac{Target\_SNR}{Original\_SNR} \right)^2 N_{Obs}$$

From the *KOI* catalog, we selected the transits, excluding the ones still labeled as "CANDIDATE", with *SNR* in the ranges $20,00 < SNR \leq 30,00$, $30,00 < SNR \leq 50,00$ and $50,00 < SNR \leq 100,00$, since these ranges offer an interesting balance of classes. By defining a target *SNR* of 5.00 and using it to calculate the number of transit observations for each *KOI* in those ranges to achieve this *SNR*, we divided each *LC* into splits containing only this amount of transit observations. These splits are then to be treated by the classifiers as different lower *SNR* transit events. The resulting data set comprises 40,828 examples, being 27,864 of the negative class, henceforth called "AFP" (False positives in Fig. 3), and 12,964 of the positive class, henceforth called "PC" (Confirmed in Fig. 3), creating a unbalanced data set with proportions 68.25% / 31.75%, respectively.

To select our classifiers, we refer to the state of art in the application of *ML* to the transit method, discussed in Sect. 2. We then chose three different classifiers, which used different *ML* techniques, to study their behavior in low *SNR* data, and how our sampling strategies might affect them. We began with those whose source code was available for download, and that we could easily make compatible to our data set. From this criterion, we selected *AstroNet* [32], a *CNN* which has two convolutional columns, one for a whole vision of the *LC* folded over 2,001 buckets, called "global view", and a second with a zoomed in vision of the transit event, with only 201 buckets, called "local view". The available implementation already contains all the preprocessing and execution framework necessary for *AstroNet* out of the box.

Since we could find no other classifiers with these characteristics, we changed our criterion to reflect easiness of implementation and code reuse. So our next choice was *SIDRA* [26], a *RF* which uses only four input features, derived directly from the *LC*. As our third classifier, we selected the *SVM* from [21], henceforth called *ExoplanetSVM*, since we can reuse in it *AstroNet*'s preprocessing by making a minor tweak, changing the original 2.048 inputs from [21] to the 2.001 inputs *AstroNet* generates. This change is expected not to cause great impacts on the classifier, since it was made only to better accommodate data treated with *DWT*, a preprocessing technique which we are not going to use. Implementations were done in *Python* 3, using the configuration described in each method's article, with *AstroNet* using the *TensorFlow* package [1], and *SIDRA* and *ExoplanetSVM* using the *Scikit-Learn* package [29].

As mentioned, we modeled the problem as a binary classification task, whereby a classifier had to tell exoplanets from false positives in the *LC* (*i.e.* Confirmed × False Positive in Fig. 3). To deal with class imbalance, we applied two strategies. The first one consisted in the *Random Undersampling (RUS)* of the majority class, where the examples of this class are chosen at random to build the training set. The second strategy, in turn, comprehends the *Random Oversampling (ROS)* of the minority class, whereby examples from the minority class are randomly chosen (with reposition) to build the training set. For the implementation of *RUS* and *ROS* we used *Python*'s *Imbalanced-Learn* [24] package.

Finally, to evaluate our classifiers, we defined four quality metrics [18]:

– *Accuracy*: The rate of correct predictions

$$A = (TP + TN)/(Pos + Neg)$$

where TP (True Positives) is the number of positive examples correctly classified as such, TN (True Negatives) is the number of negative examples correctly classified as such, FP (False Positives) is the number of negative examples misclassified as positives, Pos is the total number of positive examples and Neg is the total number of negative examples;

- *Recall*: The rate of correctly classified examples of the positive class in regarding to the number of positive examples

$$R = TP/Pos$$

- *Precision*: Ratio between the correctly classified positive examples over the total predictions of the positive class

$$P = TP/(TP + FP)$$

- *Specificity*: The rate of correctly classified examples of the negative class in relation to the number of negative examples

$$S = TN/Neg$$

5   Results

Our experiments followed a *repeated hold-out* strategy with $n = 10$ repetitions, with random stratified sampling for the train and test sets split [18]. Within this setup, one randomly separates the data set into training and testing sets, training the models in the training set and evaluating them, according to the metrics listed in Sect. 4, in the test set. This process is repeated a number of times (in our case, 10), each time separating 70% of the data for training, with the remaining 30% being left for test purposes.

Since we are following the configuration and hyperparameter values defined by the authors of each method, hyperparameter tuning was not necessary. The application of *RUS* or *ROS* was always done over the sampled training set, not affecting the test set. Through all repetitions, all models were trained and tested in the same sets. The seed for the random split of the training and test sets was the number of the repetition, *i.e.* 1 for the first, 2 for the second and so forth.

Our first round of experiments was done without the application of *RUS* or *ROS*, so as to build a baseline for each classifier. Table 1 gives the mean results for our quality metrics. Differences were found to be statistically significant, according to an ANOVA test ($F = 402.28, p \ll 0,001$, at the 95% confidence level[3]), with a pairwise confirmation by a *post hoc Tukey HSD* test run on accuracy figures[4]. This led us to believe the three classifiers actually generated distinct classification models.

With this baseline, we proceeded to experiment with *RUS*. Since our original data set had a $31,75\%/68,25\%$ proportion for Positive and Negative cases,

---

[3] The test was executed through the *f_oneway* function of the *SciPy* package [33].

[4] Calculated with the *pairwise_tukeyhsd* function of the *statsmodels* package [31], leading to *AstroNet* $\times$ *SIDRA* $t = -0,0561, p \ll 0,001$; *AstroNet* $\times$ *ExoplanetSVM* $t = -0,0908, p \ll 0,001$; *SIDRA* $\times$ *ExoplanetSVM* $t = -0,0347, p \ll 0,001$.

Table 1. Mean accuracy (A), specificity (S), recall (R) and precision (P) in the test set, over 10 executions, for the classifiers trained with so sampling strategy.

Classifier	A	S	R	P
SIDRA	75.16%	89.92%	43.88%	67.23%
AstroNet	80.77%	83.97%	73.99%	68.52%
ExoplanetSVM	71.69%	93.15%	26.20%	64.32%

respectively, and due to limitations of the *RUS* implementation, we were forced to work with the intervals between the original proportion and $50,00\%/50,00\%$. So we decided to experiment with two different proportions to better understand how undersampling can affect our classifiers: $40,00\%/60,00\%$ (closer to the original proportion) and and $50,00\%/50,00\%$ (a balanced data set).

We then retrained our classifiers in the new undersampled training sets and tested them against the original (not modified) test sets, producing the results shown in Table 2. Once again, overall differences in accuracy were found to be statistically significant ($ANOVA\ F = 406, 12, p \ll 0,001$). However, a pairwise *post hoc Tukey HSD*, run with all ten possible pairs of classifiers, showed there to be no significant difference between *AstroNet* $50,00\%/50,00\%$ and *AstroNet* $40,00\%/60,00\%$ ($p = 1.000$), and between *SIDRA* $50,00\%/50,00\%$ and *ExoplanetSVM* $40,00\%/60,00\%$ ($p = 0.798$). Interestingly, no difference could be observed between *AstroNet* $50,00\%/50,00\%$ and *AstroNet* $40,00\%/60,00\%$ in any of the adopted metrics.

Table 2. Mean accuracy (A), specificity (S), recall (R) and precision (P) on the test set, over 10 executions, for our three classifiers trained with *RUS* using the proportions $50,00\%/50,00\%$ and $40,00\%/60,00\%$.

Classifier	$50,00\%/50,00\%$				$40,00\%/60,00\%$			
	A	S	R	P	A	S	R	P
SIDRA	69.44%	64.70%	79.50%	51.54%	74.89%	83.37%	56.89%	61.76%
AstroNet	80.49%	83.50%	74.08%	68.09%	80.41%	83.74%	73.33%	68.12%
ExoplanetSVM	66.31%	64.28%	70.61%	48.25%	70.59%	79.38%	51.93%	54.30%

Next, we executed our *ROS* experiments, following the same steps as with *RUS*, with the same class proportions, as shown in Table 3. As previously, overall differences were found to be significant ($ANOVA\ F = 191.48, p \ll 0.001$). A pairwise analysis, however, pointed there to be four pairs with no significant differences in accuracy. These were *AstroNet* $50,00\%/50,00\%$ and *AstroNet* $40,00\%/60,00\%$ ($p > 0.876$); *SIDRA* $50,00\%/50,00\%$ and *ExoplanetSVM* $40,00\%/60,00\%$ ($p > 0.876$); *SIDRA* $50,00\%/50,00\%$ and *ExoplanetSVM* $40,00\%/60,00\%$ ($p > 0.074$); and *ExoplanetSVM* $50,00\%/50,00\%$ and *ExoplanetSVM* $40,00\%/60,00\%$ ($p > 0.074$). Of these, only *AstroNet* $50,00\%/ 50,00\%$ and *AstroNet* $40,00\%/60,00\%$ showed no significant difference across all metrics.

Table 3. Mean accuracy (A), specificity (S), recall (R) and precision (P) on the test set, over 10 executions, for our three classifiers trained with *ROS* using the proportions $50,00\%/50,00\%$ and $40,00\%/60,00\%$.

Classifier	$50,00\%/50,00\%$				$40,00\%/60,00\%$			
	A	S	R	P	A	S	R	P
SIDRA	70.00%	65.90%	78.70%	52.16%	74.96%	84.25%	55.27%	62.35%
AstroNet	81.02%	83.67%	75.42%	68.74%	80.45%	83.54%	73.93%	68.04%
ExoplanetSVM	70.00%	77.07%	55.02%	53.09%	71.44%	84.91%	42.88%	57.27%

6   Discussion

All classifiers were found to perform better than random guessing or zero rule classifiers, ranging from 69% to 81% accuracy, as shown in Fig. 4, even under very low *SNR* conditions (recall that our target *SNR* was only 5, whereas most of current results work with values above 20). There has been, however, some differences found across *ML* models, with *AstroNet*'s *CNN* delivering the best results (around 80% accuracy), independently of the applied sampling strategy.

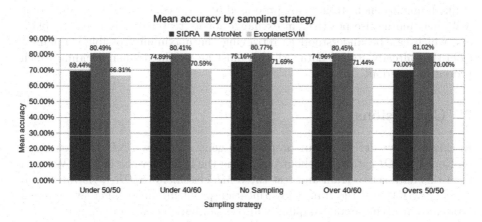

Fig. 4. Mean accuracy results for each classifier and sampling strategy.

As a matter of fact, *AstroNet*'s accuracy and precision figures outscored its counterparts in all experiments, only loosing to *SIDRA* in the context of a $50,00\%/50,00\%$ sampling strategy, with recall. The fact that *AstroNet* performed very similarly across the experiments shows this classifier to be rather stable, even when presented with an unbalanced data set, in comparison to a more balanced one. In comparison with its original results, *AstroNet*'s accuracy decreased almost 15%, which shows the difficulties in processing low SNR data.

SIDRA, in turn, was an average performing classifier, with its results being the closest ones to *AstroNet*'s, while still lying roughly 5.00% accuracy and 6.00% precision below it when no sampling is applied, and being almost 22% bellow its original results. This is nonetheless an interesting result, given this classifier's lower complexity, taking only four inputs (against *AstroNet*'s 2,202 and *ExoplanetSVM*'s 2,001). These features translate into a fluent training of multiple instances at a really low computational cost. On the other hand, *SIDRA* seems to be affected by different sampling strategies. Lastly, *ExoplanetSVM* was the worst performing classifier, being also affected, as it was the case with *SIDRA*, by the adopted sampling strategy.

All in all, accuracy does change across sampling strategies. At the same time, specificity and precision drop, whereas recall raises. The only exception to these rules is *AstroNet*, which shows little change across the tested scenarios. Despite these results, the decision on whether to use or not a sampling strategy is actually deeply tied to the goal and background of the proposed model. For instance, new missions or equipment, which are still generating lots of data, might prefer a more conservative classifier, which has a higher precision and a lower recall, so false positives might not overburden the scientists doing the post pipeline validation of the "PC" transits, and so using sampling strategies with simple classifiers might not help.

On the other hand, older missions or equipment, such as the Kepler telescope, which are generating little to no new data, and whose data was already deeply studied, might benefit from being processed by a less conservative classifier, that will allow more false positives in exchange for a greater recall of real exoplanets, which can be understood as retrieving the maximum possible exoplanets from that already stalled data. All this must be weighted when deciding which procedure to adopt.

7   Conclusions

In this work we described our proposal to generate a low *SNR* data set based on Kepler's *KOI* catalog, without tampering the light curve artificially. We then used this data set to train and test three *ML* classifiers (*SIDRA*, *AstroNet* and *ExoplanetSVM*) for the transit identification problem, under different sampling strategies involving undersampling and oversampling of the training set.

The results point that *AstroNet* was basically unaffected by the adopted sampling strategy. On the other hand, both *SIDRA* and *ExoplanetSVM* seem to be affected, in a similar way, by both undersampling and oversampling. As it tuned out, with these classifiers, the closer the proportion of both classes to each other in the training set, the higher the classifiers' recall in the test set, whereas the lower their precision, sensitivity and accuracy. As a result, although sampling strategies can be used in the context of exoplanet transit identification, more complex classifiers do not seem to benefit from them, and so the overhead caused by these strategies may not justify their use in these cases.

Obtained results are nonetheless encouraging since, despite the difference, all classifiers were found to perform rather well, ranging from 69% to 81% accuracy. As already pointed out, being able to correctly identify low *SNR* transits can lead to the identification of smaller exoplanets (which cause only small fluctuations in a star's LC), potentially leading to an explosion in the number of known exoplanets.

These results also show there to be a good chance of identifying planets with long transit periods, and which would take a long time to produce the number of folds necessary for a high SNR, but that might otherwise be detected through techniques capable of dealing with low SNR values. In addition, top accuracy techniques could be used to expand the range of stars under consideration, by including stars farther from earth which, despite being massive, produce low SNR values. Last, but not least, our findings could lead to a reduction in the time needed to capture an LC, if only by demanding fewer transit observations to be made.

As for directions for future work, we believe other techniques should be tested on the same grounds as those we report here, so that their behavior in low SNR scenarios could be assessed. Also, other techniques could be applied, such as ensemble models, that could benefit from the best each of its comprising models can deliver. This is a direction we intend to pursue in the near future.

References

1. Abadi, M., et al.: Tensorflow: a system for large-scale machine learning. In: 12th {USENIX} Symposium on Operating Systems Design and Implementation ({OSDI} 2016), pp. 265–283 (2016)
2. Alshehhi, R., Rodenbeck, K., Gizon, L., Sreenivasan, K.R.: Detection of exomoons in simulated light curves with a regularized convolutional neural network. Astron. Astrophys. **640**, A41 (2020). https://doi.org/10.1051/0004-6361/201937059
3. Amin, R.A., et al.: Detection of exoplanet systems in kepler light curves using adaptive neuro-fuzzy system. In: 2018 International Conference on Intelligent Systems (IS), pp. 66–72. IEEE (2018)
4. Ansdell, M., et al.: Scientific domain knowledge improves exoplanet transit classification with deep learning. Astrophys. J. **869**(1), L7 (2018). https://doi.org/10.3847/2041-8213/aaf23b
5. Armstrong, D.J., Gamper, J., Damoulas, T.: Exoplanet validation with machine learning: 50 new validated kepler planets (2020)
6. Armstrong, D.J.: Automatic vetting of planet candidates from ground-based surveys: machine learning with NGTS. Monthly Not. Roy. Astron. Soc. **478**(3), 4225–4237 (2018)
7. Armstrong, D.J., Pollacco, D., Santerne, A.: Transit shapes and self organising maps as a tool for ranking planetary candidates: application to kepler and k2. Monthly Not. Roy. Astron. Soc., stw2881 (2016)
8. Assembly, I.G.: Resolutions b5 and b6 on the definition of a planet in the solar system and pluto (2014)
9. Battley, M.P., Pollacco, D., Armstrong, D.J.: A search for young exoplanets in sectors 1–5 of the tess full-frame images. Monthly Not. Roy. Astron. Soc. **496**(2), 1197–1216 (2020)
10. Boss, A.P., et al.: Working group on extrasolar planets. Proc. Int. Astron. Union **1**(T26A), 183–186 (2005)
11. Bugueno, M., Mena, F., Araya, M.: Refining exoplanet detection using supervised learning and feature engineering. In: 2018 XLIV Latin American Computer Conference (CLEI), pp. 278–287. IEEE (2018)

12. Caceres, G.A., et al.: Autoregressive planet search: application to the kepler mission. Astron. J. **158**(2), 58 (2019)
13. Chaushev, A., et al.: Classifying exoplanet candidates with convolutional neural networks: application to the next generation transit survey. Monthly Not. Roy. Astron. Soc. **488**(4), 5232–5250 (2019)
14. Chintarungruangchai, P., Jiang, G.: Detecting exoplanet transits through machine-learning techniques with convolutional neural networks. Publ. Astron. Soc. Pac. **131**(1000), 064502 (2019)
15. Coughlin, J.L., et al.: Planetary candidates observed by kepler. vii. the first fully uniform catalog based on the entire 48-month data set (q1–q17 dr24). Astrophys. J. Suppl. Ser. **224**(1), 12 (2016)
16. Dattilo, A., et al.: Identifying exoplanets with deep learning. ii. two new super-earths uncovered by a neural network in k2 data. Astron. J. **157**(5), 169 (2019)
17. Grziwa, S., Pätzold, M.: Wavelet-based filter methods to detect small transiting planets in stellar light curves. arXiv preprint arXiv:1607.08417 (2016)
18. Han, J., Pei, J., Kamber, M.: Data Mining: Concepts and Techniques. Elsevier, Amsterdam (2011)
19. Hinners, T.A., Tat, K., Thorp, R.: Machine learning techniques for stellar light curve classification. Astron. J. **156**(1), 7 (2018)
20. Hippke, M., Heller, R.: Optimized transit detection algorithm to search for periodic transits of small planets. Astron. Astrophys. **623**, A39 (2019)
21. Jara-Maldonado, M., Alarcon-Aquino, V., Rosas-Romero, R., Starostenko, O., Ramirez-Cortes, J.M.: Transiting exoplanet discovery using machine learning techniques: a survey (2020)
22. Jenkins, J.M., et al.: Overview of the kepler science processing pipeline. Astrophysi. J. Lett. **713**(2), L87 (2010)
23. Jenkins, J.M., et al.: Auto-vetting transiting planet candidates identified by the kepler pipeline. Proc. Int. Astron. Union **8**(S293), 94–99 (2012)
24. Lemaître, G., Nogueira, F., Aridas, C.K.: Imbalanced-learn: a python toolbox to tackle the curse of imbalanced datasets in machine learning. J. Mach. Learn. Res. **18**(1), 559–563 (2017)
25. McCauliff, S.D., et al.: Automatic classification of kepler planetary transit candidates. Astrophys. J. **806**(1), 6 (2015)
26. Mislis, D., Bachelet, E., Alsubai, K., Bramich, D., Parley, N.: Sidra: a blind algorithm for signal detection in photometric surveys. Monthly Not. Roy. Astron. Soc. **455**(1), 626–633 (2016)
27. Osborn, H.P., et al.: Rapid classification of tess planet candidates with convolutional neural networks. Astron. Astrophys. **633**, A53 (2020)
28. Pearson, K.A., Palafox, L., Griffith, C.A.: Searching for exoplanets using artificial intelligence. Monthly Not. Roy. Astron. Soc. **474**(1), 478–491 (2018)
29. Pedregosa, F., et al.: Scikit-learn: machine learning in Python. J. Mach. Learn. Res. **12**, 2825–2830 (2011)
30. Schanche, N., et al.: Machine-learning approaches to exoplanet transit detection and candidate validation in wide-field ground-based surveys. Monthly Not. Roy. Astron. Soc. **483**(4), 5534–5547 (2019)
31. Seabold, S., Perktold, J.: Statsmodels: econometric and statistical modeling with python. In: Proceedings of the 9th Python in Science Conference, vol. 57, p. 61. Austin, TX (2010)
32. Shallue, C.J., Vanderburg, A.: Identifying exoplanets with deep learning: a five-planet resonant chain around kepler-80 and an eighth planet around kepler-90. Astron. J. **155**(2), 94 (2018)

33. Virtanen, P., et al.: Scipy 1.0: fundamental algorithms for scientific computing in python. Nat. Methods **17**(3), 261–272 (2020)
34. Weiss, L.M., Petigura, E.A.: The kepler peas in a pod pattern is astrophysical. Astrophys. J. Lett. **893**(1), L1 (2020)
35. Armstrong, D.J., Gamper, J., Damoulas, T.: Exoplanet validation with machine learning: 50 new validated kepler planets (2020)
36. Yu, L., et al.: Identifying exoplanets with deep learning. iii. automated triage and vetting of tess candidates. Astron. J. **158**(1), 25 (2019)
37. Zucker, S., Giryes, R.: Shallow transits-deep learning. i. feasibility study of deep learning to detect periodic transits of exoplanets. Astron. J. **155**(4), 147 (2018)

Estimating Bone Mineral Density Based on Age, Sex, and Anthropometric Measurements

Gabriel Maia Bezerra[1,2(✉)] [ID], Elene Firmeza Ohata[1,3] [ID],
Pedro Yuri Rodrigues Nunes[1,2] [ID], Levy dos Santos Silveira[1,2] [ID],
Luiz Lannes Loureiro[4] [ID], Victor Zaban Bittencourt[4] [ID],
Valden Luis Matos Capistrano[3] [ID], and Pedro Pedrosa Rebouças Filho[1,2,3] [ID]

[1] Laboratory of Image Processing, Signals, and Applied Computing (LAPISCO),
Fortaleza, Brazil
{gabrielmaia,elene.ohata}@lapisco.ifce.edu.br,
{pedro.yuri.rodrigues08,levy.silveira07}@aluno.ifce.edu.br
[2] Federal Institute of Ceará (IFCE), Fortaleza, Brazil
[3] Federal University of Ceará (UFC), Fortaleza, Brazil
[4] Federal University of Rio de Janeiro (UFRJ), Rio de Janeiro, Brazil

Abstract. Osteoporosis is a global health problem characterized by low bone density and deterioration of bone tissue that increases the risk of fracture. Early identification of low bone mineral density (BMD) is crucial to reducing risks by providing correct treatment or prevention methods. Dual-energy x-ray absorptiometry (DXA) is often used to measure BMD. However, it is not affordable or accessible to some patients and is rarely suggested to non-risk groups. Alternatively, information such as age, sex, weight, height, and body circumferences have shown an association with BMD and are inexpensive and easy to obtain. Thus, this paper proposes a method to estimate BMD through anthropometric measurements, age, and sex. We also introduce BMD-10, a dataset containing 911 patients with their respective BMD values and 10 other features. Our approach evaluates the performance of different types of regression algorithms through nested cross-validation. A Least-Squares Support-Vector Machine achieves the lowest Mean Absolute Error: $0.0769\,\mathrm{g/cm^2}$. Lastly, we interpret the model predictions with SHAP (SHapley Additive exPlanations), finding that weight is the most important feature for the estimation.

Keywords: Bone mineral density · Regression · Anthropometric measurements

1 Introduction

Low bone mass and calcium leading to a deterioration of bone structure are characteristics of osteoporosis, a disease that affects the skeleton [1]. Its incidence

J. C. Xavier-Junior and R. A. Rios (Eds.): BRACIS 2022, LNAI 13653, pp. 122–134, 2022.
https://doi.org/10.1007/978-3-031-21686-2_9

has been rising in developed and developing countries, thus becoming a global health concern. Approximately 2.5 million fractures reported in the USA and Europe yearly are associated with osteoporosis. Furthermore, the complication of such fractures and costs related to surgery, extended care, disability, and hospitalizations carry considerable economic expenses [25]. An example of such fractures is hip fractures. They often require hospitalization and are considered one of the most critical fracture types. In Latin America, more than 23% of patients die in the first year after a hip fracture [3].

Bone mineral density (BMD) value is a source of information about the patient's bone health that can be used to diagnose osteoporosis and determine fracture risk [22]. There are several methods to determine the BMD; among them, one of the most popular is the dual energy x-ray absorptiometry (DXA). Although the DXA is considered the most reliable method for estimating BMD [25], it presents challenges regarding its lack of portability and cost [8,21]. Therefore, people of lower income, rural communities, and populations from underdeveloped countries may have difficulty accessing or affording this procedure.

Lately, machine learning methods are becoming more popular due to their versatility and their capacity to establish convoluted connections between features and predictions [4,6]. Several papers sought to predict fracture risks in individuals with osteoporosis [16]. Other papers aimed the BMD estimation using computed tomography (CT) scan or X-ray images [11]. Many studies focused on predicting osteoporosis in postmenopausal women since this data is more readily available [15,24]. However, there is a lack of research on estimating BMD with more accessible features in a broader range of demographic groups.

Often, osteoporosis is not detected until it has reached an advanced state. Most of the time, the first symptom that manifests in a patient is a broken bone. However, bone fracture risk can be mitigated significantly with preventive measures or early detection and treatment of osteoporosis [25]. Thus, we propose a method that uses artificial intelligence (AI) to estimate BMD using age, sex at birth, and anthropometric measurements such as weight, height, and six body circumferences measured by a specialist. Our main contributions are enumerated as follows:

1. An evaluation of different methods for BMD estimation using easy to obtain features: age, sex at birth, and anthropometric measurements;
2. A new dataset for estimating BMD, consisting of 911 female and male patients (ages between 18 and 65) with a wide variety of body types;
3. An interpretation and analysis of the impact of features on the model output, which can increase the trust in the algorithm and bring insights to other researchers.

This paper is organized into enumerated sections. Section 2 introduces related works to this study. The proposed dataset, the learning process, evaluation, and interpretation steps are explained in Sect. 3. The experiment results and their discussion are exhibited in Sect. 4. Finally, in Sect. 5, we conclude our findings and present future works.

2 Related Work

This section presents related works that use machine learning to aid in osteoporosis detection or bone-related risk prediction.

Iliou and Anagnostopoulos [12] evaluated the use of machine learning and feature selection to detect osteoporosis. They used a dataset composed of 3426 patients. The authors concluded that only age and weight achieved satisfactory results in predicting osteoporosis. However, their dataset was primarily composed of women (98%). In addition, the authors did not provide the dataset and did not present the data distribution.

Kim et al. [15] proposed the use of machine learning to predict osteoporosis risk. The dataset used in the study comprised 1674 postmenopausal women, which included clinical and demographic data, such as age, weight, waist circumference, fracture history, and osteoarthritis, among other information. They reached an accuracy of 76.7% and 0.827 of area under the curve when using the Support Vector Machine (SVM) classifier. Nevertheless, the main shortcoming of this study is that using only postmenopausal women limits the demographic portion of the population in which the method can be applied.

Yang et al. [26] developed a machine learning model to predict osteoporosis. The authors evaluated several machine learning methods, achieving 0.843 as the highest area under the receiver operating characteristic curve for men and 0.811 for women. The dataset was composed of 2929 female patients and 3053 male individuals. However, this study also focuses on a strict age group and uses several hematological and biochemical tests as features.

Kilic and Hosgormez [14] proposed using six bone densitometry feature sets and the patient's age (resulting in 24 attributes from 350 postmenopausal women) to perform a 3-class classification: osteoporosis, osteopenia, and control group. The authors concluded that selecting five BMD measures and five T-Score values achieved the best result in identifying osteoporosis using the Random Subspace Method combined with Random Forest. Nonetheless, they also concentrated their study on postmenopausal women. Furthermore, the detailed data was composed of measures from high-end equipment that is already used for osteoporosis diagnoses.

3 Methodology

In this section, we describe how we built the proposed dataset and we investigate the data. We also inform which regression models are used, the configuration of the experiments, and the methodology used to train, evaluate and interpret these models.

3.1 Dataset

We have not found any public datasets for estimating BMD through anthropometric measurements. Therefore, we have built a new dataset, called BMD-10, and will make it available to other researchers. The BMD-10 contains data from 911 adults living in Brazil. More specifically, all patients reported their current age and sex at birth, and we measured their height, weight, total body BMD, and six body circumferences: arm, forearm, waist, hip, thigh, and calf. For measuring the body circumferences, we followed the protocol from the International Society for the Advancement of Kinanthropometry (ISAK). For obtaining the total body BMD, all patients underwent a DXA scan (GE Prodigy Advance).

Table 1 presents the means, standard deviations (SD), and range for each feature and the target (BMD). In addition, we compare these statistics between the subset of male patients (containing 411 people) and the subset of female patients (containing 500 people). From Table 1, we observe that all patients are adults aged between 18 and 65. Moreover, we notice that the BMD has a range of $0.9\,\mathrm{g/cm^2}$, reaching as high as $1.689\,\mathrm{g/cm^2}$ and as low as $0.789\,\mathrm{g/cm^2}$. When comparing the means of the subsets, we verify that male patients tend to have higher anthropometric measurements than female ones, with the only exception being the hip circumference.

Table 1. Measure of the mean, standard deviation (SD), minimum, and maximum values for the features and the target of all patients in the dataset. Additionally, the same statistics are shown for the male and female patients, separately.

Features	All patients (N = 911)		Male patients (N = 411)		Female patients (N = 500)	
	Mean (SD)	Range	Mean (SD)	Range	Mean (SD)	Range
age (years)	34.69 (11.90)	[18, 65]	33.66 (11.36)	[18, 65]	35.53 (12.26)	[19, 65]
height (cm)	168.2 (9.258)	[144.0, 199.0]	175.6 (6.670)	[160.5, 199.0]	162.2 (6.155)	[144.0, 182.0]
weight (kg)	73.42 (16.61)	[30.0, 143.5]	82.37 (14.69)	[53.9, 143.5]	66.06 (14.33)	[30.0, 141.6]
arm (cm)	31.93 (4.614)	[20.65, 48.55]	34.43 (3.703)	[25.30, 48.55]	29.88 (4.264)	[20.65, 46.25]
forearm (cm)	26.23 (3.293)	[7.85, 50.45]	28.67 (2.580)	[7.85, 50.45]	24.23 (2.316)	[15.95, 34.30]
waist (cm)	81.30 (12.61)	[37.40, 134.2]	86.55 (11.63)	[37.40, 134.2]	76.98 (11.71)	[55.05, 124.0]
hip (cm)	100.5 (9.427)	[47.50, 148.5]	99.96 (8.161)	[82.75, 135.1]	101.1 (10.32)	[47.50, 148.5]
thigh (cm)	54.28 (6.381)	[28.00, 98.35]	55.43 (5.730)	[28.00, 79.95]	53.34 (6.725)	[37.90, 98.35]
calf (cm)	37.26 (3.555)	[24.55, 53.70]	38.27 (3.096)	[30.00, 49.50]	36.43 (3.690)	[24.55, 53.70]
BMD (g/cm^2)	1.232 (0.136)	[0.789, 1.689]	1.313 (0.123)	[0.944, 1.689]	1.166 (0.107)	[0.789, 1.465]

Figure 1 investigates the linear association between features and target pairs, showing the matrix of Pearson's correlation coefficients. The sex feature is encoded as 1 for *female* and 0 for *male*. From the figure, we see that all anthropometric measurements positively correlate with each other, especially weight and arm circumference (coefficient of 0.89). In addition, all these measurements have a positive correlation with BMD, especially the forearm circumference (coefficient of 0.61). The only features that negatively correlate with BMD are age and especially sex (coefficient of -0.54), indicating that people identified as female at birth tend to have lower BMD.

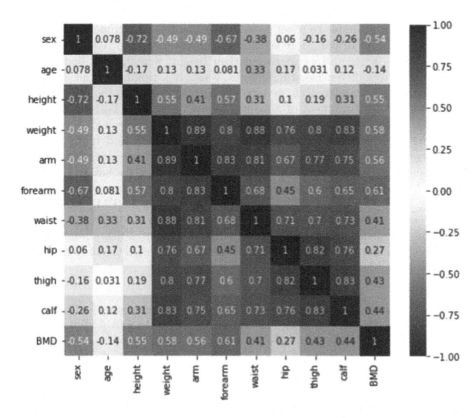

Fig. 1. Matrix of Pearson's correlation coefficients measured between every feature-feature and feature-target pairs. The features that positively correlate with bone mineral density, in descending order, are the forearm, weight, arm, height, calf, thigh, waist, and hip. The features that negatively correlate with bone mineral density, in ascending order, are sex and age.

3.2   Learning Process

For the BMD estimation task, we train and compare 11 regression models. We choose algorithms from different categories to evaluate a wider variety of learning methods. These categories and their respective models are listed below:

- Linear Models: Simple Linear Regression (LR) and Elastic Net (EN);
- Nearest Neighbors: k-Nearest Neighbors (kNN);
- Neural Networks: Multilayer Perceptron (MLP);
- Kernel-based: Support-Vector Machines with a polynomial kernel (SVM-Poly) and a sigmoid kernel (SVM-Sig); and Least-Squares Support-Vector Machines with a linear kernel (LSSVM-Linear) and a radial basis function kernel (LSSVM-RBF).
- Tree-based: Decision Trees (DT), Random Forest (RF), and Gradient-Boosted Decision Trees (GBDT).

We perform nested cross-validation (CV) for hyperparameter tuning (inner loop) and model selection (outer loop). In both loops, we execute a 10-fold CV. In the hyperparameter optimization, we perform a grid search, and the search space for each model is defined in Table 2. In every loop, we standardize the training set and transform the test set using the original mean and standard deviation values from the training set, avoiding data leakage.

Table 2. Hyperparameters tuned by Grid Search and their respective search spaces and models.

Hyperparameter	Search space	Models
alpha $(a + b)$	10^i, $i \in [-6..6]$	EN
L1 ratio $(a/(a + b))$	$\{0, 0.01, 0.1, 0.5, 0.9, 0.95, 0.99, 1\}$	EN
k (number of neighbours)	$\{1, 2, 5, 10, 25, 50, 75, 100\}$	kNN
number of hidden layers	$\{1, 2\}$	MLP
neurons in hidden layer	3^i, $i \in [0..3]$	MLP
activation function	$\{$logistic, tanh, ReLU$\}$	MLP
C	10^i, $i \in [-2..6]$	All SVM and LSSVM
gamma	10^i, $i \in [-6..2]$	SVM (RBF, Poly, and Sig), and LSSVM-RBF
degree	$\{2, 3, 4, 5\}$	SVM-Poly
criterion	$\{$MSE, MAE, Friedman's MSE$\}$	DT and RF
minimum samples at a leaf (%)	$\{1, 2.5, 5, 7.5, 10, 25\}$	DT and RF
maximum tree depth	$\{1, 2, 4, 6, 8, \text{none}\}$	DT, RF, and GBDT
number of trees	$\{10, 50, 100, 500\}$	RF and GBDT
maximum number of leaves	$\{2, 4, 16\}$	GBDT

3.3   Evaluation

Our method uses the coefficient of determination (R^2) to evaluate model performance and select the best hyperparameter configuration in the grid search. For the model selection, besides the R^2 metric, we also report the mean absolute error (MAE) of the estimations.

For a proper comparison among the models, we perform statistical tests on the models' results. First, we use the Friedman test to determine if there are any significant differences between the mean values of the metrics obtained in the outer loop of the nested CV. If we reject the null hypothesis, we conclude that at least two models have significantly different performances. If we arrive at that conclusion, we apply the post hoc Nemenyi test to determine which groups of models have different results.

3.4   Interpretation

Clinicians are responsible for analyzing patient information and recommending the proper treatments or preventive actions. Thus, the lack of transparency in an AI system reduces the level of trust from health professionals, causing it to be one of the main barriers to implementation [19].

Therefore, we use Shapley Additive Explanations (SHAP) [18] to interpret the predictions of the best-performing model. SHAP is a popular model-agnostic method that uses Shapley values (from game theory) to measure the contributions of each feature to the model's prediction [10].

4   Results and Discussion

In this section, we report and discuss the results of the nested cross-validation, the statistical tests, and the interpretation analyses. We make the dataset and the code (with clear instructions) available in a Github repository for computational reproducibility[1].

4.1   Cross-Validation Results

Table 3 presents the mean and standard deviation of the metrics (R^2 and MAE) achieved by the 11 models after the nested CV. We sort the models ascendingly according to the mean MAE and highlight in italics the model that achieved the highest mean R^2 and lowest mean MAE.

Table 3. Results of model evaluation considering the means and standard deviations of MAE and R^2.

Model	MAE (g/cm^2)	R^2
LSSVM-RBF	*0.0769 (0.0056)*	*0.4870 (0.0652)*
MLP	0.0784 (0.0038)	0.4586 (0.0519)
kNN	0.0787 (0.0053)	0.4583 (0.0731)
LSSVM-Linear	0.0788 (0.0054)	0.4540 (0.0704)
EN	0.0788 (0.0054)	0.4543 (0.0706)
LR	0.0789 (0.0053)	0.4526 (0.0667)
RF	0.0789 (0.0054)	0.4598 (0.0567)
SVM-Sig	0.0790 (0.0055)	0.4519 (0.0710)
GBDT	0.0793 (0.0057)	0.4506 (0.0667)
DT	0.0841 (0.0065)	0.3884 (0.0703)
SVM-Poly	0.0842 (0.0051)	0.2928 (0.2188)

Except for Decision Tree and SVM-Poly, all models achieved an R^2 of at least 0.45 and an MAE of less than 0.08 g/cm^2. The algorithm that obtained the best mean value in both metrics was LSSVM-RBF, achieving an R^2 of 0.4870 and an MAE of only 0.0769 g/cm^2. This MAE value represents only 8.5% of the BMD range found in the dataset.

---

[1] https://github.com/gmaiab/Estimating-Bone-Mineral-Density.

However, we cannot assume beforehand that the LSSVM-RBF had better and significantly different results than the other models. For that analysis, we apply the Friedman test to the MAE sets obtained in the ten iterations of the nested CV outer loop. We reject the null hypothesis, concluding that there are significant differences among the regression methods. Thus, we use the Nemenyi test to identify which groups of models have no significant differences. The test shows that algorithms with a mean rank greater than the critical distance (CD) of 4.774 are significantly different. In Fig. 2, we show the critical difference diagram. The image shows that LSSVM-RBF, MLP, and Linear Regression performed significantly better than SMV-Poly and the Decision Tree. We can also observe that LSSVM-RBF achieved a mean rank of 2.4, which is the closest to 1.

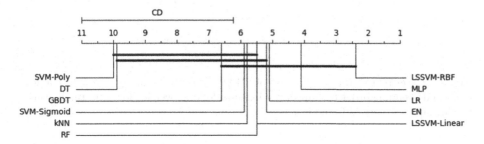

Fig. 2. Critical difference diagram showing the mean rank of each model regarding their MAE.

4.2 Interpretation

First, we calculate the SHAP values for each observation in the data. Then we aggregate these values into a swarm plot, shown in Fig. 3. This plot summarizes the distribution of the impact of each feature on the model's output. Each row shows one feature, and we sort the features by the sum of the SHAP value magnitudes for all patients (the top row has the highest sum).

In a row, each dot represents one observation. Dots further to the left represent cases where the feature contributed to a lower BMD estimation. In contrast, those to the right represent contributions to a higher BMD estimation. Dots with the same impact are distributed vertically to represent the density in the region. In addition, the dots are color-coded to represent the value of the feature: red, yellow, and blue represent high, intermediate, and low values, respectively.

From Fig. 3, we observe that weight has the highest impact on the output of the LSSVM-RBF for BMD prediction, while arm circumference had the lowest impact. The association between weight and BMD is expected and has been found by other researchers [20]. From the distributions presented, high values of weight, height, and forearm, thigh, and arm circumference positively impact the model's output. On the other hand, high values of age and waist, hip, and

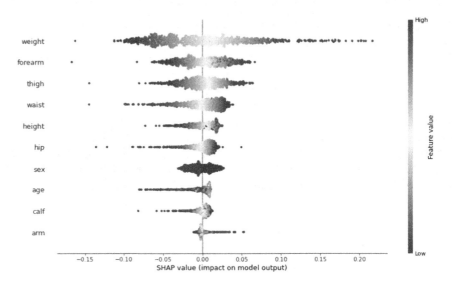

Fig. 3. Swarm plot of the distribution of SHAP values for each observation. The features in descending order by impact are listed as follows: weight, forearm, thigh, waist, height, hip, sex, age, calf, and arm.

calf circumference negatively impact the model's output. Being born female also tends to decrease the estimated BMD value.

It is worth mentioning that, in general, the lowest age values made little difference in the output or even impacted negatively. To evaluate this behavior, we plot the SHAP values for age in Fig. 4. In addition, we explore how the interaction between age and sex affects the model's output.

In Fig. 4, we notice that age has a different impact on the model's output depending on the sex of birth. Younger male patients tend to have a lower BMD estimate than female patients. The peak impact of age happens earlier for female individuals. However, starting at about 40 years of age, advancing age causes a more significant negative impact in female patients than in male patients. This behavior is consistent with studies showing that men have a peak BMD later than women [2, 23] and that after menopause, women's BMD drops considerably [13].

Other distributions that stand out are the forearm, the waist, and the hip. The forearm proved to be the second feature with the greatest impact on model output. It was also the feature with the highest correlation with BMD, as presented in Fig. 1. This relationship can be explained by the high correlation between forearm circumference and lean body mass [5], which has a high association with BMD [9]. On the other hand, the SHAP values showed that high waist or hip circumference values negatively affect the model output. This behavior could be caused by the fact that the waist and the hip have a high association with body fat percentage [7], which has been shown to impact BMD negatively [17].

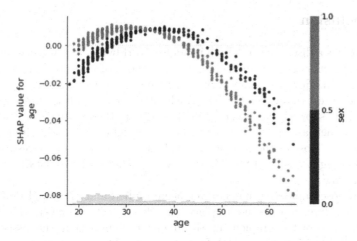

Fig. 4. Scatter plot of the SHAP value for age and its interaction with sex. Red observations represent female patients, and blue observations represent male patients. (Color figure online)

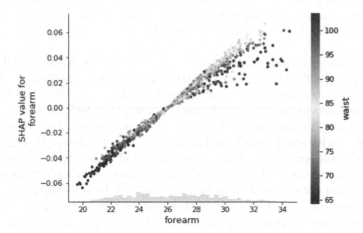

Fig. 5. Scatter plot of the SHAP value for forearm circumference and its interaction with waist circumference. Red, yellow, and blue dots represent high, intermediate, and low waist circumference values, respectively. (Color figure online)

To better analyze the impact of the forearm on model output and its relationship to waist circumference, we plot the chart in Fig. 5. The distribution shows that low forearm values cause a negative impact on the model's output, while high values cause a positive impact. In addition, the x-axis region between 26 and 32 cm indicates that, for people with similar forearm values, there is a more positive impact for patients with low or intermediate waist values than high values. This behavior is consistent with the previous discussion.

5   Conclusion

Osteoporosis is a disease with a high incidence worldwide. It generates high costs for health systems and is responsible for many cases of fracture, hospitalization, and even death. However, not every patient has easy access to DXA, the most recommended test to measure BMD (which allows the diagnosis of osteoporosis).

Our results showed that it is possible to estimate BMD using easily accessible information such as age, sex at birth, height, weight, and body circumference measurements. Most of the algorithms evaluated had satisfactory results. However, the highlight was the LSSVM-RBF, which achieved R^2 of 0.4870 and MAE of $0.0769\,\mathrm{g/cm^2}$.

The proposed approach has not yet been applied and evaluated in a clinical setting. Therefore, it does not replace an official test. However, this method can become a tool to complement the medical examination and assist in the professional's decision. A low BMD estimate might suggest that the patient needs an official test to confirm the value. A low estimate could also influence the health care provider to suggest lifestyles for osteoporosis prevention, such as dietary changes and exercise. Finally, a high BMD estimate, combined with the physician's analysis of other factors, could indicate that the patient does not need to have a DXA at that time, which would reduce costs and avoid radiation exposure.

An essential contribution of our research is the creation and availability of a new dataset for BMD estimation. BMD-10 contains the BMD value, age, sex, and anthropometric measurements of 911 adult patients. Another contribution is the interpretation of the estimates made by the model, which allows for a better understanding of its functioning. The impacts of the features also showed behavior consistent with studies conducted in the area. This consistency brings greater confidence to the specialists who may use the model. In addition, the analysis of the importance of the features and their interactions can bring insights to other researchers and reveal associations that have not yet been widely studied.

For future work, we aim to include and analyze new features, such as eating habits, frequency of sports practice, medication use, and medical history. In addition, we intend to include more patients and patients from other countries. Finally, we would like to evaluate the use of patient images for BMD estimation, which could remove the need for a professional to perform the circumference measurements.

Acknowledgments. This study was financed in part by the Coordenação de Aperfeiçoamento de Pessoal de Nível Superior - Brasil (CAPES) - Finance Code 001. Also Pedro Pedrosa Rebouças Filho acknowledges the sponsorship from the Brazilian National Council for Research and Development (CNPq) via Grants Nos. 431709/2018-1 and 311973/2018-3.

References

1. Consensus development conference: Diagnosis, prophylaxis, and treatment of osteoporosis. Am. J. Med. **94**(6), 646–650 (1993). https://doi.org/10.1016/0002-9343(93)90218-e

2. Avdagić, S., et al.: Differences in peak bone density between male and female students. Arch. Ind. Hyg. Toxicol. **60**(1), 79–86 (2009). https://doi.org/10.2478/10004-1254-60-2009-1886

3. Aziziyeh, R., et al.: A scorecard for osteoporosis in four Latin American countries: Brazil, Mexico, Colombia, and Argentina. Arch. Osteoporos. **14**(1), 1–10 (2019). https://doi.org/10.1007/s11657-019-0622-1

4. Benke, K., Benke, G.: Artificial intelligence and big data in public health. Int. J. Environ. Res. Public Health **15**(12), 2796 (2018). https://doi.org/10.3390/ijerph15122796

5. Chien, K.Y., Chen, C.N., Chen, S.C., Wang, H.H., Zhou, W.S., Chen, L.H.: A community-based approach to lean body mass and appendicular skeletal muscle mass prediction using body circumferences in community-dwelling elderly in taiwan. Asia Pac. J. Clin. Nutr. **29**(1), 94–100 (2020). https://doi.org/10.6133/apjcn.202003_29(1).0013

6. Deo, R.C.: Machine learning in medicine. Circulation **132**(20), 1920–1930 (2015). https://doi.org/10.1161/circulationaha.115.001593

7. Freedman, D.S., et al.: The body adiposity index (hip circumference ÷ height-sup1.5/sup) is not a more accurate measure of adiposity than is BMI, waist circumference, or hip circumference. Obesity **20**(12), 2438–2444 (2012). https://doi.org/10.1038/oby.2012.81

8. Hien, V.T.T., et al.: Determining the prevalence of osteoporosis and related factors using quantitative ultrasound in Vietnamese adult women. Am. J. Epidemiol. **161**(9), 824–830 (2005). https://doi.org/10.1093/aje/kwi105

9. Ho-Pham, L.T., Nguyen, U.D.T., Nguyen, T.V.: Association between lean mass, fat mass, and bone mineral density: a meta-analysis. J. Clin. Endocrinol. Metab. **99**(1), 30–38 (2014). https://doi.org/10.1210/jc.2013-3190

10. Holzinger, A., Saranti, A., Molnar, C., Biecek, P., Samek, W.: Explainable AI methods - a brief overview. In: Holzinger, A., Goebel, R., Fong, R., Moon, T., Müller, K.R., Samek, W. (eds.) xxAI - Beyond Explainable AI, pp. 13–38. Springer, Cham (2022). https://doi.org/10.1007/978-3-031-04083-2_2

11. Hwang, J.J., et al.: Strut analysis for osteoporosis detection model using dental panoramic radiography. Dentomaxillofac. Radiol. **46**(7), 20170006 (2017). https://doi.org/10.1259/dmfr.20170006

12. Iliou, T., Anagnostopoulos, C.N., Anastassopoulos, G.: Osteoporosis detection using machine learning techniques and feature selection. Int. J. Artif. Intell. Tools **23**(05), 1450014 (2014). https://doi.org/10.1142/s0218213014500146

13. Ji, M.X., Yu, Q.: Primary osteoporosis in postmenopausal women. Chronic Dis. Transl. Med. **1**(1), 9–13 (2015). https://doi.org/10.1016/j.cdtm.2015.02.006

14. Kilic, N., Hosgormez, E.: Automatic estimation of osteoporotic fracture cases by using ensemble learning approaches. J. Med. Syst. **40**(3), 1–10 (2015). https://doi.org/10.1007/s10916-015-0413-1

15. Kim, S.K., Yoo, T.K., Oh, E., Kim, D.W.: Osteoporosis risk prediction using machine learning and conventional methods. In: 2013 35th Annual International Conference of the IEEE Engineering in Medicine and Biology Society (EMBC). IEEE, July 2013. https://doi.org/10.1109/embc.2013.6609469

16. Kruse, C., Eiken, P., Vestergaard, P.: Machine learning principles can improve hip fracture prediction. Calcif. Tissue Int. **100**(4), 348–360 (2017). https://doi.org/10.1007/s00223-017-0238-7

17. Liu, Y., et al.: Association of weight-adjusted body fat and fat distribution with bone mineral density in middle-aged Chinese adults: a cross-sectional study. PLoS ONE **8**(5), e63339 (2013). https://doi.org/10.1371/journal.pone.0063339

18. Lundberg, S.M., Lee, S.I.: A unified approach to interpreting model predictions. In: Guyon, I., et al. (eds.) Advances in Neural Information Processing Systems, vol. 30. Curran Associates, Inc. (2017). https://proceedings.neurips.cc/paper/2017/file/8a20a8621978632d76c43dfd28b67767-Paper.pdf

19. Markus, A.F., Kors, J.A., Rijnbeek, P.R.: The role of explainability in creating trustworthy artificial intelligence for health care: a comprehensive survey of the terminology, design choices, and evaluation strategies. J. Biomed. Inform. **113**, 103655 (2021). https://doi.org/10.1016/j.jbi.2020.103655

20. Meybodi, H.A., et al.: Association between anthropometric measures and bone mineral density: population-based study. Iran. J. Public Health **40**(2), 18 (2011). https://www.ncbi.nlm.nih.gov/pmc/articles/PMC3481769/

21. Nayak, S., Roberts, M.S., Greenspan, S.L.: Cost-effectiveness of different screening strategies for osteoporosis in postmenopausal women. Ann. Intern. Med. **155**(11), 751 (2011). https://doi.org/10.7326/0003-4819-155-11-201112060-00007

22. NIH Osteoporosis and Related Bone Diseases - National Resource Center: Bone mass measurement: What the numbers mean, October 2018. https://www.bones.nih.gov/health-info/bone/bone-health/bone-mass-measure

23. NIH Osteoporosis and Related Bone Diseases - National Resource Center: Osteoporosis: Peak bone mass in women, October 2018. https://www.bones.nih.gov/health-info/bone/osteoporosis/bone-mass

24. Shim, J.-G., et al.: Application of machine learning approaches for osteoporosis risk prediction in postmenopausal women. Arch. Osteoporos. **15**(1), 1–9 (2020). https://doi.org/10.1007/s11657-020-00802-8

25. WHO Scientific Group on Prevention and Management of Osteoporosis, World Health Organization: Prevention and management of osteoporosis: report of a WHO scientific group. No. 921, World Health Organization (2003)

26. Yang, W.Y.O., Lai, C.C., Tsou, M.T., Hwang, L.C.: Development of machine learning models for prediction of osteoporosis from clinical health examination data. Int. J. Environ. Res. Public Health **18**(14), 7635 (2021). https://doi.org/10.3390/ijerph18147635

Feature Extraction for a Genetic Programming-Based Brain-Computer Interface

Gabriel Henrique de Souza[✉][ID], Gabriel Oliveira Faria[ID],
Luciana Paixão Motta[ID], Heder Soares Bernardino[ID], and Alex Borges Vieira[ID]

Federal University of Juiz de Fora, Juiz de Fora, Brazil
gabriel.souza@engenharia.ufjf.br

Abstract. Brain-Computer Interfaces (BCI) open a two-way communication channel between a computer and the brain: while the brain can control the computer, the computer can induce changes in the brain through feedback. This mechanism is used in post-stroke motor rehabilitation, in which a BCI provides feedback by classifying signals collected from a patient's brain. Single Feature Genetic Programming (SFGP) can create classifiers for these signals. However, the Genetic Programming (GP) step in SFGP requires a set of extracted features to generate its model. To the best of our knowledge, the LogPower function is the only initial feature extraction function used in SFGP. Nevertheless, other functions can improve the quality of the generated classifiers. Thus, we analyze new initial feature extraction functions for GP in SFGP. We test the Common Spatial Patterns, Nonlinear Energy, Average Power Spectral Density, and Curve Length methods on two datasets suitable for post-stroke rehabilitation training. The results obtained show that the analyzed functions outperform LogPower in all our experiments, with a kappa value up to 25.20% better. We further test the proposed methods on a third dataset, created with low-cost equipment. In this case, we show that the Average Power Spectral Density function outperforms LogPower by 11.39% when three electrodes are used. Thus, we demonstrate that the new approach can be used with low-cost equipment and a small number of electrodes, reducing the financial costs of treatment and improving patients' comfort.

Keywords: Brain-machine interface · Common spatial pattern · Stroke rehabilitation · Power spectral density

1 Introduction

Brain-Computer Interfaces (BCI) aim at providing a non-muscular channel for sending commands to the external world using electroencephalographic activity or other electrophysiological measurements of the brain function [5]. Any system

Supported by CNPq, Capes, FAPEMIG, and UFJF.

that takes human input to control an electronic device can use a BCI approach. Therefore, BCIs have a variety of applications, including prosthesis control [36], computer spelling [40], drone control [16], neuromarketing [42], and security [23]. In addition, there are healthcare applications such as post-stroke motor rehabilitation [4], attention deficit/hyperactivity disorder treatment [25], brain tumor detection [30], epileptic seizure detection [39], and dyslexia diagnosis [9].

BCI software usually classifies brain signals in order to perform tasks. In this process, many methods can be used, including Bandpass Filtering [29], Wavelet Transform [27], Common Spatial Pattern [28], Convolutional Neural Network [26], Support Vector Machine [28]. In addition, several evolutionary computing techniques have been used to improve the efficiency of classifying BCIs such as Genetic Algorithm [8], Particle Swarm Optimization [21], Differential Evolution [34], Genetic Programming [35], and Ant Colony Optimization [22].

The design of BCI software consists of four main phases: signal acquisition, preprocessing, feature extraction, and classification [1]. Feature extraction is an important task to predict outcomes from the raw signal [2]. Genetic Programming (GP) is an efficient way to extract features from EEG data [35]. However, GP requires the extraction of initial features for its leaves. In the literature, the LogPower function is used to generate these initial features for a post-stroke motor rehabilitation BCI [35].

We propose the use of new functions to extract the initial features for a post-stroke motor rehabilitation BCI, which help patients recover motor ability. Within this treatment, the patient imagines the movement of their body, and the BCI provides the patient with feedback about the imagined movement. We evaluate the Power Spectral Density, Common Spatial Pattern, Non-Linear Energy, and Curve Length Functions to extract these initial features for GP. The experiments show that these functions are more suitable for extracting the initial features for GP than the Logpower function.

2   Brain-Computer Interface and Post-stroke Motor Rehabilitation

BCI development is an interdisciplinary problem, involving neurobiology, psychology, engineering, mathematics, computer science, and clinical rehabilitation [41]. A BCI transforms brain signals from reflections of the central nervous system into the products of that activity: messages or commands [41].

The BCI pipeline consists of four main steps [1], as shown in Fig. 1: (i) the person imagines/performs an activity, (ii) a device collects brain signals, (iii) a BCI software classifies the signals, and (iv) an application executes the command. After performing the command, the person observes the result of the application (feedback) and the BCI pipeline restarts. A BCI software receives the recorded signal and returns its label. To that end, the signal goes through three main steps [1]: (i) a preprocessing step consisting of temporal and spatial filters, (ii) a feature extraction step, and (iii) a classification step. The preprocessing step aims to reduce signal noise. The feature extraction step aims to extract good features for the classifier. Finally, the classifier predicts the label corresponding to the input signal.

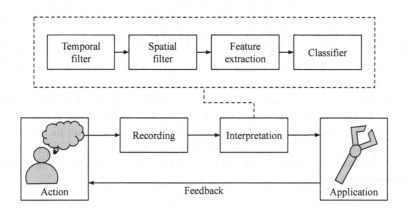

Fig. 1. Brain-Computer Interface pipeline.

In 2019, there were 113.2 million cases of stroke worldwide, which caused 6.55 million deaths and 143 million DALYs [10]. A large proportion of people who survive a stroke need treatments in stroke units. There are several types of intervention for post-stroke rehabilitation such as pharmacological, technological, and neuromodulation-based treatments [37]. Other strategies based on neuroplasticity can be used, such as robot-assisted therapy, reinforced feedback in virtual environment, and Brain-Computer Interface [31]. Most BCIs for motor rehabilitation are based on motor imagery and motor execution. The typical activities for these types of BCI are: (i) the patient imagines a movement and (ii) they receive a feedback of this movement from a screen or a robotic arm. This feedback is able to reinforce the brain circuit of the imagined movement [31]. It is recommended to use few electrodes in this application to facilitate the treatment and to increase the comfort of the patient [19,34].

3 Datasets

Several BCI datasets are available in the literature. In general, brain activity patterns are measured using electroencephalography (EEG) which is a noninvasive, easy to use, low cost method [24]. EEG uses electrodes to record the electrical potential on the scalp. We used two datasets from BCI Competition IV and a third dataset collected by our research group. We obtained and processed the two first datasets using the MOABB [15] and MNE [13] libraries.

3.1 BCI Competition IV 2a

The BCI Competition IV 2a dataset [38] is a motor imagery dataset with four motor imagery tasks (left-hand, right-hand, both-feet, and tongue). The signals were collected using twenty-two monopolar electrodes with left-mastoid as a reference. The signals were sampled 250 Hz and then preprocessed with a bandpass filter of 0.5–100 Hz, and 50 Hz notch filter. The positions of the 22 electrodes are Fz, FC3, FC1, FCz, FC2, FC4, C5, C3, C1, Cz, C2, C4, C6, CP3, CP1, CPz, CP2, CP4, P1, Pz, P2, POz. The dataset contains EEG signals from 9 subjects. Each subject has two sessions collected on different days. Each session has six runs with 48 trials in each one (12 per class), resulting in 288 trials per session. Each trial is approximately 8 s long and adhere to the following steps: (i) 0 to 2 s: a fixation cross appears in the screen; (ii) 2 to 3.25 s: a cue in the form of an arrow pointing left, right, down, or up is shown; (iii) 3 to 6 s: the subject imagines the indicated movement; and (iv) 6 to ∼8 s: a short break is provided before the next run.

3.2 BCI Competition IV 2b

The BCI Competition IV 2b dataset [20] is a motor imagery dataset with two motor imagery tasks (left-hand and right-hand). The signals were collected using three bipolar electrodes at a sampling rate 250 Hz, and then preprocessed with a bandpass filter of 0.5–100 Hz and 50 Hz notch filter. The bipolar electrodes are C3 (difference between FC3 and CP3), Cz (difference between FCz and CPz), and C4 (difference between FC4 and CP4). The dataset contains data from nine participants and each subject has five sessions. The last three sessions contain feedback and the subject can see the classification while they execute the task. Each session has a 2 weeks interval and 6 runs with 20 trials (10 per class), resulting in 120 trials per session. The sessions with feedback contain 160 trials per session. Each trial is approximately 9 s long, and comprises the following steps: (i) 0 to 2 s: a gray smiling face appears in the screen; (ii) 2 s: a beep sounds; (ii) 3 to 7.5 s: a cue is shown in the screen; (iii) 3 to 7.5 s: the subject imagines the movement; and (iv) 7.5 to ∼9 s: a short break is provided before next run.

3.3 Our Dataset

The collection of these data was approved by an ethical committee[1]. We collected the data using the OpenBCI Cyton+Daisy Biosensing Boards (16-Channels) and electrode cap equipment[2]. This is a low-cost BCI equipment that has suitable accuracy for BCI applications [11]. The signals were sampled 125 Hz and were collected through 16 electrodes: Fp1, Fz, C3, C4, T5, T6, Cz, Pz, F7, F8, F3, F4,

---

[1] Approved by the ethics committee of Federal University of Juiz de Fora under the number 47866121.5.0000.5147.

[2] https://openbci.com/.

T3, T4, P3, P4. The dataset consists of left-hand and right-hand motor imagery task data from six healthy subjects. Each subject has one session with 4 runs with 40 trials (20 per class). In this way, this dataset has 160 trials per subject. Each trial is 10 s long and comprises the following steps: (i) 0 to 3 s: a fixation cross appeared in the screen; (ii) 2 s: a beep sounds; (ii) 3 to 5 s: a cue is shown in the screen; (iii) 5 to 9 s: the subject imagine the movement; and (iv) 9 to 10 s: a short break is provided before next run.

4   Proposed Method

Feature extraction is an important step within the BCI software pipeline. These features try to represent the main signal information for the classifier. Single Feature Genetic Programming (SFGP) is a BCI pipeline that uses genetic programming (GP) to extract a feature from EEG data [35]. Figure 2 shows the SFGP pipeline, which has three main steps: (i) preprocessing, (ii) initial feature extraction, (iii) feature extraction, and (iv) classification.

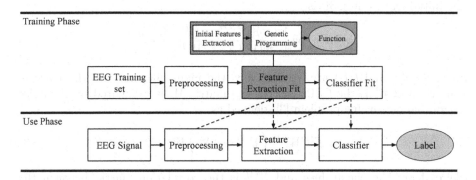

Fig. 2. Steps of training and use phases of single feature genetic programming [35].

In the first step, a preprocessing methodology must be applied to the signal in order to increase the signal-to-noise ratio [2]. Band-pass filter, wavelet transformation, and filter bank are commonly used for this purpose, with Band Pass filter yielding the best results for SFGP according to the literature [35]. The Bandpass Filter preprocessing can be represented as $Z_i = BP_{7-35}(X_i)$, where $X_i \in \mathbb{R}^{e \times t}$ is the ith trial of the training dataset, e is the number of electrodes, t is the number of time samples in the trial, $BP_{7-35}(\cdot)$ is the 7–35 Hz bandpass function, and $Z_i \in \mathbb{R}^{e \times t}$ is the filtered signal.

4.1   Initial Feature Extraction

After the preprocessing step, the signal Z_i goes through the feature extraction step. The feature extraction has 2 parts: (i) Initial feature extraction; and

(ii) Genetic Programming. The initial feature extraction transforms each electrode series in a number $I_i = E(Z_i)$, where $I_i \in \mathbb{R}^{e \times t}$ is the initial features of $Z_{c,i}$ and $E(\cdot)$ is the initial feature extraction function. The literature uses the LogPower function to extract these initial features as

$$I_{i,e}^{LogPower} = \log \left(\sum_{j}^{N} \frac{|Z_{i,e,j}|}{N} \right) \tag{1}$$

where N is the number of electrodes and $Z_{i,e}$ is the eth electrode of the trial Z_i. However, it is possible to use other initial feature extraction functions. In this work, we propose and analyze the use of four new functions to extract the initial features on SFGP: Common Spatial Patterns, Nonlinear Energy, Average Power Spectral Density, and Curve Length.

Common Spatial Pattern. Common Spatial Pattern (CSP) is a spatial filter and a feature extraction method. It is a transformation that aims to maximize the variance between two groups and minimize the variances within each group [12]. The initial feature in CSP is

$$I_{CSP} = \log \left[\frac{diag(W^T Z_i \times (W^T Z_i)^T)}{tr(W^T Z_i \times (W^T Z_i)^T)} \right] \tag{2}$$

where $diag(\cdot)$ and $tr(\cdot)$ are, respectively, the main diagonal and the trace of a matrix and W is the CSP matrix. The matrix W is calculated by solving a generalized eigenvalue problem [3], defined as

$$\Sigma^{(1)} W = \left(\Sigma^{(1)} + \Sigma^{(2)} \right) W \Lambda \tag{3}$$

where Λ are the eigenvalues and $\Sigma^{(c)}$ is the correlation matrix between the electrodes of a class c trial, defined as

$$\Sigma^{(c)} = \frac{1}{N} \sum_{i}^{N} Z_i^{(c)} \times Z_i^{(c)T} \tag{4}$$

Nonlinear Energy. The Nonlinear Energy (NE) is a extension of the concept of quadratic measure energy [6,17] and can be calculated as

$$I_{NE} = \sum_{j=1}^{N-2} (Z_i^2[j] - Z_i[j+1] \cdot Z_i[j-1]) \tag{5}$$

where Z_i is the ith electrode of a trial Z.

Average Power Spectral Density. Power Spectral Density (PSD) extracts the power intensity in the frequency spectrum. PSD is calculated by estimating the autocorrelation sequence across frequencies. Here, we use the Welch's method to estimate the autocorrelation [14, 32]. Thus, the initial feature using the average PSD (APSD) is obtained by

$$I_{APSD} = \frac{1}{K} \sum_{i=0}^{K-1} \frac{1}{MU} \left| \sum_{j=0}^{N-1} w(n) Z_{n_i + iD} e^{-j2\pi fn} \right|^2 \tag{6}$$

where N is the number of time samples of the trial, f is the sampling rate, K is the number of overlapping segments, M is the length of each segment, D_i is the overlapped segment, $j = \sqrt{-1}$, $w(n)$ is the window function, Z_t is the tth time step of a electrode Z and U is the power of the window function.

Curve Length. Curve Length (CL) is the total vertical length of the signal. CL was proposed to approximate the Katz's fractal dimension [6]. The initial feature using CL can be defined as

$$I_{CL} = \sum_{j=1}^{N-1} |Z_i[j] - Z_i[j-1]| \tag{7}$$

where Z_i is the ith electrode of a trial Z.

4.2   Genetic Programming

Genetic Programming (GP) is an evolutionary algorithm for generating programs in arbitrary languages [18]. These programs can be a function, a circuit design, a formal language, a classifier, among other structures. Like most evolutionary algorithms, GP has four fundamental steps: (i) create an initial population, (ii) reproduce the population, (iii) select the individuals for the next generation, and (iv) verify a stop criterion.

The tree structure is used to represent the candidate functions in SFGP. The operators are in the internal nodes and the initial features are in the leaves of the trees. GP creates the initial population randomly and Fig. 3 shows an example of an individual.

The GP algorithm creates new individuals by applying crossover and mutation operators on the current population. The parents are chosen by a tournament selection. The one-point crossover swaps a subtree from the selected individuals with a user-defined probability. Thereafter, the uniform mutation operator replaces a subtree of the new individual with a new subtree with a user-defined probability.

We used the cross-entropy as the fitness function. The cross-entropy of each individual is calculated as

$$G(W) = -\sum_{i=1}^{N} \sum_{j=1}^{C} \ln(1 - Y_W(X_i, j)) \cdot int(j \neq \theta_i)$$

$$+ \ln(Y_W(X_i, j)) \cdot int(j = \theta_i) \tag{8}$$

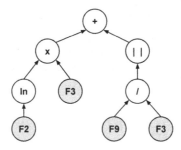

Fig. 3. Function $V(F) = (\ln(F2) \times F3) + |F9/F3|$ represented using a tree [35]

where N is the number of training samples, C is the number of classes, $Y_W(X_i, j)$ is the probability calculated by the classifier for the sample X_i being of class j when using the spatial filter W, $int(.)$ is a function that returns 1 if its argument is true, and 0 otherwise, and θ_i is the class of the i-th sample [34]. The probability calculated by the classifier $Y_W(X_i, j)$ in Eq. 8 is defined as

$$Y_W(X_i, j) = \frac{1}{1 + e^{-(\alpha + \beta V)}} \tag{9}$$

where V is the final feature calculated by GP using the initial features, α and β are estimated by the Limited-memory Broyden-Fletcher-Goldfarb-Shanno algorithm (LBFGS) [7] using the training data.

The new generation totally replaces the previous generation and the best individual in each generation is stored. The process is repeated until the maximum number of generations is reached. Finally, the best individual of all generations is used to evaluate the test data.

5   Computational Experiments

Computational experiments were performed to evaluate the performance of the feature extraction methods when compared to other approaches from the literature. We chose the datasets for the computational experiments to simulate the most suitable conditions for post-stroke motor rehabilitation. The conditions of interest in choosing the dataset were: (i) 3 electrodes or the possibility of reduction to 3 electrodes, (ii) 2 classes of motor imagery or the possibility of reduction to 2 classes, and (iii) different forms of acquisition (monopolar and bipolar). We used only right-hand and left-hand classes in the experiments as two classes are enough for post-stroke motor rehabilitation. Using few electrodes are recommended to increase comfort of the patient during treatment and preparation time. Therefore, different numbers of electrodes were evaluated with dataset 2a in Sect. 5.1 to verify if the quality of the results are affected by the reduction of the number of electrodes. We selected dataset 2b to verify if a bipolar acquisition improves the SFGP efficiency. Bipolar acquisition requires approximately twice as many electrodes to collect the same amount of signals compared to monopolar acquisition. In addition, as the 2b dataset has a 2-weeks interval between sessions, we also evaluated the cross-session result. SFGP was also compared against Single Electrode Energy (SEE) [33], the BCI method used by the winning team of the Clinical BCI Challenge 2020[3]. This competition was focused on motor imagery classification for post-stroke rehabilitation and, thus, SEE was developed for the same application addressed here and obtained results better than those found by other approaches. For all experiments, we used 5-fold cross-validation for each subject/session pair, except for the cross-session case. We use the same method for evaluating the results of the BCI Competition IV, from where we take datasets 2a and 2b. We used a 2-second window for training and testing. The training was performed with the window starting at 0.5 s after the exhibition of the cue. We calculated the kappa every 0.1 s of the test trials (the first window started when the cue is presented), and the final result is the best kappa found. We implemented SFGP using the DEAP library[4], and the source code of the proposed method and other implemented techniques are available[5]. The parameters used for GP were: (i) population size = 100; (ii) number of generation = 300; (iii) crossover probability = 0.85; (iv) mutation probability = 0.2; (v) max depth = 6; (vi) max initial population death = 3; (vii) max mutation tree depth = 2. Similar to the parameters used in the literature for the SFGP [35].

5.1   Analysis of the Number of Electrodes

We used the 2a dataset to evaluate the use of three different numbers of electrodes for monopolar acquisition. We consider the following configurations: (i) all 22 electrodes presented in Sect. 3.1; (ii) 8 electrodes (FC3, C3, CP3, FCz, CPz, FC4, C4, CP4); and (iii) 3 electrodes (C3, Cz, C4). The results for each number of electrodes can be seen in Tables 1 and 2.

Table 1 shows the results for 22 electrodes. SFGP presented better results than SEE regardless of the extraction function and the best result was found with the APSD function. The SFGP-APSD was the best method for 4 of 9 subjects. Its average result for all subjects was 74.01% better than that reached by SEE and 25.20% better than SFGP-LogPower. It is also interesting to notice that SFGP-CSP presented results similar to those found by SFGP-APSD (4.05% difference). Table 2 shows the results for 8 electrodes. SFGP presented better results than SEE regardless of the extraction function and LogPower function presented the best result for 4 out of 9 subjects. However, SFGP-CSP presented the best average result, and SFGP-LogPower was only the third-best average result. SFGP-CSP presented an average result 69.44% better than that found by SEE and 19.14% better than SFGP-LogPower.

Table 2 shows the results for 3 electrodes. Again, SFGP presented better results than SEE regardless of the extraction function used. SFGP-CL and

---

[3] https://sites.google.com/view/bci-comp-wcci.

[4] https://github.com/DEAP/deap.

[5] https://github.com/ghdesouza/bci.

Table 1. Median kappa for 5-fold cross-validation for dataset 2a. The best results are boldfaced.

ID	SEE	SFGP LogPower	SFGP CSP	SFGP NE	SFGP APSD	SFGP CL
1	0.089	0.235	**0.331**	0.281	0.301	0.301
2	0.086	0.214	0.276	0.159	**0.346**	0.173
3	0.179	0.407	0.374	0.248	**0.418**	0.313
4	0.165	0.233	0.268	0.090	**0.298**	0.157
5	**0.308**	0.202	0.244	0.207	0.266	0.249
6	0.115	**0.242**	0.213	0.207	0.165	0.159
7	0.173	0.274	0.264	0.138	**0.294**	0.132
8	0.136	0.303	0.308	**0.346**	0.278	0.302
9	0.312	0.347	0.380	0.316	0.316	**0.422**
All	0.177	0.246	0.296	0.239	**0.308**	0.248

SFGP-LogPower presented the best result for 3 out of 9 subjects each. However, SFGP-APSD presented the best average result, and SFGP-LogPower reached the third-best average result. SFGP-APSD performed an average result 63.64% better than that found by SEE and 22.89% better than SFGP-LogPower. SFGP-APSD presented the best average result using both 22 and 3 electrodes. Furthermore, the result with 22 electrodes was only 0.65% better than that obtained with 3 electrodes. This result indicates that the reduction in the number of electrodes does not significantly impact the final result for the best method found (SFGP-APSD). It is important to remember that it's recommends the use of few electrodes in this application to facilitate the treatment and to increase the comfort of the patient [19,34]. Thus, the possibility of reducing the number of electrodes without reducing the classification efficiency is an important advantage of SFGP-APSD.

Table 2. Median kappa for 5-fold cross-validation for dataset 2a using 8 and 3 electrodes. The best results are boldfaced.

	8 electrodes						3 electrodes					
ID	SEE	SFGP LogPower	SFGP CSP	SFGP NE	SFGP APSD	SFGP CL	SEE	SFGP LogPower	SFGP CSP	SFGP NE	SFGP APSD	SFGP CL
1	0.238	0.244	**0.337**	0.279	0.276	0.237	0.104	0.179	0.251	0.246	0.228	**0.266**
2	0.116	**0.270**	0.244	0.244	0.239	0.246	0.111	0.196	0.185	0.109	**0.306**	0.189
3	0.167	**0.383**	0.345	0.290	0.314	0.316	0.187	**0.449**	0.308	0.278	0.401	0.377
4	0.250	0.251	**0.325**	0.202	0.159	0.212	0.242	0.232	0.236	0.167	0.308	**0.313**
5	0.283	0.185	**0.303**	0.283	0.230	0.175	0.280	**0.301**	0.246	0.113	0.264	0.228
6	0.080	**0.269**	0.239	0.244	0.248	0.178	0.019	**0.237**	0.228	0.183	0.201	0.152
7	0.161	**0.256**	0.140	0.155	0.226	0.163	0.226	0.239	**0.248**	0.231	0.239	0.226
8	0.153	0.335	0.342	0.209	0.257	**0.414**	0.067	0.209	0.327	0.204	0.357	**0.379**
9	0.380	0.313	0.550	0.366	0.366	**0.471**	0.374	0.251	**0.415**	0.332	0.382	0.386
All	0.187	0.256	**0.305**	0.244	0.269	0.246	0.187	0.249	0.251	0.214	**0.306**	0.268

5.2 Analysis of Within and Cross-Session Training

We use the 2b dataset to evaluate within-session and cross-session experimental procedures. This dataset has five sessions with two weeks intervals between each of them. We performed two experiments considering this interval between sessions: (i) within-session: we performed 5-fold cross-validation for each session separately; and (ii) cross-session: we mixed all sessions before splitting the 5 folds. Table 3 presents the results for the within and cross-session cases. For the within-session case, the best method was SFGP-CSP. Again, it is possible to observe that SFGP presented results better than those found by SEE regardless of the extraction function used. SFGP-CSP performed a kappa average 101.20% better than SEE and 33.60% better than SFGP-LogPower. Also, the results obtained by SFGP-CL showed an average kappa difference of only 0.60% compared to SFGP-CSP. We also compared with-session monopolar and bipolar acquisitions and the results are presented, respectively, in Tables 2 and (Table 3. The results obtained using monopolar acquisition are better than those reached using bipolar acquisition. However, bipolar acquisition requires approximately twice as many electrodes compared to monopolar acquisition to record the same amount of signal. For the cross-session case, the best methods were SFGP-APSD and SFGP-CL, both with a kappa of 0.063. However, the results obtained for the cross-session case were not satisfactory. The results obtained with the within-session case was 165.08% better than the those with the cross-session case, indicating that the features found by SFGP change over the weeks. Thus, it is necessary to retrain the model periodically for its continued use.

Table 3. Median kappa for within-subject within- and cross-session 5-fold cross-validation for dataset 2b. The best results are boldfaced.

	Within-session						Cross-session					
ID	SEE	SFGP LogPower	SFGP CSP	SFGP NE	SFGP APSD	SFGP CL	SEE	SFGP LogPower	SFGP CSP	SFGP NE	SFGP APSD	SFGP CL
1	0.083	0.188	0.188	0.167	0.125	**0.250**	0.042	0.042	0.042	**0.083**	**0.083**	0.063
2	0.083	0.167	0.167	0.167	0.167	**0.188**	0.000	**0.068**	0.023	0.023	0.023	**0.068**
3	0.083	0.125	**0.250**	0.167	0.125	0.083	**0.125**	0.021	0.021	−0.021	**0.125**	0.063
4	0.083	0.143	**0.188**	0.125	0.063	0.167	0.000	0.021	0.000	0.083	−0.042	**0.104**
5	0.167	0.143	0.125	**0.188**	**0.188**	**0.188**	0.125	0.125	0.042	0.083	**0.188**	0.146
6	0.000	**0.188**	0.167	0.125	**0.188**	**0.188**	**0.063**	0.042	0.021	0.000	**0.063**	0.021
7	0.083	0.063	**0.167**	0.083	0.125	0.063	0.000	**0.083**	0.021	0.021	0.042	0.063
8	0.063	0.125	**0.188**	**0.188**	0.125	0.167	**0.125**	0.021	0.021	0.000	0.083	0.042
9	0.125	0.125	**0.188**	0.083	0.167	0.167	**0.104**	0.042	0.021	0.042	0.000	0.021
all	0.083	0.125	**0.167**	0.125	0.125	0.166	0.042	0.042	0.021	0.023	**0.063**	**0.063**

5.3 Analysis Using Data Obtained with Low-Cost Equipment

We created the protocol presented in Sect. 3.3 based on the results presented in Sects. 5.1 and 5.2. We collected data from 6 subjects with 16 electrodes, and we used an acquisition protocol similar to the other two datasets. We used within-session monopolar acquisition as this setup yielded the best results in the other datasets. In this Section, we performed the experiments using 16 and 3 electrodes. The case of 3 electrodes using OpenBCI is interesting as it is the most suitable for post-stroke motor rehabilitation. This experiment was performed with low-cost equipment, when compared to the others results, and has the advantage of using a small number of electrodes. This allows for greater diffusion of this treatment. Table 4 shows the results for 16 electrodes. SFGP-NE obtained the best result with a average kappa of 0.344. SFGP-NE average kappa was 37.60% better than SEE and 10.97% better than SFGP-LogPower.

The results for 3 electrodes are shown in Table 4. SFGP-APSD and SFGP-CL obtained the best results with an average kappa of 0.313. This result was 25.20% better than SEE and 11.39% better than SFGP-LogPower. This result is the most promising initiative for the dissemination of the use of BCIs for post-stroke motor rehabilitation.

Table 4. Median kappa for 5-fold cross-validation for our dataset with 8 and 3 electrodes. The best results are boldfaced.

ID	8 electrodes						3 electrodes					
	SEE	SFGP LogPower	SFGP CSP	SFGP NE	SFGP APSD	SFGP CL	SEE	SFGP LogPower	SFGP CSP	SFGP NE	SFGP APSD	SFGP CL
1	**0.438**	0.313	0.125	0.375	0.313	**0.438**	**0.438**	0.375	0.313	**0.438**	**0.438**	**0.438**
2	0.250	0.313	0.188	0.375	**0.438**	0.375	0.250	0.313	0.188	0.375	0.375	**0.438**
3	0.313	0.438	**0.563**	0.375	0.438	0.438	0.313	0.313	0.313	**0.375**	0.250	0.250
4	0.125	**0.250**	0.063	0.188	0.125	0.125	0.125	0.125	0.063	**0.188**	0.125	0.063
5	0.125	0.250	0.063	**0.375**	0.313	0.250	0.125	0.125	0.063	0.063	0.250	**0.313**
6	0.188	0.063	0.125	**0.313**	0.188	0.125	**0.188**	0.125	0.063	**0.188**	**0.188**	0.125
all	0.250	0.313	0.125	**0.344**	0.313	0.313	0.250	0.281	0.125	0.281	**0.313**	**0.313**

6   Concluding Remarks and Future Work

BCIs can be used in many types of applications, including post-stroke motor rehabilitation. This work introduced four new initial feature extraction methods for genetic programming in BCI training: Common Spatial Patterns, Nonlinear Energy, Average Power Spectral Density, and Curve Length. We performed the experiments with three different datasets. Each dataset experimented with has its characteristics, such as the number of electrodes, acquisition method, and data collection protocol. We tested different feature extraction functions in three relevant situations for the advancement and dissemination of BCI: (i) different number of electrodes; (ii) cross-session training; (iii) low-cost equipment. The new evaluated functions obtained better results than single electrode energy (SEE) and single feature genetic programming with the Log-Power (SFGP-LogPower) function for all experiments performed. The results obtained show that the average power spectral density (APSD) function is the best feature extraction function for SFGP. Furthermore, the results show that SFGP-APSD is robust in terms of the number of electrodes, achieving similar results with 22, 16, or 3 electrodes. Using SFGP-APSD, proposed in this work, we obtained a better result than datasets 2a and 2b using low-cost equipment with only three electrodes. This approach reduces the financial cost of treatment, improves patient comfort, shortens the time to prepare patient sessions, and increases the classification capability of the BCI. For the cross-session case, none of the evaluated functions presented satisfactory results. In addition, more experiments are needed to verify the statistical significance of the results obtained. In future work, it is possible to investigate ways to improve SFGP for the cross-session case. The possibility of cross-session training with SFGP can further improve the practicality of using BCI for post-stroke motor rehabilitation.

References

1. Abdallah, N., Khawandi, S., Daya, B., Chauvet, P.: A survey of methods for the construction of a brain computer interface. In: 2017 Sensors Networks Smart and Emerging Technologies (SENSET), pp. 1–4. IEEE (2017)
2. Alamdari, N., Haider, A., Arefin, R., Verma, A.K., Tavakolian, K., Fazel-Rezai, R.: A review of methods and applications of brain computer interface systems. In: 2016 IEEE International Conference on Electro Information Technology (EIT), pp. 0345–0350. IEEE (2016)
3. Ang, K.K., Chin, Z.Y., Wang, C., Guan, C., Zhang, H.: Filter bank common spatial pattern algorithm on BCI competition iv datasets 2a and 2b. Front. Neurosci. **6**, 39 (2012)
4. Ang, K.K., Guan, C.: Brain-computer interface in stroke rehabilitation. J. Comput. Sci. Eng. **7**(2), 139–146 (2013)
5. Bashashati, A., Fatourechi, M., Ward, R.K., Birch, G.E.: A survey of signal processing algorithms in brain-computer interfaces based on electrical brain signals. J. Neural Eng. **4**(2), R32 (2007)
6. Boonyakitanont, P., Lek-Uthai, A., Chomtho, K., Songsiri, J.: A review of feature extraction and performance evaluation in epileptic seizure detection using EEG. Biomed. Signal Process. Control **57**, 101702 (2020)
7. Broyden, C.G.: The convergence of a class of double-rank minimization algorithms 1. general considerations. IMA J. Appl. Math. **6**(1), 76–90 (1970)
8. Eslahi, S.V., Dabanloo, N.J., Maghooli, K.: A GA-based feature selection of the EEG signals by classification evaluation: application in BCI systems. arXiv preprint arXiv:1903.02081 (2019)
9. Fadzal, C., Mansor, W., Khuan, L.: Review of brain computer interface application in diagnosing dyslexia. In: 2011 IEEE Control and System Graduate Research Colloquium, pp. 124–128. IEEE (2011)
10. Feigin, V.L., et al.: Global, regional, and national burden of stroke and its risk factors, 1990–2019: a systematic analysis for the global burden of disease study 2019. Lancet Neurol. **20**(10), 795–820 (2021)
11. Frey, J.: Comparison of an open-hardware electroencephalography amplifier with medical grade device in brain-computer interface applications. arXiv preprint arXiv:1606.02438 (2016)
12. Fukunaga, K.: Introduction to Statistical Pattern Recognition. Academic Press, San Diego (1990)
13. Gramfort, A., et al.: MEG and EEG data analysis with MNE-python. Front. Neurosci. **7**, 267 (2013)

14. Gupta, A., et al.: On the utility of power spectral techniques with feature selection techniques for effective mental task classification in noninvasive BCI. IEEE Trans. Syst. Man Cybern. Syst. **51**(5), 3080–3092 (2019)
15. Jayaram, V., Barachant, A.: MOABB: trustworthy algorithm benchmarking for BCIs. J. Neural Eng. **15**(6), 066011 (2018)
16. Jeong, J.H., Lee, D.H., Ahn, H.J., Lee, S.W.: Towards brain-computer interfaces for drone swarm control. In: 2020 8th International Winter Conference on Brain-Computer Interface (BCI), pp. 1–4. IEEE (2020)
17. Kaiser, J.F.: On a simple algorithm to calculate the 'energy' of a signal. In: International Conference on Acoustics, Speech, and Signal Processing, pp. 381–384. IEEE (1990)
18. Koza, J.R.: Genetic Programming: On the Programming of Computers by Means of Natural Selection. The MIT Press, Cambridge (1992)
19. Leeb, R., Lee, F., Keinrath, C., Scherer, R., Bischof, H., Pfurtscheller, G.: Brain-computer communication: motivation, aim, and impact of exploring a virtual apartment. IEEE Trans. Neural Syst. Rehabil. Eng. **15**(4), 473–482 (2007)
20. Leeb, R., Lee, F., Keinrath, C., Scherer, R., Bischof, H., Pfurtscheller, G.: Brain-computer communication: motivation, aim, and impact of exploring a virtual apartment. IEEE Trans. Neural Syst. Rehabil. Eng. **15**(4), 473–482 (2007)
21. Li, Z., et al.: Enhancing BCI-based emotion recognition using an improved particle swarm optimization for feature selection. Sensors **20**(11), 3028 (2020)
22. Miao, M., Zhang, W., Hu, W., Wang, R.: An adaptive multi-domain feature joint optimization framework based on composite kernels and ant colony optimization for motor imagery EEG classification. Biomed. Signal Process. Control **61**, 101994 (2020)
23. Nakanishi, I., Ozaki, K., Li, S.: Evaluation of the brain wave as biometrics in a simulated driving environment. In: 2012 BIOSIG-Proceedings of the International Conference of Biometrics Special Interest Group (BIOSIG), pp. 1–5. IEEE (2012)
24. Pawar, D., Dhage, S.: Feature extraction methods for electroencephalography based brain-computer interface: a review. IAENG Int. J. Comput. Sci. **47**(3) (2020)
25. Qian, X., et al.: Brain-computer-interface-based intervention re-normalizes brain functional network topology in children with attention deficit/hyperactivity disorder. Transl. Psychiatry **8**(1), 1–11 (2018)
26. Ravi, A., Beni, N.H., Manuel, J., Jiang, N.: Comparing user-dependent and user-independent training of CNN for SSVEP BCI. J. Neural Eng. **17**(2), 026028 (2020)
27. Sadiq, M.T., Yu, X., Yuan, Z., Aziz, M.Z.: Motor imagery BCI classification based on novel two-dimensional modelling in empirical wavelet transform. Electron. Lett. **56**(25), 1367–1369 (2020)
28. Selim, S., Tantawi, M.M., Shedeed, H.A., Badr, A.: A CSP\-BA-SVM approach for motor imagery BCI system. IEEE Access **6**, 49192–49208 (2018)
29. Shajil, N., Mohan, S., Srinivasan, P., Arivudaiyanambi, J., Murrugesan, A.A.: Multiclass classification of spatially filtered motor imagery EEG signals using convolutional neural network for BCI based applications. J. Med. Biol. Eng. **40**(5), 663–672 (2020)
30. Sharanreddy, M., Kulkarni, P.: Detection of primary brain tumor present in EEG signal using wavelet transform and neural network. Int. J. Biol. Med. Res. **4**(1), 2855–9 (2013)
31. Silvoni, S., et al.: Brain-computer interface in stroke: a review of progress. Clin. EEG Neurosci. **42**(4), 245–252 (2011)
32. Solomon, O.M., Jr.: PSD computations using Welch's method. NASA STI/Recon Technical Report N, vol. 92, p. 23584 (1991)

33. de Souza, G.H., Bernardino, H.S., Vieira, A.B.: Single electrode energy on clinical brain-computer interface challenge. Biomed. Signal Process. Control **70**, 102993 (2021)
34. de Souza, G.H., Bernardino, H.S., Vieira, A.B., Barbosa, H.J.C.: Differential evolution based spatial filter optimization for brain-computer interface. In: Proceedings of the ACM Genetic and Evolutionary Computation Conference, pp. 1165–1173 (2019)
35. de Souza, G.H., Bernardino, H.S., Vieira, A.B., Barbosa, H.J.C.: Genetic programming for feature extraction in motor imagery brain-computer interface. In: Marreiros, G., Melo, F.S., Lau, N., Lopes Cardoso, H., Reis, L.P. (eds.) EPIA 2021. LNCS (LNAI), vol. 12981, pp. 227–238. Springer, Cham (2021). https://doi.org/10.1007/978-3-030-86230-5_18
36. Staffa, M., Giordano, M., Ficuciello, F.: A WiSARD network approach for a BCI-based robotic prosthetic control. Int. J. Soc. Robot. **12**(3), 749–764 (2020)
37. Stinear, C.M., Lang, C.E., Zeiler, S., Byblow, W.D.: Advances and challenges in stroke rehabilitation. Lancet Neurol. **19**(4), 348–360 (2020)
38. Tangermann, M., et al.: Review of the BCI competition IV. Front. Neurosci. **6**(55) (2012)
39. Tzallas, A.T., et al.: EEG classification and short-term epilepsy prognosis using brain computer interface software. In: 2017 IEEE 30th International Symposium on Computer-Based Medical Systems (CBMS), pp. 349–353. IEEE (2017)
40. Vansteensel, M.J., Jarosiewicz, B.: Brain-computer interfaces for communication. In: Handbook of Clinical Neurology, vol. 168, pp. 67–85. Elsevier (2020)
41. Wolpaw, J.R., Birbaumer, N., McFarland, D.J., Pfurtscheller, G., Vaughan, T.M.: Brain-computer interfaces for communication and control. Clin. Neurophysiol. **113**(6), 767–791 (2002)
42. Yoshioka, M., Inoue, T., Ozawa, J.: Brain signal pattern of engrossed subjects using near infrared spectroscopy (NIRS) and its application to tv commercial evaluation. In: The 2012 International Joint Conference on Neural Networks (IJCNN), pp. 1–6. IEEE (2012)

Selecting Optimal Trace Clustering Pipelines with Meta-learning

Gabriel Marques Tavares[1(✉)] , Sylvio Barbon Junior[2] , Ernesto Damiani[3] , and Paolo Ceravolo[1]

[1] Università degli Studi di Milano (UNIMI), Milan, Italy
{gabriel.tavares,paolo.ceravolo}@unimi.it
[2] Università degli Studi di Trieste (UniTS), Trieste, Italy
sylvio.barbonjunior@units.it
[3] Khalifa University (KUST), Abu Dhabi, UAE
ernesto.damiani@kustar.ac.ae

Abstract. Trace clustering has been extensively used to discover aspects of the data from event logs. Process Mining techniques guide the identification of sub-logs by grouping traces with similar behaviors, producing more understandable models and improving conformance indicators. Nevertheless, little attention has been posed to the relationship among event log properties, the pipeline of encoding and clustering algorithms, and the quality of the obtained outcome. The present study contributes to the understanding of the aforementioned relationships and provides an automatic selection of a proper combination of algorithms for clustering a given event log. We propose a Meta-Learning framework to recommend the most suitable pipeline for trace clustering, which encompasses the encoding method, clustering algorithm, and its hyperparameters. Our experiments were conducted using a thousand event logs, four encoding techniques, and three clustering methods. Results indicate that our framework sheds light on the trace clustering problem and can assist users in choosing the best pipeline considering their environment.

Keywords: Process mining · Trace clustering · Meta-learning · Recommendation · Pipeline design

1 Introduction

Executing business processes leaves trails of the accomplished activities, performances achieved, and resources consumed. This information is stored in event logs, which embrace the history of the process. Executions generating the same sequence of activities are observed as the same trace by Process Mining (PM) algorithms that can group multiple executions in a single representation. Often, the variability of traces is however remarkable, and traces by themselves do not offer a helpful representation of the process. This variability causes problems for existing PM techniques. For instance, business processes with high trace

J. C. Xavier-Junior and R. A. Rios (Eds.): BRACIS 2022, LNAI 13653, pp. 150–164, 2022.
https://doi.org/10.1007/978-3-031-21686-2_11

variability generate spaghetti-like models, i.e., complex models with an enormous number of relations, often unreadable [1]. Neubauer et al. [24] identified two elements that contribute primarily to the inherent complexity of business processes: (i) knowledge-intensive processes where decision-making is human-dependent, and (ii) processes from large organizations with a fast generation rate, and therefore high volume output. Therefore, it is of interest to simplify the analysis representation, thus, allowing an easier interpretation for stakeholders and leveraging efficiency. For instance, consider an event log and its sequences of activities (traces) $L = \{\langle a,b,c,d,e\rangle, \langle a,d,c,e\rangle, \langle a,b,c\rangle, \langle a,d,c\rangle\}$, it is possible to notice two groups of closely related traces, i.e., trace $\langle a,b,c,d,e\rangle$ is similar to $\langle a,b,c\rangle$ whereas trace $\langle a,d,c\rangle$ has a sequence closer to trace $\langle a,d,c,e\rangle$. Grouping these similar traces may improve the accuracy and comprehensibility of process discovery techniques [11], and at the same time, support the identification of deviating or anomalous instances [17]. Moreover, concurrency might also be a problem in some domains. For instance, traces $\langle a,b,c,e\rangle$ and $\langle a,c,b,e\rangle$ may be considered the same from a business perspective if the order of activities b and c do not affect the process outcome. This way, these trace representations should be close when projected into the feature space.

Trace clustering techniques have been adopted to solve these issues by identifying sub-logs grouped by trace similarity. This way, by detecting groups with homogeneous behavior, process discovery techniques can be executed in sub-logs, producing higher quality models, which are instead accessible for stakeholders [14]. Trace clustering has also been studied in the context of explainability for PM [20] and, more recently, adapted to incorporate expert knowledge [19]. However, selecting the appropriate clustering technique is a complex task. Many transformation methods were presented, treating traces as vectors generated from bags of activities [22], edit distance [4] or dependency spaces [12], discriminant rules [15,26] or log footprints [20]. The set of clustering algorithms applied is also ample, e.g., k-means [15], hierarchical clustering [4], spectral clustering [12], constrained clustering [19], among others. Given this large set of options to set up a clustering pipeline, a non-expert user can likely feel overwhelmed.

Considering the challenge of designing pipelines to identify the correct encoding method, clustering algorithms, and hyperparameters to use for a specific log, we propose a framework based on Meta-learning (MtL). Our framework recommends the trace clustering pipeline that best fits a specific event log. MtL is a learning process applied to meta-data representing other learning processes [31] and has been used successfully to emulate experts' recommendations, maximize performance, and improve quality metrics [16].

The problem of simultaneously recommending an algorithm and tuning its hyperparameters to optimize a task is defined as the combined algorithm selection and hyperparameter optimization problem [28]. Alternatively, it is possible to exploit similar recommending tasks, in which algorithms and hyperparameters are represented as discrete spaces, mapping possible inter-correlations between the different hyperparameters as a multi-output machine learning problem. In this work, the meta-data consists of a large set of event log features that are provided as input to the MtL workflow that outputs trace clustering pipelines described by an encoding technique, a clustering algorithm, and hyperparameters modeled using a problem transformation approach. In our scenario, MtL serves as an automated approach as it suppresses the need for expert interaction to work correctly. The relationship between event log features and the quality of PM techniques has been already pointed out in the literature [2,3]. We introduce a general framework for studying this relationship for the trace clustering task using MtL. Moreover, we instantiate this framework to provide an example of its functionality. In particular, in our experiments, we submit the method to a set of 1091 event logs described by 93 log features, four encoding techniques, and three clustering algorithms. Results show that our approach achieves considerable performance for recommending encoding and clustering techniques. We also provide a comparison with two baseline methods, highlighting the improvement supported by the MtL strategy.

The remainder of this paper is organized as follows. Section 2 gives a historical overview of trace clustering solutions, focusing on the employed transformation and clustering methods. Section 3 defines the task and its configuration steps, while Sect. 4 presents our proposed framework to solve the trace clustering recommendation problem. Section 5 presents the material used for experiments, the techniques, and quality metrics adopted. Section 6 shows the results and raises a discussion around them. Section 7 concludes the paper and Sect. 8 lists its broader impact.

2   Related Work

Trace clustering research is deeply connected to the variant analysis problem, that is, detecting groups of similar behavior within a single business process [20]. Clustering traces is partitioning an event log into groups of comparable traces such that each trace is assigned to a unique group [19]. Since its initial adoption, trace clustering has been proposed as an instrument to reduce variability. Discovering process models from clusters, for example, generally improves quality [14]. An early work in the area used a set of n-grams to encode a trace activity sequence, thus, mapping traces to a feature vector space [15]. Song et al. [26] went further by defining multiple encoding procedures, named profiles, to represent traces as vectors. Furthermore, the authors call attention to the modularity between the profiling and clustering steps. Bose and Aalst [4] represent traces as strings and apply edit distance to measure trace similarity. Delias et al. [12] proposed a measure to calculate trace distance based on dependency. However, approaches based on instance-level similarity may be applicable only to particular domains. Thaler et al. [27] highlight that bags of activities may lose critical information regarding the execution order. Delias et al. [12] show that no single optimal similarity metric is applicable for all domains and applications while Zandkarimi et al. [34] stated that trace clustering is a context-specific task. Koninck et al. [20] characterize the complexity of clustering with the assessment of the best event log splitting operations. A well-performing encoding method improves a wide range of posterior analyses without the need to tune them [3]. The authors also showed that there is no best encoding method for every scenario, that is, different event logs are encoded better, considering several quality criteria, by different encoding techniques. A similar conclusion is achieved in [27] when analyzing clustering algorithms applied to PM.

The authors stated that some techniques are suitable for particular scenarios, reinforcing the argument that process characteristics may guide the decision of the appropriate clustering technique. Besides, different from supervised approaches, unsupervised learning performance is severely affected by small changes in hyperparameters, depending heavily on user-domain knowledge [18]. This implies that the solutions proposed today are far from optimal as they are attached to a unique set of encoding and clustering algorithms.

Considering the multiple available profiling and clustering algorithms, we envision two main building blocks regulating the success of clustering techniques. The first regards the encoding method, i.e., converting the trace sequences into feature vectors, and the latter comprises the clustering techniques. The approaches currently available in the literature are strictly attached to a specific combination of these building blocks; hence, they neither offer a means to study the relationship between the different steps that compose a pipeline nor relate process behavior to optimal solutions.

3 Problem Statement

Given the plethora of configuration steps and parametrization, designing the appropriate trace clustering pipeline is a complex issue even for experts. We identified in the literature three configuration steps that highly affect the clustering results: (i) trace encoding, (ii) clustering algorithm, and (iii) hyperparameters regulating the clustering algorithm. The choice of each step is critical since slight changes deeply affect the clustering results.

PM techniques ingest event logs. An *event log* is the set of events executed in a business process. An *event* records the execution of an *activity*. It follows that each event is strictly related to a unique process instance, identified by its *case*. A unique end to end sequence of activities within a case is known as a *trace*. Let Σ be the *event universe*, i.e. the set of all possible event identifiers; Σ^* denotes the set of all sequences over Σ. A *trace* is a non-empty sequence of events $t \in \Sigma^*$ where each event appears only once and time is non-decreasing, i.e., for $1 \le i < j \le |t| : t(i) \ne t(j)$. In PM applications, encoding aims at transforming traces into vectors, mapping process instances into a feature space. Therefore, an *encoding* method is a function $E^{()}$ that maps a set of traces into a n-dimensional feature space, projecting the instances' distances according to their trace sequence.

The problem of selecting a trace clustering pipeline is different from the traditional algorithm selection, in which it is expected to recommend a tuple $\langle encoding, clustering, hyperparameters \rangle$. It is worth mentioning that the hyperparameters are continuous values with a high dimensional space that might present different inter-correlations between them. Here, they were discretized at frequent intervals to cover a wide range of promising possibilities. Further, we employed a multi-output strategy to take advantage of inter-correlations from the clustering algorithm and its hyperparameters.

We formulate the problem as a set of encoding methods $\mathcal{E} = \{E^{(1)}, ..., E^{(j)}\}$, clustering algorithms $\mathcal{C} = \{C^{(1)}, ..., C^{(l)}\}$ associated with a hyperparameter space $\mathcal{H} = \{H^{(1)}, ..., H^{(l)}\}$ and event log data mapped by meta-features and best pipeline as $\mathcal{D} = \{(x^{(i)}, Y^{(s)}|s = 2)\}$. We have j encoding methods, l combinations of clustering algorithms and hyperparameters, and s is the number of expected pipelines' steps to be recommended. It is important to mention that the clustering algorithm and hyperparameter recommendation were modeled as a single step of the pipeline. For each event log sample $(x^{(i)}, Y^{(i)})$, $x^{(i)} \in \mathcal{X}$ is a d-dimensional meta-feature vector $(x^{(i1)}, x^{(i2)}, ..., x^{(id)})$ and $Y^{(i)} \subseteq \mathcal{Y}$ is the tuple $\langle encoding, clustering, hyperparameters \rangle$ associated to $x^{(i)}$. The goal is that for any unseen event log x, the MtL model $h_E()$ recommends $h_E(x) \in \mathcal{E}$ as the proper encoding method for x and $h_{CH}()$ recommends $h_{CH}(x) \in \mathcal{L}$. In the proposed setup, we are facing a *multi-output* problem, where a set of labels $\mathcal{C} \times \mathcal{H} \subseteq L$ is associated with a single instance [29]. Following the taxonomy proposed in [29], we adopt a problem transformation approach, which converts the data into a format that can be used in conjunction with traditional techniques. More specifically, we employed the Binary Relevance (BR) problem transformation approach [33]. BR works by transforming the original data set into q data sets D_{λ_j}, where $j = [1, ..., q]$ contains all instances of the original data that are labeled according to the existence or not of single labels λ_j. Thus, BR learns q binary classifiers, one for each label L. Given a new instance, BR provides the union of the labels λ_j predicted by the q classifiers.

There are several ways to model this problem. In this paper, we followed the supervised machine learning approach to build $h_E()$ and $h_{CH}()$ towards determining a promising pipeline candidate configuration. The problem is, in nature, a multi-output problem. Therefore, we model this through the BR approach to combine outputs from both $h_E()$ and $h_{CH}()$. Alternative optimization-based modeling methods to control the trade-off between exploitation versus exploration of pipeline combinations exists, but as an initial study exploring new meta-features and meta-target selection in a new application on the PM domain, we adopted this modeling strategy for simplicity.

4   MtL-Based Solution for Trace Clustering

Trace clustering solutions must be able to adapt according to domain characteristics. We then propose a framework grounded in MtL capable of delivering suitable recommendations according to different business process behaviors. The main goal of our approach is to recommend a tuple $\langle encoding, clustering, hyperparameters \rangle$ that maximizes quality metrics for the trace clustering problem. Figure 1 shows the overview of the framework. First, an event log repository is created to represent different business scenarios. The *Meta-Feature Extraction* step mines features for each event log in the repository, creating *meta-features* according to MtL terminology. The description quality of the *meta-features* is an important constraint bounding the performance of the complete pipeline. Moreover, the *Meta-Target Definition* defines a set of encoding and clustering (coupled with its hyperparameter) techniques that are assessed by quality metrics and ranked according to a ranking function. Then, the *Meta-Database* combines the *meta-features* and *meta-targets* defined in previous steps, creating a data set populated by *meta-instances*. Using the *meta-database*, the *Meta-learning* step induces a *Meta-model* that is, then, used to recommend a pipeline for a given event log considering its *meta-features*. It is worth mentioning that multi-output machine learning modeling for the meta-model can bring important achievements in terms of performance, considering the interrelations between each step of the pipeline. In Fig. 1, green arrows indicate the steps that are used for the creation and training of the framework, while blue arrows represent a production environment where one assesses the meta-model for recommendation.

Given the adaptable setup of our framework, one can implement it using a different set of *meta-features* and *meta-targets*. The automatic aspect of this approach provides the user with recommendations based on event log behavior, considering the possible options among the configurable steps. Moreover, other aspects are adaptable, such as the adopted quality metrics and the ranking function. Nonetheless, we note that the robustness of the approach depends on the MtL structure, which must be maintained when the framework is instantiated in real scenarios.

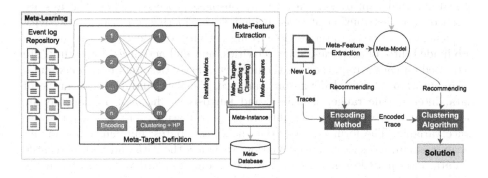

Fig. 1. Overview of MtL proposal for trace clustering.

5   Experimental Setup

In this section, we expose the details of each framework step, as seen in Fig. 1, and reveal the experiments implemented to study a possible instance of our MtL framework. The implementation is available for replication purposes[1].

---
[1] https://github.com/gbrltv/meta_trace_clustering/.

5.1 Event Logs and Featurization

MtL benefits from using a large set of instances in the meta-database. Hence, we are aiming at a heterogeneous set of business process logs representing different scenarios and behaviors. For that, we rely on the set of logs proposed in [2]. These event logs were grouped to represent a plethora of business behaviors, mapping the relationship between process characteristics and quality metrics. This set contains both real and synthetic event logs. Regarding real-life data, there are six logs from past Business Process Intelligence Challenges (BPIC)[2], the environmental permit[3], helpdesk[4] and sepsis[5] logs. For synthetic data, the authors adopted 192 logs from the Process Discovery Contest (PDC) 2020[6], an annual event organized to evaluate the efficiency of process discovery algorithms. The PDC logs are complex given the nature of employed behaviors, such as dependent tasks, loops, invisible and duplicate tasks, and noise. The next group of synthetic data contains 750 logs proposed in the context of online PM [7]. These logs are built to depict process drifts, i.e., behavior change during the business process execution, containing four drift types, five noise percentages, and three trace lengths. The final group of synthetic event logs was proposed for the evaluation of trace encoding techniques [3]. This set contains 140 logs generated from five process models, six anomaly types, and four frequency percentages.

The performance of the meta-model is directly dependent on the quality of the meta-features. Thus, the meta-features extracted from event logs must capture the process behavior and describe it from complementary perspectives. We adopted the featurization introduced in [2]. The authors presented a group of features that capture several layers of business processes. These features are based on the distribution of trace behavior, considering trace length, activity frequencies, and trace variants. Regarding activity-level features, the group is subdivided into all activities, start activities, and end activities. 12 features are extracted for each group, they are the number of activities, minimum, maximum, mean, median, standard deviation, variance, the 25th and 75th percentile of data, interquartile range, skewness, and kurtosis coefficients. To capture the behavior at the trace level, the authors propose features for trace lengths and trace variants. The former group contains 29 attributes: minimum, maximum, mean, median, mode, standard deviation, variance, the 25th and 75th percentile of data, interquartile range, geometric mean and standard variation, harmonic mean, coefficient of variation, entropy, and a histogram of 10 bins along with its skewness and kurtosis coefficients. Trace variants are captured by 11 descriptors: mean number of traces per variant, standard variation, skewness coefficient, kurtosis coefficient, the ratio of the most common variant to the number of traces, and ratios of the top 1%, 5%, 10%, 20%, 50% and 75% variants to the total number of traces. Log-level behavior is captured by: number of traces, unique

---

[2] https://www.tf-pm.org/resources/logs.

[3] https://doi.org/10.4121/uuid:26aba40d-8b2d-435b-b5af-6d4bfbd7a270.

[4] https://doi.org/10.17632/39bp3vv62t.1.

[5] https://doi.org/10.4121/uuid:915d2bfb-7e84-49ad-a286-dc35f063a460.

[6] https://doi.org/10.4121/14626020.v1.

traces, their ratio, and number of events. To describe log complexity, entropy-based measures have been adopted recently in PM literature [1] aiming at the discretization between logs that are better mined by declarative or imperative algorithms. Hence, such metrics capture the structuredness and variability of the log. The 14 entropy features we adopt are: trace, prefix, k-block difference and ratio (k values of 1, 3 and 5), global block, k-nearest neighbor (k values of 3, 5, and 7), Lempel-Ziv, and Kozachenko-Leonenko. Considering all groups, 93 meta-features were used to extract log behavior covering log structuredness and variability, statistical dispersion, probability distribution shape, and tendency.

5.2 Trace Encoding Techniques

Many PM techniques rely on encoding to transform event log-specific representations to other formats [8,25,30]. The transformation usually applies at the trace-level, that is, converting the sequence of activities respective to a unique trace into a feature vector. In [3], the authors compared ten different encoding techniques through the lens of quality metrics measuring data dispersity, representativeness, and compactedness. They concluded that there is no encoding that excels in all tasks and perspectives concomitantly. For instance, graph embeddings outperform the others in the classification task and representation quality. However, these encoding methods are costly and usually sparse, meaning that there are better encoding techniques considering space and time complexity. The trace clustering literature has already experimented with several types of encoding methods, such as one-hot encoding [15,26], edit distance [4], log footprints [20], activity profiles and n-grams [9]. Nonetheless, no trace similarity measure is general enough to be applicable in all scenarios [10].

In this work, we adopt four encoding techniques that were frequently applied in the context of trace clustering. The first one is one-hot encoding. This technique encodes activities as categorical dimensions, creating a feature vector of binary values for each trace based on the occurrence of activities in a trace. Next, we adopt n-grams, a common technique used in text mining applications. This encoding maps groups of activities of size n into a feature vector, accounting for their occurrence or not. More specifically, we apply bi-gram and tri-gram. Finally, we applied position profiles [6], an approach that relates activity frequency and position. A log profile is created by computing the activity appearances in each trace position and its respective frequency. A trace is encoded considering the frequency of its activities in their positions according to the log profile.

5.3 Trace Clustering Algorithms

We selected three clustering techniques commonly applied in data mining and trace clustering literature. These techniques are grounded in different heuristics, and with this, we aim to evaluate if a particular clustering structure outperforms the others. The choice of parameters was also guided by considering the literature on trace clustering, comprising different trace behaviors and complexities. It is important to note that we selected a range of possible values to support the exploration of the algorithmic space.

First, we adopt the Density-based Spatial Clustering of Applications with Noise (*dbscan*) algorithm [13]. The *dbscan* method guides its clustering based on the density of the feature space, hence, instances in high-density regions form a cluster while instances sitting at low-density regions are regarded as outliers. The main hyperparameter affecting the clustering results is *eps*, which regulates the maximum distance between two points for them to be considered of the same neighborhood. We explore different configurations of the *eps* hyperparameter to evaluate its impact and to recommend the best configuration in the meta-model step. For that, we apply the following *eps* values: 0.001, 0.005, 0.01, 0.05, 0.1, 0.5, 1. Moreover, we adopt *k*-means [21], a clustering technique that randomly selects centroids, which are the initial cluster points, and works by iteratively optimizing the centroid positions. The *k*-means technique requires the expected number of clusters (*k*) from a given data set as a hyperparameter. We set *k* to these values: 2, 3, 4, 5, 6, 7, 8, 9, 10. Finally, the last technique is *agglomerative* clustering [32], a type of hierarchical clustering with a bottom-up approach. The algorithm starts by considering each point as a cluster and merges the clusters as the hierarchy moves up, creating a tree-like structure depicting the cluster levels and merges. As with *k*-means, *agglomerative* clustering requires the number of clusters as input, we then adopted the same range of values for the *k* parameter.

5.4   Ranking Metrics

To complete the creation of a meta-database, meta-targets must be defined for each meta-instance. This way, a ranking strategy is required to compare the possible trace clustering pipelines. Hence, the technique sitting at the top of the ranking strategy is the one recommended for a meta-instance, i.e., it is defined as the meta-target. As pointed out in the literature [3,10], there is no unique solution for a problem that outperforms the others from all perspectives. Considering this hypothesis, we propose three complementary metrics to evaluate trace clustering solutions, this way, capturing different degrees of performance. Moreover, a user applying a trace clustering solution may expect to evaluate the results from several perspectives. Here, we support such a user by assessing clustering quality from a set of criteria.

Silhouette coefficient (*s*), the first metric we propose to measure performance, is based on the traditional clustering literature. The Silhouette score is computed at the cluster level to capture its tightness and separation, judging instances that fit their cluster or are in between different clusters. The scores of a group of clusters can be combined to assess the relative quality of the clustering technique.

To complement this evaluation with a PM-inspired metric, we propose to measure the quality of clusters concerning trace variants. This way, by computing the trace variant frequency in each

$$v = \frac{\sum_{C_i \in C} var(C_i) - 1}{\#traces} \quad (1)$$

cluster, we can evaluate if the solution provides a clear separation of variants in the feature space. For that, we compute the unique traces in a cluster, and by a weighted mean, the Variant score (v) is obtained. Consider C the group of all clusters, C_i the cluster of index i, $var(C_i)$ the number of unique traces found in cluster C_i and $\#traces$ the total number of traces in the event log, Eq. 1 depicts the Variant score calculation, 0 is the optimal value. As resource consumption is an important aspect in organizations, we also consider the clustering time (t) as a metric to assess its quality. The lower the t metric for a particular solution, the better it is ranked compared to others.

Given this set of metrics, i.e. s for silhouette coefficient, v for the variant score, and t for computational time, a *meta-target* $\langle encoding, clustering, hyperparameters \rangle$ has to successfully balance between all metrics to be considered good. This way, the app-

Table 1. Ranking trace clustering pipelines.

Log	Encoding	Clustering	s	v	t	R_s	R_v	R_t	R
L	E_1	C_1	0.9	0.5	50	1	2	3	2
L	E_2	C_2	0.3	0	10	3	1	1	1.67
L	E_3	C_3	0.8	0.7	15	2	3	2	2.33

roach rewards techniques that excel in the three metrics, such as ignoring one or more may lead to a lack of tightness, improper variant identification, and high resource consumption. Hence, we propose a ranking strategy (R) that combines all dimensions. Table 1 presents an example of the ranking strategy we propose. For each pair of encoding techniques and clustering algorithms, we apply it for a given event log (L) and measure the quality metrics (s, v, t). Following, a positional rank is built for each metric (R_s, R_v, R_t), i.e, comparing the pairs of encodings and clustering in each dimension. Finally, a rank (R) is computed by the average of the metrics ranks. For example, considering the pairs $\langle E_1, C_1 \rangle$, $\langle E_2, C_2 \rangle$ and $\langle E_3, C_3 \rangle$, their respective final ranks are 2, 1.67 and 2.33. The solution chosen as the meta-target is the one that minimizes the R function, which, in this example, is the pair $\langle E_2, C_2 \rangle$.

5.5   Meta-model

Regarding the meta-learner, we applied the Random Forest (RF) algorithm [5] due to its robustness, being less prone to overfitting. Moreover, we applied a hyperparameter tuning technique to improve performance in the recommendation task. For that, we adopted a holdout strategy where 80% of the meta-database was used for tuning and 20% as the validation set. After a grid search tuning strategy with 5-fold cross-validation, the best hyperparameters were: (i) *50* as the number of trees composing the forest, (ii) *gini* as the criterion measuring split quality, (iii) *3* as the required minimum number of samples for a node split, (iv) *1* as the minimum number of samples required to be a leaf node, and (v) *log2* as the number of considered features for a split. The results reported in Sect. 6 were extracted when applying the tuned meta-model to validation data.

6   Results and Discussion

This section explores the meta-database composition by observing the encoding techniques and clustering algorithms chosen by their performance and balancing.

Next, an overall analysis, including the comparison of the proposed strategy with the baselines, is introduced.

6.1   Meta-learning Exploratory Analysis

The rank results, considering all algorithms for setting the meta-database, including the metrics used for ranking the meta-targets, are presented in Fig. 2. The heatmap plots show the ranking of the metrics s, v, and t for encoding (Fig. 2a) and clustering (Fig. 2b) used to sort and identify promising algorithms as meta-targets. Each ranking varies from 1 to 81, in which 1 is the best-ranked algorithm for a given metric.

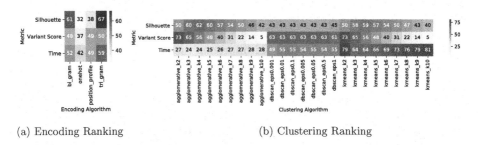

(a) Encoding Ranking   (b) Clustering Ranking

Fig. 2. Encoding and clustering rankings. Color gradient represents the ranking position variation.

Observing the encoding techniques (Fig. 2a), it is possible to identify a large discrepancy between them when evaluated by s, revealing the superiority of *one-hot* and *position profile* algorithms, whereas v score and t do not present a such prominent variation, leading to closer ranking positions. Note that the results report the average ranking position. In other words, one-hot encoding is the most well-ranked across the set of event logs, although it is not unanimous. However, when observing the clustering algorithms (Fig. 2b), it is possible to note a balance regarding s while v and t reveal discrepancies. The former (v) exposes the importance of hyperparameter definition since *agglomerative* and k-means ranged throughout the rankings when changing their hyperparameter k. Moreover, the t metric delivered an important perspective, in which each clustering algorithm is recognizable regardless of its hyperparameters. In particular, *agglomerative* and *dbscan* were superior to k-means. This superiority led to no usage of k-means as a clustering meta-target.

The meta-database was built using the combination of the top-ranked algorithms for each meta-instance (event logs). This combination leads to an imbalanced multi-output dataset. Combinations such as *one-hot* encoding with *agglomerative* clustering using 10 as k value (*onehot_agglomerative_k*10) represented 469 meta-instances. The second most frequent combination (171 meta-instances) was *position profile* with *agglomerative* clustering using 10 as k (*position_profile_agglomerative_k*10). The third was *one-hot* using *dbscan* adopting a *eps* equals 0.001 (*onehot_dbscan_eps0.001*) in 125 meta-instances. These meta-target frequencies show the evident dominance of *one-hot* and *position profile* over the other encoding methods. *Bi-gram* was the best encoding technique for 37 meta-instances while *tri-gram* was the best one, combined with *dbscan*, only with four meta-instances. When evaluating from a clustering perspective, we observe a balance between *dbscan* with a wide range of *eps* and *agglomerative* using k as 10. Different values of k for *agglomerative* did not meet many meta-instances. Conversely, *dbscan* demonstrates the necessity of hyperparameter adjustments since different values of *eps* could match particular meta-instances. The imbalance issue was addressed by removing the minority class combinations, that is, meta-targets that appear less than five times. The final meta-database was composed of 1036 samples, with fifteen different combinations of *one-hot, position profile,* and *bi-gram* with *agglomerative* (k in {8, 9, 10}) and *dbscan* (*eps* in {0.001, 0.005, 0.05, 0.01, 0.1, 0.5, 1}).

6.2   Meta-model Performance

Using RF as our meta-model built over the meta-database, we analyzed the performance for both encoding and clustering algorithm recommendations (Fig. 3). It is worth mentioning that the problem was modeled as a multi-output problem using the BR transformation approach, addressing encoding and clustering at once. Since there are no other literature references, we employed majority voting and random selection as baseline approaches for comparison reasons. Majority voting works by always indicating the most common meta-target, i.e., the majority class in the meta-database. In this setup, *one-hot* and *agglomerative_k*10 are the most common encoding technique and clustering algorithm, respectively. Although a simple baseline, majority voting is a suitable comparison in machine learning applications, clearly specifying the minimum performance threshold. The random selection approach randomly chooses one of the possible pipeline combinations (coming from the set of meta-targets). This technique simulates a PM practitioner in a scenario without the availability of experts, a common situation in real environments. This way, we situate our method's performance both in relation to the machine learning and PM landscapes, creating an initial assessment and benchmark for the trace clustering problem.

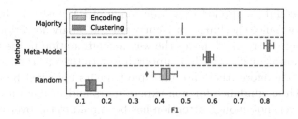

Fig. 3. Performance of the MtL framework to recommend the encoding technique and clustering algorithm in terms of accuracy and F1.

As observed in Fig. 3, our proposal obtained an F1 of 0.81 ($\pm$0.01) when recommending the encoding technique and an F1 of 0.59 ($\pm$0.01) for the recommendation of the clustering algorithm. The majority baseline for encoding obtained an F1 of 0.71 while the random baseline achieved 0.42 ($\pm$0.03). Regarding clustering, the majority obtained 0.49 of F1, and random selection reached 0.14 ($\pm$0.03). Our approach obtained a mean predictive performance of 0.7 for the whole trace clustering pipeline. The results were superior to the majority and random baselines, which averaged 0.59 and 0.28, respectively. Note that the majority voting results are boosted by the imbalanced scenario, for balanced meta-databases, the tendency is to underperform. The superior performance of our proposal confirms our hypothesis, i.e., there is a relationship between event log behavior and optimal pipelines. Since this relationship exists (and is partially captured by our proposed meta-features), our method outperforms the baselines. Given the universe of possibilities (81 combinations) and the limitations imposed by the imbalanced scenario, we consider the F1 performances suitable. Furthermore, this assessment serves as a benchmark for the area to be compared to alternative solutions proposed in the future.

7 Conclusion

This paper proposes an MtL framework to recommend the best pipeline for trace clustering based on a specific event log and its behavior. For that, we extract meta-features to describe event logs and match them with the best clustering pipeline by assessing three complementary metrics. The framework recommends a tuple ⟨*encoding, clustering, hyperparameters*⟩, making trace clustering solutions accessible for non-expert users and assisting experts with guided recommendations. Results have shown that the framework outperforms baseline approaches. In future research, we aim to extend the experimental evaluation to gather further insights into the relationship between trace clustering quality and event log behavior. Moreover, we plan to improve the modeling of the multi-output approach by testing different techniques, possibly taking advantage of the intercorrelation between different steps of the recommended pipeline.

8 Limitations and Broader Impact Statement

Our approach could be applied in a wide range of PM tasks, including process discovery, conformance checking, trace clustering, anomaly detection, and several others. This way, our research paves the way for automation in the business process domain, complemented with a supporting data-driven framework to study PM problems. Therefore, there are multiple benefits unlocked by the proposed technology, such as guidance for non-expert users and insights for experienced analysts. However, not enough attention has been paid to the over-application of automation techniques in PM. An important aspect touches the validity of the research and experimental design [23]. More specifically, the underlying behavior distribution of event logs might lead to unexpected results. This way, adopters of this tool should be careful in selecting a representative set of event logs to serve as the basis of the meta-database. Otherwise, the insights or recommendation quality might decrease. Being a technique that abstracts the pipeline for non-experts, the possibility of results misuse rises, thus requiring understanding about the possible domain risks when applying the tool.

References

1. Back, C.O., Debois, S., Slaats, T.: Entropy as a measure of log variability. J. Data Semant. **8**(2), 129–156 (2019)
2. Barbon Jr., S., Ceravolo, P., Damiani, E., Tavares, G.M.: Using meta-learning to recommend process discovery methods (2021)
3. Barbon J., Sylvio, C., Paolo, D., Marques Tavares, G.: Evaluating trace encoding methods in process mining. In: Bowles, J., Broccia, G., Nanni, M. (eds.) DataMod 2020. LNCS, vol. 12611, pp. 174–189. Springer, Cham (2021). https://doi.org/10.1007/978-3-030-70650-0_11
4. Bose, R.P.J.C., van der Aalst, W.M.: Context aware trace clustering: towards improving process mining results. In: Proceedings of the 2009 SIAM International Conference on Data Mining. Society for Industrial and Applied Mathematics (2009)
5. Breiman, L.: Random forests. Mach. Learn. **45**(1), 5–32 (2001)
6. Ceravolo, P., Damiani, E., Torabi, M., Barbon, S.: Toward a new generation of log pre-processing methods for process mining. In: Carmona, J., Engels, G., Kumar, A. (eds.) BPM 2017. LNBIP, vol. 297, pp. 55–70. Springer, Cham (2017). https://doi.org/10.1007/978-3-319-65015-9_4
7. Ceravolo, P., M. Tavares, G., Barbon Jr., S., Damiani, E.: Evaluation goals for online process mining: a concept drift perspective. IEEE Trans. Serv. Comput. 1 (2020)
8. De Koninck, P., vanden Broucke, S., De Weerdt, J.: act2vec, trace2vec, log2vec, and model2vec: representation learning for business processes. In: Weske, M., Montali, M., Weber, I., vom Brocke, J. (eds.) BPM 2018. LNCS, vol. 11080, pp. 305–321. Springer, Cham (2018). https://doi.org/10.1007/978-3-319-98648-7_18
9. De Koninck, P., De Weerdt, J.: Scalable mixed-paradigm trace clustering using super-instances. In: International Conference on Process Mining (2019)
10. de Leoni, M., van der Aalst, W.M., Dees, M.: A general process mining framework for correlating, predicting and clustering dynamic behavior based on event logs. Inf. Syst. **56**, 235–257 (2016)
11. De Weerdt, J., vanden Broucke, S., Vanthienen, J., Baesens, B.: Active trace clustering for improved process discovery. IEEE Trans. Knowl. Data Eng. **25**(12), 2708–2720 (2013)
12. Delias, P., Doumpos, M., Grigoroudis, E., Manolitzas, P., Matsatsinis, N.: Supporting healthcare management decisions via robust clustering of event logs. Knowl. Based Syst. **84**, 203–213 (2015)
13. Ester, M., Kriegel, H.P., Sander, J., Xu, X.: A density-based algorithm for discovering clusters in large spatial databases with noise. In: Proceedings of the Second International Conference on Knowledge Discovery and Data Mining, pp. 226–231. KDD 1996, AAAI Press (1996)
14. Fani Sani, M., Boltenhagen, M., van der Aalst, W.: Prototype selection using clustering and conformance metrics for process discovery. In: Del Río Ortega, A., Leopold, H., Santoro, F.M. (eds.) BPM 2020. LNBIP, vol. 397, pp. 281–294. Springer, Cham (2020). https://doi.org/10.1007/978-3-030-66498-5_21

15. Greco, G., Guzzo, A., Pontieri, L., Sacca, D.: Discovering expressive process models by clustering log traces. IEEE Trans. Knowl. Data Eng. **18**(8), 1010–1027 (2006)

16. He, X., Zhao, K., Chu, X.: Automl: a survey of the state-of-the-art. Knowl. Based Syst. **212** (2021)

17. Hompes, B., Buijs, J., van der Aalst, W., Dixit, P., Buurman, J.: Discovering deviating cases and process variants using trace clustering. In: 27th Benelux Conference on Artificial Intelligence (2015)

18. Hou, J., Gao, H., Li, X.: Dsets-dbscan: a parameter-free clustering algorithm. IEEE Trans. Image Process. **25**(7), 3182–3193 (2016)

19. Koninck, P.D., Nelissen, K., vanden Broucke, S., Baesens, B., Snoeck, M., Weerdt, J.D.: Expert-driven trace clustering with instance-level constraints. Knowl. Inf. Syst. **63**(5), 1197–1220 (2021)

20. Koninck, P.D., Weerdt, J.D., vanden Broucke, S.K.L.M.: Explaining clusterings of process instances. Data Mining Knowl. Discov. **31**(3), 774–808 (2016)

21. MacQueen, J.: Some methods for classification and analysis of multivariate observations. In: Proceedings of the fifth Berkeley Symposium on Mathematical Statistics and Probability, vol. 1, pp. 281–297. Oakland, CA, USA (1967)

22. de Medeiros, A.K.A.., et al.: Process mining based on clustering: a quest for precision. In: ter Hofstede, A., Benatallah, B., Paik, H.-Y. (eds.) BPM 2007. LNCS, vol. 4928, pp. 17–29. Springer, Heidelberg (2008). https://doi.org/10.1007/978-3-540-78238-4_4

23. Mendling, J., Depaire, B., Leopold, H.: Theory and practice of algorithm engineering (2021)

24. Neubauer, T.R., Pamponet Sobrinho, G., Fantinato, M., Peres, S.M.: Visualization for enabling human-in-the-loop in trace clustering-based process mining tasks. In: 2021 IEEE International Conference on Big Data (Big Data), pp. 3548–3556 (2021)

25. Polato, M., Sperduti, A., Burattin, A., Leoni, M.d.: Time and activity sequence prediction of business process instances. Computing **100**(9), 1005–1031 (2018)

26. Song, M., Günther, C.W., van der Aalst, W.M.P.: Trace clustering in process mining. In: Ardagna, D., Mecella, M., Yang, J. (eds.) BPM 2008. LNBIP, vol. 17, pp. 109–120. Springer, Heidelberg (2009). https://doi.org/10.1007/978-3-642-00328-8_11

27. Thaler, T., Ternis, S.F., Fettke, P., Loos, P.: A comparative analysis of process instance cluster techniques. Wirtschaftsinformatik **2015**, 423–437 (2015)

28. Thornton, C., Hutter, F., Hoos, H.H., Leyton-Brown, K.: Auto-weka: combined selection and hyperparameter optimization of classification algorithms. In: Proceedings of the 19th ACM SIGKDD International Conference on Knowledge Discovery and Data Mining, pp. 847–855 (2013)

29. Tsoumakas, G., Katakis, I., Vlahavas, I.: Mining Multi-label Data, pp. 667–685. Springer, US, Boston, MA (2010)

30. van der Aalst, W., Weijters, T., Maruster, L.: Workflow mining: discovering process models from event logs. IEEE Trans. Knowl. Data Eng. **16**(9), 1128–1142 (2004)

31. Vanschoren, J.: Meta-learning: a survey (2018). arxiv.org/abs/1810.03548

32. Ward, J.H.: Hierarchical grouping to optimize an objective function. J. Am. Statist. Assoc. **58**(301), 236–244 (1963)

33. Xu, D., Shi, Y., Tsang, I.W., Ong, Y.S., Gong, C., Shen, X.: Survey on multi-output learning. IEEE Trans. Neural Networks Learn. Syst. **31**(7), 2409–2429 (2020)

34. Zandkarimi, F., Rehse, J.R., Soudmand, P., Hoehle, H.: A generic framework for trace clustering in process mining. In: 2020 2nd International Conference on Process Mining (ICPM), pp. 177–184 (2020)

Sequential Short-Text Classification from Multiple Textual Representations with Weak Supervision

Ivan J. Reis Filho[1,2(✉)] [ID], Luiz H. D. Martins[1] [ID], Antonio R. S. Parmezan[2] [ID], Ricardo M. Marcacini[2] [ID], and Solange O. Rezende[2] [ID]

[1] Minas Gerais State University, Frutal, Brazil
[2] University of São Paulo, São Carlos, Brazil
ivan.filho@uemg.br

Abstract. The amount of news generated on the internet has increased significantly in recent years. As a trend, text data has gained attention from industry, government, academia, and the financial market. This information is potentially valuable to assist domain experts in decision making. Therefore, related applications based on machine learning have been widely available in several areas of knowledge. However, for supervised learning tasks, the availability of annotated texts in quantity and quality is a recurring problem. This work proposes a time-series-driven approach to labeling chronologically arranged documents. Our proposal categorizes short texts for a particular domain according to the level and trend patterns of a given time series. We use the obtained weak labels with the understanding that they are imperfect but still useful for building predictive text models. Documents and agribusiness commodity price series were employed to assess performance in four classification scenarios. The experimental evaluation considered nine textual representations and different learning paradigms. Neural language-based models demonstrated better classification performance than traditional ones. The results indicate that the proposed approach can be an alternative for automatically labeling a large news volume.

Keywords: Data labeling · Machine learning · Text mining · Weak supervision

1 Introduction

We have witnessed an increased interest in machine learning-based applications for industry, government, academia, and the financial market [18]. Supervised learning tasks such as classification and regression have been widely explored to assist domain experts in decision making. In this way, emerging intelligent technologies have enhanced the offer of computational resources capable of storing, analyzing, and predicting information from a large volume of data [7].

J. C. Xavier-Junior and R. A. Rios (Eds.): BRACIS 2022, LNAI 13653, pp. 165–179, 2022.
https://doi.org/10.1007/978-3-031-21686-2_12

Predictive models are generally learned from a dataset containing many training samples, each corresponding to an object or event. In this context, the performance of machine learning models depends on the availability of labeled data in sufficient quantity and quality [6]. However, annotated data for some domains can be scarce, and the typical process of obtaining labels with experts inspecting individual samples is usually expensive and time-consuming. Thus, to overcome this limitation, machine learning techniques should be able to work under weak supervision [27].

Weak supervision provides a significantly inexpensive alternative to traditional annotation, reducing the need for humans to hand label large datasets to train machine learning models [6,8,25]. Researchers have employed this technique to support many applications, including annotating and detecting fake news [12,21,25], labeling images from social media posts [9], recognizing named entities [16], and classifying texts using external sources [8,18].

In recent decades, the amount of news generated and made available on the internet has grown exponentially [14]. Text mining and natural language processing methods allowed the conversion of such documents into helpful information for experts in different domains [11]. However, due to the lack of annotated news, unsupervised and semi-supervised learning tasks have been adopted for these applications [24]. In light of this, this paper proposes a time-series-driven approach to labeling chronologically arranged documents. Time series are ordered sequences of numerical observations recorded over time. In finance, the price series represents daily records of prices practiced on the stock exchange or commodities. Sudden fluctuations in the price series can mean political, climatic and macro-economic events, as well as market supply and demand.

Interestingly, events that alter market behavior are often reported explicitly in text news. Thus, we design in this work a function that uses the price series of two Brazilian agribusiness commodities to label short texts that correspond to agricultural news. Our proposal weakly categorizes the documents according to the time series's level and trend patterns. We use the obtained weak labels with the understanding that they are imperfect but still useful for building predictive text models. An experimental evaluation estimated the efficiency of our approach in the face of nine textual representations and different learning paradigms. Furthermore, we propose a vector representation of texts based on bag-of-words that uses a distance measure between Terms and Documents through pre-trained BERT models, designated here as TD-BERT.

The remainder of this paper is structured as follows: Section 2 describes and contrasts related work. Section 3 presents our contribution. Section 4 reports the empirical evaluation and discusses the results. Finally, Sect. 5 concludes our study and lists future work.

2   Related Work

There are many strategies for labeling training data automatically. These annotation tactics generate imperfect (less accurate) labels based on domain knowledge and are commonly pervasive as weak supervision [18]. We can categorize weak supervision approaches into three types [27]: (i) incomplete supervision, where only a small subset of training data is available with annotations; (ii) inexact supervision, in which training data is only provided with coarse annotations; and (iii) inaccurate supervision, where available labels are not always ground-truth.

Many incomplete supervision approaches have been proposed to identify fake news. The studies developed techniques to annotate social media news and increase the amount of training data from various sources [12,21,25]. Inexact supervision approaches have been proposed in which some supervisory information is provided but not as accurate as desired [3,9,13]. For example, a study [6] developed a framework for weak interactive supervision where a method proposes heuristics and learns from user feedback on each heuristic. The experiments demonstrated that only a few feedback iterations are needed to train models that achieve highly competitive test performance without access to ground-truth training labels.

In this paper, we focus on inaccurate supervision procedures due to the similarity of the proposed approach. Inaccurate supervision concerns the situation in which the supervision information is not always ground-truth, and some annotations may suffer from errors [27]. Weak labeling techniques in text classification tasks are known as distant supervision [17]. Distant supervision generates training annotations by heuristically aligning data points with an external knowledge base [2,15]. In addition, heuristic rules for labeling data are also common sources of weak supervision. That is, weak supervision sources mainly contain distant supervision [5,20,26] and heuristic rules [10,19].

A study presented a practical approach for treating the identification of fake news on Twitter as a binary machine learning problem [12]. The tweets were labeled by their sources, *i.e.*, tweets issued by accounts known to spread fake news were labeled as fake, and tweets issued by accounts known as trustworthy were labeled as accurate. Two datasets and six textual representation models were considered for experimental evaluation. Two alternatives were explored to represent the tweet textual contents: a Bag-of-Words (BoW) employing TF-IDF vectors and a neural Doc2vec model trained on the corpus. Instead of creating a small but accurate hand-labeled dataset, the authors demonstrated that using a large-scale dataset with inaccurate labels yields competitive results.

A more specific study for short-text classification involving insufficient unlabeled data, data sparsity, and imbalanced classification was reported in [8]. The proposed method can generate probabilistic labels through the conditional independent model. Six pre-training models were adopted: BERT Base and Multilingual Chinese, RoBERTa Base and Large Chinese, ERNIE and ERNIE Chinese. According to experimental results on public and synthetic datasets, unlabeled imbalanced short-text classification problems can be solved effectively by multiple weak supervision. Notably, recall and $F_1 Score$ can be improved without reducing precision by adding distant supervision clustering, which can be employed to meet different application needs.

The authors [16] presented an approach to bootstrap named entity recognition models without requiring any labeled data from the target domain. Instead, the approach relies on labeling functions by automatically annotating documents with named entity labels. A Hidden Markov Model (HMM) was trained to unify the noisy labeling functions into a single (probabilistic) annotation, considering each labeling function.

The success of machine learning methods for texts is closely related to the pre-processing strategy of textual data and the characteristics of the application domains [4]. We highlight that studies on textual representations for weak supervision tasks have received much attention in the literature. However, no in-depth studies were found on text classification for supervision heuristically aligning textual and time series data. In this sense, we introduce the proposed method in Sect. 3.

3   Methods

This work investigates a short-text labeling function of the commodity market using time series data (Fig. 1). In addition, it contemplates a vector text representation model based on BoW that adopts a measure of distance between Terms and Documents from pre-trained BERT models, called TD-BERT. Thus, classification models are applied to assess the predictive performance of the proposed approaches. Figure 1 illustrates the steps performed in this study.

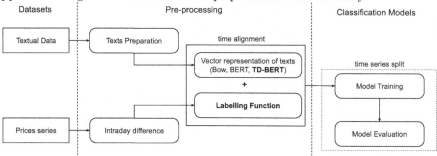

Fig. 1. Conceptual model of the proposed method.

3.1   Labeling Function

A price series S of size m is defined as an ordered sequence of observations, *i.e.*, $S = (s_1, s_2, ..., s_m)$, where s_t represents an observation s at time t. The textual documents D is also an ordered sequence $D = (d_1, d_2, ..., d_k)$, where d_t is a text d at time t, and size n. Therefore, we attribute via time alignment a label (-1, 0 or 1) to texts using the following equation:

$$d_t = \begin{cases} -1 \ if \ s_{t+lag} < (s_t + s_{t-lag})/2 \\ \ \ 1 \ if \ s_{t+lag} > (s_t + s_{t-lag})/2 \\ \ \ 0 \ \ \ \ \ \ \ \ \ \ otherwise \end{cases} \qquad (1)$$

the text d_t receives a label according to the level and trend patterns of the time series S. The constant lag corresponds to the seasonal period of the time series in number of observations. To exemplify, Fig. 2 portrays the result of Eq. 1 applied to a synthetic time series with $lag = 5$. This function aims to capture the time series' stable, increasing, and decreasing behaviors to assign labels to short texts arranged chronologically in time.

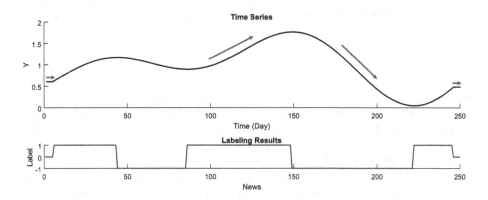

Fig. 2. Illustration of how the labeling function works.

3.2   TD-BERT

We implemented the proposed approach to obtain a new textual representation that considers the semantic features. First, we extract a collection of documents $D = [d_1, d_2, ..., d_k]$ containing k documents and a set $T = [w_1, w_2, ..., w_b]$ with b terms from D. This process is similar to the one used in BoW. However, we take into account here the sentence transformers of the pre-trained BERT models to obtain the cosine distance of each term in each document.

The textual representation D with sentence transformers is defined as $DS = ([B_1], [B_2], ...[B_k])$, where each B is a BERT vector of h positions representing a document d at time t. The representation of Terms with the sentence transformers is defined as $TS = ([W_1], [W_2], ...[W_b])$, where W_j is a BERT vector of h positions that represents a term w_j. The set of documents is represented as a document-term matrix constituted by cosine distance c from each vector k composed of b dimensions, as depicted in Fig. 3.

The matrix values correspond to the cosine distance of each term in each document, i.e., $c(B_k, W_b)$ equals the distance between vectors W_j and B_i. The vector values DS and TS are assigned according to a pre-trained BERT model. Thus, in this work, we evaluate the classification performance applying three

	W_1	W_2	...	W_{b-1}	W_b
B_1	$c(B_1,W_1)$	$c(B_1,W_2)$	...	$c(B_1,W_{b-1})$	$c(B_1,W_b)$
B_2	$c(B_2,W_1)$	$c(B_2,W_2)$	...	$c(B_2,W_{b-1})$	$c(B_2,W_b)$
$\vdots$	$\vdots$	$\vdots$	$\ddots$	$\vdots$	$\vdots$
B_{k-1}	$c(B_{k-1},W_1)$	$c(B_{k-1},W_2)$	...	$c(B_{k-1},W_{b-1})$	$c(B_{k-1},W_b)$
B_k	$c(B_k,W_1)$	$c(B_k,W_2)$	...	$c(B_k,W_{b-1})$	$c(B_k,W_b)$

Fig. 3. Illustration of the representation of document k as a document-term matrix.

pre-trained models: BERT base multilingual (TD-BERT), DistilBERT base multilingual (TD-DistilBERT), and BERT base Portuguese (TD-BERTimbau) [23].

4   Evaluation

We present a weak supervision evaluation of two agricultural commodities datasets: corn and soybean. Furthermore, we compare several predictive models for the classification task, considering distinct textual representations. We selected methods from different machine learning paradigms to better compare the investigated configurations. The K-Nearest Neighbors (KNN) method belongs to the instance-based classification paradigm. Multi-Layer Perceptron (MLP) is an algorithm from the connectionist paradigm. Gaussian Naive Bayes (GNB) and Multinomial Naive Bayes (MNB) are probabilistic methods. Support Vector Machine (SVM), in turn, is a model of the statistical learning theory paradigm.

The following subsections report the steps illustrated in Fig. 1. We present an analysis of the impact of textual representations on the classification of agricultural commodity headlines. These tasks are relevant to emerging research topics related to classifying large volumes of unlabeled text. For reproducibility purposes, we provide a GitHub repository at https://github.com/ivanfilhoreis/ws_text with the source code of the classification methods and the textual representations.

4.1   Datasets

We used texts and time series of corn and soybean. The Portuguese textual data were extracted from an agricultural news website[1]. Founded in 1997, *Notícias Agrícolas* is one of the Brazilian agribusiness's most influential media. Table 1 describes the dataset period, the number of days, and information about the text data.

Time series data were extracted from the Center for Advanced Studies in Applied Economics (CEPEA) at the University of São Paulo (USP).

---

[1] https://www.noticiasagricolas.com.br/.

Table 1. Overview of the time series and textual data used in our experimental evaluation.

Commodity	Corn and Soybean
Period	2015-01-05 to 2021-12-10
Number of days	1753
TS Attributes	Values (Open, **Close**, High, Low)
Number of headlines/News	7172 (Corn) - 8394 (Soybean)

4.2 Pre-processing

We evaluated the predictive model performances considering weak labels in positive and negative binary scenarios. The Positive Binary (PB) scenario has the labels [0, 1], and Negative Binary (NB) one has the labels [−1, 0]. Table 2 presents examples of labeled headlines in agreement with the function formalized in Sect. 3.1.

Table 2. News samples labeled using the labeling function.

Com.	Date	Headline	Lab.
Corn	2016-01-12	Dólar sobe nesta 4ª com atenção à política interna; milho acompanha	1
	2016-06-21	Preços do milho recuam até 15% no Brasil com colheita da 2ª safra	−1
	2017-03-27	Incerteza sobre a demanda por milho resulta em nova queda de preço	−1
	2018-02-27	USDA reporta a venda de 130 mil toneladas para destinos desconhecidos	1
Soybean	2016-05-10	Chuva do início de outubro ainda não foi suficiente para as lavouras no Sul do MS	1
	2017-01-30	Com queda do dólar e perspectiva de safra elevada, preço da soja cai no Brasil	−1
	2018-11-30	Soja opera estável na Bolsa de Chicago observando início da reunião do G20	−1
	2020-09-21	USDA informa nova venda de 435 mil t de soja para China e demais destinos	1

Among the 7172 corn headlines, 3209 were labeled as negative (−1), 66 as neutral (0), and 3897 as positive (1). Regarding soybean headlines, 3681 were labeled as negative, 82 as neutral, and 4631 as positive. In order to make a binary assessment, negative labels were assigned as neutral in the PB scenario, and in the NB one, positive labels were also changed to neutral.

Our work applies BoW-based representations, pre-trained NLM models, and the proposed TD-BERT model for vector representation of the texts. In the BoW modeling, we used three-term weighting techniques: Binary, TF, and TF-IDF. We considered only unigram versions of each of these weighting terms. In these models, we applied a text cleaning process to decrease the data dimensionality and increase representation quality. According to [1], this process improves the quality of the classification algorithms. The cleaning steps were: (1) converting words to lowercase and removal of accents; (2) removal of punctuation marks and alphanumeric characters; (3) removal of stopwords; and (4) word-stemming.

We used three pre-trained neural language models to assess weak supervision techniques: Multilingual (M) versions of BERT, M. DistilBERT, and the Portuguese version BERTimbau [22]. In the pre-trained models, we do not use text cleaning techniques to maintain the original text structure, which is essential for context-dependent NLMs. Thus, the sentence transformers of each trained model were employed as input for the predictive models. Also, we used the pre-trained models to build the proposed models: TD-BERT (TD-Be), TD-DilstilBERT (TD-Di), and TD-BERTimbau (TD-Ba).

4.3   Classification Models and Experimental Setup

We used five traditional classification algorithms: MLP, SVM, KNN, GNB, and MNB. The parameters of the ML algorithms we adopted in our experiments were default values of the scikit-learn library.

The time series split evaluation strategy was employed to consider temporal dependence of the textual data, *i.e.*, we train past news to evaluate a future scenario. Thus, seven splits were used for eight evaluations. In this configuration, each split represents one year of the textual dataset. Figure 4 outlines the time series split assessment strategy adopted in this study.

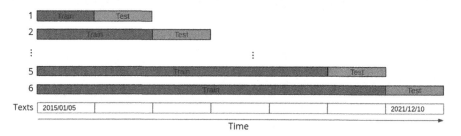

Fig. 4. Time series split used in the experimental setup.

For the evaluation step, we used the F_1 evaluation measure, which corresponds to the harmonic mean of Precision 3 and Recall 4. Equation 2 defines the F_1 index. We employed this metric because the classes are imbalanced in all evaluation splits.

$$F_1 = \frac{2 \times Prec \times Rec}{Prec + Rec},$$
(2)

$$Prec = \frac{TP}{TP + FP},$$
(3)

$$Rec = \frac{TP}{TP + FN},\qquad(4)$$

where TP (True Positive) refers to the number of documents of a class in which the algorithm has correctly classified, and FP (False) indicates the number of documents that do not belong to a class the algorithm wrongly classified as belonging. Finally, FN (False Negative) refers to the number of documents from a class that the algorithm wrongly classified as another class.

4.4 Results and Discussion

We conducted an experimental evaluation to investigate two aspects of weak supervision. In the first aspect, we sought to analyze each textual representation model's impact considering the five different classification algorithms. In the second aspect, we assessed the influence of the neural language model on two weak supervision classification tasks.

Concerning the first aspect, Tables 3 and 5 present the classification results of the MLP, SVM, KNN, GNB, and MNB algorithms on the corn and soybean datasets. This table covers an evaluation scenario PB for Corn and Soybean, which we named CPB and SPB. Each row represents the result of F_1 for a specific algorithm. In bold, we highlighted the highest values for each classification model. The underlined values reflect the best performance of the textual representation models (BoW, BERT and TD-BERT), and the value in parenthesis is the best result considering all the performances.

Table 3. Positive binary evaluation results. Comparison (macro F_1 measure) of BoW models, pre-trained neural language and the proposed TD-BERT hybrid model.

Corn - Positive Binary (CPB)									
Mod.	Bin.	TF	TFIDF	BERT	Distil.	B.Br	TD-B	TD-D	TD-Br
MLP	<u>0.496</u>	0.495	0.495	0.486	0.488	(**0.499**)	0.378	0.342	0.356
SVM	**0.456**	0.454	0.452	0.431	0.422	0.412	0.416	0.389	0.388
KNN	0.484	0.483	0.490	0.476	0.479	0.492	0.483	0.486	<u>0.495</u>
GNB	0.439	0.439	0.444	0.496	0.485	**0.497**	0.494	<u>0.495</u>	0.462
MNB	0.487	0.488	0.451	–	–	–	–	–	–
Soybean - Positive Binary (SPB)									
Mod.	Bin.	TF	TFIDF	BERT	Distil.	B.Br	TD-B	TD-D	TD-Br
MLP	<u>**0.490**</u>	0.488	0.488	0.476	0.489	0.485	0.344	0.312	0.352
SVM	0.440	0.439	**0.442**	0.398	0.371	0.387	0.381	0.355	0.357
KNN	0.483	0.481	0.484	0.478	0.474	**0.486**	0.477	0.474	0.481
GNB	0.470	0.470	0.469	0.498	0.499	(**0.500**)	<u>0.494</u>	0.481	0.469
MNB	0.472	0.472	0.436	–	–	–	–	–	–

The neural language models were not processed for MNB because it does not accept vectors with negative values. However, we considered it essential to keep the MNB results for the BoW representations in order to compare them with other results. Analyzing the highlighted values of CPB (bold) for each classification model, we observed that the representations BERTimbau (B.Br), Binary (Bin), and TD-BERTimbau (TD-Br) obtained the best values F_1. We also noticed that the BERTimbau (MLP) model had the highest value among all results (0.499). The SPB results showed Binary, TF-IDF, and BERTImbau as the best values for each F_1 ranking ratio, with BERTimbau (0.500) being the highest value among all the SPB results. Table 4 presents the best CPB and SPB results in terms of precision, recall, and accuracy.

Table 4. Evaluation metrics concerning the best SPB and CPB classification results.

	CPB: BERTimbau (MLP)				SPB: BERTimbau (GNB)			
	Prec	Recall	F1-score	Support	Prec	Recall	F1-score	Support
0	0.489	0.385	0.422	2917	0.478	0.465	0.458	3358
1	0.534	0.636	0.574	3227	0.544	0.559	0.540	3836
Accuracy	0.518				0.511			
Macro avg	0.512	0.511	**0.499**	6144	0.511	0.512	**0.500**	7194
Weighted avg	0.527	0.518	0.500	6144	0.546	0.511	0.517	7194

The CPB and SPB accuracies were 0.518 and 0.51, respectively. However, looking at the support values of Table 4, we can see that the weak labels are reasonably balanced. Therefore, by analyzing results for this type of evaluation, Macro F_1 becomes more appropriate. Regarding NB, Table 5 displays two assessment scenarios. We called them the Negative Binary classification of Corn (CNB) and Soybean (SNB) approaches. Table 6 lists the best results of CNB and SNB concerning precision, accuracy, and recall.

Table 5. Negative Binary evaluation results. Comparison (macro F_1 measure) of BoW models, pre-trained neural language, and the proposed TD-BERT hybrid model.

Corn - Negative Binary (CNB)									
Mod.	Bin.	TF	TFIDF	BERT	Distil.	B.Br	TD-B	TD-D	TD-Br
MLP	0.491	0.495	<u>0.496</u>	0.482	**0.497**	0.493	0.358	0.343	0.362
SVM	**0.452**	0.451	0.451	0.428	0.418	0.411	0.406	0.375	0.374
KNN	0.484	0.481	0.490	0.474	0.479	0.492	0.485	0.483	<u>0.494</u>
GNB	0.439	0.439	0.443	0.497	0.490	**(0.507)**	0.493	<u>0.494</u>	0.469
MNB	0.491	0.490	0.446	–	–	–	–	–	–
Soybean - Negative Binary (SNB)									
Mod.	Bin.	TF	TFIDF	BERT	Distil.	B.Br	TD-B	TD-D	TD-Br
MLP	0.48	0.487	<u>0.488</u>	0.474	0.481	0.485	0.339	0.288	0.344
SVM	**0.434**	0.432	0.432	0.389	0.363	0.378	0.376	0.358	0.364
KNN	0.471	0.471	0.472	0.471	0.468	**0.480**	0.47	0.47	0.478
GNB	0.451	0.452	0.453	0.500	**(0.501)**	0.496	0.485	<u>0.486</u>	0.461
MNB	0.474	0.473	0.429	0	0	0	0	0	0

Table 6. Evaluation metrics regarding the best CNB and SNB classification results.

	CNB: BERTimbau (GNB)				SNB: DistilBERT (GNB)			
	Prec	Recall	F1-score	Support	Prec	Recall	F1-score	Support
−1	0.489	0.520	0.492	2881	0.479	0.524	0.483	3339
0	0.554	0.520	0.520	3263	0.553	0.512	0.517	3855
Accuracy	0.517				0.506			
Macro avg	0.521	0.520	**0.507**	6144	0.516	0.518	**0.501**	7194
Weighted avg	0.534	0.517	0.511	6144	0.552	0.506	0.513	7194

Observing the CNB results, we emphasize that the DistilBERT (Diltil.), Binary, TD-BERTimbau, and BERTimbau (B.Br) representations had the highest values (bold) of F_1 for each classification algorithm, respectively. Regarding the SNB results, the TF-IDF, Binary, BERTimbau, and DiltilBERT representations achieved the best results. In this case, the values 0.517 and 0.569, in parentheses, represent the best results of CNB and SNB, respectively.

Aiming to investigate the second aspect of the experimental evaluation, we analyzed the impact of the neural language model on weak supervision. According to the underlined result of CPB and SPB in Table 3, the binary representation had the highest F_1 values, i.e., 0.496 and 0.490, respectively. The Neural language model BERTimbau performed better in the two scenarios with F_1 values of 0.499 and 0.500. Finally, TD-BERTimbau and TD-BERT representations achieved F_1 results with values of 0.495 and 0.494. Thus, representations models based on neural language had better performance than the BoW models. To illustrate the vector distribution of the texts, Fig. 5 presents a graph of the textual representations that performed better in each representation model of Table 3 (underlined values).

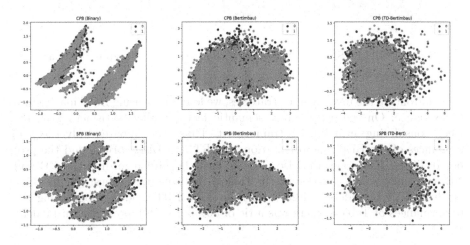

Fig. 5. CPB and SPB. PCA technique for plotting textual representations.

The PCA technique was used to reduce the dimensionality of the textual representation of the agricultural commodities dataset. We observed that headlines classified as positive (1) are more concentrated in the graph distribution, while headlines classified as neutral are a little more sparse. Furthermore, the CPB (TF-IDF) and SPB (TF) representations have smaller ranges on the axes than the BERT-based representations. In this sense, we believe that this broader spectrum can abstract more semantic information from texts.

Comparing the CNB and SNB underlined results in Table 5, the TF-IDF, BERTimbau, DistilBERT, TD-DistilBERT, and TD-BERTimbau obtained the best classification performance for the learning algorithms, respectively. The DistilBERT and BERTimbau neural language models performed better for the GNB method. In both experiments (PB and NB), we observed that the best results came from representations based on Distilbert and BERTimbau with the GNB and MLP models. Figure 6 illustrates a graph of the textual representations that performed better in each representation model of Table 5.

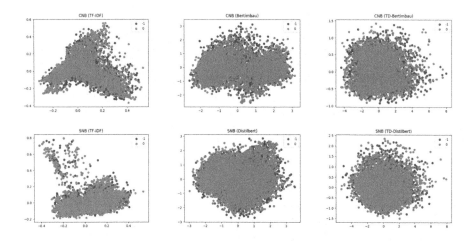

Fig. 6. CNB and SNB. PCA technique for plotting textual representations.

Confronting the investigated strategies, we found that using BoW may reduce performance. On the other hand, semantic features allow satisfactory results when considering an extensive training set. Thus, we compared only the performances of the BERT representations (BERT and Dist. B.Bau) and the proposed TD-BERT model (TD-Be, TD-Di and TD-Bau) regarding the PB and NB assessments. We can observe in Table 3 that among the best results, 75% are from BERT models and 25% are from TD-BERT models. Concerning the performances of Table 5, there was a tie of 50% for each representation model.

5   Conclusion

We introduced an automatic text labeling approach using information extracted from time series. This paper innovates by considering a weak supervision technique to label a large volume of texts. Text documents and agribusiness commodity price series were employed to assess performance in four classification scenarios (CPB, SPB, CNB and SNB). Our experimental evaluation considered nine textual representations and different learning paradigms. In addition, we proposed a text representation model that measures the distance between Terms and Documents from pre-trained BERT models (TD-BERT).

Regarding the best results of the Positive Binary and Negative Binary assessment scenarios, ten between sixteen are representations from the BERT models (62.5%). The proposed TD-BERT models performed better in some cases by analyzing neural language-based representation models. In general, neural language-based representation models outperformed BoW-based models. However, a limitation of the TD-BERT models is processing time, and future work can be conducted to reduce computational costs.

The designed labeling function can be an alternative to annotating a large volume of text documents. Automatic labeling can be imprecise but useful when many texts are not labeled. In this study, the limitation of the weak supervision analysis consisted of the class imbalance. Future research can developed strategies to propagate labels through semi-supervised learning to reinforce labeling. In addition, through connectionist approaches, other external factors can be used in the labeling function; *e.g.*, we can consider a weighting coefficient in the labeling of news.

Acknowledgements. This work was carried out at the Center for Artificial Intelligence (C4AI-USP) and partially supported by the São Paulo Research Foundation (FAPESP) (grant #2019/07665-4) and the IBM Corporation. The authors of this paper thank FAPESP (Process 2019 / 25010-5) and the National Center for Scientific and Technological Development (CNPq) (process 309575/2021-4). The corresponding author thanks the Minas Gerais State Research Support Foundation (FAPEMIG) (Process PCRH BPG-00054-210).

References

1. Aggarwal, C.C.: Machine Learning for Text. Springer, Cham (2018). https://doi.org/10.1007/978-3-319-73531-3
2. Alfonseca, E., Filippova, K., Delort, J.Y., Garrido, G.: Pattern learning for relation extraction with a hierarchical topic model. In: Annual Meeting of the Association for Computational Linguistics, vol. 2, pp. 54–59 (2012)
3. Anklin, V., et al.: Learning whole-slide segmentation from inexact and incomplete labels using tissue graphs. In: de Bruijne, M., et al. (eds.) MICCAI 2021. LNCS, vol. 12902, pp. 636–646. Springer, Cham (2021). https://doi.org/10.1007/978-3-030-87196-3_59
4. Araujo, A.F., Gôlo, M.P., Marcacini, R.M.: Opinion mining for app reviews: an analysis of textual representation and predictive models. Autom. Software Eng. **29**(1), 1–30 (2022)

5. Batista-Navarro, R., Hawkins, O.: Topic modelling vs distant supervision: a comparative evaluation based on the classification of parliamentary enquiries. In: Doucet, A., Isaac, A., Golub, K., Aalberg, T., Jatowt, A. (eds.) TPDL 2019. LNCS, vol. 11799, pp. 415–419. Springer, Cham (2019). https://doi.org/10.1007/978-3-030-30760-8_46

6. Boecking, B., Neiswanger, W., Xing, E., Dubrawski, A.: Interactive weak supervision: learning useful heuristics for data labeling. arXiv preprint arXiv:2012.06046 (2020)

7. Chatfield, C., Xing, H.: The Analysis of Time Series: An Introduction with R. CRC Press (2019)

8. Chen, L.M., Xiu, B.X., Ding, Z.Y.: Multiple weak supervision for short text classification. Appl. Intell. 1–16 (2022)

9. Dai, E., Shu, K., Sun, Y., Wang, S.: Labeled data generation with inexact supervision. In: ACM SIGKDD Conference on Knowledge Discovery and Data Mining, pp. 218–226 (2021)

10. De Sa, C., Ratner, A., Ré, C., Shin, J., Wang, F., Wu, S., Zhang, C.: Deepdive: declarative knowledge base construction. ACM SIGMOD Record **45**(1), 60–67 (2016)

11. dos Santos, B.N., Marcacini, R.M., Rezende, S.O.: Multi-domain aspect extraction using bidirectional encoder representations from transformers. IEEE Access **9**, 91604–91613 (2021)

12. Helmstetter, S., Paulheim, H.: Collecting a large scale dataset for classifying fake news tweets using weak supervision. Fut. Internet **13**(5), 114 (2021)

13. Hsieh, C.Y., Lin, W.I., Xu, M., Niu, G., Lin, H.T., Sugiyama, M.: Active refinement for multi-label learning: a pseudo-label approach. arXiv preprint arXiv:2109.14676 (2021)

14. Janev, V., Pujić, D., Jelić, M., Vidal, M.-E.: Chapter 9 survey on big data applications. In: Janev, V., Graux, D., Jabeen, H., Sallinger, E. (eds.) Knowledge Graphs and Big Data Processing. LNCS, vol. 12072, pp. 149–164. Springer, Cham (2020). https://doi.org/10.1007/978-3-030-53199-7_9

15. Krause, S., Li, H., Uszkoreit, H., Xu, F.: Large-scale learning of relation-extraction rules with distant supervision from the web. In: Cudré-Mauroux, P., et al. (eds.) ISWC 2012. LNCS, vol. 7649, pp. 263–278. Springer, Heidelberg (2012). https://doi.org/10.1007/978-3-642-35176-1_17

16. Lison, P., Hubin, A., Barnes, J., Touileb, S.: Named entity recognition without labelled data: a weak supervision approach. arXiv preprint arXiv:2004.14723 (2020)

17. Mintz, M., Bills, S., Snow, R., Jurafsky, D.: Distant supervision for relation extraction without labeled data. In: Joint Conference of the 47th Annual Meeting of the ACL and the 4th International Joint Conference on Natural Language Processing of the AFNLP, pp. 1003–1011 (2009)

18. Ratner, A., Bach, S.H., Ehrenberg, H., Fries, J., Wu, S., Ré, C.: Snorkel: rapid training data creation with weak supervision. In: International Conference on Very Large Data Bases, vol. 11, p. 269. NIH Public Access (2017)

19. Rekatsinas, T., Chu, X., Ilyas, I.F., Ré, C.: Holoclean: holistic data repairs with probabilistic inference. arXiv preprint arXiv:1702.00820 (2017)

20. Shi, Y., Xiao, Y., Niu, L.: A brief survey of relation extraction based on distant supervision. In: Rodrigues, J.M.F., et al. (eds.) ICCS 2019. LNCS, vol. 11538, pp. 293–303. Springer, Cham (2019). https://doi.org/10.1007/978-3-030-22744-9_23

21. Shu, K., et al.: Leveraging multi-source weak social supervision for early detection of fake news. arXiv preprint arXiv:2004.01732 (2020)

22. Souza, F., Nogueira, R., Lotufo, R.: Portuguese named entity recognition using bert-CRF. arXiv preprint arXiv:1909.10649 (2019)
23. Souza, F., Nogueira, R., Lotufo, R.: BERTimbau: pretrained BERT models for Brazilian Portuguese. In: Brazilian Conference on Intelligent Systems (2020)
24. de Souza, M.C., Nogueira, B.M., Rossi, R.G., Marcacini, R.M., dos Santos, B.N., Rezende, S.O.: A network-based positive and unlabeled learning approach for fake news detection. Mach. Learn. 1–44 (2021)
25. Wang, Y., et al.: Weak supervision for fake news detection via reinforcement learning. In: AAAI Conference on Artificial Intelligence, vol. 34, pp. 516–523 (2020)
26. Yao, W., Liu, J., Cai, Z.: Personal attributes extraction in chinese text based on distant-supervision and LSTM. In: Park, J.J., Loia, V., Yi, G., Sung, Y. (eds.) CUTE/CSA -2017. LNEE, vol. 474, pp. 511–515. Springer, Singapore (2018). https://doi.org/10.1007/978-981-10-7605-3_84
27. Zhou, Z.H.: A brief introduction to weakly supervised learning. Natl. Sci. Rev. 5(1), 44–53 (2018)

Towards a Better Understanding of Heuristic Approaches Applied to the Biological Motif Discovery

Jader M. Caldonazzo Garbelini[1]([⊠]) [ID], Danilo Sipoli Sanches[2][ID], and Aurora Trinidad Ramirez Pozo[1][ID]

[1] Federal University of Paraná, Curitiba, Brazil
{jmcgarbelini,aurora}@inf.ufpr.br
[2] Federal University of Technology, Cornélio Procópio, Brazil
danilosanches@utfpr.edu.br

Abstract. The detection of transcription factor binding sites (TFBS) play a important role inside bioinformatics challenges. Its correct identification in the promoter regions of co-expressed genes is a crucial step for understanding gene expression mechanisms and creating new drugs and vaccines. The problem of finding motifs consists of looking for conserved patterns in biological datasets of sequences through the use of unsupervised learning algorithms. For that reason, it is considered one of the classic problems of computational biology, which in its simplest formulation has been proven to be NP-HARD. Moreover, heuristic and meta-heuristic algorithms have been shown to be very promising in solving combinatorial problems with very large search spaces. In this work, we propose an evaluation of different heuristics and meta-heuristics approaches in order to measure its performance: Variable Neighborhood Search (VNS), Expectation Maximization (EM) and Iterated Local Search (ILS). For each of them, two sets of experiments were carried out: In the first, the heuristics were performed alone and in the second, a constructive procedure was introduced with respect to improve the quality of initial solutions. Finally, the metrics were compared with the state-of-art MEME algorithm, which is very used in biological motif discovery. The results obtained suggest that the heuristics are more efficient when used together and also, a constructive procedure was very promising, managing to improve the performance metrics of the evaluated heuristics in most experiments. Also, the combination between a constructive procedure and EM proved to be quite competitive, managing to outperform the MEME algorithm in several datasets.

Keywords: Biological motifs · Heuristics · Meta-heuristics · Unsupervised learning

Supplementary Information The online version contains supplementary material available at https://doi.org/10.1007/978-3-031-21686-2_13.

J. C. Xavier-Junior and R. A. Rios (Eds.): BRACIS 2022, LNAI 13653, pp. 180–194, 2022.
https://doi.org/10.1007/978-3-031-21686-2_13

1   Introduction

In biological sequence analysis, motifs are small fragments of conserved nucleotides that are presumed to have some biological significance. These small patterns, most of the time, appear recurrently in the promoter regions of the co-expressed genes [6]. They act as binding sites for specific proteins and play a key role in the activation and primary repression of gene expression. Although they are often found in promoter regions, they also appear in exonic sequences, within introns, or on the negative strand of genes [19]. Although it is possible to identify motifs with relative precision using experimental techniques such as DNAse footprinting, ChIPseq, gel-shift, and reporter construct assays, these approaches are often expensive and time-consuming [12]. Furthermore, in each experiment, it is possible to analyze only a small region of the gene. With the increase in available data due to the number of sequenced genomes, it was necessary to develop faster and cheaper techniques that would maintain a good level of reliability in data analysis [5]. For this reason, computational techniques have gained importance and are being widely used in the analysis of biological sequences.

The main objective of finding motifs is to identify the sites responsible for the gene transcription initiation. Through this information, it is possible to recognize which sub-sequences have over-representation in relation to the others. In recent years, many approaches have been proposed in the literature. In general, they are divided into: i) approaches that use probabilistic models and ii) approaches that use consensus sequences. Both techniques are reviewed in Sect. 2. Probabilistic methods seek to maximize some type of probabilistic model or probabilistic function created from the oligo sequences dataset, such as relative entropy or some other statistical measure. These algorithms are generally quick to run, however, depending on the optimization strategy, they can get stuck in local optima. Exact approaches generally use the consensus sequence for motif representation, employing some mathematical optimization as a search model. Often, these approaches have a high convergence time, in particular for long motif lengths.

This work aims to investigate the behavior of the VNS, EM, and ILS algorithms applied to biological motif discovery problem. In that case, two different evaluations were performed: i) first, the algorithms were executed alone on 10 datasets extracted from the Jaspar repository [17] and their results were collected, ii) after that, the same tests were performed, but now, the constructive procedure were used to initialize the solutions. Furthermore, this study contributes to better understand the relationship among the heuristics used applied to biological motif discovery, and also providing valuable information that can guide for future works, such as build integrated frameworks for the motif discovery.

The results suggest that using an initialization heuristic is a good way and contributes positively to the quality of the solutions found, and this was even more evident in the tests performed with the EM algorithm. The rest of the paper is organized as follows: The bibliographic review is presented in Sect. 2. In Sect. 3 the problem is defined and the implementation details of each algorithm are presented. In Sect. 4, the results are displayed and the discussion takes place. Finally, in Sect. 5 the conclusion and the future steps that should be followed in the continuation of this work are presented.

2   Literature Review

Due to its importance, the problem of finding motifs has already been the subject of much research. One of the first surveys published in the area was in 2006 by Sandve and Drabløs [18]. In their article, they listed more than 100 algorithms applied in motif problem. In 2018, Nung Kion Lee et al. [10] published a work whose focus was to enumerate the main Evolutionary Algorithms used in this domain. The last review published until the writing of this work was in 2020, by the authors Ying He et al. [9] whose essence was the survey of the main Deep Learning techniques successfully employed in this context.

Due to the high density of algorithms, we will review just a few of them here. Pavesi et al. published the WEEDER algorithm [16], in which motifs are modeled as strings. The main idea of the weeder is to looking for strings (with some degeneration degrees) that occur many times (as much as possible) in the input sequences. To do this, they employed a suffix tree data structure and a strategy in which a pattern-oriented search was used on the given strings. STREME [2] was published by Timothy L. Bailey and finds fixed-length ungapped motifs. This algorithm can find motifs limited to 30 columns wide in large sets of sequence data. Its input consists of one or two sets of sequences, with the control sequences having approximately the same distribution as the primary sequences. The program uses Fisher's Exact Test or Binomial Test to determine the significance of each motif found in the positive set compared to its representation in the control set. Just like WEEDER, STREME employs a suffix tree as its data structure and this gives it high speed in the optimization process.

MEME [3] is a classic approach and is currently the most used algorithm by the scientific community. Developed by Timothy L. Bailey and Charles Elkan, its purpose is to perform the alignment of sequences using weight matrices. Motifs are represented as probability matrices and optimized using a strategy based on Expectation Maximization (EM). Although it was published in 1995, MEME underwent constant updates and the last one, according to the MEME SUITE platform [4], was carried out in August 2021. MotifSampler [22] is a probabilistic motif detection tool idealized by Gert Thijs et al. The search is performed through a stochastic optimization strategy, based on the Gibbs Sampling approach, which looks for all possible sets of short DNA segments that are over-represented in the sequence dataset. MotifSampler was originally published in 2001, however, according to the authors, it was last updated in August 2020.

FMGA [13] is an approach based on Genetic Algorithms (GA) developed by Liu F.F.M. et al. According to its authors, its aims to locate motifs between −2000bp upstream and +1000bp downstream of several groups of co-expressed genes. The motif pattern length is fixed until the end of GA and a distance-based fitness function has been employed. An initial population of individuals is randomly generated and after evaluation, an elitist competition is held so that individuals with the best fitness values are automatically qualified for the next generation. MFEA [1] is an approach conceived by Faisal Bin Ashraf and Md Shafiur Raihan Shafi, based on meta-heuristics that aims to minimize the trade-off between exploration and exploitation of the search space through a mutation technique defined over a normal distribution, thus managing to efficiently measure the adequacy of a candidate motif. The authors used a reference dataset published by Tompa et al. [23] to evaluate the motifs found.

The heuristics implemented in this paper employ probabilistic scoring functions, being therefore classified as belonging to the group of approaches that use probabilistic models (group i). Furthermore, algorithms that use the constructive procedure are better described by hybridization theory, which can be consulted in article [21]. In this way, the results obtained by them were compared to the results achieved by the MEME algorithm, which is a reference in the area and one of the most used techniques by the scientific community to find biological motifs.

3   Problem Definition and Algorithms

Let $X = \{x_1, \ldots, x_N\}$, be a set of N sequences of length M, defined over an alphabet $\Sigma = \{A, C, G, T\}$. Let w be the length of the motif. We start from the premise that the relation $0 < w \ll M$ is true. Let $Y = \{y_1, \ldots y_n\}$ be a set of n sequences of length w extracted from X. Each x_i has $L = M - w + 1$ overlapping substrings. Thus, Y has $L * N$ substrings of length w. The problem consists of classification the w-$mers$[1] using only the data provided as input (*ab-initio*). Motif finding can also be defined in a combinatorial way, in which we want to find $X^* = \{x_1^*, \ldots, x_N^*\}$ and their respective initial positions in X. The choice of a given pattern is based on the definition of one or more score functions that measure the similarity or difference between motifs and their respective occurrences. Li et al. [11] proved that the canonical definition of the motif problem is NP-HARD even with the most simplified assumptions.

The solutions were implemented as integer vectors (the pseudo-code of all algorithms used in this paper are available in the supplementary material). Each vector index corresponds to a sequence in the dataset and each value reflects the estimated starting position of each motif, with $1 \leq p_i \leq L$. For example, for a dataset with $N = 4$ sequences and $L = 10$, $S_1 = \{1, 6, 5, 10\}$, $S_2 = \{2, 2, 7, 3\}$ and $S_3 = \{6, 1, 2, 4\}$ are valid solutions. The total number of valid solutions is given by L^N. First, each solution is converted into a set of oligo sequences, as shown in Fig. 1a. Also in this figure, we have the frequency matrix (PFM). Each $PFM_{i,j}$ element corresponds to the count of symbols in each column of the string set. Given the assumption that symbols are independent and identically distributed (i.i.d.), then each column of the PFM matrix can be modeled as a Multinomial distribution, as shown in Eq. 1. To the left of Fig. 1b, we have the PFM matrix with the addition of pseudo-counters. This technique is also known

---

[1] Text segment of size w.

as Rule of Succession or Laplace's rule and its objective is to add small values (normally one) to the PFM matrix when there are few available observations, preventing further calculations from tending to infinity. To the right of Fig. 1b we have the PWM matrix, which is calculated by dividing each entry of the PFM matrix by n, where n is equal to the sum of the elements of an arbitrary column of the PFM (all PFM columns add the same value). Assuming that the values of a PWM are i.i.d, then each column can be modeled as a Dirichlet distribution, as shown in Eq. 2. Finally, Fig. 1c displays the PSSM matrix. It is generated by calculating the logarithm of the division of each entry in the PWM matrix by the background probability (which can be obtained by counting the symbols belonging to the genome of the studied organism). A positive entry in this matrix implies that the probability of the symbol belonging to a specific distribution is greater than the background probability. For this reason, the PSSM matrix is also known as the log-odds matrix. Equation 4 shows how the PSSM matrix is calculated. Once the PSSM matrix is created, then we can easily score the substrings just by adding the values corresponding to each symbol. The sum of all values in the PSSM matrix is called Information Content (IC). Equation 4 shows how this value is calculated. Through the IC, it is possible to assess the suitability of each solution.

$$Pr(X_1 = x_1, X_2 = x_2, \ldots, X_k = x_k) = \frac{\Gamma(\sum_i)x_i + 1}{\prod_i \Gamma(x_i + 1)} \prod_i^k p_i^{x_i} \tag{1}$$

where x_i represents the counts and p_i represents the probability of the i-th symbol respectively.

$$Pr(P_1 = p_1, \ldots, P_k = p_k; \alpha_1, \ldots, \alpha_k) = \frac{1}{\beta(\alpha)} \prod_i^k p_i^{\alpha_i - 1} \tag{2}$$

where p_i represents the probability and α_i represents the counts of the i-th symbol respectively.

$$PSSM_{k,j} = \log_2 \left[\frac{(PWM_{k,j})}{b_k} \right] \tag{3}$$

where $PWM_{k,j}$ represents the j-th probability of the line k and b_k represents the background probability of the k-th symbol.

$$IC = \sum_{i=1}^{\Sigma} \sum_{j=1}^{w} \Theta_{(i,j)} \log_2 \left[\frac{\Theta_{(i,j)}}{\Theta_{(0,i)}} \right] \tag{4}$$

where w is the length of the motif, Σ is the number of symbols in the alphabet ($\Sigma = 4$ for nucleotides), $\Theta(i,j)$ is the matrix of relative frequencies and $\Theta(0,i)$ is the background probability vector.

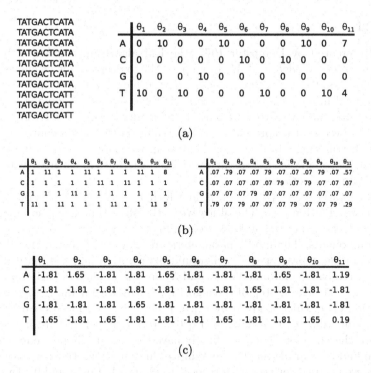

Fig. 1. Step-by-step construction of the PSSM matrix. (a) Left, motif alignment; on the right, PFM matrix (Position Frequency Matrix). (b) Left, PFM with pseudo-counters; on the right, PWM matrix (Position Weight Matrix); (c) PSSM matrix.

3.1 VNS

VNS is a meta-heuristic proposed by Mladenović and Hansen [15] whose objective is to solve complex combinatorial problems in large search spaces. Its basic operation is based on searching increasingly distant neighborhoods with the hope of finding more promising solutions. According to its authors, the VNS is based on the following principles: i) a local minimum found with the help of a certain neighborhood structure is not necessarily a local minimum for another neighborhood; ii) The global minimum is a local minimum for any neighborhood structure; iii) local minima are located close to each other.

First, the initial solution S is generated randomly. It is then converted to a PSSM matrix and its score is calculated. From this point onwards, the search for more distant neighborhoods starts. The search starts with $k = 1$ and goes to $kmax$. An intermediate solution Q is generated from a perturbation applied to S. The level of disturbance depends on the value of k, the greater the disturbance, the more intense it is. In practice, this is reflected in the search for a more distant neighborhood. If $IC(Q) > IC(S)$, then $S = Q$ and $k = 1$. Otherwise, k is incremented and a new perturbation is applied. This is until the algorithm reaches convergence or a maximum number of iterations has been reached.

3.2   EM

Originally created by Dempster 1977, the EM algorithm consists of basically two main steps, called E and M. In step E, the expectation is calculated using the log-likelihood function and in step M, the model is re-estimated with the data obtained in step E. The premise is that the samples belonging to Y have originated from at least two distributions belonging to the same family. Mixture models can be used to represent a $p(y)$ distribution through a convex combination of k distributions. In this implementation, we model the following distributions: i) background distribution and ii) motif distribution. Both are categorical, but the latter is positional.

Let a dataset $Y = \{y_1, \ldots, y_n\}$, in which y_i was extracted *i.i.d.* from an unknown distribution $p(y)$. The objective of the algorithm is to find an approximate representation of this unknown distribution through a mixture model with $k = 2$ components. The model parameters are $\Theta = \{\theta_1, \theta_2\}$ and $\Lambda = \{\lambda_1, \lambda_2\}$, where θ_1 and θ_2 represent the background distribution and the motif distribution respectively. The parameters λ_1 and λ_2 are the weights of each model, such that $\lambda_1 + \lambda_2 = 1$. The E step of the algorithm starts with defining the predictive distribution of y_i which is given by Eq. 5. From this equation, we can write the likelihood function of the Y dataset (Eq. 6). For numerical reasons, it's better to work with the odds log. Thus, we finally have the log-likelihood, represented by Eq. 7. To find the parameters that maximize this equation, we resort to calculus. For that, we need to derive 7 with respect to Θ and Λ (Eqs. 8 and 9). By solving Eqs. 8 and 9 for Θ and Λ, Eqs. 10 and 11 appear. After the expectation step, EM re-estimates the parameters of the k models through Eqs. 12 and 13 (step M). The algorithm alternates between steps E and M until convergence or by a number of iterations defined a priori. The models Θ_0 and Λ_0 are initialized randomly. In particular, to initialize θ_2 (distribution of the motifs model) the algorithm converts a set of oligo sequences into a probability matrix, as shown in Eq. 14. The parameter θ_1 is initialized using the following uniform distribution, $\theta_1 = \{0.25, 0.25, 0.25, 0.25\}$.

$$p(y_i|\Theta, \Lambda) = \sum_{k=1}^{2} \lambda_k p(y_i|\theta_k) \tag{5}$$

where $p(y_i|\Theta, \Lambda)$ represents the predictive distribution of the example y_i and $p(y_i|\theta_k)$ is the likelihood function of y_i given the model k.

$$p(Y|\Theta, \Lambda) = \prod_{i=1}^{n} p(y_i|\Theta, \Lambda) \tag{6}$$

where $p(Y|\Theta, \Lambda)$ is the probability of the dataset given the model.

$$\mathcal{L} = \sum_{i=1}^{n} \log(p(y_i|\Theta, \Lambda)) \tag{7}$$

where $\mathcal{L}$ is the log-likelihood of the dataset given the model.

$$\frac{\partial \mathcal{L}}{\partial \theta_k} = \sum_{i=1}^{n} \frac{\partial \log(p(y_i|\Theta, \Lambda))}{\partial \theta_k} \tag{8}$$

$$\frac{\partial \mathcal{L}}{\partial \lambda_k} = \sum_{i=1}^{n} \frac{\partial \log(p(y_i|\Theta, \Lambda))}{\partial \lambda_k} \tag{9}$$

$$\frac{\partial \mathcal{L}}{\partial \theta_k} = \sum_{i=1}^{n} \frac{1}{p(y_i|\Theta)} \frac{\partial p(y_i|\Theta)}{\partial \theta_k} \tag{10}$$

$$\frac{\partial \mathcal{L}}{\partial \lambda_k} = \sum_{i=1}^{n} \frac{1}{p(y_i|\Theta)} \frac{\partial p(y_i|\Theta)}{\partial \lambda_k} \tag{11}$$

where $\frac{\partial \mathcal{L}}{\partial \theta_k}$ and $\frac{\partial \mathcal{L}}{\partial \lambda_k}$ are the partial derivatives of the log-likelihood function in with respect to θ_k and λ_k.

$$\theta_{knew} = \frac{\sum_{i=1}^{n} y_i \lambda_k p(y_i|\theta_k, \lambda_k)}{\sum_{i=1}^{n} \sum_{k=1}^{2} y_i \lambda_k p(y_i|\theta_k, \lambda_k)} \tag{12}$$

$$\lambda_{knew} = \frac{\sum_{i=1}^{n} \lambda_k p(y_i|\theta_k, \lambda_k)}{n} \tag{13}$$

where θ_{knew} and λ_{knew} are the models θ_k and λ_k after the j-th iteration.

$$\theta_2 = \frac{F(x;i)}{\sum_{y \in a,c,g,t} F(y;i)}, \forall i, 1 \leq i \leq w \tag{14}$$

where $F(x, i)$ represents the absolute frequency of the symbol x in the column i.

3.3 ILS

The ILS algorithm was first published under that name in the seminal work of Thomas Stützle in 1998 [20]. Since then, several works have been done, including the recent study in [14]. At first, a solution S is generated randomly, and then the EM algorithm is used to perform a local search in S, generating S^*. After that, a perturbation is applied to S^* generating S^{**}.

The perturbation needs to be strong enough to allow the local search to explore different solutions and weak enough to prevent a random restart. An adaptive function was used for this purpose. In that case, the function can perturb up to $k = w$ components of the vector S^*. With the increase of iterations, the value of k progressively decays according to Eq. 15. This equation has the following property: $\lim_{i \to \infty} k = 1$ This means that the algorithm will explore more at the beginning and exploit more at the end. If $\delta = IC(S^{**}) - IC(S*)$ is positive, then the solution S^{**} is accepted. Otherwise, an uniform distributed variable r is drawn. If $r < g(T, \delta)$, then the solution S^{**} is accepted, otherwise S^* remains the current solution. g is an exponential decay function such that $g(\delta, T) = \exp(\delta/T)$.

$$k = h(i) = \lfloor 1 + imax/(\frac{imax}{w} + i)\rfloor \tag{15}$$

where i is the value of the i-th iteration, $imax$ is the hyperparameter represent the expected number of iterations until convergence.

3.4   Constructive Procedure

The constructive procedure used in this paper was based on GRASP meta-heuristic published in 1995 by Thomas A. Feo and Maurício GC. Resende [7]. In general, the algorithm has two stages: i) constructive phase and ii) local search. In this work, we have used only the constructive step, and all details are described as follow: the constructive step of GRASP can be carried out through several types of strategies. In this paper, the randomized greedy procedure created by Hart and Shogan [8] was used. This process is known to make a random choice from a restricted set, greedily organized.

This set is called the Restricted Candidate List (LCR). What the algorithm does is sort each solution by its cost (for this Eq. 4 is used) and add them to the LCR. LCR size is a hyper-parameter that needs to be controlled. If $|LCR| = 1$, then the algorithm becomes greedy, otherwise, if $|LCR|$ is too big, then the algorithm becomes random. In the motifs problem, an important point is to know which will be the first element to be inserted in the solution vector. If the starting position is too far from the correct position, the solution will be poor. To solve this problem, L solutions are created with GRASP and only the best ones are chosen at the end of the process. This guarantees that, $\forall l \in L, \exists S$ such that $S = G(l)$, where $G(.)$ is the GRASP constructive function. In other words, GRASP generates a valid solution for each starting position in the string dataset. To make it easier to read, from this point forward we will call our construction procedure GRASP.

4   Experiments

The following combinations of experiments were performed: i) VNS, ii) EM, iii) ILS, iv) GRASP+VNS, v) GRASP+EM, vi) GRASP+ILS, and vii) MEME. Each experiment was run 30 times on each dataset. In addition, each algorithm received 300 s of processing time for stop condition, except for the MEME that ran to completion. This was necessary so that the tests were adequate from a computational point of view and the comparison between the approaches could be carried out without bias. All algorithms and their respective parameters can be found at the following address: https://drive.google.com/drive/folders/ 1oPW0plMuD9SmQMtJuXxkO3ll9paVVoMK?usp=sharing. In addition, the MEME algorithm was downloaded from the following site: https://meme-suite. org/meme/. It is important to note that the best parameters were chosen through trial and error, since this choice is an optimization process by in itself.

4.1 Datasets

To perform the experiments, 10 datasets of organisms *H. sapiens* and *Mus. musculus* were randomly extracted from the Jaspar repository [17]. For each algorithm, the mean and standard deviation of the *f-score* was calculated. This procedure was not performed for the MEME, as it is a deterministic approach, its score does not vary over the runs. Therefore, the metrics extracted from this algorithm were obtained from only a single execution. To calculate the f-score, the initial positions found by each approach were converted into a matrix $N \times M$ of $0s$ and $1s$ in which 0 indicates absence and 1 presence of motif. For each position p, w sequences of $1s$ are generated. The same procedure is performed for the real motif locations. This method allowed a very accurate calculation of motifs positioning and, consequently, greater rigor in measuring the f-score. After that, the confusion matrix was generated and the f-scores was calculated according to: $F_1 = 2 \times \frac{precison \times recall}{precison + recall}$.

4.2 Results and Discussion

Table 1 illustrates the results achieved by each approach. In that case, we can note VNS, EM and ILS approaches had a weak performance to find good results when executed by itself (without GRASP combination). Moreover, considering the quality of results found by MEME, it is important to highlight MEME is also based on Expectation Maximization, but it also employs other helper mechanisms that guide the optimization process across the search space. For that reason, we realized the limitations found by EM when compared to MEME.

Furthermore, after GRASP inclusion in the experiments, we noted an increase of the quality of solutions by the improved initialization strategy. Also, all algorithms combined with GRASP have improved the quality of solutions when compared to its stand alone version. These results indicate the benefits of GRASP algorithm as a constructive heuristic. In addition, we realized a significant improvement with GRASP+EM in 8 (MA0036.2, MA0463.1, MA0475.1, MA0479.1, MA0497.1, MA0506.1, MA0508.1, and MA0518.1) out of the 10 evaluated datasets. The major important difference was identified in MA0036.2 dataset, where the GRASP+EM approach scored an average of 0.964 ± 0.059, followed by GRASP+ILS (0.951 ± 0.081) and MEME (0.913). GRASP+ILS algorithm had also achieved good results compared with MEME, achieving higher scores in 4 datasets (MA0036.2, MA0497.1, MA0506.1, and MA0518.1) and tying in 2 (MA0463.1 and MA0508.1). However, this approach has still some limitations to found best solutions in some datasets. A possible limitation of this approach is related to decrease process of k value or the stochastic nature of the acceptance function. Also, these limitations will be further investigated in future works. Finally, MEME was the best algorithm in 2 datasets (MA0106.2, MA0523.1) and performed better than GRASP+ILS in 4 (MA0106.2, MA0475.1, MA0479.1, and MA0523.1). In summary, GRASP+EM had 8 wins and 2 losses, with a balance of 6. GRASP+ILS had no wins, with a final balance of −10, and MEME had 2 wins and 8 losses, ending with a balance of −8.

Table 1. F-measure achieved by each algorithm in the performed experiments.

ID	VNS	EM	ILS	GRASP+VNS	GRASP+EM	GRASP+ILS	MEME
MA0036.2	0.156 ± 0.003	0.308 ± 0.002	0.809 ± 0.099	0.411 ± 0.029	**0.964 ± 0.059**	0.951 ± 0.081	0.913
MA0106.2	0.164 ± 0.001	0.141 ± 0.013	0.802 ± 0.111	0.341 ± 0.004	0.946 ± 0.067	0.942 ± 0.063	**0.952**
MA0463.1	0.156 ± 0.002	0.126 ± 0.006	0.591 ± 0.247	0.408 ± 0.052	**0.969 ± 0.009**	0.962 ± 0.008	0.962
MA0475.1	0.116 ± 0.004	0.361 ± 0.013	0.669 ± 0.102	0.343 ± 0.025	**0.905 ± 0.014**	0.898 ± 0.002	0.900
MA0479.1	0.112 ± 0.002	0.146 ± 0.072	0.841 ± 0.137	0.035 ± 0.058	**0.986 ± 0.004**	0.982 ± 0.005	0.983
MA0497.1	0.213 ± 0.008	0.425 ± 0.009	0.691 ± 0.105	0.529 ± 0.042	**0.962 ± 0.009**	0.957 ± 0.023	0.956
MA0506.1	0.111 ± 0.003	0.079 ± 0.003	0.786 ± 0.052	0.034 ± 0.024	**0.840 ± 0.014**	0.838 ± 0.034	0.832
MA0508.1	0.158 ± 0.004	0.585 ± 0.001	0.828 ± 0.034	0.556 ± 0.032	**0.956 ± 0.007**	0.951 ± 0.019	0.951
MA0518.1	0.146 ± 0.006	0.239 ± 0.002	0.619 ± 0.152	0.504 ± 0.026	**0.990 ± 0.005**	0.983 ± 0.051	0.968
MA0523.1	0.146 ± 0.005	0.352 ± 0.091	0.821 ± 0.108	0.435 ± 0.039	0.971 ± 0.007	0.971 ± 0.012	**0.984**

4.3 Statistical Analysis

Analysis by counting wins and losses is insufficient to conclude whether one approach is superior to another. For this reason, we performed a hypothesis test on the results obtained by each algorithm. The non-parametric Friedman test with post-hoc Dunn-Bonferroni was applied because the data failed the Shapiro-Wilk test.

The objective of this analysis was to compare the results obtained by the algorithms using statistical methods, in order to identify significant differences between the metrics obtained by the approaches. Statistical significance tests were performed between the methods that achieved the highest f-scores in the experiment steps: GRASP+EM, GRASP+ILS, and MEME. The tested hypotheses were:

$$\begin{cases} H_0 : & \text{median of differences} = 0 \\ H_1 : & \text{median of differences} \neq 0 \end{cases}$$

Table 2 displays the results achieved by the algorithms in the Friedman and Dunn-Bonferroni tests (Bonferroni correction is used in paired tests with the objective of repairing type-i errors). Through it, we can conclude that, for most datasets, the Friedman test showed that there is a predictor effect on the f-scores. Furthermore, Dunn-Bonferroni's post-hoc pointed out that the f-scores achieved by GRASP+EM tend to be higher than the f-scores obtained by GRASP+ILS and MEME. According to the calculated p-values, of the 10 datasets, it is possible to conclude that GRASP+EM was better in 7 of them (MA0036.2, MA0463.1, MA0475.1, MA0497.1, MA0506.1, MA0508.1 and MA0518.1). MEME achieved significantly better results in 2 (MA0106.2 and MA0523.1). With a p-value of 0.272531, it was not possible to conclude which approach did better in dataset MA0479.1. Figure 2b shows the distribution of data, considering the average of the 30 executions in the 10 datasets. It is interesting to note in Fig. 2a that, although GRASP+EM has some outliers, the dispersion (interquartile range) is the smallest all of them. This indicates that their results were more homogeneous and consequently more consistent. Furthermore, if we disregard the outliers, GRASP+EM achieved the highest f-score among the maximums and minimums obtained by the 3 approaches. Although GRASP+ILS achieved worse results than MEME in statistical tests, its metrics spread less, which indicates greater similarity. Also according to Fig. 2a, MEME was the algorithm that presented the greatest dispersion of data, suggesting a greater variation between its results across all datasets. Figure 2b shows the data distribution in stack histogram format. In that case, it is possible to notice that the metrics were concentrated 0.90 and 1.00 (space most to the right of the graph). Analyzing this region, it is evident that the GRASP+EM algorithm has a greater representation in relation to the others. In the other regions of the graph, the algorithms were visually similar, with the exception of the central region, where GRASP+EM has a smaller representation.

Figure 3a estimates kernel density (KDE) belonging to each distribution considering the average of the f-scores achieved by each approach in each dataset. In it, we can verify that the curves drawn by each algorithm escape the "bell" format, characteristic of Gaussian's distributions, exhibiting in some cases multimodality. In addition, Figure 3b shows the classic quantile-quantile plot, which represents the empirical and theoretical quantiles, assuming a normal distribution. We can note that the data on the diagonal of the graph are not well-fitted, indicating that they cannot be approximated by a Gaussian distribution. Analyzing Fig. 3 in its entirety, we can see that the pooled data cannot be approximated by a normal distribution. This analysis was necessary to provide the correct direction for the next tests performed. Figure 4a shows the critical differences plot between the approaches. This graph is interesting because it visually shows whether there is a significant difference between the examined algorithms. Through it, we can clearly see that there is a difference between GRASP+EM and GRASP+ILS. Although this diagram did not show that GRASP+EM was different from MEME, it is possible to see that a horizontal line separating them is quite long, which indicates a borderline value. To aid in the analysis of the previous figure, we plot Fig. 4b. Through it, we can directly visualize the p-value matrix constructed by making all pairwise comparisons (corrected using the Bergmann and Hommel procedure). Also in this figure, it is possible to conclude that the GRASP+EM algorithm was different (p-value ≤ 0.05) in relation to the MEME and GRASP+ILS approaches. It was not possible to conclude whether MEME and GRASP+ILS achieved different performances.

Table 2. P-values obtained in the tests of Friedman e Dunn-Bonferroni. ***(0.001),
**(0.01), *(0.05), =(no difference). G+EM (GRASP+EM), G+ILS (GRASP+ILS).

			Dunn_Bonferroni pairwise p-value			
ID	Friedman p-value	Result	G+EM vs G+ILS	G+EM vs MEME	G+ILS vs MEME	Winner
MA0036.2	2.460e–13	***	0.000901	7.797e–14	0.000188	G+EM
MA0106.2	1.094e–09	***	0.084557	0.000108	7.55e–10	MEME
MA0463.1	0.007202	**	0.029469	0.013526	1	G+EM
MA0475.1	0.017715	*	0.466741	0.013526	0.466741	G+EM
MA0479.1	0.272531	=	0.364005	1	0.735834	None
MA0497.1	5.095e–11	***	1.373e–10	0.364005	1.434e–06	G+EM
MA0506.1	5.942e–06	***	0.466741	5.346e–06	0.002367	G+EM
MA0508.1	0.003345	**	0.005837	0.020118	1	G+EM
MA0518.1	2.997e–12	***	0.013526	1.453e–12	3.410e–05	G+EM
MA0523.1	9.925e–11	***	0.905098	3.608e–07	7.554e–10	MEME

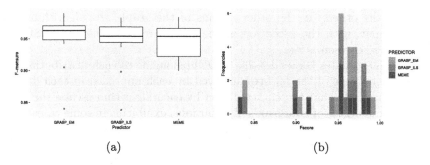

(a)   (b)

Fig. 2. Distribution of f-scores considering all datasets. (a) Boxplots of the best algorithms. (b) Histogram in stack format of the best algorithms.

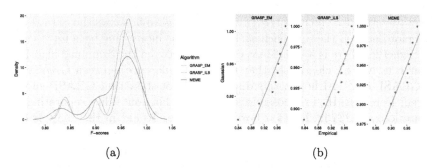

(a)   (b)

Fig. 3. (a) Kernel density estimation (KDE) of the sample distributions. (b) Quantile-quantile plots representing the empirical and theoretical quantiles of the data.

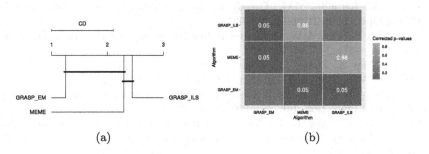

Fig. 4. (a) Critical plot differences between algorithms. (b) P-value matrix using Friedman test with Bergmann and Hommel correction.

5 Conclusion

In this paper, we present an evaluation among different heuristics approaches and highlighting the benefits when these heuristics are associated with a non-randomly initialization strategy. This was evident when all algorithms were initialized with a constructive procedure. Also, GRASP proved to be a very promising constructive heuristic, contributing to each heuristic optimize the search space. At the same time, EM algorithm proved to be an important enhancer, achieving good results when associated with a some kind of constructive methods. Although EM has a fast convergence and, some times, gets stuck into local optima, it has proven to be very effective in improving non-randomly initialized solutions. In general, EM needs good initial sampling to get satisfactory results and GRASP can provide this through its constructive mechanism. The ILS-based approach proved to be quite promising, but it still needs some adjustments in future works, especially in relation to the perturbation and acceptance function. Finally, we intend to use the results achieved in this study to explore new heuristics and, in this way, propose an unified framework that can be used in different domains of problems.

References

1. Ashraf, F.B., Shafi, M.S.R.: MFEA: an evolutionary approach for motif finding in DNA sequences. Inf. Med. Unlocked **21** (2020)
2. Bailey, T.L.: Streme: accurate and versatile sequence motif discovery. Bioinformatics **37**(18), 2834–2840 (2021)
3. Bailey, T.L., Elkan, C.: Unsupervised learning of multiple motifs in biopolymers using expectation maximization. Mach. Learn. **21**(1–2), 51–80 (1995)
4. Bailey, T.L., Johnson, J., Grant, C.E., Noble, W.S.: The meme suite. Nucleic Acids Res. **43**(W1), W39–W49 (2015)
5. D'haeseleer, P.: How does DNA sequence motif discovery work? Nature Biotechnol. **24**(8), 959–961 (2006)
6. D'haeseleer, P.: What are DNA sequence motifs? Nature Biotechnol. **24**(4), 423–425 (2006)

7. Feo, T.A., Resende, M.G.: Greedy randomized adaptive search procedures. J. Global Optimiz. **6**(2), 109–133 (1995)
8. Hart, J.P., Shogan, A.W.: Semi-greedy heuristics: an empirical study. Oper. Res. Lett. **6**(3), 107–114 (1987)
9. He, Y., Shen, Z., Zhang, Q., Wang, S., Huang, D.S.: A survey on deep learning in DNA/RNA motif mining. Brief. Bioinf. **22**(4), bbaa229 (2021)
10. Lee, N.K., Li, X., Wang, D.: A comprehensive survey on genetic algorithms for DNA motif prediction. Inf. Sci. **466**, 25–43 (2018)
11. Li, M., Ma, B., Wang, L.: Finding similar regions in many strings. In: Proceedings of The Thirty-first Annual ACM Symposium on Theory of Computing, pp. 473–482. ACM (1999)
12. Lihu, A., Holban, Ş.: A review of ensemble methods for de novo motif discovery in chip-seq data. Briefings in bioinformatics p. bbv022 (2015)
13. Liu, F.F., Tsai, J.J., Chen, R.M., Chen, S., Shih, S.: FMGA: finding motifs by genetic algorithm. In: Fourth IEEE Symposium on Bioinformatics and Bioengineering, BIBE 2004. Proceedings, pp. 459–466. IEEE (2004)
14. Lourenço, H.R., Martin, O.C., Stützle, T.: Iterated local search: framework and applications. In: Gendreau, M., Potvin, J.-Y. (eds.) Handbook of Metaheuristics. ISORMS, vol. 272, pp. 129–168. Springer, Cham (2019). https://doi.org/10.1007/978-3-319-91086-4_5
15. Mladenović, N., Hansen, P.: Variable neighborhood search. Comput. Oper. Res. **24**(11), 1097–1100 (1997)
16. Pavesi, G., Mauri, G., Pesole, G.: An algorithm for finding signals of unknown length in DNA sequences. Bioinformatics **17**(suppl 1), S207–S214 (2001)
17. Sandelin, A., Alkema, W., Engström, P., Wasserman, W.W., Lenhard, B.: Jaspar: an open-access database for eukaryotic transcription factor binding profiles. Nucleic acids Res. **32**(suppl 1), D91–D94 (2004)
18. Sandve, G.K., Drabløs, F.: A survey of motif discovery methods in an integrated framework. Biol. Direct **1**(1), 11 (2006)
19. Stormo, G.D., Hartzell, G.W.: Identifying protein-binding sites from unaligned DNA fragments. Proc. Natl. Acad. Sci. **86**(4), 1183–1187 (1989)
20. Stützle, T.: Local search algorithms for combinatorial problems. Darmstadt University of Technology PhD Thesis, p. 20 (1998)
21. Talbi, E.G.: A taxonomy of hybrid metaheuristics. J. Heurist. **8**(5), 541–564 (2002)
22. Thijs, G., et al.: A higher-order background model improves the detection of promoter regulatory elements by gibbs sampling. Bioinformatics **17**(12), 1113–1122 (2001)
23. Tompa, M., et al.: Assessing computational tools for the discovery of transcription factor binding sites. Nat. Biotechnol. **23**(1), 137–144 (2005)

Mutation Rate Analysis Using a Self-Adaptive Genetic Algorithm on the OneMax Problem

João Victor Ribeiro Ferro[(✉)], José Rubens da Silva Brito,
Roberta Vilhena Vieira Lopes, and Evandro de Barros Costa

Federal University of Alagoas, Campus A.C. Simões, Maceió, Brazil
{jvrf,jrsb,rvvl,evandro}@ic.ufal.br

Abstract. In this paper, a variation of a genetic algorithm for optimization problems is presented, focusing on the adjustment of the mutation rate parameter by fuzzifying the diversity of the population and the value of the individual's adaptation. Here, it is important to remember that this parameter directly interferes with the convergence and quality of the solution found by the genetic algorithm. To evaluate the performance of the proposed solution, experiments were conducted on the OneMax problem, analyzing aspects such as: convergence, quality of the solution, the diversity of the population, and the number of individuals evaluated. Obtained results and their impacts are presented in this paper.

Keywords: Genetic algorithm · Mutation rate · Fuzzy logic · OneMax

1 Introduction

The Genetic algorithms simulate the process of species evolution described in Darwin's theory combined with the concepts of heredity, crossover, and mutation from genetic biology [5]. With the goal of exploring the search space of complex problems behind a global optimal solution, by performing both a macro search and a micro search in the investigation space of the problem.

The macro search is performed by mutation and serves to identify the promising sub-spaces. These are the sub-spaces of the search space whose solutions have better quality than the solutions of the neighboring sub-spaces. Micro search is performed by crossover, on the other hand, it will work on the promising sub-spaces in order to find the local optimal solutions of a promising subspace, as shown in Fig. 1.

To perform macro search, genetic algorithms make use of the mutation rate, which is responsible for determining how many new search sub-spaces will be considered for construction in the next iteration of the algorithm. Finding the best value for the mutation rate is a challenge, since different problems may require different values. The choice of mutation rate is determined before the algorithm is run by the programmer, who may use a default value, a success value for a problem in the same class, or a value obtained by trial-and-error. However,

J. C. Xavier-Junior and R. A. Rios (Eds.): BRACIS 2022, LNAI 13653, pp. 195–208, 2022.
https://doi.org/10.1007/978-3-031-21686-2_14

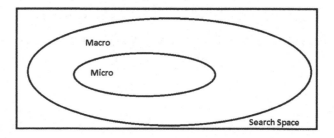

Fig. 1. Search space illustration

all of these ways of choosing the mutation rate can produce an inefficient macro search. A macro search is considered efficient when the diversity of the population worked on by the genetic algorithm is maintained throughout its execution.

One proposal for maintaining population diversity is to have the mutation rate value calculated as a function of the diversity present in the current population and the quality value assigned to an element of the population. In the literature there are several approaches for performing this calculation, such as:

– The study by FERRARI (2014) [6], used Fuzzy Logic to determine the crossover and mutation rate parameters iteratively in Differential Evolution, showing that the algorithm using Fuzzy Logic converges faster than the conventional Differential Algorithm.
– In the master's dissertation by BURDELIS (2019) [3] Fuzzy Logic was also used to decrease the number of generations and the time required for the Genetic Algorithm to converge for different problems such as Traveling Salesman, Function Minimization, thus showing a great approach to using this logic for problem solving.
– The work in CARVALHO (2017) [4], analyzed the mutation rate and the crossover rate, in relation to the convergence time of elitist GAs. These rates ranged from 0.11 to 0.9, using the 8-bit binary representation. It was applied to one-dimensional and two-dimensional problems. Hence, the author concluded that the mutation operator had the greatest influence on convergence speed relative to the crossover operator.

The present work proposes a calculation method for the value of the mutation rate based on the fuzzyfication of the diversity found in the current population and the value of the quality assigned to the individual to be mutated, where, initially was develop the Holland's genetic algorithm [8] to be compared with the algorithm proposed in the paper.

2   Background

2.1   Genetic Algorithm

Simple genetic algorithm (Simple GA) is a parallel guided random search algorithm for solutions to an NP-complete problem [9], whose idea is to assume the existence of one or more operators capable of modifying a bag of solutions of fixed size, called population, for the problem in focus into another bag of solutions with better quality.

The solutions that make up the population are called individuals. Each individual represents the encoding, in a binary vector of fixed size, of a solution belonging to the solution space of the problem. The quality of the population is the sum of the quality of all its individuals. The quality of an individual is calculated by a function defined by the programmer that relates the objective and the intention of the problem.

The operators responsible for transforming the initial population into a final population containing the individual that is the solution to the problem are selection, crossover, and mutation as shown in Fig. 2.

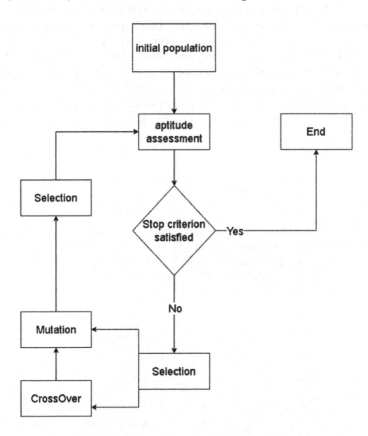

Fig. 2. Basic structure of simple GA

According to this Fig. 2, the genetic algorithm starts in the population module, which can be generated by a population according to some heuristic or captured from the environment the population. If the solution to the problem is present in this population, the algorithm will finish its execution and display the individuals with the best quality. Otherwise, the search operators will start their work.

First, the individuals in the population go through the quality assessment procedure, which calculates the fitness adaptation of each individual. Next, the selection operator uses the roulette wheel method, which consists of a roulette wheel that assigns each individual a slice with a size proportional to its score [11]. Once the roulette wheel is created, it is spun n times until a section corresponding to an individual is selected. This process is repeated until the number of parents is equal to the expected number, thus composing the population of Ancestral individuals, which will undergo both crossover and mutation by the genetic operator.

The crossover or recombination operator is inspired by sexual reproduction, with the goal of heritable transmission of adapted characteristics from the current population to future generations [13]. This operator takes as input a real value, n individuals from the Ancestry population and a set of $n - 1$ integer values. It then generates a random number belonging to the interval $[0, 100]$ and asks if it is smaller than the real number, if yes, it will generate n individuals formed by interweaving the pieces of the provided individuals from position 1 to the first provided integer value i_1, from position i_1 to the second provided integer value i_2, ..., from position i_{n-2} to the last provided integer value i_{n-1} (Fig. 3).

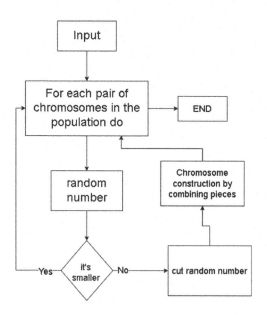

Fig. 3. Crossover

The mutation operator (Fig. 4) is intended to ensure the appearance of new features in the population [9]. Thus, it serves to drive the evolution of the population through search sub-spaces not yet investigated. This operator takes an individual from the Ancestry population and a real value. It then generates a random number belonging to the range $[0, 100]$ and asks if this value is smaller than the provided real number, if yes for each position of the provided individual a random value between $[0, 1]$ is generated, if the rounded value is 1 then the position value is changed.

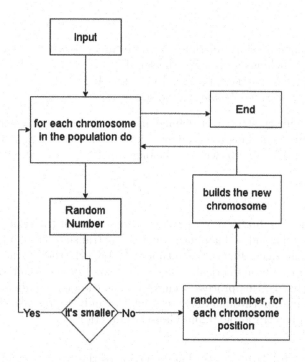

Fig. 4. Mutation operator

Finally, the fitness evaluation procedure will calculate the fitness of each individual generated by the action of the genetic crossover and mutation operators, then the selection operator will select the most valued individuals to join the best-fit chromosomes from the current population, thus forming the new population.

2.2   Fuzzy Logic

Fuzzy logic may be seen as a response to the need to perform the representation of imprecise knowledge expressed in natural language by a human, being a form of many-valued logic in which the truth-values of the propositions are interpreted as degrees of truth, typically expressed by any real number between 0 and 1 [7].

Unlike classical logic, in fuzzy logic a given proposition is evaluated by assigning degrees of truth to it, considering a degree of membership, typically expressed by any real number between 0 and 1, ranging between completely true and completely false. The information is fuzzified using a membership function.

The fuzzy sets and its subsets comes from mathematical modeling, where a subset A of a set of subsets U is considered a fuzzy subset of subsets U, and can be described as a set of ordered pairs:

$$A = \{(x, \mu_A(x)); x \in U(x) \in [0,1]) \mu_A(x) \in [0,1]\} \tag{1}$$

where:

- $\mu_A(x)$, is a membership function that determines to what extent x is in A;
- $\mu_A(x) = 1$, x belongs entirely to the set A;
- $0 < \mu_A(x) < 1$, x partially belongs to the set A;
- $\mu_A(x) = 0$, x does not belong to the set A;

The fuzzy subset A of x is defined using a function μ, which is called a membership function. This function associates each element of x with a degree $\mu_A(x)$, between 0.0 and 1.0, where $\mathbf{x}$ belongs to A:

$$\mu_A : x \rightarrow [0,1]$$

Fuzzification. The fuzzification process occurs through the relationship of the imprecision found in natural language when being translated into input values for fuzzy description. According to the author IVANQUI [1], the fuzzification process aims to transform the numerical values (non-fuzzy) of the input variables into fuzzy values, that is, this process occurs to allow the mathematical modeling of the input information through fuzzy sets, which can be represented by functions, such as triangular, trapezoidal and Gaussian functions.

Fuzzy Rules and Inference. The behavior of the fuzzy system is dynamic, being modeled by means of fuzzy rules. These rules are important in the knowledge structure of the inference system, related to the fuzzy variables associated with one of its predicates or linguistic terms, being defined as follows:

$$If < antecedents > then < consequents >$$

The antecedents and consequents of a fuzzy rule are: propositions containing linguistic variables and associated fuzzy sets:

$$If\ A\ is\ a\ and\ B\ is\ b\ then\ C\ is\ c$$

These rules are obtained by experts, in the form of linguistic sentences, and are a key aspect in the performance of a fuzzy inference system. The rule base and the data base form the knowledge base of a fuzzy system [1].

Defuzzification. In the defuzzification phase, the values inferred through fuzzy rules, will have a numeric (non-fuzzy) value as equivalence. In this phase, we look for a single discrete numeric value that can represent the values inferred through the output variables [1]

3   Methodology

3.1   Test Environment

The test environment used to make the comparisons was Google Colab (Google Computer Engine)[1] which has the following settings:

– RAM: approximately 12 GB;
– HD: approximately 108 GB;

Python version 3.7.13 was used, and some libraries were also used, such as:

– ipython-autotime version 0.3.1 to determine the runtime;
– matplotlib version 3.2.2 to show the graphs;
– scikit fuzzy version 0.4.2 to build the fuzzy system;

3.2   Problem Description

The problem worked on in this paper was OneMax, which consists of counting the one (1) bits that each chromosome has, and also represents the fitness of the individual. Thus, the optimal binary is the string where all bits are one (1). The solution space or domain of the OneMax problem depends on the length of the string, an important feature of the OneMax domain is that all bits are unrelated [8]. The simplicity of the OneMax problem makes it an excellent candidate for studying the performance evaluation of simple genetic algorithm, Fuzzy logic on mutation rate.

3.3   Genetic Algorithm

The genetic algorithm developed in this work follows the following specification [8]:

– Representation of the individual: binary vector of size m.
– Population representation: vector of individuals of size n.
– Genetic operators:
  1. **Roulette Selection:** creates a roulette wheel of individuals, where the area occupied by each individual is proportional to its adaptation, so the adapted individuals occupy a larger area [9].
  2. **Crossover:** of a cut-off point with rate in 90%, both defined in according the theoretical reference.

---

[1] https://colab.research.google.com.

3. **Mutation:** per complement defined in the theoretical reference, with mutation rate adjustable by fuzzy rules.
4. **Elitist Selection:** which chooses the best-fit individuals present in the current population and generated by genetic operators to compose the next population.

3.4   Application of Fuzzy Logic

In our application of fuzzy logic, we have considered the possibility of twenty-five rules for the OneMax problem, using the triangular membership function representation. The rules have the following antecedent values of the individual's adaptation and population diversity (PD), both with the fuzzy set 'very bad', 'bad', 'medium', 'good', 'verygood' as represented in Fig. 5 and as a consequence the individual's mutation percentage (MP), which is represented by the fuzzy set 'verylow', 'low', 'medium', 'high', 'veryhigh', which is described in Fig. 6 [0, 100] (Table 1).

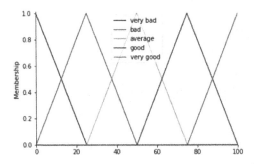

Fig. 5. Population quality and diversity

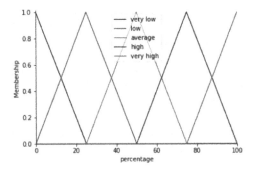

Fig. 6. Mutation percentage graph

The rules adopted in the fuzzy system were expressed as follows:

Table 1. Fuzzy rules

Rule	If	Quality	And	DP	Then	PM
1	If	very bad	And	very bad	Then	very high
2	If	very bad	And	bad	Then	very high
3	If	very bad	And	average	Then	very high
4	If	very bad	And	good	Then	very high
5	If	very bad	And	very good	Then	very high
6	If	bad	And	very bad	Then	very high
7	If	bad	And	bad	Then	high
8	If	bad	And	average	Then	average
9	If	bad	And	good	Then	low
10	If	bad	And	very good	Then	high
11	If	average	And	very bad	Then	high
12	If	average	And	bad	Then	average
13	If	average	And	average	Then	average
14	If	average	And	good	Then	low
15	If	average	And	very good	Then	low
16	If	good	And	very bad	Then	how
17	If	good	And	bad	Then	low
18	If	good	And	average	Then	low
19	If	good	And	good	Then	low
20	If	good	And	very good	Then	very low
21	If	very good	And	very bad	Then	very low
22	If	very good	And	bad	Then	very low
23	If	very good	And	average	Then	very low
24	If	very good	And	good	Then	very low
25	If	very good	And	very good	Then	very low

3.5   Evaluation Metrics

The metrics chosen in this work to perform the comparison were the number of generations, the diversity of the population, the runtime required for the genetic algorithm to find the solution to the problem, the score of the best chromosome per generation interval, and the T-Student test analysis.

4   Results and Discussions

Each algorithm was run 100 times, that is, each run means that the algorithm was started and is only stopped when it finds the solution to the problem. When it is found, the data is stored and then the system is restarted, this *looping* happens until the 100 runs are completed, and then the results are extracted, as shown in the graph in Fig. 7.

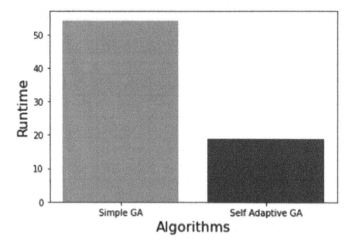

Fig. 7. Execution graph

Figure 8 shows the total time of the 100 runs that each algorithm obtained, that is, the arithmetic mean of the time that each run took to find the best solution to the OneMax problem. Given the Fig. 7, it is remarkable how much difference there is in the execution time of the two algorithms.

The simple genetic algorithm performed worse, about 54.0s, when compared to the performance of the auto adaptive genetic algorithm, with a run time of about 16.0s. Across all runs, the run time of the simple GA is about 237.5% longer than the fuzzy logic algorithm.

One of the reasons analyzed for this difference in the time of the simple GA compared to the self-adaptive GA was the delayed convergence of the population, as seen in (LINDEN, 2012), due to destruction or chromosomes with good *scores*. The destruction being generated by the action of the genetic operators and the loss of the choice of chromosomes that will generate offspring and remain in the population.

The self-adaptive GA, on the other hand, maintains a high search rate in macro regions, i.e., it efficiently searches a larger number of search spaces. This is due to the fact that it can infer an appropriate mutation rate for each chromosome in the population, taking into account the diversity of the population and the quality of the chromosome, which interferes with faster convergence.

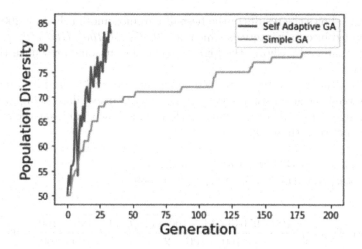

Fig. 8. Population diversity by generation

In addition, the diversity of the population present in a run was verified, in which the first 200 generations of each algorithm were extracted, as shown in Fig. 8, in which, it was observed that the proposed GA presents at the beginning a diversity equal to that of the simple GA, but throughout the generations this diversity increases, thus generating greater chances of finding the individual that satisfies the OneMax problem.

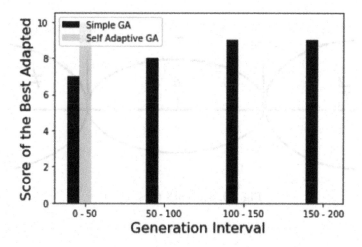

Fig. 9. Best suited by generation interval

Figure 9 shows the results of the best-fit chromosomes in the first 200 genera-tions, divided into four intervals which are $[0,50[, [50,100[, [100,150[, [150,200]$, where each interval is shown the highest *score* chromosome present in the given algorithm.

Finally, Student's T-test with 95% confidence interval is used to compare the mean between the generation of the two algorithms, these data being in the gen-eration range, where this variation will determine how significant the statistical difference between these algorithms is [10].

$$Var_{holland} = 976434.48 \ , \ Var_{fuzzy} = 1195.27$$
$$Mean_{holland} = 945.49 \ , \ Mean_{fuzzy} = 68.389 \ ,$$

$$S_{X1-X2} = \sqrt{\frac{(100 * 976434.48 + 100 * 1195.27)}{100 + 100 - 2}} * \frac{100 + 100}{100 * 100} = 99.37 \quad (2)$$

$$t_{Obs} = \frac{945.49 - 68.389}{99.37} = 8.82 \quad (3)$$

$$GL = 100 + 100 - 2 = 198 \quad (4)$$

When obtaining the degree of freedom (GL), it was verified in the t-Student distribution table that T Critical is 1.960. Thus, it is proven that there is a statistically significant difference between the algorithms, since the T observed is 8.82, in other words, it exceeds the T Critical interval, as can be seen in the Fig. 10.

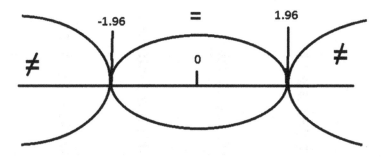

Fig. 10. Student's t test chart

5   Conclusion

In this paper, we have presented a variation of a genetic algorithm for optimization problems in order to adjust of the mutation rate parameter by using fuzzy logic to model the diversity of the population and the value of the individual's adaptation. From the previous analyses, it was observed that the genetic algorithm with the mutation rate derived from the Fuzzy rules in general presented a performance in execution time 237.5% lower than the Simple Genetic Algorithm, which corroborates the work of BARCELLOS, 2000 [2]. However, the way fuzzy logic was used in the proposed algorithm is simpler to implement than the one presented in these works, which confirms the adoption of these modifications in its implementation, as presented in the works in [4,11,12], thus showing the importance of the modification in the interactive mutation rate sensitive to the current population.

For immediate future work, it is most important, at this point, to apply the proposed GA solution to other classes of problems where the chromosome bits are not correlated and analyze how this algorithm behaves. In this way, it becomes easier for the developer to make a decision about which problems it is recommended to adopt the self-adaptive genetic algorithm.

References

1. Bai, Y., Wang, D.: Fundamentals of fuzzy logic control - fuzzy sets, fuzzy rules and defuzzifications. In: Bai, Y., Zhuang, H., Wang, D. (eds.) Advanced Fuzzy Logic Technologies in Industrial Applications. Advances in Industrial Control. Springer, London (2006). https://doi.org/10.1007/978-1-84628-469-4_2
2. Barcellos, J.C.H.: Algoritmos genéticos adaptativos: um estudo comparativo. Dissertação (Mestrado em Engenharia) - Escola Politécnica da Universidade de São Paulo (2000)
3. Burdelis, M.A.P.: Ajuste de Taxas de Mutação e de Cruzamento de Algoritmos Genéticos Utilizando-se Inferências Nebulosas. 2009. Dissertação (Mestre em Engenharia) - Departamento de Engenharia de Computação e Sistemas Digitais (PCS), [S. l.] (2009)
4. Carvalho, W.L.O.: Estudo de parâmetros ótimos em algoritmos genéticos elitistas. Dissertação (Mestrado em Matemática Aplicada e Estatística) - Universidade Federal do Rio Grande do Norte, Natal (2017)
5. Eiben, A.E., Smith, J.E.: Introduction to Evolutionary Computing. Springer, New York, NY (2003)
6. Ferrari, A.C.K., Leandro, G.V., Olieveira, G.: Evolução Diferencial com Parâmetros Ajustáveis por Lógica Fuzzy. Anais do XX Congresso Brasileiro de Automática, [S. l.], pp. 796–803 (2014)
7. Gerla, G.: Fuzzy Logic: Mathematical Tools for Approximate Reasoning, Trends in Logic, Kluwer Ac. Press (2000). https://doi.org/10.1007/978-94-015-9660-2
8. Giguere, P., Goldberg, D.E.: Population sizing for optimum sampling with genetic algorithms: a case study of the onemax problem. Genetic Program. **98**, 496–503 (1998)
9. Linden, R. Algoritmos genéticos. 3rd edn. Editora Ciência Moderna, Rio de Janeiro (2012)

10. Livingston, E.H.: Who was student and why do we care so much about his t-test?1. J. Surgical Res. **118**(1), 58–65 (2004)
11. Mitchell, M.: An Introduction to Genetic Algorithms. MIT Press (1999)
12. de Oliveira, N.J.R.R.: Avaliação de Taxas de Cruzamento e Mutação em um Algoritmo Genético Baseado em Ordem Aplicado ao Problema do Caixeiro Viajante. 2018. Monografia (Bacharel) - Curso de Bacharelado em Sistemas de Informação, [S. l.] (2018)
13. Pappa, G.L., Freitas, A.A.: Evolutionary algorithms. In: Automating the Design of Data Mining Algorithms. NCS. Springer, Heidelberg (2010). https://doi.org/10.1007/978-3-642-02541-9_3

Application of the Sugeno Integral in Fuzzy Rule-Based Classification

Jonata Wieczynski[1] , Giancarlo Lucca[2(✉)] , Eduardo Borges[2] ,
and Graçaliz Dimuro[1,2]

[1] Departamento de Estadística, Informática y Matemáticas,
Universidad Pública de Navarra, Pamplona, Spain
jonata.wieczynski@unavarra.es
[2] Centro de Ciêcias Computacionais, Universidade Federal de Rio Grande,
Rio Grande, Brazil
{giancarlo.lucca,eduardoborges,gracalizdimuro}@furg.br

Abstract. Fuzzy Rule-Based Classification System (FRBCS) is a well known technique to deal with classification problems. Recent studies have considered the usage of the Choquet integral and its generalizations to enhance the quality of such systems. Precisely, it was applied to the Fuzzy Reasoning Method (FRM) to aggregate the fired fuzzy rules when classify new data. On the other side, the Sugeno integral, another well known aggregation operator, obtained good results when applied to brain-computer interfaces. Those facts led to the present study in which we consider the Sugeno integral in classification problems. That is, the Sugeno integral is applied in the FRM of a widely used FRBCS and its performance is analyzed over 33 different datasets from the literature. In order to show the efficiency of this new approach, the obtained results are also compared to past studies involving the application of different aggregation functions. Finally, we perform a statistical analysis of the application.

Keywords: Classification problem · Fuzzy Rule-Based Classification System · Fuzzy reasoning method · Sugeno integral · Choquet integral

1 Introduction

Fuzzy Rule-Based Classification Systems (FRBCS's) [18] is a technique used to deal with classification problems [13] which has been applied to diverse problems, e.g., big data [32], image segmentation [21], health [22] and others. The Fuzzy Reasoning Method (FRM) [7,8] used by FRBCSs is a key component which is composed by four steps. One of them is the aggregation, where the information of the system's fired rules is aggregated, per class. For this step, the FRM normally uses an aggregation function [5], and by doing so, the system will have a different performance (notice that performance in this paper is related to the method accuracy, and not the runtime one) whenever one changes the function.

© The Author(s), under exclusive license to Springer Nature Switzerland AG 2022
J. C. Xavier-Junior and R. A. Rios (Eds.): BRACIS 2022, LNAI 13653, pp. 209–220, 2022.
https://doi.org/10.1007/978-3-031-21686-2_15

The work proposed by Barrenechea et al. [4] introduced a new FRM that accounts the usage of all given information by the fired fuzzy rules when classifying a new instance. To do so, they have considered the Choquet integral [10]. Moreover, they introduced a fuzzy measure [30] that is adapted for each class of the problem.

Considering the Choquet integral as basis, was introduced by Lucca et al. [24] the concept of pre-aggregation functions. One way to produce such function is by generalizing the base integral by different t-norms [19]. The generalizations where applied in the FRM to cope with classification problems and elevated the system quality. After that, also considering the Choquet integral as basis, different generalizations where provided and applied, namelY: CC-integral [27], C_F-integral [28], C_{F1F2}-integrals [23] and gC_{F1F2}-integrals [12]. Additionally, this generalizations were also applied in multi-criteria decision making problems [35,36] and image processing [29][1].

On the otter hand, the Sugeno integral [33] is another fuzzy integral which has been applied to diverse problems in the literature. More recently it was applied to a Motor-Imagery Brain-Computer Interface [20], where it obtained good results when compared to the standard Choquet integral (for more information see [20]).

Having in consideration that the Choquet integral was used as base to different generalizations and the good results that the Sugeno integral achieved in recent applications, this paper intends to analyze if the usage of the Sugeno integral as aggregation function in the FRM is able to produce a system with competitive results. To do so, we apply and analyze this new base function in the FRM of a state-of-art classifier and provided an analysis over 33 distinct datasets from the literature.

This work is organized as follows. Section 2 presents the background theory in respect to the following sections. In Sect. 3, the new framework of FRBCS using the Sugeno integral is presented. Then, Sect. 4 presents and discuss the results. Lastly, Sect. 5 is the conclusion thoughts of the work.

2 Preliminary Concepts and the Sugeno-like Generalization

In this section the theoretical background necessary to better understanding of the paper is provided. In what follows consider the following notation: $N = \{1, \ldots, n\}$, that is, the subset of the natural numbers up to n.

An aggregation function (**AF**) [14] is a function $f : [0,1]^n \rightarrow [0,1]$ such that the boundary conditions, $f(\mathbf{0}) = 0$ and $f(\mathbf{1}) = 1$, where $\mathbf{0} = (0, \ldots, 0)$ and $\mathbf{1} = (1, \ldots, 1)$, and the monotonicity properties, $\boldsymbol{x} \leq \boldsymbol{y} \implies f(\boldsymbol{x}) \leq f(\boldsymbol{y})$, $\forall \boldsymbol{x}, \boldsymbol{y} \in [0,1]^n$, hold.

A triangular norm (t-norm) is an aggregation function $T : [0,1]^2 \rightarrow [0,1]$ that satisfies, for any $x, y, z \in [0,1]$: the commutative ($T(x,y) = T(y,x)$), the

---

[1] An overview of the different generalizations of the Choquet integral is available in [11].

associative $(T(T(x, y), z) = T(x, T(y, z)))$ properties and the boundary condition.

An example of t-norm is the Hamacher t-norm, defined for $x, y \in [0, 1]$ as:

$$T_{HP}(x, y) = \begin{cases} 0 & \text{if } x = y = 0 \\ \frac{xy}{x+y-xy} & \text{otherwise .} \end{cases}$$

A fuzzy measure [33] is a function $m : 2^N \to [0, 1]$ that for all $X, Y \subseteq N$ holds the conditions: (i) $m(\emptyset) = 0$ and $m(N) = 1$; (ii) if $X \subset Y$, then $m(X) \leq m(Y)$.

In this study the Power Measure (PM) is considered as the fuzzy measure. It is defined for $X \subseteq N$ as: $m_P(X) = (|X|/n)^q$, with $q > 0$ being genetically learned.

Let m be a fuzzy measure. The standard Choquet integral [6] $\mathfrak{C}_m : [0, 1]^n \to [0, 1]$ of $x \in [0, 1]^n$ with respect to m is defined as:

$$\mathfrak{C}_m(x) = \sum_{i=1}^{n} \left(x_{(i)} - x_{(i-1)} \right) \cdot m(A_{(i)})$$

where (i) is a permutation on 2^N such that $x_{(i-1)} \leq x_{(i)}$ for all $i = 1, \ldots, n$, with $x_{(0)} = 0$ and $A_{(i)} = \{(1), \ldots, (i)\}$.

Let m be a fuzzy measure and $T : [0, 1]^2 \to [0, 1]$ a t-norm. Then a C_T-integral is defined as $\mathfrak{C}_m^T : [0, 1]^n \to [0, 1]$, given, for all $x \in [0, 1]^n$, by

$$\mathfrak{C}_m^T(x) = \sum_{i=1}^{n} T\left(x_{(i)} - x_{(i-1)}, \; m(A_{(i)}) \right)$$

where $x_{(i)}$, $A_{(i)}$ and i is defined as the standard Choquet integral.

Notice that the Choquet integral is an averaging functions [14], i.e., it always holds that for any $x \in [0, 1]$ and any fuzzy measure m, $\min(x) \leq \mathfrak{C}_m^T(x) \leq \max(x)$

Let Co be a bivariate copula [31]. The Choquet-like integral based on copula with respect to a fuzzy measure m, named CC-integral, is defined as a function $\mathfrak{C}_m^{Co} : [0, 1]^n \to [0, 1]$, for all $x \in [0, 1]^n$, by

$$\mathfrak{C}_m^{Co}(x) = \sum_{i=1}^{n} Co\left(x_{(i)}, \; m(A_{(i)}) \right) - Co\left(x_{(i-1)}, \; m(A_{(i)}) \right)$$

where $x_{(i)}$, $A_{(i)}$ and i is defined as the standard Choquet integral.

Lastly, the C_F-integral [28] is a generalization of the standard Choquet integral which uses an generic function F instead of the product operator. The definition is as follows: let $F : [0, 1]^2 \to [0, 1]$ be a function and $m : 2^N \to [0, 1]$ a fuzzy measure. Then the C_F-integral $\mathfrak{C}_m^F : [0, 1]^n \to [0, 1]$ is defined, for all $x \in [0, 1]^n$ by:

$$\mathfrak{C}_m^F(x) = \min\left\{ 1, \sum_{i=1}^{n} F\left(x_{(i)} - x_{(i-1)}, m\left(A_{(i)} \right) \right) \right\},$$

where $x_{(i)}$, $A_{(i)}$ and i is defined as the standard Choquet integral.

In this study as function F the following is considered, $F_{NA} : [0,1]^2 \rightarrow [0,1]$:

$$F_{NA}(x,y) = \begin{cases} x, \text{ if } x \leq y, \\ \min\{\frac{x}{2}, y\}, \text{ otherwise.} \end{cases}$$

The Sugeno integral is a well know operator, that have been used in many different applications. It is defined with respect to a fuzzy measure m by:

$$Su_m(\boldsymbol{x}) = \bigvee_{i=1}^{n} \left(x_{(i)} \wedge m(A_{(i)}) \right)$$

where $x_{(i)}$, $A_{(i)}$ and i is defined as the Choquet integral. Moreover, it is observable that this integral share the same averaging characteristic as the Choquet integral [14].

3 Application of the Sugeno Integral to Classification in FRBCS

In this section, the application of the Sugeno integral in a Fuzzy Rule-Based Classification System is presented. We begin presenting the new Fuzzy Reasoning Method that uses of the Sugeno integral. Thereafter, the experimental framework is described. At the end, the obtained results are shown.

3.1 The New Fuzzy Reasoning Method

In this paper, the application of the Sugeno integral take into account a fuzzy classifier that is widely used. Precisely, it considers the Fuzzy Association Rule-based Classification model for High Dimensional Problems (FARC-HD) [1].

The rules used by FARC-HD follows this structure:

$$\text{Rule } R_j : \text{If } x_1 \text{ is } A_{j1} \text{ and } \dots \text{ and } x_n \text{ is } A_{jn}$$
$$\text{then Class is } C_j \text{ with } RW_j,$$

where R_j is the label of the j-th rule, A_{ji} is a fuzzy set representing a linguistic term modeled by a triangular shaped membership function, C_j is the class label, and $RW_j \in [0,1]$ is the rule weight [17], which in this case is computed as the confidence of the fuzzy rule.

Once the fuzzy rules composing the system have been created, the FRM is responsible for classifying new examples. Specifically, let $x_p = (x_{p1}, \dots, x_{pn})$ be a new example to be classified, L being the number of rules in the rule base, and M being the number of classes of the problem. The new FRM, where the Sugeno integral is used, consist of 4 different steps:

1. To compute the *matching degree*, that is, the strength of the activation of the if-part of the rules for the example x_p, which is computed using a t-norm $T' : [0,1]^n \rightarrow [0,1]$:

$$\mu_{A_j}(x_p) = T'(\mu_{A_{j1}}(x_{p1}), \ldots, \mu_{A_{jn}}(x_{pn})),$$
$$\text{with } j = 1, \ldots, L.$$

2. *Association degree computation*, that is, for the class of each rule the matching degree is weighted with the corresponding rule weight, given by:

$$b_j^k(x_p) = \mu_{A_j}(x_p) \cdot RW_j^k,$$
$$\text{with } k = Class(R_j), \ j = 1, \ldots, L.$$

3. The *example classification soundness degree for all classes* in this step that the aggregation functions are applied to combine the association degrees obtained in the previous step. The Sugeno integral Su is used as follows:

$$Y_k(x_p) = Su\left(b_1^k(x_p), \ldots, b_L^k(x_p)\right), \tag{1}$$
$$\text{with } k = 1, \ldots, M.$$

Since, whenever $b_j^k(x_p) = 0$, it holds that:

$$Su\left(b_1^k(x_p), \ldots, b_L^k(x_p)\right) = Su\left(b_1^k(x_p), \ldots, b_{j-1}^k(x_p), b_{j+1}^k(x_p), \ldots, b_L^k(x_p)\right),$$

then, for practical reasons, only those $b_j^k > 0$ are considered in Equation (1).

4. A *Classification* decision function $C : [0,1]^M \rightarrow \{1, \ldots, M\}$ is applied over the example classification soundness degrees of all classes and thus, the class corresponding to the maximum soundness degree is determined.

$$C(Y_1, \ldots, Y_M) = \min_{k=1,\ldots,M} k \ \text{ s.t. } \ Y_k = \max_{w=1,\ldots,M}(Y_w).$$

In practical applications, it is sufficient to consider

$$C(Y_1, \ldots, Y_M) = \arg \max_{k=1,\ldots,M}(Y_k).$$

Finally, its necessary highlight that the fuzzy measure used by the Sugeno and the generalizations of the Choquet integral is the Power Measure, with the exponent q genetically learned as proposed by Barrenechea et al. [4]. This is due to the fact that this fuzzy measure achieved the superior performance in all generalizations. A comparison of the usage of the PM (applied with different generalizations in the FRM) against different fuzzy measure is done in [25].

Table 1. Summary of the datasets used in the study.

Id	Dataset	#Inst.	#Atts.	#Class	Id.	Dataset	#Inst.	#Atts.	#Class
App	Appendicitis	106	7	2	Pen	Penbased	10,992	16	10
Bal	Balance	625	4	3	Pho	Phoneme	5,404	5	2
Ban	Banana	5,300	2	2	Pim	Pima	768	8	2
Bnd	Bands	365	19	2	Rin	Ring	740	20	2
Bup	Bupa	345	6	2	Sah	Saheart	462	9	2
Cle	Cleveland	297	13	5	Sat	Satimage	6,435	36	7
Con	Contraceptive	1,473	9	3	Seg	Segment	2,310	19	7
Eco	Ecoli	336	7	8	Shu	Shuttle	58,000	9	7
Gla	Glass	214	9	6	Son	Sonar	208	60	2
Hab	Haberman	306	3	2	Spe	Spectfheart	267	44	2
Hay	Hayes-Roth	160	4	3	Tit	Titanic	2,201	3	2
Ion	Ionosphere	351	33	2	Two	Twonorm	740	20	2
Iri	Iris	150	4	3	Veh	Vehicle	846	18	4
Led	led7digit	500	7	10	Win	Wine	178	13	3
Mag	Magic	1,902	10	2	Wis	Wisconsin	683	11	2
New	Newthyroid	215	5	3	Yea	Yeast	1,484	8	10
Pag	Pageblocks	5,472	10	5					

3.2   Experimental Framework

To demonstrate the efficiency and the quality of the proposal, this study uses 33 different datasets. It is necessary to highlight that these datasets are public available in KEEL dataset repository [2]. Also, these datasets are the same used in previous studies (see [23,28] and [25]).

In Table 1, the characteristics of the datasets are summarized. Then, for each dataset, it is presented the corresponding identification (Id), the number of instances (#Inst), attributes (#Atts), and classes (#Class).

Following the idea of previous generalizations, the results are presented taking into account a 5-fold cross-validation procedure [34]. To analyze the classifier performance the accuracy [34] is used. Consequently, the results presented in this study are related to the average accuracy obtained in the five different folds.

As mentioned before, the fuzzy classifier used in this paper is the FARC-HD, therefore, the configuration of this classifier follows the original author's suggestion. That is, the product t-norm as conjunction operator, the certainty factor is the RW, with 0.05 as minimum support, the threshold for the confidence as 0.8, the depth of the tree is 3, and k_t equals 2.

In relation to the parameters used by the genetic algorithm applied to learn the fuzzy measure, it is considered the same configuration used in different studies ([23,27] and [28]). To the genetic part of the algorithm it have a population composed by 50 individuals, 30 bits per gene in the gray codification, 20.000 evaluations and the fitness is calculated in therms of the accuracy.

4   Experimental Results

This section describes the obtained results. As discussed in [28] the application of non-averaging functions in the FRM statistically outperformed all the averaging functions. Thus, considering that the proposed FRM, using the Sugeno integral as aggregation, is an averaging approach, in order to provide a fair comparison, in this study we have only performed comparisons against averaging operators.

Again, it is necessary to point out that this study intend to observe that the usage of the Sugeno integral in the FRM can produce a competitive model to deal with classification problems. We are mainly interested in observing if this function is comparable against the standard Choquet integral, since this can allow promising researches on future generalizations of the Sugeno integral, in a similar way that was done with the Choquet integral.

However, aiming at providing a more robust and complete study, comparisons of the new approach against classical FRMs are provided. Precisely, against the Winning Rule (WR) [8], the standard Choquet integral and the best generalizations of the Choquet integral. In this sense we selected the CC-integral (the Choquet integral in its expanded form and generalized by Copulas functions) [27], the best C_T-integral [24] that is based on the Hammacher t-norm and the best averaging C_F-integral that is based in the F_{NA} function.

The obtained results are shown in Table 2. In it, the rows are related to the different datasets (for more details about the dataset see Table 1), per columns different FRMs are compared. The result in each cell is related to the accuracy mean obtained in the cross-validation process. The largest obtained mean in the study, among all approaches, is highlighted in **boldface**.

By taking a general look over the obtained results one can notice that the behavior of FRMs considering the Sugeno integral and the CC-integral are similar. In fact, only in four specific datasets (Ban, Bup, Mag and Two) the achieved result are different.

The biggest obtained accuracy mean is obtained by the C_T- integral, followed closely by the C_F- integral (mean difference of 0.10), CC-integral(mean difference of 0.20) and Sugeno (mean difference of 0.20). Considering the WR and the Choquet integral the obtained mean achieved a low performance.

In a closer look, considering the specific cases where the FRM's provided the largest results (the ones highlighted in **boldface**), the C_T-integral obtained the largest accuracy in 10 of the 33 datasets. However, another interesting result is seen for both, the Sugeno and the CC-integral, where the obtained results are the biggest accuracy in 9 of the 33 datasets. For the remaining cases, the C_F-integral, the Choquet integral and the WR present 6, 4 and 4 of the 33 datasets, respectively. Notice that for the Tit dataset, the obtained means are all equal and therefore are not included in the above count.

By considering and comparing only the Sugeno and Choquet integrals we have that for 19 different datasets the obtained means are superior in favor of the Sugeno integral in comparison to the Choquet (13 cases). On the other hand, when comparing the Sugeno integral against the C_T-integral, the latter achieves

Table 2. Accuracy mean obtained in test by the application of different averaging functions in the FRM.

Dataset	WR	Choquet	CC-integral	C_T-integral	C_F-integral	Sugeno integral
App	83.03	80.13	**85.84**	82.99	82.99	**85.84**
Bal	81.92	82.40	81.60	**82.72**	82.56	81.60
Ban	83.94	**86.32**	84.30	85.96	86.09	85.26
Bnd	69.40	68.56	71.06	**72.13**	69.40	71.06
Bup	62.03	66.96	61.45	65.80	**67.83**	60.87
Cle	56.91	55.58	54.88	55.58	**57.92**	54.88
Com	52.07	51.26	52.61	**53.09**	52.27	52.61
Ecp	75.62	76.51	77.09	**80.07**	78.88	77.09
Gla	64.99	64.02	**69.17**	63.10	64.51	**69.17**
Hab	70.89	72.52	**74.17**	72.21	73.51	**74.17**
Hay	78.69	79.49	**81.74**	79.49	78.72	**81.74**
Ion	90.03	90.04	88.89	89.18	**90.60**	88.89
Iri	**94.00**	91.33	92.67	93.33	93.33	92.67
Led	**69.40**	68.20	68.40	68.60	68.60	68.40
Mag	78.60	78.86	79.81	79.76	**80.02**	79.70
New	94.88	94.88	93.95	**95.35**	93.49	93.95
Pag	94.16	94.16	93.97	**94.34**	93.97	93.97
Pen	**91.45**	90.55	91.27	90.82	**91.45**	91.27
Pho	82.29	82.98	82.94	**83.83**	82.86	82.94
Pim	74.60	73.95	74.21	74.87	**75.64**	74.21
Rin	90.00	**90.95**	87.97	88.78	90.27	87.97
Sah	68.61	69.69	**70.78**	70.77	68.61	**70.78**
Sat	79.63	79.47	79.01	**80.40**	78.54	79.01
Seg	93.03	**93.46**	92.25	93.33	92.55	92.25
Shu	96.00	97.61	**98.16**	97.20	96.78	**98.16**
Son	77.42	77.43	76.95	**79.34**	78.85	76.95
Spe	77.90	77.88	**78.99**	76.02	78.26	**78.99**
Tit	78.87	78.87	78.87	78.87	78.87	78.87
Two	**86.49**	84.46	85.14	85.27	83.92	84.86
Veh	66.67	68.44	**69.86**	68.20	67.97	**69.86**
Win	96.60	93.79	93.83	**96.63**	96.03	93.83
Wis	96.34	**97.22**	95.90	96.78	96.34	95.90
Yea	55.32	55.73	**57.01**	56.53	56.40	**57.01**
Mean	79.15	79.20	79.54	**79.74**	79.64	79.54

Table 3. Average Rankings of the algorithms by using the Aligned Friedman and the obtained APV

(Pre-)Aggregation function	Ranking	APV
C_T-integral	80.19	
C_F-integral	91.33	0.56
Sugeno integral	97.98	0.56
CC-integral	98.74	0.56
Choquet integral	114.18	<u>0.07</u>
WR	114.56	<u>0.07</u>

Table 4. Results obtained by the Wilcoxon test to pair-wise comparison among the different approaches.

		WR	Choquet	CC-integral	C_T-integral	C_F-integral
Sugeno integral	P-value	0.26	0.21	0.87	0.19	0.88
	R^+	338.5	350.5	250.5	204.5	272
	R^-	222.5	210.5	310.5	356.5	289

superior mean in more than double the number of datasets than the former. However, it is necessary to point that the C_T-integral is a generalization of the Choquet integral, and that the t-norm T was chosen because of its superior results in the FRM, and in this study only the standard Sugeno is considered. Lastly, the results of both F_{NA} and WR are quite similar.

4.1   Statistical Analysis

Making comparisons considering the obtained means is a good approach. However, in order to provide a more robust study, in this subsection, we provide a statistical analysis from the different approaches, since it is an interesting question that can enligh the efficiency of the usage of the Sugeno integral.

The statistical analysis consider a non-parametric tests [9], the first analysis is a group comparison using the Aligned Friedman rank test [15]. This test consider a reverse ranking, where the lowest one is considered as control variable and is compared against the others. The results of this test is available in Table 3, which is sorted from the lowest to the largest rank. Also, the Adjusted P-Value (APV) is provided. To calculate the APV the post-hoc Holm's test [16] is used. In this Table the cases where the null hypothesis is rejected are <u>underlined</u>, having a significance level of 90% ($\alpha = 10\%$).

It can be observed from the group test that the C_T-integral is considered as control method and present statistical differences against the standard Choquet integral and WR. However, when compared to the remaining methods no significant difference were found.

Up to this point, to clarify even more the efficiency of the usage of the Sugeno integral, we have performed a set of pairwise comparisons, with the Wilcoxon signed-rank test [37]. This allows to direct compare the Sugeno integral with the different considered approaches.

The results of the Wilcoxon's test is provided in Table 4. In this table, is shown the obtained p-value, the rank obtained by the Sugeno integral (R^+) and the ranking obtained by the compared method (R^-).

The obtained results reinforce that the Sugeno integral is equivalent to any averaging operator used in different FRMs in the literature, since no statistical difference were found. Moreover, it is observable that comparing our approach against the standard Choquet integral, the obtained ranking is superior.

5 Conclusion

The usage of Fuzzy Rule-Based Classification Systems are an interesting technique to deal with classification problems. The Fuzzy Reasoning Method is the mechanism to perform the classification of different examples. The aggregation used in the FRM is a key point to define the performance of the system.

The usage of the standard Choquet integral in the FRM have been proposed in the literature and provided satisfactory results. After that, many generalizations of this integral where provided, such as: C_T-integral, CC-integral, C_F-integral and otters.

In this paper we provided an application of the Sugeno integral in the FRM. Precisely, the Sugeno integral. This function have been applied, among otters, in the Brain-Computer Interface (BCI) and demonstrated promising results.

In the experimental results we have compared our approach against classical FRMs using the maximum and the Choquet integral and the ones composed by the generalizations of the Choquet integral. The results demonstrated that the Sugeno integral is able to provide superior results in many different datasets and also that this method is statistically equivalent to the compared ones.

Considering the satisfactory obtained results, some future works can be followed. For instance, to create generalizations of the Sugeno integral, e.g. the FG-functional [3], in the FRM and compare the results to past results from generalizations of the Choquet integral. A deep analysis on the characteristics of the datasets (by using data complexity measures for example [26]) that could affect the performance of the classifier by using the Sugeno integral, is another interesting path.

Acknowledgments. The authors would like to thank CNPq (proc. 305805/2021-5, 301618/2019-4), FAPERGS (proc. 19/2551-0001660-3) and Navarra de Servicios y Tecnologías, S.A. (NASERTIC).

References

1. Alcala-Fdez, J., Alcala, R., Herrera, F.: A fuzzy association rule-based classification model for high-dimensional problems with genetic rule selection and lateral tuning. IEEE Trans. Fuzzy Syst. **19**(5), 857–872 (2011)

2. Alcalá-Fdez, J., et al.: Keel: a software tool to assess evolutionary algorithms for data mining problems. Soft Comput. **13**(3), 307–318 (2009)
3. Bardozzo, F., et al.: Sugeno integral generalization applied to improve adaptive image binarization. Inf. Fus. **68**, 37–45 (2021)
4. Barrenechea, E., Bustince, H., Fernandez, J., Paternain, D., Sanz, J.A.: Using the Choquet integral in the fuzzy reasoning method of fuzzy rule-based classification systems. Axioms **2**(2), 208–223 (2013)
5. Beliakov, G., Pradera, A., Calvo, T.: Aggregation Functions: A Guide for Practitioners. Springer, Berlin (2007)
6. Choquet, G.: Theory of capacities. Annales de l'Institut Fourier **5**, 131–295 (1953–1954)
7. Cordón, O., del Jesus, M.J., Herrera, F.: A proposal on reasoning methods in fuzzy rule-based classification systems. Int. J. Approx. Reason. **20**(1), 21–45 (1999)
8. Cordon, O., del Jesus, M.J., Herrera, F.: Analyzing the reasoning mechanisms in fuzzy rule based classification systems. Mathware Soft Comput. **5**(2–3), 321–332 (1998)
9. Demšar, J.: Statistical comparisons of classifiers over multiple data sets. J. Mach. Learn. Res. **7**, 1–30 (2006)
10. Dias, C.A., et al.: Using the choquet integral in the pooling layer in deep learning networks. In: Barreto, G.A., Coelho, R. (eds.) NAFIPS 2018. CCIS, vol. 831, pp. 144–154. Springer, Cham (2018). https://doi.org/10.1007/978-3-319-95312-0_13
11. Dimuro, G.P., et al.: The state-of-art of the generalizations of the Choquet integral: from aggregation and pre-aggregation to ordered directionally monotone functions. Inf. Fus. **57**, 27–43 (2020)
12. Dimuro, G.P., et al.: Generalized $C_{F_1F_2}$-integrals: from choquet-like aggregation to ordered directionally monotone functions. Fuzzy Sets Syst. **378**, 44–67 (2020)
13. Duda, R.O., Hart, P.E., Stork, D.G.: Pattern Classification, 2nd edn. Wiley-Interscience (2000)
14. Grabisch, M., Marichal, J.L., Mesiar, R., Pap, E.: Aggregation Functions, p. 480 (2009)
15. Hodges, J.L., Lehmann, E.L.: Ranks methods for combination of independent experiments in analysis of variance. Ann. Math. Statist. **33**, 482–497 (1962)
16. Holm, S.: A simple sequentially rejective multiple test procedure. Scandinavian J. Statist. **6**, 65–70 (1979)
17. Ishibuchi, H., Nakashima, T.: Effect of rule weights in fuzzy rule-based classification systems. IEEE Trans. Fuzzy Syst. **9**(4), 506–515 (2001)
18. Ishibuchi, H., Nakashima, T., Nii, M.: Classification and Modeling with Linguistic Information Granules. Advanced Approaches to Linguistic Data Mining. Advanced Information Processing, Springer, Berlin (2005). https://doi.org/10.1007/b138232
19. Klement, E.P., Mesiar, R., Pap, E.: Triangular Norms. Kluwer Academic Publisher, Dordrecht (2000)
20. Ko, L., et al.: Multimodal fuzzy fusion for enhancing the motor-imagery-based brain computer interface. IEEE Comput. Intell. Magaz. **14**(1), 96–106 (2019)
21. Leon-Garza, H., Hagras, H., Peña-Rios, A., Conway, A., Owusu, G.: A fuzzy rule-based system using a patch-based approach for semantic segmentation in floor plans. In: 2021 IEEE International Conference on Fuzzy Systems (FUZZ-IEEE), pp. 1–6 (2021)
22. Lixandru-Petre, I.O.: A fuzzy system approach for diabetes classification. In: 2020 International Conference on e-Health and Bioengineering (EHB), pp. 1–4 (2020)

23. Lucca, G., Dimuro, G.P., Fernandez, J., Bustince, H., Bedregal, B., Sanz, J.A.: Improving the performance of fuzzy rule-based classification systems based on a nonaveraging generalization of CC-integrals named $C_{F_1 F_2}$-integrals. IEEE Trans. Fuzzy Syst. **27**(1), 124–134 (2019)

24. Lucca, G., et al.: Pre-aggregation functions: construction and an application. IEEE Trans. Fuzzy Syst. **24**(2), 260–272 (2016)

25. Lucca, G., Sanz, J.A., Dimuro, G.P., Borges, E.N., Santos, H., Bustince, H.: Analyzing the performance of different fuzzy measures with generalizations of the choquet integral in classification problems. In: 2019 IEEE International Conference on Fuzzy Systems (FUZZ-IEEE), pp. 1–6 (2019)

26. Lucca, G., Sanz, J., Dimuro, G.P., Bedregal, B., Bustince, H.: Analyzing the behavior of aggregation and pre-aggregation functions in fuzzy rule-based classification systems with data complexity measures. In: Kacprzyk, J., Szmidt, E., Zadrożny, S., Atanassov, K.T., Krawczak, M. (eds.) IWIFSGN/EUSFLAT -2017. AISC, vol. 642, pp. 443–455. Springer, Cham (2018). https://doi.org/10.1007/978-3-319-66824-6_39

27. Lucca, G., et al.: CC-integrals: choquet-like copula-based aggregation functions and its application in fuzzy rule-based classification systems. Knowl. Based Syst. **119**, 32–43 (2017)

28. Lucca, G., Sanz, J.A., Dimuro, G.P., Bedregal, B., Bustince, H., Mesiar, R.: CF-integrals: a new family of pre-aggregation functions with application to fuzzy rule-based classification systems. Inf. Sci. **435**, 94–110 (2018)

29. Marco-Detchart, C., Lucca, G., Lopez-Molina, C., De Miguel, L., Pereira Dimuro, G., Bustince, H.: Neuro-inspired edge feature fusion using Choquet integrals. Inf. Sci. **581**, 740–754 (2021)

30. Murofushi, T., Sugeno, M., Machida, M.: Non-monotonic fuzzy measures and the Choquet integral. Fuzzy Sets Syst. **64**(1), 73–86 (1994)

31. Nelsen, R.B.: An Introduction to Copulas. Springer Science & Business Media (2007)

32. da S. E. Tuy, P.G., Nogueira Rios, T.: Summarizer: fuzzy rule-based classification systems for vertical and horizontal big data. In: 2020 IEEE International Conference on Fuzzy Systems (FUZZ-IEEE), pp. 1–8 (2020)

33. Sugeno, M.: Theory of Fuzzy Integrals and its Applications. Ph.D. thesis, Tokyo Institute of Technology, Tokyo (1974)

34. Tan, P.N., Steinbach, M., Kumar, V.: Introduction to Data Mining, 1st edn. Addison-Wesley Longman Publishing Co., Inc, Boston, MA, USA (2005)

35. Wieczynski, J.C., et al.: Generalizing the GMC-RTOPSIS method using CT-integral pre-aggregation functions. In: 2020 IEEE International Conference on Fuzzy Systems (FUZZ-IEEE), pp. 1–8. IEEE, Los Alamitos (2020)

36. Wieczynski., J., Lucca., G., Borges., E., Dimuro., G., Lourenzutti., R., Bustince, H.: CC-separation measure applied in business group decision making. In: Proceedings of the 23rd International Conference on Enterprise Information Systems - Volume 1: ICEIS, pp. 452–462. SciTePress (2021)

37. Wilcoxon, F.: Individual comparisons by ranking methods. Biometrics **1**, 80–83 (1945)

Improving the FQF Distributional Reinforcement Learning Algorithm in MinAtar Environment

Júlio César Mendes de Resende[✉], Edimilson Batista dos Santos, and Marcos Antonio de Matos Laia

Department of Computer Science, Federal University of São João del-Rei, São João del-Rei CEP 36301-360, Brazil
julio.cmdr@gmail.com, {edimilson.santos,marcoslaia}@ufsj.edu.br

Abstract. Reinforcement learning algorithms allow agents to learn from experience, without the need for prior knowledge. For this reason, they have been widely used and the use of low and medium complexity digital games as benchmark environments has become a common practice. In 2013, a new algorithm, called DQN (Deep Q Network), caused a great impact in the academic environment by obtaining human-level results in several Atari 2600 games, using artificial neural networks. Consequently, new lines of research emerged and new derived algorithms were proposed. Among these, the FQF (Fully Parameterized Quantile Function) stands out, an algorithm that has become the state of the art among the non-distributed algorithms in the Atari 2600 domain. However, the FQF has not yet achieved results obtained by a human expert in all evaluated games, thus demonstrating that better results than current ones can still be obtained. Therefore, this work sought to combine two improvements in the FQF that brought success in algorithms proposed before the FQF, with the objective of improving it. The improvements applied to the FQF are: the use of 3 steps in the temporal difference and the application of the Munchausen approach. The FQF changed with the improvements was evaluated in 5 MinAtar games and the results obtained brought gains of approximately 147% over the original FQF in the median of agent's returns.

Keywords: Reiforcement learning · Deep learning · Distributional reinforcement learning

1 Introduction

Learning is a concept shared and explored by several areas of study, such as pedagogy, psychology, neuroscience and computer science. Although each area and author have a different formal definition of what learning is, it is common for the term to be exemplified using the development process of a child, who learns as he observes, interacts with his environment and memorizes [17]. It is also

J. C. Xavier-Junior and R. A. Rios (Eds.): BRACIS 2022, LNAI 13653, pp. 221–236, 2022.
https://doi.org/10.1007/978-3-031-21686-2_16

possible to elucidate learning through animals, such as, for example, a domestic dog, which receives a reward when it has a good attitude or a punishment when it takes an action considered improper by its owner.

In computer science, specifically in the area of artificial intelligence, such behaviors of living beings and natural phenomena are often studied in order to obtain efficient algorithms capable of teaching machines, these being called machine learning algorithms [12]. Although interaction learning is a reference on the subject, the most studied machine learning algorithms are supervised learning [18], in which a predictive model is built based on a previously labeled dataset. There are also algorithms called unsupervised, which, in turn, seek to group unlabeled data. Finally, a third paradigm encompasses reinforcement learning algorithms, which are the only ones that actually learn by interaction.

Reinforcement learning algorithms differ from others in that they do not require a database and, mainly, have a fixed objective in each task, which involves challenges of exploring environments and planning [18]. A new reinforcement learning algorithm, called DQN (Deep Q Network) [13], appeared in 2013 and revolutionized the literature by combining classical reinforcement learning algorithms with deep learning techniques, which had been obtaining important results in algorithms of supervised learning. DQN was the first algorithm to be able to play several Atari 2600 games at the same level as a human, which boosted a new line of research: that of reinforcement learning algorithms evaluated in Atari 2600 games.

In this context, it is important to emphasize that digital games are just environments used to study the algorithms, which can later be applied in other environments, such as in industry - through the control of machines [16] - or even on public roads, driving autonomous vehicles [11]. In addition, the growing worldwide technological expansion, together with the possibility of applying reinforcement learning algorithms in complex tasks, demands that they are constantly evolving, which in fact has happened.

After the emergence of DQN, several derived algorithms were proposed in the literature. Among these, the FQF (Fully Parameterized Quantile Function) [22] stands out, which has become the state of the art among the non-distributional algorithms in the Atari 2600 domain. However, the FQF has not yet achieved results obtained by a human expert in all evaluated games and thus better results can still be obtained. Therefore, considering the relevance of the contributions of several works present in the literature and the need for a constant improvement of the methods, this paper seeks to improve the FQF, inserting two improvements that were able to boost the results of algorithms proposed before the FQF. The improvements applied to the FQF are: the use of 3 steps in the temporal difference and the application of the Munchausen approach. The FQF was run with these improvements and evaluated in the MinAtar [23] domain, which is a smaller graphical environment. The results obtained with the FQF altered with the improvements tend to be superior to the original FQF in the initial experiments.

This paper is organized as follows. In Sect. 2, an introduction to reinforcement learning is presented, describing the main concepts used throughout the text. In Sect. 3, some related works are described, taking the publication of the DQN as a starting point. This section also includes a description of Distributional Reinforcement Learning and the FQF algorithm. Sections 4, 5 and 6 present the methodology, results of the experiments and the conclusion, respectively.

2 Reinforcement Learning

A reinforcement learning problem can be modeled using finite Markov Decision Process (MDP). In this type of problem, an agent interacts with an environment in a discrete sequence of time steps $t = 0, 1, 2, 3, ...$ After each action $A_t \in \mathcal{A}(s)$, in the state $S_t \in \mathcal{S}$, the agent observes the new state S_{t+1} and the reward $R_{t+1} \in \mathcal{R} \subset \mathbb{R}$, which indicates how good it is for the agent to be in S_{t+1}. This sequence of interactions gives rise to a trajectory which begins with: $S_0, A_0, R_1, S_1, A_1, R_2, S_2, A_2, ...$

At each step t, the agent's objective is to maximize the expected return of rewards received up to the final step T. The return is formulated as shown in Eq. 1. However, usually the rewards in G_t are weighted by a discount factor $\gamma \in [0, 1]$, representing how much future rewards will be considered, which makes common, the representation $G_t = R_{t+1} + \gamma G_{t+1}$.

$$G_t \doteq R_{t+1} + R_{t+2} + R_{t+3} + \cdots + R_T \tag{1}$$

The agent chooses the actions to take at each step according to its policy, which defines a probability distribution over actions for each state. For this, the policy can be based on two important functions, which are: i) $v_\pi(s) \doteq \mathbb{E}[G_t|S_t = s]$, which returns how good it is for the agent to be in a state s following a policy π, and ii) $q_\pi(s, a) \doteq \mathbb{E}[G_t|S_t = s, A_t = a]$, which tells how good it is for the agent to take an action a in a state s following a policy π afterwards.

Considering that all MDP transition probability functions are known, it can be solved using dynamic programming (DP) methods, which use tables to store the values of $q_\pi(s, a)$ or $v_\pi(s)$. These values are constantly updated. However, problems in which the probability functions of the MDP are known are rare, which makes the application of DP in the context of reinforcement learning very limited [18]. This work addresses a class of methods that is model-free and seeks to estimate state values by sampling, without the need for model equations, such as Monte Carlo (MC) and temporal difference (TD) methods.

MC methods are only applicable to tasks that are composed of episodes. The central idea of the method is based on a list that stores all the rewards received by the agent in the episode, as well as the states and actions that the agent went through. So, at the end of the episode, the algorithm goes through this list backwards, adding up the rewards and updating the algorithm table. TD methods, on the other hand, update their tables throughout the episode, through an estimate of later states. An example of an algorithm that uses TD is Sarsa [14], where the step update equation is represented in Eq. 2:

$$Q(S, A) \leftarrow Q(S, A) + \alpha[R + \gamma Q(S', A') - Q(S, A)] \tag{2}$$

$\alpha \in (0, 1]$, here, is an adjustable parameter of the learning factor and S', A' is the pair of the next state and action taken, respectively.

Note that the Sarsa equation uses as a target $R + \gamma Q(S', A')$. This type of composition, in which an explicit reward is used in the target and the rest of the return estimated by the next state, is called TD(0). TD(0) algorithms bring agility to the table update, but also have greater divergence, because they learn one guess from the next [18]. On the other hand, MC methods do not use estimates of the remainder of the return, but they have to wait for the end of the episode (which can be huge) to update the table. One way to balance the benefits of MC with TD(0) is to use n-step TD methods. This means that more rewards are explicitly represented in the equation, deferring the estimation of the next states to a lower weighted part of the return, without delaying the table update too much. A 3-step version of Sarsa, for example, would have, as a target, $R_{t+1} + \gamma R_{t+2} + \gamma^2 R_{t+3} + \gamma^3 Q(S_{t+3}, A_{t+3})$.

Besides Sarsa, another famous TD(0) algorithm is Q-learning [21]. Q-learning has, as a target, $R + \gamma max_a Q(S', a)$, which considers the estimate of the most promising action of the next state. However, this action is not necessarily the one chosen by the agent in S', which makes the algorithm considered off-policy. Off-policy algorithms have greater generalization capacity and are applicable to a greater number of tasks, as they can learn from data generated by an agent that is not learning, such as a human expert. However, they take longer to converge and have greater divergence than single-policy algorithms, called on-policy. Both Sarsa and Q-learning use a table to store function values $q_\pi(s, a)$. In applications where the number of states is very large, the use of tables can become unfeasible, and the use of function approximators is a viable alternative. Section 3 presents some works that use neural networks to obtain an estimate of $q_\pi(s, a)$.

3   Related Work

3.1   DQN

In [13], a reinforcement learning algorithm called DQN (Deep Q Network) was proposed with the aim of learning to play Atari 2600 games through ALE (Arcade Learning Environment) [2]. The DQN makes use of a convolutional neural network and therefore has an image as a representation of the state, not being necessary to perform state feature extraction. For this reason, DQN is able to learn several Atari 2600 games just adjusting the number of neurons in the last layer for the number of possible actions in each game. The authors of DQN have opened up the frontiers of deep reinforcement learning and made ALE and its Atari games an important set of benchmark environments for the algorithms, being used in several later works.

Another important contribution of the authors of DQN was to apply the technique of experience replay, together with mini-batch in reinforcement learning. The experience replay consists of keeping a record of the last n state transitions in a buffer. Thus, the update of the weights in the network is based on a batch that is formed through samples chosen in each update step in a uniformly random way among the n stored in the buffer. This is an important characteristic of the DQN, as training the network based on consecutive samples can be inefficient due to the strong correlation between samples. In DQN, the objective of the network is to approximate the values of state and action following the Q-Learning algorithm, which as already seen is an off-policy algorithm.

By using a temporal difference algorithm, the DQN would fit into the algorithms that are considered semi gradient and consequently have great variation. To alleviate this problem, the authors of the DQN rely on another neural network called the target network. The target network has the same structure as the main network, however its weights θ^- are only updated every k steps, as a copy of the weights of the original network. This auxiliary network was important to reduce the variance in the DQN.

3.2   Prioritized Experience Replay

In [15], one important improvement for the DQN algorithm has been proposed, altering the way in which the experience replay occurs. The authors proposed the creation of a priority queue, in which transition samples that have a greater temporal difference are more likely to be selected. The motivation for the use of this technique is that, many times, important transitions for the evolution of the agent are not selected and consequently learned as well.

3.3   Rainbow

After the publication of the DQN, several works appeared proposing improvements in the algorithm. However, as many were produced simultaneously, there was not a good synchronization between them. Thus, in [9], an algorithm called Rainbow was proposed, which combined several of these improvements, becoming the state of the art in the Atari 2600 domain until then.

Rainbow combined features of C51 [1], a distributional reinforcement learning algorithm; DDQN [8], which uses Double Learning techniques; the Dueling Network [20], which uses a concept of an action's advantage function over others; the Noisy Nets [6], which offer greater exploration capabilities to the agent; in addition to incorporating the prioritized repetition replay and the n-step, with $n = 3$. According to the results of the experiments, the authors observed that the components that most positively impacted individually on Rainbom have been the n steps algorithm, the prioritized repetition replay and the representation of the distribution of returns.

3.4 Distributional Reinforcement Learning and FQF

DQN, and several other algorithms reported so far, seek to learn the q_π function, and for each state entered, the network provides several outputs, one for each action. It is worth remembering that the functions q_π and v_π are the expectation of a distribution of returns G_t. However, learning the mean of a random variable can be a difficult task, due to the possibility of large sample divergence. For this reason, in [1], a discussion was started on the application of distributional reinforcement learning.

The authors presented important theoretical contributions and presented a distributional reinforcement learning algorithm called C51. C51 sought to learn a discretized distribution of returns for each action, which is composed of several atoms, each represented by an output in the network. Each atom in the network represents a range of values that a return can take, and the atom's output is an indicator of the frequency of this value range, later normalized by a softmax function. The C51 presented important results, but it has the limitation of acting in only a pre-defined distribution range, in addition to not having proof of convergence.

Therefore, in [5], the QR-DQN algorithm was proposed, which performs a quantile regression and transposes the parameterization of C51. While the C51 fixes distribution values and estimates probabilities, the QR-DQN has quantiles fixed at the network outputs and estimates the values of such quantiles. One more evolution was proposed in [4], with an implicit version of QR-DQN called IQN. In the IQN, the network has only one output per action, but it is dependent on an extra network that receives the quantile for which the value is to be estimated. Thus, to obtain a distribution of values, it is necessary to forward the network several times[1].

In the IQN, the quantiles entered for processing in the network are chosen randomly. However, as it is not feasible to use a huge number of quantiles, the choice of which quantiles will compose the distribution can significantly affect the final result. For this reason, in [22], the FQF (Fully Parameterized Quantile Function) algorithm was proposed, which adds one more flow in the network to provide quantiles whose values must be estimated, thus parameterizing the entire distribution. In Fig. 1, there is an illustration of the distributional reinforcement learning algorithms described here, in which it is observed that the FQF has one more network than the IQN, called φ. This network aims to provide 32 quantiles to be estimated by the ϕ network.

---

[1] As the quantiles are processed by a parallel network, the IQN main stream that processes the state does not need to be reprocessed.

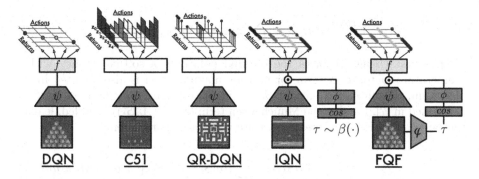

Fig. 1. Distributional reinforcement learning algorithms presented in this paper. The illustrations are extracted from [22] and [4].

The FQF training takes place in two parts: first, the φ network is adjusted by a function based on the Wassertein metric and then the ϕ and ψ networks are trained, in the same way as in the IQN : training each quantile individually, based on the Huber loss function [10]. A comparison presented in [22] shows the superiority of FQF over other algorithms discussed here when analyzing its performance in 55 Atari 2600 games, creating a new record for the Atari Learning Environment for non-distributed agents. However, the results also point out that the FQF still fails to surpass the human baseline in 9 out of 55 games.

It is important to note that although these algorithms estimate a return distribution, they use this distribution to estimate the value of q_π posteriorly and thus choose an action to take based on q_π. They also inherit DQN technical features such as the use of a target network and experience replay.

3.5   Munchausen R.L.

In 2020, a new and simple improvement applied to the DQN emerged and was named Munchausen R.L. (MRL) [19]. Based on elementary knowledge of reinforcement learning, this approach assumes that, indirectly, the policy is always being evolved and that, therefore, it can emit a reinforcement in the agent's learning. If an optimal policy is assumed, the log of the probability of taking the best action will always be 0, while for the other actions it will be $-\infty$. The MRL adds this signal to the reward, replacing R_t by $R_t + bln(\pi(A_t|S_t))$, where b is an adjustable parameter.

An initial problem with the application of this approach in DQN is the fact that the network does not learn stochastic policies, but the optimal policy, which is an obstacle for the application of the log, which would always result in 0 or $-\infty$. Before introducing the Munchausen approach, let's rewrite the DQN target (Q-Learning) in the same way as in the Expected Sarsa algorithm, but remembering that the DQN learning policy is greedy and deterministic:

$$Y_t^{DQN} = R_{t+1} + \gamma \sum_a \pi(a|S_{t+1})\widehat{q}(S_{t+1}, a; \theta_t^-) \tag{3}$$

The first change replaces the optimal policy by a softmax policy with a temperature factor τ. A term for entropy maximization is also adopted in the algorithm, by subtracting $\tau ln\pi(a|S_{t+1}))$ in the estimation of the next state, as performed in [7]. Finally, the Munchausen term $b\tau ln(\pi(A_t|S_t))$ is added, leaving the target as in Eq. 4, where b and τ are adjustable parameters:

$$Y^{M-DQN} = R_{t+1} + b\tau ln(\pi(A_t|S_t)) + \gamma \sum_a \pi(a|S_{t+1})(\widehat{q}(S_{t+1}, a; \theta_t^-) - \tau ln\pi(a|S_{t+1}))$$
$$\tag{4}$$

with $\pi = sm(\frac{\widehat{q}(S_{t+1}, a; \theta_t^-)}{\tau})$.

To avoid large variance, the authors also implemented an upper and lower bound for the value of the Munchausen term and use a stable implementation of the policy's softmax function. The adapted version of DQN was named M-DQN and achieved excellent results, which made M-DQN the first non-distributional reinforcement learning algorithm to outperform a distributional one.

3.6 MinAtar

ALE with its Atari 2600 games has established itself as one of the main benchmark environments for reinforcement learning algorithms, being used in several works. Its games bring two main challenges: learning features and learning behavior. Initially, learning features was an interesting challenge. However, it is observed that, after the emergence of DQN, all evolutions are focused on learning behavior, as the convolutional layers of networks have not received much attention. Despite this, the learning of features is the most computationally expensive part of the algorithms, and can even be considered an obstacle for those who want to explore the behavior more. The high computational cost prevents more diversified experiments from being carried out [23].

Considering that feature learning is a problem that has already been shown to be solvable with the proper adjustment of convolutional layers, in [23], a new set of environments called MinAtar is proposed. This environment is a miniaturized version of 5 Atari 2600 games (Seaquest, Breakout, Asterix, Freeway and Space invaders)[2] and seeks to simplify the game frame so that researchers can focus on reinforcement learning algorithms.

In [3], the authors performed a set of tests using the 5 MinAtar games with the same algorithms evaluated in [9], in addition to including QR-DQN and IQN (FQF was not evaluated in [3]). The results found in these environments were in agreement with those of Rainbow, presented in [9], with 57 Atari 2600 games. This highlights the relevance of MinAtar and leads us to believe that it is possible to evaluate an algorithm in this simpler environment initially, before evaluating it in a complex environment that requires a higher computational cost.

---

[2] Information on game modeling can be found at [23].

4   Methodology

According to the works discussed in Sect. 3, it is notable that there have been great evolutions since the emergence of DQN. Among these, FQF stands out, a powerful distributional reinforcement learning algorithm. However, it is also notorious that the results obtained by FQF can still be improved, as this algorithm has not yet achieved results obtained by a human expert in all rated Atari 2600 games. Considering that features from related works were able to improve the C51 algorithm considerably, producing the Rainbow, in this work, we evaluated the addition of two features in the FQF, which has already presented better results than the C51 algorithm. Initially, we chose to evaluate the addition of the n-steps method, which had a high impact on Rainbow, and also the addition of the Munchausen approach, as this was an evolution that showed promise and has not been evaluated in Rainbow. These improvements were combined and the algorithms created as follows:

- FQF: Nature FQF
- FQFn3: FQF + n-step with $n = 3$
- FQFM: FQF + MRL
- FQFn3M: FQF + 3-steps + MRL

These algorithms were evaluated in the 5 games of the MinAtar environment, which is a less complex environment, but which proved to be a good environment for tests in [3]. Since the algorithms were evaluated in a continuous process of training, it is possible to observe the evolution of their performance. In the experiments performed in this work, the algorithms were evaluated by 20 million steps, which were grouped into 20 iterations, and each iteration, having 1 million steps, represents several episodes. Note that, in this type of grouping by number of steps, the number of episodes will vary according to the agent's performance in each iteration. It is normal that the agent has a low performance at the beginning, which causes the episodes to be shorter and therefore more numerous in the same iteration. The opposite happens when the agent is already more mature. In this case, the agent is expected to survive longer in episodes, thus accumulating more rewards. For this reason, cluster measures in relation to the return of episodes represent good alternatives for agent evaluation.

Here, we use the same evaluation measures used in [1,4,5,9,22], which are the mean and median of the returns in the last iteration (LI). In the cited works, these values were normalized in relation to a human score. As in MinAtar we do not have the data from a human base and the work is focused on improvements over the FQF, we normalized the scores against the FQF. In addition, we also evaluated the area under the curve (AUC) of the iterations using the same aggregating measures, which allows us to identify those algorithms that evolved faster.

The algorithms evaluated in this work were implemented using the Python language, together with the Rljax[3], [4] framework, which is a framework for rein-

---

[3] The Rljax original source code can be found here: https://github.com/ku2482/rljax.
[4] The source code used in this work can be found here: https://github.com/julio-cmdr/rljax.

forcement learning with the implementation of several algorithms, including the FQF. All experiments were performed using GPUs.

5   Experiments and Analysis of Results

Like several of the algorithms mentioned in this work, the FQF also has a large set of hyperparameters, which were proposed based mainly on the parameterization presented for the DQN in ALE. Since we did not find evaluations of the FQF in MinAtar in the literature, the hyperparameters for it were defined here based on the work presented in [3], which evaluated several algorithms in MinAtar, with the exception of the FQF. Hyperparameters that are particular to the FQF, such as the learning rate of the fraction proposal network, were initially maintained as proposed in [22].

However, running the FQF on MinAtar in initial experiments carried out in this work, the FQF showed characteristics of catastrophic forgetting (especially in the game Space Invaders), in which the agent has a sudden drop in its performance after some training time. Although it is a classic problem of neural networks in continuous learning, this behavior has not been reported in previous works, possibly due to the use of several techniques originally proposed in DQN, such as the experience replay. For this reason, we believed this is a hyperparameter tuning issue.

5.1   Hyperparameter Tuning

Tuning a large number of hyperparameters can easily result in a combinatorial explosion, which is compounded when it comes to deep learning, which are computationally expensive algorithms. Therefore, an initial adjustment was made in 3 hyperparameters that are highly related to the learning capacity of the algorithm: the learning rate of the fraction proposal network (lr_frac_net), which was not adjusted for MinAtar, the learning rate of the main network (lr) and the batch size.

A combinational evaluation of these 3 hyperparameters was carried out, and in each hyperparameter, through changes in the exponent, an alternative value greater and another smaller than the original value was added (extracted from [3] and [22]), which resulted in 27 combinations. For the batch size, usually defined as a power of 2, the values 16, 32 and 64 were evaluated. In the case of lr, the values 2.5e−05, 2.5e−04 and 2.5e−03 were evaluated. For lr_frac_net, the values used were 2.5e−10, 2.5e−09 and 2.5e−08.

The evaluation of hyperparameters was performed sequentially with 9 seeds in the game Space Invaders, which was the game that in preliminary analyzes caused catastrophic forgetting in the FQF. As the focus of this analysis was to find a set of hyperparameters that did not suffer from catastrophic forgetting in any execution, whenever a set of hyperparameters led to a sudden drop in performance in an algorithm, it was eliminated from the next executions.

To detect when a sudden drop occurred or not, we use a stopping criterion that terminates the algorithm whenever the average of the returns in an iteration is less than or equal to 50% of the highest average obtained by the algorithm in the execution in question. The result of this analysis can be seen in Fig. 2, where the shaded area on the curve represents the 95% confidence interval with respect to aggregate returns. In the graphs, it is observed that after 5 executions, only one set of hyperparameters remained, which is the set that uses batch_size = 16, lr_frac_net = 2.5e−10 and lr = 2.5e−05, the only hyperparameter that kept the original value. These values are used in all subsequent runs in this work, in the 5 games.

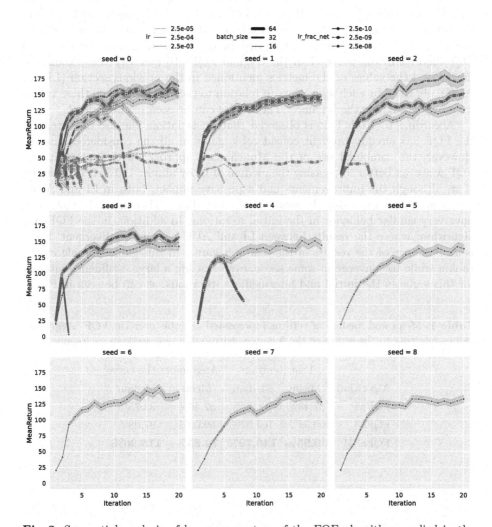

Fig. 2. Sequential analysis of hyperparameters of the FQF algorithm applied in the game Space Invaders with 9 seeds. Every time a set of hyperparameters caused an algorithm to crash, it was eliminated from subsequent runs.

5.2   Main Results

With the hyperparameters defined, the improvements were applied to the FQF. Each algorithm was executed 5 times in each game, using seeds 0, 1, 2, 3 and 4, which totaled 100 executions. The graphs in Fig. 3 allow to follow the evolution of the algorithms for each game through the average of the returns, with the shaded area representing the 95% confidence interval of this value, according to the average of each seed. In the analysis of these graphs, we consider that when there is an overlap of the shaded areas between two algorithms in an iteration, the algorithms tie, given the confidence level considered.

Analyzing the average returns in the graphs in Fig. 3, we observed that the FQFM algorithm was better than the FQF in 4 of the 5 games, with a tie in the seaquest game. The FQFn3 and FQFn3M algorithms were also better than the FQF in 4 out of 5 games, but tied in the breakout game. No improved FQF algorithm was worse than the original FQF. To quantify these gains, we normalized the values of the metrics mentioned in the previous section (LI and AUC), obtained by each algorithm, in relation to the FQF, and then these values were aggregated for the 5 games. The results can be seen in the Table 1.

According to Table 1, it is observed that, in isolation, the use of 3 steps in the FQF has already brought considerable gains in all the considered metrics. However, the combined use of the Munchausen approach with 3 steps in the FQF was even better, increasing the median of LI by almost 150%. It is observed that, although the improvements had a positive impact, both in LI and AUC, the gains in LI were slightly more expressive. This is because all algorithms have very similar behavior in the initial iterations. In addition, in the FQFn3M algorithm, where the results between LI and AUC were more divergent, there is an influence of the results of the seaquest game. In this game, the algorithm took a while to converge in some seeds, even causing a large confidence interval for this game in the initial and intermediate iterations, as can be seen in Fig. 3.

Table 1. Mean and median of returns represented as gains over the FQF. Aggregate values referring to the scores of the 5 games, with 5 different seeds in each game.

Algorithm	Last iteration		Area under the curve	
	Mean	Median	Mean	Median
FQFM	37.43%	42.20%	32.46%	32.75%
FQFn3	60.08%	103.07%	59.00%	97.08%
FQFn3M	**90.95%**	**146.79%**	**72.85%**	**115.96%**

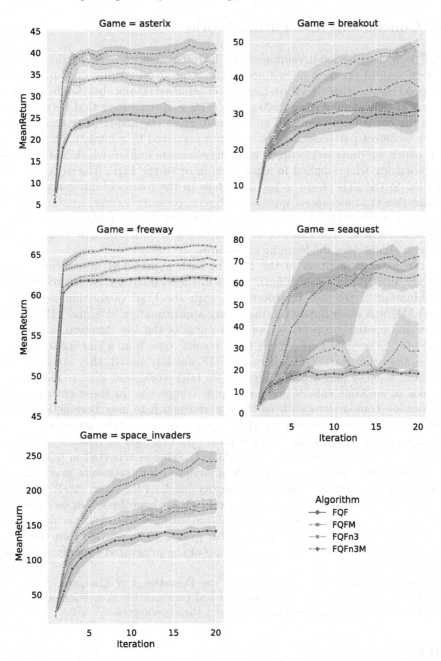

Fig. 3. Evolution of the average of returns for the 4 algorithms evaluated in the 5 games. Aggregated values referring to runs with 5 different seeds.

6   Conclusions and Future Works

FQF is a distributional reinforcement learning algorithm, which has outperformed several previous algorithms in the Atari 2600 gaming domain. Despite the good results, this algorithm has not yet achieved results obtained by a human expert in all evaluated Atari 2600 games. This leads to believe that better results than the current ones can still be obtained, given the ability of artificial intelligence to detect patterns that are often not perceived by a human. Therefore, in this paper, we proposed to apply two improvements to FQF which had shown to be promising when applied in algorithms prior to the FQF. The improvements applied to the FQF are: the use of 3 steps in the temporal difference and the application of Munchausen approach. These improvements were combined in the FQF and evaluated in the domain of MinAtar, which is a smaller graphical environment, ideal for experiments where the focus is on learning of behavior, but without eliminating the learning of features.

The isolated addition in the FQF of both approached improvements was already able to present significant gains in the evaluated games. However, the combination of the two improvements with the FQF (algorithm here called FQFn3M) was more impactful, increasing approximately 91% and 147% in the mean and median, respectively, of the returns of the last iteration, when compared to the original FQF. While these results come from a smaller-scale set of games than the Atari 2600, where the FQF was introduced, they are extremely relevant, because they act as a direction for later works that want to make applications in environments of greater graphic complexity. In these environments, performing comparisons such as those performed here may become unfeasible for many researchers, due to the high demand for computational power.

In addition, the relevant gains obtained by FQFn3M in relation to FQF in the game Seaquest, wich is the most difficult game as reported in [23], lead us to believe that the inclusion of both approached improvements may generate gains in environments of greater complexity (such as the Atari 2600). This can boost FQF gains over a human considerably. This hypothesis can be confirmed through experiments using the Atari 2600 in future works. Another proposal is the evaluation of other improvements that proved to be relevant, but that were not considered in this paper, as is the case of the prioritized experience replay.

Acknowledgement. The authors thank the Department of Computer Science at UFSJ, especially professor Diego Colombo and professor Elverton Fazzion, for their support with powerful computers to carry out the experiments.

References

1. Bellemare, M.G., Dabney, W., Munos, R.: A distributional perspective on reinforcement learning. In: Precup, D., Teh, Y.W. (eds.) Proceedings of the 34th International Conference on Machine Learning. Proceedings of Machine Learning Research, vol. 70, pp. 449–458. PMLR, 06–11 August 2017. http://proceedings.mlr.press/v70/bellemare17a.html

2. Bellemare, M.G., Naddaf, Y., Veness, J., Bowling, M.: The arcade learning environment: an evaluation platform for general agents. J. Artif. Int. Res. **47**(1), 253–279 (2013)

3. Ceron, J.S.O., Castro, P.S.: Revisiting rainbow: promoting more insightful and inclusive deep reinforcement learning research. In: Meila, M., Zhang, T. (eds.) Proceedings of the 38th International Conference on Machine Learning. Proceedings of Machine Learning Research, vol. 139, pp. 1373–1383. PMLR, 18–24 July 2021. http://proceedings.mlr.press/v139/ceron21a.html

4. Dabney, W., Ostrovski, G., Silver, D., Munos, R.: Implicit quantile networks for distributional reinforcement learning. In: Dy, J., Krause, A. (eds.) Proceedings of the 35th International Conference on Machine Learning. Proceedings of Machine Learning Research, vol. 80, pp. 1096–1105. PMLR, 10–15 July 2018. http://proceedings.mlr.press/v80/dabney18a.html

5. Dabney, W., Rowland, M., Bellemare, M.G., Munos, R.: Distributional reinforcement learning with quantile regression. In: McIlraith, S.A., Weinberger, K.Q. (eds.) Proceedings of the Thirty-Second AAAI Conference on Artificial Intelligence, (AAAI-18), The 30th Innovative Applications of Artificial Intelligence (IAAI-18), and the 8th AAAI Symposium on Educational Advances in Artificial Intelligence (EAAI-18), New Orleans, Louisiana, USA, 2–7 February 2018, pp. 2892–2901. AAAI Press (2018)

6. Fortunato, M., et al.: Noisy networks for exploration. In: International Conference on Learning Representations (2018). http://openreview.net/forum?id=rywHCPkAW

7. Haarnoja, T., Zhou, A., Abbeel, P., Levine, S.: Soft actor-critic: off-policy maximum entropy deep reinforcement learning with a stochastic actor. In: Dy, J., Krause, A. (eds.) Proceedings of the 35th International Conference on Machine Learning. Proceedings of Machine Learning Research, vol. 80, pp. 1861–1870. PMLR, 10–15 July 2018. http://proceedings.mlr.press/v80/haarnoja18b.html

8. Hasselt, H.v., Guez, A., Silver, D.: Deep reinforcement learning with double q-learning. In: Proceedings of the Thirtieth AAAI Conference on Artificial Intelligence, pp. 2094–2100. AAAI 2016, AAAI Press (2016)

9. Hessel, M., et al.: Rainbow: combining improvements in deep reinforcement learning. AAAI 2018/IAAI 2018/EAAI 2018, AAAI Press (2018)

10. Huber, P.J.: Robust estimation of a location parameter. Ann. Math. Stat. **35**(1), 73–101 (1964). https://doi.org/10.1214/aoms/1177703732

11. Kendall, A., et al.: Learning to drive in a day, pp. 8248–8254 (2019). https://doi.org/10.1109/ICRA.2019.8793742

12. Mitchell, T.M.: Machine Learning. McGraw-Hill, New York (1997)

13. Mnih, V., et al.: Human-level control through deep reinforcement learning (2015). https://doi.org/10.1038/nature14236

14. Rummery, G., Niranjan, M.: On-Line Q-learning Using Connectionist Systems. Technical report. CUED/F-INFENG/TR 166, Cambridge University, Cambridge (1994)

15. Schaul, T., Quan, J., Antonoglou, I., Silver, D.: Prioritized experience replay (2016). http://arxiv.org/abs/1511.05952. published as a conference paper at ICLR 2016

16. Schoettler, G., et al.: Deep reinforcement learning for industrial insertion tasks with visual inputs and natural rewards. In: 2020 IEEE/RSJ International Conference on Intelligent Robots and Systems (IROS), pp. 5548–5555 (2020)

17. Subramanian, A., Chitlangia, S., Baths, V.: Reinforcement learning and its connections with neuroscience and psychology. Neural Netw. **145**(C), 271–287 (2022). https://doi.org/10.1016/j.neunet.2021.10.003
18. Sutton, R.S., Barto, A.G.: Reinforcement Learning: An Introduction. The MIT Press, Cambridge second edn. (2018). http://incompleteideas.net/book/the-book-2nd.html
19. Vieillard, N., Pietquin, O., Geist, M.: Munchausen reinforcement learning. In: Larochelle, H., Ranzato, M., Hadsell, R., Balcan, M.F., Lin, H. (eds.) Advances in Neural Information Processing Systems. vol. 33, pp. 4235–4246. Curran Associates, Inc. (2020)
20. Wang, Z., Schaul, T., Hessel, M., Hasselt, H., Lanctot, M., Freitas, N.: Dueling network architectures for deep reinforcement learning. In: Balcan, M.F., Weinberger, K.Q. (eds.) Proceedings of The 33rd International Conference on Machine Learning. Proceedings of Machine Learning Research, vol. 48, pp. 1995–2003. PMLR, New York, 20–22 Jun 2016. http://proceedings.mlr.press/v48/wangf16.html
21. Watkins, C.J.C.H.: Learning from delayed rewards. Ph.D. thesis, King's College, Oxford (1989)
22. Yang, D., Zhao, L., Lin, Z., Qin, T., Bian, J., Liu, T.Y.: Fully parameterized quantile function for distributional reinforcement learning. In: Wallach, H., Larochelle, H., Beygelzimer, A., d' Alché-Buc, F., Fox, E., Garnett, R. (eds.) Advances in Neural Information Processing Systems, vol. 32. Curran Associates, Inc. (2019)
23. Young, K., Tian, T.: MinAtar: an atari-inspired testbed for thorough and reproducible reinforcement learning experiments. arXiv preprint arXiv:1903.03176 (2019). http://arxiv.org/abs/1903.03176

Glomerulosclerosis Identification Using a Modified Dense Convolutional Network

Justino Santos[1,2($\boxtimes$)], Vinicius Machado[2], Luciano Oliveira[3],
Washington Santos[4], Nayze Aldeman[5], Angelo Duarte[6], and Rodrigo Veras[2]

[1] Instituto Federal do Piauì, São Raimundo Nonato, PI, Brazil
`justinoduarte@ifpi.edu.br`
[2] Departamento de Computação, Universidade Federal do Piauí, Teresina, PI, Brazil
`{vinicius,rveras}@ufpi.edu.br`
[3] Departamento de Ciência da Computação, Universidade Federal da Bahia,
Salvador, BA, Brazil
`lrebouca@ufba.br`
[4] Centro de Pesquisas Gonçalo Moniz, Fundação Oswaldo Cruz, Salvador, BA, Brazil
`washington.santos@fiocruz.br`
[5] Curso de Medicina, Universidade Federal do Delta do Parnaíba,
Parnaíba, PI, Brazil
`nayzealdeman@ufpi.edu.br`
[6] Departamento de Tecnologia, Universidade Estadual de Feira de Santana,
Feira de Santana, BA, Brazil
`angeloduarte@uefs.br`

Abstract. Glomerulosclerosis is a common kidney disease characterized by the deposition of scar tissue, which replaces the renal parenchyma, and is quantified by renal pathologists to indicate the presence and extent of renal damage. It is paramount to guide the appropriate treatment and minimize the chances of the disease progressing to chronic stages. Thus, to identify glomerulus with sclerosis, this article proposes a convolutional neural network (CNN) inspired by convolutional blocks of DenseNet-201 but with smaller dense layers. We analyzed five CNNs - VGG-19, Inception-V3, ResNet-50, DenseNet-201, and EfficientNet-B2 - to define the best CNN model and evaluated several configurations for the fully connected layers. In total, 25 different models were analyzed. The experiments were carried out in three datasets, composed of 1,062 images, on which we applied data-augmentation techniques in the training set. These CNNs demonstrated effectiveness in the task and achieved an accuracy of 92.7% and kappa of 85.3%, considered excellent.

Keywords: Transfer learning · Kidney disease · Computer-aided diagnosis · Image analysis

1 Introduction

Glomerulosclerosis (GS) is a health condition that causes morphological alterations, including sclerosis scarring or hardening, of tiny blood vessels in the

J. C. Xavier-Junior and R. A. Rios (Eds.): BRACIS 2022, LNAI 13653, pp. 237–252, 2022.
https://doi.org/10.1007/978-3-031-21686-2_17

kidneys called glomeruli [2]. Within the glomeruli, waste and fluid are filtered from the blood and removed from the body through urine. Children and adults can be diagnosed with glomerulosclerosis, and almost 20% of adults with kidney disease will develop it[1]. The greater the delay in diagnosing it, the less chance a patient can preserve their kidney function.

A kidney biopsy is often required to diagnose kidney diseases accurately. In addition to establishing the diagnosis, the biopsy allows assessing chronicity and activity to define prognosis and choose the proper therapy [4]. However, making an accurate diagnosis is time-consuming, even for trained pathologists. Thus, there is an expectation that automated processing to support this task will improve the efficiency of renal pathology, contributing to a more objective and standardized diagnosis, especially in hospitals or countries with poor numbers of nephropathologists.

According to Yi et al. [25], the integration of artificial intelligence and clinical medicine allows the development of tools and resources to support clinical decision-making at the bedside. Several aspects of nephrology are likely to be affected by these new technologies. Renal pathology is a good example where improvements in workflows and more subtle histological phenotyping are expected, although much work remains to be done to validate these new methods [8].

Machine learning algorithms, especially those based on Convolutional Neural Networks (CNNs), have achieved excellent performance in image classification and are applied in several research areas, including medicine [26]. They have acquired considerable attention in histology and pathology [12]. Considering the importance of diagnostics and the potential of computer-aided diagnosis systems, in this work, we propose a convolutional neural network (CNN) inspired by convolutional blocks of DenseNet-201 but with smaller dense layers to detect glomerulosclerosis in glomerulus images. To establish the best model, we modified and fine-tuned the architecture of five general-purpose CNNs: VGG-19 [20], Inception-V3 [21], ResNet-50 [6], DenseNet-201 [7], and EfficientNet-B2 [23].

We organized this work as follows: in Sect. 2 we presented recent works and methodologies on the problem under study. Section 3 presents the proposed method, the image data sets, the applied techniques, and the evaluation metrics adopted to validate our results. The results and their discussion are presented in Sect. 4. Finally, conclusions and future work are drawn in Sect. 5.

2   Related Works

Several studies applied image processing techniques and artificial intelligence to renal pathology. Some have analyzed tubules, blood vessels, and interstitium [10, 27]. However, most studies have focused on the glomeruli, which have several essential histological findings for the diagnosis [17,28].

The first step in the glomerulosclerosis diagnostic procedure is the detection of a glomerulus on a whole-slide image of kidney tissue samples. This problem is also the subject of recent studies. The proposed solutions use methods to define various characteristics [18] or use CNNs [5].

---

[1] www.davita.com/education/kidney-disease/symptoms/what-is-glomerulosclerosis.

With the same purpose as our methodology, Araujo et al. [1], Marsh et al. [14], Kannan et al. [9], and Pesce et al. [15] reported solutions to distinguish between sclerotic and non-sclerotic glomeruli.

Araujo et al. [1] used images of single glomeruli to detect segmental glomerulosclerosis. Their architecture had the typical structure, consisting of a digital image processing and pattern recognition system. Three feature vectors were extracted and supplied to four classifiers: KNN, SVM, a neural network, and Naive Bayes.

Marsh et al. [14] described the development of a deep learning model that identifies and classifies glomeruli with and without sclerosis in 48 images of donor kidney biopsy entire sections. This differentiation is meaningful because the criterion for accepting or rejecting the donor's kidneys relies heavily on the pathologist's determination of the percentages of glomeruli that are normal and sclerotic. According to the authors, the model achieved a precision of 81.28% in identifying non-sclerosed glomeruli.

Kannan et al. [9] proposed a CNN capable of classifying portions of slides into three classes: no glomerulus, normal or partially sclerosed (NPS) glomerulus, and globally sclerosed (GS) glomerulus. According to the authors, the CNN model could accurately discriminate non-glomerular images from NPS and GS images with an accuracy of 92.67%.

Pesce et al. [15] designed, tested, and compared two artificial neural networks (ANN) classifiers. The former implements a shallow ANN classifying hand-crafted features extracted from Regions of Interest (ROIs) employing image-processing procedures. The latter, instead, employs the IBM Watson Visual Recognition System, which uses a deep artificial neural network to make decisions taking the images as input. The input dataset consisted of 428 sclerotic glomeruli and 2,344 non-sclerotic glomeruli derived from images of kidney biopsies scanned by the Aperio ScanScope System. According to the authors, the two approaches allowed accurately distinguishing between sclerotic and non-sclerotic glomeruli with a mean accuracy of 99%.

A common factor among the works that distinguish between sclerotic and non-sclerotic glomeruli was the creation of models based on deep learning. On the other hand, several researchers in medical image processing demonstrate that, despite the proposal of numerous CNNs, for implementation in a natural system to aid the diagnosis, the consolidated architectures are the ones that tend to present better effectiveness. The authors achieved promising results with consolidated state-of-the-art CNNs, such as VGGs, DenseNet, ResNet, and Inception. In this way, we investigated the use of five well-known CNNs, evaluated the shallow and deep fine-tuning, and proposed modifications in the fully connected layers.

3   Materials and Methods

This work presents a method based on modified CNN architectures to detect glomerulosclerosis in renal biopsy images. To reach the proposed solution, we used the methodology shown in Fig. 1. As an initial step, we select pre-trained CNN and extract the convolutional layers with their weights. Then, we add new fully connected layers and perform the fine-tuning in two steps: in shallow fine-tuning (SFT), we freeze the convolutional weights and train only the new layers; in Deep Fine-Tuning (DFT), we retrain the entire CNN. After that, we conduct the performance analyses and choose the proposed method by analyzing the performances obtained. The following subsections describe the proposed methodology, the image dataset, and the techniques and metrics adopted to assess the solution.

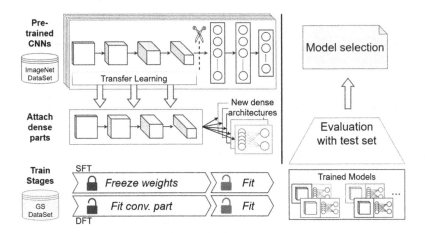

Fig. 1. Summary of the experimental set. We took the convolutional layers of pre-trained CNNs and added new dense architectures. Then we fine-tuned the models in two stages, SFT (train only the dense part) and DFT (train the entire network) with our dataset. Finally, we evaluate the trained models performance.

3.1   Proposed Methodology

Following the methodology illustrated in Fig. 1 and analyzing the achieved results with each CNNs, we reached the proposed approach shown in Fig. 2.

The input image goes through a pre-processing step, aiming to adapt its dimensions and pixel values range to the CNN's inputs. The image is classified by a DenseNet architecture fine-tuned to perform this task.

3.2   Image Dataset

The dataset contains 1,062 optical microscopy images of renal biopsies with one glomerulus per image. Biological material samples were treated with three chemical stains: Hematoxylin-Eosin (HE), Periodic Acid-Schiff (PAS), and Periodic Schiff-Methenamine Silver (PASM).

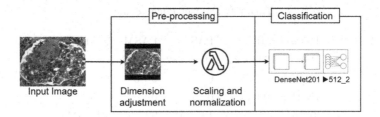

Fig. 2. Proposed methodology. The image undergoes dimensional and scale adjustments, after which it goes to CNN model for final classification.

We used images from three different datasets. Most of them are from the dataset (here called PSK1) built by pathologists from the Gonçalo Moniz Institute from Oswaldo Cruz Foundation, and made available in the PathoSpotter[2] project scope. The others were extracted from the DME and INetDB databases found in the work of Santos et al. [17].

In Fig. 3 we show dataset samples. We verified that the tissue is complex with various colors (mainly due to staining), textures, and glomerulus shapes. In addition, there are visually similar images belonging to different classes and vice versa. Such characteristics make the classification task more challenging.

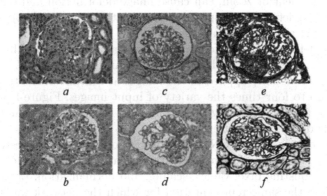

Fig. 3. Samples of images from the used dataset. *a* and *b*: HE stained; *c* and *d*: PAS stained; *e* and *f*: PASM stained. The first row has glomeruli with glomerulosclerosis and the second row has healthy glomeruli.

The images from the healthy class were randomly selected from a larger dataset provided by the authors of the PSK1 dataset. We selected images that contained a single glomerulus and were treated with the same stain as sample images with glomerulosclerosis. It was also sought to balance classes in the number of samples. Table 1 shows the distribution of the final number of images by staining and class.

---

[2] https://pathospotter.bahia.fiocruz.br.

Table 1. Dataset image count.

Dataset	HE	PAS	PASM	Total
PSK1	244 | 240	195 | 182	92 | 92	531 | 514
DME	0 | 3	0 | 6	0 | 0	0 | 9
INetDB	0 | 1	0 | 7	0 | 0	0 | 8
Total	244 | 244	195 | 195	92 | 92	531 | 531

Each pair of values means healthy and Glomeru-losclerosis image count respectively.

3.3 Pre-processing and Data Augmentation

The input dimensions (height and width) of the CNNs evaluated are different in absolute values and proportional terms from the original image's dimensions. Therefore, we added black borders (padding) to the images to make their appearance square before resizing. This strategy avoids deformations in the format of the objects present in the image and an eventual loss of data when performing procedures to cut specific regions.

To increase the number of images to improve the generalization performance [19], we apply transformations to the training input images. Some operations, such as shear or zoom, can cause image deformation and data loss. We decided, then, to define simple spatial transformations that can happen naturally when the expert analyzes the sample.

Thus, we defined two transformations: vertical and horizontal flip, since, for the present dataset, they are label-preserving transformations [19]. These transformations are randomly applied to the images at runtime of the training, multiplying by up to four times the variety of input images. Figure 4 presents one image sample after the padding operation (from pre-processing) and shows the results of the data-augmentation operations.

Before being taken to the CNNs, the images undergo a pre-processing process, where specific operations for each architecture are performed. Such operations may include adjusting pixel values, scaling, and normalization. In this way, the images acquire the same representation for which the network was pre-trained in ImageNet.

3.4 Evaluated Convolutional Neural Networks

CNNs have been widely used in machine learning, especially in medical imaging. With their deep architectures, CNNs can map image features at different levels of abstraction and have been successfully used in the development of medical diagnostics tools, often surpassing in accuracy conventional feature extraction methods [22].

The typical architecture of a CNN can be divided into two parts: (1) in the convolutional part, there are convolution operations with matrix weight filters,

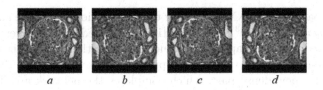

Fig. 4. Pre-processing and data augmentation operations. a is an image with padding addiction; b, c and d are generated from a by data-agumentation transformations, respectively vertical flip, horizontal flip and in both directions.

which extract feature maps, and pooling operations, which reduce the dimensionality of the maps, preserving the more essential features. After a succession of these operations, the map features are taken to the dense part (2), similar to a fully connected multi-layer perceptron, producing the desired output (classification). Strictly speaking, there are other topologies with residual flows and normalization layers, for example. For convenience, we will call the initial part of the network up to the point where the maps are linearized as convolutional.

We use pre-trained CNNs on the ImageNet [16] dataset, which contains over 1.2 million images and 1,000 classes. We only took the convolutional part (with the weights) of five architectures already established in the literature and with high generalization capacity: VGG-19 [20], Inception-V3 [21], ResNet-50 [6], DenseNet-201 [7] and EfficientNet-B2 [23]. Table 2 provides details of these networks. According to Kornblith et al. [11], the better the architecture performs on the ImageNet dataset, the better the transfer to other natural image datasets.

Table 2. Evaluated convolutional architectures summary.

Name	Abbr.	Parameters (original)	Parameters (without dense)	Top-5 acc (ImageNet)
VGG-19	v19	143,667,240	20,024,384	90.0%
Inception-V3	iv3	23,851,784	21,802,784	93.7%
Resnet-50	rsn	25,636,712	23,587,712	92.1%
DenseNet-201	dsn	20,242,984	18,321,984	93.6%
EfficientNet-B2	ef2	9,177,569	7,768,569	94.9%

3.5   Transfer Learning and Fine-Tunning

Training a CNN from scratch is computationally expensive and requires a large amount of training data to achieve good generalization power. An alternative is the use of transfer of learning (TL) techniques, which allow reusing knowledge learned in one field and applying it in another related field [22].

We took each of the CNNs, and after the convolutional part, we added a layer of *Global Average Pooling*[3]. The operation performed on this layer synthesizes each generated feature map into a value based on the average, in this way it produces one-dimensional vectors to supply the dense part. After this layer, we concatenated a new dense part, which completes the network architecture and provides the classification.

For the dense part, we evaluated five architectural alternatives. The simplest of them only consists of a classification layer with two neurons. The other four include a hidden layer containing 64, 256, 512, or 1024 neurons. We found that adding more hidden layers did not bring benefits in initial tests. We will use the abbreviations from Table 2 followed by the representation of its dense layers, separated by "_" to name a specific architecture. Ex. rsn_64_2 represents the Resnet-50 with a hidden layer of 64 neurons and the output layer with two neurons.

The training strategy occurred in two stages. We applied shallow fine-tuning (SFT) in the first step, where only the dense layers were trained. Moreover, in the second step, deep fine-tuning (DFT), the entire network is trained at a very low learning rate.

The network training process was followed and monitored through the loss function during 80 epochs in the SFT stage, as illustrated in Fig. 5. For the second training stage, each network was taken when it presented the lowest loss in the validation set, and the training process continued from this state for another 80 epochs or until stagnation or an increase in loss for 20 consecutive epochs (early stopping) was observed. We used the RMSprop algorithm optimizer [24], a learning rate of 10^{-3} in SFT and 10^{-6} in the DFT, and the categorical crossentropy as the loss function.

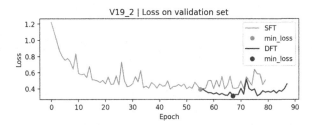

Fig. 5. Loss monitoring on the validation set during the training process. Dot marks show epoch with minimal loss. DFT stage starts over the green dot marked epoch. (Color figure online)

3.6 Evaluation Metrics

We employed the stratified k-fold cross-validation. This technique randomly distributed the dataset instances into k mutually exclusive subsets (folds) of approx-

---

[3] https://keras.io/api/layers/pooling_layers/global_average_pooling2d/.

imately equal size and in the same proportion observed in the original dataset. In this way, the CNN is fine-tuned and tested k times. Each round uses a different subset for evaluation, and the remaining $k - 1$ subsets are used for fine-tuning.

We splited the image dataset into five folds ($k = 5$), so 20% of the instances make up the test subset, which is not used during the training of the network, but in the evaluation of the final performance of the classifiers. The images from the remaining folds were splited into two subsets: the training subset (70% from total), and the network weights are adjusted based on this subset's loss. Furthermore, the validation subset (10% from total) was used to monitor training and detect overfitting.

The confusion matrix confronts the predicted and actual results for the same set of tests. There are four values in this matrix: the true positive (TP), which indicates the number of images correctly classified as GS; the true negative (TN), corresponding to the number of correct healthy classifications; the false positive (FP), representing the number of images classified as GS, but which are healthy; and finally, the false negative (FN), which refers to the number of images erroneously classified as healthy. From the confusion matrix, we evaluated the classification performance in the test set by the mean values of four metrics: accuracy (A), precision (P), recall (R), and kappa (K) [3].

The kappa (Eq. 1) allows measuring the degree of agreement between the classifier and the expert. We used kappa as the main evaluation metric because it is more challenging than accuracy, as it takes into account the expected probability (P_e) of the evaluators agreeing on the classification. This metric is also calculated based on the entire confusion matrix, i.e., correct and wrong results of both classes are considered.

$$K = \frac{A - P_e}{1 - P_e}. \tag{1}$$

The maximum kappa value (100%) indicates the perfect agreement among the evaluators; other ranges of values are categorized according to to the k value as follows: k $\leq$ 20: Bad; 20 < k $\leq$ 40: Fair; 40 < k $\leq$ 60: Good; 60 < k $\leq$ 80: Very Good and k > 80: Excellent [13].

4   Results and Discussion

Here we detail the individual results obtained with the 25 CNNs architectures evaluated. The metrics presented were obtained in the test subsets and represent the average of the values measured in the five folds of the cross-validation.

Table 3 presents the results observed in architectures based on VGG-19. Comparing the results obtained in the training stages, we verified that the continuity of training to adjust the weights of the convolutional layers (at the DFT stage) had positive effects on all architectures based on VGG-19.

The topology of VGG-19 is simple (sequential type), and compared to the other architectures evaluated, it originally had more parameters in the dense part (Table 2). We believe that when replacing the original dense part with a much

Table 3. Results from VGG-19 based architectures.

	Architecture	A (%)	P (%)	R (%)	K (%)
SFT	v19_2	83.6 ±4.5	85.1 ±5.0	81.9 ±9.0	67.2 ±9.0
	v19_64_2	86.2 ±4.1	**86.7 ±4.9**	85.7 ±6.0	72.3 ±8.2
	v19_256_2	**86.4 ±5.0**	85.2 ±6.1	**88.5 ±4.5**	72.9 ±10.1
	v19_512_2	86.3 ±4.9	85.6 ±6.0	87.7 ±5.5	72.7 ±9.7
	v19_1024_2	85.0 ±4.6	83.3 ±4.6	87.8 ±5.3	70.0 ±9.3
DFT	v19_2	88.5 ±3.7	88.8 ±5.8	88.7 ±6.3	77.0 ±7.4
	v19_64_2	**90.9 ±3.3**	90.0 ±3.6	**92.1 ±4.3**	**81.7 ±6.5**
	v19_256_2	**90.9 ±2.4**	**90.2 ±2.1**	91.7 ±3.5	**81.7 ±4.9**
	v19_512_2	89.3 ±3.7	89.4 ±3.6	89.4 ±6.5	78.7 ±7.4
	v19_1024_2	89.2 ±4.0	88.5 ±3.8	90.0 ±5.4	78.3 ±8.0

- Bold are the best values.

smaller one with fewer layers and parameters, it was not able to perform well the necessary transformations in the attributes received from the convolutional layer. So the general adjustment of the weights (at the DFT) made possible a better accord between the convolutional and dense parts in the network and allowed more adaptation to the new data.

Table 4 presents the results observed in architectures based on Inception-V3. The original architecture of this CNN contains only one dense layer (the output layer), which shows that the convolutional part performs almost all the processing and already delivers a simpler task to the final layers. The DFT also positively affected the results in Inception-V3, but the improvement was smaller than that observed in VGG-19.

Table 4. Results from Inception-V3 based architectures.

	Architecture	A (%)	P (%)	R (%)	K (%)
SFT	iv3_2	89.2 ±3.1	88.4 ±3.6	**90.2 ±2.7**	78.4 ±6.2
	iv3_64_2	**89.5 ±1.8**	88.9 ±1.9	**90.2 ±4.1**	**78.9 ±3.6**
	iv3_256_2	**89.5 ±1.2**	**89.6 ±1.7**	89.3 ±1.5	**78.9 ±2.5**
	iv3_512_2	88.9 ±2.8	88.6 ±3.3	89.3 ±2.7	77.8 ±5.6
	iv3_1024_2	88.8 ±2.2	88.0 ±2.3	89.8 ±3.4	77.6 ±4.4
DFT	iv3_2	90.0 ±2.1	88.7 ±3.7	**91.9 ±1.1**	80.0 ±4.1
	iv3_64_2	90.5 ±2.2	89.9 ±3.3	91.3 ±1.2	81.0 ±4.3
	iv3_256_2	89.9 ±1.7	90.0 ±2.2	89.8 ±2.3	79.8 ±3.5
	iv3_512_2	90.3 ±1.8	90.4 ±2.7	90.2 ±1.1	80.6 ±3.7
	iv3_1024_2	**90.8 ±1.5**	**90.9 ±3.3**	90.8 ±1.3	**81.5 ±3.0**

- Bold are the best values.

Table 5 presents the results observed in architectures based on Resnet-50. The continuity of training in the DFT, in some scenarios, had diffuse and small-magnitude effects on the measured metrics. We verified the early stagnation of the loss (in the validation set) in the ninth epoch on average, so minor effects are expected, considering the low learning rate inherent to the DFT stage. The rsn_256_2 architecture achieved the best kappa, reaching 82.9%.

Table 5. Results from Resnet-50 based architectures.

	Architecture	A (%)	P (%)	R (%)	K (%)
SFT	rsn_2	89.3 ±2.3	89.2 ±2.3	89.5 ±4.6	78.5 ±4.6
	rsn_64_2	90.6 ±2.2	91.0 ±2.9	90.2 ±2.5	81.2 ±4.4
	rsn_256_2	90.3 ±2.8	90.6 ±4.2	90.2 ±4.7	80.6 ±5.5
	rsn_512_2	**91.1 ±2.6**	91.1 ±4.5	**91.3 ±4.2**	**82.1 ±5.2**
	rsn_1024_2	90.3 ±2.0	**92.1 ±3.0**	88.3 ±5.6	80.6 ±3.9
DFT	rsn_2	89.8 ±3.2	90.0 ±3.2	89.6 ±4.9	79.7 ±6.4
	rsn_64_2	91.0 ±2.3	**91.0 ±2.8**	91.0 ±3.5	81.9 ±4.6
	rsn_256_2	**91.4 ±2.2**	90.3 ±3.5	**93.0 ±2.1**	**82.9 ±4.3**
	rsn_512_2	90.5 ±2.3	89.6 ±2.7	91.7 ±2.3	81.0 ±4.6
	rsn_1024_2	90.9 ±1.9	90.3 ±2.8	91.7 ±4.2	81.7 ±3.8

- Bold are the best values.

Table 6 presents the results observed in architectures based on DenseNet-201. Similar to Inception-V3, we see a positive impact generated by the DFT; all metrics improved after DFT training stage. We also verify the DFT duration in training epochs, the loss stagnation occurred at the 17th epoch on average. The dsn_512_2 architecture got the best results at DFT, reaching 85.3% of kappa.

Table 6. Results from DenseNet-201 based architectures.

	Architecture	A (%)	P (%)	R (%)	K (%)
SFT	dsn_2	**91.9 ±2.0**	91.8 ±1.8	92.1 ±2.8	**83.8 ±4.1**
	dsn_64_2	**91.9 ±2.0**	92.1 ±1.4	91.7 ±3.5	**83.8 ±4.0**
	dsn_256_2	90.8 ±1.8	91.4 ±3.2	90.2 ±1.1	81.6 ±3.6
	dsn_512_2	91.1 ±2.9	90.8 ±4.1	91.5 ±2.5	82.1 ±5.8
	dsn_1024_2	90.6 ±2.5	90.4 ±3.7	91.0 ±2.4	81.2 ±5.0
DFT	dsn_2	92.4 ±2.4	92.2 ±3.0	92.6 ±2.7	84.7 ±4.9
	dsn_64_2	92.6 ±1.9	92.6 ±2.1	92.5 ±1.9	85.1 ±3.9
	dsn_256_2	92.3 ±1.9	92.5 ±2.4	92.1 ±1.3	84.6 ±3.8
	dsn_512_2	**92.7 ±2.7**	92.6 ±3.5	**92.8 ±2.2**	**85.3 ±5.3**
	dsn_1024_2	92.6 ±2.7	**92.7 ±3.0**	92.5 ±2.3	85.1 ±5.4

- Bold are the best values.

Table 7 presents the results observed in the architectures based on EfficientNet-B2. Training this network in the DFT stage promoted accuracy, Precision, and Kappa improvements. However, the recall was reduced, which indicates that the DFT decreased the classifier's ability to find the positive samples.

The part used in TL (convolutional only) from the EfficientNet-B2 is about 1/3 of the other networks, considering the number of parameters. It indicates that their few weights were well refined to the original problem (ImageNet). So, in this case, the DFT fulfilled its function more intensely. It is still worth reporting that the networks based on EfficientNet-B2 had a more extended training at the DFT stage (the loss stagnation happened about the 42th epoch in average).

Table 7. Results from EfficientNet-B2 based architectures.

	Architecture	A (%)	P (%)	R (%)	K (%)
SFT	ef2_2	90.0 ±2.2	89.1 ±2.0	91.1 ±2.6	80.0 ±4.5
	ef2_64_2	90.5 ±1.7	89.3 ±2.8	**92.1 ±2.2**	81.0 ±3.5
	ef2_256_2	89.8 ±2.2	88.5 ±3.7	91.7 ±2.3	79.7 ±4.5
	ef2_512_2	**90.6 ±2.1**	**89.5 ±3.4**	**92.1 ±2.2**	**81.2 ±4.2**
	ef2_1024_2	89.7 ±2.7	88.2 ±3.7	91.9 ±3.2	79.5 ±5.4
DFT	ef2_2	**91.6 ±1.6**	**92.3 ±1.7**	90.8 ±2.0	**83.2 ±3.2**
	ef2_64_2	90.9 ±2.1	91.8 ±3.3	89.8 ±2.1	81.7 ±4.2
	ef2_256_2	90.7 ±2.0	91.1 ±1.9	90.2 ±2.9	81.4 ±4.0
	ef2_512_2	90.9 ±1.5	91.5 ±1.8	90.2 ±3.4	81.7 ±3.1
	ef2_1024_2	89.9 ±2.1	90.0 ±2.3	89.8 ±2.7	79.9 ±4.1

- Bold are the best values.

We reach the best results using the Densenet-201 convolutional part, folowed by two dense layers having 512 and 2 neurons (dsn_512_2). The accuracy achieved was 92.7%, and the kappa was 85.3%, which characterizes almost perfect agreement with the classification performed by the specialist. The other evaluation metrics also reached similar levels: The precision of 92.6% and the recall of 92.8% point to low FP and FN ratings, respectively. The graph in Fig. 6 allows comparing the overall performance of the best of each CNN based architecture.

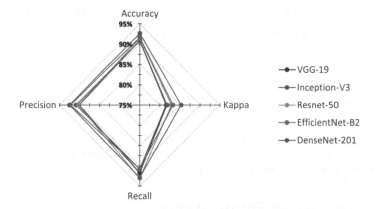

Fig. 6. Performance comparison between each best CNN-based architecture.

Figure 7 presents the dsn_512_2 predictions for some images from the test set; we include in the figure the percentage of each category from the confusion matrix (TP, TN, FP, and FN) separately for each staining. Considering the images treated with HE, for example, in the True Positive quadrant we have $TP_{HE}/N_{HE} = 47.5\%$, and in the True Negative we have $TN_{HE}/N_{HE} = 46.1\%$, totaling 93.6% success. Since there were 3.9% FP, and 2.5% of FN, 6.4% of errors completes 100% of the images of this staining. The accuracy among the images treated with PAS is 92.8%, in the PASM stained images, 89.7%.

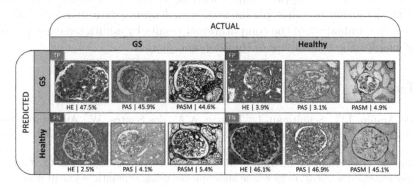

Fig. 7. Sample results from classifications made by dsn_512_2.

Figure 8 presents the accuracy and loss ratio of the training and validation sets over the dsn_512_2 training epochs. We can observe that, although there are some peaks, there is a relationship between the curves (training and validation), and there are no inflection points, which characterizes a good generalization capacity [19]. From the results, it is possible to conclude that there was no overfitting during training. We attribute this fact to the decrease in complexity provided by the reduction in fully connected layers and the application of data augmentation techniques.

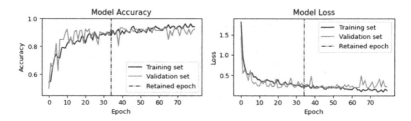

Fig. 8. Accuracy and loss of DenseNet based-model in training and validation sets by epochs of SFT stage. These plots show no inflection point, and accuracy increases for both sets together, indicating no overfitting.

5   Conclusion and Future Works

This work presented a DenseNet-based CNN architecture and training strategy to identify glomerulus with sclerosis. Several architectures, fine-tuning schemes, and other convolutional architectures were studied to define the proposed model. The experiments showed that the deep fine-tuning was more efficient than the shallow fine-tuning.

The results obtained were promising, but they can be improved. Future work may also investigate the use of generative adversarial networks in increasing data availability; notably, these networks can generate heterogeneous images that adequately represent the original distribution. Furthermore, we also intend to improve the models by applying meta-learning to adjust the architecture design and the definition of other parameters, e.g., learning rate and loss function. Finally, the evaluation of the computational results by additional experts would be crucial for the routine use of the proposed model.

References

1. de Araújo, I.C., Schnitman, L., Duarte, A.A., dos Santos, W.: Automated detection of segmental glomerulosclerosis in kidney histopathology. In: XIII Brazilian Congress on Computational Intelligence, p. 12 (2017)
2. Bueno, G., Fernandez-Carrobles, M.M., Gonzalez-Lopez, L., Deniz, O.: Glomerulosclerosis identification in whole slide images using semantic segmentation. Comput. Methods Programs Biomed. **184**, 105273 (2020)
3. Cohen, J.: A coefficient of agreement for nominal scales. Educ. Psychol. Measur. **20**, 37–46 (1960)
4. Dhaun, N., Bellamy, C., Cattran, D., Kluth, D.: Utility of renal biopsy in the clinical management of renal disease: hematuria should not be missed reply. Kidney Int. **86**(6), 1269–1269 (2014)
5. Ginley, B., et al.: Computational segmentation and classification of diabetic glomerulosclerosis. J. Am. Soc. Nephrol.: JASN **30**(10), 1953-1967(2019)
6. He, K., Zhang, X., Ren, S., Sun, J.: Deep residual learning for image recognition. In: Proceedings of the IEEE Conference on Computer Vision and Pattern Recognition, pp. 770–778 (2015)

7. Huang, G., Liu, Z., Weinberger, K.Q.: Densely connected convolutional networks. CoRR abs/1608.06993 (2016)
8. Huo, Y., Deng, R., Liu, Q., Fogo, A.B., Yang, H.: AI applications in renal pathology. Kidney Int. **99**(6), 1309–1320 (2021)
9. Kannan, S., et al.: Segmentation of glomeruli within trichrome images using deep learning. Kidney Int. Rep. **4**(7), 955–962 (2019)
10. Kolachalama, V.B.: Association of pathological fibrosis with renal survival using deep neural networks. Kidney Int. Rep. **3**(2), 464–475 (2018)
11. Kornblith, S., Shlens, J., Le, Q.V.: Do better imagenet models transfer better? In: Proceedings of the IEEE/CVF Conference on Computer Vision and Pattern Recognition, pp. 2661–2671 (2019)
12. van der Laak, J., Litjens, G., Ciompi, F.: Deep learning in histopathology: the path to the clinic. Nat. Med. **27**(5), 775–784 (2021)
13. Landis, J.R., Koch, G.G.: The measurement of observer agreement for categorical data. Biometrics, 159–174 (1977)
14. Marsh, J.N., et al.: Deep learning global glomerulosclerosis in transplant kidney frozen sections. IEEE Trans. Med. Imag. **37**(12), 2718–2728 (2018)
15. Pesce, F., et al.: Identification of glomerulosclerosis using IBM Watson and shallow neural networks. J. Nephrol. **35**(4), 1235–1242 (2022). https://doi.org/10.1007/s40620-021-01200-0
16. Russakovsky, O., et al.: ImageNet large scale visual recognition challenge. Int. J. Comput. Vision **115**, 211–252 (2015)
17. Santos, J.D., et al.: A hybrid of deep and textural features to differentiate glomerulosclerosis and minimal change disease from glomerulus biopsy images. Biomed. Signal Process. Control **70**, 103020 (2021)
18. Sheehan, S.M., Korstanje, R.: Automatic glomerular identification and quantification of histological phenotypes using image analysis and machine learning. Am. J. Physiol.-Ren. Physiol. **315**(6), F1644–F1651 (2018)
19. Shorten, C., Khoshgoftaar, T.M.: A survey on image data augmentation for deep learning. J. Big Data **6**(1), 1–48 (2019)
20. Simonyan, K., Zisserman, A.: Very deep convolutional networks for large-scale image recognition. In: Bengio, Y., LeCun, Y. (eds.) 3rd International Conference on Learning Representations, ICLR 2015, San Diego, CA, USA, 7–9 May 2015, Conference Track Proceedings (2015)
21. Szegedy, C., Vanhoucke, V., Ioffe, S., Shlens, J., Wojna, Z.: Rethinking the inception architecture for computer vision. In: Proceedings of the IEEE Conference on Computer Vision and Pattern Recognition (CVPR), pp. 2818–2826 (2016)
22. Tajbakhsh, N., et al.: Convolutional neural networks for medical image analysis: full training or fine tuning? IEEE Trans. Med. Imaging **35**, 1299–1312 (2016)
23. Tan, M., Le, Q.V.: EfficientNet: rethinking model scaling for convolutional neural networks. In: International Conference on Machine Learning (2019)
24. Tieleman, T., Hinton, G., et al.: Lecture 6.5-rmsprop: Divide the gradient by a running average of its recent magnitude. COURSERA: Neural Netw. Mach. Learn. **4**, 26–31 (2012)
25. Yi, T.W., et al.: Digital health and artificial intelligence in kidney research: a report from the 2020 Kidney Disease Clinical Trialists (KDCT) meeting. Nephrol. Dial. Transplant. **37**(4), 620–627 (2021)
26. Yu, H., Yang, L.T., Zhang, Q., Armstrong, D., Deen, M.J.: Convolutional neural networks for medical image analysis: State-of-the-art, comparisons, improvement and perspectives. Neurocomputing **444**, 92–110 (2021)

27. Zheng, Y., et al.: Deep-learning-driven quantification of interstitial fibrosis in digitized kidney biopsies. Am. J. Pathol. **191**(8), 1442–1453 (2021)
28. Zheng, Z., et al.: Deep learning-based artificial intelligence system for automatic assessment of glomerular pathological findings in lupus nephritis. Diagnostics **11**(11) (2021)

Diffusion-Based Approach to Style Modeling in Expressive TTS

Leonardo B. de M. M. Marques[1]([✉]), Lucas H. Ueda[1,2], Flávio O. Simões[2],
Mário Uliani Neto[2], Fernando O. Runstein[2], Edson J. Nagle[2], Bianca Dal Bó[2],
and Paula D. P. Costa[1]

[1] Department of Computer Engineering and Automation, School of Electrical and
Computer Engineering, University of Campinas (UNICAMP), Campinas, Brazil
1218479@dac.unicamp.br, paulad@unicamp.br
[2] CPQD, Campinas, Brazil
{lhueda,simoes,uliani,runstein,nagle,bdalbo}@cpqd.com.br

Abstract. In this article, we propose an aggregation of denoising diffusion probabilistic models (DDPMs) onto an end-to-end text-to-speech system to learn a distribution of reference speaking styles in an unsupervised manner. By applying a few steps of a forward noising process to an embedding extracted from a reference mel spectrogram, we make profit of its information to reduce the diffusion chain and reconstruct an improved style embedding with only a few reverse steps, performing style transfer. Additionally, a proposed combination of spectrogram reconstruction and denoising losses allows for conditioning of the acoustic model on the synthesized style embeddings. A subjective perceptual evaluation is conducted to evaluate naturalness and style transfer capability of the proposed approach. The results show a 5-point increment on the mean of naturalness ratings and a preference of the raters (43%) of our proposed approach over state-of-the-art models (29%) in the style transfer scenario.

Keywords: Expressive speech synthesis · Style modeling · Diffusion models

1 Introduction

Among the techniques used in text-to-speech (TTS) systems, the neural approach has received much attention due to its state-of-the-art, natural-sounding generated speech that is almost indistinguishable from humans' [11,22]. However, besides naturalness, one crucial factor that still differs human speech from synthesized speech is its expressiveness. Current TTS systems only learn an average of the prosody distribution, producing a monotonous and tedious speech [8]. In this context, speech prosody, which is the information in speech not conveyed through its phonetic content, becomes determinant to make speech spontaneous [17].

Prosody can be defined as "the variation in speech signals that remains after accounting for variation due to phonetics, speaker identity, and channel

© The Author(s), under exclusive license to Springer Nature Switzerland AG 2022
J. C. Xavier-Junior and R. A. Rios (Eds.): BRACIS 2022, LNAI 13653, pp. 253–267, 2022.
https://doi.org/10.1007/978-3-031-21686-2_18

effects" [23]. Prosody is also viewed as a confluence of several factors: linguistic (modality, discourse, semantics, syntactics), paralinguistic (attitude, emotion, pragmatics) and extra-linguistic (physiological, idiolectal, geographical, sociological, situational and temporal aspects) [20], that influence the interplay between low-level acoustic characteristics, such as pitch, stress, breaks, rhythm, etc. [25]. Hence, it is an arduous task to develop explicit labels for prosody.

Speaking style, on the other hand, whilst being an ill-defined concept, can be intuitively understood as higher-level affective characteristics of speech, such as emotional valence ("good"-ness/"bad"-ness) and arousal (excitement level) [25]. Even though the speaking styles, referred here as "styles", influence prosody directly, they can provide a better comprehension of expressiveness in speech.

In the context of TTS, style modeling has been explored in a broader perspective when compared to the traditional approach of categorical emotion labels, such as the "Big-Six" emotions [6]. A few examples of modeled styles are: narrative, mean, whispering and depressive in the context of storytelling [27].

This alternative method aims at providing artificial socially interactive agents with more common and natural styles of communication that are typically observed among humans in their daily interactions. Expressive speech synthesis is a key technology to equip avatars and social robots to inform empathy in assistance tasks and show typical human social behaviors in applications such as cognitive therapy for children, elderly care, and learning tutoring [2,10]. Also, through expressive speech, voice cloning systems can create personalized voice assistants and empower individuals with speech loss [19].

Most of the successful work present in literature attempts to model style via a latent representation, which was shown to be able to modify the synthesized speech's prosody to comprise the target style without the need of explicitly controlling the acoustic parameters [29]. In these architectures on which the varying information is modeled implicitly in an unsupervised fashion, an style encoder neural network is the module responsible for extracting the style embedding, that is itself added or concatenated to condition a regular TTS acoustic model (generates Mel-spectrograms from text). On the context of implicit models, in which this work is inserted, there exists several approaches to style modeling using mostly different classes of deep generative models, such as Generative Adversarial Networks (GANs) [18], Variational Autoencoders (VAEs) [35] and Flow-based models [1].

When considering state-of-the-art generative neural networks, diffusion models [24] have been addressed a lot in recent research. This is largely due to the report that these models exhibited state-of-the-art performance by beating GANs on the image synthesis task, both on image sample quality, and in the Fréchet Inception Distance (FID) score [5]. The diffusion model consists in a forward diffusion process that converts any complex distribution into a simple and tractable distribution (Gaussian). And then in order to generate samples, the reverse process is learned, taking the Gaussian distribution from the data distribution with several denoising steps [24].

Diffusion models have already been exploited in the context of TTS. There are several approaches to a vocoder model, which converts the Mel-spectrograms generated by a TTS acoustic model into the audio speech waveforms [3,4,14]. There has also been approaches to the acoustic model itself using diffusion models, as in [11,15]. A notable property of these models is their capability of acquiring high quality audio in the case of vocoders, and high mean opinion scores (MOS) for the neural acoustic models, with the drawback of being slow, due its reverse process that requires several model passes. As a consequence, a lot of research is being conducted on how to speed up sampling in diffusion models [13].

In spite of that, to the best of the authors' knowledge, diffusion models have not been exploited in the case of style modeling. From this perspective, we explore inserting these models into a style encoder, hypothesizing that the generative power of these models can learn to reconstruct samples that better condition the acoustic model to induce the desired style.

Our contributions are the following:

1. We introduce a new technique of modeling style in TTS with a shallow diffusion mechanism, that allows the use of a reference spectrogram as input for style transfer (rather than starting from a noise distribution), and a smaller diffusion chain, speeding up training and inference.
2. We propose a new manner of training diffusion models: jointly optimizing the denoiser with a diffusion and spectrogram reconstruction losses. By going through the whole reverse diffusion chain, instead of only a single denoising step during training, we are able to generate style embeddings during training that condition the acoustic model.
3. We conduct the experiments in an expressive dataset in Brazilian Portuguese, a low resource language.

This work is divided as follows: Sect. 2 compiles the most noted works in neural TTS style modeling and the current trends; Sect. 3 details the theoretical background of the diffusion models; Sect. 4 details the proposed model architecture and the procedures of training and inference; Sect. 5 describes in detail both the setup and the experiments performed; Sect. 6 contains a brief discussion concerning the results obtained and Sect. 7 ends with the conclusion and some future work to be done.

2   Related Works

In terms of technique, there are several relevant approaches in the literature on how to model either prosody or style. In our work we consider solely approaches that use a reference audio to transfer its style to the given text. This choice is due to the fact that theoretically any style/prosody of a given reference could be transferred. However, no style extrapolation/generalization is possible from those seen in training, and current techniques still have limited performance, meaning that the problem is still being strongly tackled.

- **Reference Encoder** [23]: The reference encoder consists of a neural network with six convolutional and batch normalization layers followed by a Gated Recurrent Unit (GRU) network. It was proposed to map a reference audio spectrogram to a fixed size embedding, the style embedding, which is shared across all decoding timesteps. With this, the authors aim to define a "prosody space". During training, the target audio is used as input to the reference encoder, driven by the spectrogram reconstruction loss, and on inference, any utterance could be used to perform prosodic transfer to the inserted text. One major drawback of this approach is that there is not a guarantee that all styles drawn from the space will be meaningful, since the space itself is not compact. Practically almost all works after this use the reference encoder to consider an audio input together with some other improvement.
- **Style Tokens** [31,32]: The attention-based style tokens approach was the first relevant for style modeling in neural TTS. It consists in a bank of embeddings learned in an unsupervised manner that could decompose the reference audio into latent interpretable factors, that combined in a weighted sum, outputted a style embedding. One notable drawback of this approach is the fact that, with the bank of embeddings, there is no predictable way to manipulate either prosody or style, since there is no direct relation between a token and a prosodic attribute. Also, the tokens could represent anything in particular, not only factors related to expressiveness, such as noise.
- **VAE** [35]: The VAE-based style encoder consists in a reference encoder, as detailed previously, to obtain a reference embedding which passes through two separate fully connected layers in order to predict the mean and standard deviation of a style embedding distribution (assumed Gaussian). The VAE style encoder is trained to maximize the variational lower bound, through a combination of the acoustic models' reconstruction loss and an annealed Kullback-Leibler (KL) divergence loss. Due to the insertion of the VAE after the reference encoder, new style embeddings can now be sampled by using the reparameterization trick. This way, the authors manage to sample meaningful embeddings more often and perform style control.
- **VAE+Flow** [28]: This model aims to improve the posterior distribution of the VAE style encoder: since it is modeled as a Gaussian with learned mean and diagonal covariance matrix, the approach is not flexible enough. Thus, Householder flows [28] were proposed to enrich the model. The flow starts with a simple Gaussian distribution and then applies a series of volume-preserving transformations to obtain a more flexible distribution for the variational posterior. Thus, [1] enriches the VAE-based style encoder with the householder flows. They reported better KL and reconstruction losses and an improvement on the perceived naturalness and expressiveness in the one-shot text to speech scenario, in which the model is finetuned with only one sample from a new unseen style.

All these techniques are based on an augmentation of the acoustic Tacotron 2 [22] model. There are also other approaches in the literature for style and prosody modeling rather than those mentioned, such as: hierarchical [26] and

fine-grained [12] approaches, that consider different levels of variation information (phone-level, word-level, etc.); approaches that use low-level prosodic reference information [33] instead of the mel-spectrograms, such as pitch, duration
and energy; and style strength quantization approaches [9].

3   Background

3.1   Denoising Diffusion Probabilistic Models

Diffusion models [24] are a type of generative model. They consist on a Markov
chain that destroys data by gradually adding noise (forward process) and then
learns the reverse process to generate a new sample from noise. The forward
trajectory taking the data $z_0 \sim q(z_0)$ to its latent noisy version is given by:

$$q(z_{1:T}|z_0) := \prod_{t=1}^{T} q(z_t|z_{t-1}) \tag{1}$$

in which T is the number of steps in the chain and q is the transition probability, modeled as a multivariate Gaussian according to a predefined (or learned)
variance schedule β_t:

$$q(z_t|z_{t-1}) := \mathcal{N}(z_t; \sqrt{1 - \beta_t} z_{t-1}, \beta_t I) \tag{2}$$

One property of this forward noising process is that it admits sampling z_t at an
arbitrary noise level t in closed form, given the original sample, according to:

$$q(z_t|z_0) = \mathcal{N}(z_t; \sqrt{\bar{\alpha}_t} z_0, (1 - \bar{\alpha}_t) I) \tag{3}$$

being $\alpha_t := 1 - \beta_t$ and $\bar{\alpha}_t := \prod_{s=1}^{t} \alpha_s$. Then, the noisy version z_t is obtained by
reparameterization, with $\epsilon \sim \mathcal{N}(0, I)$:

$$z_t(z_0, \epsilon) = \sqrt{\bar{\alpha}_t} z_0 + \sqrt{1 - \bar{\alpha}_t} \epsilon \tag{4}$$

Assuming T is large enough, $q(z_T)$ is nearly an isotropic Gaussian distribution [7], thus the reverse process is given by:

$$p_\theta(z_{0:T}) := p(z_T) \prod_{t=1}^{T} p_\theta(z_{t-1}|z_t) \tag{5}$$

in which $p(z_T) = \mathcal{N}(z_T; 0, I)$, and θ is the model's parameters. Each reverse
transition is normally intractable, so they are modeled as multivariate Gaussians
whose mean's and covariance's can be learned by a neural network:

$$p_\theta(z_{t-1}|z_t) := \mathcal{N}(z_{t-1}; \mu_\theta(z_t, t), \Sigma_\theta(z_t, t)) \tag{6}$$

On training, the negative log-likelihood is indirectly optimized through the Variational Lower Bound:

$$\mathbb{E}_q[-\log p_\theta(z_0)] \geq \mathbb{E}_q\left[-\log \frac{p_\theta(z_{0:T})}{q(z_{1:T}|z_0)}\right] := \mathbb{L} \tag{7}$$

$\mathbb{L}$ can be decomposed in combination of losses each corresponding to a time step. In this context, [7] showed that efficient training can be achieved by optimizing with stochastic gradient descent separate random terms of $\mathbb{L}$, each given by:

$$\mathbb{L}_{t-1} := D_{KL}(q(z_{t-1}|z_t, z_0)||p_\theta(z_{t-1}|z_t)) \tag{8}$$

Developing each loss term, fixing $\Sigma_\theta(z_t, t) = \sigma_t^2 I$, and choosing the neural network to predict the noise ϵ in z_t given t (thus denoising diffusion probabilistic models), we obtain the following objective:

$$\mathbb{E}_{z_0,\epsilon}\left[\frac{\beta_t^2}{2\sigma_t^2\alpha_t(1-\bar{\alpha}_t)}||\epsilon - \epsilon_\theta(\sqrt{\bar{\alpha}_t}z_0 + \sqrt{1-\bar{\alpha}_t}\epsilon, t)||^2\right] \tag{9}$$

With this parameterization, sampling $z_{t-1} \sim p_\theta(z_{t-1}|z_t)$ corresponds to:

$$z_{t-1} = \frac{1}{\sqrt{\alpha_t}}\left(z_t - \frac{\beta_t}{\sqrt{1-\bar{\alpha}_t}}\epsilon_\theta(z_t, t)\right) + \sigma_t\eta \tag{10}$$

in which $\eta \sim \mathcal{N}(0, I)$.

Finally, [7] found that the following simplified variant (ignoring the scaling term) of the objective was better for both simplification and sample quality:

$$\mathbb{L}_{simple} := \mathbb{E}_{t,z_0,\epsilon}\left[||\epsilon - \epsilon_\theta(\sqrt{\bar{\alpha}_t}z_0 + \sqrt{1-\bar{\alpha}_t}\epsilon, t)||^2\right] \tag{11}$$

3.2 Shallow Diffusion Mechanism

The shallow diffusion mechanism [15], instead of starting the reverse diffusion process from the noise Gaussian distribution, it takes profit of prior knowledge to assist the chain in the synthesis of a new sample. On inference, the authors take over-smoothed mel-spectrograms outputted by a simple decoder trained with L1 loss directly on mel-specs and apply k steps of noise. Then, the decoder is able to synthesize a sample by taking this noisy over-smoothed mel-spectrogram and performing the reverse denoising process with only k steps, in which $k < T$ (the diffusion chain size when starting from noise).

The noise level (number of noising steps) k is set when both ground-truth and over-smoothed mel-spectrograms are indistinguishable (both manifolds intersect), thus allowing the model to generate mel-spectrograms with rich details between neighboring harmonics whilst starting from the noisy over-smoothed version. This procedure was used in a TTS acoustic model and was found to accelerate inference and improve the quality of the synthesized audio.

4   Model

4.1   Model Architecture

Our architecture, shown in Fig. 1, consists of an acoustic TTS model, chosen to be a Tacotron 2 network with the same hyperparameters as described in [22], augmented with a style encoder module based on denoising diffusion probabilistic models. The style encoder consists of a reference encoder with the same hyperparameters described in [23], and a denoiser which, for generality, was configured as a sequence of five Feed-Forward Transformer blocks [21], although other architectures are possible as well.

Commonly, the diffusion-based generation starts by sampling random noise from a standard Gaussian distribution that goes through an iterative denoising process, resulting in the synthesis of a sample, as detailed in Eq. 5. However, since our objective is to transfer style from a reference mel-spectrogram to the desired input phoneme sequence, we apply a mechanism similar to the shallow diffusion process [15], described in Sect. 3.2, which allows the generation of the style embedding from the reference embedding. So, we use a diffusion chain with only up to 25 steps.

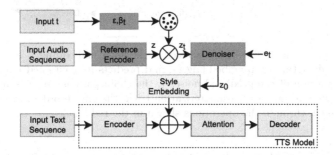

Fig. 1. Block diagram of the proposed architecture. The green blocks represent the required entries, the blue blocks compose the style encoder that generates the style embedding, used to condition the acoustic model, comprised by the gray blocks. (Color figure online)

4.2 Training and Inference

As shown in Fig. 1, in our training approach, the model receives the text sequence; the target mel-spectrogram, which is also passed as reference to the style encoder and converted to a reference embedding z by the reference encoder, and t, the parameter that controls number of noise-adding steps on the extracted reference embedding, taking it to the noise level z_t. After this noising process of the reference embedding, the reconstruction (reverse process) takes place.

In each denoising step, presented in Fig. 2, the denoiser takes a concatenation of the reference embedding at noise level t (obtained with the noising process), z_t, and a time-step embedding e_t, which is a Transformer-like sinusoidal positional embedding to indicate which step of the chain is being executed, and outputs the predicted noise contained in the reference embedding, ϵ_t. With these parameters combined and a reparameterization trick, the denoised embedding at noise level $t - 1$, z_{t-1}, is obtained.

To condition the acoustic model, we propose a new training procedure. Commonly, the diffusion model does not synthesize any sample during training, since each weight update is done only on isolated random steps of the diffusion chain, not requiring going through the whole reverse process. This would not enable to condition the Tacotron on the learned distribution given that no style embedding would be outputted during training.

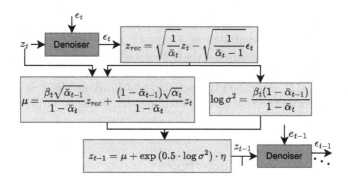

Fig. 2. A denoising step. The denoiser receives the noisy embedding z_t at noise level t, the correspondent time-step embedding e_t, and predicts its contained noise ϵ_t. Then, with a combination of the predicted noise ϵ_t, the chain parameters (α's and β's), the noise embedding z_t, and a reparameterization trick, the one-step denoised embedding z_{t-1} is obtained, through the operations shown in the red blocks.

To solve this problem, instead of only performing a single denoising step to z_{t-1}, shown in Fig. 2, we reconstruct the style embedding z_0, by going through the whole reverse chain (t denoising steps). The style embedding is then concatenated to the encoder states of the Tacotron 2 to condition the decoder with style information, as exhibited in Fig. 1.

This approach allows us to train the denoiser simultaneously with a combined loss composed with both standard simple diffusion loss presented in Eq. 11 between the noise levels t and $t - 1$ and also the Tacotron 2 loss, given that we are now able to generated a style embedding and concatenate it to the encoder outputs. The extra denoising steps needed could lead to slower training, but this is compensated by the shallow diffusion mechanism since it allows a smaller diffusion chain. Our loss is thus given as:

$$\mathcal{L}_{comb} = \mathbb{L}_{simple} + \mathbb{L}_{Mel} + \mathbb{L}_{Gate}, \tag{12}$$

in which the mel loss is an MSE between the outputs of the Tacotron 2 pre-net and post-net with the target spectrogram (reconstruction loss); the gate loss is a simple BCE (generation stop loss). Both the acoustic model and the reference encoder are updated with the gradients from these losses.

The denoiser is also trained with the mel and gate loss, whose gradients update the model t times through all denoising steps in a single backward pass, and additionally the simple diffusion loss solely between the steps t and $t - 1$. So, the denoiser learns simultaneously how to reconstruct the style embedding such that it improves the mel reconstruction, and also the reverse conditional distribution of the diffusion process to synthesize style embeddings $(q(z_{t-1}|z_t, z_0) \approx p_\theta(z_{t-1}|z_t) := \mathcal{N}(z_{t-1}; \mu_\theta(z_t, t), \Sigma_\theta(z_t, t)))$.

During inference, the model is fed with the text sequence, a reference spectrogram whose style is to be captured, and an input t corresponding to the noise level that the reference embedding will be submitted to and reconstructed from. We hypothesize that, the smaller the t, the more the output style will be similar to the input style, and the greater the t the more information of the reference embedding will be lost to noise and a more different style will be generated.

While previous approaches such as the VAE and the VAE+Flow based style encoders focus directly on constructing a disentangled space of styles, diffusion models' latent variables are naturally defined on noisy spaces. Thus, the disentangling concept cannot be easily applied. However, in this specific case, our latent variables contains prior information from the noisy reference embedding vector. Therefore, we hypothesize that the diffusion chain is able to synthesize styles conditioned on the reference input by learning a reconstruction in a way that better guides the acoustic model while pondering useful information from the noisy reference embedding, as the whole chain is driven by both reconstruction and denoising losses.

5 Experiments

5.1 Experimental Setup

An internal Brazilian Portuguese single speaker dataset was used in all experiments. It consists of 15 h of speech, with 6 h of expressive content, spoken by a professional voice actress. The expressive styles present in the dataset are labeled as "lively", "welcoming", and "harsh", and were projected in a non-archetypal manner, such that all of them are applicable to real-life customer-based services. There is a total of 12400 neutral, 1307 lively, 1308 welcoming, and 1256 harsh utterances. For each label, 90% of the sentences were used for training and 10% was used for validation and testing.

Phonemes are used as inputs to the model. To extract the audio features, the 22 kHz audios are converted to 80-dimensional mel-spectrograms with Hanning windowing, a window length of $\approx$46 ms, a hop length of $\approx$11 ms and a 1024-point Fourier transform.

All models were trained using the Coqui TTS[1] framework for 1000 epochs on a single NVIDIA T4 GPU, with a batch size of 32. The whole training process took up to 4 days[2]. The optimizer used was the RAdam [16] with $\beta_1 = 0.9$, $\beta_2 = 0.998$, and $\epsilon = 10^{-6}$. A scheduler [30] for the learning rate initiating in 10^{-4} was used. The Parallel WaveGAN [34] vocoder trained in a proprietary database with 30 h of Brazilian Portuguese (2 male and 4 female speakers) was used to synthesize all waveforms from the mel-spectrograms generated by the models.

---

[1] https://github.com/coqui-ai/TTS.
[2] Our code is available at https://github.com/AI-Unicamp/TTS.

In order to evaluate the developed diffusion-based approach to style modeling, we ran a subjective perceptual evaluation.[4] In our experiments, VAETacotron [35] and the VAE+Flow [1] model were used as baselines to be compared.

The subjective experiments were performed with 30 randomly selected native Brazilian Portuguese speakers (15 male and 15 female volunteers) ranging from 18 to 61 years old. The first experiment was a test in which the listeners were given instructions to rate the naturalness of each sample from 0 meaning "completely artificial" to 100 meaning "completely natural", whilst ignoring any audio quality issues. The second was a comparison test to analyze style transfer capability between the proposed approach and the VAE+Flow model. In this test, the listeners were given instructions on what consists a speaking style, and were asked to select which style of the synthesized utterances was closer to the reference, with the option of equally close also being considered.

5.2   Naturalness

In order to evaluate the naturalness of the proposed approach, 40 utterances (10 of each style) from the test set were used. All models synthesized the audios by receiving the unseen text, their corresponding unseen audios, and with the t parameter manually adjusted to 5 noising/denoising steps, which was shown to yield better results. For each utterance, four audios were rated: a re-synthesis of the GT mel-spectrogram with the vocoder (GT+Vocoder), and the synthesis with the three considered models: VAE [35], VAE+Flow [1] and our proposed approach (Diffusion). The results are shown in Fig. 3.

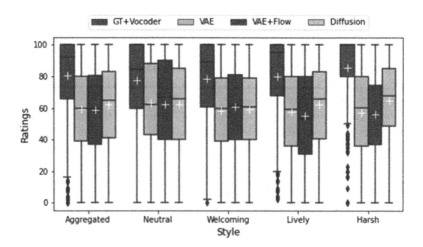

Fig. 3. Results for the naturalness subjective evaluation. The ratings of the listeners are shown in box plots grouped by style. The black stripe represents each median and each white cross the mean.

---

[4] Listening samples are available in https://ai-unicamp.github.io/publications/tts/diffusion_for_style/.

5.3   Style Transfer

We evaluate both parallel (when the text content of the reference audio is the same as the input) and non-parallel (when the content of the reference audio is different from the input) style transfer. Taken from the test set, 16 parallel and 16 non-parallel (four of each style) utterances that were conditioned on other unseen utterances of the test set were used for rating. The raters had to make a choice between the synthesis of the model A (our diffusion-based approach), the model B (the VAE+Flow [1]) and also a third option stating that both models' audios style were equally distant to the reference's style (Neutral). The results are presented in Fig. 4.

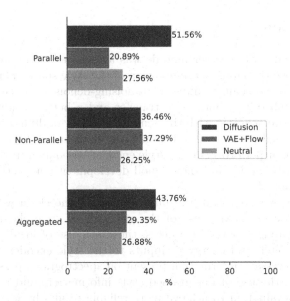

Fig. 4. Results for the style transfer test. We plot each model's score in percentage grouped by the style transfer type.

6   Discussion

From a general perspective, our proposed approach to model style based on denoising diffusion models was shown to improve the performance on both naturalness and style transfer on the scenario of expressive TTS. On the naturalness case, our approach resulted in an improvement in both the aggregated styles ($\approx$5 points in the mean), the lively ($\approx$8 points in the mean) and harsh ($\approx$9 points in the mean). A better rating means the raters found the model's samples to be closer to "completely natural". On the other styles, neutral and welcoming, our approach performed only slightly worse than the baseline models (≈-1 point in the mean from both the neutral and welcoming styles). We hypothesize this is due to the fact that the lively and harsh are styles have more emotional strength and arousal than the neutral and welcoming, showing that our approach behaves better for the more expressive cases while performing similar to baseline in the more close-to-neutral examples.

On the style transfer experiment, our model had a greater percentage of choices among the examples on both the parallel, with a gain of 30% more choices than the baseline VAE+Flow model, and on the aggregated transfer (ensemble of the parallel and non-parallel cases) with a gain of 14.41 in the percentage of choices. On the non-parallel case our approach was outperformed by the baseline with only 0.83% of the choices, meaning a similar performance. This behavior was associated to the struggle of both models on the difficult non-parallel style transfer task.

7   Conclusion

A diffusion-based style encoder module was aggregated in an end-to-end TTS model to improve the latent representation of the style space. Through a mechanism similar to the shallow diffusion (a noising-denoising process) applied on the reference embedding, both style transfer and a better conditioning of the acoustic model are acquired with the generative power of diffusion models. This is verified with a subjective evaluation on naturalness and style transfer: the proposed approach outperforms current style encoder based architectures especially on more expressive styles and shows good development perspectives to become a competitive model.

Future work will focus on: the improvement of the chain with the guided diffusion [5] process, making use of the gradients of style classifiers to guide the style embedding generation process; the direct use of low-level features of audio, such as pitch and energy as inputs to the style encoder, to reduce the amount of unnecessary information present on spectrograms passed to the reference encoder; the use of fine-grained style information; and a refinement of the subjective evaluation to achieve more reliable results by focusing more on better explaining the experiment to the raters rather than having more samples to evaluate.

Acknowledgment. The authors would like to thank the Research and Development Institute CPQD and the Ministry of Science, Technology and Innovations for supporting and funding this project. This work is supported by the BI0S - Brazilian Institute of Data Science, grant #2020/09838-0, São Paulo Research Foundation (FAPESP).

References

1. Aggarwal, V., Cotescu, M., Prateek, N., Lorenzo-Trueba, J., Barra-Chicote, R.: Using VAEs and normalizing flows for one-shot text-to-speech synthesis of expressive speech. In: ICASSP 2020–2020 IEEE International Conference on Acoustics, Speech and Signal Processing (ICASSP), pp. 6179–6183 (2020). https://doi.org/10.1109/ICASSP40776.2020.9053678

2. Aylett, M.P., Clark, L., Cowan, B.R., Torre, I.: Building and designing expressive speech synthesis. In: The Handbook on Socially Interactive Agents: 20 years of Research on Embodied Conversational Agents, Intelligent Virtual Agents, and Social Robotics Volume 1: Methods, Behavior, Cognition, pp. 173–212. Association for Computing Machinery, New York (2021). https://doi.org/10.1145/3477322

3. Chen, N., Zhang, Y., Zen, H., Weiss, R.J., Norouzi, M., Chan, W.: WAVEGRAD: estimating gradients for waveform generation (2020). https://doi.org/10.48550/ARXIV.2009.00713

4. Chen, Z., et al.: InferGrad: improving diffusion models for vocoder by considering inference in training (2022). https://doi.org/10.48550/ARXIV.2202.03751

5. Dhariwal, P., Nichol, A.: Diffusion models beat GANs on image synthesis. In: Ranzato, M., Beygelzimer, A., Dauphin, Y., Liang, P., Vaughan, J.W. (eds.) Advances in Neural Information Processing Systems, vol. 34, pp. 8780–8794. Curran Associates, Inc. (2021). https://proceedings.neurips.cc/paper/2021/file/49ad23d1ec9fa4bd8d77d02681df5cfa-Paper.pdf

6. Ekman, P., Friesen, W.V.: Constants across cultures in the face and emotion. J. Pers. Soc. Psychol. **17**(2), 124 (1971). https://doi.org/10.1037/h0030377

7. Ho, J., Jain, A., Abbeel, P.: Denoising diffusion probabilistic models. In: Larochelle, H., Ranzato, M., Hadsell, R., Balcan, M.F., Lin, H. (eds.) Advances in Neural Information Processing Systems, vol. 33, pp. 6840–6851. Curran Associates, Inc. (2020). https://proceedings.neurips.cc/paper/2020/file/4c5bcfec8584af0d967f1ab10179ca4b-Paper.pdf

8. Hodari, Z., Lai, C., King, S.: Perception of prosodic variation for speech synthesis using an unsupervised discrete representation of f0. In: Proceedings of Speech Prosody 2020, pp. 965–969 (2020). https://doi.org/10.21437/SpeechProsody.2020-197. Published 24 May 2020; Speech Prosody 2020; Conference date: 24-05-2020 Through 28-05-2020

9. Im, C.B., Lee, S.H., Kim, S.B., Lee, S.W.: EMOQ-TTS: emotion intensity quantization for fine-grained controllable emotional text-to-speech. In: ICASSP 2022–2022 IEEE International Conference on Acoustics, Speech and Signal Processing (ICASSP), pp. 6317–6321 (2022). https://doi.org/10.1109/ICASSP43922.2022.9747098

10. James, J., Balamurali, B.T., Watson, C.I., MacDonald, B.: Empathetic speech synthesis and testing for healthcare robots. Int. J. Soc. Robot. **13**(8), 2119–2137 (2020). https://doi.org/10.1007/s12369-020-00691-4

11. Jeong, M., Kim, H., Cheon, S.J., Choi, B.J., Kim, N.S.: DIFF-TTS: a denoising diffusion model for text-to-speech (2021). https://doi.org/10.48550/ARXIV.2104.01409

12. Klimkov, V., Ronanki, S., Rohnke, J., Drugman, T.: Fine-grained robust prosody transfer for single-speaker neural text-to-speech. In: 2019 Proceedings of the Interspeech, pp. 4440–4444 (2019). https://doi.org/10.21437/Interspeech.2019-2571

13. Kong, Z., Ping, W.: On fast sampling of diffusion probabilistic models. In: ICML Workshop on Invertible Neural Networks, Normalizing Flows, and Explicit Likelihood Models (2021). https://openreview.net/forum?id=agj4cdOfrAP

14. Kong, Z., Ping, W., Huang, J., Zhao, K., Catanzaro, B.: DiffWave: a versatile diffusion model for audio synthesis (2020). https://doi.org/10.48550/ARXIV.2009.09761

15. Liu, J., Li, C., Ren, Y., Chen, F., Zhao, Z.: DiffSinger: singing voice synthesis via shallow diffusion mechanism (2021). https://doi.org/10.48550/ARXIV.2105.02446

16. Liu, L., et al.: On the variance of the adaptive learning rate and beyond. In: International Conference on Learning Representations (2020). https://openreview.net/forum?id=rkgz2aEKDr

17. Liu, R., Sisman, B., Gao, G., Li, H.: Expressive TTS training with frame and style reconstruction loss. IEEE/ACM Trans. Audio Speech Lang. Proc. **29**, 1806–1818 (2021). https://doi.org/10.1109/TASLP.2021.3076369

18. Ma, S., McDuff, D., Song, Y.: Neural TTS stylization with adversarial and collaborative games. In: International Conference on Learning Representations (ICLR) (2019). https://www.microsoft.com/en-us/research/publication/neural-tts-stylization-with-adversarial-and-collaborative-games/

19. Neekhara, P., Hussain, S., Dubnov, S., Koushanfar, F., McAuley, J.: Expressive neural voice cloning. In: Balasubramanian, V.N., Tsang, I. (eds.) Proceedings of The 13th Asian Conference on Machine Learning. Proceedings of Machine Learning Research, vol. 157, pp. 252–267. PMLR, 17–19 November 2021. https://proceedings.mlr.press/v157/neekhara21a.html

20. Obin, N.: MeLos: analysis and modelling of speech prosody and speaking style. Ph.D. thesis, Ecole Doctorale Informatique, Télécommunications et Electronique (EDITE) (2011). https://tel.archives-ouvertes.fr/tel-00694687v2/document

21. Ren, Y., et al.: FastSpeech: fast, robust and controllable text to speech. In: Wallach, H., Larochelle, H., Beygelzimer, A., d' Alché-Buc, F., Fox, E., Garnett, R. (eds.) Advances in Neural Information Processing Systems, vol. 32. Curran Associates, Inc. (2019). https://proceedings.neurips.cc/paper/2019/file/f63f65b503e22cb970527f23c9ad7db1-Paper.pdf

22. Shen, J., et al.: Natural TTS synthesis by conditioning waveNet on MEL spectrogram predictions. In: 2018 IEEE International Conference on Acoustics, Speech and Signal Processing (ICASSP), pp. 4779–4783 (2018). https://doi.org/10.1109/ICASSP.2018.8461368

23. Skerry-Ryan, R., et al.: Towards end-to-end prosody transfer for expressive speech synthesis with tacotron. In: Dy, J., Krause, A. (eds.) Proceedings of the 35th International Conference on Machine Learning. Proceedings of Machine Learning Research, vol. 80, pp. 4693–4702. PMLR, 10–15 July 2018. https://proceedings.mlr.press/v80/skerry-ryan18a.html

24. Sohl-Dickstein, J., Weiss, E., Maheswaranathan, N., Ganguli, S.: Deep unsupervised learning using nonequilibrium thermodynamics. In: Bach, F., Blei, D. (eds.) Proceedings of the 32nd International Conference on Machine Learning. Proceedings of Machine Learning Research, vol. 37, pp. 2256–2265. PMLR, Lille, France, 07–09 July 2015. https://proceedings.mlr.press/v37/sohl-dickstein15.html

25. Stanton, D., Wang, Y., Skerry-Ryan, R.: Predicting expressive speaking style from text in end-to-end speech synthesis. In: 2018 IEEE Spoken Language Technology Workshop (SLT), pp. 595–602 (2018). https://doi.org/10.1109/SLT.2018.8639682

26. Sun, G., Zhang, Y., Weiss, R.J., Cao, Y., Zen, H., Wu, Y.: Fully-hierarchical fine-grained prosody modeling for interpretable speech synthesis. In: ICASSP 2020–2020 IEEE International Conference on Acoustics, Speech and Signal Processing (ICASSP), pp. 6264–6268. IEEE (2020). https://doi.org/10.1109/ICASSP40776.2020.9053520

27. Tits, N., Wang, F., Haddad, K.E., Pagel, V., Dutoit, T.: Visualization and interpretation of latent spaces for controlling expressive speech synthesis through audio analysis (2019). https://doi.org/10.48550/ARXIV.1903.11570

28. Tomczak, J.M., Welling, M.: Improving variational auto-encoders using householder flow (2016). https://doi.org/10.48550/ARXIV.1611.09630

29. Ueda, L.H., Costa, P.D.P., Simoes, F.O., Neto, M.U.: Are we truly modeling expressiveness? a study on expressive TTS in Brazilian Portuguese for real-life application styles. In: Proceedings of the 11th ISCA Speech Synthesis Workshop (SSW 2011), pp. 84–89 (2021). https://doi.org/10.21437/SSW.2021-15
30. Vaswani, A., et al.: Attention is all you need. In: Guyon, I., Luxburg, U.V., Bengio, S., Wallach, H., Fergus, R., Vishwanathan, S., Garnett, R. (eds.) Advances in Neural Information Processing Systems, vol. 30. Curran Associates, Inc. (2017). https://proceedings.neurips.cc/paper/2017/file/3f5ee243547dee91fbd053c1c4a845aa-Paper.pdf
31. Wang, Y., et al.: Uncovering latent style factors for expressive speech synthesis. In: NIPS Workshop on Machine Learning for Audio Signal Processing (ML4Audio) (2017)
32. Wang, Y., et al.: Style tokens: unsupervised style modeling, control and transfer in end-to-end speech synthesis. In: Dy, J., Krause, A. (eds.) Proceedings of the 35th International Conference on Machine Learning. Proceedings of Machine Learning Research, vol. 80, pp. 5180–5189. PMLR, 10–15 July 2018. https://proceedings.mlr.press/v80/wang18h.html
33. Wu, N.Q., Liu, Z.C., Ling, Z.H.: Discourse-level prosody modeling with a variational autoencoder for non-autoregressive expressive speech synthesis. In: ICASSP 2022–2022 IEEE International Conference on Acoustics, Speech and Signal Processing (ICASSP), pp. 7592–7596 (2022). https://doi.org/10.1109/ICASSP43922.2022.9746238
34. Yamamoto, R., Song, E., Kim, J.M.: Parallel WaveGan: a fast waveform generation model based on generative adversarial networks with multi-resolution spectrogram. In: ICASSP 2020–2020 IEEE International Conference on Acoustics, Speech and Signal Processing (ICASSP), pp. 6199–6203 (2020). https://doi.org/10.1109/ICASSP40776.2020.9053795
35. Zhang, Y.J., Pan, S., He, L., Ling, Z.H.: Learning latent representations for style control and transfer in end-to-end speech synthesis. In: ICASSP 2019–2019 IEEE International Conference on Acoustics, Speech and Signal Processing (ICASSP), pp. 6945–6949 (2019). https://doi.org/10.1109/ICASSP.2019.8683623

Automatic Rule Generation for Cellular Automata Using Fuzzy Times Series Methods

Lucas Malacarne Astore[1(✉)], Frederico Gadelha Guimarães[1],
and Carlos Alberto Severiano Junior[2]

[1] Machine Intelligence and Data Science (MINDS) Laboratory, Graduate Program in
Electrical Engineering, Universidade Federal de Minas Gerais, Av. Antônio Carlos
6627, Belo Horizonte, MG 31270-901, Brazil
astore.lucas@gmail.com, fredericoguimaraes@ufmg.br
[2] Federal Institute of Minas Gerais, Campus Sabará, IFMG, Belo Horizonte, Brazil
carlos.junior@ifmg.edu.br
https://minds.eng.ufmg.br/

Abstract. Computer simulation of land dynamics have been widely
used for several proposes, for example in epidemiological models. Cellular Automata (CA) is one of the strategies capable of predicting future
land states over time based on a set of transitional rules. Building this
set is not a straightforward task. It may require technical knowledge
about the process, through years of scientific research. If machine learning techniques are applied, there is still the challenge of finding the best
set of hyperparameters. In this context, the main goal of this paper is
presenting a different approach of CA transitional rules set construction,
based exclusively on historical data of a phenomenon. A multivariate
Fuzzy Time Series (FTS) model is applied to learn and represent the
local rules of the automaton. Therefore, we combine FTS and CA into
an integrated modeling technique. The proposed approach was able to
predict future behavior of a CA, with errors around 12%, confirming the
potential of FTS transitional rules for CA.

Keywords: Cellular automata · Fuzzy time series · Land cover land
usage · Dynamics modeling

1 Introduction

Modeling land dynamics over time through computer simulations has been considered a relevant approach in order to study and evaluate different types of real world
scenarios [9]. Comprehending and analyzing the mechanisms of land changes can
help define effective and efficient public policies and strategic planning.

This study was financed in part by the Coordenação de Aperfeiçoamento de Pessoal
de Nível Superior - Brasil (CAPES) - Finance Code 001. Supported by CNPq Grant
312991/2020-7 and FAPEMIG Grant no. APQ-01779-21.

J. C. Xavier-Junior and R. A. Rios (Eds.): BRACIS 2022, LNAI 13653, pp. 268–282, 2022.
https://doi.org/10.1007/978-3-031-21686-2_19

On this basis, in the past decades, Cellular Automata (CA) and Markov Chain models have been widely used in geographic and spatial applications [8]. CA-based models offer ways to predict and understand the land behavior in a variety of study areas such as forest cover, urban sprawl and spread of diseases such as the pandemic of Coronavirus Disease-2019 (COVID-19) [5], Chagas [6] and Dengue fever [7]. Its intrinsic discrete representation of space with a lattice of cells allows the development of models capable of simulating complex dynamic systems instead of using differential equations, which usually demand expensive computer processing for their solution.

The basic principle of a typical CA model is building a set of transition rules that describe the future cell states over time based on the neighboring cell states [1]. This can be characterized as a spatio-temporal forecasting method and comes from the idea that local changes are affected by the states of nearby cells. Determining this set of rules might require expert knowledge, which is laborious and takes time and effort to formulate. It can be found in the literature some works towards reducing this effort, for example by using shape grammar [2], genetic algorithms and genetic programming [3,4].

Another challenging task is the adjustment of several parameters that a multivariate model can hold, increasing the complexity of the modeling process. In terms of achieving the right parameters values, optimization methods and machine learning strategies have been applied to build a forecasting model, for example using multistage evolutionary strategy based on genetic algorithm [10].

In machine learning models, various methods use labeled data set in order to train the algorithms to predict a target output variable. This class of algorithms are known as supervised learning and has been popular among Artificial Intelligence (AI) techniques. It is notable that the advances in this field were possible due to new technologies of data capturing and storage in the past decades, leading to the emergence of the Big Data phenomenon [11]. A large set of historical data can be powerful in order to understand patterns and predicting variables. In this sense, Fuzzy Time Series (FTS) have been drawing attention in time series modeling and forecasting as they are computationally cheap and readable models [12].

In this context, the main goal of this paper is to employ the FTS approach to learn and generate the rules of a given Cellular Automaton from the historical data set. The rule base will then represent the transition rules of cells and the underlying dynamics of the phenomenon that generated the data. This is the first attempt in the literature to combine Fuzzy Time Series and Cellular Automata into an integrated modeling technique, to the best of our knowledge. The great advantage of the proposed approach is that the transition rules governing the CA dynamics are completely induced by the data without the need to formulate tailored rules and to calibrate specific parameters of these rules, making the whole process more automated and almost effortless.

In the results, we considered a simulation model of CA of Chagas disease in order to validate the proposed approach. The predicted values were compared with the original data using graphic analysis and the Evaluation Metrics used produced errors within 12% when comparing both data.

2   Background

2.1   Cellular Automata

Modeling phenomena using CA was first conceptualized by John Von Neumann and Stanislaw Ulam, in the late 1940s [13]. Cellular Automata can be defined as a mathematical abstraction of the real world in a discrete universe, composed with a structure of spatially grouped cells as lattice, which evolve from state in time, conditioned by a set of transition rules. The lattice of cells are mostly found in 1D or 2D, but they can have 3D structure as well. The transition rules are used in order to determine the future cell state $q(t+1)$ based on the neighbor cell states and the current state $q(t)$. As described in [8], CA is composed of five elements that form the tuple $<A, S, t, \nu, \delta>$:

1. Regular discrete space A of the set of cells (lattice of cells), which can have different configuration formats;
2. Set S of possible states $(q(t))$ of cells;
3. State update cycle t (*time steps*), i.e., in each time step the space A is updated;
4. Definition of size and layout of the neighborhood ν to be considered for state update.
5. Transition rules functions δ, that determine the states dynamics that an elementary cell can be in, i.e., $q(t + 1) = \delta(q(t), a(t))$;

Figure 1(a) presents possible configurations applicable to CA-based models. Regarding the neighborhood, it is possible to highlight the two most traditional ones found in the literature of radius 1, that of **(I) Von Neumann** in which the four cells to the north, south, east and west of the target cell are considered neighbors, and **(II) Moore** that considers all cells around the target one, and can be expanded to more cells around it by increasing the radius. Figure 1(b) shows the arrangement of the aforementioned neighborhoods, where $P(x, y)$ can be considered as an element of a matrix of discrete cells [8]. It is worth mentioning that there are several other types of neighborhood dispositions since it is a configuration that can be adjusted according to the system to be modeled [1].

In terms of transitional rules, CA models can be either deterministic or probabilistic. For a given transitional rule, if there is a probability associated to a decision of the next cell stage, thus it is a probabilistic CA. Most of natural phenomena are probabilistic and their values and distribution are defined for each study case.

2.2   Fuzzy Time Series

As Singh [14] states, the increase in information storage capacity must be accompanied by the development and improvement of processing and analysis techniques. The goals of gathering large data packages should help classify, identify

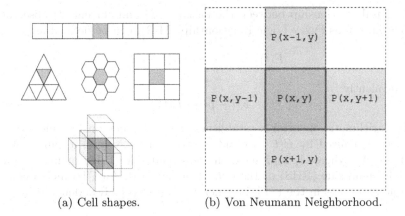

(a) Cell shapes.   (b) Von Neumann Neighborhood.

Fig. 1. Differents configurations of a CA.

and predict events and assist decision making. In this sense, forecasting models using Machine learning, statistics and soft computing (neural networks, fuzzy theory and genetic algorithms) constitute a vast area of studies today.

FTS method was first presented by Song and Chissom in the 1990s decade and its central principle is based on building a forecasting model of a given time series data through fuzzy set based representation [30]. FTS is a soft computing and data-driven technique and it has been widely applied due to its flexibility, affording a variety of data types (not normally distributed data, for instance), readability, since it is interpretable, and scalability, dealing with uncertainties of real-world data. Since its conceptualization, FTS was applied in a myriad of studies such as seasonal time series [31], stock index prices [32], electric load [33] and others [12,40].

Time series can be defined as a set of successive observations of the behavior of one or more variables over time. Such observations should not be analyzed individually or randomly, but considering the historical temporal dependence of the data. Time series forecasting is a methodology that integrates several pattern recognition techniques and creates models based on the data past behavior of a given phenomenon or process capable of conjecturing future scenarios. Examples of applications include observations of seasonality of rainfall and temperature variation in a given region, population growth, electroencephalogram examination patterns [15] and more current topics such as social media sentiment analysis [16].

As [34] describes, a Fuzzy Time Series model can be defined as: Let Y_t, $t \in \mathbb{Z}$, a conventional time series in a subset of real numbers. Let the universe of discourse be divided as $U = u_1, u_2, ..., u_n$ and the fuzzy sets A_i, $i \in \mathbb{Z}$ defined over the U intervals with corresponding membership functions f_{A_i}. Thus, $F(t)$ is a collection of f_{A_i} and is considered an FTS on Y_t.

A causal relationship between the past $(t - p)$ and current (t) observations in FTS, known as Fuzzy Logic Relationship (FLR), can be defined as:

$$F(t) = F(t - p) \circ R(t - p, t) \tag{1}$$

or, equivalently:

$$F(t - p) \rightarrow F(t) \tag{2}$$

The arithmetic operator $\circ$ establishes the fuzzy relationship and (2) shows that $F(t)$ is caused by $F(t - p)$ and in terms of fuzzy sets, it can be written as: $A_i \rightarrow A_j$, where A_i represents the left-hand side (LHS) or inputs and A_j the right-hand side (RHS) or the output of the FLR. An FTS model can have multiple fuzzy sets in the LHS, in general an order-p FLR is denoted by:

$$F(t - 1), F(t - 2), ..., F(t - p) \rightarrow F(t) \tag{3}$$

Thus, Eq. (3) shows that the weight of each $F(t - p)$ for obtaining the fuzzy forecast at time t, i.e. $F(t)$, is equal to one.

The set of rules is grouped by the precedents, creating Fuzzy Logical Relationship Groups (FLRGs). In other words, given a conventional time series, it is determined which fuzzy sets can be the resultant (RHS) of left-hand sets (LHS), forming, for example:

$$A_i \rightarrow A_1, A_3, ..., A_j \tag{4}$$

Another relevant information in FTS modeling is the number of variables that a simulation requires. For example, in deforestation forecasting, the amplitude, the wind speed, the weather and other parameters are pertinent in order to predict forest cover future states. For a multivariate FTS model with dimensionality d and order p, the FLR is expressed by:

$$
\begin{aligned}
(F_1(t - p), F_2(t - p), &\ldots, F_d(t - p)), \\
&\vdots \\
(F_1(t - 2), F_2(t - 2), &\ldots, F_d(t - 2)), \\
(F_1(t - 1), F_2(t - 1), &\ldots, F_d(t - 1)) \rightarrow F_1(t), F_2(t), ..., F_d(t)
\end{aligned}
\tag{5}
$$

Some FTS methods use weights, based on the frequency of the patterns in the data, known as Weighted FTS model (WFTS). It was first announced by [35] and followed by some improvements such as Trend WFTS [38], Improved WFTS [36], Exponentially WFTS [37] and Probabilistic WFTS [39]. The Weighted Fuzzy Logical Relationship Group (WFLRG) includes a weight matrix on FLRGs, giving greater importance to recent data in the forecast [41], satisfying the same condition $\sum_{h=1}^{k} w'_h = 1$.

Thus, the matrix is standardized $W(t)$ and the final forecast value $Y_{(t+1)}$ is equal to the product of the defuzzified matrix and the transpose of the weight matrix [35]:

$$Y_{(t+1)} = M(t) \times W(t)^T \tag{6}$$

where $M(t)$ is the defuzzified matrix forecast of $F(t)$ and $\times$ is the matrix product operator.

3   Related Work

Cellular Automata applications can be found in several areas, such as ecological models of succession in vegetation [17], urban growth [18,19], deforestation and fire propagation [20,21], fluid dynamics simulation and physical systems [22,23], urban traffic simulation [24], scattering study epidemics [7,25]. Techniques are applied in CA, either to achieve better forecasting results or searching for better model parameters via optimization algorithms.

The use of Fuzzy logic in CA models (so called Fuzzy Cellular Automata – FCA) was first proposed by [26] and it allows greater flexibility in the consideration of factors in the transition rules of states and modeling vagueness in real-world scenarios. For example, fuzzy constrained CA model was used to simulate forest insect infestations [27], or to understand logistic trends of urban development process [28]. A common usage of Fuzzy logic into CA models are related to simulate states gradient, since this logic works based on fuzzification of sharp and hard values by means of membership functions $mf_{A_i} : \mathbb{R} \rightarrow [0,1]$. Other FCA usage example is a fuzzy neighborhood CA [29], where instead of using traditional CA neighborhood configuration, a function is used to evaluate the influence of the neighboring cells, modeling it as fuzzy sets [8].

Although it can be found in the literature a variety of approaches for combining fuzzy sets into CA, none of them has used Fuzzy Time Series strategy as a forecast model. Here, the fuzzy sets are applied to the historical land dynamics dataset, in order to build a multivariate FTS model and then apply it as a transitional CA rules to predict future land states.

4   Proposed Method

As mentioned in the previous sections, the proposed method is essentially based on the development of a multivariate FTS forecasting model formed from historical data of a geospatial phenomenon and, from that, apply the model as transition rules in a simulation of a cellular automaton. In other words, for each cell in a CA grid, the FTS model is applied and thus the future state of the cells is determined. To evaluate the method, simulations were carried out using the Python programming language, due to its wide diversity of libraries, especially pyFTS (Fuzzy Time Series for Python) [42].

4.1   Training Procedure

Starting from a collection of time series historical data Y, the training procedure are summarized in the construction of a multivariate FTS model. Each sample of the collection Y is a frame over the unit of time, representing the geospatial states of the phenomena, i.e. the collection groups the data of each time unit. The training steps are described:

1) Data Preprocessing: The system variables are determined, as well as the variable of interest, and the values of their states are stored following the data collection. The system variables rely on the neighborhood chosen for the model, for example taking Moore neighborhood the variables would be all the cells surrounding the center cell. In terms of CA simulation, the variable of interest usually is the state of the center cell in the future. Each variable represents the historical states pattern over the unit time.

2) Dataset Split: The dataset is divided into training and testing data sets. The training data is then used to build the model, using the pyFTS library, and the test data is used during the model validation through the evaluation metrics.

3) Universe of Discourse Partitioning: The pyFTS library offers different types of partitioning in order to build the membership functions. Here, it was used the Grid Partitioning which divides the universe into n overlapping equal length intervals with triangular membership functions. All the variables have the same partitioning and number of fuzzy sets.

4) Data Fuzzification: The process of converting the numerical values into fuzzy linguistic variables. The crisp value from the time series $Y(t)$ is now represented by the maximum membership value fuzzy set: $F(t) = \arg\max_{A_i} \mu_{A_i}(Y(t))$, since the fuzzy sets are overlapped.

5) Weighted MVFTS Model Training: In Sect. 2.2, Eq. (5) was presented for a generic case of a multivariate FTS forecast model. From a CA perspective, considering the Von Neumann neighborhood shown in Fig. 1(b), the proposed equation becomes:

$$(F_N(t-p), F_S(t-p), F_W(t-p), F_E(t-p), F_C(t-p)),$$

$$\vdots \qquad (7)$$

$$(F_N(t-2), F_S(t-2), F_W(t-2), F_E(t-2), F_C(t-2)),$$
$$(F_N(t-1), F_S(t-1), F_W(t-1), F_E(t-1), F_C(t-1)) \rightarrow F_C(t)$$

where $F_i(t-p)$ now considers the space from the lattice, for $i \in \{North, South, West, East, Center\}$. Thus, the temporal patterns and rules $LHS \rightarrow RHS$ are created considering the matrix weights, forming the WFLRG model.

4.2   Forecast Procedure

The pyFTS model trained from the historical data is applied to the CA. The algorithm consists of scanning the CA cells at time (t) and building it at time $(t + 1)$. That is, the trained pyFTS model is used as CA state transition rules. Figure 2 shows the summary of the steps through a diagram.

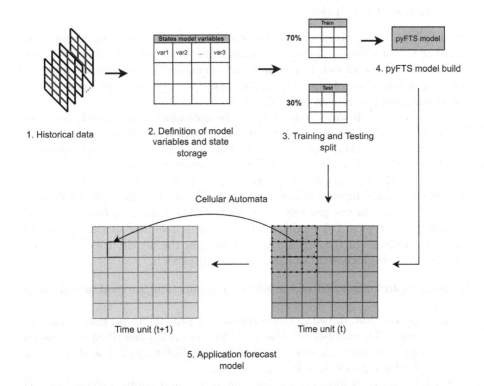

Fig. 2. Illustration of the proposed method steps.

Finally, the test data is used to compare with the period determined by the model. As a qualitative metric, a comparative graph was made between such results and a visual graphical simulation of state changes in the CA. For quantitative metrics, we chose the normalized Root Mean Square Error (nRMSE) and the Mean Absolute Error (MAE). The formulas are described as follows:

$$nRMSE = \frac{\sqrt{\frac{1}{n}\sum_{i=1}^{n}(y_i - \hat{y}_i)^2}}{A} \tag{8}$$

$$MAE = \frac{1}{n}\sum_{i=1}^{n}|y_i - \hat{y}_i| \tag{9}$$

where n is the length of the test dataset, y_i is the observed value index i, $\hat{y}_i$ denotes the predicted value and A amplitude range of the data, considering both observed and predicted values.

5 Computational Experiments

5.1 Dataset Description

For the purpose of applying the methodology described in this work, a simulation model of CA of Chagas disease [6] was reproduced to build the historical data and to validate the model. It is important to emphasize that this work started with a CA model to obtain the original data set, but any historical data of spatio-temporal models can be used.

In [6], a research was carried out on the spreading dynamics of the transmitting agent of Chagas disease, in the adult stage and larvae, in a village in the Yucatan peninsula region, in Mexico. The model parameters were calculated based on real data and the probabilistic CA simulation was built considering a 30 × 30 lattice of cells. Each cell is composed of a vector with the number of larvae and adult insects, respectively (ny, na). In addition, a maximum limit of 5 adults and 5 larvae per cell is established. The transition rules were built according to a typical CA model, that is, bringing together specific knowledge and characteristics of the region. As described in the article, these rules were divided into two processes: Demography and Dispersion.

1. Demography: composed of the phases reproduction, survival and development.
   (a) Reproduction: the adults reproduces with probability p_r, generating $F = 1$ larvae. The number of larvae generated per cell is described as a random variable following a binomial distribution $B(F.n_a, p_r)$. The state of the cell after reproduction is: $(n_y + n_{ry}, n_a)$
   (b) Survival: corresponding the amount larvae and adults survived at each time-step. It is also given by a binomial distribution $B(n*, p*)$. The probabilities are p_{sy} and p_{sa} for larvae and adults respectively. The final state after survival phase is: $(n_{ry} + n_{sy}, n_{sa})$
   (c) Development: the processes of larvae becoming adults insects. Each larva has the probability p_d to develop and the binomial distribution takes the survival larvae $B(n_{sy}, p_d)$. After the development the final states is: $(n_{sy} + n_{ry} - n_{da}, n_{sa} + n_{da})$.
2. Dispersion: Only adults insects are able to move along the cells. The CA boundary condition is related to the migration of the insects from the forest to the village, with the probability or dispersal coefficient $D_f = \frac{Q_f}{4M}$, where $Q_f = 50$ insects/day and M the lattice dimension.
   The migration happens only in the infestation period (from April to June, or 90 days). Concerning the movement through the village cells in the lattice, each adult cells have the same probability to enter or leave the cell, given by $p = \frac{D}{(2r-1)^2-1}$. where D is the dispersal coefficient in the village equals to

0.9 during the infestation period and 0.1 during the non-infestation, r is the infestation radius, equals to 4 in infestation and 1 in the non-infestation.

All the parameters of the model had to be optimized (calibrated) by using a Genetic Algorithm. The rules parameters of the CA model are: $p_r = 0.004111$; $p_{sy} = 0.90272518$; $p_{sa} = 0.9828095$; $p_d = 0.004158$;

The neighborhood considered was Moore neighborhood. For the analysis of the model, the target variable is determined by the total sum of adult individuals and larvae in the village, that is, in the lattice of cells L.

$$N_l^{(t)} = \sum_{c \in L} n_l^{(t)}(c) \qquad N_a^{(t)} = \sum_{c \in L} n_a^{(t)}(c) \qquad (10)$$

where c is a cell, n_l is the number of a larvae and n_a is the number of adults.

5.2   CA-FTS Modeling

The CA model described in the previous section was used to generate realizations of the time series of number of adults $N_a^{(t)}$. This is the only input data to the proposed CA-FTS model. The same strategy was used for larvae $N_l^{(t)}$. The idea here is to show that starting from the historical time series data, the CA-FTS model can automatically generate the transition rules and reconstruct the original data.

The next step is the development of the proposed model. For each time-step, in the daily case, the grid was traversed, extracting the variables of interest from the model. Thus, a table was created where each row refers to the configuration of states at an instant (t) that conditioned the state of the central cell at $(t+1)$. The data set was split into train and test, about 70% and 30% respectively. Then, the WMVFTS model from the pyFTS library was used, which builds the model from the historical data of the variables of the constructed table. In terms of parameters used to evaluate the model, it was analyzed two different configurations for model variables: Moore's and Von Neumann's neighborhood, and four different values for the number of fuzzy sets: 5, 10, 20 and 30.

Once the model is built, the next step is to apply it to the CA simulation. At this point, the test dataset was used, and for each time (t) in each cell of the lattice, the FTS model (represented by the learned rule base) was used to determine the state of the cell at $(t+1)$.

The graphs in Figs. 3 and 4 illustrate the simulation results. In green, the test data for adults (AA), in blue the forecasting results using the FTS model for adults (PA), in orange the test data for larvae (AL) and in red the predicted values for larvae (PL). In other words, the continuous lines are the model predicted values and the dashed lines are the actual values.

The metrics were obtained comparing the original test data with the results from the CA-FTS model simulation. Table 1 presents the respective values of nRMSE and MAE according with the FTS model order and number of fuzzy sets used. In general the results were around 12% and the best configuration was with the FTS model using Von Neumann neighborhood with 30 fuzzy sets, for adults and Moore neighborhood with 30 fuzzy sets for larvae.

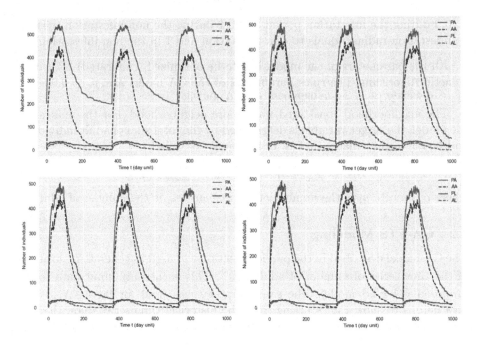

Fig. 3. Simulation results considering Von Neumann Neighborhood with 5, 10, 20 and 30 fuzzy sets. (Color figure online)

5.3 Discussion

The simulation results, as observed in Figs. 3 and 4, demonstrate the model capacity to capture and predict spatio-temporal behavior, based exclusively on a set of a phenomenon historical data. This alternative strategy for CA modeling and simulation has a very low computational cost, since most of the functions used were already developed and it is available in the pyFTS library.

It is relevant to highlight that the complexity of a multivariate model increases with the number of variables due to the number of rules combination [31]. Nevertheless, it takes less than 10 min to build an FTS model using the pyFTS library. This represents a fraction of the total cost of applying a genetic algorithm (or any other metaheuristic search algorithm) to calibrate a usual CA model, which can take hours, given that fitness calculation would require simulating the CA. As discussed in the earlier sections, optimization algorithms have to be used for tuning the parameters of CA rules. Furthermore, the proposed method can be used to model a natural phenomenon that does not have advanced studies or important information in order to build a set of transitional rules based on technical knowledge, once the method is fully data-driven.

For the specific study case, the best FTS model was obtained with increasing the number of fuzzy sets (30). Additionally, in terms of metrics, it can be noticed that using Von Neumann's neighborhood seems a better option, because the number of variables in the model is reduced compared to Moore's neighbor-

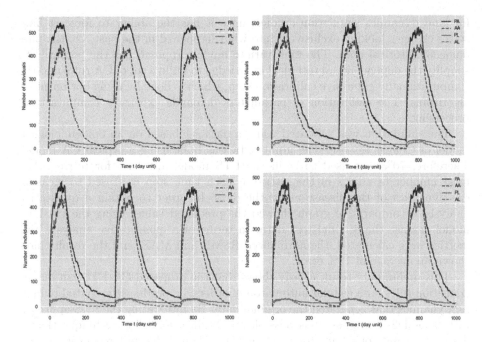

Fig. 4. Simulation results considering Moore with 5, 10, 20 and 30 fuzzy sets. (Color figure online)

Table 1. Evaluation metrics for Adults and Larvae.

Neighborhood	N fuzzy sets	Insect	nRSME	MAE	Insect	nRSME	MAE
Von Neumann	5	Adults	42,41%	179,11	Larvae	38,73%	12,91
	10		12,93%	53,07		38,51%	12,80
	20		12,69%	52,14		31,59%	9,88
	30		11,71%	48,13		29,50%	8,97
Moore	5		44,83%	189,94		38,04%	12,98
	10		13,01%	48,81		37,79%	12,88
	20		12,79%	48,06		30,53%	9,94
	30		11,81%	44,71		28,31%	8,96

hood. The flexibility of the FTS method is positive once it allows to adapt the model parameters, meeting the problem specifications, in order to select the best metrics results.

6 Conclusions and Future Work

The proposed method was able to automatically build a set of CA transitional rules using Fuzzy Times Series model. This characterizes an alternative strategy for land dynamics simulation studies, once it has the ability to forecast land states changes looking exclusively into what happened in the past, i.e. the phenomena historical data. The multivariate model proposed has the potential to simplify the dependence of technical knowledge to build a set of CA rules and the computationally expensive calibration step, which is done by using optimization algorithms. The calibration of parameters demands expensive computational processing.

The simulation was validated using a data-set based on an epidemiological model of Chagas disease. The input data was built using a typical CA model. Therefore, 70% was used to train the FTS model and 30% for testing. The total number of adults in the lattice at each time step was used to measure the dynamics of the disease along the days. In order to obtain a qualitative analysis, a comparative graph showed the predicted values using the CA-FTS model against the original test data. In terms of a quantitative analysis, two metrics were calculated, the normalized RSME and MAE, and the results have shown an average of 12% error.

This preliminary result clearly shows that the proposed CA-FTS can induce transition rules for a cellular automaton in order to reproduce the spatial temporal dynamics that generated the original data. By integrating the FTS approach to CA modeling process, it is possible to obtain a modeling technique that is data driven and not relying too much on expert knowledge and expensive parameter calibration. The discovered rules are interpretable and can potentially help understanding the underlying dynamics of the problem. Future refinements are needed in order to understand the combination of both methods. This alternative approach can be validated for other types of data-set, such as land images from drones or satellites. Another future work is implementing the method in the pyFTS library such that it can be made available to the community at large.

References

1. Kari, J.: Theory of cellular automata: a survey. Theor. Comput. Sci. **334**(1), 3–33 (2005). https://doi.org/10.1016/j.tcs.2004.11.021. ISSN 0304-3975
2. Speller, T., Whitney, D., Crawley, E.: Using shape grammar to derive cellular automata rule patterns. Complex Syst. **17**, 79 (2007)
3. Hajela, P., Kim, B.: On the use of energy minimization for CA based analysis in elasticity. In: 2000 Proceedings of the 41st AIAA/ASME/ASCE/AHS SDM Meeting, Atlanta, GA (2000)
4. Koza, J., et al.: Genetic Programming III: Darwinian Invention and Problem Solving. Morgan Kaufmann Publishers, San Francisco (1999)
5. Schimit, P.H.T.: A model based on cellular automata to estimate the social isolation impact on COVID-19 spreading in Brazil. Comput. Methods Prog. Biomed. **200**, 105832 (2021). https://doi.org/10.1016/j.cmpb.2020.105832. ISSN 0169-2607
6. Slimi, R., El Yacoubi, S., Dumonteil, E., Gourbière, S.: A cellular automata model for Chagas disease. Appl. Math. Model. **33**(2), 1072–1085 (2009). https://doi.org/10.1016/j.apm.2007.12.028. ISSN 0307-904X
7. Massahud, R.A.T.: Dengue propagation model using cellular automata. Master's thesis. Federal University of Lavras, Lavras (2011)

8. Ghosh, P., et al.: Application of cellular automata and Markov-chain model in geospatial environmental modeling - a review. Remote Sens. Appl. Soc. Environ. **5**, 64–77 (2017). https://doi.org/10.1016/j.rsase.2017.01.005. ISSN 2352-9385

9. Dawn, P., Steven, M., Marco, J., Matthew, H., Peter, D.: Multi-agent systems for the simulation of land-use and land-cover change: a review. Ann. Assoc. Am. Geogr. **93**, 314–337 (2003). https://doi.org/10.1111/1467-8306.9302004

10. Larissa, F., Gina, O., Luiz, M.: Multistage evolutionary strategies for adjusting a cellular automata-based epidemiological model, 466–473 (2021). https://doi.org/ 10.1109/CEC45853.2021.9504738

11. Lynch, C.: Big data: how do your data grow? Nature **455**(7209), 28–29 (2008)

12. Bose M., Mali, K.: Designing fuzzy time series forecasting models: a survey. Int. J. Approx. Reason. **111**, 78–99 (2019). https://doi.org/10.1016/j.ijar.2019.05.002. ISSN 0888-613X

13. Burks, A.W.: Von neumann's self-reproducing automata. Essay on Cellular Automata, pp. 3–64 (1966)

14. Singh, P.: A brief review of modeling approaches based on fuzzy time series. Inter. J. Mach. Learn. Cybern. **8**(2), 397–420 (2015). https://doi.org/10.1007/s13042-015-0332-y

15. Morettin, P.A., Toloi, C.M.D.C.: Time Series Analysis. Edgard Blucher (2004)

16. Ibrahim, N.F., Wang, X.: Decoding the sentiment dynamics of online retailing customers: time series analysis of social media. Comput. Human Behav. **96**, 32–45 (2019). https://doi.org/10.1016/j.chb.2019.02.004. ISSN 0747-5632

17. Balzter, H., Braun, P.W., Kohler, W.: Cellular automata models for vegetation dynamics. Ecol. Model. **107**(2), 113–125 (1998). https://doi.org/10.1016/S0304-3800(97)00202-0. ISSN 0304-3800

18. Liu, Y., Phinn, S.: Modeling urban development with cellular automata incorporating fuzzy-set approaches. Comput. Environ. Urban Syst. **27**, 637–658 (2003). https://doi.org/10.1016/S0198-9715(02)00069-8

19. Mantelas, L., Prastacos, P., Hatzichristos, T., Koutsopoulos, K.: Using fuzzy cellular automata to access and simulate urban growth. GeoJournal **77**, 13–28 (2012). https://doi.org/10.1007/s10708-010-9372-8

20. Zheng, Z., Huang, W., Li, S., Zeng, Y.: Forest fire spread simulating model using cellular automaton with extreme learning machine. Ecol. Model. **348**, 33–43 (2017). https://doi.org/10.1016/j.ecolmodel.2016.12.022

21. Czerniak, J., Zarzycki, H., Apiecionek, L., Palczewski, W., Kardasz, P.: A cellular automata-based simulation tool for real fire accident prevention. Math. Probl. Eng. **1–12**(02), 2018 (2018). https://doi.org/10.1155/2018/3058241

22. Chopard, B.: Cellular automata and lattice Boltzmann modeling of physical systems, pp. 287–331 (2012). https://doi.org/10.1007/978-3-540-92910-9_9. ISBN 978-3-540-92909-3

23. Chopard, B., Dupuis, A., Masselot, A., Luthi, P.: Cellular automata and lattice Boltzmann techniques: an approach to model and simulate complex systems. Adv. Comp. Syst. (ACS) **05**, 103–246 (2002). https://doi.org/10.1142/S0219525902000602

24. Tavares, L.D.: An urban traffic simulator based on cellular automata. Master's thesis, Federal University of Minas Gerais, Minas Gerais, Brazil (2010)

25. Melotti, G.: Application of cellular automata in complex systems: a study of case in spreading epidemics. Master's thesis, Federal University of Minas Gerais, Minas Gerais, Brazil (2009)

26. Cattaneo, G., Flocchini, P., Mauri, G., Vogliotti, C., Santoro, N.: Cellular automata in fuzzy backgrounds. Physica D: Nonlinear Phenom. **105**(1), 105–120 (1997). https://doi.org/10.1016/S0167-2789(96)00233-3. ISSN 0167-2789

27. Bone, C., Dragicevic, S., Roberts, A.: A fuzzy-constrained cellular automata model of forest insect infestations. Ecol. Model. **192**(1), 107–125 (2006)

28. Liu, Y., Phinn, S.R.: Modelling urban development with cellular automata incorporating fuzzy-set approaches. Comput. Environ. Urban Syst. **27**(6), 637–658 (2003)

29. Praba, B., Saranya, R.: Fuzzy graph cellular automaton and it's applications in parking recommendations. Math. Nat. Comput. **18**(1), 147–162 (2022). https://doi.org/10.1142/S1793005722500089

30. Song, Q., Chissom, B.S.: Forecasting enrollments with fuzzy time series - Part I. Fuzzy Sets Syst. **54**(1), 1–9 (1993). https://doi.org/10.1016/0165-0114(93)90355-L. ISSN 0165-0114

31. Song, Q.: Seasonal forecasting in fuzzy time series. Fuzzy Sets Syst. **107**, 235–236 (1999). https://doi.org/10.1016/S0165-0114(98)00266-8

32. Huarng, K.H., Yu, H.K.: A type 2 fuzzy time series model for stock index forecasting. Phys. A **353**, 445–462 (2005)

33. Efendi, R., Ismail, Z., Deris, M.M.: A new linguistic out-sample approach of fuzzy time series for daily forecasting of Malaysian electricity load demand. Appl. Soft Comput. **28**, 422–430 (2015)

34. Efendi, R., Deris, M.M., Ismail, Z.: Implementation of fuzzy time series in forecasting of the non-stationary data. Int. J. Comput. Intel. Appl. **15**, 1650009 (2016). https://doi.org/10.1142/S1469026816500097

35. Yu, H.-K.: Weighted fuzzy time series models for TAIEX forecasting. Physica A: Stat. Mech. Appl. **349**(3–4), 609–624 (2005). https://doi.org/10.1016/j.physa.2004.11.006. ISSN 0378-4371

36. Ismail, Z., Efendi, R.: Enrollment forecasting based on modified weight fuzzy time series. J. Artif. Intell. **4**(1), 110–118 (2011)

37. Sadaei, H.J.: Improved models in fuzzy time series for forecasting. Ph.D. thesis, Universiti Teknologi Malaysia (2013)

38. Cheng, C.-H., Chen, T.-L., Chiang, C.-H.: Trend-weighted fuzzy time-series model for TAIEX forecasting. In: King, I., Wang, J., Chan, L.-W., Wang, D.L. (eds.) ICONIP 2006. LNCS, vol. 4234, pp. 469–477. Springer, Heidelberg (2006). https://doi.org/10.1007/11893295_52

39. Silva, P.C., Sadaei, H.J., Ballini, R., Guimarães, F.G.: Probabilistic forecasting with fuzzy time series. IEEE Trans. Fuzzy Syst. **28**, 1771–1784 (2019)

40. Severiano, C.A., de Lima e Silva, P.C., Cohen, M.W., Guimarães, F.G.: Evolving fuzzy time series for spatio-temporal forecasting in renewable energy systems. Renew. Energy **171**, 764–783 (2021). https://doi.org/10.1016/j.renene.2021.02.117. ISSN 0960-1481

41. Silva, P., Lucas, P., Sadaei, H., Guimarães, F.: Distributed evolutionary hyperparameter optimization for fuzzy time series. IEEE Trans. Netw. Ser. Manage. 1. https://doi.org/10.1109/TNSM.2020.2980289

42. Silva, P.C.L. et al.: pyFTS: fuzzy time series for python. Belo Horizonte (2018). https://doi.org/10.5281/zenodo.597359

Explanation-by-Example Based on Item Response Theory

Lucas F. F. Cardoso[1,5(✉)], José de S. Ribeiro[1,2], Vitor Cirilo Araujo Santos[1,5], Raíssa L. Silva[3], Marcelle P. Mota[1], Ricardo B. C. Prudêncio[4], and Ronnie C. O. Alves[5]

[1] ICEN, Universidade Federal do Pará, Belém, Brazil
lucas.cardoso@icen.ufpa.br, mpmota@ufpa.br
[2] IFPA, Instituto Federal do Pará, Belém, Brazil
jose.ribeiro@ifpa.edu.br
[3] IRMB, Université Montpellier, Montpellier, France
[4] CIn, Universidade Federal de Pernambuco, Recife, Brazil
rbcp@cin.ufpe.br
[5] ITV, Instituto Tecnológico Vale, Belém, Brazil
{vitor.cirilo.santos,ronnie.alves}@itv.org

Abstract. Intelligent systems that use Machine Learning classification algorithms are increasingly common in everyday society. However, many systems use black-box models that do not have characteristics that allow for self-explanation of their predictions. This situation leads researchers in the field and society to the following question: How can I trust the prediction of a model I cannot understand? In this sense, XAI emerges as a field of AI that aims to create techniques capable of explaining the decisions of the classifier to the end-user. As a result, several techniques have emerged, such as Explanation-by-Example, which has a few initiatives consolidated by the community currently working with XAI. This research explores the Item Response Theory (IRT) as a tool to explaining the models and measuring the level of reliability of the Explanation-by-Example approach. To this end, four datasets with different levels of complexity were used, and the Random Forest model was used as a hypothesis test. From the test set, 83.8% of the errors are from instances in which the IRT points out the model as unreliable.

Keywords: Explainable Artificial Intelligence (XAI) · Machine Learning (ML) · Item Response Theory (IRT) · Classification

1 Introduction

The expansion and increasing use of Artificial Intelligence (AI) systems creates advances that enable these systems to learn and make decisions on their own [11]. Thus, AI becomes increasingly common in everyday society by providing for simple or complex decisions in people's lives to be taken via intelligent systems. Such decisions range from recommending movies based on the user's preferences to diagnosing a disease based on patient's exams [15].

© The Author(s), under exclusive license to Springer Nature Switzerland AG 2022
J. C. Xavier-Junior and R. A. Rios (Eds.): BRACIS 2022, LNAI 13653, pp. 283–297, 2022.
https://doi.org/10.1007/978-3-031-21686-2_20

The question "Can the decision made by a black-box model be trusted for a context-sensitive problem?" has been asked not only by the scientific community, but also by the society as a whole. For example, in 2018 the General Data Protection Regulation was implemented in the European Union. It is geared at securing anyone the right to an explanation as to why an intelligent system made a given decision [20]. In this sense, for a continuous advance in AI applications, the entire community is faced with the barrier of model explainability [9,11]. To address this issue, a new field of study is growing rapidly: Explained Artificial Intelligence (XAI). Developed by AI and Human Computer Interaction (HCI) researchers, XAI is a user-centric field of study aimed at developing techniques to make the functioning of these systems and models more transparent and consequently more reliable [2]. Recent research shows that the trust calibration on the models' decision is very important, since exaggerated or measured confidence can lead to critical problems depending on the context [19].

The models that have high success rates to solve real-world problems are usually of the black-box type. In other words, they are not easily explained and, therefore, applying XAI techniques is required so that they can be explained and then interpreted by the end user [2,9]. The emergence of XAI techniques based on different methodologies is a real fact today, but there are still many gaps in literatute. For example, XAI methods based on Explanation-by-Example in a model-agnostic fashion[1] are still underexplored by the scientific community [8,10,18]. Techniques based on Explanation-by-Example use previously known ou model-generated data instances to explain them, thus providing for a good understanding of this model and decisions thereof. This is a technique that may be natural for human beings, since humans seek to explain certain decisions they themselves make based on previously known examples and experiences [2].

This research explores a new measure of XAI based on the working principles of Item Response Theory (IRT), which is commonly used in psychometric tests to assess the performance of individuals on a set of items (e.g., questions) with different levels of difficulty [3]. To this end, the IRT was adapted for Machine Learning (ML) evaluation, treating classifiers as individuals and test instances as items [16]. In previous works [5,16] IRT was used to evaluate ML models and datasets for classification problems. By applying IRT concepts, the authors were able to provide new information about the data and the performance of the models in order to grant more robustness to the preexisting evaluation techniques. In addition, the IRT's main feature is to explore the individual's performance on a specific item and then compute the information about the individual's ability and item complexity in order to explain why a respondent got an item right or wrong. Thus, it is understood that IRT can be used as a means to comprehend the relationship between the performance of a model and the data, thus helping in explaining models and understanding the model's predictions at a local level.

Given the intrinsic characteristics of the IRT, it is understood that it can be fitted within the universe of techniques based on Explanation-by-Example. At the same time, the IRT also has concepts that allow to explain and interpret

---

[1] Model-Agnostic: it does not depend on the type of model to be explained [18].

the model in general and to shed light on details not yet explored by other XAI techniques. Based on this motivation, this research work proposes the use of IRT as a new Explanation-by-Example approach, in a model-agnostic way, aiming at greater reliability on the model's decisions by the end user. For the experiment, 4 datasets were selected with different levels of complexity indicated by [22] with the Random Forest algorithm acting as the target of the explanation. The objective of this research is to explore how the concepts from the IRT can help to open the black-box and indicate the confidence of the model's prediction.

The remainder of this paper is divided into the following sections: Sect. 2 provides a contextualization about XAI and IRT; Sect. 3 explains how IRT is applied to ML and then to XAI; Sect. 4 provides the results and discussions of the proposal presented herein; Sect. 5 carries the conclusion of the herein research and final considerations related thereof.

2 Background

2.1 Explainable Artificial Intelligence - XAI

Based on the growing need to gain confidence in black-box models, the XAI community has proposed different methodologies, techniques and tools to explain these models. It is argued that, based on the creation of model explanation layers, a human user can create their interpretations and thus better understand how the model's decisions were generated, therefore obtaining greater confidence [2,17]. One of the most popular categories of XAI techniques currently available is the so-called post-hoc explanations. The main particularity of these post-hoc explanations is the fact that they only use training data, test data, model output data and the model itself, already properly trained to generate the explanations [2]. One of the most current and necessary characteristics that an XAI technique can feature is the fact that it is applicable to computational models of independent structural natures (neural network, tree, vector of weights etc., ...). This feature is called model-agnostic [17].

Among the current post-hoc XAI techniques, the following stand out: Text Explanations, Visual Explanations, Local Explanations, Explanations-by-Example, Explanations-by-Simplification and Feature Relevance Explanations. Out of these, this research highlights the Explanation-by-Example as a poorly explored technique by the XAI community. In fact, there is a smaller number of research works that present a clear proposal or tool that can be used in a replicable way for different real-world problems [8,10,17,18].

Example-based explanation methods select specific instances of the dataset in order to explain the behavior of models or to explain the underlying data distribution [17]. Explanations based on examples are mostly model-agnostic, since they make any model more interpretable. The most popular tool proposals for example-based explanations are: Counterfactual explanations [25], Adversarial examples [4], Prototypes [13] and Influential instances [14]. Each of these proposals seeks to carry out the process of identifying relevant instances of the dataset, which directly, or even indirectly, explain and justify the model's output [17].

It should be clear that the aforementioned tools feature individual differences in terms of their ability to point to meaningful instances to explain an ML model. Therefore, they may provide different results even on the same dataset. This is directly linked to the base algorithm or function on which each tool is based, as well as the complexity of the model (dataset and algorithm) analyzed [11,22]. Thus, it is understood that the proposed study of using the IRT to explain the ML model may generate merely different results when compared to other techniques mentioned previously, so it would be difficult to make an objective comparison. Furthermore, this research aims to apply the IRT to actually explore different details from the interpretation of the IRT estimators.

2.2   Item Response Theory - IRT

Traditionally, the number of correct answers is used to evaluate the performance of individuals in a test. However, this approach has limitations to assess the real ability of an individual. On the other hand, the IRT allows for evaluating the latent characteristics of an individual that cannot be directly observed, and it aims to present the relationship between the likelihood of an individual responding correctly to an item and their ability. One of the main characteristics of the IRT is that the core elements are the items and not the test as a whole, that is, an individual's performance is evaluated based on their ability to get certain items right in a test and not how many items they get right [3].

The IRT is a set of mathematical models that seek to represent the probability of an individual correctly responding an item as a function of the item parameters and the respondent's skill, as the greater the individual's skill, the greater the chance of getting the item right. Dichotomous items are the most used, as it is only considered whether the item was answered correctly or not [3]. The IRT allows for simultaneous assessment of both the items and the respondents. In order to characterize the items, the following parameters are commonly considered by IRT models: Discrimination (a_i), which represents how much the item i differentiates between good and bad respondents. The higher its value, the more discriminating the item; Difficulty (b_i), which represents how difficult an item is to be answered correctly and the higher its value, the more difficult the item; Guessing (c_i) represents the probability of a random hit or also the probability of a low-skill respondent hit the item.

To estimate item parameters, the response set of all individuals for all items to be evaluated is used. Respondents are evaluated based on the estimated ability (θ_j) and the probability of a correct answer calculated as a function of an individual's ability and the parameters of item i. The logistic IRT model that uses the three parameters (the 3PL IRT model) calculates the probability of a correct answer U_{ij} by the following equation:

$$P(U_{ij} = 1|\theta_j) = c_i + (1 - c_i)\frac{1}{1 + e^{-a_i(\theta_j - b_i)}} \tag{1}$$

Both item and individual parameters are simultaneously estimated using the response set, usually by maximizing the likelihood of the model given the response

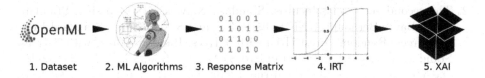

Fig. 1. Methodology for the application of IRT in ML and XAI.

data. The IRT can then be understood as a "magnifying glass" that allows for observing the individual's performance in a specific way on each item and for estimating a probable skill level in the area being evaluated.

3   Methodology

The IRT is generally applied for educational purposes, where the respondents are students and the items are test questions. To analyze datasets and learning algorithms through IRT in the herein research, instances of a dataset were used, with items and classifiers being assumed as respondents. The 3PL-IRT model was used because it is the most complete and consistent to fit responses [16].

Figure 1 illustrates the proposed methodology for applying IRT to open the box and then help explain ML models through the following steps:

1. A supervised learning dataset or benchmark is chosen and divided into training and testing;
2. Several ML models are built by using the training set and are adopted to predict the instances in the test dataset;
3. The response from these classifiers is collected in matrix form. Each row is associated with the classifier and each column is associated with a test instance. Each matrix entry represents whether an instance was correctly classified or not by a model (a 0|1 indicator);
4. The response matrix is used to build the IRT model and thus to estimate the item parameters of the test instances;
5. Finally, IRT estimators are used to open the box and assess the reliability of the model's predictions.

3.1   ML and IRT

Initially, given a dataset of interest and a pool of ML algorithms, steps 1, 2 and 3 in the proposed methodology result in a matrix of responses given as input to IRT (step 4). By default, the dataset is divided in a stratified manner, being 70% for training and 30% for validation (step 1). In order to generate a large number of responses, three sets of classifiers were built (step 2):

1. The first set is composed of 120 Random Forests models, where the number of trees gradually increases from 1 to 120;

2. The second set of classifiers is: Standard Naive Bayes Gaussian, Standard Naive Bayes Bernoulli, KNN (with 2, 3, 5 and 8 Neighbors), Standard Decision Trees, Random Forest (RF) with 3 Trees, Random Forest with 5 Trees, Standard Random Forests, Standard SVM and Standard MLP. The models classified as standard mean that the standard hyperparameters of Scikit-learn [21] were used. All models are trained using 10-fold cross-validation;
3. The third set is composed of 7 artificial classifiers advised in [16] to provide limit performance indicators of real classifiers and provide greater variability in the responses: an optimal classifier (classifies all instances correctly), a pessimal classifier (misses all classifications), a majority classifier (classifies all instances with the majority class), a minority classifier (classifies with the minority class) and three random classifiers (classifies randomly).

A matrix of responses is generated based on the predictions provided by the classifiers (step 3). The decodIRT tool [5], which automates from step 1 to step 4, was adopted in the herein paper. By definition, the tool generates 120 MLP models as the first set of classifiers. However, as one of the objectives of this study is to explain the Random Forest model, the tool was modified to suit the research objectives. To calculate the IRT estimators (step 4), the tool depends on the item parameters calculated from the model responses, the classifiers ability and the probability of success derived from the IRT logistic model.

This research proposes that the models can be explained based on the interpretation of the IRT estimators generated in the experiments (step 5). At first, it analyzes the item parameters generated for each dataset more generically to generate a general interpretation without a specific model. Then the item parameters are analyzed considering particular characteristics of the datasets. To this end, 3D graphs and histograms are generated to understand the relationship between data and item parameters. The probability of success and the ability of the models are used to measure the confidence of the classification result by comparing it to classic ML metrics. In addition, this research intends to explore the instances from the correlation analysis between the item parameters and the vector of features that make up the data to analyze and what examples are more interesting to explain and interpret a model decision. At same time exposing the models' confidence on its decision (Is the model basically guessing?).

3.2   Evaluated Datasets

As a case study, 4 binary datasets with different levels of complexity were chosen: Credit-g, Sonar, PC1 and Heart-Statlog. These datasets were selected from a total of 41 datasets, referring to binary classification problems extracted from OpenML [24]. These datasets were selected by relying on a clustering processes that identified groups of datasets in OpenML with distinct properties. Then more varied datasets across clusters were selected.

In the clustering process, *K-means* algorithm followed by a Multiple Correspondence Analysis - MCA [22] were adopted to cluster the datasets described by 15 different properties. The clustering process resulted in three main clusters, with respectively 21, 17 and 3 datasets. It is worth mentioning that this number of clusters was found from silhouette coefficient values, as recommended by the literature [12]. The MCA analysis also took into account the 15 different properties used in the clustering process, but with the addition of the label indicating the cluster to which each dataset belonged. Thus, as a result of the MCA, a graph was obtained with the spatial arrangement of all the analyzed datasets in relation to their 15 properties [1].

Thus, by inspecting the graph resulting from the MCA, it was possible to choose the 4 datasets mentioned previously in this topic, while taking due care to select 2 datasets from each cluster that exhibited considerable distances from one another, since datasets are therefore obtained with the most distinct properties possible. The cluster with 3 datasets was disregarded for being too small and for not showing sufficient separation from the other clusters according to the visual inspection of the MCA graph. It should be noted that the Credit-g and Heart-Statlog datasets belong to the most complex dataset cluster, while the Sonar and PC1 datasets belong to the simplest dataset cluster, as seen in [22].

The Heart-Statlog is a heart disease dataset, where each instance represents a diagnosed individual whether or not you have a heart disease. The dataset has 270 instances and 13 features. The dataset also has a slight class imbalance, with 55.56% of the instances being the majority. Sonar is a dataset of sonar signals, where each instance represents a sonar signal that has been reflected by a cylindrical rock or a metal cylinder. With 208 instances and 60 features, being 53.36% of the majority instances. Credit-g is a dataset for credit analysis that classifies the credit risk of individuals as good or bad. It is composed of 1000 instances, with 20 features, this dataset being more unbalanced with 70% of instances of the majority class. The PC1 dataset is a dataset and defects of the NASA Metrics Data Program, it is composed of data from the flight software for a satellite in Earth's orbit, where each instance informs whether or not the module has a defect. It has 1109 instances, with 21 features, 93% belonging to the majority class, thus configuring a very unbalanced dataset.

4   Results and Discussion

The evaluation of the use of IRT, regarding the Explanation-by-Example process, was split in two stages[2]:

1. The first focuses on the dataset and what explanations the item parameters can reveal about the data;
2. The second is about the specific model generated and how the IRT estimators can act in the explanation process at the local level.

4.1   Datasets Through the Lens of IRT

First, only the item parameters that were estimated for the test instances of the datasets will be evaluated. In IRT, discrimination and difficulty values can

---

[2] All results can be accessed at: https://github.com/LucasFerraroCardoso/IRT_XAI.

range from $-\infty$ to $+\infty$. Thus, in order to consider whether the items have high values of discrimination and difficulty, the established assessment value was zero (0). Thus, instances are considered very difficult and very discriminative if their respective values are greater than 0. For the guessing parameter, the limit presented by [5] was used, which considers that instances with high guessing values are those with values greater than or equal to 0.2. Despite the difficulty and discrimination parameters being the most directly linked to the data, due to their characteristics, it is understood that the guessing parameter is important to consider for an indirect evaluation of the model. In view of this, the following data were computed.

As can be seen in Table 1, all datasets can be considered as being very discriminative because they feature a high percentage of instances with discrimination above 0. This means that the datasets can discriminate high and low skill classifiers. Therefore, models that feature a high hit rate for these datasets, indeed, can be considered skillful.

Table 1. Table with the percentage of test instances with high values of discrimination, difficulty and guessing.

Dataset	Discrimination	Difficulty	Gessing
Sonar	87.30%	4.76%	14.29%
PC1	93.99%	2.1%	3.9%
Heart-Statlog	85.19%	3.7%	25.93%
Credit-g	77.67%	6%	15%

The difficulty parameter can also reveal important information. In this case, all datasets have few instances with high difficulty values, this can mean that the datasets themselves are easy to classify and are not a challenge. To assess model confidence, this information can be interpreted in two ways: first, considering that a dataset represents the real world very well, with more than 90% of the instances being considered easy, skilled models trained with that dataset have high chances of being reliable and correctly hitting new cases. However, if the dataset is not a reliable representation of the real world, this could also mean that few truly challenging cases are addressed by the dataset and thus the resulting model would only be prepared to correctly classify the easier cases. Thus, one can explain the model as being of high precision, but only for easy cases.

The guessing parameter is still difficult to assess. Regarding the application of IRT in ML, no research work was found that has deeply explored the impact of the guessing parameter; but, by using the IRT concepts, it is possible to raise some hypotheses. In IRT, high guesswork values usually mean that there is something in the item itself that gives a "hint" to the low-skill respondent on what the correct answer is. In ML, this may mean that within the dataset the data may have some bias that facilitates its correct classification. This may be related to the concept of "shortcut learning" [7] that happens during training when the model finds a gven characteristic in the data that correlates with the correct class and then the model starts using this shortcut instead of evaluating the entire data. Thus, if the model has low skill, then its correct classification may be biased by the data and this may not be repeated in the real world as the model would not have generalized properly. An unskilled model for high-guess data would be unreliable. Future research would involve exploring this condition through the purposeful insertion of such biases and then evaluate with the IRT.

In order to deepen the explanation of the datasets by the IRT, the specific characteristics of the datasets will be considered. It is noted that the Credit-g and Heart-Statlog datasets, when compared to the Sonar and PC1 datasets, are on average less discriminative, more difficult and have a greater chance of casual accuracy. Even if by little difference, this corroborates the classification of [22] as being more complex. However, it is clear that the Sonar dataset, considered less complex, has the second highest percentage of difficulty and this may be related to the high dimensionality of the dataset.

Furthermore, it is understood that other metadata can also help explain a model. Although all datasets have high discrimination values, the reason these values can be different for each dataset and metadata can help reveal this difference. For the very unbalanced Credit-g and PC1 datasets, it can be seen that the percentage of very discriminative instances is very close to the percentage of the majority class, with PC1 having 93% of instances of the majority class and 93.99% of discrimination and Credit-g, which is composed of 70% of the majority class and has 77.67% of very discriminating instances.

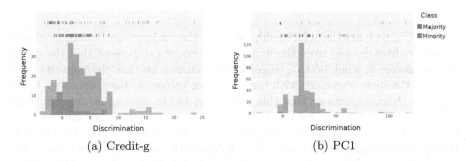

(a) Credit-g   (b) PC1

Fig. 2. Discrimination Histogram separated by majority and minority class.

It is common in IRT that items with high discrimination also have low difficulty, as it is understood that if a respondent makes a mistake with an item considered easy, then the ability must be low. In ML, a classifier can be considered unskilled if it cannot hit the instances referring to the majority class, as they are more recurrent in the dataset. So, it is correct to imagine that for very unbalanced datasets the percentage of very discriminative and easy instances may coincide with the majority class. As can be seen in Fig. 2 for the Credit-g

(a) and PC1 (b) datasets, note that the histogram shows the highest number of instances with discrimination values above 0 for the majority class, while the minority class has more instances with negative discrimination. On the other hand, it is interesting that the highest discrimination values are for the minority class, this may occur due to the lower number of items and because some instances may have a strong characteristic that links them to the minority class. This can then reveal which instances of the minority class are most representative of the group and may be the most informative instances to explain the model. Besides, this information can also be useful in selecting the most suitable instances to feed oversampling techniques.

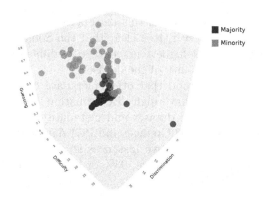

Fig. 3. Credit-g instances arranged over item parameters.

The Fig. 3 shows the relationship between the majority and minority classes of the Credit-g dataset from the item parameters, it can be seen that the guessing parameter is what best distinguishes the classes, so that the minority class exhibits the more instances with high guessing values, so they are easier to hit casually. Despite assuming that this may be related to class imbalance, the same behavior is not repeated for the PC1 dataset (see supplementary material (see footnote 2)), which is the most imbalanced one. This condition may be unique to the dataset's characteristics and would reinforce the assumption that the model may not be reliable for minority class instances if it does not have a high ability. The item parameter and model relationship will be explored in the next section.

4.2   Random Forest Through the Lens of IRT

In addition to evaluating the data in general, IRT also allows for evaluating the classifier's ability to correctly classify a specific instance. Thus, this second part of the section addresses how the IRT explains the decisions made by the evaluated model, a Random Forest with 100 trees.

Table 2 shows the results of the test set classification of datasets by Random Forest. Due to the accuracy, a model is in place with good performance in almost all cases, only for Credit-g the model showed a lower hit rate. Considering the existing imbalance in the datasets, the Matthews Correlation Coefficient (MCC) [6] of each model was also calculated and even for the least unbalanced datasets (Sonar and Heart-Statlog) the highest value of MCC was 0.71, indicating low correlation between classes and reinforcing the imbalance problem when considering all test instances. IRT, in turn, points to Random Forest as a skillful model, as the skill value is greater than the difficulty value in more than 90% of the instances in all datasets. In the IRT, the respondent's skill and the item's difficulty are measured on the same scale, so that if the skill value (θ) is equal to the item's difficulty, the chance of hitting must be equal to 50%.

Table 2. Random Forest performance for all test instances and for instances without negative discrimination.

Dataset	Acc total	MCC total	Ability θ	Acc WNG*	MCC WNG*
Sonar	86%	0.71	1.40	94%	0.88
PC1	94%	0.36	3.76	99%	0.92
Heart-Statlog	84%	0.68	1.20	97%	0.94
Credit-g	76%	0.38	2.07	96%	0.88

*Without Negative Discrimination.

When the difficulty limit is changed to the model's ability, it is noticed that the difficulty of the datasets has decreased considerably, reaching zero in the case of the Heart-Statlog dataset. In the specific case of this dataset, the IRT states that the generated model has a confidence of more than 50% of success in all test instances, at least. Furthermore, the difficulty is practically zero for PC1 as well, with 0.3% difficulty. The exception was the Sonar dataset, which kept exactly the same level of difficulty as before (4.76%), by IRT this means that the model has less than 50% confidence of success for 3 instances of the test set. For Credit-g the new percentage of difficult instances is 2.33%, which means that the model has less than a 50% hit chance for 7 instances of the test set. Such instances become very interesting to explain in what cases the model does not have a reliable prediction.

But if the model has such a high estimated skill and the datasets showed difficulty below the skill level of the model, then why were there still more errors and the MCC value was low? The answer to this question may also lie in the IRT discrimination parameter. In the IRT, negative discrimination values are not expected, despite being possible. The reason is that negative discrimination constitutes a situation where the less skilled respondents have the highest chance of getting it right, while the most skilled respondent has the least chance. Taking as an example two instances of Credit-g with very close values of difficulty and guessing, the probability of success of Random Forest can be completely different in both cases if the discrimination is negative. For the first instance with 1.59 discrimination the chance of success is almost 100%, while for the instance with -1.57 discrimination the chance of success is less than 40% (see Fig. 4).

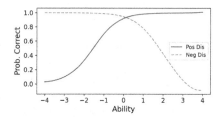

Fig. 4. Comparison between items with positive and negative discrimination.

It was observed that, on average, 83.8% of the instances that Random Forest missed in the four datasets have negative discrimination. Commonly, negative discrimination values usually mean something wrong with the item that makes it difficult to answer correctly. For the ML field, this could mean the existence of noise or an outlier in the instance. Such values may also arise when the item is different from the others, so the respondent does not have sufficient prior knowledge to answer the item correctly. For ML, this could mean that there was not enough data in training for the classifier to learn how to classify the item, as in unbalanced data correctly. Martinez et al. [16] already pointed to discrimination as a more exciting parameter than the difficulty itself.

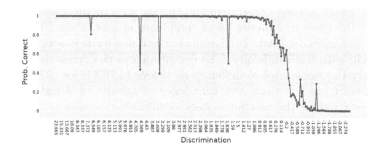

Fig. 5. Probability of correct answer for each Credit-g test instance.

Figure 5 displays the drop in the probability of success of Random Forest for the Credit-g dataset as the discrimination value of the test instances decreases; this exact configuration is repeated for the other datasets. A priori, this situation can be analyzed as follows: the model has an accuracy level above the obtained one because the instances with negative discrimination have some inconsistency and therefore can be disregarded, or the accuracy is correct, and these instances may represent types of data that the model is not able to classify correctly, as these data were not explored adequately during the training stage. In any case, this means that the model is not reliable for instances of negative discrimination.

Table 2 reveals that the impact of negative discrimination is inherent to the type of data, since datasets of different complexity and characteristics were chosen and even so, all models showed improved performance when considering only instances with positive discrimination, for both the accuracy and the MCC value that more than doubled for the unbalanced datasets. The other model errors are usually related to the difficulty of the instance. As can be seen in Fig. 5, even for instances with positive discrimination, in some cases the model has a low probability of success and this occurs when the difficulty of the instance exceeds the model's ability. It is interesting, then, to open the instances and study how the features relate to the item parameters.

Using the Heart-Statlog as an example because it is a context-sensitive dataset, a correlation analysis was performed between the values of the features and the item parameters of each instance. Out of all the test instances, only the "chest" feature has a slightly higher correlation with the difficulty parameter at 0.2485. However, when filtering the instances only for those that the model missed, it is already possible to notice greater correlations as can be seen in Table 3.

Table 3. Features with correlation above 0.4 for some item parameters of the Heart-statlog dataset.

Dataset	Discrimination	Difficulty	Guessing
chest	0.0285	0.6395	−0.3330
fasting_blood_sugar	−0.4560	0.1320	−0.1815
resting_electrocardiographic	−0.0911	−0.3875	0.4592
number_of_major_vessels	−0.5391	0.2031	−0.2924

For misclassified instances, the "chest" feature has the highest correlation, with 0.6395 for difficulty, this means that this is the feature that makes it the most difficult to classify these instances. The "chest" feature is about the type of chest pain that a patient may have and when the specific context of the dataset is explored, it is noted that this evaluation by the IRT makes sense, as other studies have already pointed out that it is difficult to identify if chest pain is a sign of heart disease [23]. In addition, the "resting_electrocardiographic_results" feature was identified as the feature that most correlates with guesswork, at 0.4592. Therefore, this would be the feature that "gives' the most clues, so that an unskilled model can get the classification of an instance right. The "fasting_blood_sugar" and "number_of_major_vessels" features are the ones that have the highest correlation with discrimination, so these may be the instances that can best be used to discriminate poorly defined models. Trusted of the most trusted. Thus, a model that has these features as the most important to classify an instance, and the model misses, means that the model has low skill. And as expected, these same features have a high correlation with negative discrimination, as it is already known that they have a high correlation with the discrimination parameter itself. However, the "oldpeak" feature also presented a high correlation, at 0.4495 for negative discrimination, which indicates that this feature may be responsible for the composition of poorly formulated instances that impair the performance of the model. When performing a percentile analysis, it was seen that 90% of the test instances have less than half of the maximum possible value of this feature, both for the majority and minority classes, this situation indicates that any value above half the maximum value can turn out to be an outlier, thus bringing about an inconsistency in the instances and resulting in negative discrimination. This does not mean that this or the other features indicated by the IRT should be removed from the dataset, but that it is important to be aware of their values and aware of their impact on the model's confidence.

5   Final Considerations

This research paper presented how the IRT can be used in the Explanation-by-Example process, aiming to assist in the process of explaining a black-box model with a focus on explaining the decision made by the model and thus greater reliability for the end user. To this end, four binary datasets of different complexity were used: Heart-statlog, Credit-g, Sonar and PC1. Along with the Random Forest black-box classifier as a case study. It was observed that the IRT is able to provide new pertinent information about the classifier and data relationship, where the item parameters can be used to evaluate if the dataset concerned really encompasses all types of cases and if its own data composition can make or break a model's classification. It was also observed that the calculation of the IRT success probability can be used to measure the level of reliability that one can have on a classifier, when the model is faced with a specific instance, and thus indicating in what specific cases the model is or is not reliable, where in 83.8% of the wrongly classified instances the IRT points out that the model is not reliable. Future research would further explore what conditions within the instances make them have a higher or lower difficulty and discrimination, in order to create conditional rules that can predict how the model will behave in view of a new instance.

References

1. Abdi, H., Valentin, D.: Multiple correspondence analysis. Encycl. Meas. Stat. **2**(4), 651–657 (2007)
2. Arrieta, A.B., et al.: Explainable artificial intelligence (XAI): concepts, taxonomies, opportunities and challenges toward responsible AI. Inf. Fusion **58**, 82–115 (2020)
3. Baker, F.B.: The basics of item response theory (2001). http://ericae.net/irt/baker
4. Biggio, B., Roli, F.: Wild patterns: ten years after the rise of adversarial machine learning. Pattern Recogn. **84**, 317–331 (2018)

5. Cardoso, L.F.F., Santos, V.C.A., Francês, R.S.K., Prudêncio, R.B.C., Alves, R.C.O.: Decoding machine learning benchmarks. In: Cerri, R., Prati, R.C. (eds.) BRACIS 2020. LNCS (LNAI), vol. 12320, pp. 412–425. Springer, Cham (2020). https://doi.org/10.1007/978-3-030-61380-8_28

6. Chicco, D., Jurman, G.: The advantages of the Matthews correlation coefficient (MCC) over F1 score and accuracy in binary classification evaluation. BMC Genomics 21(1), 1–13 (2020)

7. Geirhos, R., et al.: Shortcut learning in deep neural networks. Nat. Mach. Intel. 2(11), 665–673 (2020)

8. Gilpin, L.H., et al.: Explaining explanations: an overview of interpretability of machine learning. In: 2018 IEEE 5th International Conference on Data Science and Advanced Analytics (DSAA). IEEE (2018)

9. Gohel, P., Singh, P., Mohanty, M.: Explainable AI: current status and future directions. arXiv preprint arXiv:2107.07045 (2021)

10. Guidotti, R., et al.: A survey of methods for explaining black box models. ACM Comput. Sur. (CSUR) 51(5), 1–42 (2018)

11. Gunning, D., Aha, D.: DARPA's explainable artificial intelligence (XAI) program. AI Mag. 40(2), 44–58 (2019)

12. Rousseeuw, P.J.: Silhouettes: a graphical aid to the interpretation and validation of cluster analysis. J. Comput. Appl. Math. 20, 53–65 (1987)

13. Kim, B., Rajiv K., Koyejo, O.O.: Examples are not enough, learn to criticize! criticism for interpretability. In: Advances in Neural Information Processing Systems 29 (2016)

14. Koh, P.W., Liang, P.: Understanding black-box predictions via influence functions. In: International Conference on Machine Learning. PMLR (2017)

15. Linardatos, P., Papastefanopoulos, V., Kotsiantis, S.: Explainable AI: a review of machine learning interpretability methods. Entropy 23(1), 18 (2020)

16. Martínez-Plumed, F., et al.: Item response theory in AI: Analysing machine learning classifiers at the instance level. Artifi. Intel. 271, 18–42 (2019)

17. Molnar, C.: Interpretable machine learning (2020). Lulu.com

18. Molnar, C., Casalicchio, G., Bischl, B.: Interpretable machine learning – a brief history, state-of-the-art and challenges. In: Koprinska, I., et al. (eds.) ECML PKDD 2020. CCIS, vol. 1323, pp. 417–431. Springer, Cham (2020). https://doi.org/10.1007/978-3-030-65965-3_28

19. Naiseh, M., et al.: Explainable recommendation: when design meets trust calibration. World Wide Web 24(5), 1857–1884 (2021)

20. Regulation, P.: General data protection regulation (GDPR). Intersoft Consulting. Accessed October 24 Jan 2018

21. Pedregosa, F., et al.: Scikit-learn: machine learning in python. J. Mach. Learn. Res. 12, 2825–2830 (2011)

22. Ribeiro, J., et al.: Does dataset complexity matters for model explainers?. In: 2021 IEEE International Conference on Big Data (Big Data). IEEE (2021)

23. Sabatine, M.S., Cannon, C.P.: Approach to the patient with chest pain. In: Braunwald's Heart Disease: A Textbook of Cardiovascular Medicine. 9th edn., pp. 1076–1086. Elsevier/Saunders, Philadelphia (2012)

24. Vanschoren, J., et al.: OpenML: networked science in machine learning. ACM SIGKDD Explor. Newsl. 15(2), 49–60 (2014)

25. Wachter, S., Mittelstadt, B., Russell, C.: Counterfactual explanations without opening the black box: automated decisions and the GDPR. Harv. JL Tech. 31, 841 (2017)

Short-and-Long-Term Impact of Initialization Functions in NeuroEvolution

Lucas Gabriel Coimbra Evangelista and Rafael Giusti[(✉)] [iD]

Institute of Computing, Federal University of Amazonas, Manaus, Brazil
{lucas.evangelista,rgiusti}@icomp.ufam.edu.br

Abstract. Neural evolutionary computation has risen as a promising approach to propose neural network architectures without human interference. However, the often high computational cost of these approaches is a serious challenge for their application and research. In this work, we empirically analyse standard practices with Coevolution of Deep NeuroEvolution of Augmenting Topologies (CoDeepNEAT) and the effect that different initialization functions have when experiments are tuned for quick evolving networks on a small number of generations and small populations. We compare networks initialized with the He, Glorot, and Random initializations on different settings of population size, number of generations, training epochs, etc. Our results suggest that properly setting hyperparameters for short training sessions in each generation may be sufficient to produce competitive neural networks. We also observed that the He initialization, when associated with neural evolution, has a tendency to create architectures with multiple residual connections, while the Glorot initializer has the opposite effect.

Keywords: Deep NeuralEvolution · Genetic algorithms · Weight initialization

1 Introduction

Deep Neural Networks (DNNs) are among the most used machine learning methods nowadays. They can be applied in multiple scenarios and are able to approximate functions that are often considered too complex for "classic" models, such as Support Vector Machines and shallow Neural Networks. However, DNNs tend to be complex, so their training usually requires very large datasets and they are computationally expensive. This is particularly challenging for the task of fine-turning hyperparameters, since their validation may take considerable time.

We thank Coordination for the Improvement of Higher Education Personnel - CAPES/PROAP and Amazonas State Research Support Foundation - FAPEAM/POSGRAD 2021. This research was partially supported by CAPES via student support grant #88887.498437/2020-00.

J. C. Xavier-Junior and R. A. Rios (Eds.): BRACIS 2022, LNAI 13653, pp. 298–312, 2022.
https://doi.org/10.1007/978-3-031-21686-2_21

When designing a DNN, several factors must be considered, especially the number and design of the layers. Once the architecture has been chosen, training a DNN for a specific task requires defining several hyperparameters, such as the number of epochs, learning rate, optimization function, batch size, etc. Considering the advances in the last decade in the development of Deep Learning models to deal with challenging tasks and the remarkable effort involved in designing these models "by hand", methods capable of automatically finding ideal DNN architectures without human intervention have been growing increasingly relevant. Many of these advances were possible with the birth of the field of Deep Learning and Bio-Inspired Algorithms, which created a new area of study: Deep Evolutionary Neural Networks, also called Deep NeuroEvolution [1].

Deep NeuroEvolution (DNE) deals with using evolutionary algorithms with the specific purpose of optimizing the architecture of a DNN to solve difficult problems. Like "classical" evolutionary algorithms, the DNE approach employs a population of sub-optimal solutions, called individuals, and the goal is to identify and combine the best elements of the individuals to find solutions that are as close to the optimal as possible. This usually happens in cycles called generations. In the context of DNE, it means combining the most promising elements of neural networks (e.g., layers, neurons, or blocks of layers) at the end of each generation, creating architectures that are increasingly more suitable for some specific task, such as classification, anomaly detection, time series forecasting, and others.

One important issue that has received relatively little attention in NeuroEvolution is the weight initialization. At each generation, the evolutionary algorithm produces neural networks whose weights may need to be optimized from scratch. Different initialization functions may be used to set the initial neuron parameters, but not all of them are equally suitable. A bad initialization choice can lead to some undesirable behaviors such as the vanishing gradient problem and the gradient explosion problem [2,3]. Generally, these phenomena occur when the parameters tend to zero or to infinity, respectively, making it impossible for the machine learning model to converge during the learning process [3,4]. Furthermore, different initialization functions may lead to different architectures.

In the past decade, we have witnessed many alternative optimizations and studies on parameter initializers. These studies draw primarily on two groundbreaking researches that introduced the most commonly used functions at present time: the Glorot initialization [5] and the He initialization [6].

However, these functions were proposed to suit DNNs with a fixed architecture, which is designed "by hand" after meticulous choices on how many layers the network should have and on the layout of those layers. In NeuroEvolution, neural network components are combined in often unexpected ways, and in a sense we can think of the individuals as networks that evolve over time. Furthermore, while in "traditional" application a DNN architecture is defined once and then trained over a reasonable period of time for some specific application, this is not as easily done in NeuroEvolution, where a large number of networks have to be quickly trained at each generation, and the best elements of a DNN must be identified and combined with others to produce better individuals. With these issues in mind, we ask the following question: how do standard weight initialization functions help to rapidly converge a neural network and find the ideal DNN elements in a scenario where the network architecture is constantly changing? This question is crucial for evolutionary algorithms to become, in fact, competitive to non-evolutionary DNNs, which represent the current state of the art.

The remained of this paper is as follows. Section 2 presents related works that address weight initiatlization or maintenace in DNE. Section 3 explains the theoretical foundation behind this work. Section 4 describes and discusses our results. Finally, Sect. 5 draws final remarks.

2   Related Work

Much of the research effort in DNE is directed towards finding better ways to create ideal topological structures for the target problem, somewhat neglecting the potential benefits of better initializing or updating the weights of the networks during the evolutionary process. In this section, we give attention to some works that tackle the latter issue.

Focusing on *updating* weights, Koutnik et al. [7] created a new method to encode the weights of neural networks using Fourier coefficients. This allows exploring the spatial relationship between the weights and reducing the dimensionality of the target problem. As a main result, Koutnik et al. managed to reduce the total number of iterations to obtain the best individual in three benchmark problems (*pole-balancing*, *ball throwing*, and *octopusarm control*). Togelius et al. [8] decided to focus efforts on the crossover stage during the evolution of architectures, inspired by the workings of memetic algorithms, to find the best combination of weights between individuals rather than a random one. As benchmark they used the Race Car problem, comparing five different algorithms: Hill-Climber, Simultaneous Climber, Memetic Climber, Constrained Memetic Climber, and Inverse Memetic Climber. The authors report that the more features in the input layer (dimensions), the better results the evolutionary versions of the algorithms obtained, always outperforming the non-evolutionary versions. Neither works, however, address the total time to carry out the experiments.

Specifically considering the weight *initialization* problem, Okada et al. [9] proposed to represent the weights of neural networks not as scalars but as intervals, an extension of Evolutionary Strategy for neurevolution of intervalued neural networks, deciding not to evolve the topology of the architectures, and to work only with the prediction of values for sinusoidal functions, considering the intervals of such functions as genotypes. Also in this aspect, Desell [10], applied a new NeuroEvolutionary algorithm, EXACT (*Evolutionary eXploration of Augmenting Convolutional Topologies*), in order to address three possible fronts during the evolutionary process: (i) node-level mutation operations; (ii) epigenetic weight initialization; and (iii) pooling connections. When coining the expression *epigenetic weight initialization*, Desell aimed to investigate a better representation of the combination of the parents' genomes (weights), without necessarily changing the initial combination; in this way, the new individuals have their weights directly inherited from this epigenetic combination, similarly to [8]. As benchmark, the results were compared on the MNIST database, where the best individual obtained 99.46% accuracy. In all, Desell evolved over 225,000 neural networks with the support of 3,500 volunteers who served as hosts to run the experiments. None of these works comment on the total execution time.

In 2021, Lyu et al. [11] compared the performance of the NeuroEvolutive EXAMM (*Evolutionary eXploration of Augmenting Memory Models*), a gradient-based algorithm for recurrent neural networks, using four distinct functions to initialize the weights of the neural networks: Glorot, He, Uniform Random, and a new method called Lamarckian weight inheritence. To test the performance, EXAMM was applied in time series forecasting on real-world databases. The experiments were performed with 2,304 processing cores (Intel Xeon Gold 6150@2.70 GHz CPU) and a whopping 24 TB of RAM. This is the only work in our review that directly mentions the problem of the exploding and vanishing gradients. However, again the authors do not report the required processing time for their experiments.

3   Theoretical Foundation

3.1   CoDeepNEAT

In this paper we intend to work with the *coevolution of Deep NeuroEvolution of Augmenting Topologies* (CoDeepNEAT) algorithm [12]. As pointed out by Papavasileiou et al. [13], CoDeepNEAT is among some of the most innovative techniques that combine evolutionary algorithms based on non-gradient descent and algorithms based on gradient descent. In CoDeepNEAT, chunks of layers are developed as modules, which are merged together to create a blueprint, which in turn is used to create multiple architectures (Fig. 1).

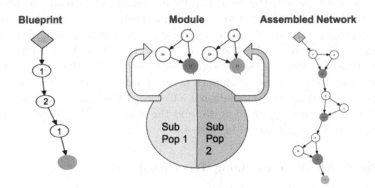

Fig. 1. Assembly of a neural network from a blueprint individual and a set of module individuals [12]. The number inside each node in the blueprint represents species in the modules population. (Color figure online)

The algorithm works with a population of modules and a population of blueprints, which are evolved separately. The blueprint chromosome is a graph where each node contains a pointer to a particular module species, while each module chromosome is a graph that represents a chunk of DNN layers.

Before the first generation, the initial population of modules is generated according to the hyperparameters of the CoDeepNEAT algorithm. A module may contain convolution layers or fully-connected layers, and the probability of getting either one is specified by the user (we keep both at 50%). If the module contains convolution layers, its hyperparameters, such as the number of filters and the kernel size, are chosen randomly from a set of possible values that are also user-specified. In our experiments, the number of filters ranges from $[32, 256]$ and the kernel size is either 1, 3, or 5.

At the beginning of each generation, a number of blueprints are randomly chosen from the blueprint population. Each node in a blueprint is replaced with a module that is randomly chosen from the module population. This results in a set of assembled neural network that comprise the individuals of that generation. Those neural networks are untrained, and their weights must be initialized, even if they come from modules that did not change since the previous generation. Each individual is then trained on a small number of epochs and evaluated on a hold-out partition of the training dataset.

Since CoDeepNEAT works with two populations, there are two fitness values to be calculated. The fitness of each blueprint individual is the average metric (loss, accuracy, F1-score etc.) of all neural networks assembled from that individual. If the blueprint was not chosen to sprout neural networks in this generation, then its fitness value is zero. Similarly, the fitness of each module is the average metric of all networks that employed that module as part of their architecture.

At the end of each generation, except for the last one, new blueprints are generated from the crossover of the best individuals, producing a new population of blueprints. The same is done to the population of modules. After the last generation is completed, the NeuroEvolution algorithm is ready to present to the user the indicated DNN for some particular task. This may be the individual with best fitness value from the final generation, or the best individual from all generations. This individual is then trained with the entire dataset for a suitable number of epochs, which is usually significantly larger than the number of epochs used to calculate the fitness of blueprints and modules during evolution.

The size of the modules population is a CoDeepNEAT hyperparameter. In our experiments, they are either 30 or 45, and the size of the blueprints population is either 10 or 25, as explained in Sect. 4.

3.2 Short, Medium and Long Term Analyses

A major drawback of bioinspired deep learning models is that they are very time consuming. In spite of their very competitive results, the time required for the evolution of architectures can make reproducibility difficult and is a challenge for researchers without access to high computational power. To put in perspective, Bohrer et al. [14] give an estimated 480+ h to reproduce the CIFAR-10 dataset experiments from the inaugural work of the CoDeepNEAT [12]. They reach that figure from the assumption that each epoch takes at least 30 s, and each of the 100 DNNs in one generation is trained for 8 epochs over a total of 72 generations.

The obvious way to reduce the required time is to reduce the complexity of the neural networks. In [14], the hyperparameter space is drastically reduced. To begin with, they evolved the networks for only 40 generations, instead of the 70 initially employed in [12]. The number of neural networks, blueprints, and modules was also reduced, as well as the complexity of the modules (fewer filters, larger possible kernels, and no max pooling). With a smaller search space, they also reduced the amount of training data. The CIFAR-10 dataset contains 50,000 training instances. While [12] uses all of them on a hold-out scheme to train and validate the DNNs, [14] employs only 40% of that data.

In this work, we consider training the neural networks during merely four generations, and we employ the same population sizes and use as much data for training as [14]. This is what we refer to as "short-term" evolution of neural networks. The main motivation behind this experiment is to verify whether an initialization function presents advantage over others when individuals are evaluated after very short training episodes. We also verify whether ReLU or hyperbolic tangent is more suitable an activation function in the short-term.

Subsequently, we also analyze the behavior of CoDeepNEAT in the medium and long term. In both cases, we use the entirety of the training dataset, with a train/validation hold-out partition, to evolve neural networks. And we increase the resources available to CoDeepNEAT in both instances.

Another argument worth mentioning to justify larger evolutionary hyperparameters is the emergence of skip connections with output summation, which resemble residual architectures, according to Miikkulainen et al. [12]. It is known that the loss function tends to be chaotic in very deep architectures, which is unfavorable for trainability and hinders generalization. Residual architectures are very popular in deep learning [15–17] because they tend to simplify that search space. This was experimentally validated in [18], which shows that the loss landscape changes significantly when skip connections are introduced, as illustrated in Fig. 2.

The residual connections implementation presents an uninterrupted flow of gradient from a given layer to the one closest to the output, avoiding the problem of vanishing gradient. Consequently, multiple skip connections are an alternative to ensure the reuse of resources of the same dimensionality as the previous layers. On the other hand, these connections are also useful for recovering spatial information lost during downsampling, and seem to stabilize gradient updates in very deep architectures, ensuring rapid convergence. Such structure is so important that it was the main motivator used by Miikkulainen et al. to introduce mutation of connections between neurons in the original CoDeepNEAT [12,19].

3.3 Initialization and Activation Functions

This work focuses on the analysis of the results obtained with coevolutionary algorithms when different initialization functions are employed. In spite of its

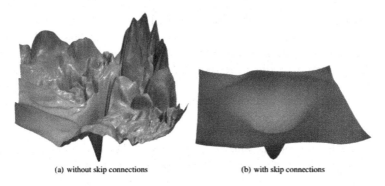

(a) without skip connections   (b) with skip connections

Fig. 2. Loss surface of a ResNet-56 without (left) and with (right) skip connections [18].

apparently limited impact on the algorithm, the choice of an initialization function can lead the algorithm to find a mostly linear neural network, or one with multiple skip connections, as illustrated in Fig. 3.

The first initialization considered is Glorot, proposed in 2010 and designed for DNNs with symmetric activation functions such as *tanh* and *softsign* [5,11]. It draws weights from a random uniform distribution such as:

$$W \sim \mathcal{U}(-\sqrt{\frac{6}{f_{in} + f_{out}}}, \sqrt{\frac{6}{f_{in} + f_{out}}}), \tag{1}$$

where f_{in} and f_{out} are the input and output sizes of the layer [5,20]. It also has a normal form, with N(0, std^2) [21], where:

$$std = \sqrt{\frac{2}{f_{in} + f_{out}}}. \tag{2}$$

As pointed out by Goodfellow et al. [4], the formula is derived on the assumption that the network consists only of a chain of matrix multiplications, with no nonlinear activations.

He initialization, in contrast, is designed for non-symmetric activation functions such as ReLU. The weights in each layer are generated to approximate the derivative of the activation function from 0 to 1 [6,11]. Its uniform formula is as follows [6,22]:

$$W \sim \mathcal{U}(-\sqrt{\frac{6}{f_{in}}}, \sqrt{\frac{6}{f_{in}}}). \tag{3}$$

Similar to Glorot, the He function also contains a normal distribution form, $\mathcal{N}(0, std^2)$ [23], where:

$$std = \sqrt{\frac{2}{f_{in}}}. \tag{4}$$

Although we have seen the emergence of non-monotonic activation functions as alternatives to new directions with the rise of functions such as swish [24] and mish [25], studies are still placed considering Glorot and He initializers. For the scope of this work, let us consider the normal and uniform versions of both Glorot and He.

As both functions are the top choice used for initiate a DNN, those functions are tested with CoDeepNEAT, alongside Random initialization. In the end, six activations are used in this paper: Glorot Normal, Glorot Uniform, He Normal, He Uniform, Random Normal and Random Uniform.

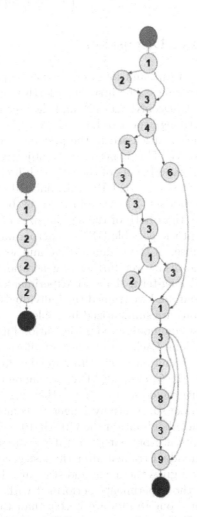

Fig. 3. Overall structure of two different networks evolved with Glorot initialization (left) and He (right). Each node represents a layer, and the different nodes represent different modules. Notice the linearity of the network on the left and the presence of skip connections on the right.

The performance between the weights initialization functions with different activation functions also needs to be addressed. Here the Hyperbolic Tangent (TanH, or tanh) and *Rectified Linear Unit* (ReLU, or relu) functions were chosen, both non-linear and commonly used as standard functions.

$$\tanh(x) = 2\sigma(2x) - 1 = \frac{2}{1 + e^{-2x}} - 1 \tag{5}$$

$$\text{relu}(x) = \max(0, x) = \begin{cases} 0, & \text{if } x < 0 \\ x, & \text{if } x \geqslant 0 \end{cases} \tag{6}$$

4   Experiments and Discussion

Our short-term, medium-term, and long-term experiments were based on and compared against two baselines. The first, henceforth named B1, is the original work that presented the CoDeepNEAT [12], and the second (B2), is the work of Boher et al. [14]. The striking aspect of B1 is the very large search space of the coevolutionary algorithm, as discussed in the previous section, whereas B2 has a much smaller search space. This is because the objective of Bohrer et al. was to reproduce the original experiments of the [12] with limited resources. Instead of multiple GPUs, they used a single CPU with only 30 GB RAM.

Our short-term experiments (ST) were inspired by both baselines, but with shorter training episodes than both of them. The sizes of the populations were similar to B2, but we kept the possible DNNs hyperparameters closer to B1. We also employed only a subset of the training data, and we reduced the duration of the coevolutionary algorithm even further (4 generations rather than 40). For medium-term (MT) and long-term (LT), we experimented with increasing the duration of the experiment, and we trained the individuals for more epochs. The experiments configurations are summarized in Table 1.

We compare the most commonly used initialization functions in Deep Learning: He, Glorot and Random. Each of them has a uniform and a normal variant, which determines the distribution it will draw weights from. For the sort-term experiments, we evaluate the CoDeepNEAT algorithm on two data sets (CIFAR-10 and MNIST) with five hold-out iterations. However, as medium-term and long-term experiments take substantially longer to complete, we had to limit the experiments to a single hold-out on the CIFAR-10 data set.

The short-term results are presented in Tables 2 and 3. Table 2 gives the accuracy of the best individual trained after the last generation on each of the five hold-out runs. Table 3 gives the mean and standard deviations.

First, let us discuss the experiments performed with the ReLU activation. We note that, in all runs, our results are better than the second baseline for the MNIST data set. The worst short-term result for MNIST was 99%, whereas B2 reported 92%. On CIFAR-10, the short-term results had a larger standard deviation. On average, the networks found with He initialization were similar to the baseline, while the individuals evolved with Glorot performed slightly

Table 1. Evolutionary and topological hyperparameters. B1 and B2 are baselines. ST, MT, and LT stand for short, medium, and long terms. B1 and B2 accuracies come from their sources. For ST, MT, and LT we report the lowest average obtained with ReLU and either Glorot or He in our experiments.

Parameters	B1 [12]	B2 [14]	ST	MT	LT
Generations	75	40	4	25	75
DNNs population	100	10	10	100	100
Blueprints population	25	10	10	25	25
Modules population	45	30	30	45	45
Epochs during evolution	8	4	5	10	10
Epochs in final training	300	40	100	150	150
Data used during evolution	100%	40%	40%	100%	100%
Filters	[32, 256]	[16, 48]	[32, 256]		
Kernel size	{1, 3}	{1, 3, 5}	{1, 3, 5}		
Dropout rate	[0, 0.7]	[0, 0.5]	[0, 0.5]		
Max pooling	{Yes, No}	{No}	{Yes, No}		
Batch normalization	{No}	{No}	{Yes, No}		
MNIST accuracy	–	92%	98.9%	–	–
CIFAR-10 accuracy	92.7%	77%	74.9%	80.1%	84.5%

better. Notice that the main difference from our short-term experiment to B2 was the larger search space for number of filters, and the possibility of employing max pooling and batch normalization. This is suggestive that running the DNE algorithm for longer generations is less important than providing the algorithm with more flexible modules.

Next, we repeated the short-term experiments with the hyperbolic tangent activation function (tanh). Again, the experiments were performed on CIFAR-10 and MNIST, and repeated a hold-out partitioning scheme five times. In both cases, the experiments were executed on single-CPU system with 15.5 GiB of RAM and a 4 GB GPU (GeForce GTX 1650/PCIe/SSE2). The total run time was 67 h. Thus, considering 2 activation functions, 4 initialization methods and 2 data sets, the total combinations add up to 16 distinct experimental settings, each repeated five times, so each experiment took on average 50 min to complete.

The average accuracy and its standard deviation are in Table 3. The performance of the individuals found with tanh was severely hindered when compared to the individuals evolved with ReLU. Still, with the exception of Glorot Normal, the results were better than B2. Once again, this suggests that it is possible to achieve decent results when the DNE is tuned to find "fast-learners", DNNs which can be trained with few epochs during the evolutionary process.

For the remaining experiments, we focused only on the CIFAR-10 data set and ReLU activation function.

Table 2. Short-term experiments with Glorot and He initialization functions (ReLU activation).

CIFAR-10			MNIST		
Initialization	*Run*	Accuracy	Initialization	*Run*	Accuracy
Glorot normal	1	0.7681	Glorot normal	1	0.9872
Glorot normal	2	0.7861	Glorot normal	2	0.9893
Glorot normal	3	**0.8057**	Glorot normal	3	0.9900
Glorot normal	4	0.8034	Glorot normal	4	0.9893
Glorot normal	5	0.7995	Glorot normal	5	**0.9903**
Glorot uniform	1	**0.8462**	Glorot uniform	1	0.9890
Glorot uniform	2	0.7188	Glorot uniform	2	0.9908
Glorot uniform	3	0.8118	Glorot uniform	3	0.9876
Glorot uniform	4	0.7654	Glorot uniform	4	**0.9914**
Glorot uniform	5	0.8180	Glorot uniform	5	0.9881
He normal	1	0.7798	He normal	1	0.9879
He normal	2	0.6894	He normal	2	0.9910
He normal	3	0.7159	He normal	3	0.9870
He normal	4	0.7706	He normal	4	**0.9921**
He normal	5	**0.7919**	He normal	5	0.9899
He uniform	1	0.7743	He uniform	1	0.9902
He uniform	2	0.7767	He uniform	2	**0.9923**
He uniform	3	0.7324	He uniform	3	0.9903
He uniform	4	0.7695	He uniform	4	0.9900
He uniform	5	**0.7998**	He uniform	5	0.9870

Table 3. Short-term accuracy with Glorot and He initialization functions with linear and non-linear activation functions.

Data set	Initialization	Activation	Mean accuracy	Std. dev.
MNIST	Glorot normal	tanh	0.9089	±0.1116
MNIST	Glorot uniform	tanh	0.9723	±0.0046
MNIST	He normal	tanh	0.9698	±0.0045
MNIST	He uniform	tanh	0.9733	±0.0026
CIFAR-10	Glorot normal	tanh	0.4532	±0.0547
CIFAR-10	Glorot uniform	tanh	0.5487	±0.0658
CIFAR-10	He normal	tanh	0.5143	±0.1237
CIFAR-10	He uniform	tanh	0.5172	±0.0777
MNIST	Glorot normal	ReLU	0.9892	±0.0010
MNIST	Glorot uniform	ReLU	0.9893	±0.0014
MNIST	He normal	ReLU	0.9895	±0.0018
MNIST	He uniform	ReLU	0.9899	±0.0016
CIFAR-10	Glorot normal	ReLU	0.7925	±0.0139
CIFAR-10	Glorot uniform	ReLU	0.7920	±0.0448
CIFAR-10	He normal	ReLU	0.7495	±0.0397
CIFAR-10	He uniform	ReLU	0.7705	±0.0217

The medium-term in this section were carried out in order to contemplate a compromise proposal between the two *baselines* used, since Boher et al. [14] used a very small amount of computational power when compared to what was used in the original CoDeepNEAT paper (2,000,000 CPUs and 5,000 GPUs). We ran the medium-term experiments on the same architecture as the short-term (a single 4 GB GPU). This time, however, each experiment took approximately 32 h in lieu of the average 50 min observed in the short-term experiments. Therefore, we limited the medium-term experiments to a single hold-out experiment.

The initialization functions were both the uniform and normal variants of He and Glorot, and we also consider a random initialization, which assigns weight values to the neurons without taking into consideration neither the activation function, nor the sizes of the input or the output. The results are shown in Table 4. In addition to the accuracy, we also report the number of trainable parameters of the best individual and whether that individual contains residual connections or not.

Table 4. Best individuals from experiments in medium-term for each evolutionary process.

Data set	Initialization	Final individual	Accuracy	Parameters
CIFAR-10	He uniform	Residual	0.8787	≈3.33 M
CIFAR-10	He normal	Residual	0.8336	≈1.25 M
CIFAR-10	Glorot uniform	Sequencial	0.8011	≈2.25 M
CIFAR-10	Glorot normal	Sequencial	0.8581	≈1.15 M
CIFAR-10	Random uniform	Sequencial	0.7380	≈1.06 M
CIFAR-10	Random normal	Sequencial	0.7271	≈2.03 M

Considering the greater number of generations, the DNNs assembled by CoDeepNEAT may obtain some residual connection due to mutations in their blueprints. However, we noticed that only the He-like initialization were able to produce the best individual with some residual connection, while all the others converged to an individual whose architecture was fully sequential (total absence of skip connections). It is important to mention that, during the evolutionary process of all, individuals with some degree of residuality were found in all cases, but only the architectures initialized with He Uniform or He Normal managed to create individuals with residual architectures that surpassed the performance of individuals with non-residual architectures during their respective evolutionary processes.

Although one of the He initializations had the best result, as seen in Table 4, with 87.87% accuracy, the Glorot Normal initialization achieved a similar result, 85.81%, with approximately 34.72% fewer floating-point operations and 34.53% less total parameters, which is equivalent to the total number of parameters to perform an inference on the model. So He Uniform got the best result, but the

Glorot Normal initializer is just as competitive with less computing resources required. A reasonable explanation for this fact is that the high frequency of mutations allows the development of residual connections, but a large number of generations may be required to produce individuals that are competitive. So much so that Miikkulainen et al. state in [12] that it was only after the 70th generation that the accuracy of the individuals in their experiments stabilized.

In order to explore this possibility, we executed long-term experiments, with three times as many generations as the medium-term. However, as the neural networks become more complex over the generations, the total time to finish was substantially greater: 40 days per experiment ($\sim$920 h), when compared to the 32 h required to finish each medium-term experiment. Only 1 hold-out was considered for Glorot Normal and He Uniform, which were the most accurate initializations in the medium-term experiments. The results are presented in Table 5.

Table 5. Best individuals from experiments in long-term for each evolutionary process.

Data set	Initialization	Final individual	Accuracy	Parameters
CIFAR-10	He uniform	Residual	0.8991	$\approx$10.23 M
CIFAR-10	Glorot normal	Sequencial	0.8454	$\approx$0.47 M

Similarly to the previous results, the best individual evolved with He initialization achieved better accuracy (89.91%) than the individual evolved with Glorot (85.54%). However, the DNN found with Glorot was completely linear, as shown in Fig. 3 (left). In comparison, the DNN found with He was rich in skip connections.

5   Conclusion

In this paper we considered the influence of different experimental settings of a coevolutionary algorithm We performed experiments with coevolution of Deep NeuroEvolution of Augmenting Topologies (CoDeepNEAT) on two benchmark data sets, on three different scenarios that involve increasingly more complex search spaces. We compared the results obtained with two popular initialization functions, He and Glorot, as well as a random-initialization strategy serving as baseline. In addition, for short-term experiments we also compared linear and non-linear activation functions.

In the initial experiments, focusing in short-term parameters, ReLU activation outperformed the results obtained by tanh in all experiments. While the results are somewhat below start-of-the-art performances, they were above 99.20% in the MNIST dataset and 89.91% in the CIFAR-10 dataset, outperforming the baseline study of [14] which attempted to run the CoDeepNEAT with limited computational resources. Our result shows that it may be more important to fine-tune the coevolutionary algorithm hyperparameters and attempt to evolve "fast-learners" in fewer generations than perform longer experiments.

It is important to note that the best result obtained for CIFAR-10 (89.91%) was derived from long-term experiments. Furthermore, we draw attention to a curious fact: considering the universe of medium and long-term experiments, all the best individuals evolved with Glorot (both Uniform and Normal variants) did not contain residual connections, while all of the best He individuals (Uniform or Normal) had multiple residual connections and were deeper. Considering the importance of residuality in neural architecture design, there seems to be evidence to suggest an investigation into this greater capacity of He initializations to facilitate the development of efficient residual architectures in short-term evolution processes.

References

1. Ma, Y., Xie, Y.: Evolutionary neural networks for deep learning: a review. Int. J. Mach. Learn. Cybern. (2022). https://doi.org/10.1007/s13042-022-01578-8
2. Kumar, S. K.: On weight initialization in deep neural networks. In: arXiv preprint arXiv:1704.08863 (2017)
3. Initializing neural networks. https://www.deeplearning.ai/ai-notes/initialization/. Accessed 12 June 2022
4. Goodfellow, I.J., Bengio, Y., Courville, A.: Deep Learning, 1st edn. Cambridge (2016)
5. Glorot, X., Bengio, Y.: Understanding the difficulty of training deep feedforward neural networks. In: Proceedings of the Thirteenth International Conference on Artificial Intelligence and Statistics, pp. 249–256. JMLR Workshop and Conference Proceedings, Sardinia (2010)
6. He, K., Zhang, X., Ren, S., Sun, J.: Delving deep into rectifiers: surpassing human-level performance on ImageNet classification. In: Proceedings of the IEEE International Conference on Computer Vision, pp. 1026–1034. IEEE, Santiago (2010)
7. Koutnik, J., Gomez, F., Schmidhuber, J.: Evolving neural networks in compressed weight space. In: Proceedings of the 12th Annual Conference on Genetic and Evolutionary Computation, pp. 619–626. ACM, Portland (2010)
8. Togelius, J., Gomez, F., Schmidhuber, J.: Learning what to ignore: memetic climbing in topology and weight space. In: Proceedings of the 2008 IEEE Congress on Evolutionary Computation (IEEE World Congress on Computational Intelligence), pp. 3274–3281. IEEE, Hong Kong (2008)
9. Okada, H., Wada, T., Yamashita, A., Matsue, T.: Interval-valued evolution strategy for evolving neural networks with interval weights and biases. In: Proceedings of the International Conference on Soft Computing and Intelligent Systems, and the 13th International Symposium on Advanced Intelligence Systems, pp. 2056–2060. IEEE, Kobe (2012)
10. Desell, T.: Accelerating the evolution of convolutional neural networks with node-level mutations and epigenetic weight initialization. In: Proceedings of the Genetic and Evolutionary Computation Conference Companion, pp. 157–158. IEEE, Kyoto (2018)

11. Lyu, Z., ElSaid, A., Karns, J., Mkaouer, M., Desell, T.: An experimental study of weight initialization and weight inheritance effects on neuroevolution. In: Proceedings of Applications of Evolutionary Computation: 24th International Conference. ACM, Seville (2021)

12. Miikkulainen, R., Liang, J., Meyerson, E., Rawal, A.: Evolving deep neural networks. In: Artificial Intelligence in the Age of Neural Networks and Brain Computing, pp. 293–312 (2019)

13. Papavasileiou, E., Cornelis, J., Jansen, B.: A systematic literature review of the successors of 'NeuroEvolution of augmenting topologies'. Evol. Comput. **29**, 1–73 (2020)

14. Bohrer, J.S., Grisci, B.I., Dorn, M.: Neuroevolution of neural network architectures using CoDeepNEAT and Keras. In: arXiv preprint arXiv:2002.04634 (2020)

15. Zhou, X., Li, X., Hu, K., Zhang, Y., Chen, Z., Gao, X.: ERV-Net: an efficient 3D residual neural network for brain tumor segmentation. Expert Syst. Appl. **170**, 114566 (2021)

16. Dogan, S, et al.: Automated accurate fire detection system using ensemble pretrained residual network. Expert Syst. Appl. **203**, 117407 (2022)

17. Hoorali, F., Khosravi, H., Moradi, B.: IRUNet for medical image segmentation. Expert Syst. Appl. **191**, 116399 (2022)

18. Li, H., Xu, Z., Tyalor, G., Studer, C., Goldstein, T.: Visualizing the loss landscape of neural nets. In: Advances in Neural Information Processing Systems (2018)

19. Intuitive Explanation of Skip Connections in Deep Learning. https://theaisummer.com/skip-connections/. Accessed 12 June 2022

20. Keras Documentation - Glorot Uniform. https://www.tensorflow.org/api_docs/python/tf/keras/initializers/GlorotUniform. Accessed 10 July 2022

21. Keras Documentation - Glorot Normal. https://www.tensorflow.org/api_docs/python/tf/keras/initializers/GlorotNormal. Accessed 10 July 2022

22. Keras Documentation - He Uniform. https://www.tensorflow.org/api_docs/python/tf/keras/initializers/HeUniform. Accessed 10 July 2022

23. Keras Documentation - He Normal. https://www.tensorflow.org/api_docs/python/tf/keras/initializers/HeNormal. Accessed 10 July 2022

24. Searching for activation functions. https://arxiv.org/abs/1710.05941. Accessed 12 June 2022

25. Mish: A self regularized non-monotonic neural activation function. https://arxiv.org/abs/1908.08681. Accessed 12 June 2022

Analysis of the Influence of the MVDR Filter Parameters on the Performance of SSVEP-Based BCI

Lucas Brazzarola Lima[1], Ramon Fernandes Viana[1], José Martins Rosa-Jr.[1], Harlei Miguel Arruda Leite[1(✉)], Guilherme Vettorazzi Vargas[2], and Sarah Negreiros Carvalho[1]

[1] Department of Electrical Engineering, Federal University of Ouro Preto, Ouro Preto, Brazil
{lucas.brazzarola,ramon.viana,jose.rosa}@aluno.ufop.edu.br,
{harlei,sarah}@ufop.edu.br
[2] School of Electrical and Computer Engineering, University of Campinas, Campinas, Brazil
g229960@dac.unicamp.br

Abstract. Brain-Computer Interface (BCI) is a communication method based on brain signals analysis. The interface enables controlling applications such as a wheelchair with minimal muscle effort, making BCI systems attractive in assistive technology development. Currently, Steady-State Visually Evoked Potential (SSVEP) represents one of the most promising BCI paradigms, since a specific physiological brain response is evoked when a subject is exposed to continuously flickering visual stimuli. In this study, we evaluated how the parameters of the Minimum Variance Distortionless Response (MVDR) filter impact the performance of the SSVEP-based BCI. Three parameters were analyzed: filter order, number of EEG signals combined at the filter input, and number of electrodes employed for filtering. Our results show that it is convenient to employ fewer electrodes, as they are closer to the visual cortex region, and to combine them spatially, using low filter orders. The best performance, among the tested configurations, was 80.20 ± 6.65%, obtained with filter order nine, employing nine EEG signals and spatially combining the inputs with eight signals at a time.

Keywords: Brain-computer interface · Steady-state visually evoked potential · Minimum variance distortionless response · Spatiotemporal filtering

1 Introduction

Brain-Computer Interface (BCI) is a technology that involves the acquisition, processing and translation of brain signals to control an external device, bypassing the conventional neuromuscular channel. Nowadays, the Steady-State Visually Evoked Potential (SSVEP) represents one of the most promising BCI

Supported by FAPEMIG and UFOP.

J. C. Xavier-Junior and R. A. Rios (Eds.): BRACIS 2022, LNAI 13653, pp. 313–324, 2022.
https://doi.org/10.1007/978-3-031-21686-2_22

paradigms [17]. In an SSVEP-based BCI, the possible commands are exclusively associated with visual stimuli. Thus, the subject focuses their gaze on the stimulus that corresponds to the desired command to be operated by the application [2,15].

There are two techniques used to acquire brain signals, one being invasive and the other not. The invasive recording methods, such as electrocorticogram, provide spatial resolution and a high signal-to-noise ratio (SNR). However, this approach is expensive and presents health risks as surgical procedures are needed [8]. Meanwhile, non-invasive methods, such as electroencephalogram (EEG), tend to be advantageous for the development of BCI systems since it is a safer, more practical and less expensive technique [7], given that the electrodes are positioned directly on the user's scalp. In contrast, the brain signal acquired presents more noise components and lower spatial resolution.

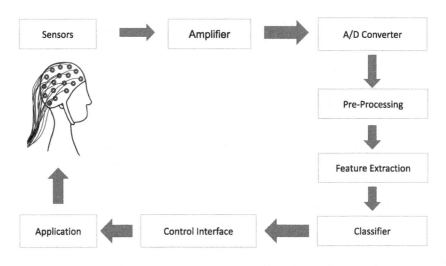

Fig. 1. Stages of a SSVEP-based BCI.

The steps of an SSVEP-based BCI System are presented in Fig. 1. Initially, brain signals are acquired by EEG, amplified and digitalized. Subsequently, the steps of pre-processing, feature extraction and classification are performed. The classifier output indicates which signal the subject was focused on and therefore generates the control signal that must be sent to the application. So, the user receives feedback on the application execution and can choose the next command to be executed.

This research focuses on the pre-processing stage, more specifically on digital filtering. The EEG signal is full of noise and interference and filtering allows for improving the SNR to generate more accurate outcomes. In this study, the digital filtering process was executed using the Common Average Reference (CAR) spatial filter, followed by a Minimum Variance Distortionless Response (MVDR) spatiotemporal filter. We evaluated how the MVDR filter order and the combination of electrodes impact the performance of the SSVEP-based BCI. This work deepens the order analysis carried out in our previous studies [1,13].

This paper is organized as follows: Sect. 2 presents an overview of the dataset and describes the signal processing techniques applied to conceive an SSVEP-based BCI at all stages: MVDR and CAR filters, the feature extraction via Fast Fourier Transform (FFT) and the linear classifier. Section 3 outlines the results and discussion, and Sect. 4 provides the conclusions.

2 Methodology

2.1 Database Description

In this study, a public database with EEG data from 35 healthy subjects (17 females), with average age of 22 years (17–34 years) was used [16]. During the experimental protocol, subjects were exposed to 40 visual stimuli, flickering at different frequencies ranging from 8 to 15.8 Hz with a regular interval of 0.2 Hz. Six samples of 5 s of valid data were collected for each visual stimulus using 64 electrodes placed according to the extended 10–20 pattern, as shown in Fig. 2. The sampling frequency 250 Hz. In our tests, we consider four visual stimuli, at the frequencies of 8, 10, 12 15 Hz. Each trial of 5 s was segmented into windows of 1 s, without overlapping. This public database was chosen with the aim of maintaining parallelism and comparing the results with our previous works [1,13] and, works by other authors [6,9].

2.2 CAR

The Common Average Reference is a filter technique that grants input signals a neutral reference value. This technique consists of obtaining the average of the input signals considering all electrodes and then subtracting the content of each electrode from the calculated average [11]. This procedure reduces the noise present in the brain signals. Mathematically, we can calculate the output of the CAR filter as:

$$x_i^{CAR} = x_i - \frac{1}{N} \sum_{j=1}^{N} x_j^{electrode} \tag{1}$$

where x_i is the signal collected between the i-th electrode and the reference and N is the total number of electrodes.

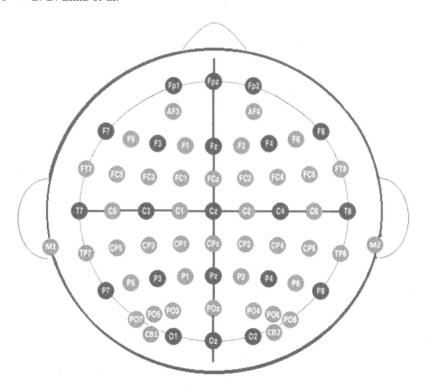

Fig. 2. Layout employed for positioning the 64 electrodes based on the 10–20 pattern.

2.3   MVDR Filter

The Minimum Variance Distortionless Response is a classic spatiotemporal filtering technique in the field of antenna positioning. It is a non-parametric [5] method which is adapted to SSVEP-based BCI systems.

The MVDR filter can process evoked signals at different frequencies by combining the input data, k by k, and producing a single output, referred to here as a channel. The objective is to apply the MVDR filter to highlight the desired frequencies and attenuate the interfering frequencies that decrease the quality of the brain signals.

The filter structure can be understood through Fig. 3, in which k from the total number of electrodes are jointly processed by finite impulse response (FIR) filters with length m, whose coefficients are mathematically represented by $\mathbf{w}_j$. This way, a brain signal $\mathbf{x}_j(n)$, representing a signal from the j-th electrode at time n is filtered as follows:

$$y_j(n) = \mathbf{w}_j^T \mathbf{x}_j(n) \tag{2}$$

Each filtered signal is summed so that a general response (channel) y(n) is obtained. The number of channels is defined by the combination of the total number of electrodes (L) taken k by k at a time, $C_{L,k}$, and it is given by the equation below:

$$C_{L,k} = \frac{L!}{k!(L-k)!} \tag{3}$$

The complete formulation of the filter is discussed by [1] and presents a closed-form solution according to the constrained problem that arises from the considered restrictions.

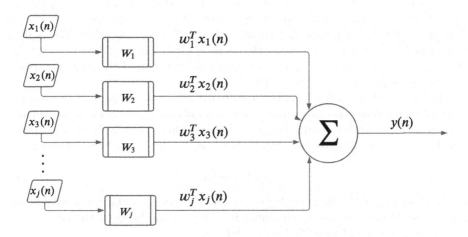

Fig. 3. Example of a scheme of an MVDR filter.

An important detail about the MVDR application for BCI-SSVEP systems is that the input signals are subtracted from an average value considering all electrodes, as is provided by the CAR [14] filtering method.

2.4 Feature Extraction

The feature extraction step consists of applying signal processing methodologies to obtain relevant attributes for the discrimination of classes associated with user intentions [4].

In this study, the Discrete Fourier Transform (DFT) was used to extract information on how the input signal is distributed in the frequency spectrum. The DFT is able to satisfactorily map the frequency domain and extract, through the magnitude of the signal, characteristics attributed to the evoked frequencies. Computationally, the DFT is quick and efficiently calculated through the algorithm of Fast Fourier Transform (FFT) [10].

2.5   Linear Classifier

Classification is a process of categorizing an input dataset into classes. This can be achieved by creating linear decision boundaries that categorize the input signals into a discrete class, based on a linear combination of their features, grouping the signals when they have the same class and separating them when they have different classes.

The built model defines the decision surfaces in the data. There are several types of linear classifiers, which vary in the decision frontier definition methods. The linear classifier based on Least Squares is used in this work, in which the objective is to find an optimal set of weights that minimize the result of the squared error of the classifier's estimation for the evoked potentials. The higher the value of the residuals, the worse the decision frontier is established concerning the input signals, the least squares method will iterate until the sum of squared residuals is equal to an established limit.

For an input $\mathbf{x}$ coming from the feature extraction stage, properly separated into test and validation data, an output $y(n)$ deciding which class the signals belong to is established, and it is still necessary to find which weights $\mathbf{w}$ promote the least error of the classifier output. Mathematically it can be written as:

$$y(n) = \mathbf{w}^T \mathbf{x} \tag{4}$$

The weight vector $\mathbf{w}$ indicates where the decision boundary will be located at. Classifier performance is measured by the hit ratio between the BCI user's intent and the actual identified command [12].

3   Results and Discussion

For each of the subjects in the dataset, different filter orders and electrodes combinations were evaluated, in order to find out the optimal set of parameters that provides the best accuracy for the BCI-SSVEP using the MVDR filter. To ensure the precision of the results in all scenarios, a 20-cross fold validation scheme was realized with data of each subject, then the average accuracy was calculated. The k-cross fold validation approach was adopted since the system is subject-dependent, i.e., 35 SSVEP-based BCI systems (one for each subject) were designed.

It was computationally impracticable to run all possible combinations considering the 64 electrodes, therefore four scenarios were chosen. The electrode arrangement in each follows the schematic shown in Fig. 2.

– **Scenario 1:** nine electrodes placed at O1, O2, Oz, PO3, PO4, POz, P1, P2 and Pz. Combination of electrodes k by k, with k ranging from 2 to 8. MVDR filter orders tested: $m = 5, 7, 9, 10, 11, 13, 15, 20, 25, 30, 35, 40, 45$ and 50. The 35-subject average accuracy was considered as a performance metric.

- **Scenario 2:** 16 electrodes placed at O1, O2, Oz, PO3, PO4, PO7, PO8, POz, C1, C2, Cz, P1, P2, Pz, CPz and FCz. Combination of electrodes k by k, with $k = 13, 14$ and 15. MVDR filter orders tested: $m = 5$ and 10. The accuracy of SSVEP-based BCI for subjects 11, 20 and 32 were evaluated.
- **Scenario 3:** similar to Scenario 2, except for the location of the 16 electrodes, which are: O1, O2, Oz, PO3, PO4, PO5, PO6, PO7, PO8, POz, C1, C2, Cz, P1, P2 and Pz.
- **Scenario 4:** is also similar to Scenario 2, except that now, all the 64 electrodes were employed, combined 63 by 63 ($k = 63$).

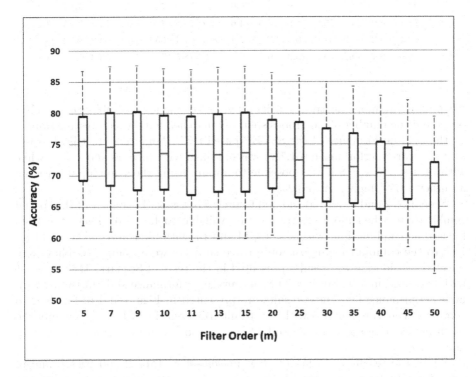

Fig. 4. Scenario 1 - SSVEP-based BCI average performance for each MVDR filter order evaluated.

Figure 4 presents the overall average accuracy of the tests for each filter order considering the average accuracy obtained for all k in Scenario 1. As it shows, lower filter orders provided better results, being order $m = 5$ the only where the average accuracy surpassed 75%.

Table 1 details the average accuracy of the 35 subjects for each combination of k and m tested in Scenario 1. Combinations of $k = 8$ electrodes resulted in more accurate outcomes for all tested m, except for $m = 50$. This fact indicates that the spatial combination, operated by the MVDR filter, becomes more effective

Table 1. SSVEP-based BCI performance considering filter order variations (m) and the number of combined signals at the MVDR filter input (k) in Scenario 1. The values highlighted in blue are the best accuracies considering different m, in green are the best accuracies for the variations of k, in cyan the best overall case and in red the worst overall case.

m \ k	5	7	9	10	11	13	15	20	25	30	35	40	45	50
2	69.3	68.4	67.7	67.8	67.5	67.4	67.9	68.2	66.5	65.8	65.6	64.6	66.2	64.4
3	77.6	76.3	75.3	75.9	74.3	75.3	75.5	73.9	73.9	72.5	73.4	71.7	73.3	71.0
4	77.0	75.2	74.9	74.3	74.7	73.9	74.7	74.1	73.4	72.6	72.5	71.6	73.8	71.4
5	77.1	75.8	74.5	74.3	74.6	74.2	75.0	74.1	73.9	72.8	72.5	72.3	73.7	72.0
6	78.1	76.9	75.0	75.0	74.8	74.5	74.6	74.2	73.8	72.8	72.9	72.5	73.5	70.0
7	70.1	69.3	67.9	67.8	66.9	67.7	67.4	68.0	67.1	66.6	66.0	64.8	66.3	61.7
8	79.5	80.1	80.2	79.7	79.5	79.9	80.1	78.9	78.5	77.5	76.8	75.3	74.4	70,0

when there are more electrodes at its input. Also, the ideal combination of filter order and the number of electrodes spatially matched have an interesting impact on the final performance of the SSVEP-based BCI. From Table 1, it is observed that the worst performance (61.7%) was obtained with $m = 50$ and $k = 7$, while with $m = 9$ and $k = 8$ it was possible to obtain, using the same input data, an accuracy 18.5% higher, of (80.2%).

Scenarios 2 and 3 were considered to precisely assess the impact of the spatial combination, when more electrodes are available to be used. Then, the tests focused on combinations with $k = 13$, 14 and 15, with filter order $m = 5$ and 10. Due to the computational cost, only three subjects were evaluated. They were selected considering the results presented in the tests of Scenario 1: Subject 11 with low performance, Subject 20 with average performance and Subject 32 with high performance [3]. These results are presented in Figs. 5 and 6. Further, to compare Scenarios 2 and 3 with the results of Scenario 1, Table 2 presents the performance of Subjects 11, 20 and 32 in Scenario 1, considering $m = 5$, $m = 10$ and $k = 8$.

Thus, the results of Scenarios 1–3, presented in Table 2 and Figs. 5 and 6, show that the best case varied between $m = 5$ and 10 according to the subject in Scenarios 1 and 2, but it was always obtained with filter order $m = 5$ for Scenario 3. Still, it is interesting to note that Subject 32, who performs well, was quite sensitive to $k = 13$ and 14, as seen in Figs. 5 and 6, with a performance difference of about 30%. This performance difference was smaller for the other two subjects tested. Also, we observed that in Scenarios 2 and 3, the best performance for all tested subjects was obtained with $k = 14$, and not with $k = 15$, which would be the largest possible combination of electrodes at the input. In scenario 1, the maximum value of combinations $k = 8$ tended to present the best performance. This fact can be better understood by looking at the results of Table 3, which presents the average SSVEP-based BCI performance for these

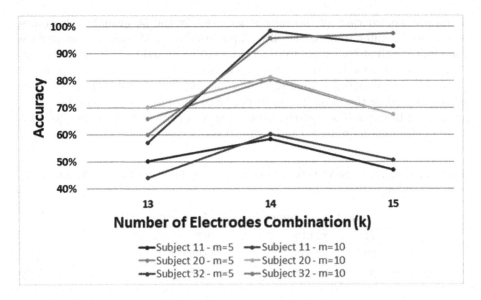

Fig. 5. Scenario 2 - SSVEP-based BCI average performance with variations of MVDR filter order and the number of entries matched in its input.

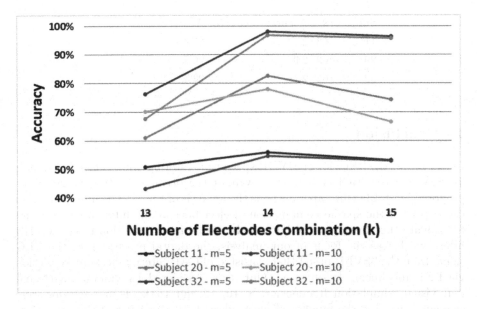

Fig. 6. Scenario 3 - SSVEP-based BCI average performance with variations of MVDR filter order and the number of entries matched in its input.

3 subjects considering all the signals acquired by the 64 available electrodes. The spatial combination at the filter input used $k = 63$ and, for comparison, the same MVDR filter orders were tested, i.e. $m = 5$ and $m = 10$. Results show that, by including other electrodes, the accuracy was reduced. This probably occurred because, when adding electrodes far from the visual cortex, the information regarding the visually evoked potential was masked by noise during the spatial combination of the filter, resulting in a reduction in the accuracy of the system, although more tests can be run to verify this effect.

Table 2. Scenario 1 - SSVEP-based BCI average performance for MVDR filter order of $m = 5$ and $m = 10$.

Subject	k	Accuracy ($m = 5$)	Accuracy ($m = 10$)
11	8	61%	67%
20	8	85%	81%
32	8	98%	99%

Table 3. Scenario 4 - SSVEP-based BCI average performance for MVDR filter order of $m = 5$ and $m = 10$.

Subject	k	Accuracy ($m = 5$)	Accuracy ($m = 10$)
11	63	24%	25%
20	63	24%	26%
32	63	24%	23%

4   Conclusion

In this study, the behavior of MVDR space-time filtering applied to EEG signals for SSVEP-based BCI systems was evaluated. Motivated by the contributions proposed by [1,13], this paper deepens the analysis of the MVDR filter order and the impact of the spatial combination of electrodes at the filter input, in order to evaluate the filtering potential and also its limits. In addition to the MVDR filter, which was the focus of our analysis, the digital processing of the EEG signal from the SSVEP-based BCI consisted of CAR filtering, feature extraction via FFT and linear classification. Our tests considered 1 s window-length and four visual stimulation frequencies: 8, 10, 12 and 15 Hz. Four scenarios were evaluated, varying the number of electrodes (9, 16 and 64) and their position. The metric used to evaluate the parameters and define the best configuration was the average accuracy of the SSVEP-based BCI, considering the 20-fold cross-validation scheme for each subject.

The best average accuracy of 80.2% was obtained in Scenario 1 (with 9 electrodes), combining the inputs with $k = 8$ and using the filter order $m = 9$.

Our analyses, considering three subjects, show that by adding more electrodes, the average accuracy decreases slightly, and in Scenario 4 it was greatly reduced. The best average performance for Scenario 1 was 82.33%, with $k = 8$ and $m = 10$; for Scenario 2 was 79.02%, obtained with $k = 14$ for both tested orders ($m = 5$ and $m = 10$); for Scenario 3, 78.96% with $k = 14$ and $m = 5$; and for Scenario 4, it was 24.67% with $k = 63$ and $m = 10$. Although further analysis is needed, our initial hypothesis is that this is due to the fact that when inserting combinations with electrodes far from the visual cortex, spatial filtering ends up reducing the amplitudes of the FFT at visually stimulated frequencies, negatively impacting the performance of the system.

In general, we can say that MVDR filtering achieves high success rates for an SSVEP-based BCI. However, the main disadvantage of this technique is the high computational cost that increases with the expansion of the number of electrodes combined at the filter inputs and the number of visual stimuli. Future work will focus on evaluating the MVDR space-time filter considering convolutional neural networks in the classification process, as the feature space tends to increase with the application of this filter, this approach seems promising.

Conflict of Interest. The authors declare that they have no conflict of interest.

References

1. Carvalho, S.N., et al.: Space-time filter for SSVEP brain-computer interface based on the minimum variance distortionless response. Med. Biol. Eng. Comput. **59**(5), 1133–1150 (2021). https://doi.org/10.1007/s11517-021-02345-7
2. Graimann, B., Allison, B., Pfurtscheller, G.: Brain-computer interfaces: a gentle introduction. In: Graimann, B., Pfurtscheller, G., Allison, B. (eds.) Brain-Computer Interfaces. The Frontiers Collection, pp. 1–27. Springer, Heidelberg (2009). https://doi.org/10.1007/978-3-642-02091-9_1
3. Guger, C., et al.: How many people could use an SSVEP BCI? Front. Neurosci. **6**, 169 (2012)
4. Haykin, S.: Neural Networks and Learning Machines, 3 edn. Pearson Education India (2010)
5. Benesty, J., Chen, J., Huang, Y.: A generalized MVDR spectrum. IEEE Trans. Audio Electroacoust. **12**(8673104), 827–830 (2005)
6. Liu, B., Chen, X., Shi, N., Wang, Y., Gao, S., Gao, X.: Improving the performance of individually calibrated SSEVP-BCI by task-discriminant component analysis. IEEE Trans. Neural Syst. Rehabil. Eng. **29**, 1998–2007 (2021)
7. Liu, Z., Shore, J., Wang, M., Yuan, F., Buss, A., Zhao, X.: A systematic review on hybrid EEG/fNIRS in brain-computer interface. Biomed. Signal Process. Control **68**, 102595 (2021)
8. Neumann, W.J., et al.: The sensitivity of ECG contamination to surgical implantation site in brain computer interfaces. Brain Stimul. **14**(5), 1301–1306 (2021)
9. Oikonomou, V.P., Nikolopoulos, S., Kompatsiaris, I.: A Bayesian multiple kernel learning algorithm for SSVEP BCI detection. IEEE J. Biomed. Health Inform. **23**(5), 1990–2001 (2018)
10. Oppenheim, A.V.: Discrete-Time Signal Processing. Pearson Education India (1999)

11. Sözer, A.T., Fidan, C.B.: Novel spatial filter for SSVEP-based BCI: a generated reference filter approach. Comput. Biol. Med. **96**, 98–105 (2018)
12. Theodoridis, S., Koutroumbas, K.: Pattern Recognition, 3rd edn. Academic Press, San Diego (2006)
13. Vargas, G.V., Carvalho, S.N., Boccato, L.: Analysis of the spatiotemporal MVDR filter applied to BCI-SSVEP and a filter bank extension. Biomed. Signal Process. Control **73**, 103459 (2022)
14. Vargas, G.V.: Filtragem espaço-temporal baseada no princípio MVDR aplicada a interfaces cérebro-computador sob o paradigma SSVEP. Master's thesis, Universidade Estadual de Campinas (2021)
15. Vialatte, F.B., Maurice, M., Dauwels, J., Cichocki, A.: Steady-state visually evoked potentials: focus on essential paradigms and future perspectives. Prog. Neurobiol. **90**(4), 418–438 (2010)
16. Wang, Y., Chen, X., Gao, X., Gao, S.: A benchmark dataset for SSVEP-based brain-computer interfaces. IEEE Trans. Neural Syst. Rehabil. Eng. **25**(10), 1746–1752 (2016)
17. Wolpaw, J., Birbaumer, N., McFarland, D., Pfurtscheller, G., Vaughan, T.: Brain-computer interfaces for communication and control. Clin. Neurophysiol. **113**(6), 767–791 (2002)

A Novel Multi-objective Decomposition Formulation for Per-Instance Configuration

Lucas Marcondes Pavelski[1]([✉]) [ID], Myriam Regattieri Delgado[1] [ID],
and Marie-Éléonore Kessaci[2] [ID]

[1] Graduate Program in Electrical and Computer Engineering (CPGEI), Federal
University of Technology – Paraná (UTFPR), Curitiba, Brazil
lpavelski@alunos.utfpr.edu.br, myriamdelg@utfpr.edu.br
[2] CRIStAL – Univ. Lille, CNRS - Centrale Lille - UMR 9189, 59000 Lille, France
marie-eleonore.kessaci@univ-lille.fr

Abstract. Per-instance algorithm configuration (PIAC) is an important
task in which, given a base problem instance, a recommendation model
indicates the best configuration to solve it. Whereas typical Automated
Algorithm Configuration (AAC) prescribes configurations for a fixed set
of instances, in PIAC, a model is trained using problem features and
past experience in solving the base problem and is tested individually on
new instances. This work proposes a novel formulation for PIAC called
Multi-Objective Automated Algorithm Configuration based on Decomposition (MOAAC/D). Unlike other PIAC approaches, it decomposes
the problem space by associating different instance sets with each objective. Using a particular implementation called iMOEA/D, we show an
efficient search with irace as a local search of the Multi-objective Algorithm based on Decomposition (MOEA/D). Experiments on 6,480 base
problem instances show that our proposal is general and performs well
on flowshop, a well-studied combinatorial optimization problem with
many variants addressed as real-world applications. During the training phase, iMOEA/D searches for good configurations by tuning Iterated Local Search with Iterated Greedy operators. The testing phase
uses the generated Pareto front to recommend parameters for new flowshop instances. The results show that the proposed approach outperforms a random selection baseline, a generalist solution provided by irace,
and is an alternative to a meta-learning-based approach. We believe the
proposed MOAAC/D formulation has the potential to open up a novel
research area: multi-objective tuners capable of providing specialist and
generalist configurations simultaneously.

Keywords: Algorithm configuration · Multi-objective optimization ·
Flowshop problem

This work has been financed in part by the Edital 4k (from Universidade Tecnológica
Federal do Paraná - UTFPR), and National Council for Scientific and Technological -
CNPq (grants 314699/2020-1 and 439226/2018-0).

J. C. Xavier-Junior and R. A. Rios (Eds.): BRACIS 2022, LNAI 13653, pp. 325–339, 2022.
https://doi.org/10.1007/978-3-031-21686-2_23

1   Introduction

Algorithm Selection (AS) [15] aims to find mappings between base problems and the best algorithms to solve them. AS is related to the Automated Algorithm Configuration (AAC) [17] problem, where search algorithms find the best configuration given a set of base problems. AACs like irace [8] can efficiently find a single best algorithm (generalist) but usually fail to yield configurations that focus on small subsets of problems (specialist). On the other hand, AS can train models that, given descriptive problem instances and well-explored performance data, map each problem to specialized algorithms. The generalist versus specialist is a relevant discussion in parameter configuration literature [18], and both have advantages and disadvantages depending on the base problem application.

The Per-Instance Algorithm Configuration (PIAC) [7] problem is a generalization of AS and AAC since it focuses on learning the mapping between each base problem instance and a well-fit configuration. PIAC proposals are less common, mainly due to PIAC formulation that can be hard to solve for diverse sets of base problem instances, particularly considering the optimization context. In some scenarios, quite different algorithms and configurations can efficiently solve the same subset of problems.

In this work, we approach PIAC through a novel formulation – MOAAC/D (Multi-Objective Automated Algorithm Configuration based on Decomposition), which uses a decomposition-based framework to provide generalist and specialist configurations at the same time. Given the configuration space as the decision space, small subsets of base problem instances are associated with each objective and the final Pareto Set ought to contain specialist (extremes) and generalist (knees) configurations. We can also select appropriated configurations from the Pareto set for new base problems. In the experiments we test this framework for tuning stochastic local search parameters (decision space) for different flowshop instances (base problems).

The remainder of the paper is organized as follows. Section 2 presents related works and Sect. 3 discusses our novel PIAC formulation as well as the proposed implementation to solve it. Section 4 presents the experiments performed, whereas results are discussed in Sect. 5. Finally, Sect. 6 concludes the paper.

2   Related Works

There are many AAC tools proposed to optimize the parameter values for a given set of problem instances. Despite that, not all AAC proposals can deal with unseen problem instances, most neither make use of problem features nor give insights into why a given configuration is chosen. Some works address the traditional AAC task under multiple criteria, by considering, for example, multiple base-problem objectives [2] or different budgets for solving the base-problem [3,5]. Hydra proposes an automatic way of building algorithm portfolios [21]. Similar to our proposal, it also uses an internal AAC procedure and iteratively samples and compares solvers to cover a heterogeneous instance set.

The main difference is that it needs an explicit portfolio builder strategy, i.e., a predefined decision-maker. The framework proposed here provides a decision-maker built in the partition method and uses the mapping from problem features to well-suited configurations. Another issue in Hydra is that its internal AACs can face difficulties updating the portfolio, which encompasses the whole set of instances. We overcome this issue by running AAC with a localized set of problems to promote specialization of the configurations.

Although very important, PIAC remains an open problem due to the complexity of recommendation space [7] and previous proposals on PIAC are limited by the problem or algorithm space. Recent applications use random forest models to learn configurations of a modular version of CMA-ES for continuous problems [13]. Also related to our proposal, Instance Space Analysis [6,19] investigates regions of the problem space in terms of algorithms' performance. None of them decompose the problem space like in MOAAC/D.

Despite the contribution of recent approaches, to the best of our knowledge, it is the first time a multi-objective formulation for PIAC based on problem space decomposition is proposed. Moreover, the proposed formulation contributes to the per-instance algorithm configuration for flowshop problems.

3   MOAAC/D: Multi-objective Automated Algorithm Configuration Based on Problem Space Decomposition

The proposed MOAAC/D formulation associates every objective with a different set of instances of the base problem – flowshop in the present work. The main idea of the framework is to build a Pareto set of configurations that can efficiently solve problems from all partitions. On the recommendation phase, we select well-suited configurations from the Pareto set. Figure 1 shows the proposed framework, including the mapping from problem features to the decomposed space that will support the Pareto set of configurations tuned to solve the base flowshop problems.

As Fig. 1 shows, the core component (in gray) of the proposed framework encompasses irace as local search and MOEA/D as the multi-objective algorithm based on decomposition used to find configurations. The diagram also highlights the principal components (flowshop features and partitioning method - PCA) used in the multi-objective decomposition, and the heuristics analyzed in the present paper, ILS and IG.

From the flowshop problem space we can extract features like the number of jobs, fitness landscape roughness, etc. In the building phase, this feature space can help group similar problems and divide the problem space. The partition could, for example, simply separate flowshop instances by their number of jobs (small, medium, or big base problems), their objective (makespan or flowtime), budget constraints, etc. Alternatively, all features can be clustered automatically using Principal Component Analysis (PCA)[1].

---

[1] Although we could use other techniques, like clustering (e.g., c-means), such a comparison is out of the scope of the paper, and we used PCA as a proof of concept.

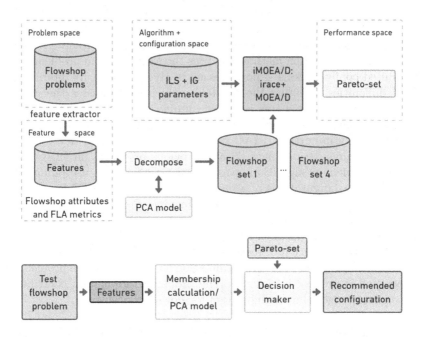

Fig. 1. The proposed framework for MOAAC/D applied to flowshop problems.

Popular metaheuristics for solving flowshop problems follow Iterated Local Search and Iterated Greedy frameworks (initialization, local search, and perturbation/destruction). The parameters of the metaheuristics form the configuration space, treated as the decision space for our multi-objective algorithm iMOEA/D (MOEA/D with irace-based local search). The aggregated performance on each partition set is used as the high-level objective function of the search. The generated Pareto front contains configurations specialized on each partition set (extreme points) and compromise solutions (knee points).

Finally, as depicted in Fig. 1(b), a decision-maker can choose from the Pareto set the best configuration for an unseen problem during the recommendation phase. Given the problem features, the proposed approach computes how close a solution is to each of the partition sets using the stored PCA model. With this information, the decision-maker determines which region of the Pareto best fits the new problem and chooses the closest configuration. Although this final phase could be considered *per se* a kind of per-instance algorithm selection since we select one from a set of configurations, this set has not been fixed *a priori* as occurs in the case of classic AS problems. Instead, it has been updated from a large set of possible configurations during the search, and the individual recommendation for a particular testing instance can be different depending on the stage of the evolutionary process.

3.1 Decomposing the Problem Space: From AAC Toward MOAAC/D

Our proposed decomposition of the problem space transforms a mono-objective AAC problem into a multi-objective one. According to the literature, AAC consists in minimizing the configuration utility value $u(\boldsymbol{\theta})$ for a set of instances of the base optimization problem (e.g., the mean objective function value of flowshop instances). In the present work, we propose a novel formulation for AAC: a multi-objective version (MOAAC/D), addressed here by partitioning or decomposing the base problem space P into non-overlapping sets of instances $\{P_1, \ldots, P_{N_{obj}}\}$ such that $P = P_1 \cup \cdots \cup P_{N_{obj}}$, where N_{obj} is the number of objectives in the MOAAC/D formulation or the total number of problem space partition subsets in our case. Formally, we have MOAAC/D defined as:

$$\text{Minimize } \mathbf{u}(\boldsymbol{\theta}) = (u_1(\boldsymbol{\theta}), \ldots, u_k(\boldsymbol{\theta}), \ldots, u_{N_{obj}}(\boldsymbol{\theta}))$$
$$\text{subject to } \boldsymbol{\theta} \in \Theta_\alpha^f \tag{1}$$

where $u_k(\boldsymbol{\theta}) = u(\boldsymbol{\theta} \mid P_k, C_{\boldsymbol{\theta}}, t), k = 1, \ldots, N_{obj}$. That is, each objective $u_k(\boldsymbol{\theta})$, associated with a problem partition subset P_k, represents a cost (drawn from a cost distribution $C_{\boldsymbol{\theta}}$) to be minimized within time t, considering as decision variables the configuration vector $\boldsymbol{\theta} \in \Theta_\alpha^f$, with $\Theta_\alpha^f \subseteq \Theta_\alpha$ as the space of feasible configurations of algorithm α. The Pareto Set resulting from this high-level optimization problem (MOAAC/D) contains specialist configurations of each partition set on the extremes of the Pareto front and compromise, generalist configurations at the Pareto front's knee.

3.2 iMOEAD: Solving the Decomposed Problem

Solving the multi-objective AAC can be difficult since the configuration space Θ contains discrete, continuous, ordinal variables and several constraints. We can leverage existing MOEAs using mono-objective AAC algorithms to perform a local search using the decomposed space.

In the present paper, we consider the well-known MOEA/D [22] to solve the problem in Eq. 1. We also embed irace as a local search operator capable of exploring well the decision space. We name this MOEA/D+irace hybrid iMOEA/D, and Algorithm 1 shows the implementation.

As usual, iMOEA/D associates each solution $\boldsymbol{\theta}_i$ in the population $\{\boldsymbol{\theta}_i\}$ with a weight vector called $\mathbf{w}_i$, according to its objective function. All the weight vectors are distributed uniformly on the simplex plane in the objective space, i.e., $\|\mathbf{w}_i\| = 1$, and remain constant through the search. The solution $\boldsymbol{\theta}_i$ associated with vector $\mathbf{w}_i$ has neighbors B_i that are N_{ngh} solutions with closest associated weights. New solutions are generated (through reproduction and/or local search), and during selection, solutions with better aggregation function values replace the previous ones. A well-known aggregation function is the Tchebycheff function [22]. Notice that original MOEA/D for continuous problems generates new individuals, i.e. new configurations, using simulated binary crossover and polynomial mutation in Line 7, whereas iMOEA/D includes the irace-based local search in Line 10 of Algorithm 1.

Algorithm 1. iMOEAD

Require: N_{pop}: MOEA/D population size
Require: N_{gen}: MOEA/D number of generations
Require: N_{ngh}: MOEA/D neighborhood size
Require: P: problem space
Require: N_{spl}: irace local search number of problems samples
Require: N_{evs}: irace local search number of configuration's evaluations
1: $\{\boldsymbol{\theta}_i\} \leftarrow N_{pop}$ random configurations from Θ_α^f
2: Calculate $\mathbf{u}(\boldsymbol{\theta}_i), i = 1, \ldots, N_{pop}$
3: $\{\mathbf{w}_i\} \leftarrow$ UNIFORMWEIGHTS(N_{pop})
4: $\{B_i\} \leftarrow$ ASSIGNNEIGHBORS($\{\mathbf{u}(\boldsymbol{\theta}_i)\}, \{\mathbf{w}_i\}, N_{ngh}$)
5: $\{P_1, \ldots, P_{N_{obj}}\} \leftarrow$ DECOMPOSE(P, N_{obj})
6: **for** $gen = 1$ to N_{gen} **do**
7:      $\{\boldsymbol{\theta}_i\}^{\text{off}} \leftarrow$ REPRODUCTION($\{\boldsymbol{\theta}_i\}$)   ▷ original MOEA/D's reproduction
8:      Calculate $\mathbf{u}(\boldsymbol{\theta}_i), i = 1, \ldots, N_{pop}$
9:      **for** $\boldsymbol{\theta}_i \in \{\boldsymbol{\theta}_i\}^{\text{off}}$ **do**
10:          $\boldsymbol{\theta}_i^n \leftarrow$ IRACELS($\boldsymbol{\theta}_i, B_i, \mathbf{w}_i, \{P_k\}, N_{spl}, N_{evs}$)
11:          Calculate $\mathbf{u}(\boldsymbol{\theta}_i^n)$
12:          Update non-dominated archive $\mathcal{A}$
13:          $\vartheta_k^* \leftarrow \min\{\vartheta_k^*, u_k(\boldsymbol{\theta}_i^n)\}, k = 1, \ldots, N_{obj}$
14:          **for** $\boldsymbol{\theta}_j \mid j \in B_i$ **do**
15:              **if** $g(\mathbf{u}(\boldsymbol{\theta}_i^n)|\mathbf{w}_i, \vartheta^*) \leq g(\mathbf{u}(\boldsymbol{\theta}_j)|\mathbf{w}_i, \vartheta^*)$ **then**
16:                  $\boldsymbol{\theta}_j \leftarrow \boldsymbol{\theta}_i^n$
17:                  $\mathbf{u}(\boldsymbol{\theta}_j) \leftarrow \mathbf{u}(\boldsymbol{\theta}_i^n)$
18:              **end if**
19:          **end for**
20:      **end for**
21: **end for**
22: **return** Non-dominated archive $\mathcal{A}$

Algorithm 2 details the proposed irace-based local search for iMOEA/D. Given the i-th incumbent solution (ie. configuration) and its associated weight vector $\mathbf{w}_i$, the local search samples instances from the problem partitions and applies irace to search on its focused instance set (P_{irace}). The sample is performed as follows. Given $\mathbf{w}_i \in \mathbb{R}^{N_{obj}}$ - the i-th weight vector that contains elements w_{ik} corresponding to the weight given by the k-th objective, the sampling process uses w_{ik} to dictate how many base problems to take from P_k. For example, $\mathbf{w} = (0.9, 0.1)$ implies that the configuration specializes most (90%) in problems like P_1 and less (10%) in problems like the ones in P_2. Therefore, in lines 1–5, Algorithm 2 samples a given number of N_{spl} problems used in irace search according to the rates given by $\mathbf{w}_i$ components ($w_{i1}, \ldots, w_{ik}, \ldots w_{iN_{obj}}$).

In addition, the proposal also uses the neighboring solutions plus the incumbent one as initial configurations for the irace local search. It has the advantage of using knowledge from good but similar configurations in the population. Irace search iteratively looks for configurations and updates the parameter value distributions. Configurations are compared against each other by solving the sampled

Algorithm 2. irace Local Search procedure for the MOEA/D framework.

Require: θ_i: the i-th incumbent solution
Require: $\mathbf{w}_i$: weight vector associated with the incumbent solution
Require: B_i^+: neighbor/closest configurations including the incumbent solution
Require: $\{P_1, \ldots, P_k, \ldots, P_{N_{obj}}\}$: problem space partition sets
Require: N_{spl}: total number of problems sampled to compose irace problem space
Require: N_{evs}: total number of configuration's evaluations
1: $P_{irace} \leftarrow \emptyset$
2: **for** $k = 1, \ldots, N_{obj}$ **do**
3:     $P_{aux} \leftarrow$ sample $\lfloor w_{ik} \times N_{spl} \rfloor$ instances from P_k
4:     $P_{irace} \leftarrow P_{irace} \bigcup P_{aux}$
5: **end for**
6: $\theta \leftarrow \text{IRACE}(\Theta_\alpha, B_i^+, P_{irace}, N_{evs})$
7: **return** θ

base problem and using non-parametric Friedman tests to select the best ones. The search stops after a maximum of N_{evs} configuration evaluations.

Figure 2 shows an example of a bi-objective space on which iMOEA/D could run. Algorithm 2 uses the knowledge from the neighborhood (green areas) to initialize the search. As shown in Fig. 2, the irace local search described in Algorithm 2 should find good configurations focused on the sampled problems, improving each solution θ_i on its search direction $\mathbf{w}_i$ of the decomposed space.

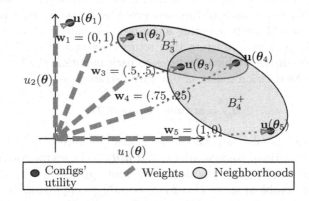

Fig. 2. An example of problem space decomposition: bi-objective formulation $(u_1(\theta), u_2(\theta))$ with five individuals in the population $(\theta_1, \ldots, \theta_5)$, five corresponding weights $(\mathbf{w}_1, \ldots, \mathbf{w}_5)$, and also $\mathbf{B}_3^+ = \{\theta_2, \theta_3, \theta_4\}$, $\mathbf{B}_4^+ = \{\theta_3, \theta_4, \theta_5\}$, ie. θ_3 and θ_4 expanded neighborhoods, respectively. (Color figure online)

Figure 2 shows 5 candidate configurations, i.e. individuals in the population of MOEA/D. In this example of a MOAAC/D formulation with $N_{obj} = 2$ objectives, every feasible configuration $\theta \in \Theta_\alpha^f$ is associated in the objective space with a 2D coordinate:

$$\mathbf{u}(\boldsymbol{\theta}) = (u_1(\boldsymbol{\theta}|P_1, C_{\boldsymbol{\theta}}, t), u_2(\boldsymbol{\theta}|P_2, C_{\boldsymbol{\theta}}, t))$$

which represents the utility of α algorithm's configuration $\boldsymbol{\theta}$ on the problem partition P_1 and P_2, respectively. In the iMOEA/D population, every candidate configuration $\boldsymbol{\theta}_i$ has also a corresponding weight $\mathbf{w}_i$. In the example shown in Fig. 2, to provide P_{irace}, the proposed irace-based local search (Algorithm 2) when performed on $\boldsymbol{\theta}_4$ would sample 75% problems from partition P_1 and 25% problems from P_2. It would also use the neighbors $\{\boldsymbol{\theta}_3, \boldsymbol{\theta}_4, \boldsymbol{\theta}_5\}$ as the initial configurations for irace.

3.3   Decision Maker: Recommending Configurations

Given the approximated Pareto set obtained by iMOEA/D, we need to provide a way to select the best fit configuration for new and unseen problems. The PIAC formulation generalizes the AS problem due to its mapping from problem features to configurations in addition to the algorithms [7]. Formally, we have $h^* = \arg\min_{h:F \to \Theta_\alpha} ||\mathbf{u}_h||$, $\mathbf{u}_h = (u(h(f(p))), \forall p \in P)$, where $f : P \to F$ extracts relevant features from instances p in the problem space P, h maps the feature space F into the configuration space Θ_α for an algorithm α, and $||\mathbf{u}_h||$ is the norm of h mapping performance to be minimized, measured by an utility function u, over all the problems or instances p in the problem space P.

A PIAC strategy can be obtained from the above formulation considering that, for a new instance, we can find the appropriate configuration by choosing one from the Pareto Set $\mathcal{PS}$. For example, if we have two partitions P_1 and P_2 for small and large instances of the base problem, respectively, a new instance of medium size could imply selecting a compromise configuration.

Formally, if we have a function that returns a vector $\boldsymbol{\psi} \in \mathbb{R}^{N_{obj}}$, indicating the membership of the problem features to each partition P_k, $k = 1, \ldots, N_{obj}$, we can define the decision maker of our PIAC strategy as:

$$\boldsymbol{\theta}_{p'}^* = \arg\min_{\boldsymbol{\theta} \in \mathcal{PS}} ||\boldsymbol{\psi}_{p'} - \hat{\mathbf{u}}(\boldsymbol{\theta})||, \tag{2}$$

where $\boldsymbol{\psi}_{p'}$ is the membership vector for a new instance p', $\mathcal{PS}$ is the Pareto Set of all (near) optimal configurations, and $\hat{\mathbf{u}}(\boldsymbol{\theta})$ is the vector corresponding to the utility vector $\mathbf{u}(\boldsymbol{\theta}) \in \mathbb{R}^{N_{obj}}$ scaled to the unit space $[0,1]^{N_{obj}}$. Therefore, by partitioning the problem space $P = P_1 \cup \cdots \cup P_{N_{obj}}$; approximating the Pareto Set of configurations (Problem in Eq. 1); defining a decision maker to select the best fit configuration (Eq. 2); and considering $h^* : F \to \Theta_\alpha$ as the best mapping from problem features to configurations of algorithm α, we can reformulate the PIAC problem as:

$$h^* = h(f(p')|\mathcal{PS}^*), \quad \mathcal{PS}^* = \arg\min_{\mathcal{PS} \subseteq \Theta} ||\mathbf{u}_{\mathcal{PS}}||$$

$$\mathbf{u}_{\mathcal{PS}} = (||\boldsymbol{\psi}_p - \hat{\mathbf{u}}(\boldsymbol{\theta})||, ||\hat{\mathbf{u}}(\boldsymbol{\theta})||), \text{ for } p \in P, \boldsymbol{\theta} \in \mathcal{PS} \tag{3}$$

where $||\mathbf{u}_{\mathcal{PS}}||$ is the utility's norm of the Pareto-Set, regarding the distance from each membership vector $\boldsymbol{\psi}_p$, $p \in P$, to every point in the Pareto-Front. As we might expect, finding a good Pareto-Front is equivalent to finding small distances between $\boldsymbol{\psi}_p$ and $\hat{\mathbf{u}}(\boldsymbol{\theta})$ (spread) and reducing the norm $||\hat{\mathbf{u}}(\boldsymbol{\theta})||$ (convergence).

Therefore, by partition the problem space, the final Pareto-Front yields a good mapping between the problems' features and the non-dominated configurations. Partitions $\{P_1, \ldots, P_{N_{obj}}\}$ can be found automatically using the feature space F and a dimensionality reduction technique, like PCA. A set of representative instances of the base problem in each axis might form the partition sets. The cosine squared metric for PCA could be used to obtain the membership vector $\boldsymbol{\psi}$ as it measures the relationship between a new point and the PCA dimensions.

4   Experiments on Flowshop Problems

To evaluate the proposal, we perform a series of experiments using the Flowshop Problem (FSP) as the base problem scenario. FSPs model a production line in which a set of machines perform operations on a set of jobs. The goal is to find a good job schedule that minimizes the total processing time (makespan) or the sum of processing times (total flowtime). The usual FSP formulation has the following conditions [1]: a set of J unrelated, multiple-operation jobs is available for processing at time zero; each job requires M operations, and each operation requires a different machine; setup times for the operations are sequence-independent and included in processing times; job times on each machine are known in advance; all machines are continuously available; once the operation begins, it proceeds without interruption.

The most studied variant of this problem is the permutation FSP, where the goal is to find the best job permutation that minimizes a given measure, like makespan (time to process all jobs) and total flowtime (sum of all job completion times). Some of the current best methods include the Nawaz-Encore-Ham (NEH) heuristic [9] and the Iterated Greedy algorithm [4,16].

In the present paper, we generate 6,480 FSPs and use 5-fold cross-validation to train and test the proposal. The problems processing times are generated with 5 random seeds from all combinations of $J = \{10, 20, 30, 50\}$ jobs, and $M = \{5, 10, 20\}$ machines, {uniform, exponential, binomial} distributions with mean 50, {random, job-correlated, machine-correlated} with 0.95 correlation for correlated instances. We also explore two different base objectives, makespan and total flowtime, and stopping criteria based on time ($2J^2 M N_{tB}$ milliseconds), or number of evaluations ($J M N_{eB}$ evaluations), where the time budget N_{tB} and evaluations budget N_{eB} are respectively, $N_{tB} \in \{10^{-4}, 10^{-3}, 10^{-2}\}$ and $N_{eB} \in \{10, 100, 1000\}$.

In the experiments, configuration space Θ is based on the ILS framework; initialization, local search, perturbation, acceptance criterion, as well as their associated parameters, are described in Table 1. Initialization can be a random or the NEH heuristic. The local search can be based on First Improvement (FI), Best Improvement (BI), or Best Insertion (BI) on a given percentage of the neighborhood, with one or unlimited steps. The perturbation procedure could be based on Ruiz-Stutzle (RS) destruction-construction [16] with a given destruction size

Table 1. The ILS+IG configuration space Θ.

Parameter	Domain
Initialization	(`random`, `NEH`)
Local search	(`none`, `FI`, `BI`, `RBI`, `BIns`)
Neighbor comparison operator	(`>`, `>=`)
Neighborhood size	$(0.0, 1.0]$
Single step local search	(`true`, `false`)
Perturbation	(`RS`, `LSPS`, `swap`)
RS or LSPS job insertion	(`randomBest`, `firstBest`, `lastBest`)
LSPS local search	(`none`, `FI`, `BI`, `RBI`, `BIns`)
LSPS single step local search	(`true`, `false`)
RS or LSPS destruction size	$[2, 3, \ldots, 8]$
Swap: number of swaps	$[1, 2, \ldots, 8]$
Acceptance criterion	(`always`, `better`, `metropolisHastings`)
Metropolis-Hastings temp. scale	$(0.0, 5.0]$

d, or include partial solution local search (LSPS) [4], including a parameterized inner local search. Finally, the acceptance criterion can be set to accept all, only improvements or based on Metropolis-Hastings [20] with a given temperature scale.

The utility function component $u_k(\boldsymbol{\theta})$ is the one commonly used to compare flowshop metaheuristics – Average Relative Percentage Deviation (ARPD) [10]. ARPD considers the difference to a reference fitness value, obtained here by running irace for a long evolution on each instance (instance-based best $\boldsymbol{\theta}^r$) and taking the average of 30 runs of the best configuration, $u_k(\boldsymbol{\theta}) = 1/30 \sum_{j=1}^{30} u_{jk}(\boldsymbol{\theta}^r)$.

The feature space F contains simple features, like the number of jobs, number of machines, budget, objective, stopping criterion, and jobs/machines ratio. It includes processing time statistics, like standard deviation, correlation per job, and correlation per machine. We also use fitness landscape metrics [12] like random walk autocorrelation, entropy, partial information, information stability, and fitness-distance correlation using different distances (all five from [14]), and local search procedures (first improvement, best improvement, and best insertion).

To investigate the MOAAC/D proposal under a PIAC perspective, we automatically decomposed the FSP problem space P into four partition subsets P_k ($k = 1, 2, 3, 4$), one for each principal component of the PCA model.

We perform a space reduction in the pre-processed F data by means of PCA, providing an $\mathbb{R}^4$ feature space. Notice that the choice of four dimensions seems a good compromise between not degenerating iMOEA/D performance and also allowing a suitable partition in the problem space. We further perform the decision making process to choose the best configuration of each test instance.

In the experiments, we compare MOEA/D (Algorithm 1 without local search in Line 10) with iMOEA/D (Algorithm 1 without reproduction in Lines 7 and 8). First, we explore partition *versus* performance, and in the sequence we analyze FSP features aspects. Then we compare iMOEA/D in terms of recommended configuration quality with three other PIAC approaches: (i) a randomly generated configuration from the same space (randPIAC); (ii) the global-best configuration found by running irace on the entire problem space; and (iii) a meta-learning-based approach that uses instance-best performance data to train multi-label random-forest models (MetaL) [11]. All experiments consider a 5-fold cross-validation procedure where the strategies use training problems $(P_{tr}^1, \ldots, P_{tr}^5)$ and have the same budget ($1000|P_{tr}^{fd}|$ configurations evaluations) in each fold fd, which took about 30 CPU days. The ARPD performance results from $(P_{ts}^1, \ldots, P_{ts}^5)$ are averaged through all test sets.

5 Results

As described in Sect. 3, we can use a PCA model of the feature space F to automatically decompose the problem space P. PCA models of the feature space provide the contribution measure for the problem features to each partition. Figure 3 shows the top six features that most contribute to dimensionality reduction.

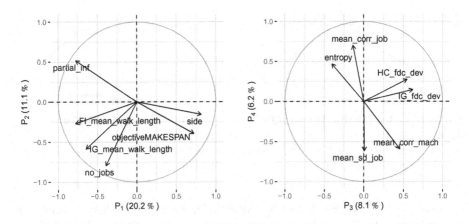

Fig. 3. PCA variable plot for flowshop features that most contribute to each partition (axis). For better visualization, we only show the top six features.

In the performed experiments partitioning explains 45,6% of the data variability as follows:

- Partition P_1: instances with many neutral (side) edges and a low number of steps to reach local optima using first improvement (FI mean walk length);
- Partition P_2: instances with a low number of jobs and makespan objective;

- Partition P_3: instances with most variations on the correlation between fitness and number of steps to reach a local minimum (fitness-distance correlation, FDC) using best insertion and best improvement local searches (HC's and IG's FDC with deviation distances).
- Partition P_4: job-correlated instances with high entropy on random walk fitness and greater variation of job times;

Figure 4a shows the ARPD performance of iMOEA/D × MOEA/D through a parallel coordinates plot for the final 20 configurations found by both. We notice that both cand find good solutions for partition P_1 and P_4, but iMOEA/D dominates MOEA/D on P_2 and P_3. This shows the effectiveness of the proposed irace-based local search when providing new individuals. The iMOEA/D non-dominated front is explored in Fig. 4b by highlighting the destruction sizes of each configuration. We notice that most configurations can find good solutions (low ARPD values) on partitions P_1 and P_4. The instances from P_3 seem much harder, and the best configurations have low destruction sizes (darker lines). The analyses emphasize how the proposed MOAAC/D adds interpretability to a black-box AAC problem by investigating the Pareto front.

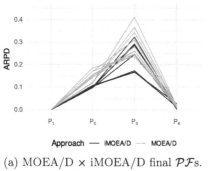

(a) MOEA/D × iMOEA/D final $\mathcal{PF}$s.

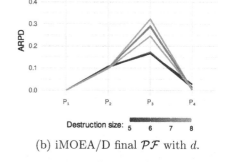

(b) iMOEA/D final $\mathcal{PF}$ with d.

Fig. 4. Final ARPD for ILS+IG configurations on FSPs, whose instances are decomposed automatically by PCA into $\{P_1, P_2, P_3, P_4\}$ partitions.

Using the Pareto Set found by iMOEA/D, we test our PIAC formulation by comparing it with randomly chosen parameters (randPIAC), the global-best configuration found by irace (G-IRACE), and a meta-learning-based approach (MetaL) [11], all using the entire problem space with the same budget of configuration evaluations. An *instance-best* reference ($\boldsymbol{\theta}^r$) is set as the best configuration found by irace for each particular instance[2]. Table 2 summarizes the percentage of times that the recommended configuration found by each approach is better, equal, or worse than the reference one.

We notice from Table 2 that, for smaller instances, iMOEA/D outperforms the remaining ones in all criteria. Considering the harder instances (larger

---

[2] $\boldsymbol{\theta}^r$, a kind of upper bound, can be surpassed due to the stochastic nature of irace.

Table 2. Percentage of times θ^*, recommended configuration by each approach, is better, equal, or worse than the reference θ^r, on test flowshop instances p'.

Number of jobs	Approach	Better $u(\theta^*_{p'}) < u(\theta^r_{p'})$	Equal $u(\theta^*_{p'}) = u(\theta^r_{p'})$	Worse $u(\theta^r_{p'}) > u(\theta^*_{p'})$
$J < 50$	randPIAC	0.59	30.14	69.27
	G-IRACE	3.30	54.49	42.21
	MetaL	6.94	53.75	39.31
	iMOEA/D	**15.56**	**60.65**	**23.80**
$J = 50$	randPIAC	0.37	17.16	82.47
	G-IRACE	3.95	35.86	60.19
	MetaL	11.54	**37.28**	**51.17**
	iMOEA/D	**27.79**	10.19	62.04

u represents a cost (drawn from a cost distribution C_{θ^*} resulting from different seeds)

instances with random processing times), although iMOEA/D is surpassed by metalearning in the second and third criteria, it outperforms the others in the first criterion, i.e. it provides almost three times better solutions than MetaL.

Table 3 shows the comparison of ARPD values of the generated configurations. We see that G-IRACE and iMOEA/D provide very similar results in terms of ARPD mean. However, iMOEA/D has the lowest mean ARPD rank indicating that, overall, it has better configurations. These conclusions are supported by the blocked Friedman test with Nemenyi posthoc, considering 30 different runs for each configuration.

Table 3. Mean and standard deviations of ARPDs and ARPD ranks for each approach.

Approach	ARPD	ARPD Rank
randPIAC	2.712 ± 10.302	3.400 ± 0.772
G-IRACE	0.082 ± 0.230	2.314 ± 0.600
MetaL	0.094 ± 1.793	2.197 ± 0.710
iMOEA/D	0.085 ± 0.466	$\mathbf{2.089 \pm 0.757}$

6 Conclusion

We have investigated the Per-Instance Algorithm Configuration problem from a completely different perspective. The proposal includes a novel multi-objective formulation of the Automated Algorithm Configuration problem, its generalization to the per-instance configuration, and an efficient local search that uses the problem space decomposition structure to find good configurations. Experiments explored the space partition and the use of PCA to provide insights into the problem characteristics and performance. The results indicated that the proposed approach is effective in finding good per-instance configurations. Overall, iMOEA/D outperformed the comparison approaches, except for the metalearning-based one for more complex instances.

Besides relieving the pressure on users that comes from the hard task of algorithm configuration on complex problems, the proposal also contributes to multi and many-objective optimization communities by providing novel applications for established MOEAs as powerful tuner algorithms. The work can be expanded in many different ways, like including new problems and testing other inner algorithms, for both local and multi-objective searches. The proposal could also use different automatic partition techniques like clustering algorithms.

References

1. Baker, K.R., Trietsch, D.: Appendix A: practical processing time distributions. In: Principles of Sequencing and Scheduling, pp. 445–458. Wiley (2009). https://doi.org/10.1002/9780470451793.app1
2. Blot, A., Hoos, H.H., Jourdan, L., Kessaci-Marmion, M.É., Trautmann, H.: MO-ParamILS: a multi-objective automatic algorithm configuration framework. In: Festa, P., Sellmann, M., Vanschoren, J. (eds.) LION 2016. LNCS, vol. 10079, pp. 32–47. Springer, Cham (2016). https://doi.org/10.1007/978-3-319-50349-3_3
3. Dréo, J.: Using performance fronts for parameter setting of stochastic metaheuristics. In: Proceedings of the 11th Annual Conference Companion on Genetic and Evolutionary Computation Conference - GECCO 2009, p. 2197. ACM Press, Montreal (2009). https://doi.org/10.1145/1570256.1570301
4. Dubois-Lacoste, J., Pagnozzi, F., Stützle, T.: An iterated greedy algorithm with optimization of partial solutions for the makespan permutation flowshop problem. Comput. Oper. Res. **81**, 160–166 (2017). https://doi.org/10.1016/j.cor.2016.12.021
5. Dymond, A.S., Kok, S., Heyns, P.S.: MOTA: a many-objective tuning algorithm specialized for tuning under multiple objective function evaluation budgets. Evol. Comput. **25**(1), 113–141 (2017). https://doi.org/10.1162/EVCO_a_00163
6. Fernandes, L.H.d.S., Lorena, A.C., Smith-Miles, K.: Towards understanding clustering problems and algorithms: an instance space analysis. Algorithms **14**(3) (2021). https://doi.org/10.3390/a14030095
7. Kerschke, P., Hoos, H.H., Neumann, F., Trautmann, H.: Automated algorithm selection: survey and perspectives. Evol. Comput. **27**(1), 3–45 (2019)
8. López-Ibáñez, M., Dubois-Lacoste, J., Cáceres, L.P., Stützle, T., Birattari, M.: The irace package: iterated racing for automatic algorithm configuration. Oper. Res. Perspect. **3**, 43–58 (2016). https://doi.org/10.1016/j.orp.2016.09.002
9. Nawaz, M., Enscore, E.E., Ham, I.: A heuristic algorithm for the m-machine, n-job flow-shop sequencing problem. Omega **11**(1), 91–95 (1983). https://doi.org/10.1016/0305-0483(83)90088-9
10. Osman, I.H., Potts, C.: Simulated annealing for permutation flow-shop scheduling. Omega **17**(6), 551–557 (1989)
11. Pavelski, L.M., Delgado, M., Kessaci, M.É., Freitas, A.A.: Stochastic local search and parameters recommendation: a case study on flowshop problems. Int. Trans. Oper. Res. itor.12922 (2020). https://doi.org/10.1111/itor.12922
12. Pitzer, E., Affenzeller, M.: A Comprehensive Survey on Fitness Landscape Analysis. In: Fodor, J., Klempous, R., Suárez Araujo, C.P. (eds.) Recent Advances in Intelligent Engineering Systems. Studies in Computational Intelligence, vol. 378, pp. 161–186. Springer, Heidelberg (2012). https://doi.org/10.1007/978-3-642-23229-9_8

13. Prager, R.P., Trautmann, H., Wang, H., Bäck, T.H.W., Kerschke, P.: Per-instance configuration of the modularized CMA-ES by means of classifier chains and exploratory landscape analysis. In: 2020 IEEE Symposium Series on Computational Intelligence (SSCI), pp. 996–1003 (2020). https://doi.org/10.1109/SSCI47803.2020.9308510

14. Reeves, C.: Landscapes, operators and heuristic search. Ann. Oper. Res. **86**, 473–490 (1999). https://doi.org/10.1023/A:1018983524911

15. Rice, J.R.: The Algorithm Selection Problem. In: Rubinoff, M., Yovits, M.C. (eds.) Advances in Computers. Advances in Computers, vol. 15, pp. 65–118. Elsevier, Washington (1976). https://doi.org/10.1016/S0065-2458(08)60520-3

16. Ruiz, R., Stützle, T.: A simple and effective iterated greedy algorithm for the permutation flowshop scheduling problem. Eur. J. Oper. Res. **177**(3), 2033–2049 (2007). https://doi.org/10.1016/j.ejor.2005.12.009

17. Schede, E., et al.: A survey of methods for automated algorithm configuration (2022). https://doi.org/10.48550/ARXIV.2202.01651

18. Smit, S.K., Eiben, A.E.: Parameter tuning of evolutionary algorithms: generalist vs. specialist. In: Di Chio, C., et al. (eds.) EvoApplications 2010. LNCS, vol. 6024, pp. 542–551. Springer, Heidelberg (2010). https://doi.org/10.1007/978-3-642-12239-2_56

19. Smith-Miles, K., Baatar, D., Wreford, B., Lewis, R.: Towards objective measures of algorithm performance across instance space. Comput. Oper. Res. **45**, 12–24 (2014). https://doi.org/10.1016/j.cor.2013.11.015

20. Stützle, T.: Applying iterated local search to the permutation flow shop problem. Technical report, FG Intellektik, TU Darmstadt, Darmstadt, Germany (1998)

21. Xu, L., Hoos, H., Leyton-Brown, K.: Hydra: automatically configuring algorithms for portfolio-based selection. In: Twenty-Fourth AAAI Conference on Artificial Intelligence (2010)

22. Zhang, Q., Li, H.: MOEA/D: a multiobjective evolutionary algorithm based on decomposition. IEEE Trans. Evol. Comput. **11**(6), 712–731 (2007). https://doi.org/10.1109/TEVC.2007.892759

Improving Group Search Optimization for Automatic Data Clustering Using Merge and Split Operators

Luciano D. S. Pacifico[1]([⊠])[iD] and Teresa B. Ludermir[2][iD]

[1] Departamento de Computação (DC), Universidade Federal Rural de Pernambuco, Recife, PE, Brazil
luciano.pacifico@ufrpe.br
[2] Centro de Informática, Universidade Federal de Pernambuco, Recife, PE, Brazil
tbl@cin.ufpe.br

Abstract. The amount of digital data produced daily has increased considerably in the last years. The need for fast and reliable information in real-world applications demands ever more precise algorithms and Data Mining tools, once most of the systems in our daily lives are executed in real-time. Data clustering is one of the most important and primitive activities in Unsupervised Machine Learning, consisting in a fundamental mechanism for exploratory data analysis. Given the complexity of data clustering task, standard clustering methods, such as the partitional algorithms, are easily trapped in local optima solutions, due to their lack of good global searching operators. In this work, three improved Group Search Optimization-based approaches are proposed, based on merge and split heuristics, in the context of Automatic Clustering Analysis: MGSO, SGSO and MSGSO. Group Search Optimization (GSO) is a natural-inspired meta-heuristic, known for its good global search abilities, and mechanisms to escape from local optima points from the problem space. The proposed models attempt to perform both cluster optimization and the determination of the best number of clusters for each dataset, overcoming the limitations of traditional partitional clustering algorithms. The proposed GSO-based models are evaluated through a testing bed composed of nine real-world problems, and compared to six state-of-the-art partitional automatic clustering approaches, include standard GSO. The experimental evaluation has been performed considering five clustering metrics, and both empirical and statistical analysis. The results showed that the proposed MGSO, SGSO and MSGSO algorithms are very promising and reliable while tackling clustering problems.

Keywords: Group search optimization · Automatic data clustering · Evolutionary operators · Merge and split heuristics · Machine learning

1 Introduction

Data Clustering is one of the most important fields in Pattern Recognition and Machine Learning, represented by an unsupervised attempt to divide a collection

© The Author(s), under exclusive license to Springer Nature Switzerland AG 2022
J. C. Xavier-Junior and R. A. Rios (Eds.): BRACIS 2022, LNAI 13653, pp. 340–354, 2022.
https://doi.org/10.1007/978-3-031-21686-2_24

of samples (observations) into groups (clusters) according to their similarities and dissimilarities. Clustering algorithms are important and fundamental tools for exploratory data analysis and Knowledge Discovery in Databases (KDD), finding applications in many fields, such as Bioinformatics, Computer Vision, Text Mining and Big Data Analysis.

The most popular clustering technique are the partitional models, which promote a partition of the dataset in evaluation into a pre-defined number of clusters (a parameter for such models), implementing their efforts in an attempt to optimize a criterion function iteratively. But partitional methods are known for their sensibility to the initial status of the search (their initial position on the problem space), what may lead to weak solutions, due to the fact that most partitional methods only perform local searches, being easily trapped in local optima points if the algorithm starts in a poor region of the problem space. Also, it is quite difficult to determine the final desired number of data clusters in applications in new research fields or problems, once few information may be available (if any) on current research subject, leading to poor clustering performances, what may result in miscomprehension of the phenomenon in study.

Natural-inspired models, such as Evolutionary Algorithms (EAs) and Swarm Intelligence (SIs) methods, are known as general-purpose meta-heuristics, usually employed in hard and complex optimization problems, once such models present good global search capabilities and mechanisms to escape from local optima points. EAs such as Genetic Algorithm (GA) [21], Differential Evolution (DE) [38] and Backtracking Search Optimization (BSA) [7], simulate biological processes like mutation, recombination and selection, to perform their searches. SIs, like Ant Colony Optimization (ACO) [14], Particle Swarm Optimization (PSO) [26] and Group Search Optimization (GSO) [19], implement their evolutionary operators as simulations to the self-organizing and collective behavior of social animals, like swarming, flocking and herding [5]. Both EAs and SIs searching strategies are guided in an attempt to optimize a criterion function, the fitness function, by improving a population of candidate solutions of the problem at hand by means of their evolutionary operators.

In the last decades, many natural-inspired meta-heuristics have been adopted to tackle data clustering problems, and, in many cases, such models are adapted as partitional clustering approaches that estimate both the best partition that would represent a given collection of sample observations, and the adequate number of final clusters, at the same time, in a field also known as Dynamic Data Clustering [31] or, more recently, Automatic Data Clustering [9]. Interesting and recent research surveys on the subject of automatic clustering using natural-inspired meta-heuristics can be found in [16, 23, 25].

In this work, we proposed three improved GSO algorithms based on the application of merge and split operators in GSO population: MGSO, SGSO and MSGSO. The proposed mechanisms seek to promote a faster exploration and exploitation of the problem search space by improving GSO global search abilities with local search heuristics adapted to the context of automatic clustering analysis.

This work is organized as follows. Section 2 discusses standard GSO algorithm. Next (Sect. 3), the proposed merge and split mechanisms, as much as the proposed GSO-based models are presented in details. Experimental results are shown in Sect. 4, followed by some conclusions and leads to future works (Sect. 5).

2   Group Search Optimization (GSO)

Group search optimization is inspired by animal social searching behavior and group living theory. GSO employs the Producer-Scrounger (PS) model as a framework. The PS model was firstly proposed by Barnard and Sibly [4] to analyze social foraging strategies of group living animals. PS model assumes that there are two foraging strategies within groups: *producing* (*e.g.*, searching for food); and joining (*scrounging*, *e.g.*, joining resources uncovered by others). Foragers are assumed to use producing or joining strategies exclusively. Under this framework, concepts of resource searching from animal visual scanning mechanism are used to design optimum searching strategies in GSO algorithm [19].

In GSO, the population G of S individuals is called *group*, and each individual is called a *member*. In an n-dimensional search space, the i-th member at the t-th searching iteration (*generation*) has a current position $\mathbf{X}_i^t \in \Re^n$ and a head angle $\alpha_i^t \in \Re^{n-1}$. The search direction of the i-th member, which is a vector $\mathbf{D}_i^t(\alpha_i^t) = (d_{i1}^t, \ldots, d_{in}^t)$ can be calculated from α_i^t via a polar to Cartesian coordinate transformation:

$$d_{i1}^t = \prod_{q=1}^{n-1} \cos(\alpha_{iq}^t), \tag{1}$$

$$d_{ij}^t = \sin(\alpha_{i(j-1)}^t) \prod_{q=1}^{n-1} \cos(\alpha_{iq}^t)(j = 1, \ldots, n-1),$$

$$d_{in}^t = \sin(\alpha_{i(n-1)}^t)$$

A group in GSO consists of three types of members: producers, scroungers and dispersed members (or *rangers*) [19]. The rangers are introduced by GSO model, extending standard PS framework.

During each GSO search generation, a group member which has found the best fitness value so far (most promising area form the problem search space) is chosen as the producer ($\mathbf{X}_p$) [8], and the remaining members are scroungers or rangers.

The producer employs a scanning strategy (*producing*) based on its vision field, generalized to a n-dimensi-onal space, which is characterized by maximum pursuit angle $\theta_{max} \in \Re^{n-1}$ and maximum pursuit distance $l_{max} \in \Re$, given by Eq. (2).

$$l_{max} = \|\mathbf{U} - \mathbf{L}\| = \sqrt{\sum_{k=1}^{n} (U_k - L_k)^2} \tag{2}$$

where U_k and L_k denote the upper bound and lower bound of the k-th dimension from the problem space, respectively.

In GSO, at the t-th generation, the producer $\mathbf{X}_p^t$ will scan laterally by randomly sampling three points in the scanning field: one at zero degree (Eq. (3)), one in the right hand side hypercube (Eq. (4)) and one in the left hand side hypercube (Eq. (5)).

$$\mathbf{X}_z = \mathbf{X}_p^t + r_1 l_{max} \mathbf{D}_p^t(\alpha_p^t) \tag{3}$$

$$\mathbf{X}_r = \mathbf{X}_p^t + r_1 l_{max} \mathbf{D}_p^t(\alpha_p^t + \tfrac{r_2 \theta_{max}}{2}) \tag{4}$$

$$\mathbf{X}_l = \mathbf{X}_p^t + r_1 l_{max} \mathbf{D}_p^t(\alpha_p^t - \tfrac{r_2 \theta_{max}}{2}) \tag{5}$$

where $r_1 \in \Re$ is a normally distributed random number (mean 0 and standard deviation 1) and $\mathbf{r}_2 \in \Re^{n-1}$ is a uniformly distributed random sequence in the range $U(0, 1)$.

If the producer is able to find a better resource than its current position, it will fly to this point; if no better point is found, the producer will stay in its current position, then it will turn its head to a new generated angle (Eq. (6)).

$$\alpha_p^{t+1} = \alpha_p^t + \mathbf{r}_2 \alpha_{max} \tag{6}$$

where $\alpha_{max} \in \Re$ is the maximum turning angle.

If after $a \in \Re$ generations the producer cannot find a better area, it will turn its head back to zero degree (Eq. (7)).

$$\alpha_p^{k+a} = \alpha_p^k \tag{7}$$

All scroungers will join the resource found by the producer, performing *scrounging* strategy according to Eq. (8).

$$\mathbf{X}_i^{t+1} = \mathbf{X}_i^t + \mathbf{r}_3 \circ (\mathbf{X}_p^t - \mathbf{X}_i^t) \tag{8}$$

where $\mathbf{r}_3 \in \Re^n$ is a uniform random sequence in the range $U(0, 1)$ and $\circ$ is the Hadamard product or the Schur product, which calculates the entrywise product of two vectors.

The rangers will perform random walks through the problem space [20], according to Eq. (9).

$$\mathbf{X}_i^{t+1} = \mathbf{X}_i^t + l_i \mathbf{D}_i^t(\alpha_i^{t+1}) \tag{9}$$

where

$$l_i = a r_1 l_{max} \tag{10}$$

In GSO, when a member escapes from the search space bounds, it will turn back to its previous position inside the search space [13]. GSO algorithm is presented in Algorithm 1.

Algorithm 1. Group Search Optimization

$t \leftarrow 0$.

Initialize randomly position $\mathbf{X}_i^{(0)}$ and head angles $\alpha_i^{(0)}$ of all members $\mathbf{X}_i^{(0)} \in G$.

Calculate the fitness value ($fitness(\mathbf{X}_i^{(0)})$) for each member $\mathbf{X}_i^{(0)}$.

while (termination conditions are not met) **do**

  Pick the best group member as the $\mathbf{X}_p^t$ for the current generation.

  Execute producing ($\mathbf{X}_p^t$ only) by evaluating three random points in its visual scanning field, $\mathbf{X}_z^t$ (Eq. (3)), $\mathbf{X}_r^t$ (Eq. (4)) and $\mathbf{X}_l^t$ (Eq. (5)).

  Choose a percentage from the members (but the $\mathbf{X}_p^t$) to perform scrounging (Eq. (8)).

  Ranging: The remaining members will perform *ranging* through random walks (Eq. (9)).

  Calculate the new fitness value $fitness(\mathbf{X}_i^t)$ for each group member $\mathbf{X}_i^t$.

  $t \leftarrow t + 1$.

end while

Return $\mathbf{X}_p^{t_{max}}$.

GSO scrounging operator focuses the search performed by the group in the most promising areas from the problem space, corresponding to the main exploration/exploitation strategy employed by many EAs (like *crossover* strategy in Genetic Algorithms and, *particle movement* in Particle Swarm Optimization).

Producing and ranging are the main mechanisms employed by GSO for escaping local optima points. When the producer of one generation is trapped in a local optimum point, all scroungers would follow that producer into that local optimum point, after executing scrounging operator, leading to a premature convergence of the group. In GSO, producing operator is proposed as a solution to the premature convergence problem, giving the producer opportunities to escape from local optima points. Rangers also provide local optima escaping mechanisms to GSO, once rangers will always keep performing random walks through the problem search space that do not depend on the results found by the producer, which may lead to most promising areas than the ones found so far by the whole group, evading local optima points.

3   Proposed Approaches: MGSO, SGSO and MSGSO

This section introduces the proposed GSO-based models: MGSO, SGSO and MSGSO. The proposed methods are introduced as partitional automatic clustering approaches, which combine the good global search capabilities of standard GSO algorithm with merge and split operators, inspired by mutation operators presented in [29], as a manner to reinforce and improve GSO mechanisms to escape from local optima points.

In the context of data clustering problems, consider a partition P_C of a dataset with N data objects $\mathbf{x}_j \in \Re^m$ ($j = 1, 2, ..., N$) in at most C_{max} clusters. Each cluster is represented by its centroid vector $\mathbf{g}_c \in \Re^m$ ($c = 1, 2, ..., C_{max}$). Each member $\mathbf{X}_i \in \Re^n$ (where $n = C_{max} + C_{max} \times m$) in group G represents

C_{max} activation threshold values and C_{max} cluster centroids at the same time, one for each candidate cluster [9,25], as illustrated in Fig. 1.

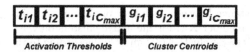

Activation Thresholds   Cluster Centroids

Fig. 1. Member representation: the first C_{max} features represent activation thresholds for each candidate cluster, while the following $C_{max} \times m$ are the C_{max} m-dimensional candidate cluster centroids.

At the t-th generation, the $\mathbf{X}_i^t$ individual will be evaluated by considering only its cluster centroids that are **active**, that is, cluster centroids with a threshold value such that $t_{ic}^t \geq 0.5$. Many functions are commonly adopted as the fitness function in Automatic Clustering applications [25], such as Calinski-Harabasz Index [6], Davies-Bouldin Index [10], Dunn Index [15], and Silhouette Index [37]. Such measures seek out the optimization of both the number of clusters and the cluster centroids themselves at the same time.

The initialization process starts with the random choice of C_{max} data objects from the currently evaluated dataset to compose the initial cluster centroids, for each individual $\mathbf{X}_i^{(0)}$, as much as the random determination of each activation threshold $t_{ic}^{(0)}$ (where $c = 1, 2, \ldots, C_{max}$), by picking a value from a uniformly distributed random sequence in the range $U(0, 1)$.

After the initialization and the evaluation of $G^{(0)}$ according to the selected fitness function, the generational process begins.

All proposed GSO-based models will execute evolutionary operators just like in standard GSO (i.e., producing, scrounging and ranging) in each generation t, in an attempt to improve their current groups. After the execution of GSO operators, each group is sorted according to the adopted fitness function, from the worst member to the best member. For each model, a percentage s_r of the current worst members is selected to perform the alternative merge and split operators, according to the GSO-based approach (as follows):

- **MGSO**: each selected member will perform Merge Operator (Algorithm 2). The merge operator will only consider members that contain at least **three active clusters** in t-th generation.
- **SGSO**: each selected member will perform Split Operator (Algorithm 3).
- **MSGSO**: each selected member will execute either Merge or Split operations randomly, with a 50% probability.

The proposed merge and split operators are local search heuristics adapted to the context of data clustering applications that seek out to promote faster recovery mechanisms from local optima points from the problem search space, once standard GSO evolutionary operators would take too many generations to promote good recovering strategies (through producing and ranging). These

---

Algorithm 2. Merge Operator

Pick all $\mathbf{T}_i^t = \{t_{i1}^t, \ldots, t_{iC_{max}}^t\}$ cluster activation thresholds and all active clusters C_k (represented by their cluster centroid $\mathbf{g}_k^t$, where $k = 1, \ldots, C_{max}$) from current member $\mathbf{X}_i^t \in G^t$.

Calculate the distance between each pair of cluster centroids $\mathbf{g}_{k_1}^t$ and $\mathbf{g}_{k_2}^t$ ($\mathbf{g}_{k_1}^t \neq \mathbf{g}_{k_2}^t$) from current member $\mathbf{X}_i^t$, for the active clusters only.

Select the pair of closest clusters C_{k_1} and C_{k_2}, according to their centroid distances.

Replace cluster C_{k_1} for the new cluster $C_{k_{new}}$, formed by allocating all data objects from both C_{k_1} and C_{k_2} into $C_{k_{new}}$, and compute the its new cluster centroid $\mathbf{g}_{k_{new}}$ as the mean value of all data objects in $C_{k_{new}}$, keeping its activation threshold $t_{ik_1}^t$.

Deactivate cluster C_{k_2}, by making its activation threshold $t_{ik_2}^t < 0.5$.

Pick a random data object from current dataset to reinitialize cluster centroid $\mathbf{g}_{k_2}^t$.

Update $\mathbf{X}_i^t$ with new activation thresholds and cluster centroids.

Algorithm 3. Split Operator

Pick all $\mathbf{T}_i^t = \{t_{i1}^t, \ldots, t_{iC_{max}}^t\}$ cluster activation thresholds and all active clusters C_k (represented by their cluster centroid $\mathbf{g}_k^t$, where $k = 1, \ldots, C_{max}$) from current member $\mathbf{X}_i^t \in G^t$.

Calculate the average distance between each active cluster centroid $\mathbf{g}_k^t$ from $\mathbf{X}_i^t$ and their corresponding data objects ($\mathbf{x}_j \in C_k$).

Pick the cluster C_k with the highest average distance (highest average dispersion value) and that contains at least two data objects ($—C_k— ¿= 2$).

Determine the new cluster $C_{k1_{new}}$ and $C_{k2_{new}}$ by picking a random data object $\mathbf{x}_j \in C_k$ as the new cluster centroid $\mathbf{g}_{k1_{new}}$ and the farthest data object $\mathbf{x}_d \in C_k$ from $\mathbf{x}_j$ as the new cluster centroid $\mathbf{g}_{k2_{new}}$.

Determine $C_{k1_{new}}$ and $C_{k2_{new}}$, by allocating each data object $\mathbf{x}_j \in C_k$ to the closest cluster, according to their cluster centroids $\mathbf{g}_{k1_{new}}$ and $\mathbf{g}_{k2_{new}}$.

Update $\mathbf{g}_{k1_{new}}$ and $\mathbf{g}_{k2_{new}}$ as the mean value of all data objects in their corresponding clusters ($C_{k1_{new}}$ and $C_{k2_{new}}$, respectively).

Replace cluster centroid $\mathbf{g}_k^t$ by $\mathbf{g}_{k1_{new}}$.

Find the cluster centroid $\mathbf{g}_l^t \in \mathbf{X}_i^t$ with the lowest activation threshold value $t_{il}^t = min\{t_{im} \in \mathbf{T}_i^t\}$.

Replace $\mathbf{g}_l^t$ with $\mathbf{g}_{k2_{new}}$.

Activate the new cluster $C_{k2_{new}}$, by making $t_{il}^t > 0.5$.

Update $\mathbf{X}_i^t$ with new activation thresholds and cluster centroids.

---

operators may be useful to speedup the exploration and exploitation capabilities of standard GSO, once group members that may have been caught in local optima points are improved by means of perturbations that may lead to more promising regions from the problem space than the areas found so far by the group.

All proposed GSO-based Automatic Clustering approaches are presented in Algorithm 4.

Algorithm 4. Merge and Split GSO (MGSO, SGSO and MSGSO)

$t \leftarrow 0$.

Initialization: For each member $\mathbf{X}_i^{(0)} \in G^{(0)}$, pick C_{max} data objects randomly as the initial cluster centroids $\mathbf{g}_{ic}(c = 1, 2, \ldots, C_{max})$. Randomly determine the cluster activation thresholds $t_{ic}^{(0)}$ and head angles $\alpha_i^{(0)}$ of all members $\mathbf{X}_i^{(0)} \in G^{(0)}$. After that, assign each data object $\mathbf{x}_j$ to its closest active cluster.

Calculate the fitness value $(fitness(\mathbf{X}_i^{(0)}))$ for each member $\mathbf{X}_i^{(0)}$.

while (termination conditions are not met) **do**

    Pick the best group member as the $\mathbf{X}_p^t$ for the current generation.

    Execute producing ($\mathbf{X}_p^t$ only) by evaluating three random points in its visual scanning field, $\mathbf{X}_z^t$ (Eq. (3)), $\mathbf{X}_r^t$ (Eq. (4)) and $\mathbf{X}_l^t$ (Eq. (5)). For each evaluated point ($\mathbf{X}_z^t$, $\mathbf{X}_r^t$ and $\mathbf{X}_l^t$), determine its partition by assigning each data object to the active cluster with the nearest centroid.

    Choose a percentage from the members (but the $\mathbf{X}_p^t$) to perform scrounging (Eq. (8)).

    Ranging: The remaining members will perform ranging through random walks (Eq. (9)).

    Apply GSO's boundary control mechanism to the out-bounded members in G^{t+1}.

    Reinitialize all members in G^{t+1} presenting less than two active clusters.

    Calculate the fitness value for each member $fitness(\mathbf{X}_i^{t+1})$ in G^{t+1}.

    Sort all members in G^{t+1} according to their fitness value, from the worst to the best.

    for (<a selected percentage s_r of the current worst members $\mathbf{X}_w^{t+1} \in G^{t+1}$>) **do**

        if (<MGSO>) **then**

            Execute Merge Operator (Algorithm 2) for current member $\mathbf{X}_w^{t+1}$.

        else if (<SGSO>) **then**

            Execute Split Operator (Algorithm 3) for current member $\mathbf{X}_w^{t+1}$.

        else // MSGSO

            Determine (with a 50% probability) whether to execute either Merge or Split operator for current member $\mathbf{X}_w^{t+1}$.

        end if

    end for

    $t \leftarrow t + 1$.

end while

Return $\mathbf{X}_p^{t_{max}}$.

---

4   Experimental Results

In this section, the proposed GSO-based Automatic Clustering models are evaluated, in comparison to six other evolutionary and swarm intelligence algorithms from the literature, by means of nine real-world datasets: Banknote Authentication, Breast Cancer Wisconsin, Pima Indians Diabetes, Heart (Statlog), Ionosphere, Iris, Page Blocks Classification, Seeds and Waveform. All real-world datasets are benchmark classification and clustering problems acquired from UCI Machine Learning Repository [3]. The selected real dataset features are shown in Table 1, presenting different degrees of difficulties, such as unbalanced and overlapping classes, different number of classes and features, and so on.

Table 1. Real-world dataset features.

Dataset	Instances	Features	Classes
Banknote Authentication	1372	4	2
Cancer	699	9	2
Diabetes	768	8	2
Heart	270	13	2
Ionosphere	351	34	2
Iris	150	4	3
Page Blocks Classification	5473	10	5
Seeds	210	7	3
Waveform	5000	21	3

Five well-known clustering measures are adopted, for comparison purposes: the Calinski-Harabasz Index (CH) [6], the Rand Index (RI) [36], the Corrected Rand Index (CR) [22], the Davies-Bouldin Index (DB) [10], and the Jaccard Index (JI) [18].

The selected comparison evolutionary and swarm intelligence models are: Genetic Algorithm, Differential Evolution, Particle Swarm Optimization, standard Group Search Optimization and standard Backtracking Search Optimization. The selected approaches are state-of-the-art models from evolutionary computing and data clustering literature, being successfully applied in many applications [12,24,27,32,34,35,39–41]. All EAs and SIs have been adapted to the context of partitional automatic clustering, using the same approach adopted by the proposed models (see Sect. 3 and Algorithm 4). Also, a hybrid GSO and BSA automatic clustering approach is employed for comparison purposes [33]. All algorithms use Calinski-Harabasz Index as their fitness function, running in a MATLAB 7.6 environment. Thirty independent tests have been executed for each dataset, and all methods have started with the same initial population in each test, obtained by a random process, as explained in Sect. 3. Each algorithm has been run and tested in a computer with an i7-7700K CPU, NVIDIA GeForce GTX 1060 6 GB GPU and 32 GB RAM, independently (one algorithm each time), and no other programs, but the Operating System, were executed during the tests, granting the same environmental conditions to each method.

Table 2 presents the hyperparameters for each EA and SI models. The selected values for each hyperparameter are adquired from the literature [1,2,7,9,19,28,33]. The only exception is the s_r for current proposed models, which was determined by a trial-and-error evaluation.

Table 2. Hyperparameters for each EA.

Algorithm	Parameter	Value
All EAs and SIs	t_{max}	200
	S	100
	C_{max}	20
GA	crossover rate	0.8
	mutation rate	0.1
	selection rate	0.8
DE	F	0.8
	crossover rate	0.9
PSO	c_1	2.0
	c_2	2.0
	w	0.9 to 0.4
	scroungers rate	0.8
GSO, BGSO,	θ_{max}	π/a^2
MGSO, SGSO and MSGSO	α_0	$\pi/4$
	α_{max}	$\theta_{max}/2$
BSA and BGSO	$mixrate$	1
	F	$3N(0,1)$
MGSO, SGSO and MSGSO	s_r	0.4*

*Hyperparameter determined by a trial-and-error approach.

The evaluation criterion includes an empirical analysis and a rank system employed through the application of Friedman test [17] for all the selected clustering measures. The Friedman test is a non-parametric hypothesis test that ranks all algorithms for each dataset separately. If the null-hypothesis (all ranks are not significantly different) is rejected, Nemenyi test [30] is adopted as the *post-hoc* test. According to Nemenyi test, the performance of two algorithms are considered significantly different if the corresponding average ranks differ by at least the critical difference

$$CD = q_a \sqrt{\frac{n_{alg}(n_{alg}+1)}{6n_{data}}} \tag{11}$$

where n_{data} represents the number of datasets, n_{alg} represents the number of compared algorithms and q_a are critical values based on a Studentized range statistic divided by $\sqrt{2}$ [11]. Since CH, RI, CR and JI are *maximization metrics* (indicated by ↑), the best methods will obtain higher ranks for the Friedman test, while for DB (a *minimization metric*, indicated by ↓), the best methods will find lower average ranks for the Friedman test.

The experimental results are presented in Table 3. The empirical analysis shows that, for the selected datasets, the proposed MGSO, SGSO and MSGSO were able to find the best values for the fitness function (CH) for most of the

Table 3. Experimental results for the real-world datasets (average ± standard deviation).

Dataset	Algorithm	$CH\uparrow$	$CR\uparrow$	$DB\downarrow$	$JI\uparrow$	$RI\uparrow$	C
Banknote Authentication	GA	1423.4 ± 2.2024	0.0487 ± 0.0015	0.8709 ± 0.0012	0.3803 ± 0.0008	0.5249 ± 0.0007	**2.000 ± 0**
	DE	1423.6 ± 0.1534	0.0485 ± 0.0006	0.8704 ± 0.0009	0.3804 ± 0.0006	0.5249 ± 0.0003	**2.000 ± 0**
	PSO	1387.5 ± 107.6	**0.0647 ± 0.0420**	0.8863 ± 0.0378	0.3573 ± 0.0522	**0.5323 ± 0.0198**	2.7667 ± 2.0625
	BSA	1423.5 ± 0.3278	0.0489 ± 0.0013	0.8707 ± 0.0009	0.3805 ± 0.0008	0.5251 ± 0.0007	**2.000 ± 0**
	GSO	1423.6 ± 0.2776	0.0486 ± 0.0008	0.8701 ± 0.0004	0.3805 ± 0.0003	0.5249 ± 0.0004	**2.000 ± 0**
	BGSO	**1423.7 ± 0.0498**	0.0486 ± 0.0003	0.8702 ± 0.0004	0.3805 ± 0.0002	0.5249 ± 0.0001	**2.000 ± 0**
	MGSO	**1423.7 ± 0.0155**	0.0486 ± 0.0002	**0.8701 ± 0.0002**	0.3805 ± 0.0001	0.5249 ± 0.00008	**2.000 ± 0**
	SGSO	**1423.7 ± 0.0012**	0.0486 ± 0.0002	**0.8701 ± 0.0002**	**0.3806 ± 0.0002**	0.5249 ± 0.0001	**2.000 ± 0**
	MSGSO	**1423.7 ± 0.0081**	0.4858 ± 0.0002	**0.8701 ± 0.0001**	0.3805 ± 0.0001	0.5249 ± 0.00008	**2.000 ± 0**
Cancer	GA	1038.9 ± 1.9792	0.8320 ± 0.0090	0.7618 ± 0.0006	0.8599 ± 0.0067	0.9169 ± 0.0044	**2.000 ± 0**
	DE	1038.9 ± 2.5719	0.8344 ± 0.0084	0.7618 ± 0.0006	0.8618 ± 0.0062	0.9181 ± 0.0041	**2.000 ± 0**
	PSO	1029.3 ± 65.950	0.8372 ± 0.0121	0.7873 ± 0.1429	0.8633 ± 0.0116	0.9194 ± 0.0063	2.0333 ± 0.1826
	BSA	1038.9 ± 2.0758	0.8337 ± 0.0105	0.7618 ± 0.0006	0.8613 ± 0.0079	0.9177 ± 0.0052	**2.000 ± 0**
	GSO	1041.3 ± 0.2277	**0.8396 ± 0.0026**	0.7612 ± 0.0001	**0.8655 ± 0.0019**	**0.9206 ± 0.0013**	**2.000 ± 0**
	BGSO	**1041.4 ± 0.0691**	0.8391 ± 0.0024	**0.7612 ± 0.00005**	0.8651 ± 0.0018	0.9204 ± 0.0012	**2.000 ± 0**
	MGSO	**1041.4 ± 0**	0.8391 ± 0	**0.7612 ± 0**	0.8651 ± 0	0.9204 ± 0	**2.000 ± 0**
	SGSO	**1041.4 ± 0.0187**	0.8392 ± 0.0010	**0.7612 ± 0.00001**	0.8653 ± 0.0007	0.9204 ± 0.0005	**2.000 ± 0**
	MSGSO	**1041.4 ± 0**	0.8391 ± 0	**0.7612 ± 0**	0.8651 ± 0	0.9204 ± 0	**2.000 ± 0**
Diabetes	GA	1139.1 ± 2.102	0.0443 ± 0.0036	0.6646 ± 0.0042	0.3789 ± 0.0041	0.5233 ± 0.0022	3 ± 0
	DE	1140.0 ± 2.251	0.0450 ± 0.0025	0.6651 ± 0.0032	0.3793 ± 0.0026	0.5236 ± 0.0014	3 ± 0
	PSO	996.17 ± 187.6	**0.0501 ± 0.0164**	0.8084 ± 0.2226	0.3277 ± 0.0969	0.5188 ± 0.0161	4.6667 ± 2.928
	BSA	1136.5 ± 3.586	0.0453 ± 0.0046	**0.6638 ± 0.0037**	**0.3806 ± 0.0050**	**0.5242 ± 0.0030**	3 ± 0
	GSO	1141.8 ± 2.930	0.0451 ± 0.0010	0.6673 ± 0.0044	0.3783 ± 0.0017	0.5233 ± 0.0004	3 ± 0
	BGSO	1142.1 ± 0.9429	0.0446 ± 0.0014	0.6679 ± 0.0018	0.3777 ± 0.0015	0.5232 ± 0.0008	3 ± 0
	MGSO	1142.3 ± 1.0431	0.0451 ± 0.0004	0.6682 ± 0.0018	0.3781 ± 0.0004	0.5235 ± 0.0002	3 ± 0
	SGSO	1142.6 ± 0	0.0452 ± 0	0.6681 ± 0	0.3781 ± 0	0.5235 ± 0	3 ± 0
	MSGSO	**1142.6 ± 0.0148**	0.0452 ± 0.0002	0.6681 ± 0.00005	0.3781 ± 0.00002	0.5235 ± 0.0001	3 ± 0
Heart	GA	**206.95 ± 0.0036**	0.0295 ± 0.0012	0.9875 ± 0.0006	0.3606 ± 0.0009	0.5150 ± 0.0006	2 ± 0
	DE	**206.95 ± 0**	**0.0302 ± 0**	**0.9871 ± 0**	**0.3611 ± 0**	**0.5154 ± 0**	2 ± 0
	PSO	206.84 ± 0.0995	0.0250 ± 0.0037	**0.9871 ± 0.0014**	0.3591 ± 0.0012	0.5128 ± 0.0018	2 ± 0
	BSA	**206.95 ± 0**	**0.0302 ± 0**	**0.9871 ± 0**	**0.3611 ± 0**	0.5153 ± 0.0003	2 ± 0
	GSO	**206.95 ± 0.0041**	0.0301 ± 0.0005	0.9873 ± 0.0007	0.3610 ± 0.0006	**0.5154 ± 0**	2 ± 0
	BGSO	**206.95 ± 0.0015**	0.0301 ± 0.0005	0.9872 ± 0.0003	0.3610 ± 0.0004	0.5153 ± 0.0002	2 ± 0
	MGSO	**206.95 ± 0**	**0.0302 ± 0**	**0.9871 ± 0**	**0.3611 ± 0**	**0.5154 ± 0**	2 ± 0
	SGSO	**206.95 ± 0**	**0.0302 ± 0**	**0.9871 ± 0**	**0.3611 ± 0**	**0.5154 ± 0**	2 ± 0
	MSGSO	**206.95 ± 0**	**0.0302 ± 0**	**0.9871 ± 0**	**0.3611 ± 0**	**0.5154 ± 0**	2 ± 0
Ionosphere	GA	115.65 ± 1.198	0.1464 ± 0.0132	1.5341 ± 0.0111	0.4190 ± 0.0064	0.5734 ± 0.0066	2 ± 0
	DE	115.48 ± 1.601	0.1427 ± 0.0158	1.5367 ± 0.0143	0.4175 ± 0.0074	0.5716 ± 0.0078	2 ± 0
	PSO	116.13 ± 9.484	**0.1791 ± 0.0214**	1.5375 ± 0.0895	0.4317 ± 0.0084	**0.5893 ± 0.0099**	2.0667 ± 0.258
	BSA	117.27 ± 0.9134	0.1564 ± 0.0151	1.5206 ± 0.0094	0.4233 ± 0.0075	0.5807 ± 0.0073	2 ± 0
	GSO	118.43 ± 0.3889	0.1697 ± 0.0091	1.5158 ± 0.0052	0.4298 ± 0.0043	0.5845 ± 0.0055	2 ± 0
	BGSO	118.60 ± 0.2825	0.1716 ± 0.0062	1.5137 ± 0.0035	0.4306 ± 0.0031	0.5859 ± 0.0031	2 ± 0
	MGSO	118.82 ± 0.0351	0.1771 ± 0.0015	**1.5134 ± 0.0007**	**0.4334 ± 0.0007**	0.5887 ± 0.0007	2 ± 0
	SGSO	118.82 ± 0.0473	0.1773 ± 0.0018	**1.5134 ± 0.0007**	**0.4334 ± 0.0009**	0.5888 ± 0.0009	2 ± 0
	MSGSO	**118.83 ± 0.0048**	0.1773 ± 0.0012	1.5135 ± 0.0006	**0.4334 ± 0.0006**	0.5888 ± 0.0006	2 ± 0
Iris	GA	561.58 ± 0.256	0.7302 ± 0.0001	0.6622 ± 0.0013	0.6958 ± 0.0003	0.8797 ± 0	3 ± 0
	DE	**561.63 ± 0**	0.7302 ± 0	**0.6620 ± 0**	0.6959 ± 0	0.8797 ± 0	3 ± 0
	PSO	560.80 ± 2.540	0.7301 ± 0.0004	0.6636 ± 0.0047	0.6956 ± 0.0007	0.8797 ± 0	3 ± 0
	BSA	**561.63 ± 0**	0.7302 ± 0	**0.6620 ± 0**	0.6959 ± 0	0.8797 ± 0	3 ± 0
	GSO	561.37 ± 0.8113	**0.7316 ± 0.0040**	0.6627 ± 0.0023	**0.6971 ± 0.0037**	**0.8803 ± 0.0019**	3 ± 0
	BGSO	**561.63 ± 0**	0.7302 ± 0	**0.6620 ± 0**	0.6959 ± 0	0.8797 ± 0	3 ± 0
	MGSO	**561.63 ± 0**	0.7302 ± 0	**0.6620 ± 0**	0.6959 ± 0	0.8797 ± 0	3 ± 0
	SGSO	**561.63 ± 0**	0.7302 ± 0	**0.6620 ± 0**	0.6959 ± 0	0.8797 ± 0	3 ± 0
	MSGSO	**561.63 ± 0**	0.7302 ± 0	**0.6620 ± 0**	0.6959 ± 0	0.8797 ± 0	3 ± 0
Page Blocks Classification	GA	14395.2 ± 567.3	0.0070 ± 0.0154	**0.5250 ± 0.0342**	0.6044 ± 0.0893	0.6287 ± 0.0757	5.5 ± 0.509
	DE	16343.2 ± 778.1	0.0003 ± 0.0129	0.6159 ± 0.0311	0.5195 ± 0.0745	0.5576 ± 0.0619	7.5 ± 0.861
	PSO	13372.5 ± 1920.9	0.0109 ± 0.0059	0.5307 ± 0.0318	0.6634 ± 0.0300	0.6782 ± 0.0260	**4.7000 ± 0.8769**
	BSA	15007.2 ± 529.4	0.0057 ± 0.0156	0.5626 ± 0.0523	0.6031 ± 0.0735	0.6196 ± 0.0675	5.9667 ± 0.8087
	GSO	12456.9 ± 1436.9	0.0110 ± 0.0108	0.5364 ± 0.0267	0.6667 ± 0.0232	0.6874 ± 0.0548	4.3667 ± 0.5561
	BGSO	12583.7 ± 1668.9	0.0125 ± 0.0181	0.5306 ± 0.0262	0.6711 ± 0.0371	0.6849 ± 0.0333	4.4 ± 0.6747
	MGSO	11841.5 ± 1082.9	**0.0202 ± 0.0252**	0.5365 ± 0.0179	**0.6860 ± 0.0448**	**0.6984 ± 0.0412**	4.0667 ± 0.5208
	SGSO	20292.6 ± 1141.9	-0.0103 ± 0.0058	0.5821 ± 0.0445	0.5021 ± 0.0589	0.5418 ± 0.0477	10.3667 ± 1.0981
	MSGSO	19083.0 ± 1250.3	-0.0079 ± 0.0069	0.6215 ± 0.0462	0.5364 ± 0.0536	0.5364 ± 0.0536	8.9333 ± 1.1725
Seeds	GA	375.31 ± 0.7548	**0.7178 ± 0.0086**	0.7535 ± 0.0010	**0.6827 ± 0.0081**	**0.8749 ± 0.0039**	3 ± 0
	DE	372.38 ± 2.3840	0.7106 ± 0.0209	0.7564 ± 0.0041	0.6763 ± 0.0194	0.8716 ± 0.0094	3 ± 0
	PSO	375.73 ± 0.2892	0.7159 ± 0.0028	0.7535 ± 0.0007	0.6808 ± 0.0026	0.8740 ± 0.0012	3 ± 0
	BSA	370.66 ± 5.5973	0.6988 ± 0.0274	0.7603 ± 0.0081	0.6656 ± 0.0243	0.8627 ± 0.0261	3 ± 0
	GSO	375.68 ± 0.3881	0.7153 ± 0.0040	0.7535 ± 0.0007	0.6803 ± 0.0037	0.8745 ± 0.0009	3 ± 0
	BGSO	374.30 ± 4.9807	0.7095 ± 0.0452	0.7514 ± 0.0121	0.6772 ± 0.0300	0.8701 ± 0.0261	2.9667 ± 0.1826
	MGSO	**375.81 ± 0**	0.7166 ± 0	**0.7533 ± 0**	0.6815 ± 0	0.8744 ± 0	3 ± 0
	SGSO	**375.81 ± 0**	0.7166 ± 0	**0.7533 ± 0**	0.6815 ± 0	0.8744 ± 0	3 ± 0
	MSGSO	**375.81 ± 0**	0.7166 ± 0	**0.7533 ± 0**	0.6815 ± 0	0.8744 ± 0	3 ± 0
Waveform	GA	2518.7 ± 11.88	0.3473 ± 0.0112	1.3783 ± 0.0036	0.4374 ± 0.0067	0.6734 ± 0.0058	2 ± 0
	DE	2544.2 ± 8.190	0.3597 ± 0.0057	1.3734 ± 0.0021	0.4450 ± 0.0035	0.6798 ± 0.0029	2 ± 0
	PSO	2552.6 ± 58.11	0.3669 ± 0.0213	1.3705 ± 0.0047	0.4480 ± 0.0209	0.6848 ± 0.0033	2.0333 ± 0.1826
	BSA	2536.2 ± 7.4291	0.3537 ± 0.0093	1.3745 ± 0.0027	0.4413 ± 0.0056	0.6780 ± 0.0038	2 ± 0
	GSO	2546.3 ± 10.52	0.3608 ± 0.0065	1.3733 ± 0.0027	0.4456 ± 0.0040	0.6795 ± 0.0037	2 ± 0
	BGSO	2558.0 ± 3.6889	0.3668 ± 0.0037	1.3704 ± 0.0009	0.4494 ± 0.0023	0.6834 ± 0.0019	2 ± 0
	MGSO	**2563.0 ± 0.6503**	0.3705 ± 0.0014	**1.3696 ± 0.0001**	0.4517 ± 0.0008	0.6852 ± 0.0007	2 ± 0
	SGSO	**2563.0 ± 0.2500**	0.3704 ± 0.0016	1.3670 ± 0.0001	0.4516 ± 0.0007	0.6852 ± 0.0006	2 ± 0
	MSGSO	**2563.2 ± 0.2599**	0.3711 ± 0.0011	**1.3696 ± 0.0001**	0.4520 ± 0.0007	0.6855 ± 0.0006	2 ± 0

cases. SGSO and MSGSO have been able to outperform standard GSO in eight out of nine datasets, and MGSO was outperformed by GSO only in one dataset (Page Blocks Classification), while the same model has been able to outperform GSO in four other datasets. The proposed models also have presented the highest degree of stability, showing their reliability. Almost all algorithms (except for PSO) have been able to predict the exact estimated number of final clusters for six out of nine datasets, which is a good result, compatible with many works from the literature [39]. Even for Diabetes, Waveform and Page Blocks Classification, the best number of clusters found by the EAs and SIs is not very much distant from the expected values, what is quite acceptable, given the different degrees of separability among the original classes in such datasets.

Table 4. Overall Evaluation: Average Ranks for the Friedman Test for each metric, with $CD = 4.0043$, and algorithm's score (in parenthesis).

Algorithm	$CH^\uparrow$	$CR^\uparrow$	$DB^\downarrow$	$JI^\uparrow$	$RI^\uparrow$	Average Score
GA	80.3167(9)	105.3352(9)	160.0667(7)	110.4907(8)	104.9519 (9)	8.4
DE	105.2611(7)	112.0278(8)	170.8370(9)	109.8352(9)	106.8704(8)	8.2
PSO	132.2093(5)	144.0722(4)	118.6148(2)	137.3185(6)	142.7407(5)	4.4
BSA	83.4074(8)	118.9593(7)	166.5926(8)	121.7389(7)	119.4537(7)	7.4
GSO	132.0704(6)	143.3074(5)	130.7759(6)	140.6796(5)	142.0648(6)	5.6
BGSO	144.1111(4)	142.2259(6)	119.9611(4)	145.1019(3)	144.8389(4)	4.2
MGSO	166.6296(3)	**159.6833**(1)	**109.6889**(1)	**161.1296**(1)	**161.8204**(1)	**1.4**
SGSO	185.3926(2)	145.5241(3)	118.6481(3)	144.7389(4)	146.5407(3)	3.0
MSGSO	**190.1019**(1)	148.3648(2)	124.3148(5)	148.4667(2)	150.2185(2)	2.4

Table 4 presents the results for the overall evaluation performed through the ranking system obtained through the application Friedman-Nemenyi hypothesis tests for each clustering metric. We also included an average score system, considering the position reached by each method in relation to each metric in the Friedman-Nemenyi ranking system, from the best method (score "1") to the worst (a score "9"). Considering the overall evaluation, the proposed models have obtained the best ranks for all evaluation metrics (except for MSGSO for Davies-Bouldin Index, where the model only reached the fifth best rank). According to the overall ranking and scoring systems, the best models for the evaluated scenarios are MGSO, MSGSO and SGSO, respectively, showing a slight preference towards Merge Operator than the Split Operator, when all evaluation metrics are taken into consideration the same way. But when the fitness function is evaluated only, the opposite situation occurs, once MSGSO and SGSO have reached better ranks in Friedman-Nemenyi tests than MGSO.

5   Conclusions

In this work, three improved GSO-based models are presented to tackle the Automatic Data Clustering problem: MGSO, SGSO and MSGSO. The proposed approaches speedup the exploration and exploitation operators of GSO, by providing perturbations on the worst members of GSO group that would lead to better local optima points escaping and recovering mechanisms.

To evaluate the proposed GSO-based Automatic Clustering models, six state-of-the-art partitional automatic clustering algorithms are adopted from the literature for comparison purposes: GA, DE, PSO, BSA, GSO and BGSO. Nine real-world datasets are employed, and five clustering metrics are used in the evaluation. The experimental analysis included an empirical method and a ranking system obtained from a hypothesis test (Friedman test).

The experiments showed that MGSO, SGSO and MSGSO are able to find better solutions than standard GSO model in most cases, and in an overall evaluation, all proposed models have been able to outperform all comparison approaches in relation to the selected clustering indices.

As future works, we intend to extend our analysis on the behavior of MGSO, SGSO and MSGSO by employing controlled scenarios obtained through the use of synthetic datasets, so we can understand the best features and limitations of the proposed models on different clustering problems. Also, we intend to evaluate the influence of the fitness function on the behavior and performance of the proposed approaches. In future researches, new evolutionary operators and local search heuristics will be employed and hybridized to improve GSO search capabilities in Automatic Data Clustering applications (in both real-world and simulated scenarios).

Acknowledgements. The authors would like to thank FACEPE, CNPq and CAPES (Brazilian Research Agencies) for their financial support.

References

1. Abdel-Kader, R.F.: Genetically improved PSO algorithm for efficient data clustering. In: 2010 Second International Conference on Machine Learning and Computing, pp. 71–75. IEEE (2010)
2. Ahmadyfard, A., Modares, H.: Combining PSO and k-means to enhance data clustering. In: International Symposium on Telecommunications, IST 2008, pp. 688–691. IEEE (2008)
3. Asuncion, A., Newman, D.: UCI machine learning repository (2007)
4. Barnard, C., Sibly, R.: Producers and scroungers: a general model and its application to captive flocks of house sparrows. Anim. Behav. **29**(2), 543–550 (1981)
5. Bonabeau, E., Dorigo, M., Theraulaz, G.: Swarm intelligence: from natural to artificial systems, vol. 4. Oxford University Press, New York (1999)
6. Caliński, T., Harabasz, J.: A dendrite method for cluster analysis. Commun. Stat.-Theory Methods **3**(1), 1–27 (1974)
7. Civicioglu, P.: Backtracking search optimization algorithm for numerical optimization problems. Appl. Math. Comput. **219**(15), 8121–8144 (2013)

8. Couzin, I.D., Krause, J., Franks, N.R., Levin, S.A.: Effective leadership and decision-making in animal groups on the move. Nature **433**(7025), 513–516 (2005)
9. Das, S., Abraham, A., Konar, A.: Automatic clustering using an improved differential evolution algorithm. IEEE Trans. Syst. Man Cybern.-Part A: Syst. Humans **38**(1), 218–237 (2007)
10. Davies, D.L., Bouldin, D.W.: A cluster separation measure. IEEE Trans. Pattern Anal. Mach. Intell. PAMI **1**(2), 224–227 (1979)
11. Demšar, J.: Statistical comparisons of classifiers over multiple data sets. J. Mach. Learn. Res. **7**, 1–30 (2006)
12. Dey, A., Dey, S., Bhattacharyya, S., Platos, J., Snasel, V.: Novel quantum inspired approaches for automatic clustering of gray level images using particle swarm optimization, spider monkey optimization and ageist spider monkey optimization algorithms. Appl. Soft Comput. **88**, 106040 (2020)
13. Dixon, A.: An experimental study of the searching behaviour of the predatory coccinellid beetle adalia decempunctata (l.). J. Animal Ecol. **28**, 259–281 (1959)
14. Dorigo, M., Maniezzo, V., Colorni, A.: Ant system: optimization by a colony of cooperating agents. IEEE Transactions Syst. Man Cybern. Part B: Cybern **26**(1), 29–41 (1996)
15. Dunn, J.C.: A fuzzy relative of the isodata process and its use in detecting compact well-separated clusters. J. Cybern. **3**(3), 32–57 (1973)
16. Ezugwu, A.E., Shukla, A.K., Agbaje, M.B., Oyelade, O.N., José-García, A., Agushaka, J.O.: Automatic clustering algorithms: a systematic review and bibliometric analysis of relevant literature. Neural Comput. Appl. **33**(11), 6247–6306 (2021)
17. Friedman, M.: The use of ranks to avoid the assumption of normality implicit in the analysis of variance. J. Am. Stat. Assoc. **32**(200), 675–701 (1937)
18. Halkidi, M., Batistakis, Y., Vazirgiannis, M.: Cluster validity methods: part I. ACM SIGMOD Rec. **31**(2), 40–45 (2002)
19. He, S., Wu, Q.H., Saunders, J.R.: Group search optimizer: an optimization algorithm inspired by animal searching behavior. IEEE Trans. Evol. Comput. **13**(5), 973–990 (2009)
20. Higgins, C.L., Strauss, R.E.: Discrimination and classification of foraging paths produced by search-tactic models. Behav. Ecol. **15**(2), 248–254 (2004)
21. Holland, J.H.: Genetic algorithms. Scientific Am. **267**(1), 66–72 (1992)
22. Hubert, L., Arabie, P.: Comparing partitions. J. Classif. **2**(1), 193–218 (1985)
23. Ikotun, A.M., Almutari, M.S., Ezugwu, A.E.: K-means-based nature-inspired metaheuristic algorithms for automatic data clustering problems: recent advances and future directions. Appl. Sci. **11**(23), 11246 (2021)
24. Jin, Y.F., Yin, Z.Y.: Enhancement of backtracking search algorithm for identifying soil parameters. Int. J. Numer. Anal. Meth. Geomech. **44**(9), 1239–1261 (2020)
25. José-García, A., Gómez-Flores, W.: Automatic clustering using nature-inspired metaheuristics: a survey. Appl. Soft Comput. **41**, 192–213 (2016)
26. Kennedy, J., Eberhart, R.: Particle swarm optimization. In: International Conference on Neural Networks, vol. 4, pp. 1942–1948. IEEE (1995)
27. Latiff, N.A., Malik, N.N.A., Idoumghar, L.: Hybrid backtracking search optimization algorithm and k-means for clustering in wireless sensor networks. In: 2016 IEEE 14th International Conference on Dependable, Autonomic and Secure Computing, 14th Intl Conference on Pervasive Intelligence and Computing, 2nd International Conference on Big Data Intelligence and Computing and Cyber Science and Technology Congress (DASC/PiCom/DataCom/CyberSciTech), pp. 558–564. IEEE (2016)

28. Liu, Y., Wu, X., Shen, Y.: Automatic clustering using genetic algorithms. Appl. Math. Comput. **218**(4), 1267–1279 (2011)
29. Naldi, M.C., Campello, R.J., Hruschka, E.R., Carvalho, A.: Efficiency issues of evolutionary k-means. Appl. Soft Comput. **11**(2), 1938–1952 (2011)
30. Nemenyi, P.B.: Distribution-free multiple comparisons. Princeton University (1962)
31. Omran, M., Salman, A., Engelbrecht, A.: Dynamic clustering using particle swarm optimization with application in unsupervised image classification. In: Fifth World Enformatika Conference (ICCI 2005), Prague, Czech Republic, pp. 199–204 (2005)
32. Pacífico, L.: Agrupamento de imagens baseado em uma abordagem híbrida entre a otimização por busca em grupo e k-means para a segmentação automática de doenças em plantas. In: Anais do XVII Encontro Nacional de Inteligência Artificial e Computacional, pp. 152–163. SBC (2020)
33. Pacifico, L., Ludermir, T.: Backtracking group search optimization: a hybrid approach for automatic data clustering. In: Cerri, R., Prati, R.C. (eds.) BRACIS 2020. LNCS (LNAI), vol. 12319, pp. 64–78. Springer, Cham (2020). https://doi.org/10.1007/978-3-030-61377-8_5
34. Pacifico, L.D., Ludermir, T.B.: An evaluation of k-means as a local search operator in hybrid memetic group search optimization for data clustering. Nat. Comput. **20**(3), 611–636 (2021)
35. Preetha, V.: Data analysis on student's performance based on health status using genetic algorithm and clustering algorithms. In: 2021 5th International Conference on Computing Methodologies and Communication (ICCMC), pp. 836–842. IEEE (2021)
36. Rand, W.M.: Objective criteria for the evaluation of clustering methods. J. Am. Stat. Assoc. **66**(336), 846–850 (1971)
37. Rousseeuw, P.J.: Silhouettes: a graphical aid to the interpretation and validation of cluster analysis. J. Comput. Appl. Math. **20**, 53–65 (1987)
38. Storn, R., Price, K.: Differential evolution-a simple and efficient adaptive scheme for global optimization over continuous spaces. International Computer Science Institute, Berkeley. Tech. Rep., CA, 1995, Tech. Rep. TR-95-012 (1995)
39. Tam, H.H., Ng, S.C., Lui, A.K., Leung, M.F.: Improved activation schema on automatic clustering using differential evolution algorithm. In: 2017 IEEE Congress on Evolutionary Computation (CEC), pp. 1749–1756. IEEE (2017)
40. Vali, M., Zare, M., Razavi, S.: Automatic clustering-based surrogate-assisted genetic algorithm for groundwater remediation system design. J. Hydrol. **598**, 125752 (2021)
41. Ye, L., Zheng, D.: Stable grasping control of robot based on particle swarm optimization. In: 2021 IEEE 2nd International Conference on Big Data, Artificial Intelligence and Internet of Things Engineering (ICBAIE), pp. 1020–1024. IEEE (2021)

Leveraging Textual Descriptions for House Price Valuation

Luís Fernando Bittencourt, Otávio Parraga$^{(\boxtimes)}$, Duncan D. Ruiz,
Isabel H. Manssour, Soraia Raupp Musse, and Rodrigo C. Barros

School of Technology, Pontifícia Universidade Católica do Rio Grande do Sul,
Av. Ipiranga, 6681, 90619-900 Porto Alegre, RS, Brazil
{luis.fernando86,otavio.parraga}@edu.pucrs.br
{duncan.ruiz,isabel.manssour,soraia.musse,rodrigo.barros}@pucrs.br

Abstract. Real estate valuation has been vastly studied by the research
community, with several articles proposing Automated Valuation Models
(AVM). However, most of those models base their estimates only on geographic location and structural characteristics of the property, disregarding several factors that influence prices, such as the need for repairs and
sun exposure. To support decision making, an AVM needs to "look" for
the same type of information a person would when valuating a property,
including photos and textual descriptions. In this work, we show that the
usage of textual data can significantly increase the performance of house
price-prediction models. Our experiments explore different combinations
of learning algorithms and methods to extract relevant information from
textual descriptions, with some surprising conclusions regarding the best
combination of approaches. Overall, we shed some light on how textual
features can be leveraged by the models, explaining the paths that lead
to predictions that end up resulting in performance gains.

Keywords: House price valuation · Information retrieval · Textual
data

1 Introduction

Real estate valuation is a key component of capitalism and the structure of our
society. Being able to assess a price that properly represents the physical characteristics and location amenities (both tangible and non-tangible) of a property
is an essential task in buy and sell operations. More than that, house valuation
is critical for insurance companies so they can firm fair contracts with clients. It
is also important for legal and mortgage purposes, and for a correct definition
of municipal property taxes [19].

Not surprisingly, there is a large number of papers proposing methods for
predicting the list price or trade price of real estate properties [19]. These models
vary in terms of both technique and coverage: some of them propose multivariate
linear regression models over a few of manually-collected samples, while others

© The Author(s), under exclusive license to Springer Nature Switzerland AG 2022
J. C. Xavier-Junior and R. A. Rios (Eds.): BRACIS 2022, LNAI 13653, pp. 355–369, 2022.
https://doi.org/10.1007/978-3-031-21686-2_25

design complex neural networks trained over millions of records provided by multiple listing services.

Most of the related work on the subject design models for predicting prices based solely on basic property features, such as size and number of rooms, and also location amenities, such as distance to green areas and commute time. Recent work [17, 21, 28] propose multimodal models, mixing these basic features with visual and geographic data, for instance. However, to the best of our knowledge, widely-available property textual data remains pretty much unexplored.

In this context, this paper aims to evaluate the impact of adding textual information to house price valuation models. For that, we collect two datasets from very different contexts—distinct cities, countries, and overall characteristics —, and we evaluate a large number of models for structuring textual information. Each experiment is a unique combination of property structural features (structured data such as size and number of rooms) and a vectorized (structured) representation of its description in natural language. The foundation of this work is to confirm whether natural-language textual information extracted from real estate listing descriptions can improve house price valuation. Assuming our hypothesis is confirmed, we also explore which terms/words and entities contribute most to this improvement. Finally, we evaluate what are the most promising methods for extracting (structuring) relevant knowledge from textual information written in natural language.

The results we present in this paper confirm that textual information can indeed improve house price valuation, while also indicating that binary term frequency-inverse document frequency (TF-IDF) is the most promising extraction (structuring) method, outperforming much more modern and trending algorithms based on deep neural networks. Finally, this work also contributes with the research community by providing a broad evaluation of how the property textual data is actually employed to make better predictions. We perform interpretability experiments that allow us to point out which terms are correlated to house pricing, helping the domain specialists when using machine learning models for house price valuation.

The rest of this paper is organized as follows. Section 2 presents related work. Section 3 describes the research methodology, including data collection, pre-processing, extraction methods, learning algorithms, model evaluation, and interpretability setup. Section 4 reports the observed quantitative and qualitative results. Finally, Sect. 5 presents the conclusions of this work.

2   Related Work

Using multimodal data to optimize house price valuation models is a valuable and promising approach. Some studies explore the usage of visual, geographic, environmental, and economic information fused with basic property attributes and demonstrate that such a strategy improves the results of the model [6, 8, 15]. Despite the fact that using multimodal data seems to be always beneficial, textual information remains a virtually-unexplored field when it comes to house price valuation[1], cited as a possibility [17] but never really actually explored.

Peng et al. [20] demonstrate the potential of textual data applied to a similar problem: they work on a case study that makes use of user reviews from Airbnb, a world-renowned lodging company, to improve rent price predictions. By mixing basic property features, geographic information, and textual data, the authors state that using multimodal data had an outstanding contribution to increasing all the evaluated metrics.

Since textual information is an unstructured data format, a crucial step for any text-based model pipeline is choosing a method to extract (structure) relevant knowledge from it. In the lack of previous works exploring this modality for house price valuation, there are a couple of papers that demonstrate extraction methods for similar problems. For instance, [20] uses a classifier to transform the user review text into a single integer value representing its sentiment. In a different approach, [22] searches the text for points of interest—relevant facilities near the property, such as schools, hospitals, and malls —, and then use these points of interest as house price predictors. Instead of restricting the search for these points, we anticipate that the listing descriptions contain a broader set of relevant aspects capable of improving the performance of AVMs.

3   Methodology

In order to assess the potential of textual data applied to house pricing predictions, we tested many distinct combinations of textual information extraction methods and learning algorithms. By methodically doing so, we cover a large number of experiments so we can validate hypotheses on whether a specific technique or algorithm is better than another. We consider an experiment as the combination between extraction method and learning algorithm[2].

We followed the same protocol for every executed experiment. After a preprocessing step, we employ an extraction method to generate a vectorized representation from the property textual data. Then we apply the Truncated Singular Value Decomposition algorithm [13] to reduce the dimensions of this vector, otherwise the high number of textual features could make the few original houses features almost insignificant[3]. Finally, we append this vector to the original house features and use the resulting data to train and evaluate the model. The pipeline steps can be seen in Fig. 1 and we detail them in the next sections.

---

[1] Some articles [1,27] use the term "textual features" to actually refer to *numerical* house attributes, such as size and number of rooms. In this paper, when we mention textual information we are actually referring to unstructured listing descriptions written in natural language.

[2] https://github.com/Otavio-Parraga/textual-house-pricing.

[3] We use scikit-learn's TruncatedSVD class. In this paper, all mentions of scikit-learn refer to version 0.24.2 (https://scikit-learn.org/0.24/).

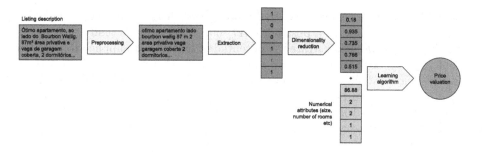

Fig. 1. The general pipeline for training the learning models (best viewed in colors).

3.1 Data Collection

Our main dataset was obtained by scraping the website of a big real estate agency from Porto Alegre, Brazil. We loaded properties for sale ads containing all the variables shown in Table 1. To perform a more controlled experiment, we removed all properties that were not apartments. In addition, we removed properties with blank descriptions, reducing the dataset to 17,406 samples.

Table 1. Dataset variables, types, and descriptive statistics.

Variable	Area	Bedrooms	Bathrooms	Suites	Parking slots	Descriptions	Price (R$)
Type	Real	Integer	Integer	Integer	Integer	Text	Real
Mean	75.56	2.13	1.56	0.47	0.93	—	483,234.72
Std Dev	42.29	0.74	0.90	0.68	0.80	—	585,227.94
Min	0.01	0.00	0.00	0.00	0.00	—	49,000.00
25%	50.00	2.00	1.00	0.00	0.00	—	220,000.00
50%	65.52	2.00	1.00	0.00	1.00	—	340,000.00
75%	86.15	3.00	2.00	1.00	1.00	—	550,000.00
Max	509.68	7.00	8.00	4.00	6.00	—	20,208,788.00

In order to assess the generalization capability of the entire learning pipeline, we also collected a second dataset from a multiple listing service in the United States (hereby simply called US dataset). We narrowed our search to the state of Florida, discarding samples with missing values or related to land lots. This dataset contains data from 22,108 houses and has some differences from our Porto Alegre dataset. First, its properties are houses, not apartments. Second, it does not contain variables for suites nor parking slots. Finally, it introduces a new variable, property age (in years).

3.2 Pre-processing and Extraction Methods

Each property listing description has gone through standard pre-processing. We converted all text to lower case and removed diacritics, punctuation, and stopwords[4]. However, even after the preprocessing stage, these descriptions are unstructured textual information that are not used directly as input for many machine learning models, so we explored a few extraction methods for turning text into numerical vectors. These methods are detailed below:

- Bag-of-words (BoW): creates a matrix representation of word frequency in a group of documents. Each word becomes a column and each line represents the presence of a word in a document. In this paper, we use two different strategies to generate the representations: binary BoW and counting BoW;
- TF-IDF: uses BoW to create a new representation that employs statistical rules to identify the importance of a specific term in a document related to a group of documents;
- Word embeddings: vectorized local representations generated by training a model to predict words in different contexts. These representations are located within a latent space where similar words are close to each other. We generate word embeddings from FastText[5] [14] and BERT [9], and we use embeddings concatenation to generate a document representation.

Since BERT is a trending representation used for a plethora of tasks [23], we tried to optimize our results by fine-tuning it to our data. For that, we trained two models on the Masked Language Modeling objective for two epochs using the AdamW optimizer [18]. We use the "bert-base-cased" model for the US dataset and its corresponding Portuguese version [24,25] for the Porto Alegre dataset.

3.3 Learning Algorithms

Our selected learning algorithms are based on ensemble strategies such as bagging [4] and boosting [10]. They are state-of-the-art approaches for structured datasets and have several advantages over neural networks, such as reduced time complexity for both training and performing inference. We choose five algorithms that are widely used in house price valuation tasks: Extremely Randomized Trees [12], Gradient Boosting [11], Random Forest [5], XGBoost [7], and Light-GBM [16].

For LightGBM and XGBoost we used the eponymous libraries, versions 3.3.0[6] and 1.5.0[7], respectively. For the other algorithms, we used scikit-learn's ExtraTreesRegressor, GradientBoostingRegressor, and RandomForestRegressor classes. In all cases, we kept the default hyperparameters values of each library.

3.4 Model Evaluation

Since our goal is to determine the impact of adding textual features to AVMs, we first define a baseline by running our models without the textual features. In

---

[4] We make use of the stopwords provided by the NLTK library [3].
[5] FastText is used directly as distributed, without any kind of fine-tuning.
[6] Available documentation at https://lightgbm.readthedocs.io/en/v3.3.0/.
[7] Available documentation at https://xgboost.readthedocs.io/en/release_1.5.0/.

this stage, the pipeline is reduced to running the learning algorithm using price as the output dependent variable and the remaining numerical attributes as input independent variables. After setting the baseline, we define six extraction methods to be combined with the learning algorithms:

1. Binary: binary bag-of-words;
2. Count: bag-of-words with counts;
3. TF-IDF: based on the eponymous extraction method;
4. Binary TF-IDF: a combination of TF-IDF with the binary strategy;
5. Word: word embedding using FastText;
6. BERT: word embedding using BERT.

All models follow two basic premises. First, we discard all words with a document frequency lower than 0.1% (about 17 samples for the Porto Alegre dataset). Second, we reduce the dimension of the textual features to 30. We use a 10-fold cross-validation protocol to compute the estimated generalization error. More specifically, we compute the MAE, RMSE, MAPE, MdAPE, and R^2, all widely-used AVM metrics presented in Eqs. 1, 2, 3, 4, and 5, respectively, where y is the vector of observed values, $\overline{y}$ is the mean of observed values, $\hat{y}$ is the vector of predicted values, and n is the number of samples.

$$MAE = \frac{1}{n} \sum_{i=1}^{n} |y_i - \hat{y}_i| \tag{1}$$

$$RMSE = \sqrt{\frac{1}{n} \sum_{i=1}^{n} (y_i - \hat{y}_i)^2} \tag{2}$$

$$MAPE = \frac{1}{n} \sum_{i=1}^{n} \frac{|y_i - \hat{y}_i|}{y_i} \tag{3}$$

$$MdAPE = median \left(\frac{|y_1 - \hat{y}_1|}{y_1}, \frac{|y_2 - \hat{y}_2|}{y_2}, \ldots, \frac{|y_n - \hat{y}_n|}{y_n} \right) \tag{4}$$

$$R^2 = 1 - \frac{\sum_{i=1}^{n} (y_i - \hat{y}_i)}{\sum_{i=1}^{n} (\hat{y}_i - \overline{y})} \tag{5}$$

3.5   Interpretability Setup

We have also performed interpretability experiments to provide qualitative assessments of the predictive models. Unlike the standard pipeline, we execute these models using just textual features as input data, without reducing their dimensions in any way. We assume that the algorithms behavior on how they handle textual data does not change significantly when fused with structured data. During those analyses, we also restrict our extraction methods to BoW and TF-IDF, leaving out word embeddings.

We used ELI5 framework[8] permutation tests to understand the overall models behavior and how they make isolated predictions. In permutation tests, we analyze the importance of a given feature by shuffling its values in the test set, running the model, and comparing its results with the ones obtained with the original test set (without feature permutation). In both cases, the results are given by the same score function (R^2 in our case). The largest the difference between them, the more important that specific feature is [2].

4 Results

As shown in Table 2, Random Forest and LightGBM provide the best results for the baseline experiments. Since Random Forest scored better for most of the scale-independent metrics (MAPE and MdAPE), we decided to use its results as the overall baseline[9]. Next, we execute each text-based model combined with each learning algorithm, surpassing the baseline on almost every metric across all experiments. The Extremely Randomized Trees (ERT) learning algorithm was the one that consistently achieved the best metrics, which was a surprise considering its average performance during the baseline experiments.

Table 2. Baseline results for the Porto Alegre dataset. Best scores in bold.

Learning algorithm	MAE	RMSE	MAPE	MdAPE	R^2
Extremely Randomized Trees	116,608	301,216	0.245	0.161	0.750
Gradient Boosting	117,910	320,622	0.268	0.196	0.724
Random Forest	**111,931**	304,317	**0.240**	**0.158**	0.747
XGBoost	112,182	298,465	0.248	0.179	0.757
LightGBM	115,316	**295,572**	0.251	0.184	**0.764**

The best result for all metrics among ERT-powered models was achieved by the binary TF-IDF extraction method (see Fig. 2). With this combination, leveraging the 30 textual features extracted from listing descriptions increased R^2 by 16.06% when compared to the baseline. It also reduced MAPE and MdAPE by 19.58% and 12.66%, respectively. Although smaller, the MdAPE reduction is significant as this metric is more immune to outliers and thus harder to improve.

4.1 Generalization Analysis

To verify the ability of the models when leveraging textual data to generalize to previously unseen data, we applied the pipeline described in Sect. 3 to the

---

[8] We used version 0.12.0 (http://eli5.readthedocs.io/).

[9] Scale-independent metrics are useful for comparing results from different studies. They are also used by AVM services such as Zillow's Zestimate (https://www.zillow.com/z/zestimate/).

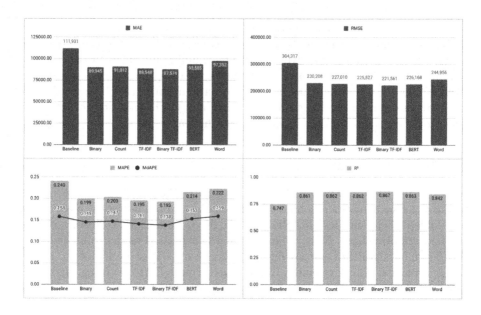

Fig. 2. Metrics for baseline and ERT models applied to the Porto Alegre dataset.

Table 3. Baseline results for the US dataset. Best scores in bold.

Learning algorithm	MAE	RMSE	MAPE	MdAPE	R^2
Extremely Randomized Trees	705,659	2,190,527	0.831	**0.341**	0.464
Gradient Boosting	689,598	**2,062,933**	1.046	0.467	**0.535**
Random Forest	683,200	2,088,720	**0.823**	0.346	0.517
XGBoost	695,317	2,181,905	0.876	0.400	0.474
LightGBM	**681,482**	2,121,460	0.929	0.416	0.504

Table 4. Best extraction method and learning algorithm for every US dataset metric.

Metric	MAE	RMSE	MAPE	MdAPE	R^2
Combination	Binary TF-IDF, XGBoost	Binary TF-IDF, XGBoost	Binary TF-IDF, Random Forest	Binary TF-IDF, Random Forest	Binary TF-IDF, XGBoost
Result	446,416	1,415,748	0.518	0.235	0.778
Improvement	34.49%	26.10%	37.06%	31.09%	45.42%

generalization (US) dataset. The idea is to check whether the very same pipeline of data preprocessing and model building would also work in a distinct dataset, albeit in the same application domain. Differently than what was observed with the Porto Alegre dataset, the generalization baseline experiments did not state a clear "best" learning algorithm. On the contrary, Table 3 shows that only

XGBoost does not achieve a best result for any given metric. For that reason, we decide not to choose a single learning algorithm as the overall baseline and we conduct our analysis by comparing experiment by experiment.

In this context, Table 4 shows the best combination of a text-based model and a learning algorithm for every metric, as well as the corresponding result and improvement over the best baseline for the same metric. Although the best learning algorithms (Random Forest and XGBoost) were different from those observed during the experiments on the Porto Alegre dataset (ERT), the best text-based model was unanimously binary TF-IDF, which is consistent with the previously observed results.

The general improvement over the baseline was significantly greater over the US dataset. This can be explained by the fact that the baseline performance was worse, as its properties are spread over a much larger area with many different price-affecting factors. As it is naturally harder to make good predictions with such a small feature set, the listing description can carry part of these factors, hence the greater improvement.

Finally, the fact that the best extraction methods were based on the relatively simple and well-known bag-of-words [26], especially its binary version, led us to think that the presence or lack of some terms in the listing description is more important for its valuation than the semantics of the text. For that reason, we conducted several qualitative analyses in order to explore which terms contributed most to improve the overall predictions. We present them next.

4.2   Qualitative Analysis

In this subsection, we detail the interpretability experiments performed to better understand how the trained models leveraged textual data and to search for interesting patterns that could explain the improvements previously presented in the quantitative analysis.

Within the same dataset, permutation analysis showed that the most significant words were quite comparable, indicating that the models discovered approximate paths on how to use the textual information. The results were very similar among the different extraction methods, which seems to point out that there is no absolute best match between a representation technique and a specific model or context. As presented in Fig. 3, similar behavior was observed among learning algorithms, with them sharing a relevant amount of best-ranked words (up to 70% for the first 50 terms). The only exception was XGBoost, with few overall patterns in its results. Evidence of its low stability is the fact that combining it with different extraction methods produced considerably distinct results, as exemplified in Fig. 4.

On the other hand, the most significant words between datasets were notably different, although some specific words such as "elevator", or adjectives such as

Random Forest

LightGBM

Extremely Randomized Trees

Gradient Boosting

Fig. 3. Each word cloud represents the 20 most relevant words for four learning algorithms and among all extraction methods for the Porto Alegre dataset.

"masterpiece" or "unique", were identified as relevant for both. This finding indicates that the method used by the models to interpret textual data is very dependent on the context in which they are used, a fact that is not surprising *per se* but that shows that there are subtle differences in distinct problems from the same application domain.

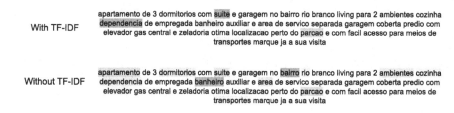

With TF-IDF

apartamento de 3 dormitorios com suite e garagem no bairro rio branco living para 2 ambientes cozinha dependencia de empregada banheiro auxiliar e area de servico separada garagem coberta predio com elevador gas central e zeladoria otima localizacao perto do parcao e com facil acesso para meios de transportes marque ja a sua visita

Without TF-IDF

apartamento de 3 dormitorios com suite e garagem no bairro rio branco living para 2 ambientes cozinha dependencia de empregada banheiro auxiliar e area de servico separada garagem coberta predio com elevador gas central e zeladoria otima localizacao perto do parcao e com facil acesso para meios de transportes marque ja a sua visita

Fig. 4. Comparison between two different models of the XGBoost learning algorithm, with notable interpretation differences. The green color is used to indicate terms related to price increases. The red color is used for the opposite. (Color figure online)

Porto Alegre Dataset. For the Porto Alegre dataset, most of the relevant words refer to amenities or places. Terms such as "suite", "pool", "fireplace", or "parking space" are unanimously listed as relevant, while adjectives such as "quiet" or "sophisticated" also appears many times. We can also extract some interesting patterns by looking at terms that are not seen in in the majority of the experiments. We found streets, neighborhoods, and city sights, which indicates that text can be used to leverage location information in the absence of precise geographic information.

US Dataset. Unlike the Porto Alegre dataset, whose properties are spread over a much smaller area, the US dataset has a much more diverse context that makes it harder to search for patterns. Although the results only converged to a sparse group of common words, we extracted some important information that confirms that all models use relevant words as key to improving predictions, even through different paths. The three most frequent words listed as relevant for the models were "ocean", "oceanfront", and "estate". A list comparing the most frequent words among learning algorithms is presented in Table 5.

Table 5. The 20 most relevant words for four learning algorithms following the count-based BoW extraction method for the US dataset. Underlined words are those that appear in multiple models.

	Random Forest	LightGBM	Gradient Boosting	ERT
1	estate	estate	estate	estate
2	oceanfront	ocean	ocean	oceanfront
3	ocean	crestron	oceanfront	ocean
4	views	oceanfront	compound	views
5	skyline	home	crestron	firm
6	firm	compound	masterpiece	residence
7	intracoastal	summer	wine	crestron
8	terraces	lifts	designed	skyline
9	community	dock	sf	elevator
10	residence	sited	firm	wine
11	crestron	architects	residence	compound
12	elevator	residence	outdoor	intracoastal
13	penthouse	design	summer	community
14	masterpiece	elevator	elevator	feet
15	camp	architect	almost	terraces
16	wolf	staff	architect	wolf
17	architects	completion	waterfrontage	searching
18	55	unit	frontage	camp
19	almost	feet	searching	grand
20	completion	finishes	sited	frontage

Finally, except for the XGBoost experiments, we see for both datasets that textual information can be a proxy for relevant aspects of the property that may not be available in other modalities. As an example of this, Fig. 5 shows that the terms considered relevant by the learning algorithms refer to a wide range of property aspects, including amenities and location.

Random Forest	viva vida club iguatemi excelente apartamento de 3 dormitorios sendo uma suite living para dois ambientes cozinha americana com churrasqueira area de servico e uma vaga de garagem escriturada condominio com infraestrutura completa de laser piscina playground portaria 24 hs quadra de esportes quiosque salao de festas salao de fitness sala de jogos jardim zelador seguranca proximo ao iguatemi
LightGBM	viva vida club iguatemi excelente apartamento de 3 dormitorios sendo uma suite living para dois ambientes cozinha americana com churrasqueira area de servico e uma vaga de garagem escriturada condominio com infraestrutura completa de laser piscina playground portaria 24 hs quadra de esportes quiosque salao de festas salao de fitness sala de jogos jardim zelador seguranca proximo ao iguatemi
Gradient Boosting	viva vida club iguatemi excelente apartamento de 3 dormitorios sendo uma suite living para dois ambientes cozinha americana com churrasqueira area de servico e uma vaga de garagem escriturada condominio com infraestrutura completa de laser piscina playground portaria 24 hs quadra de esportes quiosque salao de festas salao de fitness sala de jogos jardim zelador seguranca proximo ao iguatemi
Extremely Randomized Trees	viva vida club iguatemi excelente apartamento de 3 dormitorios sendo uma suite living para dois ambientes cozinha americana com churrasqueira area de servico e uma vaga de garagem escriturada condominio com infraestrutura completa de laser piscina playground portaria 24 hs quadra de esportes quiosque salao de festas salao de fitness sala de jogos jardim zelador seguranca proximo ao iguatemi

Fig. 5. Comparison between four learning algorithms for one description of the Porto Alegre dataset, following the count-based BoW extraction method. The green color is used to indicate terms related to price increases. The red color is used for the opposite. (Color figure online)

5   Conclusion

In this paper, we show that textual descriptions, much like geographic and visual data, may be used to reliably improve predictions in house price valuation tasks. In addition, we examined and demonstrated how models leverage textual data in their decision-making process. After running a thorough set of experiments, we confirmed that textual information extracted from real estate listing descriptions *does* improve the outcomes of house price valuation tasks, since all text-based models performed better than their baseline references (that did not employ unstructured textual data). In addition, we notice that for each dataset (Porto Alegre and US), the models reliably selected a group of different terms, most of which refer to what humans would consider relevant aspects while pricing a property. Surprisingly, binary TF-IDF is the most promising method for extracting relevant knowledge from textual information (transforming unstructured data into structured data). On the other hand, word embeddings are the worst extraction method (also quite surprisingly).

Regarding extraction methods, the best extraction methods in our case study are relatively simple BoW methods that have been employed for several years. Because of their simplicity, these methods have a lower computational cost when compared to modern methods, including deep vectorized representations such as BERT. This conclusion allows us to ensure that multimodal AVMs can save computational resources by choosing a simpler strategy for leveraging textual descriptions written in natural language.

During our experiments, binary versions of BoW performed better than those with a counting mechanism. For all cases, binary TF-IDF was the best performing approach, indicating that what drives pricing is more related to the presence or lack of certain terms than full-sentence semantics, as supported by our qualitative demonstrations. This could explain why BoW representations outperform newer approaches like BERT.

On the other hand, the best learning algorithm varied both between datasets and between baseline and text-based models, so we suggest not choosing a single one upfront. For the Porto Alegre dataset, Extremely Randomized Trees consistently achieved the best results. Combined with binary TF-IDF, it improved R^2 by 16.06% and also reduced MAPE and MdAPE by 19.58% and 12.66%, respectively. For the US dataset, the best text-based models improved R^2 (+45.42%), MAPE (−37.06%), and MdAPE (−31.09%) as well.

The fact that street and neighborhood names are considered important features by the models may indicate that incorporating structured geographic information, such as latitude and longitude, may reduce the improvements brought by textual information. At the same time, it shows that descriptions can capture some of the important factors contained in geographic data. Hence, in the lack of a large amount of structured data, which is a very common scenario in real-world applications, textual information such as listing descriptions can be used as a proxy that encodes important price-influencing features, leading to a more accurate price valuation.

In future work, it would be an interesting challenge to explore different fusion methods instead of the simple concatenation approach employed here. Moreover, we could analyze how to incorporate additional data modalities (e.g., images) to achieve even better results. As we mentioned before, combining geographic, visual, and textual data may have unexpected effects on how each modality will act in the final prediction, so we look forward to exploring more data modalities.

As an important limitation, we emphasize that our approach is very dependent on the geographic cut being analyzed, something that was demonstrated especially in our generalization experiment with the US dataset, which uses more geographically-diverse data. Designing less context-dependent approaches could amplify the model generalization across datasets, and thus it is an important research direction in this area.

Acknowledgements. Isabel H. Manssour would like to thank the financial support of the CNPq Scholarship - Brazil (308456/2020–3).

References

1. Ahmed, E.H., Moustafa, M.: House price estimation from visual and textual features. In: Guervós, J.J.M., et al. (eds.) Proceedings of the 8th International Joint Conference on Computational Intelligence, IJCCI, vol. 3: NCTA, Porto, Portugal, 9–11 Nov 2016, pp. 62–68. SciTePress (2016). https://doi.org/10.5220/0006040700620068

2. Altmann, A., Toloşi, L., Sander, O., Lengauer, T.: Permutation importance: a corrected feature importance measure. Bioinformatics **26**(10), 1340–1347 (2010). https://doi.org/10.1093/bioinformatics/btq134

3. Bird, S., Klein, E., Loper, E.: Natural language processing with python: analyzing text with the natural language toolkit. O'Reilly, Beijing (2009). http://my.safaribooksonline.com/9780596516499

4. Breiman, L.: Bagging predictors. Machine Learn. **24**(2), 123–140 (1996)

5. Breiman, L.: Random forests. Machine Learn. **45**(1), 5–32 (2001)

6. Chen, M., Liu, Y., Arribas-Bel, D., Singleton, A.: Assessing the value of user-generated images of urban surroundings for house price estimation. Landsc. Urban Plan. **226**, 104486 (2022)

7. Chen, T., Guestrin, C.: Xgboost: a scalable tree boosting system. In: Proceedings of the 22nd ACM SIGKDD International Conference on Knowledge Discovery and Data Mining, pp. 785–794 (2016)

8. Coleman, W., Johann, B., Pasternak, N., Vellayan, J., Foutz, N., Shakeri, H.: Using machine learning to evaluate real estate prices using location big data. arXiv preprint arXiv:2205.01180 (2022)

9. Devlin, J., Chang, M.W., Lee, K., Toutanova, K.: BERT: pre-training of deep bidirectional transformers for language understanding. arXiv preprint arXiv:1810.04805 (2018)

10. Freund, Y., et al.: Experiments with a new boosting algorithm. In: ICML, vol. 96, pp. 148–156. CiteSeer (1996)

11. Friedman, J.H.: Greedy function approximation: a gradient boosting machine. Annals Statist. **29**(5), 1189–1232 (2001)

12. Geurts, P., Ernst, D., Wehenkel, L.: Extremely randomized trees. Machine Learn. **63**(1), 3–42 (2006)

13. Halko, N., Martinsson, P.G., Tropp, J.A.: Finding structure with randomness: probabilistic algorithms for constructing approximate matrix decompositions (2009)

14. Joulin, A., Grave, E., Bojanowski, P., Mikolov, T.: Bag of tricks for efficient text classification. arXiv preprint arXiv:1607.01759 (2016)

15. Kang, Y., Zhang, F., Gao, S., Peng, W., Ratti, C.: Human settlement value assessment from a place perspective: considering human dynamics and perceptions in house price modeling. Cities **118**, 103333 (2021)

16. Ke, G., et al.: LightGBM: a highly efficient gradient boosting decision tree. Adv. Neural. Inf. Process. Syst. **30**, 3146–3154 (2017)

17. Law, S., Paige, B., Russell, C.: Take a look around: using street view and satellite images to estimate house prices. ACM Trans. Intell. Syst. Technol. (TIST) **10**(5), 1–19 (2019)

18. Loshchilov, I., Hutter, F.: Decoupled weight decay regularization (2017)

19. Pagourtzi, E., Assimakopoulos, V., Hatzichristos, T., French, N.: Real estate appraisal: a review of valuation methods. J. Prop. Invest. Fin. **21**(4), 383–401 (2003)

20. Peng, N., Li, K., Qin, Y.: Leveraging multi-modality data to airbnb price prediction. In: 2nd International Conference on Economic Management and Model Engineering (ICEMME), pp. 1066–1071. IEEE (2020)

21. Poursaeed, O., Matera, T., Belongie, S.: Vision-based real estate price estimation. Mach. Vis. Appl. **29**(4), 667–676 (2018). https://doi.org/10.1007/s00138-018-0922-2

22. Rae, A., Murdock, V., Popescu, A., Bouchard, H.: Mining the web for points of interest. In: Proceedings of the 35th International ACM SIGIR Conference on Research and Development in Information Retrieval, pp. 711–720 (2012)
23. Rogers, A., Kovaleva, O., Rumshisky, A.: A primer in bertology: what we know about how BERT works. Trans. Assoc. Comput. Linguist. **8**, 842–866 (2020)
24. Souza, F., Nogueira, R., Lotufo, R.: BERTimbau: pretrained BERT models for Brazilian Portuguese. In: Cerri, R., Prati, R.C. (eds.) BRACIS 2020. LNCS (LNAI), vol. 12319, pp. 403–417. Springer, Cham (2020). https://doi.org/10.1007/978-3-030-61377-8_28
25. Souza, F., Nogueira, R.F., de Alencar Lotufo, R.: Portuguese named entity recognition using BERT-CRF. CoRR abs/1909.10649 (2019)
26. Sparck Jones, K.: A statistical interpretation of term specificity and its application in retrieval. J. Doc. **28**(1), 11–21 (1972)
27. Wu, Y., Zhang, Y.: Mixing deep visual and textual features for image regression. In: Arai, K., Kapoor, S., Bhatia, R. (eds.) IntelliSys 2020. AISC, vol. 1250, pp. 747–760. Springer, Cham (2021). https://doi.org/10.1007/978-3-030-55180-3_57
28. Zhang, Y., Dong, R.: Impacts of street-visible greenery on housing prices: evidence from a hedonic price model and a massive street view image dataset in Beijing. ISPRS Int. J. Geoinf. **7**(3), 104 (2018)

Measuring Ethics in AI with AI: A Methodology and Dataset Construction

Pedro H. C. Avelar, Rafael Baldasso Audibert, and Luís C. Lamb[(✉)]

UFRGS, Instituto de Informática, Porto Alegre, RS, Brazil
{pedro.avelar,rbaudibert,lamb}@inf.ufrgs.br

Abstract. Recently, sound measures and metrics in Artificial Intelligence have become a focus of research and development in academia, government, and industry. Efforts towards measuring different phenomena have gained traction in the AI community, as illustrated by several influential field reports and policy documents. These metrics are designed to help decision-makers inform themselves about the fast-moving and impacting influences of key advances in Artificial Intelligence in general and Machine Learning in particular. In this paper, we propose to use such newfound capabilities of AI technologies to augment our AI measuring capabilities. We do so by training a model to classify publications related to ethical issues and concerns. Our methodology uses an expert, manually curated dataset as the training set and then evaluates an extensive collection of research papers. Finally, we highlight the implications of AI metrics, particularly their contribution towards developing trustful and fair AI-based tools and technologies.

1 Introduction

Recently, the use of sound measures and metrics in Artificial Intelligence (AI) has become the subject of interest of academia, government, and industry [15,20,22,32]. The widespread impact of Artificial Intelligence and Machine Learning has implied a paradigm shift in several fields of computing research, including natural language processing, machine translation, computer vision, and image recognition [4,17,27]. Leading scientists, public leaders, and entrepreneurs, including Bill Gates, Elon Musk, and the late Stephen Hawking, have raised concerns about the impact of AI on every aspect of human life. These happenings have led to increasing societal concerns about the fairness, accountability, explainability, and interpretability of AI systems and technologies [6,7,10,12]. In addition, several scientific and political organizations, the United Nations, European Union, OECD, and national governments have now invested in the development of ethical guidelines and national or multilateral AI policies, regulations, and strategies [5,22,30].

Therefore, the development of AI systems requires use of appropriate metrics for AI ethics and policies. The Stanford's AI Index Report [22], in particular, addresses these and related issues. To better understand the societal impact of AI technologies, one has to hold several useful metrics to decision-makers, including policymakers, business and technology executives, journalists, researchers and, most importantly, the general public. Educators also have an increasing responsibility as AI becomes a tool in several scientific domains, with widespread applications in every economic activity [4,6,7].

J. C. Xavier-Junior and R. A. Rios (Eds.): BRACIS 2022, LNAI 13653, pp. 370–384, 2022.
https://doi.org/10.1007/978-3-031-21686-2_26

In the development of ethically bounded AI technologies, one is typically confronted with a number of challenges, including key questions on how to embed ethical principles in AI systems [25,26]. Building intelligent agent systems, or systems that interact and work with humans in real world settings poses several challenges. Intelligent agents, the key components of any AI system, as argued by [26] must also conciliate their subjective preferences with moral and ethical values. Thus, when specifying the ethical behaviours, boundaries, and the AI agents goals one has to seek a balance between an agent's subjective preferences and ethical boundaries, which reflect in the overall AI system behaviours [19,26].

These, of course, demand the development of clearly defined methodologies and metrics in the domain of AI ethics [22]. The point we make here is that - in a nutshell - AI systems must be endowed with, and subject to, clear metrics based on data driven approaches that improve the quality, fairness, explainability, and accountability of AI systems and technologies [12]. These will certainly contribute to mitigate one of the most concerning characteristic of (unfortunately) more than a handful of AI systems: algorithmic biases. In turn, such biases coupled with data biases will lead systems that show undesired behaviours and prejudices [3,9,13]. In this paper, we shall contribute toward the aim of developing metrics for AI ethics. In order to do so, we propose to use the newfound capabilities of AI technologies to augment our AI-measuring capabilities. We train an AI model to classify if a paper is ethics-related from its title and abstract descriptions. We also use expert knowledge by means of a manually curated dataset, which is used as a training set. We then evaluate a set of papers made available in previous works, and compare the accuracy and results of our work.

Thus, our main contributions are as follows:(1) We provide a manually curated dataset of papers that present ethics-related content, which can be extracted from http:// arXiv.org through their unique identifiers. We choose arXiv.org since it has become one of a *de facto* open repository for most AI papers that will be published at mainstream AI conferences and venues. (2) We provide a trained model to evaluate whether a paper is ethics-related or not, which can be used as a tool by ethicists (and other scientists) to help them do a primary analysis of the growing amount of AI and CS-related papers that is useful to their research. (3) We evaluate our results by comparing them with an earlier methodology for measuring the role of ethics in AI research. We do so by running our classifier through the same dataset of papers abstracts from a previous study [23], and we run their methodology on our test set to compare the results.

The remainder of the paper is organised as follows. In Sect. 2 we highlight recent research which tackles key questions in developing metrics for AI ethics. Next, we describe a dataset for identifying ethics in AI research which shall be used in our methods and experiments. We then introduce a new AI-Based index that outputs the recent impact of Ethics in key AI conferences and venues. Finally, we conclude and point out directions for further research.

2   Background and Related Work

Several publications have reported efforts towards measuring different phenomena in the AI community. Some have gained widespread attention, such as the Stanford

Human-Centered AI Institute's AI Index efforts [22, 32]. These AI metrics are designed to help policy-makers and researchers to inform themselves about this fast-moving and impacting field. In this paper, we propose to use the newfound capabilities of AI technologies to augment our measuring ability, training a model to classify if an article is ethics-related from the information provided in its abstract contents.

Ethical concerns on the implications of the data-driven scientific paradigm, reinforced by the prominence of AI technologies and methods, has also raised several societal concerns. For instance, Green [16] has recently urged those who apply data-driven artificial intelligence and machine learning to social and political contexts to acknowledge the impacts of their products and take a more firm stance as policy actors. They discuss three of the main excuses used by engineers to avoid taking stances with regards to how their products are used: the first, that one is "simply an engineer" and does not dictate how the technology they produce will be used; second, that it is not the data scientist's "job" to take a political stance and that remaining neutral during the development is the best course of action; third, that perfectly managing a technology's impact is unfeasible and that this should not be an impediment to create new products that improve society incrementally. Green [16] opposes these arguments defending an apolitical stance and then proceeds to discuss one possible path to incorporating principles into data science to strive for social justice.

In [14] the authors call for data scientists to recognise themselves as a group and discusses issues raised as early as 2017 by France's Commission Nationale de l'Informatique et des Libertés [8] regarding the technologies produced by data scientists. They further discuss the various already-available ethical frameworks that data scientists can use as an azimuth, most of which had been updated in 2018, such as the codes of conduct from the American Statistical Association [2], Association for Computing Machinery [1], and German Informatics Society [15], while also stating that uniting data scientists does not imply into forming a new society and build a code of conduct from the ground up, but rather that these already existing codes could have Data-Science-specific guidelines added to them. Further, [31] raises the topics brought about in such codes of conduct and proposes a course to help professionals in the area of AI, as well as regulators and policymakers. It invites them to come to terms with the many and shifting ethical issues brought about by the AI paradigm change implications, as well as the ensuing legal and regulatory debates and constraints. In this social and scientific context, our work is situated: AI has now gone well beyond the realm of technology and has reached ubiquitous use. Therefore, measuring the social consequences of AI use is paramount to guarantee that its tools and technologies will positively impact human life.

A Note on Dataset Classification

To carry out our investigation, we use a manually curated dataset as a training set and evaluate a collection of papers made available in previous work, comparing our results with them. We chose not to use crowdsourcing or mechanical turk to classify our data. Our decision is due to both ethical and technical reasons (see, e.g. [24] for a deeper discussion on using such techniques to classify specific kinds of data). Further, the recent work of [29] analyses several ethical implications of using mechanical turk in natural language processing, AI and data analyses. Considering ethical implications and the fact that the field of AI ethics and fairness is novel, we opted to use expert knowledge to classify the papers in our dataset. We claim that to better understand whether a specific piece of research is qualified for our analyses, one has to refer to expert knowledge. As we are analysing a body of technical work that demands professional expertise of AI, an expert-curated dataset allows, in principle, a better understanding if a certain piece of research is related or not to ethics. In addition, an expert can possibly better evaluate if the AI research, tool, method, and technology described in a paper might have ethical or social consequences over third parties.

Along these lines, we highlight that an earlier study provided a metric for ethics in AI research [23] that has been used as a data source in the last two Stanford's AI Index editions [22,32]. In such work, the authors measured and analysed the use of ethics-related terms in flagship AI conferences and journals over the last fifty years. In a nutshell, their results show that, although AI was seen as a field that potentially impacted human life in the last decades, technical research papers typically do not explicitly analyse the ethical implications of their research results, tools, technologies, and achievements. One has also to mention that our work contributes towards disseminating a culture of principles Principled AI research, as defended recently by researchers, corporations, public, governments, and social organisations [11,21,30]. Only very recently, since its 2020 edition, conferences such as the Neural Information Processing Systems conference (NeurIPS) have asked authors to describe the ethical consequences of their technical work.

3 On Building a Dataset for Measuring Ethics in AI

We collected a total of 238,806 papers from Arxiv, ranging from 1989-12-31 up until 2019-10-23. These papers' metadata contained a list field "category", which unfiltered amounted to about 10441 categories. Filtering for only those papers which had an "abs" identified, there were 9839 categories. In the end, we filtered papers which had either "cs.cy" or "Computers and Society" (to filter for ethics-related papers) and had any of "cs.AI", "Artificial Intelligence", "cs.CL", "Computation and Language", "cs.CV", "Computer Vision", "Pattern Recognition", "cs.MA", "Multiagent Systems", "cs.LG", "Learning", "cs.NE", "Neural and Evolutionary Computing", "stat.ML", "Machine Learning" (to filter for AI-related papers) inside one of their category identifiers. Filtering for these we were left with 1425 papers to be annotated as related to the field or had contents associated with AI ethics.

With this subset of 1425 papers, the authors then proceeded to manually annotate and curate 200 of the papers on whether they were AI-related or ethics-related, based solely on their titles and abstracts. The final decision was done via a majority vote, where the paper would take on the label that the majority of the researchers assigned to it. The vast majority of papers were already in the AI category due to how the data was collected, however, of the 200 annotated papers there were only 54 papers which were considered to be about ethics, the rest 146 being annotated as not ethics-related.

3.1 Active Learning

To increase the number of papers available in the dataset we used active learning [18,28] to augment the dataset with machine-labelld examples. To do so, we did two rounds of machine labelling, with the first also serving as a model validation step. To keep closer to the index produced with [23], we abstained from using complex NLP models such as transformers or recurrent neural networks, depending on simpler models depending instead on a Bag-of-Words or Term-Frequency representation.

We evaluated hyperparameter combinations for models using a 4-fold cross validation, testing models such as Logistic Regression, Adaptive Boosting, Gradient Boosting, Decision Trees, Random Forests, and Multilayer Preceptrons, the combinations which were tested can be seen in Table 1.

Table 1. Models and Hyperparameters used in the experiments. LR stands for Logistic Regression, AB for Adaptive Boosting, GB for Gradient Boosting, RF for Random Forest, DT for Decision Tree, MLP for Multilayer Perceptron. For the MLP model we used the Adam optimiser.

Model	Hyperparameter	Options
LR	Inv. Reg. Strength	0.25, 0.5, 1, 2, 4
	Regularisation	$l1, l2$
AB	Num. of Estimators	8, 32, 128, 512
	Learning Rate	0.125, 0.25, 0.5, 1
GB	Num. of Estimators	8, 32, 128, 512
	Max. Tree Depth	1, 2 , 4, 8
RF	Num. of Estimators	8, 32, 128, 512
	Max. Tree Depth	1, 2 , 4, 8
DT	Optimisation Criterion	Gini, Entropy
	Splitting Method	Best, Random
	Max. Tree Depth	1, 2 , 4, 8
	Activation function	TanH, ReLU
MLP	Learning Rate	$10^{-3}, 10^{-4}, 10^{-5}$
	Learning Rate Technique	Adaptative, Constant
	Max. Training Iterations	32, 128, 512

Due to the highly imbalanced nature of the dataset, we used oversampling to create a balanced version of the dataset. We avoided using oversampling techniques that generated synthetic value such as SMOTE and ADASYN since the input values do nor represent numeric values, and we did not want to perform generative sampling without a generative model that worked on the original data format. We also tested the model without handling the imbalance on the dataset, but it produced models with a significantly lower ROC-AUC score, and a higher tendency to reject papers as not being AI-related.

Given the initial model choices, we ended up with using a random forest classifier with 512 estimators and a maximum depth of 8. We used L1 norm on the term frequencies and used IDF scaling. Given this, we proceeded with two rounds of model training and classification. On the first round the best classifier obtained an average ROC-AUC score of 0.98 on the 4-fold cross validation step, leaving us with the final model being a Random Forest with 512 estimators and a maximum tree depth of 8. The model when trained on the entirety of the 200 samples misclassified two ethics-related paper as not being on ethics. Furthermore, one of the human classifiers classified 300 more papers with which the model's results were compared, where it agreed 83.33% with the human labeler.

After this step, the human classifiers classified another 79 papers where the model was unsure of its predictions [18], where we considered it sure of its prediction if the label probability was lower than 1/3 or higher than 2/3. Given this new labelled set, containing 83 ethics-related papers on a total of 543 labelled samples, we did a 4-fold cross validation step on model with the same hyperparameter settings, obtaining an average ROC-AUC score of .99. We then re-trained the model on the all the labelled data and labelled the rest of the dataset with the second model's label whenever a human-annotated label was not available. These entries are specified in the dataset we made available so that future users can know which entries are machine-labelled if they wish to avoid them.

3.2 Dataset Analysis

The final dataset we make available with this paper has 290 hand-labelled examples, with the other 1136 being machine-labelled examples using a bag-of-words interpretation of the document and a random first classifier. Of these, 21.61% were considered to be ethics-related.

Finally, we can use the methodology proposed in [23] and used in the AI-Index [22] to assess their technique for identifying ethics-related material from abstracts and titles in our dataset. First testing on the human-labelled sampled their model achieved low scores, having a ROC AUC score of 0.68, with 68% precision and 45% recall, both severely underestimating the number of ethics-related papers while also producing some false positives.

When ran on the entirety of the dataset, the model had similar results, woth a ROC AUC score of 0.86, 54% precision and 48% recall. This result *further motivates our approach*, which tries to build a more robust method for pinpointing where the discussion about ethics in AI is, using machine learning models to help identify these papers.

4   The Construction of the AI-Based AI-Index

In order to construct an AI-based index, We shall use the same datasets as made available by the authors of [23] to perform experiments on generating an AI-based AI-Index. The available data were paper abstracts from flagship conferences (including e.g. AAAI and NeurIPS) as well as paper titles from flagship AI and robotics conferences and journals, selected by the authors. Such data shall then be analysed through a model trained on the aforementioned data.

4.1 A Logistic Regression Model

As a first study on how an alternative AI-based AI-Index model would work, we trained a logistic regression model with $l1$ normalisation. Due to how $l1$ normalisation works, the model would have weights lying close to the unit square, serving either as a positive input, which would serve to classify a paper as being on ethics-related topics, or a negative input, which would harm a paper's likelihood to be classified as being on an ethics-related topic. These keywords can give us insight on the composition of the dataset as well as being able to assess likely failings that a more complex AI-based model might associate with these papers.

The list of keywords used as ethics-related on [23] were: *accountability, accountable, employment, ethic, ethical, ethics, fool, fooled, fooling, humane, humanity, law, machine bias, moral, morality, privacy, racism, racist, responsibility, rights, secure, security, sentience, sentient, society, sustainability, unemployment, workforce.* However, since we have been using lemmatisation in our models so far, through which the list of ethic-related lemmas would be: *accountability, accountable, employment, ethic, ethical, fool, humane, humanity, law, machine bias, moral, morality, privacy, racism, racist, responsibility, right, secure, security, sentience, sentient, society, sustainability, unemployment, workforce.*

Thus, training a Logistic Regression model as specified above on our dataset, we obtained a model that weighted positively the following lemmatised keywords: *ai, bias, discrimination, ethical, fair, fairness, how, human, machine, may, social, these, trust. And weighted negatively the following lemmas: by, datum, information, method, model, network, propose, student, time, use.* The model had an intercept of -0.59, even though we used oversampling to balance the dataset.

The list of keywords learned by the model seemed to have a similar vein to that presented by [23], however it was not without its failings. First of all, some reasonably generic keywords were introduced in the model, such as "ai", "how", "human", "machine", "may" and "these", with the first and fourth ones most likely being added due to the bias our dataset has towards papers that talk about ethics in AI. An interesting note to be made, however, is that the model seemed to balance some of these generic keywords with other generic keywords on the negative part, such as *by, datum, method, model, network, propose* – all of which could be interpreted as pointing more towards an AI/ML model and further away from a paper discussing ethics in AI.

Another large failing from the AI-generated list is that it lacks keywords representing some important topics, such as AI accountability, the impact of AI in employment, AI's reinforcement of biases with regards to race, the relation of AI with law, and questions about its security (although one may argue that this last topic might've been slightly touched by the "trust" keyword). This shows a great gap that still needs to be bridged with regards to building a dataset that encompasses all of these topics.

4.2 Re-Analysing the AI-Index

The analyses presented in [23] and used in the last two Stanford's AI-Index Reports [22,32] analysed the frequency of keywords considered as ethics-related in flagship AI and robotics conferences as well as top AI and Robotics journals. Here we perform the analyses along the lines of the one done by [23], but using the random forest model trained with the dataset we made available. The first part of this study follows closely what we have done so far; their study analysed the frequency of these keywords in paper abstracts for two conferences, namely AAAI and NeurIPS. The second part, however only used paper titles.

To improve on this issue, for this part of the analysis we train our model with a dataset containing both titles and abstracts, so the model has to learn to predict if a paper is ethics-related both using its abstract and using only its title. From a preliminary study training on paper abstracts and testing on paper titles we noticed a significant drop in the model's performance, and thus decided to continue using a model trained on both alternatives. As we can see in Fig. 1, the model we produced disagreed slightly with the work of [23], that is a keyword-based classification, even though it seemed to maintain some of the peaks in Subfigure 1a in 1990, 1991, 1994, 1997, and the year 2000; on the other years, the model seemed to estimate a larger amount of ethics-related papers than the keyword-based model, and showed a decreasing tendency in the last years.

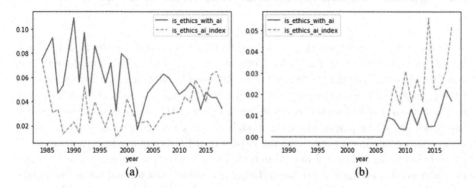

Fig. 1. Number of documents classified as ethics related in AAAI (Subfigure 1a) and NeurIPS (Subfigure 1b) abstracts by our model "is_ethics_with_ai" and by following [23] "is_ethics_ai_index".

In Subfigure 1b, however, the model estimated a much smaller amount of papers as being ethics-related, than what the keyword-based model estimated. Nonetheless, there still seemed to be an updward trend in more recent conferences, and it predicted a dip in the years of 2010, 2012 and 2014, where the keyword-based model estimated a peak instead.

In Figs. 2 and 3, however, is where we see the biggest disagreements between the two approaches, with both having very different scales.

In the conference titles (Fig. 2) we can see that the AI-based model seemed to predict more papers as being ethics-related in older conferences, where the keyword-based model predicted very few. Another big difference is that the AI-based model has a very significant peak in the AAAI conference of the year 2000, where the keyword-based one only had a much smaller peak in 1994. The 2000's peak is caused due to the fact that the dataset only had 4 paper titles for AAAI in the year 2000, among which the title

"Artificial Intelligence-Based Computer Modeling Tools for Controlling Slag Foaming in Electric Furnaces." was (we believe, wrongly) classified as being ethics-related.

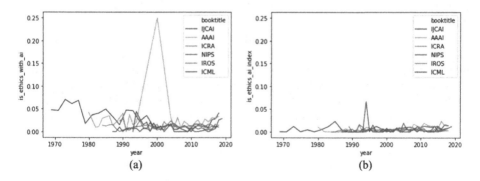

Fig. 2. Number of documents classified as ethics related in the selected flagship conferences' titles. The graph on 2a is our method and on 2b is using the keywords presented in [23]

Now, in the journal titles (Fig. 3) we see that the keyword-based model detects and upward trend in for the journal "IEEE Computer", where the AI-based model only predicts a peak in the last year for the Journal of Artificial Intelligence Research, seemingly having a stable state in the other years and journals. Looking closely at the titles classified as ethics-related by either model in the last year (2019) for both these journals we saw that many of the places where the models disagreed were on titles we believed not to be ethics-related, with only 2 of the papers in IEEE Computer being unanimously classified as such, and only 5 in total in both journals having a majority of us considering them as ethics-related. Also, the spike present in the JAIR journal for the year 2019 was mostly because of how few papers were available in the dataset used in [23].

We also present, like in [23], tables with the percentage of papers that are considered to be AI-related by the algorithm, which serves a double purpose of being another way to identify a conference's or journal's ethics in AI participation, as well as an anomaly detection for the provided algorithm.

For example, in Table 2 we can see the proportion of papers which the model considered to be ethics-related. From looking at these numbers we can quickly detect some anomalies in the model, for example the model classified 1 of the 4 papers present in the dataset for AAAI 2000 as being ethics-related, which we can easily see to not be true by looking at the paper itself, whose name was "Artificial Intelligence-Based Computer Modeling Tools for Controlling Slag Foaming in Electric Furnaces" – clearly not ethics related. However, this over-reporting is less self-evident in venues which have more occurrences, since the model still tends to predict more papers as non-ethics related. In Table 3, we can see a similar table for the analysed journal venues.

We also take 5 random examples from the conference titles and another 5 for the journal titles to see on what cases one model or the other might fail in discerning the values correctly. In the following paragraphs we will present the paper title between double quotes and an indication of the proportion of authors that thought the paper was

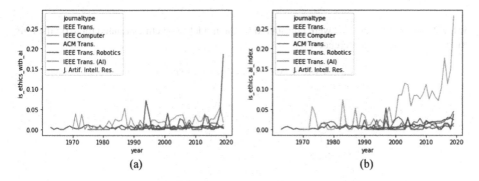

Fig. 3. Number of documents classified as ethics related in the selected journal venues' titles. The graph on 2a is our method and on 2b is using the keywords presented in [23]

ethics-related. So, if one third of the authors agreed that the paper was ethics-related, a paper would appear as "title" (1/3). If none of the authors thought the paper to be ethics, related it would only appear as "title".

In the journal titles dataset, the keyword-based method predicted that the following four titles were titles of ethics-related papers: "Secure and Efficient Handover Authentication Based on Bilinear Pairing Functions.", "When Does Relay Transmission Give a More Secure Connection in Wireless Ad Hoc Networks?", "Performance of the biased square-law sequential detector in the absence of signal.", and "Secure program partitioning.". We believe that none of these four could be considered ethics-related by looking only at the title. However, the only title the Random Forest classified as ethics-related – "Virtual Character Facial Expressions Influence Human Brain and Facial EMG Activity in a Decision-Making Game" (1/3), might be considered as ethics-related by some, although most of us who reviewed it believed it not to be so.

In the conference titles, we had that the AI-based method judged the titles "Human-machine skill transfer extended by a scaffolding framework.", "Enhanced manipulator's safety with artificial pneumatic muscle", and "Logic Programing in Artificial Intelligence" were considered to be ethics-related, while the keyword-based method classified as such the following two titles: "Coherence of Laws" (2/3), "Efficient Methods for Privacy Preserving Face Detection" (2/3). Here, the lack of relevant papers discussing law and privacy in our training set – made apparent by the lack of these keywords as discussed in Subsect. 4.1 – is clear, with the two titles that our method did not classify as ethics-related, where the keyword-based model did, were exactly about these topics.

There is one important aspect to note about the more in-depth analysis of the disagreements between the two models. Only in very few of the classifications, there was a unanimous agreement about the paper being ethics-related. Thus, even when the models theoretically should have voted them in as ethics-related, expert human labellers did not agree. The only cases where an agreement was reached were in titles such as "Codes of Ethics in a Post-Truth World" and "Algorithms: Law and Regulation", one of which was captured by the catch-all "ethics" keyword and the other touched aspects of law in algorithms, but did not necessarily inform us in the title whether it was about artificial intelligence or not.

Table 2. Ratio of papers that were considered by the model to be ethics-related topic on the four AI and two Robotics conferences.

Year	AAAI	ICML	ICRA	IJCAI	IROS	NeurIPS
1980	0.043	–	–	–	–	–
1981	–	–	–	0.037	–	–
1982	0.010	–	–	–	–	–
1983	0.011	–	–	0.041	–	–
1984	0.029	–	0.014	–	–	–
1985	–	–	0.012	0.050	–	–
1986	0.035	–	0.015	–	–	–
1987	0.014	–	0.015	0.033	–	0.000
1988	0.020	–	0.006	–	–	0.000
1989	–	–	0.000	0.015	0.023	0.010
1990	0.040	–	0.003	–	0.029	0.000
1991	0.000	–	0.005	0.047	0.004	0.007
1992	0.038	–	0.007	–	0.003	0.008
1993	0.015	0.000	–	0.046	0.010	0.006
1994	0.020	0.044	0.011	–	0.011	0.007
1995	–	0.029	0.024	0.023	–	0.000
1996	–	0.015	0.009	–	0.013	0.007
1997	–	0.021	0.007	0.036	0.014	0.006
1998	–	0.000	0.002	–	0.013	0.020
1999	–	0.019	0.012	0.010	0.010	0.007
2000	0.250	0.007	0.003	–	0.016	0.007
2001	–	0.013	0.006	0.010	0.003	0.010
2002	–	0.000	0.004	–	0.022	0.010
2003	–	0.009	0.016	0.014	0.013	0.015
2004	0.031	0.017	0.024	–	0.012	0.015
2005	0.009	0.000	0.014	0.017	0.012	0.005
2006	0.031	0.000	0.007	–	0.011	0.005
2007	0.008	0.000	0.009	0.006	0.009	0.009
2008	0.023	0.000	0.014	–	0.010	0.012
2009	–	0.000	0.008	0.000	0.012	0.008
2010	0.010	0.006	0.007	–	0.013	0.000
2011	0.019	0.000	0.008	0.016	0.011	0.010
2012	0.026	0.000	0.014	–	0.013	0.005
2013	0.012	–	0.012	0.016	0.011	0.008
2014	0.028	0.003	0.010	–	0.011	0.005
2015	0.021	0.000	0.008	0.009	0.017	0.005
2016	0.031	0.003	0.012	0.026	0.015	0.005
2017	0.030	0.007	0.011	0.028	0.008	0.016

5   Limitations

We acknowledge here that our model is not without its limitations. First of all, the use of active learning as a strategy to classify the papers was done so due to a lack of resources. If the ethics community could in turn produce a large amount of reliably classified papers, one could certainly build a probably more robust model than we have achieved. Nonetheless, our model is a step in the right direction of using a data-based solution to provide such information, and the dataset we produced is undoubtedly a contribution that can be built upon to improve this area of research.

Another issue of our work is that what defines a paper as being "ethics-related" is based on our experience in the subject. Our working group did not let each others' classification influence each other (we did a double-blind classification), and probably even if we did discuss each of the controversial classifications, it is unlikely that we could reach an agreement in all cases. We believe, again, that a large (however, ethic) paper classification could help iron out these discrepancies and provide classifications that are aligned with what is "commonsense" in the community.

Another issue one might have with our model is that we used a bag-of-words representation, and a logistic regression model to classify this paper. We justify this as to keep the model as interpretable as possible while being aligned with previous work regarding the use of a bag-of-words to classify a paper as being ethics-related or not [23].

6   Discussion

In this paper, we provided an AI-powered tool for classifying research papers as being ethics-related from their own abstract. We offer a first use of this for measuring the AI community's engagement with ethics-related research in the main tracks of flagship venues proceedings. As a consequence, we were able to identify both its characteristics and the keyword-based model's failings, providing some insight on these disagreements with the keyword-based model. This allows us to show that many of the previously reported papers as ethics-related might be wrongly classified as so. Although imperfect due to the limited scale of labelled data used for its training, our model helps alleviate this in some cases. However, one must be aware that one can work towards improving the proposed techniques. First, the data regime used to train this model is thinner than the one common to most machine learning approaches. Second, natural language models are known to exhibit biases contained in the data they have been trained, so one should be mindful that the model we provide here is not without its flaws, and should still be improved to classify correctly all the papers provided to it. Primarily, we pointed out the lack of papers discussing law and race in our training set, which might hinder the performance of our model in detecting papers on these topics, but this, in turn, is a feature of most AI flagship venues so fat, which typically have published a limited number of research papers on the social implications of AI tools and methods.

Datasets and code of this paper are available at: https://github.com/phcavelar/arxiv-ethics

Table 3. Ratio of papers that were considered by the model to be on ethics-related topics on the analysed Journal Venues; respectively, Communications of the ACM, IEEE Computer, IEEE Transactions, IEEE Transactions (AI), IEEE Transactions on Robotics, and Journal of Artificial Intelligence Research.

Year	CACM	Computer	Trans	Trans. AI	Trans. Robotics	JAIR
1980	0.000	0.000	0.002	–	–	–
1981	0.000	0.010	0.008	–	–	–
1982	0.000	0.021	0.003	–	–	–
1983	0.007	0.009	0.003	–	–	–
1984	0.000	0.025	0.003	–	–	–
1985	0.007	0.021	0.007	–	–	–
1986	0.000	0.023	0.010	–	–	–
1987	0.000	0.052	0.006	–	–	–
1988	0.008	0.000	0.003	–	–	–
1989	0.000	0.030	0.009	–	0.000	–
1990	0.000	0.012	0.004	0.000	0.000	–
1991	0.000	0.000	0.004	0.000	0.010	–
1992	0.013	0.023	0.003	0.000	0.000	–
1993	0.000	0.013	0.003	0.000	0.000	0.000
1994	0.008	0.011	0.003	0.008	0.014	0.071
1995	0.008	0.006	0.003	0.000	0.000	0.037
1996	0.012	0.023	0.002	0.005	0.011	0.000
1997	0.008	0.023	0.002	0.000	0.000	0.000
1998	0.008	0.017	0.002	0.000	0.010	0.000
1999	0.004	0.017	0.002	0.000	0.002	0.000
2000	0.007	0.020	0.004	0.004	0.002	0.000
2001	0.004	0.006	0.003	0.004	0.005	0.040
2002	0.005	0.023	0.004	0.000	0.002	0.000
2003	0.009	0.027	0.003	0.000	0.007	0.000
2004	0.004	0.033	0.005	0.004	0.002	0.000
2005	0.010	0.035	0.003	0.000	0.002	0.026
2006	0.002	0.036	0.003	0.000	0.005	0.023
2007	0.007	0.011	0.006	0.004	0.010	0.000
2008	0.006	0.024	0.006	0.002	0.010	0.017
2009	0.008	0.013	0.005	0.008	0.007	0.000
2010	0.005	0.034	0.004	0.003	0.005	0.000
2011	0.005	0.020	0.007	0.003	0.006	0.000
2012	0.010	0.020	0.004	0.010	0.005	0.000
2013	0.008	0.038	0.005	0.003	0.003	0.033
2014	0.008	0.014	0.007	0.002	0.002	0.030
2015	0.011	0.026	0.004	0.016	0.003	0.000
2016	0.007	0.025	0.006	0.015	0.001	0.000
2017	0.006	0.039	0.006	0.015	0.003	0.015

Acknowledgements. This work is partly sponsored by the Brazilian Research Council CNPq and the CAPES Foundation, Finance Code 001.

References

1. ACM: ACM Code of Ethics and Professional Conduct (2018). https://www.acm.org/code-of-ethics
2. ASA: Ethical Guidelines for Statistical Practice (2018). https://www.amstat.org/ASA/Your-Career/Ethical-Guidelines-for-Statistical-Practice.aspx
3. Bender, E.M., Gebru, T., McMillan-Major, A., Shmitchell, S.: On the dangers of stochastic parrots: can language models be too big? In: Proceedings of the 2021 ACM Conference on Fairness, Accountability, and Transparency, FAccT 2021, pp. 610–623. Association for Computing Machinery, New York (2021). DOIurlhttps://doi.org/10.1145/3442188.3445922
4. Bengio, Y., LeCun, Y., Hinton, G.: Deep learning for AI. Comm. ACM **64**(7), 58–65 (2021)
5. Bostrom, N., Yudkowsky, E.: The ethics of artificial intelligence. In: Frankish, K., Ramsey, W. (eds.) The Cambridge Handbook of Artificial Intelligence, pp. 316–334. Cambridge Univ. Press (2014)
6. Burton, E., Goldsmith, J., Koenig, S., Kuipers, B., Mattei, N., Walsh, T.: Ethical considerations in artificial intelligence courses. AI Mag. **38**(2), 22–34 (2017)
7. Chouldechova, A., Roth, A.: A snapshot of the frontiers of fairness in machine learning. Commun. ACM **63**(5), 82–89 (2020). https://doi.org/10.1145/3376898
8. CNIL: Algorithms and artificial intelligence: CNIL's report on the ethical issues. Tech. rep., CNIL (2018). https://www.cnil.fr/en/algorithms-and-artificial-intelligence-cnils-report-ethical-issues
9. Crawford, K.: The hidden biases in big data. Harvard Bus. Rev. **1**(4) (2013)
10. Doran, D., Schulz, S., Besold, T.: What does explainable AI really mean? A new conceptualization of perspectives. arXiv:1710.00794 (2017)
11. Fjeld, J., Achten, N., Hilligoss, H., Nagy, A., Srikumar, M.: Principled artificial intelligence: mapping consensus in ethical and rights-based approaches to principles for AI. Berkman Klein Center Research Publication (2020–1) (2020). Available at SSRN: https://ssrn.com/abstract=3518482 or http://dx.doi.org/10.2139/ssrn.3518482
12. Floridi, L., et al.: AI4People - an ethical framework for a good AI society: opportunities, risks, principles, and recommendations. Minds Mach. **28**(4), 689–707 (2018)
13. Garcia, M.: Racist in the machine: the disturbing implications of algorithmic bias. World Pol. J. **33**(4), 111–117 (2016)
14. Garzcarek, U., Steuer, D.: Approaching ethical guidelines for data scientists. In: Bauer, N., Ickstadt, K., Lübke, K., Szepannek, G., Trautmann, H., Vichi, M. (eds.) Applications in Statistical Computing. SCDAKO, pp. 151–169. Springer, Cham (2019). https://doi.org/10.1007/978-3-030-25147-5_10
15. GI: Ethical Guidelines of the German Informatics Society (2018). https://gi.de/ethicalguidelines/
16. Green, B.: Data science as political action: grounding data science in a politics of justice. Available at SSRN 3658431 (2020)
17. LeCun, Y., Bengio, Y., Hinton, G.: Deep learning. Nature **521**(7553), 436–444 (2015)
18. Lewis, D.D., Gale, W.A.: A sequential algorithm for training text classifiers. In: W.B. Croft, C.J. van Rijsbergen (eds.) Proceedings of ACM-SIGIR, pp. 3–12. ACM/Springer (1994)
19. Mehrabi, N., Morstatter, F., Saxena, N., Lerman, K., Galstyan, A.: A survey on bias and fairness in machine learning. CoRR abs/1908.09635 (2019). http://arxiv.org/abs/1908.09635
20. Mishra, S., Clark, J., Perrault, C.R.: Measurement in AI policy: opportunities and challenges. CoRR abs/2009.09071 (2020). https://arxiv.org/abs/2009.09071

21. OECD: OECD principles on AI (2019). https://www.oecd.org/going-digital/ai/principles/
22. Perrault, R,et al.: The ai index 2019 annual report. Tech. rep, AI Index Steering Committee, Human-Centered AI Institute (2019)
23. Prates, M.O.R., Avelar, P.H.C., Lamb, L.C.: On quantifying and understanding the role of ethics in AI research: a historical account of flagship conferences and journals. In: GCAI, *EPiC Series in Computing*, vol. 55, pp. 188–201. Easy Chair (2018)
24. Ratcliff, R., Hendrickson, A.T.: Do data from mechanical turk subjects replicate accuracy, response time, and diffusion modeling results? Behavior Research Methods, pp. 1–24 (2021)
25. Rossi, F.: Safety constraints and ethical principles in collective decision making systems. In: KI 2015, pp. 3–15 (2015)
26. Rossi, F., Mattei, N.: Building ethically bounded AI. In: The Thirty-Third AAAI Conference on Artificial Intelligence, pp. 9785–9789. AAAI Press (2019)
27. Schmidhuber, J.: Deep learning in neural networks: an overview. Neural Netw. **61**, 85–117 (2015)
28. Settles, B.: Active learning literature survey, Tech. Rep. University of Wisconsin-Madison Department of Computer Sciences (2009)
29. Shmueli, B., Fell, J., Ray, S., Ku, L.W.: Beyond fair pay: ethical implications of NLP crowd-sourcing. arXiv preprint arXiv:2104.10097 (2021)
30. United Nations: The Age of Digital Interdependence - Report of the UN Secretary-General's High-level Panel on Digital Cooperation (2019)
31. Wilk, A.: Teaching AI, Ethics, Law and Policy. arXiv preprint arXiv:1904.12470 (2019)
32. Zhang, D., et al.: The AI Index 2021 Annual Report. AI Index Steering Committee, Human-Centered AI Institute, Stanford University, Stanford, CA (2021). https://aiindex.stanford.edu/wp-content/uploads/2021/03/2021-AI-Index-Report_Master.pdf

Time Robust Trees: Using Temporal Invariance to Improve Generalization

Luis Moneda$^{(\boxtimes)}$ and Denis Mauá

University of Sao Paulo, Sao Paulo, Brazil
{luis.moneda,denis.maua}@usp.br

Abstract. As time passes by, the performance of real-world predictive models degrades due to distributional shifts and learned spurious correlations. Typical countermeasures, such as retraining and online learning, can be costly and challenging in production, especially when accounting for business constraints and culture. Causality-based approaches aim to identify invariant mechanisms from data, thus leading to more robust predictors at the possible expense of decreasing short-term performance. However, most such approaches scale poorly to high dimensions or require extra knowledge such as data segmentation in representative environments. In this work, we develop the Time Robust Trees, a new algorithm for inducing decision trees with an inductive bias towards learning time-invariant rules. The algorithm's main innovation is to replace the usual information-gain split criterion (or similar) with a new criterion that examines the imbalance among classes induced by the split through time. Experiments with real data show that our approach improves long-term generalization, thus offering an exciting alternative for classification problems under distributional shift.

Keywords: Invariance · Generalization · Distributional shift · Inductive bias

1 Introduction

Machine learning techniques are mainly evaluated by their ability to generalize, that is, to find valuable patterns from a training data sample that satisfactory apply to unseen instances [4]. Typically, that process involves a time dimension: training data refers to the past, while unseen instances come from the future. This temporal characteristic is usually dismissed by a time-stationary assumption of the data generating distribution. In practice, sampling distributions are seldom stationary, which causes spurious correlations to be learned and performance to degrade quickly.[1] A stereotypical anecdotal example is that of learning to classify an image of a husky dog as a wolf due to the presence of snow [1]. By blindly minimizing training error (or empirical risk), machine learning models

---

[1] There is often an inductive bias in learning algorithms towards estimating simpler accurate models. For complex tasks, it is often the case that spurious correlations are often simpler than non-spurious ones [1,32].

J. C. Xavier-Junior and R. A. Rios (Eds.): BRACIS 2022, LNAI 13653, pp. 385–397, 2022.
https://doi.org/10.1007/978-3-031-21686-2_27

absorb such relationships [1] and fail to generalize, even when a generalization promise from the validation stage is observed [8,25].

A quick and dirty solution often employed is regularly retraining predictive models as new data arrives. However, this is unsatisfying from a business perspective, as labeling new data is often costly, and recurrently deploying new models into production can introduce inadvertent behavior and lead to significant harm. Also, business culture often is conservative towards adding or modifying existing systems, and such a constant update can decrease trust in machine learning models.

Spurious correlations can be defined as the non-causal statistical relationships between the target and non-target (covariate) variables [20,21]. Thus, the impact of spurious correlations can be alleviated by incorporating causal reasoning into the learning process. However, performing causal inference in the absence of interventional data can be detrimental to predictive performance (hence to generalization) and is generally avoided unless the end task involves causal reasoning (such as producing counterfactuals). Recently, researchers have started advocating the benefits of ensuring some of the properties of causal inference for purely predictive problems without going through a full causal analysis [11].

One interesting property is invariance: causal relationships are invariant to change of environments, which are external settings of the covariates [6,23]. By enforcing invariance in a learning algorithm, we regularize against spurious correlations and decrease the generalization error [1]. While samples from multiple environments are available in specific circumstances (e.g., clinical data collected at different health care centers), this type of data is missing and difficult to generate for most prediction tasks in the real world. Instead, a different type of information about the environment is present in the form of the temporal order in which observations are collected, often spanning a significant time period.

To circumvent the shortcomings highlighted and make use of the often abundant temporal information available in real industry datasets, in this work, we develop the Time Robust Trees (TRT), a new decision tree-inducing algorithm with a strong learning bias towards time-invariant predictive models. A TRT is obtained by modifying the standard recursive partitioning algorithm used to induce decision trees, replacing typical split criteria such as information gain or standard deviation with a new criterion that measures impurity across different time periods. We thus assume that the data is temporally ordered and the training set is segmented by, e.g., yearly data. A hyper-parameter defines the minimum number of examples by segment the model should keep as it learns new rules. This ensures that predictions on unseen data (which do not need temporal information) are more robust to spurious correlations in training data without requiring specific information about environments, causal relationships, or retraining.

Our experiments with seven real-world datasets show that when domain shift is significant, as measured by a domain classifier [24], there is a benefit to using TRTs as a base estimator for an ensemble instead of Decision Trees. The higher the change between the training period and future data, the higher the benefit of signaling to the model via the TRT design that we prefer to learn stable relationships.

The rest of the paper is organized as follows. We start in Sect. 2 reviewing related work in invariance learning, robust learning, and causal analysis. We then present our proposal in Sect. 3. We explain the experimental setup and present the empirical results with real data in Sect. 4. We conclude the paper in Sect. 5 with a discussion about the limitations of the proposed method and some possible improvements for the future.

2   Related Work

As discussed in the introduction, one way of hedging against spurious correlations is to explicitly consider causal relationships in model building. While there are many ways that can be achieved in the literature, a common approach is to resort to the principle of Independent Causal Mechanisms (ICM), which states that the causal generative process of a phenomenon is composed of autonomous modules that do not inform or influence each other [26]. In the probabilistic case, this means that the conditional distribution of each variable given its direct causes (i.e., its causal mechanism) does not inform or influence the other conditional distributions (mechanisms). Recall that an environment is defined as an external setting of the covariates and target variables. As such, data from different environments can be used to identify and learn the causal mechanisms and avoid learning spurious correlations. The hypothesis is that such causal mechanisms are time-invariant, hence improving the generalization ability of the model.

The Invariant Causal Prediction (ICP) [22] is a feature selection algorithm that finds the subset of causal features by testing if the error in the residual on this subset follows a property only found on the target variable's parents under the needed assumptions. ICP requires some mild conditions to be met and scales poorly to high-dimensional data.

The Invariant Risk Minimization (IRM) [1] exploits invariance without explicitly modeling causal relationships. Instead, it modifies the objective function to iterate in training environments and penalizes the lack of invariance across environments. A penalization term is derived for the case of linear classifiers (possibly after the input has been modified by a feature extractor), and the more general case of nonlinear (e.g., neural net) classifiers is left open. ICP requires data to be collected from different environments and annotated accordingly.

There are many other approaches designed to take advantage of the ICM principle. In the Recurrent Independent Mechanisms (RIM) network [10], attention [30] is used to activate different modules composed by RIMs. These modules learn different aspects of the problem, and it is expected that the invariant aspects will be useful when it needs to predict data that differ from the training distribution. Neural Causal Models [15] leverage known or unknown interventions in the observational data to learn. The weakly supervised disentanglement approach [16] learns how to disentangle components from high-dimensional feature spaces, like images, using the hypothesis that they are composed of a small number of relevant factors that can change, thus exploring the modularity of ICMs. The do-calculus in the presence of interventional data [3] is also used to tackle the problem of learning from available data from a few environments and generalizing to unseen environments.

In Robust Supervised-Learning (RSL) [2], the concept of an environment is explicitly missing, but the learning algorithm assigns weights to the training data and optimizes a worst-case scenario for such weights, hoping that such a scenario will also protect the performance at future unseen examples [12]. The first difference with respect to the model we propose here is that RSL considers an adversarial optimization problem while our approach considers worst-case optimization by segmenting the training data.

Our proposed learning algorithm is heavily inspired by the Causal Forest algorithm [31], which induces a decision tree from interventional data segmented into treatment and control groups. The algorithm enforces invariance by requiring a minimum number of examples from treatment and control groups at each split to contrast them in the leaves and reveal heterogeneous causal effects. In contrast, our proposed method assumes purely observational data and regularizes against spurious correlations and distributional shifts by requiring a minimum number of examples from every time period in every node of the decision tree.

There are also approaches that exploit temporal information as an environment proxy like ours in order to mitigate the effects of distributional shifts. The Temporal Decision Trees [13,14] use timestamped data to induce a decision tree, targeting the construction of sequential predictions; as is the case with sequential prediction, the algorithm assumes time-dependence among examples with a stationary generating process. Our method instead makes the common assumption of independent and identically distributed data points.

3   Learning Time Robust Trees

Before formally describing the proposed algorithm, we will first motivate the necessity of time-robust learning methods and explain the limitations of current approaches with a toy example.

3.1   Motivational Example

Consider a setting with two finite-valued input variables X_1 and X_2, a binary target variable Y, and a time period variable T_{period}, used to segment the data into different diverse environments. We use three time periods to illustrate, thus $T_{period} = \{1, 2, 3\}$. Suppose we collect the data shown in Fig. 1, where the data segments for $t = 1, 2$ consist of the available training set, and the data segment for $t = 3$ is observed after model deployment. We will call it the holdout set (note: this is not the typical validation dataset since we assume it is taken from a different distribution, arising possibly from a different environment and certainly from a different time period in the future). According to the example, X_1 is mildly predictive and stable for Y, while X_2 is a perfect predictor at $t = 1$ but

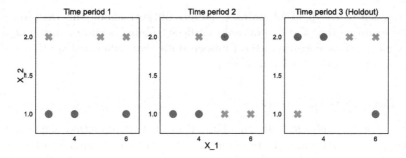

Fig. 1. An artificial example of data with spurious correlations. In the time period 1, there is an pure split on X_2 that leads to maximum accuracy in that data, but does much poorer performance for the time periods 2 and 3. The Time Robust Tree prefers the impure split on X_1, which provides more stable performance across different time periods.

irrelevant for $t = 2$ and $t = 3$. Thus, X_2 can be considered a spurious correlation or a non-static causal relation that shifted.

If the modeler uses all the available training data, a typical Decision Tree (DT) inducing algorithm will combine the data from periods 1 and 2 into a single training data set to evaluate the possible splits. In contrast, in the Time Robust Tree (TRT), as long as the modeler sets the period information as the environment, we consider the split performance separately when looking at every period. To illustrate it, we prune the example tree to have a single split in both cases. We use the Gini impurity (GI) minimization process in Table 1.

Table 1. Split evaluation process for the Decision Tree and the Time Robust Tree for the motivating example

		DT	TRT		
Variable	Split value	GI	GI at t = 1	GI at t = 2	Max. GI value
X_1	3	0.49	0.50	0.40	0.50
X_1	4	0.44	0.44	0.44	**0.44**
X_1	5	0.49	0.50	0.40	0.50
X_2	1	**0.27**	0.00	0.50	0.50

We use the Area Under the Curve (AUC) to evaluate the prediction quality. The measure goes from 0 to 1, and the higher, the better. By learning these splits, the Decision Tree achieves a 0.83 AUC on training but a poor result on holdout data of 0.50 AUC. The Time Robust Tree performs significantly worse in training, achieving an AUC of 0.67; however, it maintains that same performance in the holdout dataset, largely outperforming the Decision Tree. As this example shows, our proposal sacrifices training accuracy in the hopes of achieving superior performance on unseen data that suffer distributional shifts due to the presence of spurious correlations.

3.2   Time Robust Forests

We can now formally describe the Time Robust Tree induction algorithm. We denote an arbitrary impurity function used to evaluate the quality of a dataset split as L. The algorithm work as follows. Consider a timestamp column T_{stamp} representing the data point's capture time with the exact dimension of the random variables vectors $(X_1, ..., X_d, Y)$, where the X variables represent inputs and Y the variable of interest, that is, the target. The time period T_{period} is an aggregation of sequential examples when ordered by T_{stamp} using a human-centered concept, like hourly, daily, weekly, monthly, yearly, or simply putting together a fixed number of examples and reducing T_{stamp} granularity.

Given n time periods $T_{period} = t_1, t_2, \ldots, t_n$ in the training set, we find the best split s^* to divide the examples in X_{node} using the rule $X_f \leq v_f$ where f is a feature from all available features F at a certain value v_f from all possible values for the feature f in the training set V_f by applying recursively to every node data X_{node} until the constraints are not satisfied, being the first node the root containing all the training set:

$$s^* = \min_{\forall f \in F, \forall v \in V_f} \max_{t \in T_{period}} L(X_{node}),$$

$$\text{subject to } |X_{right,t}| \geq \rho \text{ and } |X_{left,t}| \geq \rho, \forall t \in T_{period}. \tag{1}$$

The ρ is a scalar representing the minimum number of examples in every time period to perform a split. The model also accepts the average loss criteria.

$$s^* = \min_{\forall f \in F, \forall v \in V_f} \frac{1}{|T_{period}|} \sum_{t=1}^{T_{period}} L(X_t),$$

$$\text{subject to } |X_{right,t}| > \rho \text{ and } |X_{left,t}| > \rho, \forall t \in T_{period}. \tag{2}$$

For the predictions $\hat{Y}$, the average from the leaf is taken without any consideration about the time period it belongs, $\hat{Y} = \frac{1}{|Y|} \sum y_i$.

It is worth isolating in the Eq. 3 one of the differences from TRT. This period-wise score considers how the model performs in the different periods defined by the user to decide the optimal split. The other difference is the hyper-parameter ρ. It interacts a lot with this part of the process-higher ρ guarantees a higher sample in each period for their evaluation regarding the split.

$$\frac{1}{|T_{period}|} \sum_{t=1}^{T_{period}} L(X_t). \tag{3}$$

There is nothing particularly different in the step from Time Robust Tree to Time Robust Forest (TRF) in comparison to the one from a Decision Tree to a Random Forest [5]. Considering M trees, the final prediction $\hat{Y}$ becomes $\frac{1}{M}\sum_{m=1}^{M}\hat{Y}_m$, a random proportion of the input features F is considered when finding the best split for a node on Eq. 1, and bootstrapping is performed in the training data before learning every tree.

3.3 Synthetic Example

In order to see how TRT prevents spurious correlations from a causal perspective, consider the following artificial example. Once again, we include a spurious feature X_2 in the data generating process that makes the prediction non-stable in the training data. The example is extreme, since X_2 mimics Y in $t = 1$, while it is random in $t = 2$, both of them available for training. The X_2 keeps random in the following periods, consisting of the holdout set. It emulates the hypothesis that unstable properties are less likely to persist.

$$X_1 \sim \mathbb{N}(0,1)$$
$$Y \sim X_1 + \mathbb{N}(0,1) \tag{4}$$
$$X_2 \sim f(e)$$

where e is the time period variable, which is our environment. In the training, we have two training environments $\epsilon_{train} = \{1,2\}$. The $f(e)$ defines X_2 following:

$$f(e) = \begin{cases} Y, \text{ if } e = 1 \\ \mathbb{N}(0,1), \text{ if } e \neq 1 \end{cases} \tag{5}$$

We make it a binary classification task by converting y to a positive class when greater than 0.5 and to the negative one otherwise. The holdout is composed of the following periods, starting at $t = 3$.

At first, we apply the TRT and the DT using similar hyper-parameters: 30 as maximum depth, 0.01 as minimum impurity decrease, 10 as a minimum sample by period for the TRT, and 20 as a minimum sample to split for the DT since we have two periods. The TRT presents an AUC of 0.83 in train and 0.81 in the holdout, while the DT performs around 0.92 AUC in training and 0.64 in the holdout. It shows how the TRT avoids learning from the spurious variable X_2, which lowers its training performance but makes it succeed in the holdout, while the DT goes in the opposite direction. However, we need to define the hyper-parameters following a process and objective criteria in a real-world case. In the following subsection, we show how to execute this step when using the TRT.

3.4 Hyper-parameter Optimization

When selecting hyper-parameters, a common strategy is to use the K-fold validation design [29]. However, during the hyper-parameter selection, this design pools

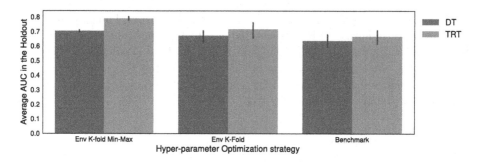

Fig. 2. A hyper-parameter optimization design that keeps the period-wise evaluation from the TRT algorithm is important to make the model keep its purpose of learning stable relationships.

the data from the periods and then select the set of parameters in which the performance is the highest. This process does not favor the period-wise design from TRT. We use a K-fold that generates folds containing just one environment, used as test folds to overcome it. We identify this approach as Environment K-Folds (Env K-Folds). Similar to what we use to learn the best split in the TRT. Besides taking the average performance in the folds to decide the hyper-parameters, we evaluate a second strategy when using the Env K-Folds. First, we average the performance in all folds consisting of the same environment and hyper-parameter set, then we group by only hyper-parameters sets and select the minimum performance, which is the worst environment case. Finally, we take the set with the highest performance among the worst cases to determine the model using the best worst case. We identify this approach as Env K-folds Min-Max.

We bootstrap the data and repeat the process ten times to evaluate these different designs. The results are the average of these ten best models following each approach. As seen in Fig. 2, the TRT performs significantly better than the DT in the holdout set when using the Env K-folds Min-Max, while in the other two strategies, they are very similar.

4   Experiments

To validate the approach, seven public datasets in which a timestamp information and a reasonable time range are available were selected [7,9,17–19,27,28].

We split every dataset into two time periods: training and holdout. Then training period data is split randomly between training and test. For both benchmark and challenger, we use the Time Robust Forest python package.[2] The

---

[2] The source code and datasets used and install instructions are available on GitHub at (https://github.com/lgmoneda/time-robust-tree-paper).

benchmark has all training examples with the same T_{period}, which is a special case the TRF becomes a regular Random Forest. The challenger uses yearly or year-monthly segments.

In Table 2, it is possible to verify that the cases where TRF is an exciting challenger are the ones in which the benchmark has problems performing in the holdout as well as it does in the test. We train a domain classifier using the holdout as the target to clarify the evidence under scenarios the future data changes the most. The higher the AUC, the more significant the difference between test and holdout in that dataset. As seen in Fig. 3, the results show the TRF performed better in the datasets with a more remarkable shift between training data and holdout data.

Table 2. Performance results. When comparing the AUC in the holdout from the TRF to the RF, the benchmark gets better performance on three cases. However, the difference between challenger and benchmark in the holdout always drops compared to the same difference in the test.

Dataset	Data split	Volume	Time range	RF	TRF	Δ TRF-RF
Kickstarter	Train	98k	2010–2013	**.736**	.717	−.019
	Test	24k	2010–2013	**.705**	.701	−.004
	Holdout	254k	2014–2017	.647	**.661**	.014
GE News	Train	21k	2015–2018	**.927**	.865	−.062
	Test	5k	2015–2018	**.879**	.839	−.040
	Holdout	58k	2019–2021	.805	**.821**	.017
20 News	Train	8k	–	**.939**	.869	−.070
	Test	2k	–	**.867**	.828	−.039
	Holdout	8k	–	.768	**.774**	.006
Animal Shelter	Train	75k	2014–2017	**.814**	.803	−.011
	Test	19k	2014–2017	**.792**	.790	−.002
	Holdout	61k	2018–2021	**.791**	.791	.000
Olist	Train	41k	2017	**.799**	.695	−.104
	Test	10k	2017	**.664**	.641	−.023
	Holdout	62k	2018	**.635**	.635	.000
Chicago Crime	Train	100k	2001–2010	**.936**	.909	−.027
	Test	61k	2001–2010	**.904**	.899	−.005
	Holdout	90k	2011–2017	**.905**	.902	−.003
Building Permits	Train	90k	2013–2015	**.990**	.984	−.006
	Test	22k	2013–2015	**.974**	.972	−.002
	Holdout	193k	2016–2017	**.977**	.973	−.004

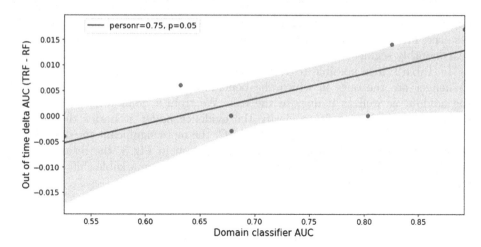

Fig. 3. Domain classifier performance by the delta improvement in the TRF. The greater the difference between the source and target data, translated by a high AUC for the domain classifier, the greater the benefit of learning invariant relationships to generalize to future unseen data.

5   Discussion and Conclusion

Ultimately, machine learning models are evaluated by their ability to generalize observed patterns to unseen data. In realistic scenarios, this often involves using the model under different conditions than those observed in the training stage, causing models learned by standard empirical risk minimization to perform unsatisfactorily when deployed. Common solutions such as constantly retraining are costly, unwanted from a business perspective, and may introduce inadvertent behavior in the system.

Typical real-world datasets are often collected during a significant period and contain temporal order information (i.e., timestamps) that is most often ignored during model construction. In this work, we proposed the Time Robust Trees, a new decision tree induction algorithm that uses temporal order to improve the generalization ability of predictions on unseen, future data. Our method segments data according to time and minimizes the variance of predictions across different time segments, delivering more time-stable models. Experiments with real-world data showing varying degrees of distributional shift suggest that Time Robust Forests are a promising alternative for applications where it is not possible to update the model continuously. Beyond the immediate practical purpose, the experiments show that exploiting time invariance as an inductive learning bias is attractive for non-sequential predictive tasks.

A main limitation of the proposed method is the requirements of temporal order (that is, a timestamp column in the dataset), a reasonable time range, and overlapping empirical distribution support for every period regarding the input features. Timestamp information is most commonly available in real-world datasets since every data is generated in time, and standard practice stores such information. Data collected during long time periods is not uncommon since (unlabeled), as businesses often store data continuously for significant amounts of time. Therefore, the most severe limitation is the need for overlapping empirical distribution support of the input features in every period. We can mitigate such a shortcoming by considering different time scales for the time periods while evaluating model performance and overlap. For example, consider an application that predicts customer acquisition success rate and has customer age as one of the input features. If the time period scale is too small, such as in days, it is very likely that certain age ranges be present in one time period and not in others. In the limit, each period would consist of a single example, causing the learned model to ignore age as a relevant predictive feature. By considering increasingly larger periods (say, of months or years), we can ensure that every data segment contains enough examples for every age range.

The proposed method identifies time periods and environments which are prone to problems. For example, considering two time periods as different environments while they were generated under the same environment will not degrade performance if data is sufficiently abundant but might do so if data is less abundant since it will require more data to meet the cut-off level in the splits. Instead, suppose we place two different environments in the same time period segment. In that case, we are losing an opportunity to offer the model two cases we want to keep relationships invariant and potentially enable the algorithm to create splits that are good only for one of the environments in the same period. However, we still want invariance between this period with the two environments and other periods in the training data. While fewer segments provides a higher volume of data in every period and enables learning more complex rules, it will also make it more likely that two different environments share the same period, compromising the inductive bias for invariant rules.

Real-world data with a good time range should offer enough flexibility to enable a period segmentation to overcome the requirement of overlapping input distribution support. For the future, we plan on exploring different automatic segmentation strategies, representation learning schema that satisfies the overlapping support requirement, combining boosting while respecting the invariance preference, and ensembles of differently regularized Time Robust Forests.

References

1. Arjovsky, M., Bottou, L., Gulrajani, I., Lopez-Paz, D.: Invariant risk minimization (2019)
2. Bagnell, J.A.: Robust supervised learning. In: AAAI, pp. 714–719 (2005)
3. Bareinboim, E., Pearl, J.: Transportability from multiple environments with limited experiments: completeness results. Adv. Neural. Inf. Process. Syst. **27**, 280–288 (2014)
4. Bishop, C.M.: Pattern recognition and machine learning. springer (2006)
5. Breiman, L.: Random forest. Mach. Learn. **45**(1), 5–32 (2001)

6. Cartwright, N.: Two theorems on invariance and causality. Philos. Sci. **70**(1), 203–224 (2003)
7. City of Chicago : Chicago crime - bigquery dataset (2021), version 1. Accessed 13 Mar 2021. https://www.kaggle.com/chicago/chicago-crime
8. D'Amour, A., et al.: Underspecification presents challenges for credibility in modern machine learning. CoRR (2020). http://arxiv.org/abs/2011.03395v1
9. Daoud, J.: Animal shelter dataset (2021), version 1. Accessed 13 Mar 2021. https://www.kaggle.com/jackdaoud/animal-shelter-analytics
10. Goyal, A., et al.: Recurrent independent mechanisms. arXiv preprint arXiv:1909.10893 (2019)
11. Gulrajani, I., Lopez-Paz, D.: In search of lost domain generalization. arXiv preprint arXiv:2007.01434 (2020)
12. Hu, W., Niu, G., Sato, I., Sugiyama, M.: Does distributionally robust supervised learning give robust classifiers? In: International Conference on Machine Learning, pp. 2029–2037. PMLR (2018)
13. Karimi, K., Hamilton, H.J.: Generation and interpretation of temporal decision rules. arXiv preprint arXiv:1004.3334 (2010)
14. Karimi, K., Hamilton, H.J.: Temporal rules and temporal decision trees: A C4. 5 approach. Department of Computer Science, University of Regina Regina, Saskatchewan . . . (2001)
15. Ke, N.R., et al.: Learning neural causal models from unknown interventions. arXiv preprint arXiv:1910.01075 (2019)
16. Locatello, F., Poole, B., Rätsch, G., Schölkopf, B., Bachem, O., Tschannen, M.: Weakly-supervised disentanglement without compromises. In: International Conference on Machine Learning, pp. 6348–6359. PMLR (2020)
17. Mitchell, T.M., et al.: Machine learning (1997)
18. Moneda, L.: Globo esporte news dataset (2020), version 11. Accessed 31 Mar 2021. https://www.kaggle.com/lgmoneda/ge-soccer-clubs-news
19. Mouillé, M.: Kickstarter projects dataset (2018), version 7. Accessed 13 Mar 2021. https://www.kaggle.com/kemical/kickstarter-projects?select=ks-projects-201612.csv
20. Pearl, J.: Causality. Cambridge University Press, Cambridge, UK, 2nd edn. (2009). https://doi.org/10.1017/CBO9780511803161
21. Pearson, K.: On a form of spurious correlation which may arise when indices are useed in the measurement of organs. In: Royal Society of London Proceedings, vol. 60, pp. 489–502 (1897)
22. Peters, J., Bühlmann, P., Meinshausen, N.: Causal inference using invariant prediction: identification and confidence intervals. arXiv preprint arXiv:1501.01332 (2015)
23. Peters, J., Janzing, D., Schlkopf, B.: Elements of causal inference: foundations and learning algorithms. The MIT Press (2017)
24. Rabanser, S., Günnemann, S., Lipton, Z.C.: Failing loudly: an empirical study of methods for detecting dataset shift (2018)
25. Ribeiro, M.T., Singh, S., Guestrin, C.: why should i trust you? explaining the predictions of any classifier. In: Proceedings of the 22nd ACM SIGKDD International Conference on Knowledge Discovery and Data Mining, pp. 1135–1144 (2016)
26. Schölkopf, B., Janzing, D., Peters, J., Sgouritsa, E., Zhang, K., Mooij, J.: On causal and anticausal learning. arXiv preprint arXiv:1206.6471 (2012)
27. Shastry, A.: San francisco building permits dataset (2018), version 1. Accessed 13 Mar 2021. https://www.kaggle.com/aparnashastry/building-permit-applications-data

28. Sionek, A.: Brazilian e-commerce public dataset by olist (2019), version 7. Accessed 13 Mar 2021. https://www.kaggle.com/olistbr/brazilian-ecommerce
29. Stone, M.: Cross-validatory choice and assessment of statistical predictions. J. Roy. Stat. Soc.: Ser. B (Methodol.) **36**(2), 111–133 (1974)
30. Vaswani, A., et al.: Attention is all you need. arXiv preprint arXiv:1706.03762 (2017)
31. Wager, S., Athey, S.: Estimation and inference of heterogeneous treatment effects using random forests. J. Am. Stat. Assoc. **113**(523), 1228–1242 (2018)
32. Wilson, A.C., Roelofs, R., Stern, M., Srebro, N., Recht, B.: The marginal value of adaptive gradient methods in machine learning. arXiv preprint arXiv:1705.08292 (2017)

Generating Diverse Clustering Datasets with Targeted Characteristics

Luiz Henrique dos Santos Fernandes[1]([⊠]) [ID], Kate Smith-Miles[2] [ID],
and Ana Carolina Lorena[1] [ID]

[1] Instituto Tecnológico de Aeronáutica, São José Dos Campos/SP, Brazil
{lhsf,aclorena}@ita.br
[2] The University of Melbourne, Melbourne, Australia
smith-miles@unimelb.edu.au

Abstract. When evaluating clustering algorithms, it is important to assess their performance in retrieving clusters of datasets with known structures. Nonetheless, generating and choosing diverse datasets to compose such test benchmarks is non-trivial. The datasets must present a large variety of structures and characteristics so that the algorithms can be challenged and their strengths and weaknesses can be revealed. The use of generators currently available in the literature relies on trial and error procedures that can be quite costly and inaccurate. Taking advantage of an Instance Space Analysis of popular clustering benchmarks, where datasets are projected into a 2-D embedding with linear trends according to different characteristics, we use a genetic algorithm to produce new datasets at targeted locations in the instance space. This is a natural extension of the Instance Space Analysis framework, and as a result, we are able to produce diverse datasets for composing test benchmarks for clustering.

Keywords: Meta-learning · Instance space analysis · Clustering

1 Introduction

It is commonplace to observe that the performance of an algorithm depends critically on the choice of test problem, and this statement is true whether the algorithm is for clustering, classification, optimization, or many other tasks. The challenge is to learn how the characteristics of the test problems affect the performance of algorithms in order to select the most suitable algorithm for a given test problem instance. The algorithm selection problem presented by Rice in [14] makes up one of the pioneering frameworks in Meta-learning (MtL) for automated algorithm selection. Since then, several methodologies for the meta-analysis of a large variety of problems have emerged. One of them is Instance Space Analysis (ISA), which has successfully been applied to various

Supported by FAPESP (grant 2021/06870-3), CNPq and the Australian Research Council (Laureate Fellowship scheme FL140100012).

problems such as optimization [8,16], classification [9], regression [10], forecasting [6], anomaly detection [5] and related topics. ISA allows the construction of a 2-D embedding of the datasets, organized to highlight linear trends of different meta-features describing the characteristics of the datasets, as well as directions of hardness indicated by algorithm performance metrics. By inspecting the projection of the datasets in this space, it is possible to not only gain insights into the relationships between dataset characteristics and algorithm performance, but also to visually assess the diversity and sufficiency of the datasets for rigorous algorithm testing conclusions.

There are multiple natural challenges associated with clustering, mainly due to the lack of an expected output (ground truth), and the possible presence of multiple valid grouping structures within the data. These challenges are also reflected in the construction of a suitable instance space for clustering problems, which must consider appropriate datasets, meta-features, algorithms and evaluation measures. The ISA for clustering problems built in [2] considers more than 500 datasets, but it still has gaps and empty regions to be filled by datasets that could challenge the clustering algorithms differently and expand the current knowledge on their capabilities and limitations. While there are synthetic dataset generators available in the literature, when used alone they may not be sufficient to fill this gap. Specifically, it is hard to tune these tools to produce datasets with targeted characteristics that make it possible to evaluate the performance of different algorithms under controlled or desired conditions.

In this work we present an ISA-based method that employs Genetic Algorithms (GA) in the search for datasets that target particular locations of the instance space. The GA adopts a synthetic dataset generator from the literature and guides it towards specific regions of interest in the instance space, such as gaps in the instance space corresponding to datasets with previously unstudied combinations of meta-features. The generated datasets provide an opportunity to evaluate different capabilities of clustering algorithms such that their strengths and weaknesses can be better understood. In addition, they have distinct characteristics that allow us to push the boundaries of the current analysis of clustering problems and algorithms. Similar efforts to evolve new test problems have been demonstrated in machine learning [9], optimization [17], and time series forecasting [6], but each time the idea is applied to a new problem domain, considerable thought must go into the problem encoding and tailoring the search to create valid and useful test problems. This approach of using ISA to guide the search for new and diverse clustering problems has not previously been attempted.

The remainder of this paper is organized as follows: In Sect. 2, we recall a theoretical summary of ISA. Section 3 describes the preparation of the meta-dataset and the ISA results for this study. In Sect. 4 we describe the experimental methodology adopted to generate new clustering datasets at targeted regions of the instance space. Final considerations are presented in Sect. 5.

2   Instance Space Analysis

ISA is an MtL methodology that enables visual insights about the strengths and weaknesses of a portfolio of algorithms for different problem instances. Figure 1 summarizes the ISA framework [9]. In the center, I contains a subset of instances of the problem space P for which computational results are available. In Machine Learning (ML), this subset is usually composed of datasets collected from benchmark repositories. The feature space F contains multiple measures used to characterize the properties of the instances in I. They are also referred as meta-features in the MtL literature. The algorithm space A is composed of a portfolio of algorithms that can be used to solve the instances in I. The performance space Y measures the performance of the algorithms in A, when evaluated on the solution of the instances in I. Through a computational process, for all instances in I and all algorithms in A, a meta-dataset containing an ordered quadruple (I, F, A, Y) is composed. One can then learn, through an appropriate supervised learning method, the relationship between the features in F and the performance Y of the algorithms so that algorithms can be recommended for new problems with similar characteristics.

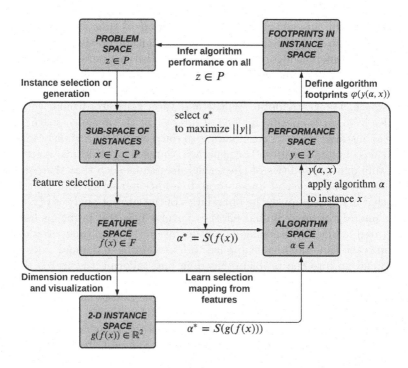

Fig. 1. Instance Space Analysis (ISA) framework, building upon Rice's algorithm selection framework, highlighted in the center [9].

ISA extends beyond algorithm selection and builds a 2-D projection of the instances, known as an instance space (IS), where (meta-)features and algorithmic performance values show linear trends to visualize relationships and insights. While algorithm recommendation can also be performed in this space, other important meta-analysis are made possible. One of them is generating footprints of the algorithms in the IS, which define regions for which each algorithm has a consistently good performance. One can also assess the diversity and sufficiency of currently available benchmark instances to challenge the algorithms, which is our interest here. The main step for the generation of the IS is to build the projection model, for which an optimization problem is formulated and solved to achieve dimension reduction with linear trends. For such, we solve the optimization problem described as follows. According to [9], given a (meta-)feature matrix $\mathbf{F} = [\mathbf{f}_1\ \mathbf{f}_2\ \cdots\ \mathbf{f}_m] \in \mathbb{R}^{m \times n}$ and algorithm performance vector $\mathbf{y} \in \mathbb{R}^n$ (which can also be generalized to a matrix containing multiple evaluations), where m is the number of meta-features and n is the number of problem instances, we achieve an ideal projection of the instances if we can find the matrices $\mathbf{A_r} \in \mathbb{R}^{2 \times m}$, $\mathbf{B_r} \in \mathbb{R}^{m \times 2}$ and the vector $\mathbf{c_r} \in \mathbb{R}^2$ which minimize the approximation error $||\mathbf{F} - \hat{\mathbf{F}}||_F^2 + ||\mathbf{y}^\top - \hat{\mathbf{y}}^\top||_F^2$, such that $\mathbf{Z} = \mathbf{A_r}\mathbf{F}$, $\hat{\mathbf{F}} = \mathbf{B_r}\mathbf{Z}$ and $\hat{\mathbf{y}}^\top = \mathbf{c_r}^\top\mathbf{Z}$.

A Matlab toolkit named MATILDA (acronym for Melbourne Algorithm Test Instance Library with Data Analytics) implements this procedure. The MATILDA tool consists of a data pipeline that integrates some steps. In a preprocessing stage, the metadata is normalized using the Box-Cox and Z transformations. Next, a feature selection process is performed in order to find a subset of meta-features which best explains how the properties of the instances affect the performance of the algorithms. The resulting meta-dataset is then projected to 2-D using the aforementioned optimization model to dimensionality reduction. Once the location of each dataset/instance is project into the 2-D coordinate system defined by $\mathbf{Z}$, the footprints of the algorithms can also be defined as areas in the IS where an algorithm is expected to perform well, given a threshold of good performance. These areas are calculated by Delaunay triangulation of the instances where the algorithm shows a (user-defined) good performance, considering the removal of any contradictory evidence. MATILDA also trains a series of Support Vector Machine (SVM) meta-models for algorithm selection, aimed to predict whether an algorithm will perform well or not for future instances at any location of the instance space. More details on MATILDA and the ISA framework can be found at [9,10].

3   Meta-dataset and ISA Results

This section presents how ISA was framed for the analysis of clustering problems in this work. We used MATILDA for obtaining the IS. All codes used in this experiment were executed on an Intel Core i7-7500U CPU with 2.70 GHz, 16 GB RAM, Microsoft Windows 10 operating system and can be found at https://github.com/ml-research-clustering.

3.1 Clustering Datasets

A total of 553 datasets composed our instance set I. Among them, 336 are synthetic clustering datasets collected from repositories and previously used as benchmarks. This subset contains 80 Gaussian datasets of low dimension and 80 ellipsoidal datasets of high dimension [3]. Another 176 synthetic datasets are from different sources and have different shapes, number of observations (examples), attributes and clusters. In the selected datasets, the number of examples ranges from 100 to 5000, the number of attributes ranges from 2 to 100 and the number of embedded clusters ranges from 1 to 40. Another 217 datasets were collected from the OpenML repository and used as benchmark for clustering algorithms in [12]. In these datasets, the number of examples ranges from 100 to 5000, the number of attributes ranges from 2 to 100, and the number of embedded clusters ranges from 2 to 30. One must notice that these OpenML datasets are originally representatives of classification problems. Although the usage of classification datasets in the evaluation of clustering algorithms is largely employed, some care must be taken in the interpretation of the obtained results as the problems considered have notable differences, as pointed out in [2].

Figure 2 presents the IS obtained, with instances colored by: (a) type of dataset among ellipsoidal, Gaussian, multiple shapes and OpenML datasets; (b) number of examples; (c) number of attributes and (d) number of clusters represented by ground truth. There is a clear distinction between the different types of datasets. The ellipsoidal and multiply shaped datasets are spread in different regions. Gaussian datasets occupy a region between them, with some degree of overlap. The real (OpenML) datasets have a large dispersion towards the upper region of the IS. We can see from the figure that the number of examples, together with the number of attributes, seem to be decisive in the distribution of datasets. The number of examples increases from top to bottom, while the number of attributes increases from left to right. These variables are implicitly represented by one of the selected meta-features that deals with the ratio between the number of attributes and the number of examples, as presented next.

3.2 Meta-features

We used a total of 25 meta-features to describe the clustering datasets. They were divided into seven categories, according to the main properties they extract from data: Distribution, Neighborhood, Density, Dimensionality, Network centrality, Distance and Entropy. The measures had been used in [1,2]. Table 1 presents all the meta-features used in this experiment. Distribution-based measures quantify if the data distribution roughly approximates to a normal distribution. If this is the case, the dataset might have compact hyper-spherical clusters. Neighbourhood-based measures quantify the local nearest neighbour influence in clustering, being a rough indicator of the presence of connected clusters. Density measures quantify whether there are dense regions of data in the input space, a surrogate for the presence of dense clusters. Dimensionality measures regard on the dimensions of the dataset concerning the number

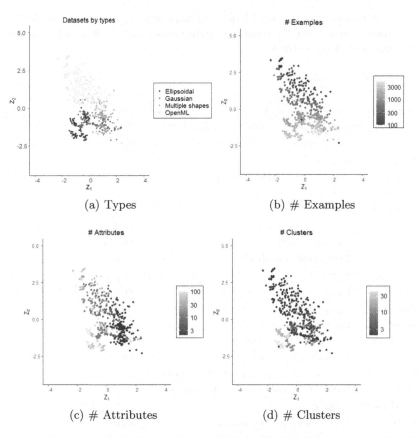

Fig. 2. Characteristics of the datasets projected in the IS, color coded according to (a) distribution of the datasets by type, (b) number of examples, (c) number of attributes and (d) number of clusters represented by ground-truth. The characteristics have been log_{10}-scaled for better visualization.

of examples and input attributes. Some quantify whether the data is sparsely distributed. Network centrality measures quantify whether there are connected structures on data. Distance-based measures quantify the relative differences of distances between the dataset observations. They resemble the neighborhood-based measures, but while the former considers all distances in the dataset, the later takes local distance-based information only. Entropy measures quantify the statistical dependence between random variables and the presence of some structure on data.

From that initial set, six meta-features were selected by the feature selection step of MATILDA to be employed in the ISA analysis: avg_abs_cor, avg_nnd, avg_pca, cop_entropy, ratio_ftr_ex and sd_dist. Equation 1 presents the projection matrix that allows the transformation of instances from the 6-D feature space to the 2-D instance space.

Table 1. Meta-features classified by the main properties they measure. In the Asymptotic column, n stands for the number of data items a dataset has and d corresponds to its number of input features.

Meta-feature	Description	Asymptotic
1) Distribution:		
Multi_norm	Multivariate normality	$O(d \cdot n + n^2)$
Skewness	Multivariate normality skewness	$O(d \cdot n + n^2)$
Kurtosis	Multivariate normality kurtosis	$O(d \cdot n + n^2)$
2) Neighbourhood:		
avg_nnd	Avg. nearest neighbour degree	$O(d \cdot n^2)$
contrast	Contrast	$O(n^2)$
3) Density:		
clust_coef	Clustering coefficient	$O(d \cdot n^2)$
net_dens	Network density	$O(d \cdot n^2)$
perc_out	Percentage of outliers	$O(n^2)$
4) Dimensionality:		
number_ex	$\log_{10}$ number of examples	$O(n)$
number_ftr	$\log_{10}$ number of attributes	$O(d)$
ratio_ftr_ex	Ratio number of attributes to examples	$O(d + n)$
avg_abs_cor	Avg. absolute correlation	$O(n)$
intr_dim	Intrinsic dimensionality	$O(n^2)$
avg_pca	Avg. number of points per PCA dimension	$O(d^2 \cdot n + d^3)$
ratio_pca	Ratio PCA to the original dimension	$O(d^2 \cdot n + d^3)$
5) Network Centrality:		
power_cent	Bonacich's power centrality	$O(n^3)$
eigen_cent	Eigenvalue centrality of MST	$O(n^2)$
hub_score	Kleinberg's hub centrality	$O(n^3)$
6) Distance:		
mean_dist	Mean distance	$O(n^2)$
var_dist	Variance of distances	$O(n^2)$
sd_dist	Standard deviation of distances	$O(n^2)$
high_dist	Percentage of points of high distance	$O(n^2)$
low_dist	Percentage of points of low distance	$O(n^2)$
7) Entropy:		
cop_entropy	Copula entropy	$O(n^2 + n \cdot \log n + n \cdot \sqrt{k \cdot n})$
knn_entropy	k-NN method entropy	$O(n^2 + n \cdot \sqrt{k \cdot n})$

$$
\begin{bmatrix} Z_1 \\ Z_2 \end{bmatrix} = \begin{bmatrix} -0.360 & 0.419 \\ -0.022 & 0.063 \\ 0.478 & -0.195 \\ 0.081 & -0.808 \\ 0.397 & 0.055 \\ -0.134 & -0.446 \end{bmatrix}^{\top} \begin{bmatrix} \text{ratio\_ftr\_ex} \\ \text{avg\_pca} \\ \text{avg\_abs\_cor} \\ \text{avg\_nnd} \\ \text{sd\_dist} \\ \text{cop\_entropy} \end{bmatrix} \tag{1}
$$

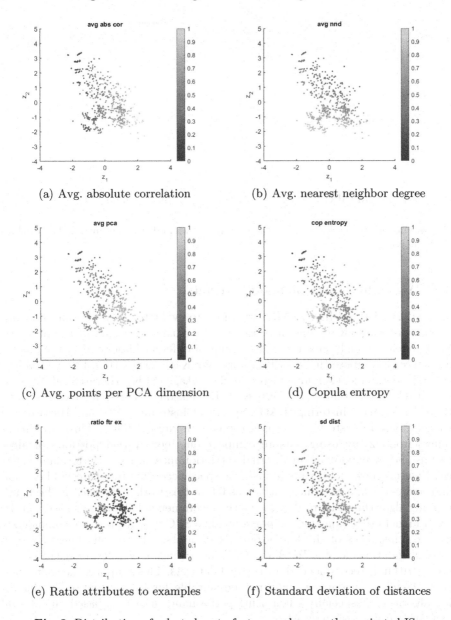

(a) Avg. absolute correlation   (b) Avg. nearest neighbor degree

(c) Avg. points per PCA dimension   (d) Copula entropy

(e) Ratio attributes to examples   (f) Standard deviation of distances

Fig. 3. Distribution of selected meta-features values on the projected IS.

We can notice the presence of meta-features from these categories: Dimensionality, Neighbourhood, Distance and Entropy, with a predominance of dimensionality measures. The distribution of the selected features in the IS, each scaled to $[0, 1]$, is shown in Fig. 3. Some meta-features values decrease from the bottom to the top of the IS (avg_nnd, cop_entropy), others from the right to the left of the IS (avg_abs_cor and sd_dist) and avg_pca and ratio_ftr_ex have mixed behaviors.

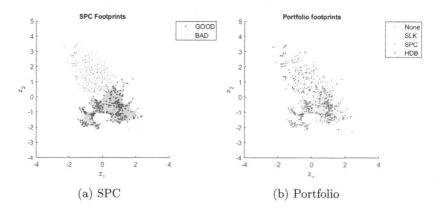

(a) SPC   (b) Portfolio

Fig. 4. Footprints highlighting (a) the best algorithm and (b) recommended algorithms in the portfolio.

3.3   Algorithm Portfolio and Footprints

The Adjusted Rand Index (ARI) was used to evaluate the performance of different clustering algorithms in retrieving the cluster structure of the datasets. We have chosen eight clustering algorithms of different biases widely used in the literature to compose the algorithm portfolio: K-Means (KME), Fuzzy C-Means (FCM), Hierarchical Agglomerative Single Linkage (SLK), Hierarchical Agglomerative Complete Linkage (CLK), High Dimensional Gaussian Mixture Model (GMM), Bagged Clustering (BAG), Spectral Clustering (SPC) and Hierarquical Density Based Clustering of Applications with Noise (HDB). This experiment differs from [2] by using a more compact but more varied portfolio of algorithms and by employing an external validation measure as a performance metric. Figure 4 shows footprints in instance space according to a threshold of good and bad performance, which was set as 0.5 in this work. SPC was the best performing algorithm. Figure 4b presents the recommended algorithms based on the SVM model predictions for the portfolio. The SPC algorithm was recommended for most instances of the four groups of datasets, but not all. There are also recommendations for the HDB and CLK algorithms. One relevant result is that no algorithm is recommended for the real datasets. This supports the contention that classification datasets may not be suitable for the evaluation of clustering algorithms, especially when using performance measures based on external validation criteria such as ARI.

3.4   Analysis of ISA Results

More analysis is possible by relating algorithmic performance along the IS (as shown in Fig. 4) to the meta-characteristics of the datasets (plotted in Fig. 2

and 3). Albeit interesting, our focus in this experiment is on the analysis of the diversity of the datasets included in the IS. In this case, although the set of OpenML datasets has occupied a considerable portion of the IS, they are not really appropriate for evaluating clustering algorithms, as previously discussed. The synthetic benchmarks, on the other hand, are concentrated in the center of the IS (with coordinates z_1 between -2 and 2.2 and z_2 between -2.5 and 1.5). This already indicates the need for more diverse datasets suitable for clustering. Specifically, the synthetic benchmarks lack datasets with features corresponding to a lower average nearest neighbor degree (Fig. 3b), a lower average number of examples per PCA dimension (Fig. 3c), lower Copula entropy values (Fig. 3d) and higher ratios of number of attributes to the number of examples (Fig. 3e). It will be very challenging to try to generate clustering datasets that have these exact characteristics in order to increase the diversity of the instance space. Next section will address how new synthetic datasets with these specific meta-characteristics can be evolved to occupy targeted regions of the instance space and achieve such diversity in a controllable manner.

4 Generation of Artificial Problem Instances

The generation of clustering datasets at controlled locations of the IS is a natural next step when ISA has revealed gaps. For instance, an algorithm based on a mixture of Gaussians was proposed for generating new classification datasets in the related work [9]. Clustering problems need dedicated solutions, as evidenced by our ISA results.

4.1 Proposed Method

Our solution uses a Genetic Algorithm (GA) to guide a clustering dataset generator towards occupying target coordinates of the IS in an ISA-targeted method. Its stages are shown in Fig. 5 and are described next.

The synthetic dataset generator embedded in our solution is MDCGen (Multidimensional Dataset Generator for Clustering), developed by [4]. MDCGen is an open source tool implemented in MATLAB and Python. According to the authors, MDCGen has some advantageous features when compared to other clustering dataset generators available in the literature [3,7,11,13,15,18], such as the ability to generate datasets of both low and high dimensionality, using diverse distributions and shapes for the clusters, controlling cluster overlapping and different clustering properties (e.g., size, number of examples, shape, orientation, cluster inter-distances), among others.

Initially we import the MATILDA ISA results: coordinates, boundaries, raw meta-features, preprocessed meta-features and projection matrix. We also extract the additional dimensionality features from the original datasets: number of examples, attributes and clusters. We then generate a set of regression models using k-Nearest Neighbours (k-NN) for supporting our method and narrowing the search space of the GA. These k-NN models aim to take best advantage of

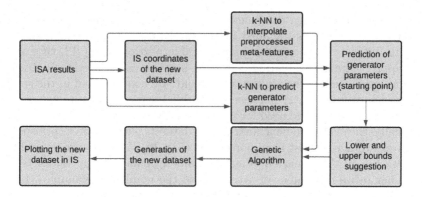

Fig. 5. Generation of new clustering datasets in IS.

the ISA results for guiding the GA solutions and making the entire search process less computationally intensive. A total of nine k-NN models are considered. A 10-fold cross-validation procedure was employed for their tuning, where the k value leading to minimal Root Mean Square Error (RMSE) in each subproblem is chosen. Three of the k-NN models predict the dimensionality features of a dataset, namely number of examples, number of attributes and number of clusters, using the coordinates of the datasets in the IS as inputs. The idea is to narrow the possible values for these features in the GA search. The remaining six models predict the preprocessed values of the six selected meta-features using raw values as inputs. They were necessary to estimate the preprocessed meta-features values of the new datasets required by ISA.

Next we define the target IS coordinates of the new dataset we want to produce, fed to the k-NN models as input, which output the expected dimensions of the new dataset and define suggestions of starting points for the GA search, including lower and upper bounds on the search variables. Based on our previous results, we opted to vary MDCGen parameters related to the number of examples, attributes, clusters, cluster distribution, absolute correlation between the variables, cluster compactness and degree of cluster overlap. The choice of parameters took into consideration dimensionality factors and direct or indirect relationship to the meta-features selected in the ISA. The initial parameters are defined as follows: (i) number of examples, attributes and clusters are obtained from the k-NN predictive models; (ii) distribution is chosen from a list (Gaussian, normal, triangular, gap, logistic, uniform and gamma), starting with the Gaussian distribution; (iii) absolute correlation between the variables, cluster compactness and degree of cluster overlap vary in the range $(0, 1)$ and start with value 0.5. Other parameters included vary over their full range. The number of outliers and the number of noise variables in MDCGen were nullified.

The GA is then tasked with generating a new dataset approximating the given target coordinates of the IS, which is finally plotted on the IS. In the GA implementation we used the R package gramEvol, GeneticAlg.int function, which enables the use of integer decision variables and has allowed managing the search problem more effectively. Each chromosome is encoded as a vector $\mathbf{p} \in \mathbb{Z}_+$ defining parameters of the MDCGen tool. A gene p_i corresponds to one of the following parameters: number of examples, number of attributes, number of clusters, cluster distribution, absolute correlation between the variables, cluster compactness and degree of cluster overlap. Real-valued variables were coded between 0 and 10 and are divided by 10 to obtain their corresponding real values, with experiments confirming that higher precision has little impact on search accuracy. Therefore, each chromosome encodes parameter values of the MDCGen tool, which can be run for generating a new dataset. A population of 10 chromosomes and a maximum number of 500 iterations were used. The mutation probability was set to 0.125 and the number of top ranking chromosomes (elitism) was set to 1. Classic mutation and single-point crossover operators were used.

Let (z_1, z_2) be a point that represents the coordinates of one of the generated datasets when projected onto the IS and (z_1^t, z_2^t) be a target point. The objective of the GA is to minimize a fitness function defined by $f = |z_1 - z_1^t| + |z_2 - z_2^t|$, that is, at each iteration of the GA the Manhattan distance between the coordinates of the generated point and the target point is reduced to an acceptable tolerance level, set by the user (default of 0.1). There is a natural trade-off between the computational cost of the GA and the degree of precision achievable in such approximation, such that stricter tolerance values imply a larger processing time. The steps of the GA fitness function evaluation for each individual are summarized in Fig. 6.

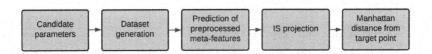

Fig. 6. Steps of the GA fitness function in each generation.

Given a GA individual, it is first decoded so the MDCGen parameters are defined and can be run using these input values. Next, the meta-features of the generated dataset are calculated. k-NN models are used to predict the preprocessed values of such meta-features necessary for ISA. These values are used as input to the projection equation used by ISA to calculate the coordinates of the new dataset and the Manhattan distance to the target point is calculated.

4.2 Generated Datasets and Discussion

Figure 7 illustrates the results of applying our method for generating eight new datasets located at distributed target points (targets shown as crosses, solutions obtained shown as red points). Their characteristics are detailed in Table 2.

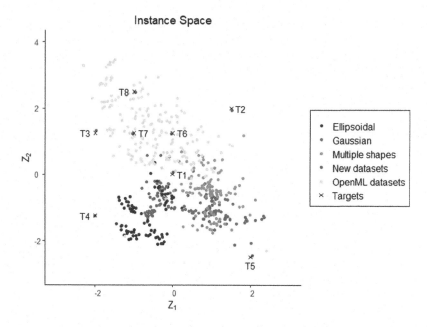

Fig. 7. Examples of new datasets in instance space generated by our method.

Table 2. Characteristics of the new datasets, where i is the number of iterations of GA required, n is the number of examples, d is the number of attributes, k is the number of clusters, *Corr.* is the correlation between variables, *Comp.* is the compactness factor and *Over.* is the degree of overlap.

Target	(z_1, z_2)	i	n	d	k	Distribution	Correlation	Compact	Overlap
T1	(0.00, 0.00)	14	572	18	5	Gap	0.6	0.6	0.7
T2	(1.50, 2.00)	124	36	3	2	Gaussian	0.4	0.2	0.5
T3	(−2.00, 1.25)	18	249	34	13	Logistic	0.5	0.5	0.2
T4	(−2.00, −1.25)	5	2110	98	31	Gap	0.5	0.5	0.5
T5	(2.00, −2.50)	89	3336	34	2	Triangular	0.6	0.2	0.8
T6	(0.00, 1.25)	42	158	18	6	Gap	0.5	0.5	1.0
T7	(−1.00, 1.25)	14	255	19	6	Gap	0.5	0.5	0.3
T8	(−1.00, 2.50)	26	98	40	4	Gap	0.7	0.2	0.3

Interestingly, we are also able to produce clustering datasets in regions occupied originally by the classification datasets only.

Table 3 presents the ARI results of all clustering algorithms in our portfolio for the new datasets. The best and worst performances per dataset are highlighted in bold and italics, respectively. While on average SPC and SLK remained as the best and worst performing algorithms, there are some interesting subtleties in the results. SLK had the largest variation on the average ARI results in the generated datasets (mean ARI of 0.58 with std of 0.48). It performed poorly

Table 3. ARI performance of the clustering algorithms on the new generated datasets. Best and worst ARI results per dataset are highlighted in bold and italics, respectively.

Target	KME	FCM	SLK	CLK	GMM	BAG	SPC	HDB	Mean	Std
T1	**0.99**	0.78	*0.72*	0.91	0.98	**0.99**	0.93	0.91	0.90	0.10
T2	*0.59*	**1.00**	**1.00**	**1.00**	0.96	**1.00**	**1.00**	**1.00**	**0.94**	0.14
T3	0.43	0.11	*0.00*	0.06	0.22	*0.00*	**0.56**	0.11	*0.19*	0.21
T4	0.85	*0.34*	0.89	**0.99**	0.67	**0.99**	0.96	**0.99**	0.83	0.23
T5	0.72	0.87	**1.00**	0.91	*0.61*	0.74	**1.00**	**1.00**	0.86	0.15
T6	0.75	0.74	**1.00**	**1.00**	*0.54*	**1.00**	**1.00**	**1.00**	0.88	0.18
T7	0.76	0.67	*0.00*	0.54	0.52	0.41	**0.95**	0.39	0.53	0.28
T8	0.54	**1.00**	*0.02*	0.19	0.55	0.53	**1.00**	0.93	0.60	0.37
Mean	0.70	0.69	*0.58*	0.70	0.63	0.71	**0.93**	0.79	–	–
Std	0.18	0.31	0.48	0.38	0.25	0.37	0.15	0.34	–	–

on datasets T3, T7 and T8, generated in the OpenML region, with ARI results indicating a total discordance of the clustering assignments to the ground truth of the data. In contrast, a perfect agreement between the ground truth and the SLK cluster's assignments is verified in datasets T2, T5 and T6. While T2 and T5 are targeted at underrepresented regions of the IS that no benchmark dataset currently occupies, T6 is in a region previously occupied by OpenML datasets, as other challenging datasets for SLK. In fact, horizontal shifts in the IS seem to impact more the results of the SLK algorithm, something that can also be observed in the SLK footprint presented in Fig. 4a. Meta-features such as avg_abs_cor (Fig. 3a) and sd_dist (Fig. 3f) are more explanatory of the horizontal placement of the datasets in the IS, with datasets with lower values for these measures placed in the left of the IS, such as T3, T7 and T8. Dataset T3 was very challenging for all clustering algorithms (even SPC), with an average ARI value of 0.19. T7 was also challenging for most of the algorithms (average ARI value of 0.53), with an exception of SPC. T8 allowed to stress a lot of differences between the algorithms, whose ARI performances varied more (mean ARI of 0.60 with std of 0.37). Some of the generated datasets therefore provide greater challenge for some clustering techniques, and highlight better the differences in competence among them. Meanwhile, we also found that it is possible to generate new clustering datasets with a similar degree of difficulty when compared to datasets collected from repositories such as OpenML and datasets extrapolating the current boundaries of the benchmarks included in our ISA.

5   Final Considerations

In this paper we proposed a methodology for generating novel clustering datasets with targeted characteristics. The results show that it is possible to generate test datasets suitable for clustering algorithms and allowing to differentiate subtleties in their performance. Furthermore, it is feasible to diversify the datasets located in other regions of the instance space.

References

1. Fernandes, L.H.S., de Souto, M.C.P., Lorena, A.C.: Evaluating data characterization measures for clustering problems in meta-learning. In: Mantoro, T., Lee, M., Ayu, M.A., Wong, K.W., Hidayanto, A.N. (eds.) ICONIP 2021. LNCS, vol. 13108, pp. 621–632. Springer, Cham (2021). https://doi.org/10.1007/978-3-030-92185-9_51
2. Fernandes, L.H.d.S., Lorena, A.C., Smith-Miles, K.: Towards understanding clustering problems and algorithms: An instance space analysis. Algorithms **14**(3), 95 (2021)
3. Handl, J., Knowles, J.: Cluster generators for large high-dimensional data sets with large numbers of clusters. Dimension **2**, 20 (2005)
4. Iglesias, F., Zseby, T., Ferreira, D., Zimek, A.: Mdcgen: Multidimensional dataset generator for clustering. J. Classification **36**(3), 599–618 (2019)
5. Kandanaarachchi, S., Muñoz, M.A., Hyndman, R.J., Smith-Miles, K.: On normalization and algorithm selection for unsupervised outlier detection. Data Mining Knowl. Disc. **34**(2), 309–354 (2020)
6. Kang, Y., Hyndman, R.J., Smith-Miles, K.: Visualising forecasting algorithm performance using time series instance spaces. Int. J. Forecast. **33**(2), 345–358 (2017)
7. Milligan, G.W., Cooper, M.C.: A study of the comparability of external criteria for hierarchical cluster analysis. Multivariate Behav. Res. **21**(4), 441–458 (1986)
8. Muñoz, M.A., Smith-Miles, K.A.: Performance analysis of continuous black-box optimization algorithms via footprints in instance space. Evol. Comput. **25**(4), 529–554 (2017)
9. Munoz, M.A., Villanova, L., Baatar, D., Smith-Miles, K.: Instance spaces for machine learning classification. Mach. Learn. **107**(1), 109–147 (2018)
10. Muñoz, M.A., et al.: An instance space analysis of regression problems. ACM Trans. Knowl. Discovery Data (TKDD) **15**(2), 1–25 (2021)
11. Pei, Y., Zaïane, O.: A synthetic data generator for clustering and outlier analysis. Tech. rep., Department of Computing Science, University of Alberta Edmonton, AB, Canada (2006). https://era.library.ualberta.ca/items/63beb6a7-cc50-4ffd-990b-64723b1e4bf9
12. Pimentel, B.A., de Carvalho, A.C.: A new data characterization for selecting clustering algorithms using meta-learning. Inform. Sci. **477**, 203–219 (2019)
13. Qiu, W., Joe, H.: Generation of random clusters with specified degree of separation. J. Classification **23**(2), 315–334 (2006)
14. Rice, J.R.: The algorithm selection problem. In: Advances in Computers, vol. 15, pp. 65–118. Elsevier (1976)
15. Schubert, E., Koos, A., Emrich, T., Züfle, A., Schmid, K.A., Zimek, A.: A framework for clustering uncertain data. Proc. VLDB Endowment **8**(12), 1976–1979 (2015)
16. Smith-Miles, K., Baatar, D., Wreford, B., Lewis, R.: Towards objective measures of algorithm performance across instance space. Comput. Op. Res. **45**, 12–24 (2014)
17. Smith-Miles, K., Bowly, S.: Generating new test instances by evolving in instance space. Comput. Oper. Res. **63**, 102–113 (2015)
18. Steinley, D., Henson, R.: Oclus: an analytic method for generating clusters with known overlap. J. Classification **22**(2), 221–250 (2005)

On AGM Belief Revision
for Computational Tree Logic

Marlo Souza$^{(\boxtimes)}$ (iD)

Federal University of Bahia, Salvador, Brazil
`msouza1@ufba.br`

Abstract. Computation Tree Logic is a temporal logic proposed as a tool for formal design and verification of concurrent systems in which execution flows may have many possible branches. A specification of a concurrent system may thus be constructed using a relational model for this logic, and the formal verification of the system's properties may be carried out through model checking. While CTL has obtained a great deal of success as a specification tool, there was a lack of systematic processes for reasoning about the revision and evolution of such specifications. Recently, some literature has proposed applying Belief Change techniques to the problem of revising CTL theories, with implications for revision of formal specifications of systems. This work expands this idea by proposing a Temporal Preference Logic, a multimodal logic with which we can reason about Belief Change in CTL. We show general results regarding the expressivity and decidability of some AGM Belief Revision operators for CTL.

Keywords: Belief revision · Computational tree logic · Preference logic

1 Introduction

Computation Tree Logic (CTL) is a temporal logic proposed by Clarke and Emerson [7] as a tool for formal design and verification of concurrent systems. CTL is built on an interpretation of multiple futures, where several time-flows can succeed the same instant of time. CTL is especially useful to specify properties of systems in which execution flows may have many possible branches.

Belief Change, on the other hand, is the study of how an agent comes to change their mind after acquiring new information. The most influential approach to Belief Change in the literature is the AGM paradigm [1]. While the seminal work of AGM focus on Belief Change for Classical Logic, recently, several works investigated its applications to a wide range of non-classical logics of interest in areas such as Artificial Intelligence, Knowledge Representation, Normative Systems, etc. [9,16,22].

Particularly, Belief Change operations on temporal logics have been studied before as an approach to the problem of model repair, i.e., the problem of how to modify a system model in order to satisfy desired properties [12,16,23]. These

© The Author(s), under exclusive license to Springer Nature Switzerland AG 2022
J. C. Xavier-Junior and R. A. Rios (Eds.): BRACIS 2022, LNAI 13653, pp. 413–428, 2022.
https://doi.org/10.1007/978-3-031-21686-2_29

studies aim to find suitable modifications to the system specification (encoded as a model for temporal logic) that generate admissible models, i.e., representing the intended design for the system.

Recently, authors such as Girard and Rott [10] and Souza et al. [21] propose the use of Dynamic Preference Logic (DPL), a dynamic logic concerning changes in comparative attitudes, to study belief change *a la* AGM. These works, however, have limited their analysis to applying of such logic to belief change in a classical propositional language.

In this work, we investigate the application of Temporal Preference Logic, a temporal extension of Souza et al.'s [21] Dynamic Preference Logic, to study Belief Change in Computational Tree Logic. Within this formalism, we establish the connection between Temporal Preference Logic and AGM Belief Revision [1] and Booth et al.'s Credibility-limited Revision [4], showing that our logic is a general framework to reason about prioritized and non-prioritized belief change. More yet, based on the results about the characterization of finite Kripke models [6], we can show the decidability of belief change for a class of AGM Belief Revision operators for CTL.

This work is structured as follows: in Sect. 2, we introduce the fundamental notions related to AGM Belief Revision Theory that will be used in this work; in Sect. 3, we present Computation Tree Logic (CTL), a temporal logic for which we will investigate belief change operations, and the main results about this logic that we will employ in this work; in Chap. 4, we introduce our Temporal Preference Logic, an extension of CTL and Preference Logic [10], which we employ to define and investigate prioritized and non-prioritized belief change operations for CTL; finally, in our Final Considerations, we present future developments and possible applications of our results.

2   Preliminaries

Given a logic $\mathcal{L} = \langle L, Cn \rangle$, where L is a logical language and $Cn : 2^L \to 2^L$ is a consequence operator, we define the set of $\mathcal{L}$-theories as $Th(\mathcal{L}) = \{\Gamma \subseteq L \mid \Gamma = Cn(\Gamma)\}$.

In AGM's approach, a belief change operator is any operation $\star : Th(\mathcal{L}) \times L \to Th(\mathcal{L})$ that, given a belief set $B \in Th(\mathcal{L})$ and some information φ, changes B in some way. AGM investigated three basic belief change operators: expansion, contraction, and revision.

Belief expansion + blindly integrates a new piece of information into the agent's beliefs. Belief contraction removes a currently believed sentence from the agent's set of beliefs with minimal alterations. Finally, belief revision is the operation of integrating new information into an agent's beliefs while maintaining consistency.

Among these basic operations, only expansion can be univocally defined, $B + \varphi = Cn(B \cup \{\varphi\})$. The other two operations are characterized by a set of postulates, which define a class of suitable change operators representing different rational ways of changing the agent's beliefs.

Let $B \subseteq L$ be a belief set $\mathcal{L}$-formulas. We say that an operation $\star$ is an AGM belief revision function on B if for any formulas $\varphi, \psi \in L$, it satisfies the following postulates:

(R1) $B \star \varphi = Cn(B \star \varphi)$
(R2) $\varphi \in B \star \varphi$
(R3) $B \star \varphi \subseteq Cn(B \cup \{\varphi\})$
(R4) If $\neg\varphi \notin B$, then $B \star \varphi = Cn(B \cup \{\varphi\})$
(R5) $B \star \varphi = Cn(\{\bot\})$ iff $\neg\varphi \notin Cn(\varnothing)$
(R6) If $Cn(\varphi) = Cn(\psi)$ then $B \star \varphi = B \star \psi$
(R7) $B \star (\varphi \wedge \psi) \subseteq Cn((B \star \varphi) \cup \{\psi\})$
(R8) If $\neg\psi \notin B \star \varphi$, then $Cn((B \star \varphi) \cup \{\psi\}) \subseteq B \star (\varphi \wedge \psi)$

While the AGM approach is independent of the supporting logic's syntax, meaning it does not assume an underlying language, it lacks a clear semantic interpretation for its operations. Grove [11] provided one such interpretation using possible world semantics. A Grove system of spheres (SOS) is a pair $\mathcal{S} = \langle W, \leqslant \rangle$ where W is the set of models for the logic $\mathcal{L}$ and $\leqslant \subseteq W \times W$ satisfies the following conditions: (i) $\leqslant$ is reflexive (ii) $\leqslant$ is connected, (iii) $\leqslant$ is transitive, and (iv) for any $S \subseteq W$, if $S \neq \varnothing$, then exists $x \in S$ minimal in $\leqslant$ in regards to S.

Given a system of spheres $\mathcal{S} = \langle W, \leqslant \rangle$, Grove defines the set $Min_{\leqslant} X = \{w \in X \mid \not\exists w' \in X \text{ s.t. } w' \leqslant w \wedge w \not\leqslant w'\}$ and shows that for any belief revision operator $\star$ satisfying the AGM postulates (R1) - (R8) and any belief set B, there is a system of spheres $\mathcal{S}_B = \langle W, \leqslant \rangle$ such that $w \in Min_{\leqslant} W$ iff $w \models B$ and $[\![B \star \varphi]\!] = Min_{\leqslant} [\![\varphi]\!]$.

Investigating the definability of AGM belief change operators in non-classical logics, Flouris et al. [9] show that any logic in which for any set B and non-tautological formula $\varphi \in Cn(B)$, the set

$$\varphi^-(B) = \{K' \subseteq K : Cn(K' \cup \{\varphi\}) = Cn(K)\}$$

is not empty, then an AGM belief contraction (and, thus, also a revision) is definable within that logic.

As CTL satisfies the separation property above, given that it is a Tarskian and boolean logic, we can conclude that AGM Belief Revision operators are definable in CTL, but no general construction of such operators had been proposed. Later, studying the definability of AGM belief change in non-compact logics, such as CTL, Ribeiro et al. [17] show how to construct AGM belief change operators based on a generalization of AGM's partial meet operators, which is closely related to the Grove's Systems of Spheres.

Recently, it has been proposed in the literature that integrating of belief change operations within the object language used to represent the agent's beliefs may have important expressive consequences, as it allows reasoning about change and introspection. It also allows exploring established results from Modal Logic to construct applications of the AGM belief change theory. Work on dynamic logics of belief and information change has been proposed since at least the work of Segerberg, Lindström, and Rabinowicz on DDL [14,19], with important results regarding the generalization of AGM's postulates for introspective agents. More yet, Girard and Rott [10] propose Dynamic Preference Logic (DPL) as a logic to study and reason about Belief Change for Classical Propositional Logic, which Souza et al. [21] have demonstrated to be expressive enough to encode belief change postulates. In doing so, Souza et al. [21] show we can use DPL as a language to reason about classes of Belief Change operators for Classical Propositional Logic.

In this work, we explore how a Dynamic Logic of this kind can be employed to study AGM belief change for CTL.

3   Computational Tree Logic

Computation Tree Logic (CTL) is a branching-time temporal logic proposed by Clarke and Emerson [7] as a tool for formal design and verification of concurrent systems. It has been largely studied for its application in systems specification and model checking [8]. Let us introduce the language of CTL.

Definition 1. *Let P be a set of propositional symbols. We define the language $\mathcal{L}_{CTL}(P)$ by the following grammar (where $p \in P$):*

$$\varphi ::= p \mid \neg\varphi \mid \varphi \wedge \varphi \mid EX\varphi \mid EG\varphi \mid E(\varphi \; U \; \varphi).$$

We call a CTL formula any formula $\varphi \in \mathcal{L}_{CTL}$.

As usual, we define $AF\varphi = \neg EG\neg\varphi$, $AG\varphi = \neg EF\neg\varphi$, $AX\varphi = \neg EX\neg\varphi$ and $A(\varphi \; U \; \psi) = \neg(E(\neg\psi \; U \; \neg(\varphi \vee \psi)) \vee EG\neg\psi)$. CTL formulas are interpreted by means of transition systems, or Kripke models, in which the accessibility relation is understood as temporal possibility, i.e., as possible progressions of time. As such, time in CTL is discrete and non-deterministic.

Definition 2. *We call a branching time model (or CTL model) over P any tuple $M = \langle W, R, v \rangle$, where W is a non-empty set of possible worlds, R is a serial accessibility relation over W and $v : P \to 2^W$ is a valuation function. We denote by $Mod(CTL(P))$ the class of CTL models over P - when P is clear, we will only say $Mod(CTL)$.*

Given a CTL model, a point of evaluation determines the current state of affairs, or the notion of 'present', through which we evaluate the progression of time. To capture this notion, let us introduce the notion of pointed model.

Definition 3. *We call a pointed CTL model over P any tuple $\mathcal{M} = \langle M, w \rangle$ where $M = \langle W, R, v \rangle W$ is a CTL model and $w \in W$ is a possible world.*

We will omit the rules of the interpretation of a temporal formula in a CTL model, as they will be presented in Sect. 4 for Temporal Preference Logic. We will denote that a world w in a model M satisfies the formula φ by $M, w \vDash \varphi$, as usual. Similarly, if $\mathcal{M} = \langle M, w \rangle$ is a pointed CTL model, we say $\mathcal{M} \models \varphi$ if $M, w \models \varphi$.

Browne et al. [6] define the notion of equivalence up to k-steps between two pointed CTL models M and M' as a k-bissimulation.

Definition 4. *[6] Let $M = \langle W, R, v \rangle$ and $M' = \langle W', R', v' \rangle$ be CTL models, and $w \in W$ and $w' \in W'$ be a possible worlds of M and M', respectively. We define a sequence of equivalence relations $E_0, E_1, \cdots, E_k$ on $W \times W'$ as follows:*

- *$w \, E_0 \, w'$ if $\forall p \in P$: $w \in v(p) \Leftrightarrow w' \in v'(p)$*
- *$w \, E_k \, w'$ if*
    - *$w \, E_0 \, w'$ and*
    - *$\forall w_1 \in W[w \, R \, w_1 \Rightarrow \exists w_1' \in W'[w' \, R' \, w_1' \text{ and } w_1 \, E_{k-1} \, w_1']]$*
    - *$\forall w_1' \in W'[w' \, R' \, w_1' \Rightarrow \exists w_1 \in W[w \, R \, w_1 \text{ and } w_1 \, E_{k-1} \, w_1']]$*

Further, we define the relation $w \, E \, w'$ iff $w E_k w'$ for all $k \geqslant 0$.

We say the pointed models $\langle M, w \rangle$ and $\langle M', w' \rangle$ are equivalent up to k-steps if $w \, E_k \, w'$, and that $\langle M, w \rangle$ and $\langle M', w' \rangle$ are equivalent if $w \, E \, w'$

It is easy to see that, since k-step equivalence corresponds to k-bissimulations, these relations are, in fact, equivalence relations.

Lemma 5. *E and E_n are an equivalence relation, for each $n \geqslant 0$.*

We can characterize the k-step equivalence symbolically by means of characteristic formulas obtained through unravelling the model, as studied by Browne et al. [6].

Lemma 6. *[6] Let P be a finite set of propositional symbols, and $\langle M, w \rangle$ and $\langle M', w' \rangle$ be finite pointed CTL models, there is a CTL formula $tr_i(M, w)$ s.t. $w \, E_n \, w'$ iff $M', w' \models tr_n(M, w)$.*

As state equivalence corresponds to bissimulation, it is easy to see that a state equivalence implies a modal equivalence between two possible worlds.

Lemma 7. *[6] Let M and M' be two CTL models, and w, w' be possible worlds of M and M', respectively. If $w \, E \, w'$, then for any $\varphi \in \mathcal{L}_{CTL}$, it holds that $M, w \models \varphi$ iff $M', w' \models \varphi$.*

Surprisingly, however, for any pair of models, Browne et al. [6] show that there is a maximal number of steps that we need to consult in order to decide whether two CTL models are equivalent.

Lemma 8. *[6] Let M and M' be two finite CTL models and w, w' be possible worlds of M and M', respectively. There is some $k \in \mathbb{N}$ s.t. $w \, E \, w'$ iff $w \, E_k \, w'$.*

We can thus define a characteristic number of a model as the smallest number for which we can differentiate between all non-modally equivalent worlds of the model.

Definition 9. *Let $M = \langle W, R, v \rangle$ be a CTL model. We say the characteristic number of M is the smallest c s.t. for any $w, w' \in W$, $w \mathrel{E} w'$ iff $w \mathrel{E_c} w'$.*

Notice that the characteristic number of a model M can be obtained with standard machinery of automata theory, as the characteristic number c describes the greatest size of acyclic paths in M. With these ingredients, we can construct a characteristic formula for the model, i.e., a CTL formula that encodes all the information in the model.

In fact, the characteristic formula of a model M completely characterizes its structure, as any pointed model satisfying it must be modally equivalent to M.

Proposition 10. *[6] Let P be a finite set of propositional symbols, and $\langle M, w \rangle$ and $\langle M', w' \rangle$ be finite pointed CTL models. There is a CTL formula $\mathcal{C}(M, w)$ s.t. if $M', w' \models \mathcal{C}(M, w)$, then for any $\varphi \in \mathcal{L}_{CTL}(P)$ its holds that $M, w \models \varphi \Leftrightarrow M', w' \models \varphi$.*

4   Temporal Preference Logic

We extend the language of CTL to include preference modalities. Preference modalities come from the work on Preference Logic [10,13,21], a logic to reason about comparative attitudes. Preferences have a deep connection to a variety of phenomena in Deontic Logic [22], Logics of Belief [2], and others.

Particularly, authors such as Baltag and Smets [2], Girard and Rott [10] and Souza et al. [21] employ Preference Logic as a language to reason about Belief Change, connecting the work of the AGM-inspired literature to the work on Dynamic Logic, particularly on Dynamic Epistemic Logic. In this work, we aim to extend Preference Logic with temporal modalities in order to reason about AGM Belief Change in CTL. Let us introduce the language of our supporting logic.

Definition 11. *Let P be a set of propositional symbols, we define the language $\mathcal{L}_{\leqslant}^{CTL}(P)$ by the following grammar (where $p \in P$):*

$$\varphi ::= p \mid \neg\varphi \mid \varphi \wedge \varphi \mid \langle\sim\rangle\varphi \mid \langle\leqslant\rangle\varphi \mid \langle<\rangle\varphi \mid EX\varphi \mid EG\varphi \mid E(\varphi U \varphi)$$

We will often refer to the language $\mathcal{L}_{\leqslant}^{CTL}(P)$ simply as $\mathcal{L}_{\leqslant}^{CTL}$, by supposing the set P is fixed. In this language, a formula $\langle\sim\rangle\varphi$ can be read as "*in a conceivable state of affairs φ holds*". Similarly, $\langle\leqslant\rangle\varphi$ ($\langle<\rangle\varphi$) can be read as "*in a state of affairs at least as preferable as (strictly preferable than) the current one, φ holds*." To provide the semantics of this language, we will use a special kind of multimodal frames, called *temporal preference frame*. As usual, we define the dual box modalities $[\leqslant]\varphi$, $[<]\varphi$, $[\sim]\varphi$ as $\neg\langle\leqslant\rangle\neg\varphi$, $\neg\langle<\rangle\neg\varphi$, and $\neg\langle\sim\rangle\neg\varphi$, respectively

Definition 12. *A temporal preference model is a tuple $M = \langle W, R, \leqslant, \sim, v \rangle$, where W is a non-empty set of possible worlds, $R \subseteq W \times W$ is a serial transition*

relation over possible worlds, $\leqslant$ is a reflexive, transitive relation over W with no infinite descending chains, $\sim$ is an equivalence relation on W s.t. $\leqslant \subseteq \sim$ and $v : P \to 2^W$ is a valuation function. We denote by $Mod(\mathcal{L}_{\leqslant}^{CTL}(P))$ the class of temporal preference models over P.

The accessibility relation $\leqslant$ represents an ordering of the possible worlds according to the preferences of a given agent - here understood as a relation of epistemic comparability, i.e. evaluation of plausibility of states of affairs, as commonly done in Epistemic Logic [2]. As such, given two possible worlds $w, w' \in W$, we say that w is at least as plausible as w' if, and only if, $w \leqslant w'$.

We can, thus, define the relation of satisfaction of a formula in a pointed model $\langle M, w \rangle$, i.e., a temporal preference model M and a possible world w.

Definition 13. *Let $M = \langle W, R, \leqslant, \sim, v \rangle$ be a temporal preference model, $w \in W$ be a possible world of M, and $\varphi \in \mathcal{L}_{\leqslant}^{CTL}(P)$ a temporal preference formula. We define the satisfaction relation $M, w \models \varphi$ as follows:*

1. *$M, w \models p$ iff $w \in v(p)$.*
2. *$M, w \models \neg\varphi$ iff $M, w \not\models \varphi$.*
3. *$M, w \models \varphi_1 \wedge \varphi_2$ iff $M, w \models \varphi_1$ and $M, w \models \varphi_2$*
4. *$M, w \models \langle\sim\rangle\varphi$ iff there is some $w' \in W$ s.t. $w \sim w'$ and $M, w' \models \varphi$.*
5. *$M, w \models \langle\leqslant\rangle\varphi$ iff there is some $w' \in W$ s.t. $w' \leqslant w$ and $M, w' \models \varphi$.*
6. *$M, w \models \langle<\rangle\varphi$ iff there is some $w' \in W$ s.t. $w' \leqslant w$, $w' \not\leqslant w$ and $M, w' \models \varphi$.*
7. *$M, w \models EX\varphi$ iff there is some $w' \in W$ s.t. wRw' and $M, w' \models \varphi$.*
8. *$M, w \models EG\varphi$ iff there is a path $\pi = \langle w_0, w_1, w_2, \cdots \rangle$ s.t. $w_0 = w$ and $M, w_i \models \varphi$ for $i \geqslant 0$.*
9. *$M, w \models E(\varphi U \psi)$ iff there is a path $\pi = \langle w_0, w_1, w_2, \cdots \rangle$ s.t. $w_0 = w$ and there is some $i \geqslant 0$ s.t. $M, w_i \models \psi$ and for any $0 \leqslant j < i$, it holds that $M, w_j \models \varphi$.*

As usual, we say φ is valid in M, denoted $M \models \varphi$, if for any $w \in W$, $M, w \models \varphi$ and that φ is valid, denoted $\models \varphi$, if for any temporal preference model M, it holds that $M \models \varphi$.

Given a temporal preference model M and a formula φ, we use the notation $[\![\varphi]\!]_M$ to denote the set of all the worlds in M satisfying φ, or only $[\![\varphi]\!]$ when the model is clear from the context. We assume in the rest of this work that the propositional symbol set P is fixed and will omit it in the presentation unless necessary.

Clearly, temporal preference models are fusion models [3] of preference models, which have the form $M = \langle W, \leqslant, \sim, v \rangle$ and, and CTL models, of the form $M = \langle W, R, v \rangle$, as such Temporal Preference Logic can be characterized as the fusion[1] of Preference Logic [21] and CTL [7]. From this, we conclude from basic results on combining modal logics [3, Chapter 15, Theorem 5] that we can obtain a weakly-complete axiomatization for the logic from the axiomatizations of base logics.

---

[1] Notice that both Preference Logic and CTL are normal modal logics and the class of preference frames and of CTL frames are closed under disjoint unions and isomorphic copies, thus TPL is a fusion logic of CTL, and Preference Logic [3, Chapter 15, Theorem 3].

Theorem 14. *Temporal Preference Logic is weakly-complete axiomatized by the axioms depicted in Fig. 1 along with the substitution and modus ponens rule, the necessitation rules for the box operators $[\leqslant]$, $[<]$, $[\sim]$, AF, AG, AU, AX and EX, as well as the rules below.*

$$\frac{\vdash r \to (\neg q \wedge EXr)}{\vdash r \to \neg A(p \ U \ q)} \qquad \frac{\vdash r \to (\neg q \wedge AX(r \vee \neg E(p \ U \ q)))}{\vdash r \to \neg E(p \ U \ q)}$$

CP   All axioms from Classical Propositional Logic

$\mathbf{K}_{\leqslant}$: $\langle\leqslant\rangle(p \vee q) \to (\langle\leqslant\rangle p \vee \langle\leqslant\rangle q)$

$\mathbf{T}_{\leqslant}$: $p \to \langle\leqslant\rangle p$

$\mathbf{4}_{\leqslant}$: $\langle\leqslant\rangle\langle\leqslant\rangle p \to \langle\leqslant\rangle p$

$\mathbf{K}_{<}$: $\langle<\rangle(p \vee q) \to (\langle<\rangle p \vee \langle<\rangle q)$

$\mathbf{W}_{<}$: $\langle<\rangle p \to \langle<\rangle(\neg\langle<\rangle p \wedge p)$

$<\leqslant_1$: $\langle<\rangle p \to \langle\leqslant\rangle p$

$<\leqslant_2$: $\langle\leqslant\rangle\langle<\rangle p \to \langle<\rangle p$

$<\leqslant_3$: $\langle<\rangle\langle\leqslant\rangle p \to \langle<\rangle p$

$<\leqslant_4$: $p \wedge \langle\leqslant\rangle q \to (\langle<\rangle q \vee \langle\leqslant\rangle(q \wedge \langle\leqslant\rangle p))$

$\mathbf{K}_{\sim}$: $\langle\sim\rangle(p \vee q) \to (\langle\sim\rangle p \vee \langle\sim\rangle q)$

$\mathbf{T}_{\sim}$: $p \to \langle\sim\rangle p$

$\mathbf{4}_{\sim}$: $\langle\sim\rangle\langle\sim\rangle p \to \langle\sim\rangle$

$\mathbf{B}_{\sim}$: $p \to [\sim]\langle\sim\rangle p$

$\sim\leqslant$: $\langle\leqslant\rangle \sim p \to \langle\sim\rangle p$

$\mathbf{K_X}$ $EX(p \to q) \leftrightarrow (EXp \to EXq)$

$\mathbf{D_X}$ $\neg X\bot$

$\mathbf{K_G}$ $EG(p \to q) \leftrightarrow (EGp \to EGq)$

$\mu_{\mathbf{G}}$ $EGp \leftrightarrow (p \wedge EXEGp)$

$\mu_{\mathbf{F}}$ $EFp \leftrightarrow (p \vee EXEFp)$

$\mu_{\mathbf{U}}$ $E(p \ U \ q) \leftrightarrow (q \vee (p \wedge EXE(p \ U \ q)))$

$\mathbf{XF}$ $EXp \leftrightarrow EFp$

Fig. 1. Axiomatization for Temporal Preference Logic

More yet, as satisfiability and model checking are decidable for both CTL and Preference Logic, they also are for Temporal Preference Logic.

4.1   Belief Change in CTL

Now, we focus our attention on using Temporal Preference Logic to study AGM revisions for CTL. Notice that, as Ribeiro et al. [17] show, AGM Belief Revision in CTL can be characterized through Grove-like models or preferences over the class of CTL theories. Such models are, however, very similar to temporal preference models. In fact, Ribeiro et al.'s construction can be translated into temporal preference models, and, thus, we can apply our Temporal Preference Logic to reason about AGM belief Change for CTL.

Let us establish the connection between Belief Change and our logic. Given a temporal preference model M, we can interpret its preference relation $\leqslant$ as some plausibility relation that orders the space of CTL theories (i.e., models). As Grove [11], we can characterize the result of a revision as the set of minimal models according to this plausibility relation.

Definition 15. *Let* $M = \langle W, R, \leqslant, \sim, v \rangle$ *be a temporal preference model and* $S \subseteq W$, *we define the set of minimal worlds in* S *according to* $\leqslant$ *as*

$$Min_{\leqslant} S = \{w \in S \mid \nexists w' \in S \text{ s.t. } w' \leqslant w\}$$

Notice that as the relation $\leqslant$ in temporal preference models is well-founded, i.e., it does not contain infinite descending chains, for any $\varnothing \neq S \subseteq W$, it holds that $Min_{\leqslant} S \neq \varnothing$. We can encode that notion in our language, as previously done for the propositional case by several authors [5,10,21,22].

Definition 16. *Let* $\varphi, \psi \in \mathcal{L}_{\leqslant}^{CTL}(P)$ *be temporal preference formulas, we define the minimization of* φ, *the formula*

$$\mu\varphi \doteq (\varphi \wedge \neg\langle < \rangle\varphi)$$

It is easy to see that this formula captures the notion of minimal conceivable worlds in our language.

Lemma 17. *Let* M *be a temporal preference model and* $w \in W$ *a possible world of* M. *Let yet* $\varphi \in \mathcal{L}_{\leqslant}^{CTL}(P)$ *be temporal preference formula.*

$$M, w \models \mu\varphi \text{ iff } w \in Min_{\leqslant}[\![\varphi]\!]_M$$

We can, thus, represent a counterfactual conditional $\varphi \Rightarrow \psi$ in our language as the notion of satisfying ψ in the most plausible (or typical) φ-worlds.

Definition 18. *Let* $\varphi, \psi \in \mathcal{L}_{\leqslant}^{CTL}(P)$ *be temporal preference formulas, we define the belief in* ψ *conditioned by* φ, *the formula*

$$B(\psi \mid \varphi) \doteq [\sim](\mu\varphi \to \psi)$$

It is not difficult to see that this formula encodes exactly the notion of revisions as conditionalization in which we are interested - as done in several logical interpretations of AGM operations. In fact, given Lemma 17, $B(\psi \mid \varphi)$ can be interpreted as encoding the notion that '*in the most typical/preferred φ-worlds, ψ holds*', similar to the semantics of revision as conditionalization observed in works such as that of Grove [11], Boutilier [5], and Segerberg [19]. Notice that formulas φ and ψ can be any temporal preference formula, meaning that the logic allows us to reason about beliefs regarding temporal information.

Lemma 19. *Let M be a temporal preference model and $w \in W$ a possible world of M. Let yet $\varphi, \psi \in \mathcal{L}_{\leqslant}^{CTL}(P)$ be temporal preference formulas.*

$$M, w \vDash B(\psi \mid \varphi) \text{ iff } Min_{\leqslant}[\![\varphi]\!] \cap [w]_{\sim} \subseteq [\![\psi]\!],$$

where $[w]_{\sim} = \{w' \in W : w \sim w'\}$.

Most importantly, our temporal preference logics are flexible enough to represent any preference over maximal CTL theories, i.e., any partial pre-order over pointed CTL models.

Proposition 20. *Let $\mathcal{M} \subseteq Mod(CTL)$ be a set of pointed CTL models and $\preceq \subseteq \mathcal{M} \times \mathcal{M}$ be a well-founded preference relation on $\mathcal{M}$, i.e., a well-founded reflexive and transitive relation on $\mathcal{M}$. There is a temporal preference model $M = \langle W, R, \leqslant, \sim, v \rangle$ s.t. for any $\langle M_1, w_1 \rangle, \langle M_2, w_2 \rangle \in \mathcal{M}$, $\langle M_1, w_1 \rangle \preceq \langle M_2, w_2 \rangle$ iff there are $w, w' \in M$ s.t. for any temporal formula $\varphi \in \mathcal{L}_{CTL} M, w \vDash \varphi$ iff $\langle M_1, w_1 \rangle \vDash \varphi$ and $M, w' \vDash \varphi$ iff $\langle M_2, w_2 \rangle \vDash \varphi$ and $w \leqslant w'$.*

Proof. Let $\mathcal{M} = \{\langle M_i, w_i \rangle \mid i \in I\}$ for some index set I and $M_i = \langle W_i, R_i, \leqslant_i, \sim_i, v_i \rangle$.

Lets define $M = \langle W, R, \leqslant, \sim, v \rangle$, with:

- $W = \biguplus_{i \in I} W_i$;
- $w \leqslant w'$ if there are $i, j \in I$ s.t. $w = w_i$, $w' = w_j$ and $\langle M_i, w_i \rangle \preceq \langle M_j, w_j \rangle$;
- $w \sim w'$ if there are $i, j \in I$ s.t. $\langle M_i, w_i \rangle \preceq \langle M_j, w_j \rangle$, and $w = w_i$ and $w' = w_j$, or $w' = w_i$ and $w = w_j$;
- wRw' if there is some $i \in I$ s.t. $w, w' \in W_i$ and $wR_i w'$;
- $w \in v(p)$ if there is some $i \in I$ s.t. $w \in W_i$ and $w_i \in v_i(p)$.

Clearly, by construction, $\leqslant$ is a well-founded preference, $\sim$ is the symmetric closure of $\leqslant$ (thus equivalence relation), and R is serial. Clearly, for any $\langle M_i, w_i \rangle \in \mathcal{M}$ there is a $w \in W$ s.t. $M, w \vDash \varphi$ iff $M_i, w_i \vDash \varphi$ and $\leqslant$ reproduces $\preceq$. ☐

This indicates that we have all the resources in our logic to reason about AGM Belief Revision within Temporal Preference Logic, and, indeed, it can be used to investigate properties of belief change operators for CTL. In our construction, the relation $\leqslant$ in model M represents the qualitative epistemic value, plausibility, or credence an agent attributes to a state of affairs. As such, we can understand the most plausible states of affairs, i.e., the minimal φ-elements of the relation $\leqslant$, as representing the most plausible counterfactual scenarios in which φ holds.

Corollary 21. *Let $\star$ be an AGM Belief Revision, for any CTL theory $\Gamma \in Th(CTL)$, there is temporal preference model $M_\Gamma = \langle W, R, \leqslant, \sim, v \rangle$ s.t. for any formulas $\psi, \varphi \in \mathcal{L}_{CTL}$ it holds that*

$$\psi \in \Gamma \star \varphi \text{ iff } M_\Gamma \vDash B(\psi \mid \varphi)$$

Proof. The corollary follows from Ribeiro's [17] characterization of AGM revisions for boolean and non-compact logics as a selection of minimal elements given a preference over maximal theories.   □

Not only do AGM belief change operations induce temporal preference models, but they are characterized by them, as is the case for the propositional case [11].

Corollary 22. *Let $M = \langle W, R, \leqslant, \sim, v \rangle$ be a temporal preference model. If there is a world $w \in W$ s.t. for any satisfiable formula $\varphi \in \mathcal{L}_{CTL}$ there is $w' \in W$ s.t. $w \sim w'$ and $M, w' \models \varphi$, then for some satisfiable set of formulas $\Gamma_M \in Th(CTL)$ there is an AGM belief revision function $\star_w$ on Γ_M, s.t. for any $\varphi, \psi \in \mathcal{L}_{CTL}$, it holds that*

$$\psi \in (\Gamma_M \star_w \varphi) \ \textit{iff} \ M, w \models B(\psi \mid \varphi)$$

As such, we can see that Temporal Preference Logic can be used to reason about AGM-like belief change on the logic CTL. We call temporal preference models that satisfy the conditions of Corollary 22, i.e., in which there is a world that can "see" all temporally satisfiable possibilities, an AGM-like preference model.

For any temporal preference model M, we define the k-step equivalence between the worlds of M similarly as in Definition 4. Let us also define the temporal submodels defined by a temporal preference model.

Definition 23. *Let $M = \langle W, R, \leqslant, \sim, v \rangle$ be a temporal preference model and $w \in W$ a possible world. The temporal restriction of M relative to w is the CTL model $\downharpoonright (M, w) = \langle W_w, R_w, v_w \rangle$ s.t.*

- *$W_w = \{w' \in W \mid wR^*w'\}$ where R^* is the reflexive transitive closure of R;*
- *$R_w = R \cap (W_w \times W_w)$*
- *For all $p \in P$, $v_w(p) = v(p) \cap W_w$*

As for preference models over propositional valuations, which can be represented by orders over propositional formulas [15,20], some temporal preference models can be represented by means of partial orders over temporal formulas. For that to hold, however, we need to impose some restrictions on the interplay between preference relations and temporal relations.

Definition 24. *Let $M = \langle W, R, \leqslant, \sim, v \rangle$ be a temporal preference model and $\equiv = \leqslant \cap \leqslant^{-1}$ the equivalence relation obtained from $\leqslant$, we say M syntactically representable if (i) for any $w \in W$, there is a temporal formula φ_w s.t. for any $w' \in W$, $M, w' \models \varphi_w$ if, and only if, $w \equiv w'$, and (ii) for any $w, w' \in W$ if $\langle \downharpoonright (M, w), w \rangle \ E \ \langle \downharpoonright (M, w'), w' \rangle$, then $w \equiv w'$.*

Proposition 25. *Let P be a finite set of propositional symbols and let $M = \langle W, R, \leqslant, \sim, v \rangle$ be an AGM-like temporal preference model. If M is syntactically representable and $\equiv$ has a finite support, then it is decidable whether $\psi \in \Gamma \star_w \varphi$ for any $\varphi, \psi \in \mathcal{L}_{CTL}$ and $\Gamma \subseteq \mathcal{L}_{CTL}$*

Proof. (Sketch of the proof). From $\equiv$ having finite support, we conclude that there are finitely many equivalence classes $[w]_\equiv$ in $W_{/\equiv}$. As such, there is some finite temporal preference model $M' = \langle W', R', \leqslant', \sim', v' \rangle$ s.t. for any $w \in W$ there is $w' \in W'$ s.t. for any $\varphi \in \mathcal{L}_{\leqslant}^{CTL}(P)$, $M, w \vDash \varphi$ iff $M', w' \vDash \varphi$ (it suffices to take $W' = W_{/\equiv}$). More yet, as M is syntactically representable, it means that for any $\varphi \in \mathcal{L}_{CTL}$, $M, w \vDash [\leqslant]\varphi$ (similarly $[<]\varphi$, or $[\sim]\varphi$) iff for all $w' \in W$ s.t. $w' \leqslant w$ ($w' < w$, or $w' \sim w$) it holds that $\varphi_{w'} \to \varphi$ is CTL-valid[2].

From that and model checking being decidable in Temporal Preference Logic, we conclude decidability. $\qquad\qquad\qquad\qquad\qquad\qquad\qquad\qquad\qquad\qquad\qquad\qquad\quad\square$

Notice that the connection between temporal preference models and AGM belief revisions is established only for AGM-like temporal preference models. The reason for this restriction is that AGM belief revisions allow non-trivial revisions for any formula of the object language. In other words, it is a prioritized operation. Let us formalize this notion using the concept of the scope of a belief change operation.

Definition 26. *[18] Let $\star : Th(CTL) \times \mathcal{L}_{CTL} \to Th(CTL)$ be a belief change operator. We define the scope of $\star$ relative to the set K as the set*

$$Scp^\star(K) = \{\varphi \in \mathcal{L}_{CTL} \mid \varphi \in K \star \varphi\}$$

As a prioritized revision, AGM revision has the maximal possible scope.

Lemma 27. *[18] If $\star$ is an AGM revision, then for any $K \subseteq \mathcal{L}_{CTL}$, $Scp^\star(K) = \mathcal{L}_{CTL}$.*

It is easy to see that for non-AGM-like temporal preference models, a belief change operation definable as in Corollary 22 has a limited scope. This indicates that temporal preference models are a general framework for investigating prioritized and non-prioritized belief revision. In fact, we will establish the connection between temporal preference models and Booth et al.'s [4] credibility-limited revisions.

Definition 28. *[18] Let $\star : Th(CTL) \times \mathcal{L}_{CTL} \to Th(CTL)$ and $K \in Th(CTL)$, we say $\star$ is a credibility-limited revision operator on K, if it satisfies the following postulates:*

(CL1) $\varphi \in K \star \varphi$ or $K \star \varphi = K$
(CL2) *If* $K + \varphi \nvdash \bot$ *then* $K \star \varphi = K + \varphi$
(CL3) $K \star \varphi \nvdash \bot$
(CL4) *If* $\vDash \varphi \leftrightarrow \psi$ *then* $K \star \varphi = K \star \psi$
(CL5) *If* $\varphi \in K \star \varphi$ *and* $\varphi \vDash \psi$, *then* $\psi \in K \star \psi$
(CL6) $K \star (\varphi \vee \psi) = \begin{cases} K \star \varphi \text{ or} \\ K \star \psi \text{ or} \\ K \star \varphi \cap K \star \psi \end{cases}$

---

[2] Notice that there is only finitely many that we need to consider, as $\equiv$ has finite support.

To characterize credibility-limited revisions, we must guarantee that the scope of the operation satisfies the following conditions.

Definition 29. *Let* $X \subseteq \mathcal{L}_{CTL}$, *we say* X *satisfies:*

- *disjunction completeness if for any* $\varphi \vee \psi \in X$, *then* $\varphi \in X$ *or* $\psi \in X$
- *single-sentence closure if for* $\varphi \in X$ *and* $\varphi \models \psi$, *then* $\psi \in X$

We can, then, characterize credibility-limited revisions for CTL.

Proposition 30. *Let* $K \subseteq Th(CTL)$ *be a CTL theory, and* $\star : Th(CTL) \times \mathcal{L}_{CTL} \rightarrow Th(CTL)$ *a belief change operator, the following statements hold:*

1. *If* $\star$ *is a credibility-limited revision operator on* K, *then* $K \subseteq Scp^{\star}(K)$ *and* $Scp^{\star}(K)$ *satisfies single sentence-closure and disjunction completeness;*
2. *For each* $X \subseteq \mathcal{L}_{CTL}$ *s.t. any consistent extension* K' *of* K *is contained in* X, *i.e., for any* $K' \subseteq \mathcal{L}_{CTL}$ *s.t.* $\perp \notin Cn(K')$ *and* $K \subseteq K'$ *it holds that* $K' \subseteq X$, *and* X *satisfies single-sentence closure and disjunction completeness, there exists a credibility-limited revision operator s.t.* $Scp^{\star}(K) = X$

Proof. The proof of item 1 is immediate from Definition 28, with $K \subseteq Scp^{\star}(K)$ following from (**CL2**), single sentence-closure from (**CL5**) and disjunction completeness from (**CL1**), (**CL2**), and (**CL6**);

To prove of item 2, we construct the set of K-thories in X as

$$\overline{K}_X = \{\Gamma \in Th(CTL) \mid K \subseteq \Gamma \subseteq X\}.$$

Further, we define a preference relation $\leqslant_K$ on $\overline{K}_X$, as

$$\Gamma \leqslant_K \Gamma' \text{ iff } \Gamma' \subseteq \Gamma$$

For any formula $\varphi \in \mathcal{L}_{CTL}$, we define the set $[\![\varphi]\!]_X = \{\Gamma \in \overline{K}_X \mid \varphi \in \Gamma\}$. We construct an operation $\star_M : Th(CTL) \times \mathcal{L}_{CTL} \rightarrow Th(CTL)$ s.t.

$$K \star \varphi = \begin{cases} \bigcap Min_{\leqslant_K}[\![\varphi]\!]_X & \text{if } [\![\varphi]\!]_X \neq \varnothing \\ K & \text{if } [\![\varphi]\!]_X = \varnothing \end{cases}$$

The proof that this operator satisfies the (**CL1**)−(**CL6**) is similar to that presented by Booth et al. [4] for the characterization of credibility-limited revision for Classical Propositional Logic.

To prove that $Scp(\star) = X$, it suffices to see that if $\varphi \notin X$, then $[\![\varphi]\!]_X = \varnothing$. □

Now, we can define the notion of belief revision operations being induced by a temporal preference model.

Definition 31. *Let* $M = \langle W, R, \leqslant, \sim, v \rangle$ *be a temporal preference model and* $w \in W$ *be a possible world. We say a belief change operator* $\star_M$ *is a revision induced by* M *and* w *if for* $K = Bel_w(M) = \{\varphi \in \mathcal{L}_{CTL} \mid \forall w \in Min_{\leqslant}[w]_{\sim} : M, w \models \varphi\}$, *it holds that*

$$K \star_M \varphi = \{\psi \in \mathcal{L}_{CTL} \mid \forall w \in Min_{\leqslant}[\![\varphi]\!] : M, w \models \psi\}$$

Clearly, the set of valid formulas in any temporal preference model satisfies single-sentence closure and disjunction.

Lemma 32. *Let $M = \langle W, R, \leqslant, \sim, v \rangle$ be a temporal preference model and $w \in W$. $\Diamond Know_w(M) = \{\varphi \in \mathcal{L}_{CTL} \mid \exists w' \in [w]_\sim : M, w' \models \varphi\}$ satisfies single-sentence closure and disjunction completeness.*

We can, thus, show that temporal preference models induce credibility-limited revisions.

Corollary 33. *Let $M = \langle W, R, \leqslant, \sim, v \rangle$ be a temporal preference model, $w \in W$, and $\star_M$ a revision induced by M and w, then $\star_M$ is a credibility-limited operator on $Bel_w(M)$ with scope $\Diamond Know_w(M)$.*

5   Conclusions

In this work, we investigated the application of Temporal Preference Logic to study Belief Change in Computational Tree Logic. We established the connection between Temporal Preference Logic with prioritized and non-prioritized Belief Revision through change operations induced by models. We show that all temporal preference models induce a credibility-limited revision [4] and, when the model satisfies some structural conditions, such an operation is an AGM belief revision. More yet, based on the results about the characterization of finite Kripke models [6], we showed the decidability of belief change for a class of AGM Belief Revision operators for CTL.

Future work includes the study of iterated belief change operators for CTL, connecting our logic to the axiomatic characterization of iterated belief change postulates previously provided by Souza et al. [21]. Also, our results on the decidability of some AGM belief revisions for CTL indicate connections to the belief change operations defined by transformations of syntactic representations of preferences, such as graph transformations studied by Souza and Moreira [20] for Classical Propositional Logic. From Proposition 25, we can infer that Souza and Moreira's method for defining belief change operations can be extended for CTL, obtaining a class of decidable CTL belief change operations.

Acknowledgments. This study was financed in part by the Coordenação de Aperfeiçoamento de Pessoal de Nível Superior - Brasil (CAPES) - Finance Code 001.

References

1. Alchourrón, C.E., Gärdenfors, P., Makinson, D.: On the logic of theory change: Partial meet contraction and revision functions. J. Symbolic Logic **50**(2), 510–530 (1985)

2. Baltag, A., Smets, S.: A qualitative theory of dynamic interactive belief revision. In: Arló-Costa, H., Hendricks, V.F., van Benthem, J. (eds.) Readings in Formal Epistemology. SGTP, vol. 1, pp. 813–858. Springer, Cham (2016). https://doi.org/10.1007/978-3-319-20451-2_39

3. Blackburn, P., van Benthem, J.F., Wolter, F.: Handbook of modal logic, vol. 3. Elsevier, Amsterdam, NL (2006)

4. Booth, R., Fermé, E., Konieczny, S., Pérez, R.P.: Credibility-limited revision operators in propositional logic. In: Thirteenth International Conference on the Principles of Knowledge Representation and Reasoning (2012)

5. Boutilier, C.: Conditional logics for default reasoning and belief revision. University of British Columbia, Tech. rep. (1992)

6. Browne, M.C., Clarke, E.M., Grümberg, O.: Characterizing finite kripke structures in propositional temporal logic. Theor. Comput. Sci. **59**(1–2), 115–131 (1988)

7. Clarke, E.M., Emerson, E.A.: Design and synthesis of synchronization skeletons using branching time temporal logic. In: Kozen, D. (ed.) Logic of Programs 1981. LNCS, vol. 131, pp. 52–71. Springer, Heidelberg (1982). https://doi.org/10.1007/BFb0025774

8. Clarke Jr, E.M., Grumberg, O., Kroening, D., Peled, D., Veith, H.: Model checking. MIT press (2018)

9. Flouris, G., Plexousakis, D., Antoniou, G.: On applying the AGM theory to DLs and OWL. In: ISWC, pp. 216–231 (2005)

10. Girard, P., Rott, H.: Belief revision and dynamic logic. In: Baltag, A., Smets, S. (eds.) Johan van Benthem on Logic and Information Dynamics. OCL, vol. 5, pp. 203–233. Springer, Cham (2014). https://doi.org/10.1007/978-3-319-06025-5_8

11. Grove, A.: Two modelings for theory change. J. Philosophical Logic **17**(2), 157–170 (1988)

12. Guerra, P.T., Wassermann, R.: Two agm-style characterizations of model repair. Ann. Math Artif. Intell. **87**(3), 233–257 (2019)

13. Hansson, S.O.: Preference logic. In: Handbook of philosophical logic, pp. 319–393. Springer (2001). https://doi.org/10.1007/978-94-017-0456-4

14. Lindström, S., Rabinowicz, W.: DDL unlimited: dynamic doxastic logic for introspective agents. Erkenntnis **50**(2), 353–385 (1999)

15. Liu, F.: Reasoning about preference dynamics, vol. 354. Springer, New York, US (2011)

16. Ribeiro, J.S., Andrade, A.: A 3-Valued contraction model checking game: deciding on the world of partial information. In: Butler, M., Conchon, S., Zaïdi, F. (eds.) ICFEM 2015. LNCS, vol. 9407, pp. 84–99. Springer, Cham (2015). https://doi.org/10.1007/978-3-319-25423-4_6

17. Ribeiro, J.S., Nayak, A., Wassermann, R.: Towards belief contraction without compactness. In: Sixteenth International Conference on Principles of Knowledge Representation and Reasoning (2018)

18. Sauerwald, K., Kern-Isberner, G., Beierle, C.: On limited non-prioritised belief revision operators with dynamic scope. arXiv preprint arXiv:2108.07769 (2021)

19. Segerberg, K.: Two traditions in the logic of belief: bringing them together. In: Ohlbach, H.J., Reyle, U. (eds.) Logic, Language and Reasoning. TL, vol. 5, pp. 135–147. Springer, Dordrecht (1999). https://doi.org/10.1007/978-94-011-4574-9_8

20. Souza, M., Moreira, Á.: Belief base change as priority change: a study based on dynamic epistemic logic. J. Logical Algebraic Methods Programm. **122**, 100689 (2021)

21. Souza, M., Vieira, R., Moreira, Á.: Dynamic preference logic meets iterated belief change: Representation results and postulates characterization. Theor. Comput. Sci. **872**, 15–40(2020)
22. Van Benthem, J., Grossi, D., Liu, F.: Priority structures in deontic logic. Theoria **80**(2), 116–152 (2014)
23. Zhang, Y., Ding, Y.: Ctl model update for system modifications. J. Artif. Intell. Res. **31**, 113–155 (2008)

Hyperintensional Models and Belief Change

Marlo Souza[1](✉) ⓘ and Renata Wassermann[2] ⓘ

[1] Federal University of Bahia, Salvador, Brazil
`msouza1@ufba.br`
[2] University of São Paulo, São Paulo, Brazil
`renata@ime.usp.br`

Abstract. Formal frameworks for Epistemology need to have enough logical structure to enable interesting conclusions regarding epistemic phenomena and to be expressive enough to model competing positions in the philosophical and logical literature. While beliefs are commonly accepted as hyperintensional attitudes, most work on standard epistemic logic has relied on idealised and intensional agents. This is particularly true in the area of AGM-inspired Belief Change. In this work, we investigate hyperintensional belief change operations providing a semantic framework based on impossible worlds semantics to hyperintensional variants of belief change operations. In doing so, we provide the basis for deepening the connection between AGM-inspired Belief Change literature and current discussions on Formal Epistemology and Metaphysics.

Keywords: Hyperintensional logic · Impossible worlds semantics · Belief revision

1 Introduction

Belief Change is the area that studies how doxastic agents change their minds after acquiring new information. One of the most influential approaches in the literature, namely the AGM framework [2], studies rational constrains, or postulates, that characterise rational ways of changing beliefs, given a representation of one's doxastic state.

Since its seminal work, the AGM-inspired literature has traditionally relied on a highly idealised notion of belief and representation of an agent's reasoning power. For example, AGM admits as a representation of an agent's belief state a consequentially-closed set of formulas, requiring thus that an agent believes in all consequences of their beliefs. While works, such as that of Hansson [20], advocate for more realistic representations of the agent's epistemic state, these works still admit an intentional treatment of belief, namely that equivalent sentences have equivalent results in changing one's beliefs and thus that agents are logically omniscient.

It is well recognised in the literature [41], however, that beliefs and other mental attitudes are sensitive to hyperintensional distinctions. We call, after

J. C. Xavier-Junior and R. A. Rios (Eds.): BRACIS 2022, LNAI 13653, pp. 429–443, 2022.
https://doi.org/10.1007/978-3-031-21686-2_30

Cresswell [13], *hyperintensional attitudes* those which can draw distinctions between necessarily equivalent contents. For example, while the sentences "*3 is a prime number*" and "*3068 is divisible by 13*" have the same intension as mathematical necessities, they certainly cannot be transparently substituted for the other in the sentence "Alice believes that *3 is a prime number.*"

As discussed by Berto and Hawke [6], Formal Epistemology frameworks should, at the same time, be sufficiently powerful to derive proofs about a broad class of doxastic agents and epistemic phenomena but flexible enough to model different philosophical positions in the field. Aiming to obtain a compromise between the logical power of standard epistemic logic and the lack of expressiveness to encode current positions in philosophical debate [6], some recent work on hyperintensional belief change has risen in the literature to deal with this limitation.

Most prominent, Berto [7] proposes a hyperintensional logic of conditional beliefs and investigates hyperintensional belief revision operations interpreted as conditional beliefs. On the other hand, Souza [32] and Souza and Wassermann [34,35] investigate hyperintensional belief change operations using tools similar to that of the AGM framework, based on abstract logics. These authors propose connections between hyperintensional belief change and belief change in non-classical logics.

The main drawback of the work of Souza and Wassermann [34,35] is that hyperintensional belief change is investigated through the framework of abstract logic. While the use of abstract logic to encode reasoning processes allows a clear connection between their results and those in AGM-inspired literature, it obscures their framework's connections to those proposed for hyperintensional logics - usually reliant on model-theoretic approaches. Moreso, as we shall provide evidence in this work, it obscures the connection between different notions of belief change and thus, we argue, the true nature of *rational* belief change encoded in the area.

In this work, we extend Souza and Wassermann's [35] investigation on hyperintensional belief change by proposing a semantic analysis of different hyperintensional operations in the literature and unifying these notions in a single semantic framework. We employ an impossible worlds semantics, a framework with deep connections to the study of hyperintensional phenomena [12,22,23,29], to reason about hyperintensional differences between sentences and show that, depending on the properties of the logic, both partial meet contractions and AGM contractions coincide with our proposed notion of selection operator. Further, our results point to important connections between results in Abstract Model Theory [26] and the definability of belief change operations for certain logics.

This work is structured as follows: in Sect. 2 we discuss some of the related literature focusing on hyperintensional phenomena in belief change; in Sect. 3, we present the basic concepts and notations employed in this work; in Sect. 4, we study Hyperintensional Belief Contractions, as studied by Souza and Wassermann [34,35], extending their operations to define and characterise a notion of hyperintensional AGM contraction; Sect. 5 discusses our semantic framework and its application to study intensional and hyperintensional belief change operations. In that section, we provide connections between our semantic (hyperintensional) contraction operations with both Hansson's and Souza and Wasserman's partial meet contractions and AGM's contractions, showing that, despite earlier negative results, these notions share a similar underlying structure. Finally, in Sect. 6, we present our final considerations and future directions of our work.

2   Related Work

Work on hyperintensional phenomena in representations of beliefs and other mental attitudes has a long standing tradition on epistemic logic at least since the work of Cresswell [12–14], c.f. also [15,29,38,39].

Regarding hyperintensional phenomena in Belief Change, work has mainly focused on the representation of explicit doxastic commitments of an agent and syntactic representations of their belief state [1,20,31,40].

On the other hand, work on belief change for non-classical logics, such as that of Hansson and Wassermann [21], Flouris [16], Tennant [37], or Girard and Tanaka [19] give clues of how hyperintensional phenomena can interact with Belief Change and what restrictions they impose on the results and tools of the area. Nevertheless, these works do not propose an explicit investigation of the connection of hyperintensional modelling of belief and their effect on the definability of belief change operations. Thus they do not provide theoretical and philosophical connections that allow us to understand their underlying commitments.

Work on genuinely hyperintensional models for belief change, i.e. explicitly considering hyperintensional differences between formulas both in the agent's beliefs and in the input, is far more recent in the literature. To our knowledge, Berto [7] was the first to propose a hyperintensional notion of belief change, applying his mereological theory of propositional contents to study conditional beliefs, understood as belief change in his work. That work was extended by Özgün and Berto [41], who propose a dynamic logic of hyperintensional belief change connected to the tradition of Dynamic Epistemic Logic for Formal Epistemology. Unlike their work, however, ours investigates how a general notion of hyperintensional belief change can be defined, based on the AGM approach, that can be connected to different semantic frameworks for hyperintensional reasoning.

Similarly, Bozdag [11] proposes a hyperintensional doxastic logic, based on the HYPE framework [24], in which belief base revision can also be thought of as a form of conditionalisation. As before, it is not completely clear how we can compare her proposal with competing notions of belief change in the AGM-inspired literature since, as observed by Lindström and Rabinowicz [25], Baltag and Smets [4], or even Souza et al. [33], modal-based semantic approaches to AGM Belief Change often surpass the expressive power of the original frameworks and, thus, postulates need to be generalised for these settings.

Our work follows the line delineated in [32] and [34], proposing a semantic framework with which we can understand, define and study the connection between different notions of hyperintensional belief change.

3   Preliminaries

In this work, we employ the tools from Abstract Logic and Model Theory to study classes of belief change operations and their definability in non-classical logics.

We will call a logic any pair $\mathcal{L} = \langle L, Cn \rangle$, where L is a non-empty set, called the logical language, and $Cn : 2^L \to 2^L$ is a function called a consequence operator satisfying the following properties[1]

- **inclusion:** $\Gamma \subseteq Cn(\Gamma)$.
- **idempotence:** $Cn(\Gamma) = Cn(Cn(\Gamma))$.
- **monotonicity:** If $\Gamma \subseteq \Gamma'$ then $Cn(\Gamma) \subseteq Cn(\Gamma')$.

Aside from the three basic Tarskian properties above, a consequence relation may satisfy some important properties present in some logics of interest for Philosophy and Artificial Intelligence.

- **compactness:** for any $\varphi \in Cn(\Gamma)$, there is some finite $\Gamma' \subseteq \Gamma$ s.t. $\varphi \in Cn(\Gamma')$.
- **distributivity:** for any $\Gamma \subseteq L$ and any finitely representable[2] $\Gamma', \Gamma'' \subseteq L$, it holds that $Cn(\Gamma \cup (Cn(\Gamma') \cap Cn(\Gamma''))) = Cn(\Gamma \cup \Gamma') \cap Cn(\Gamma \cup \Gamma'')$.
- **closure under negation:** for any $\Gamma \subseteq L$ finitely representable, there is $\Gamma' \subseteq L$ finitely representable s.t. $Cn(\Gamma \cup \Gamma') = L$ and $Cn(\Gamma \cap \Gamma') = Cn(\varnothing)$.
- **booleanicity:** Cn is tarskian, distributive and closed under negation.

Given a logic $\mathcal{L}$ as above, we call a *belief change operation* and function $\star : 2^L \times L \to 2^L$, which maps pairs of sets of sentences and a sentence, called a set of beliefs and a piece of input information, to a set of sentences, the resulting beliefs.

AGM investigate three basic belief change operations: expansions, contractions and revisions. Belief expansion blindly integrates a new piece of information into the agent's beliefs. Belief contraction removes a currently held belief from the agent's set of beliefs, with minimal alterations. Finally, belief revision is the operation of integrating new information into an agent's beliefs while maintaining consistency.

---

[1] Notice that, differently than Souza [32] or Souza and Wassermann [34,35], we require our logics to be Tarskian. The reason for this is to make clearer the presentation of the key aspects of our semantic framework. Our results could be reproduced without requiring tarskianicity - although with a rather more involved way, since the relationship between intensions and inferences becomes less well-behaved.

[2] We say a set $\Gamma \subseteq L$ is finitely representable if there is some finite $X \subseteq L$ s.t. $Cn(X) = Cn(\Gamma)$.

Among these basic operations, only expansion can be univocally defined. The other two are defined by a set of rational constraints or postulates, usually referred to as the AGM postulates. These postulates define a class of suitable change operators representing different rational ways in which an agent can change their beliefs. Given a closed set of beliefs[3], i.e. $K \subseteq L$ s.t. $K = Cn(K)$, we say a belief change operation $\dot{-}$ is an AGM contraction on K if for any $\varphi, \psi \in L$, it satisfies:

(closure) $K\dot{-}\varphi = Cn(K\dot{-}\varphi)$
(success) If $\varphi \notin Cn(\varnothing)$ then $\varphi \notin K\dot{-}\varphi$
(inclusion) $K\dot{-}\varphi \subseteq K$
(vacuity) If $\varphi \notin K$ then $K\dot{-}\varphi = K$
(recovery) $K \subseteq Cn(K\dot{-}\varphi \cup \{\varphi\})$
(extensionality) If $Cn(\varphi) = Cn(\psi)$ then $K\dot{-}\varphi = K\dot{-}\psi$

To characterise their rational contractions, AGM propose the notion of partial meet belief contraction, an operation that preserves a maximal amount of "safe" information from the agent's beliefs, i.e., information that cannot be used to derive what the agent has ceased to believe. To formalise this notion, Alchourrón and Makinson [3] propose the notion of remainder set.

Definition 1. *Let $B \subseteq L$ be a set of formulas and $\varphi \in L$ be a formula of L, the remainder set $B\perp_{\mathcal{L}}\varphi$ is the set of sets B' satisfying:*

- $B' \subseteq B$
- $\varphi \notin Cn(B')$
- $B' \subset B'' \subseteq B$ *implies* $\varphi \in Cn(B'')$.

When it is clear to which logic $\mathcal{L}$ we are referring, we will denote $B\perp_{\mathcal{L}}\varphi$ by $B\perp\varphi$.

A partial meet contraction $\dot{-}$ is an operation for which there is a selection function γ, that characterises this operation. By selection function, we mean that the function γ satisfies (i) $\varnothing \neq \gamma(B\perp\varphi) \subseteq B\perp\varphi$ if $B\perp\varphi \neq \varnothing$ and (ii) $\gamma(B\perp\varphi) = \{B\}$ otherwise.

Definition 2. *We say a belief base change operator $\dot{-}$ is a belief base contraction on a set $B \subseteq L$ if there is a selection function γ, s.t. for any φ*

$$B\dot{-}\varphi = \gamma(B\perp\varphi).$$

The authors show that for any boolean and compact logic, an AGM contraction on a closed set K is a partial meet contraction on K and vice versa. Further, Hansson and Wassermann [21] show that for any monotonic and compact logic, an operation $\dot{-}$ is a partial meet contraction on a set B of beliefs if and only if it satisfies the following postulates:

(**success**) If $\varphi \notin Cn(\varnothing)$, then $\varphi \notin Cn(B\dot{-}\varphi)$

---

[3] In the following we will often refer to closed sets by the letter K (K',K'', etc.), arbitrary sets of formulas by the letter B (B',B'', etc.).

(**inclusion**) $B \dot{-} \varphi \subseteq B$

(**uniformity**) If for any $B' \subseteq B$ it holds that $\varphi \in Cn(B')$ iff $\psi \in Cn(B')$, then it holds that $B \dot{-} \varphi = B \dot{-} \psi$

(**relevance**) If $\psi \in B \backslash B \dot{-} \varphi$, then there is some $B' \subseteq B$ s.t. $B \dot{-} \varphi \subseteq B'$, $\varphi \notin Cn(B')$, and $\varphi \in Cn(B' \cup \{\psi\})$

On the other hand, Flouris [16] studied the definability of AGM contraction operations, i.e. operations satisfying the original AGM postulates, in Tarskian logics, obtaining sufficient and necessary conditions for such definability. Let us introduce these notions in order to compare our hyperintensional belief change operations and AGM contractions.

Definition 3. *Let $\mathcal{L} = \langle L, Cn \rangle$ be a logic, a set $B \subseteq L$ is said to be decomposable in $\mathcal{L}$, if for any $\varphi \in L$, with $Cn(\varnothing) \subset Cn(\varphi) \subseteq Cn(B)$, the set*

$$\varphi^-(B) = \{B' \subseteq B \mid \varphi \notin Cn(B') \wedge Cn(B) = Cn(B' \cup \{\varphi\})\}$$

is not empty. A logic is said to be decomposable if every $B \subseteq L$ is decomposable.

Flouris [16] show that decomposability is a necessary and sufficient condition for the definability of AGM contraction operations.

Proposition 4 (Adapted from [16]). *Let $\mathcal{L} = \langle L, Cn \rangle$ be a logic, $K \subseteq L$ be a closed set of formulas, which is decomposable in $\mathcal{L}$, and $\dot{-}$ be a belief change operation, then $\dot{-}$ is an AGM contraction on K iff for any $\varphi \in L$ it holds that (i) $K \dot{-} \varphi = Cn(K')$ for some $K' \in \varphi^-(K)$, if $Cn(\varnothing) \subset Cn(\varphi) \subseteq Cn(K)$, and (ii) $K \dot{-} \varphi = K$, otherwise.*

4   Hyperintensional Belief Contractions

The term hyperintensionality describes phenomena in which it is possible to draw distinctions between necessarily equivalent formulas - or those having the same intension. As such, hyperintensionality is commonly explained through the relation between the contents of a sentence and its intension with respect to a standard semantics. In the remainder of this work, we will represent hyperintensional reasoning by means of the relationship between two consequence operators over a given language. Let us define this notion formally.

Definition 5. *We call a sound hyperintensional logic a tuple $\mathcal{L} = \langle L, Cn, C \rangle$, where L is a logical language and $Cn, C : 2^L \to 2^L$ are logical consequence operators s.t. for any $\Gamma \subseteq 2^L$, it holds that $C(\Gamma) \subseteq Cn(\Gamma)$. We say that Cn is the intensional consequence of $\mathcal{L}$ and that C is the hyperintensional consequence of $\mathcal{L}$.*

In the following, we will often refer to a sound hyperintensional logic simply as a logic unless such terminological abuse may induce confusion. Similarly, we will often omit the definition of the tuple $\mathcal{L} = \langle L, Cn, C \rangle$ and will always refer to Cn and C as the consequences of some abstract logic $\mathcal{L}$, unless when necessary to explicit the definition of the logic. We say that a logic $\mathcal{L}$ satisfies some property, e.g. compactness, if Cn does, and that $\mathcal{L}$ hyperintensionally satisfies some property if C does.

We begin our exposition by studying Souza and Wassermann's [35] notion of hyperintensional partial meet operations. These authors propose a generalisation of Hansson's notion of partial meet belief contraction to hyperintensional logics, showing that we can use a more structured foundational logic to construct hyperintensional belief change operations for its sub-logics. To do that, they propose generalising the notion of remainder set as follows.

Definition 6 [35]. *Let $\mathcal{L}$ be compact logic, $B \subseteq L$ be a set of formulas and $\varphi \in L$ a formula of L. The hyperintensional remainder set of B by φ is the set:*

$$B \perp^C \varphi = \{B' \subseteq B | \varphi \notin C(B') \text{ and } \exists B'' \in B \perp \varphi \text{ s.t. } B'' \subseteq B'\}$$

The hyperintensional remainder set of B by φ, relative to C, contains all parts of B that do not imply φ, while maintaining a maximal amount of "safe" information in B. With this notion, the authors define their hyperintensional partial meet belief contractions.

Definition 7 [35]. *Let $\mathcal{L}$ be a compact logic, and $B \subseteq L$ be a set of formulas. We say a belief base change operator $\dot{-} : 2^L \times L \rightarrow 2^L$ is a hyperintensional belief contraction on B iff there is a selection function γ s.t. for any $\varphi \in L$:*

$$B \dot{-} \varphi = \gamma(B \perp^C \varphi).$$

To characterise this operation Souza and Wassermann [35] propose the following postulates, based on Hasson's [20] postulates for Belief Base Contraction.

(inclusion) $B \dot{-} \varphi \subseteq B$
(C-success) If $\varphi \notin C(\varnothing)$, then $\varphi \notin C(B \dot{-} \varphi)$
(hyperintensional uniformity) If for any $B', B'' \subseteq B$ it holds that
   1. $\varphi \in Cn(B')$ iff $\psi \in Cn(B')$
   2. $\varphi \notin Cn(B')$ and $\varphi \in C(B' \cup B'')$ implies that $\psi \in C(B' \cup B'')$
   3. $\psi \notin Cn(B')$ and $\psi \in C(B' \cup B'')$ implies that $\varphi \in C(B' \cup B'')$
   then $B \dot{-} \varphi = B \dot{-} \psi$
(hyperintensional relevance) If $\psi \in B \backslash B \dot{-} \varphi$, there is some $B' \subseteq B$ s.t. $B \dot{-} \varphi \subseteq B'$, $\varphi \notin C(B')$ but $\psi \notin B'$ and $\varphi \in Cn(B' \cup \{\psi\})$. Furthermore, there is some $B'' \subseteq B'$ s.t. $\varphi \notin Cn(B'')$ but $\varphi \in Cn(B'' \cup \{\xi\})$ for any $\xi \in B \backslash B''$.

With these postulates, the authors prove the characterisation of hyperintensional partial meet belief contraction for any monotonic hyperintensional consequence C.

Theorem 8 [35]. *Let $\mathcal{L}$ be a compact logic, and $B \subseteq L$ be a set of formulas. An operator $\dot{-}$ is a hyperintensional belief contraction on B iff $\dot{-}$ satisfies (inclusion), (C-success), (hyperintensional uniformity) and (hyperintensional relevance).*

In a similar fashion, we can extend Souza and Wassermann's [35] notion of hyperintensional contractions for other types of contraction operations, not based on partial meet contraction. Following the same idea as those authors, we propose an extension of AGM's notion of rational contraction to a hyperintensional setting by providing adequate generalisations of AGM's postulates.

Definition 9. *Let $\mathcal{L}$ be a logic and $B \subseteq L$ be a set of formulas. We say a belief change operator $\dot{-} : 2^L \times L \to 2^L$ is a hyperintensional AGM contraction if it satisfies the following postulates.*

(hyperintensional closure) $B \dot{-} \varphi = C(B \dot{-} \varphi)$
(inclusion) $B \dot{-} \varphi \subseteq B$
(C-success) If $\varphi \notin C(\varnothing)$, then $\varphi \notin C(B \dot{-} \varphi)$
(vacuity) If $\varphi \notin C(B)$, then $B \subseteq B \dot{-} \varphi$
(hyperintensional extension) If $Cn(\varphi) = Cn(\psi)$ and for any $B' \subseteq B$ it holds that $\varphi \notin C(B')$ and $\varphi \in Cn(B')$ iff $\psi \notin C(B')$ and $\psi \in Cn(B')$, then $B \dot{-} \varphi = B \dot{-} \psi$
(recovery) $Cn(B) = Cn(B \dot{-} \varphi \cup \{\varphi\})$.

The postulate of *(hyperintensional extension)*, similar to Souza and Wassermann's *(hyperintensional uniformity)* with Hansson's *(uniformity)* postulate, extends AGM's extensionality requiring that the contraction operation coincides for any two formulas when they behave similarly with respect to the logic Cn, but also to C in the subsets of the set of beliefs B.

Definition 10. *Let $B \subseteq L$ be a set of logical formulas and $\varphi \in L$ a logical formula. We define the hyperintensional set of complements of B with respect to φ, the set*

$$\varphi_C^-(B) = \{B' \subseteq B \mid \varphi \notin C(B') \text{ and } \exists B'' \in \varphi^-(B) : B'' \subseteq B'\}$$

Following Flouris' [16] characterisation of AGM base contractions, we can fully characterise hyperintensional AGM contractions on decomposable logics.

Theorem 11. *Let $\mathcal{L}$ be a logic, and $K \subseteq L$ a hyperintensionally closed set of formulas, i.e. $K = C(K)$, with some hyperintensionally closed subset $K' \subseteq K$ decomposable in $\mathcal{L}$. A belief change operation $\dot{-}$ is a hyperintensional AGM contraction on K iff for any $\varphi \in L$ it holds that (i) if $\varphi \in Cn(K) \backslash Cn(\varnothing)$, then $K \dot{-} \varphi = C(K')$ for some $K' \in \varphi_C^-(K)$, (ii) if $K \in \varphi_C^-(K)$, then $K \dot{-} \varphi = K$, and (iii) $K \dot{-} \varphi = K$, otherwise.*

5   A Model of Hyperintensional Belief Change

We turn our attention to the search for a proper semantic characterisation of hyperintensional belief change operations. As stated before, hyperintensional logics have, traditionally, a deep connection to model-theoretic approaches, as the crucial question any framework for hyperintensionality must answer is the nature of propositional contents and their relation to meaning. Providing a semantic interpretation of hyperintensional belief change allows us to establish a bridge between the questions and results of both areas.

Let us first introduce the basic framework and connect it to hyperintensional logics. In this work, we focus on an impossible world semantics for propositions, as this is a rich framework with a vast philosophical tradition [12, 14, 22, 23, 27], although their adoption is not without controversy [8].

Definition 12. *Let L be a logical language, we call an impossible worlds model (IWM) on L any tuple $M = \langle W, N, v \rangle$, where*

- *W is a non-empty set of possible worlds;*
- *$N \subseteq W$ is a set of normal worlds;*
- *$v : L \to 2^W$ is a valuation function.*

Any impossible worlds model M induces intensional and hyperintensional consequence operators associated with the valuations at the normal and non-normal worlds of M. We can, thus, construct such operators by examining the interpretations of formulas and sets of formulas in a given IWM. Let us define these notions formally.

Definition 13. *Let L be a logical language and $M = \langle W, N, v \rangle$ be an IWM on L. For any $\varphi \in L$ and $\Gamma \subseteq L$, we define:*

$$[\![\varphi]\!]_N = \{w \in N \mid w \in v(\varphi)\}$$
$$[\![\Gamma]\!]_N = \bigcap_{\varphi \in \Gamma} [\![\varphi]\!]_N$$
$$[\![\varphi]\!] = \{w \in W \mid w \in v(\varphi)\}$$
$$[\![\Gamma]\!] = \bigcap_{\varphi \in \Gamma} [\![\varphi]\!]$$

Further, let $X \subseteq W$ be a set of possible worlds, we define:

$$Th(X) = \{\varphi \in L \mid \forall w \in X : w \in v(\varphi)\}$$

With that, it is easy to construct the logic induced by an impossible worlds model M.

Definition 14. *Let L be a logical language and $M = \langle W, N, v \rangle$ be an IWM on L. We define the logic induced by M, the hyperintensional logic $\mathcal{L}_M = \langle L, Cn, C \rangle$ s.t. for any $\Gamma \subseteq L$: $Cn(\Gamma) = Th([\![\Gamma]\!]_N)$ and $C(\Gamma) = Th([\![\Gamma]\!])$.*

More yet, it is easy to see that any sound hyperintensional logic is induced by some impossible worlds model.

Lemma 15. *Let L be a logical language and $Cn, C : 2^L \to 2^L$ be tasrkian consequence operators. $\mathcal{L} = \langle L, Cn, C \rangle$ is a sound hyperintensional logic iff there is an IWM on L, $M = \langle W, N, v \rangle$ s.t. $\mathcal{L}_M = \mathcal{L}$.*

To prove Lemma 15, it suffices to take the set of hyperintensional theories of $\mathcal{L}$, with N the set of intensional theories of $\mathcal{L}$, as possible worlds and the trivial valuation $v(\varphi) = \{\Gamma \mid \varphi \in \Gamma\}$. It is trivial to show that the logic induced by the model coincides with $\mathcal{L}$.

As we will use impossible worlds models as a foundational framework to define our hyperintensional logics, we expand our models to include the elements we will employ to encode belief change operations.

The study of belief change has a deep connection and cross-contribution with the study of conditional beliefs, counterfactual conditionals and non-monotonic reasoning [2,10,17,18,36]. In Conditional Logic [28], we can obtain modal conditional structures to interpret conditional implications of the form $\varphi \Rightarrow \psi$ by augmenting a modal structure with a selection function $f : L \to 2^W$, which can be used to define a notion of minimality according to the desired conditional interpretation, e.g. 'most typical worlds' when modelling conditional beliefs. Similarly, we will augment our impossible worlds models with a generalisation of such functions, and eliminating the function's dependence on the language's syntax.

Definition 16. *Let L be a logical language, we call selection impossible worlds model (SIWM) on L any tuple $M = \langle W, N, f, v \rangle$, where*

- *$\langle W, N, v \rangle$ is an IWM;*
- *$f : 2^W \times 2^W \to 2^W$ is a selection function on possible worlds, i.e. a function satisfying the following conditions for all $X, Y, Z \subseteq W$:*
  1. *$X \subseteq f(X, Y) \subseteq X \cup Y$*
  2. *If $Y \neq \varnothing$, then $f(X, Y) \cap Y \neq \varnothing$*
  3. *If $f(X, Y) \cap Z \neq \varnothing$ and $f(X, Z) \cap Y \neq \varnothing$, then $f(X, Y) = f(X, Z)$*

For any sound hyperintensional logic $\mathcal{L} = \langle L, Cn, C \rangle$, we say M is a SIWM for $\mathcal{L}$ if $\mathcal{L}$ is induced by $\langle W, N, v \rangle$ - in that case we also say $\mathcal{L}$ is a logic induced by M.

Conditions 1–3 on the selection function f, in Definition 16, are standard restrictions to ensure the appropriate behaviour of f. Condition 1 is similar to the postulate of *(inclusion)*, and states that the selection on the set Y, based on the set X of possible worlds, denoted by $f(X, Y)$, must contain all X-worlds and, possibly, some Y worlds, and none more. Condition 2 ensures that if Y is not empty, some Y-worlds will be selected. Finally, Condition 3 states that there are Z-worlds among the selected Y-worlds, based on X, and vice-versa, it must be the case that they coincide. This condition encodes the notion of 'minimality' imbued in conditional logics *a la* Stalnaker [36]. We can, thus, define our semantic contraction operations based on the interpretation of conditionals in such models, as usual.

Definition 17. *Let $\dot{-} : 2^L \times L \to 2^L$ a belief change operator on a logic $\mathcal{L}$, and $B \subseteq L$ be a set of formulas. We say $\dot{-}$ is a normal selection contraction operator on $\mathcal{L}$ iff there is some SIWM $M = \langle W, N, f, v \rangle$ s.t. for any $\varphi \in L$, it holds that*

$$B \dot{-} \varphi = (Th(\llbracket B \rrbracket_N \cup (N \backslash \llbracket \varphi \rrbracket))) \cap B.$$

Similarly, we say $\dot{-}$ is a non-normal selection contraction operator on B iff there is some SIWM $M = \langle W, N, f, v \rangle$ s.t. for any $\varphi \in L$, it holds that

$$B \dot{-} \varphi = (Th(\llbracket B \rrbracket \cup (W \backslash \llbracket \varphi \rrbracket))) \cap B.$$

To establish the connection between partial meet contractions and selection contractions, let us see that remainder sets can be defined through our semantics.

Proposition 18. *Let $\mathcal{L}$ be a logic and $M = \langle W, N, v \rangle$ be an IWM on L s.t. $\mathcal{L}$ is induced by M. Let yet $B \subseteq L$ be a set of formulas s.t. $B \bot \varphi \neq \varnothing$ and $\varphi \in L$ be a formula of L, then the following hold:*

(i) If $B' \in B \bot \varphi$, then there is some $w \in N \backslash \llbracket \varphi \rrbracket$ s.t. $B' = Th(\llbracket B \rrbracket_N \cup \{w\}) \cap B$.
(ii) If $B' \in B \bot^C \varphi$, then there is some $X \subseteq W \backslash \llbracket \varphi \rrbracket$ s.t. $B' \subseteq Th(\llbracket B \rrbracket \cup X) \cap B$.

Proof (Sketch of the proof). To prove (i), take $w \in \llbracket B' \rrbracket_N \backslash \llbracket B \rrbracket_N$, which must exist if $\varphi \notin Cn(\varnothing)$ (otherwise $B \bot \varphi = \varnothing$). Then $B' \subseteq B'' = Th(\llbracket B \rrbracket_N \cup \{w\}) \cap B$ and $\varphi \notin Cn(B'')$. By maximality of B', then $B' = B''$.

To prove (ii), take $X = \llbracket B' \rrbracket \backslash \llbracket B \rrbracket$, which must not be empty if $\varphi \notin C(\varnothing)$ (otherwise $B \bot^C \varphi = \varnothing$). Then $B' \subseteq C(B') \cap B = Th(\llbracket B' \rrbracket) \cap B = Th(\llbracket B \rrbracket \cup X) \cap B$.

From Proposition 18, it is easy to see that any (hyperintensional) partial meet contraction on a set B is also a (non-)normal selection contraction on B.

Corollary 19. *Let $\mathcal{L}$ be compact logic, $B \subseteq 2^L$ be set of formulas, $\varphi \in L$ be a logical formula, and $\dot{-}$ be a belief change operator. The following hold:*

- *if $\dot{-}$ is a partial meet contraction on B then it is a normal selection contraction operator on B.*
- *if $\dot{-}$ is a hyperintensional partial meet contraction on B then there is a non-normal selection contraction operator $-$ on B, s.t. for any $\varphi \in L$, it holds that $B \dot{-} \varphi \subseteq B - \varphi$.*

Notice that Corollary 19 only establishes a one-way connection between partial meet contractions and selection contractions. In fact, we will see that selection contractions are more general than partial meet and can be used to unify different competing notions of 'minimality' or 'rationality of choice' in the area in a single framework with roots in similar ideas in Conditional Logic and Nonmonotonic reasoning. To show this, let us examine how our hyperintensional AGM contractions can be interpreted in this framework.

Proposition 20. *Let $\mathcal{L}$ be a logic and $M = \langle W, N, v \rangle$ be an IWM on L s.t. $\mathcal{L}$ is induced by M. Let yet $K \subseteq L$ be a set of formulas and $\varphi \in L$ be a formula of L s.t. $\varphi \in Cn(K) \backslash Cn(\varnothing)$ and $\varphi^-(K) \neq \varnothing$, then the following hold:*

(i) for any $K' \in \varphi^-(K)$ s.t. $Cn(K') = K'$, there is some $X \subseteq N \backslash \llbracket \varphi \rrbracket$ s.t. $K' = Th(\llbracket K \rrbracket \cup X)$;
(ii) for any $K' \in \varphi_C^-(K)$ s.t. $C(K') = K'$, there is some $X \subseteq W \backslash \llbracket \varphi \rrbracket$ s.t. $K' \subseteq Th(\llbracket K \rrbracket \cup X)$

Proof (Sketch of the proof). To prove (i), take $X = [\![K']\!]_N \backslash [\![K]\!]$, which must be not be empty since $\varphi \notin Cn(K')$. Since $K' \in \varphi^-(K)$, then $[\![K']\!]_N = [\![K]\!]_N \cap [\![\varphi]\!]_N$, then $[\![K']\!]_N = [\![K]\!]_N \cup X$. As $K' = Cn(K')$, it must hold that $K' = Th([\![K']\!]_N) = Th([\![K]\!]_N \cup X)$.

To prove (ii), take $X = [\![K']\!] \backslash [\![\varphi]\!]$, which must be not be empty since $\varphi \notin K'$. Well, $[\![K']\!] = [\![K]\!] \cup ([\![K']\!] \backslash [\![K]\!]) \supseteq [\![K]\!] \cup ([\![K']\!] \backslash [\![\varphi]\!]) = [\![K]\!] \cup X$. Then $K' \subseteq Th([\![K]\!] \cup X) = Th([\![K']\!])$.

With that, it is easy to see that any (hyperintensional) AGM contraction on a closed set K can be obtained through a (non-)normal selection contraction for K decomposable.

Corollary 21. *Let $\mathcal{L}$ logic, $K \subseteq L$ be a set of formulas s.t. $\mathcal{L}$ is decomposable on K. The following hold:*

- *If the operator $\dot{-}$ is an AGM contraction on K iff it is a normal selection contraction operator on K.*
- *The operator $\dot{-}$ is an hyperintensional AGM contraction on K iff there is a non-normal selection contraction operator $-$ on K s.t. for any $\varphi \in L$ it holds that $K \dot{-} \varphi \subseteq K - \varphi$.*

As with Corollary 19, we can see that the connection between hyperintensional contractions and non-normal selection operators is not straightforward, meaning that non-normal selection operators provide possible upper bounds for the result of a contraction. The reason for this is that hyperintensional contractions, as studied by [32,34], are syntactic in nature, while selection operators are semantic - in the sense that they select hyperintensional theories, not formulas. Notice that this is a feature of our modelling, as the notion of limited reasoning in our approach is already encoded within the hyperintensional consequence operator of the logic.

It has been noted in the literature that, while in some logics the operations of AGM contraction and partial meet contraction coincide [2,30], they may differ in others, even when both are definable [30,35]. In fact, the following example shows two logics in which not all AGM contractions are partial meet or not all partial meet contractions are AGM.

Example 22. *Consider the logics over the logical language $L = \{a, b, c\}$ in Fig. 1. In Fig. 1a, there is an AGM contraction for $K = \{a, b, c\}$, with $K \dot{-} b = \{a\}$ which is not partial meet. In Fig. 1b, on the other hand, there is a partial meet contraction for K, with $K \dot{-} b = \{c\}$, which is not AGM. Regardless, these operations are all normal selection contractions - it suffices to see that these lattices can be converted into selection models (excluding the $\varnothing$ node).*

Example 22 shows that our notion of selection contraction is more general than both AGM contraction and partial meet contraction while still maintaining cognitive plausibility for encoding a notion of *'minimality of change'* or *'rationality'*, which can be easily connected to previous work on both conditional and non-monotonic reasoning, and hyperintensional logics.

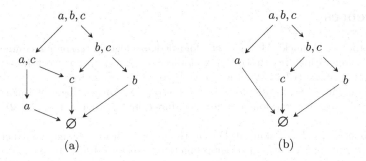

Fig. 1. Logics in which AGM and Partial meet contractions are definable and may not coincide.

Corollary 23. *There is a logic $\mathcal{L}$ s.t. there is a (non-)normal selection contraction which is not a (hyperintensional) partial meet contraction or an (hyperintensional) AGM contraction.*

6 Conclusions

In this work, we propose hyperintensional belief contraction operations based on a generalisation of previous operations in the literature and provide a semantic framework to reason about these operations. In doing so, we provide the basis for deepening the connection between AGM-inspired Belief Change literature and current discussions on Formal Epistemology and Metaphysics [5,6,8,9,12, 22,23,29]. More yet, we show that our belief change operators are more general than those studies in the literature, be them traditional intensional ones, as AGM's and Hansson's, or hyperintensional ones, as Berto's [7] or Souza and Wassermann's [34,35], while still maintaining cognitive plausibility.

By showing that different notions of minimality arise as the reflection of topological properties of the model space, encoded in the logical consequences, our results point to an interesting connection between Belief Change and Abstract Model Theory [26]. It particularly points to possible general answers to definability (and construction) of belief change operations in non-classical and infinitary logics, based on topological and categorical properties of these model spaces and of selection functions over them.

Acknowledgments. This study was financed in part by the Coordenação de Aperfeiçoamento de Pessoal de Nível Superior - Brasil (CAPES) - Finance Code 001 and in part by the Center for Artificial Intelligence (C4AI-USP), with support by the São Paulo Research Foundation (FAPESP grant #2019/07665-4) and by the IBM Corporation.

References

1. Ågotnes, T., Walicki, M.: Syntactic knowledge: a logic of reasoning, communication and cooperation. In: Proceedings of the Second European Workshop on Multi-Agent Systems (EUMAS), Barcelona, Spain. Citeseer (2004)
2. Alchourrón, C.E., Gärdenfors, P., Makinson, D.: On the logic of theory change: partial meet contraction and revision functions. J. Symb. Log. **50**(2), 510–530 (1985)
3. Alchourrón, C.E., Makinson, D.: On the logic of theory change: contraction functions and their associated revision functions. Theoria **48**(1), 14–37 (1982)
4. Baltag, A., Smets, S.: A qualitative theory of dynamic interactive belief revision. Texts Logic Games **3**, 9–58 (2008)
5. Berto, F.: Impossible worlds and the logic of imagination. Erkenntnis **82**(6), 1277–1297 (2017)
6. Berto, F., Hawke, P.: Knowability relative to information. Mind **130**(517), 1–33 (2021)
7. Berto, F.: Simple hyperintensional belief revision. Erkenntnis **84**(3), 559–575 (2019)
8. Bjerring, J.C.: Impossible worlds and logical omniscience: an impossibility result. Synthese **190**(13), 2505–2524 (2013)
9. Bjerring, J.C., Skipper, M.: A dynamic solution to the problem of logical omniscience. J. Philos. Log. **48**(3), 501–521 (2019)
10. Boutilier, C.: Conditional logics for default reasoning and belief revision. Technical report, University of British Columbia (1992)
11. Bozdag, S.: A semantics for hyperintensional belief revision based on information bases. Studia Logica 1–38 (2021)
12. Cresswell, M.J.: Intensional logics and logical truth. J. Philos. Log. **1**(1), 2–15 (1972)
13. Cresswell, M.J.: Hyperintensional logic. Studia Logica: Int. J. Symb. Logic **34**(1), 25–38 (1975)
14. Cresswell, M.J.: Classical intensional logics. Theoria **36**(3), 347–372 (1970)
15. Fagin, R., Halpern, J.Y.: Belief, awareness, and limited reasoning. Artif. Intell. **34**(1), 39–76 (1987)
16. Flouris, G.: On belief change and ontology evolution. Ph.D. thesis, University of Crete (2006)
17. Gärdenfors, P.: Belief revisions and the Ramsey test for conditionals. Philos. Rev. **95**(1), 81–93 (1986)
18. Gärdenfors, P.: Belief revision and nonmonotonic logic: two sides of the same coin? In: van Eijck, J. (ed.) JELIA 1990. LNCS, vol. 478, pp. 52–54. Springer, Heidelberg (1991). https://doi.org/10.1007/BFb0018432
19. Girard, P., Tanaka, K.: Paraconsistent dynamics. Synthese **193**(1), 1–14 (2016)
20. Hansson, S.O.: In defense of base contraction. Synthese **91**(3), 239–245 (1992)
21. Hansson, S.O., Wassermann, R.: Local change. Stud. Logica **70**(1), 49–76 (2002)
22. Jago, M.: The Impossible: An Essay on Hyperintensionality. OUP, Oxford (2014)
23. Jago, M.: Hyperintensional propositions. Synthese **192**(3), 585–601 (2015)
24. Leitgeb, H.: HYPE: a system of hyperintensional logic (with an application to semantic paradoxes). J. Philos. Log. **48**(2), 305–405 (2019)
25. Lindström, S., Rabinowicz, W.: DDL unlimited: dynamic doxastic logic for introspective agents. Erkenntnis **50**(2), 353–385 (1999)

26. Makowsky, J.A., Shelah, S.: Positive results in abstract model theory: a theory of compact logics. Ann. Pure Appl. Logic **25**(3), 263–299 (1983)
27. Nolan, D.: Hyperintensional metaphysics. Philos. Stud. **171**(1), 149–160 (2014)
28. Nute, D., Cross, C.B.: Conditional logic. In: Gabbay, D.M., Guenthner, F. (eds.) Handbook of Philosophical Logic, pp. 1–98. Springer, Dordrecht (2001). https://doi.org/10.1007/978-94-017-0456-4_1
29. Rantala, V.: Impossible worlds semantics and logical omniscience. Acta Philosophica Fennica **35**, 106–115 (1982)
30. Ribeiro, M.M.: Belief Revision in Non-classical Logics. Springer, Heidelberg (2013)
31. Rott, H.: 'Just because': taking belief bases seriously. In: Buss, S.R. (ed.) Lecture Notes in Logic, vol. 13, pp. 387–408. Association for Symbolic Logic, Urbana (1998)
32. Souza, M.: Towards a theory of hyperintensional belief change. In: Cerri, R., Prati, R.C. (eds.) BRACIS 2020. LNCS (LNAI), vol. 12320, pp. 272–287. Springer, Cham (2020). https://doi.org/10.1007/978-3-030-61380-8_19
33. Souza, M., Vieira, R., Moreira, Á.: Dynamic preference logic meets iterated belief change: representation results and postulates characterization. Theor. Comput. Sci. **872**, 15–40 (2021)
34. Souza, M., Wassermann, R.: Belief contraction in non-classical logics as hyperintensional belief change. In: Proceedings of the International Conference on Principles of Knowledge Representation and Reasoning, vol. 18, pp. 588–598 (2021)
35. Souza, M., Wassermann, R.: Hyperintensional partial meet contractions. In: Proceedings of the International Conference on Principles of Knowledge Representation and Reasoning (2022)
36. Stalnaker, R.C.: A theory of conditionals. In: Harper, W.L., Stalnaker, R., Pearce, G. (eds.) IFS, pp. 41–55. Springer, Dordrecht (1968). https://doi.org/10.1007/978-94-009-9117-0_2
37. Tennant, N.: Contracting intuitionistic theories. Stud. Logica **80**(2), 369–391 (2005)
38. Vardi, M.Y.: On epistemic logic and logical omniscience. In: Theoretical Aspects of Reasoning About Knowledge, pp. 293–305. Elsevier (1986)
39. Wansing, H.: A general possible worlds framework for reasoning about knowledge and belief. Stud. Logica **49**(4), 523–539 (1990)
40. Williams, M.A.: Iterated theory base change: a computational model. In: Proceedings of the 14th International Joint Conference on Artificial intelligence, pp. 1541–1547 (1995)
41. Özgün, A., Berto, F.: Dynamic hyperintensional belief revision. Rev. Symb. Logic **14**, 1–46 (2020). https://doi.org/10.1017/S1755020319000686

A Multi-population Schema Designed for Biased Random-Key Genetic Algorithms on Continuous Optimisation Problems

Mateus Boiani[1]([envelope]) [iD], Rafael Stubs Parpinelli[2] [iD], and Márcio Dorn[1] [iD]

[1] Institute of Informatics, Federal University of Rio Grande do Sul, Porto Alegre, Rio Grande do Sul, Brazil
{mboiani,mdorn}@inf.ufrgs.br
[2] Graduate Program in Applied Computing, Santa Catarina State University, Joinville, Santa Catarina, Brazil
rafael.parpinelli@udesc.br

Abstract. In Evolutionary Algorithms, population diversity is a determinant factor for the quality of the final solutions. Due to diverse problem characteristics, many techniques face difficulties and converge prematurely in local optima. The maintenance of diversity allows the algorithm to explore the search space and efficiently achieve better results. Parallel models are well-known techniques to maintain population diversity; however, design choices lead to different characteristics for the optimization process. For instance, the migration policy on the Island model can control how fast the algorithm converges. This work proposes a new migration policy designed for the Biased Random-Key Genetic Algorithm (BRKGA). Also, the proposal is compared with two traditional strategies and evaluates its performance in continuous search spaces. The results show that the proposal can improve the BRKGA optimization capability with suitable parameters.

Keywords: Genetic algorithms · Parallel metaheuristics · Island model

1 Introduction

Optimization problems are common in many areas and domains, most of them concerned with efficiently allocating limited resources to meet desired objectives. Over the last decades, metaheuristics have been used as an alternative to achieve good results in a reasonable time. Several techniques have been developed, such as Particle Swarm Optimization (PSO), Tabu Search (TS), Genetic Algorithm (GA), Simulated Annealing (SA), and Ant Colony (ACO) [5,8,14,17,20]. However, such methods often find reasonable solutions in a relatively short execution time, but optimal solutions can not be guaranteed. GAs are well-known population metaheuristics based on Charles Darwin's theory of evolution by natural

J. C. Xavier-Junior and R. A. Rios (Eds.): BRACIS 2022, LNAI 13653, pp. 444–457, 2022.
https://doi.org/10.1007/978-3-031-21686-2_31

selection. In a nutshell, the individuals in the population represent potential solutions to an optimization problem. Then, the population evolves through recombination operators, seeking convergence for the best results [9,13,22].

Some problems present dynamic landscapes or multiple local optima, and this can lead the algorithm to be trapped in specific regions along the search process, resulting in premature convergence. One of the most critical factors that determine the performance of a GA is population diversity. Undiversified populations accelerate the diffusion of genetic material from the elite solutions among the other individuals, resulting in premature convergence. Maintaining a certain level of diversity allows the population to explore new regions of the search space and improve the found results. Some strategies have been developed over the last years to achieve and maintain population diversity [11,24,26]. Gonçalves and Resende [10] proposed a GA version called Biased Random-Key Genetic Algorithm (BRKGA). On BRKGA, the population is structured into groups, and genetic operators are applied to use individuals from all groups, proving to be efficient and reach good levels of diversity. Another commonly used strategy is the Island Model (IM) on Distributed Genetic Algorithms (DGAs), where individuals are divided into smaller sub-populations (called demes or islands) [11]. Each subpopulation evolves in isolation and periodically carries out an exchange of individuals (*migration*) under specific criteria (*migration policy*). In this way, the global population (composed of all demes) has its convergence decelerated, keeping it diversified. This is mainly due to the migration policy, which controls the frequency of migration, the number of migrated individuals, how these individuals are selected and replaced, and the subpopulation communication topology [2,11,23].

This paper investigates the development of a new migration policy based on the BRKGA's structured population on IM-DGA. The Fitness-based Migration Policy (FBMP) designed takes advantage of the population structure to promote maintenance of diversity through a mechanism that combines groups of individuals to alternate between exploration and exploitation.

An essential aspect of GAs, and most metaheuristics, refers to its parameterization. Because such approaches use stochastic components that define their behavior and guide the search, the ideal parameters are problem-dependent, so the user must perform a parameter tuning for each problem that wishes to apply the optimizer [21]. To find the most appropriate set of parameters, we adopt offline parameter tuning through an iterated racing procedure provided by the irace package [18].

The paper is organized as follows. Section 2 presents the theoretical background. Section 3 presents the proposed method. The design of the experiments is described in Sect. 4, with the results and analysis in Sect. 5. Finally, the conclusion and future research directions are shown in Sect. 6.

2   Background

2.1   Biased Random-Key Genetic Algorithms

A Genetic Algorithm (GA) is a population-based metaheuristic that runs for many iterations, called generations [12]. Individuals are combined through a *crossover* operator during each generation to generate individuals for the next generation. In the GA context, a solution is represented as an individual, a set of individuals forms a population, and each solution value is an allele. Each population individual is evaluated through an objective function (also called the fitness function). A GA selects well-evaluated individuals to *crossover*, aiming to improve the population's quality from one generation to the next. Therefore, pairs of them are chosen to participate in Recombination and Mutation. According to the *Roulette Selection* scheme, each individual has a probability of selection proportional to their fitness value [9,12,25]. Equation 1 presents the individual selection probability according to the *Roulette Selection* scheme where i represents the i-th individual and NP represents the size of the population.

$$Probability(i) = \frac{Fitness(i)}{\sum_{j=1}^{NP} Fitness(j)} \tag{1}$$

Once selected, the pair of individuals (*offsprings*) are generated from the parents' matting of genetic material. A *cut-off* point k is randomly defined between 1 and $l-1$, where l is the individual's number of genes (dimensions). Then, the first *offspring* receives genetic material from 1 until k from parent one, and from $k+1$ until l from parent two, the second *offspring* is generated inversely. Additionally, under a probability m, these *offsprings* may be submitted to mutation process, where a gene is randomly selected and altered. The whole process is repeated until NP new individuals are generated, forming the new population. This cycle, called generation, repeats until a stop criterion is met [22].

A common variation consists in promote the best individual (elite) to the next population, thus ensuring that the best solution found by GA throughout its execution will always be maintained [6]. For a complete description of Evolutionary Computation and Genetic Algorithms, please refer to [17] and [7].

A *Biased Random-Key Genetic Algorithm* (BRKGA) is an evolutionary algorithm mainly designed for discrete and global optimization problems [10]. Basically, each solution is encoded as a vector of random keys, where each random key is a real number, randomly generated, in the continuous interval [0, 1). A decoder maps each vector of random keys to a solution of the optimization problem being solved and computes its cost. This normalization makes it independent of the application and, when necessary, the decoding operator is applied to the solution found, bringing it back to the problem domain [4,10].

A particularity of BRKGA is the way used to structure its population, where individuals are organized according to their fitness value, which aims to preserve the diversity of the population. After creating and ordering the initial population, the individuals are divided into two groups: *Elite* and *Non-elite*, as illustrated in Fig. 1. In the next generation, the *Elite* is preserved and entirely copied to the new population (Algorithm 1, line 4). Then, P_m mutant (randomly) new individuals are generated uniformly at random in the interval [0, 1) and aims to help escape local optima (Algorithm 1, line 5). Lastly, the crossover between *Elite* and *Non-elite* is performed (Algorithm 1, lines 6–11). On crossover, the BRKGA guarantees that the crossover is performed with individuals from different groups to explore the population diversity. The first parent is randomly selected from the *Elite* group, while the second parent is randomly chosen from the *Non-elite* groups (which includes the mutants). Differently from canonical GAs, each crossover operation results in only one offspring. For each gene, the crossover operator flips a biased coin to choose which parent passes genetic material to the child. Generally, the bias tends to favor the *Elite* parent; however, a parameter ρ_e is defined by the algorithm designer to specify the *Elite* gene inheritance probability. BRKGA's authors [10] suggest a value between 50% and 70% to ρ_e since parent one is guaranteed a better fitness than parent two.

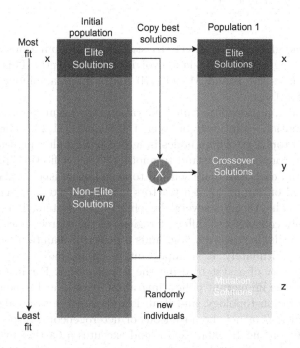

Fig. 1. BRKGA's population structure, transition from initial population to first generation. Adapted from [10].

2.2   Distributed Genetic Algorithms

Distributed Genetic Algorithms (DGAs) are one of the widely known techniques to improve the exploration of the search space and maintain diversity on

---

Algorithm 1. BRKGA – A population P with P_e *Elite*, $\bar{P}_e$ *Non-Elite*, and P_m *Mutant* groups.

Require: $P, P_e, \bar{P}_e, P_m$
Ensure: $\bar{P}_e = P \setminus P_e$
 1: Create initial population P
 2: **while** stopping criteria not met **do**
 3:    Sort population
 4:    Copy P_e from population k to $k+1$
 5:    Add P_m mutants individuals to population $k+1$
 6:    **while** $(k+1)$-th population $< P$ **do**
 7:       select a random individual from P_e
 8:       select a random individual from $\bar{P}_e$
 9:       produce offspring with a probability ρ_e
10:       add new offspring to population $k+1$
11:    **end while**
12: **end while**
13: **return** fittest individual

GAs [2,11]. The global population's division into subpopulations gives rise to a kind of archipelago, with each island evolving independently. This technique has become known as the Island Model (IM) and is popular among Evolutionary Algorithms (EAs).

In this model, a procedure called migration may occur periodically to interchange information (solutions) between the islands [11,23]. Migrations occur according to a topology, where nodes in a directed graph represent each island, and each edge connects one island to another. At specific time points, selected individuals from each island are sent off to neighboring islands [23]. As a result, islands trapped on strong local attractors may be affected by successful island migrants [23]. The IM adds several benefits to the system; it coordinates the search, possibly converges to different regions of the search space, improves the usage of the available resources, and adds a powerful maintaining diversity feature [2,11,23]. Ultimately, it is important to know that many design choices affect the behavior of a system using the island model. For instance, the emigration policy, the immigrant policy, migration interval or frequency, number of migrants, migration topology, and if all islands run the same algorithm under identical conditions or not, homogeneous or heterogeneous, respectively.

Specifically, on the IM-DGA, each island executes a GA that evolves independently. Nonetheless, the IM design choices may result in different characteristics for the optimization process. Regarding the migration topology, the most common include *Ring*, *Star*, and *Fully-Connected* [1]. For a complete description about Distributed Evolutionary Algorithms, please refer to [11].

This work uses *Ring* and *Fully-Connected* topologies as baseline approaches to validate and compare the proposal. The best/replace worst fashion is adopted in the *Ring*, sending the elite individual to the right neighborhood island.

Furthermore, the *Fully-Connected* topology consists of each island broadcasting η elite individuals; we call this policy (ηBest).

3   Proposed Model

This section describes a new migration policy focused on maintaining population diversity through a mechanism for sharing individuals with similar positions on fitness ranking in their populations. Like BRKGA, the proposed approach takes advantage of the fact that the population is structured, thus exploring the existing diversity and sharing it with other demes in a fully-connected topology. The proposed migration is based only on fitness values, avoiding the computation of similarity among individuals, significantly reducing the complexity compared with policies that use similarity information. The proposal's main idea is that the fitness and the structured population provide enough information during the optimization process, evolving the global population while maintaining the diversity among demes.

With the individuals ranked according to their adaptability to the problem under optimization, the population is sliced in equal parts of size ω computed according to Eq. 2, where NP represents the population size, and ι represents the number of demes. This means that NP and ι are expected to be multiple. Then, the computed sliced size that we call window defines the permutation of individuals. Basically, each deme is formed by individuals from the i-th window of each deme. For illustration, the first deme receives the first ω individuals from all demes, the second deme receives the next window of individuals from all demes, and so on.

$$\omega = \frac{NP}{\iota} \tag{2}$$

In the proposed migration policy, the exchange of individuals configures a fully connected communication. Each population has genetic material available from all demes, allowing to explore combinations that might otherwise be unreachable due to local convergence. However, to avoid a significant disparity after migration, the new population is formed by individuals in the same fitness range (structured population) in their respective deme but not necessarily having similar fitness values. This ensures that, although the initial algorithm behavior is related to the exploitation capacity, it will oscillate between exploitation and exploration during the optimization process. Figure 2 exemplifies the approach with three demes, each with six individuals. The $\lambda_{i,j}$ value represents the fitness difference between the best and the worst individual from a population i in a migration step j.

Due to the fact that individuals with similar fitness are sorted according to their adaptability and maintained in the same population - the first group is formed by the fittest individuals, followed by the second group, whose contain individuals with average fitness, and finally, the third one compiles the worst-performing individuals -, we can observe a significant decrease in the values of λ

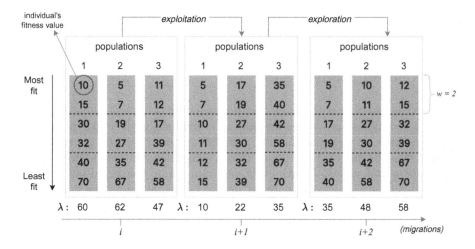

Fig. 2. The behavior of the approach over time with $NP = 6$, $K = 3$ and $W = 2$. From i to $i + 1$, an exploitation behavior is observed, i.e., similar individuals are gathered in the same *deme* (indicated by λ values). Then, a new migration is applied, resulting in an exploration behavior (from $i + 1$ to $i + 2$), i.e., similar individuals are spread over the *demes*.

from the migration step i to the step $i + 1$, which characterizes an *exploitation* process. The transition between steps $i + 1$ and $i + 2$ characterize an *exploration* process, where similar individuals of each *deme* are separated and sent to all others, resulting in a more diversified new population. It is essential to point out that Fig. 2 illustrates the migration policy behavior in an advanced-convergence scenario, where no improvements have been achieved in the last generations. Thus, when successively applied, the proposed policy results in demes with populations sometimes diversified, sometimes with similar individuals. Despite the stochastic factor of the genetic operators, this migration policy is characterized mainly by its ability to alternate between *exploration* and *exploitation* behavior according to the convergence level.

4   Design of Experiments

This section presents details regarding the design of experiments: hardware and software details, statistical tests, parameter setup, problem instances, convergence, and diversity analysis.

Experiments were conducted on a single machine using the same hardware throughout the complete experimentation set. It consists of a computing node equipped with 2 Intel Xeon Silver 4216 at 2.1 GHz, 32 cores, and 64 threads. The development environment is made of Ubuntu 20.04 operating system. The programming languages are C and Python.

In this experimentation, we have used the benchmark proposed for the "Special Session on Single Objective Real Parameter Numeric Optimization" held at the IEEE CEC 2017 [3]. This benchmark defines 29 continuous optimization functions varying from 10 up to 100 dimensions structured in four groups: unimodal, simple multimodal, hybrid, and composition. A more detailed description of the functions and their characteristics can be found in [3].

The test results are evaluated by the *Error* metric, as used in the CEC 2017 Benchmark [3] presented by Eq. 3, where F_i^* is the optimal value and F_i is the obtained result, both relative to the i-th function.

$$Error(F_i) = F_i - F_i^* \tag{3}$$

An analysis methodology proposed by LaTorre, Muelas, and Peña [15] was adopted for statistical assessment. We report the best, median, worst, mean and standard deviation of the error for each benchmark on every function. For brevity purposes, this information is only available in the supplementary material[1]. Furthermore, following the methodology, we have used the Friedman test (with a degree of confidence of 95%) for multiple comparison to check if there are significant differences among the considered algorithms [15,16]. If the difference exists, it is computed the following values for each algorithm:

- **Overall ranking:** relatively to the Friedman test, it computes the relative ranking of each algorithm according to its mean performance on each function and reports the average ranking computed through all the functions. For instance, given the following mean performance in a benchmark of three functions for algorithms A and B: A = (0.21, 3.45, 1.20), B = (2.25, 1.33, 0.80); their relative ranking would be: Rank(A) = (1, 2, 2), Rank(B) = (2, 1, 1); and thus their corresponding average rankings are 1.67 and 1.33, respectively.
- **# Best:** refers to the number of functions in which each algorithm obtains the best results compared to other algorithms.
- **nWins:** refers to the number of other algorithms for which each algorithm is statistically better minus the number of algorithms for which each algorithm is statistically worse according to the *Wilcoxon Signed Rank Test* in a pairwise comparison (with a degree of confidence of 95%).

4.1 Parameter Settings

In the experiments, some parameters were fixed and empirically defined. Others were defined using an automatic algorithm configuration package that varies its values and tries to determine the most suitable set of values. The algorithms run 31 times starting from different random seeds with population size (NP) of 100. The BRKGA parameters are balanced according to the author's recommendation [10]. The population (P) incorporates an elite group (P_e) of 10% of the total population, the crossover solutions comprise 70%, and 20% are reserved

---

[1] Available at https://shorturl.at/cg238.

for mutation solutions. The elite allele inheritance probability (ρ_e) was fixed at 0.70.

To fairly assess the performance of each migration policy, the `irace`[2] automatic algorithm configuration package [18] was applied to the parameters listed in Table 1. In the table, each parameter is described with its type and value range.

Table 1. Description and ranges of parameters for automatic parameter tuning with irace [18].

Parameter	Description	Type	Value range
τ	Migration frequency	Categorical	$\{32, 64, 128, 256, 512, 1024\}$
ι	Number of islands/*demes*	Categorical	$\{2, 4, 5, 10\}$
η	Number of migrant individuals on ηBest policy	Integer	$[1, 10]$

We try to find a reasonable threshold between experimentation time and result quality concerning offline tuning. For this reason, instead of searching for the most suitable set of parameter values for the 29 functions of the test suite, we dig for the most appropriate set of parameters for each group of functions (unimodal, simple multimodal, hybrid, and composition). The maximum number of runs (tuning budget) for `irace` is set according to the number of instances (functions) variations, i.e., a budget of 500 runs per function is available. For instance, the first group has two functions and a budget of 1000. Table 2 summarizes the 48 experiment results conducted to find the most suitable parameters for each group of problems varying the migration policy and problem dimensionality.

Lastly, each execution has $D \cdot 10^4$ functions evaluations, where D refers to the problem dimensionality and $D = 10, 30, 50, 100$.

5 Results and Analysis

Table 3 presents the results obtained by varying the migration policy and applying the suitable parameters found on the offline tuning with `irace`. From the table, it is possible to observe that for problems of up to 50 dimensions, `FBMP` proved to be the best choice migration strategy. Furthermore, the values of #Best and nWins indicate the superiority of the results obtained by the proposed migration policy. On the other hand, for 100 dimensions, the result obtained by the `FBMP` deteriorated. This may indicate some points of attention related to the proposal. For instance, the demes had not yet converged to local or global attractors at migration, so the optimization's progress could be compromised, and the search stopped abruptly. In addition, there is a possibility that the parameters

---

[2] Available at https://cran.r-project.org/web/packages/irace version 3.4.1.

Table 2. Parameters found by irace tuning: migration frequency (τ), number of islands (ι), and ηBest (η).

	D_{10}			D_{30}			D_{50}			D_{100}		
	τ	ι	η	τ	ι	η	τ	ι	η	τ	ι	η
Unimodal Functions $F_{1,3}$												
Ring	32	10	–	32	10	–	64	10	–	32	10	–
ηBest	32	10	6	32	10	4	64	10	2	32	10	6
FBMP	64	10	–	32	10	–	32	10	–	128	10	–
Simple Multimodal Functions F_{4-10}												
Ring	32	10	–	64	10	–	128	10	–	32	10	–
ηBest	256	10	5	128	10	1	32	10	4	32	10	4
FBMP	256	10	–	64	10	–	64	10	–	128	10	–
Hybrid Functions F_{11-20}												
Ring	32	10	–	128	10	–	32	10	–	256	10	–
ηBest	256	10	8	256	4	2	128	10	1	32	10	1
FBMP	512	10	–	128	10	–	256	10	–	64	10	–
Composition Functions F_{21-30}												
Ring	32	10	–	32	10	–	256	10	–	32	10	–
ηBest	64	10	4	256	10	1	256	10	2	64	10	10
FBMP	64	10	–	256	10	–	512	10	–	1024	10	–

found by the offline tuning are not ideal, and in this case, a more extended experimentation period and a higher budget for irace are necessary.

A new experiment was carried out to identify whether the set of parameters found for the FBMP in 100 dimensions is not ideal. Hence, the irace budget was expanded. The budget was defined by the product between the number of different parameter combinations, the number of functions in the group, and the number of executions planned. For instance, for FBMP, there are 24 combinations for the parameters, 2 functions, and 30 runs intended, so the budget is $24 \cdot 2 \cdot 30 = 1440$. Regarding the parameters, irace found a new set that increased the migration frequency to 256 for the unimodal, hybrid, and composition function groups. Table 4 presents the results of this experiment. The table shows improved performance for FBMP. The ranking reduced from 2.34 to 2.08, the #Best advanced from 7 to 14, and nWins from -11 to -6. It significantly improved, yet, it was not enough to overcome the Ring policy for $D = 100$.

From the conducted experiments, it is possible to imply that FBMP is sensitive to the problem dimensionality and its parameters. Moreover, the budget increase results prove that there is room to search for suitable parameters that are clearly still not found. On the other hand, results of up to 50 dimensions show that FBMP is a promising migration policy with potential application in different scenarios and experiments.

Table 3. Average ranking, number of functions for which the algorithm obtains the best results and number of wins in pair-wise comparisons on the CEC 2017 benchmark.

D	Migration policy	Ranking	#Best	nWins
10	FBMP	1.80	13	3
	Ring	2.00	11	5
	ηBest	2.20	5	−8
30	FBMP	1.62	16	12
	Ring	1.81	11	13
	ηBest	2.57	2	−25
50	FBMP	1.44	17	18
	Ring	2.17	8	2
	ηBest	2.39	4	−20
100	Ring	1.55	15	10
	ηBest	2.10	7	1
	FBMP	2.35	7	−11

Table 4. Offline tuning increased budget results.

D	Migration policy	Ranking	# Best	nWins
100	Ring	1.71	9	8
	FBMP	2.08	14	−6
	ηBest	2.21	6	−2

Convergence and diversity charts were generated to better understand the effects of the proposed migration policies. It is well known that convergence and diversity analysis are fundamental tools to inspect the algorithm working process. The convergence chart represents the distance from the optimal solution at each iteration from which it is possible to visualize the algorithm's convergence. The population's diversity is obtained using the momentum of inertia proposed by Morrison and De Jong [19]. The diversity chart represents how spread the solutions are in the search space in a given iteration. This makes it possible to draw whether the algorithm is performing a local or global search at each iteration.

Figure 3 presents the behavior obtained concerning convergence and diversity when applying the FBMP during optimization. In the figure, it is possible to observe the characteristics of the proposed method. Observing the convergence of Island #1 (Fig. 3a), it is possible to observe that during the optimization process, the exchange of individuals between islands favors convergence since Island #1 is composed predominantly of elite individuals; naturally, the BRKGA dynamic introduces diversity to this island. On the other hand, on Island #10 (last island, Fig. 3c), we observe how the removal of elite individuals has an

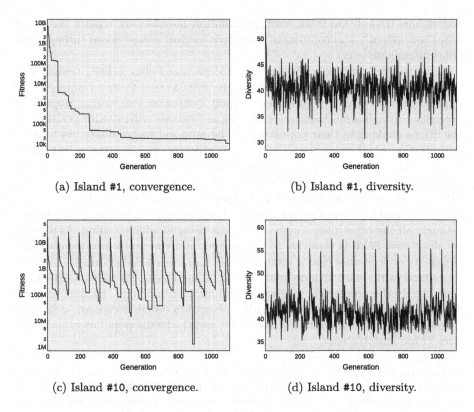

(a) Island #1, convergence.

(b) Island #1, diversity.

(c) Island #10, convergence.

(d) Island #10, diversity.

Fig. 3. Example of FBMP's convergence and diversity for island #1 and island #10.

impact similar to a population reset, allowing the method with high diversity (Fig. 3d) to restart the search for promising regions of the search space. It is interesting to highlight that the FBMP's fully-connected migration allows the information of new promising regions of the search space to reach all islands, especially elite islands. According to convergence charts, we can observe the dynamic of population diversity. When the FBMP is applied, the diversity of Island #10 undergoes a sudden increase (*exploration*). Conversely, on Island #1, we can observe that the constant diversity is interrupted in the migration movement giving rise to an *exploitation* behavior. Again, the mechanisms of BRKGA control back the diversity.

6   Conclusion and Future Works

Metaheuristics have been successfully applied to solve real-world continuous optimization problems. Over the years, scientists worldwide have been studying strategies to deal with premature convergence, which compromises metaheuristics accuracy. Specifically, metaheuristics success is strongly related to balancing exploration and exploitation during the search process. A well-known strategy to mitigate this balance is using parallel models, such as the Distributed Genetic

Algorithms (DGAs). On DGAs, multiple solution populations explore the search space concurrently and coordinate the search for promising regions through periodic communications.

This paper proposes a Fitness-based Migration Policy (FBMP) designed for the Biased Random-Key Genetic Algorithm (BRKGA) that takes advantage of the population structure. The proposed method focuses on maintaining population diversity through a mechanism for sharing individuals with similar positions on the fitness ranking in their populations. The proposal was evaluated in terms of optimization accuracy on the CEC'17 single objective real-parameter benchmark against two well-known migration topologies, *Ring* and *Fully-Connected*. The analysis points out that our migration policy is highly competitive, presenting better results up to 50 dimensions. Moreover, we observe that the parameters (frequency and number of islands) substantially affect the methods' performance. As such, all the experiments were carried out using an offline tuning. As a result, we show that FBMP is sensitive to parameters and the problem dimensionality. Hence, its performance can be improved with a more detailed parameter analysis.

Future works may include analysis and offline tuning for independent problems. Apply the proposal to real-world optimization problems. Explore methods to make dynamic or self-adaptive critical algorithm design decisions, such as the number of islands, the migration frequency, and the BRKGA population and group sizes (*Elite* and *Mutant*).

Acknowledgment. This work was supported by grants from the Fundação de Amparo à Pesquisa do Estado do Rio Grande do Sul (FAPERGS) [19/2551-0001906-8], Conselho Nacional de Desenvolvimento Científico e Tecnológico (CNPq) and was financed, in part, by the Coordenação de Aperfeiçoamento de Pessoal de Nível Superior (CAPES) [*Finance Code 001*] - Brazil.

References

1. Alba, E.: Parallel Metaheuristics: A New Class of Algorithms, vol. 47, 1st edn. Wiley-Interscience, New York (2005)
2. Alba, E., Luque, G., Nesmachnow, S.: Parallel metaheuristics: recent advances and new trends. Int. Trans. Oper. Res. **20**(1), 1–48 (2013)
3. Awad, N.H., Ali, M.Z., Suganthan, P.N., Liang, J.J., Qu, B.Y.: Problem definitions and evaluation criteria for the CEC 2017 special session and competition on single objective real-parameter numerical optimization. Technical report, Nanyang Technological University, Singapore (2016)
4. Bean, J.C.: Genetic algorithms and random keys for sequencing and optimization. ORSA J. Comput. **6**(2), 154–160 (1994). https://doi.org/10.1287/ijoc.6.2.154
5. Chopard, B., Tomassini, M.: An Introduction to Metaheuristics for Optimization. Springer, Cham (2018). https://doi.org/10.1007/978-3-319-93073-2
6. Deb, K., Pratap, A., Agarwal, S., Meyarivan, T.: A fast and elitist multiobjective genetic algorithm: NSGA-II. IEEE Trans. Evol. Comput. **6**(2), 182–197 (2002). https://doi.org/10.1109/4235.996017
7. Eiben, A.E., Smith, J.E.: Introduction to Evolutionary Computing, 2nd edn. Springer, Heidelberg (2015). https://doi.org/10.1007/978-3-662-44874-8

8. Fister, I., Yang, X.S., Fister, I., Brest, J., Fister, D.: A Brief Review of Nature-Inspired Algorithms for Optimization (2013). https://doi.org/10.48550/ARXIV.1307.4186

9. Goldberg, D.E.: Genetic Algorithms in Search, Optimization and Machine Learning, 13th edn. Addison-Wesley Professional (1989)

10. Gonçalves, J.F., Resende, M.G.: Biased random-key genetic algorithms for combinatorial optimization. J. Heuristics **17**(5), 487–525 (2011)

11. Gong, Y.J., et al.: Distributed evolutionary algorithms and their models: a survey of the state-of-the-art. Appl. Soft Comput. **34**, 286–300 (2015)

12. Holland, J.H.: Adaptation in Natural and Artificial Systems: An Introductory Analysis with Applications to Biology, Control, and Artificial Intelligence, 1st edn. MIT Press (1992)

13. Holland, J.H.: Genetic algorithms. Sci. Am. **267**(1), 66–73 (1992)

14. Kar, A.K.: Bio inspired computing - a review of algorithms and scope of applications. Expert Syst. Appl. **59**, 20–32 (2016). https://doi.org/10.1016/j.eswa.2016.04.018. https://www.sciencedirect.com/science/article/pii/S095741741630183X

15. LaTorre, A., Muelas, S., Peña, J.M.: A comprehensive comparison of large scale global optimizers. Inf. Sci. **316**, 517–549 (2015). https://doi.org/10.1016/j.ins.2014.09.031

16. Lilja, D.J.: Measuring Computer Performance: A Practitioner's Guide. Cambridge University Press, Minneapolis (2000). https://doi.org/10.1017/CBO9780511612398

17. Luke, S.: Essentials of Metaheuristics, 2nd edn. Lulu (2013). http://cs.gmu.edu/~sean/book/metaheuristics/

18. López-Ibáñez, M., Dubois-Lacoste, J., Pérez Cáceres, L., Birattari, M., Stützle, T.: The irace package: iterated racing for automatic algorithm configuration. Oper. Res. Perspect. **3**, 43–58 (2016). https://doi.org/10.1016/j.orp.2016.09.002

19. Morrison, R.W., De Jong, K.A.: Measurement of population diversity. In: Collet, P., Fonlupt, C., Hao, J.-K., Lutton, E., Schoenauer, M. (eds.) EA 2001. LNCS, vol. 2310, pp. 31–41. Springer, Heidelberg (2002). https://doi.org/10.1007/3-540-46033-0_3

20. Parpinelli, R.S., Lopes, H.S.: New inspirations in swarm intelligence: a survey. Int. J. Bio-Inspired Comput. **3**(1), 1–16 (2011). https://doi.org/10.1504/IJBIC.2011.038700

21. Parpinelli, R., Plichoski, G., Silva, R., Narloch, P.: A review of techniques for online control of parameters in swarm intelligence and evolutionary computation algorithms. Int. J. Bio-Inspired Comput. **13**(1), 1–20 (2019). https://doi.org/10.1504/IJBIC.2019.10018955

22. Srinivas, M., Patnaik, L.: Genetic algorithms: a survey. Computer **27**(6), 17–26 (1994). https://doi.org/10.1109/2.294849

23. Sudholt, D.: Parallel evolutionary algorithms. In: Kacprzyk, J., Pedrycz, W. (eds.) Springer Handbook of Computational Intelligence, pp. 929–959. Springer, Heidelberg (2015). https://doi.org/10.1007/978-3-662-43505-2_46

24. Črepinšek, M., Liu, S.H., Mernik, M.: Exploration and exploitation in evolutionary algorithms: a survey. ACM Comput. Surv. **45**(3) (2013). https://doi.org/10.1145/2480741.2480752

25. Whitley, D.: A genetic algorithm tutorial. Stat. Comput. **4**(2), 65–85 (1994)

26. Xu, J., Zhang, J.: Exploration-exploitation tradeoffs in metaheuristics: survey and analysis. In: Proceedings of the 33rd Chinese Control Conference, Nanjing, China, pp. 8633–8638. IEEE (2014)

Answering Questions About COVID-19 Vaccines Using ChatBot Technologies

Matheus Letzov Pelozo, Marcelo Custódio, and Alison R. Panisson[(⊠)]

Department of Computing (DEC), Federal University of Santa Catarina (UFSC),
Araranguá, Brazil
`alison.panisson@ufsc.br`

Abstract. Chatbots are a powerful tool to design and implement sophisticated computer systems able to interact with human users through natural language. Chatbots are considered more friendly to users than other sources of information, and consequently, they have been largely applied to various domains. In this work, we propose a chatbot application aimed at answering questions about COVID-19 vaccines. Besides the interesting application domain and the knowledge engineering behind this development, we also introduce a modular chatbot architecture based on an easy-to-update database and natural language templates in which new information (for example, new vaccines) can be added without the need for retraining the chatbot. Furthermore, in this paper, we provide an empirical evaluation of the proposed chatbot application.

Keywords: Artificial intelligence · Chatbots · COVID-19 pandemic

1 Introduction

Chatbots are considered artificial intelligence programs (agents) with sophisticated Human-Computer Interaction (HCI) models [3]. They are equipped with Natural Language Processing (NLP) units, which allow them to communicate in human language by text or oral speech with human users [10]. Consequently, considering their capability of imitating human interaction, chatbots have been applied to various domains, such as education, business and e-commerce, health, and entertainment [18]. Furthermore, chatbots are considered more friendly to users than other sources of information, for example, the static content search in frequently asked questions (FAQs) lists [1]. This is because they offer efficient assistance when communicating to users, also providing more comfortable interactions through more engaging answers, directly responding to users' problems, etc. [4,15].

The rise of the Internet and the view of user-driven content have provided a venue for quick and broad dissemination of information. Nowadays, the world is more connected than ever. However, false information also can be disseminated in a very fast way, and every day is more difficult to prevent misinformation

© The Author(s), under exclusive license to Springer Nature Switzerland AG 2022
J. C. Xavier-Junior and R. A. Rios (Eds.): BRACIS 2022, LNAI 13653, pp. 458–472, 2022.
https://doi.org/10.1007/978-3-031-21686-2_32

(false or inaccurate information) [5]. During the COVID-19 pandemic, it was not different, there was several inaccurate information widespread on the Internet, mainly about the COVID-19 vaccines[1]. In this context, we propose a chatbot application aimed at answering questions about COVID-19 vaccines, which could be used to prevent (combat) the misinformation phenomenon, considering the trustworthy sources of information used to build the chatbot database, and the capability of these technologies to provide information in a very friendly way, which inspire trust from the users [1].

Further, in our approach, considering that information about COVID-19 vaccines is constantly being updated, for example, new vaccines are being developed, we propose a modular and easy-to-update chatbot extended architecture. The proposed extensions to the usual chatbot architecture comprise two modular components: (i) a database containing the information about the COVID-19 vaccines; and (ii) natural language templates the chatbot uses to build answers to users, using the information from the database. This extended architecture allows us to update information in the database, for example, adding new vaccines, without the need for retraining the proposed chatbot (for some case it does not require even updating the natural language templates).

This paper is structured as follows. First, in Sect. 2, we describe the background for this work, introducing chatbot technologies and contextualising the COVID-19 pandemic. After, in Sect. 3, we introduce the proposed chatbot extended architecture. Also, we describe how we have identified the most common questions about COVID-19 vaccines in Subsect. 3.1 and collected the data set of natural language expressions to train the natural language unit model we present in Subsect. 3.2. After, in Sect. 4, we describe the process of knowledge engineering to implement the proposed chatbot application, including the search for trustworthy sources of information, the modelling of the database and natural language templates the chatbot uses to build answers for users, and the proposed dialogue strategies in Subsect. 4.1. After, in Sect. 5, we describe an empirical evaluation for both: (i) the natural language unit model; (ii) and the proposed dialogue strategies. After in Sect. 6, we discuss the related work, and in Sect. 7, we conclude the work by also pointing out future work.

2 Background

2.1 Chatbot Technologies

Chatbot technologies have grown worldwide. They are considered easy to use, even if users have never used a chatbot before, because they simulate a conversation as they were talking to other humans. Also, chatbots can answer multiple users instantaneously and they are 24/7 available.

Currently, chatbot technologies are available not only on multinational corporation, thanks to the variety of frameworks to develop such technologies, many

---

[1] https://www.who.int/docs/default-source/coronaviruse/vaccine-misinformation-toolkit_desktop1.pdf.

of them open-source. Also, thanks to the portability with text messages apps, chatbot technologies are being more explored by small companies, in which they take advantage of such technologies by interacting with the customer whenever necessary [17]. Many companies are replacing customer services by chatbots, because chatbots are faster, cheaper and always friendly. Chatbot technologies provide benefits not only to e-commerce and client support, they are also popular technology in other domains, and they will be even more popular in the future.

There are different chatbot technologies, for example, DialogFlow[2], IBM Watson[3], and Rasa[4]. In this work, we will use Rasa Framework, one of the most used frameworks to develop chatbots [17]. Rasa has been chosen because it is open source, allows us to import the chatbot to websites and message apps such as Telegram and Whatsapp, it allows incorporate natural language models, for example, those provided by spaCy[5], it is highly customisable, and it has a vast documentation.

2.2   COVID-19 Pandemic

The COVID-19 pandemic became part of the history of humanity for many factors. It has caused millions of deaths[6]. Immeasurable global financial losses also were consequences of the pandemic. Global events like the Olympics and the 2020 Eurocup were delayed due to the pandemic. Indisputably the COVID-19 pandemic changed the world and the way of life around the planet.

To combat the pandemic, vaccines were produced in record time, some of them exploring new technologies. No matter the reason, (political ideology, religion, misinformation, bigotry or skepticism), a large group of people on the whole planet were not convinced about the efficiency of the vaccines, for example. At the moment people had access to the vaccines, fake news about them started to spread. Consequently, COVID-19 vaccines became a target of misinformation.

As a tool to combat misinformation about COVID-19 vaccines, we propose a chatbot aimed at answering questions about the COVID-19 vaccines, using information that comes from trustworthy sources. In order to develop a trustworthy source of information, the chatbot was built with the information provided by the vaccine manufacturers and surveys published on the most regarded medicine portals, like the Lancet. Then, the many doubts and questions about the vaccines were properly answered.

---

[2] https://dialogflow.cloud.google.com/.
[3] https://www.ibm.com/br-pt/products/watson-assistant.
[4] https://rasa.com/.
[5] https://spacy.io/.
[6] 6.475.346 deaths were caused by COVID-19, according to https://covid19.who.int/ August 2022.

3   A Chatbot Extended Architecture

To develop a chatbot aimed at answering questions about the COVID-19 vaccines domain that could be easily updated according to new information (or even new vaccines) become available, we proposed a chatbot extended architecture. The proposed architecture, shown in Fig. 1, besides the usual chatbot components, includes: (i) an independent module containing the database with information used to build answers by the chatbot, and (ii) a module with natural language templates used to present the data from the database.

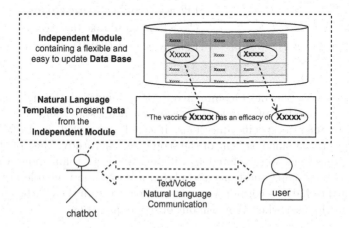

Fig. 1. Chatbot extended architecture.

To create the database to feed the independent module for the proposed chatbot architecture, and collect data to train the Natural Language Unit (NLU), we made available on the Internet an anonymous survey to ask people which kind of information they would like to know about the COVID-19 vaccines. In addition to answering that, we asked the respondents to describe different ways of questioning that particular information they were interested to know about COVID-19 vaccines. Thus, we were able not only to evaluate the most common questions people had about COVID-19 vaccines, which we will present in Sect. 3.1, but also to collect a data set with a variety of natural language expressions for those questions, which we used to train the NLU module we will present in Sect. 3.2.

3.1   Most Common Questions About COVID-19 Vaccines

During the period the survey was available, 19 people answered the survey. In the survey, the respondents were able to include how many questions they wanted, resulting in a total of 39 questions. From the total of questions, we grouped the questions according to their main subject, resulting in 9 questions (D1-D9) shown in Table 1.

Table 1. Most common questions about COVID-19 vaccines.

Code	Number of people	Information about vaccines
D1	8	Efficacy (first and second shot)
D2	7	Collateral effects
D3	5	Vaccines available in Brazil
D4	5	Out of scope questions
D5	4	How each vaccine works
D6	3	Interval between doses
D7	3	Efficacy according to age groups
D8	2	Close location to take the vaccine
D9	2	Efficacy against COVID-19 variants

Table 1 shows the most common questions identified through the survey, including the number of people that asked each question, and a code we will use later to refer to specific questions. It can be observed, in Table 1, that 5 questions were classified as "out of the scope" for this study. Most of those "out of scope" questions were related to different subjects than about COVID-19 vaccines. Out of scope questions also played a very important role in the development of the chatbot, considering we were able to train the chatbot to identify and inform the user when they ask out of scope questions.

3.2 NLU Training

When respondents were answering the survey we used to build Table 1, we also asked them to express five different ways they would use to ask each one of their questions to a chatbot (or another person). Thus, we were able to build a data set with about 200 sentences, representing different ways to ask the questions from Table 1, using natural language.

Using the data set with questions in natural language, we trained the chatbot NLU to recognise the users' intents, and whether the user provides specific entities or not during their questions (according to questions from Table 1, and more specific interactions we will present later). For example, when asking about the efficacy of vaccines, without informing a specific vaccine or dose, the chatbot will recognise that the user has the intent to know about the efficacy of COVID-19 vaccines but the user did not provide any specific details about their question:

```
- intent: non_immediate_efficacy
  example: |
    - What is the efficacy of COVID-19 vaccines?
    - What efficacy of the vaccines?
    - I am wondering about the efficacy of COVID-19 vaccines.
    ...
    - I would like to know more about the efficacy of vaccines.
```

Also, the chatbot NLU was trained to recognise more specific intents used to collect data from the user (and more specific interactions). That information is used during the chatbot dialogue strategy, as we will present in Sect. 4.1. For example, when the user asks a question about the efficacy of COVID-19 vaccines according to its age group, without telling their age during the first interaction, then the chatbot will ask the user to provide their age, and then the chatbot will be able to provide the requested information to the user. Those more specific interactions can occur in different forms. For example, the user can inform their age to the chatbot using different natural language expressions, such as:

- "19',
- "I am 19"
- "I am 19 years old"
- "19 years old"
- "nineteen"
- "nineteen year old"

We manually specified examples of natural language expressions for those specific cases and used them to train the chatbot NLU, enabling the chatbot to recognise specific entities. Recognising entities results from the process of annotating those entities in the natural language examples used during training, for example, ``I am [19](age) years old'' is annotated to recognise the entity **age** from that sentence.

4   Knowledge Engineering

After evaluating the most common questions through the survey, we have searched for trustworthy sources of information to answer those questions (D1-D9), and built the database and natural language templates shown in Fig. 1. In total, we used 32 sources of information[7], including the webpage of the vaccine manufacturing companies, recognised public health organisations, and scientific studies published by recognised institutions (research institutions and universities) to build a modular and easy-to-update database, as shown in Fig. 1.

In the database built, each line corresponds to one COVID-19 vaccine, and each column[8] corresponds to one piece of information about that particular vaccine. Thus, when the chatbot agent needs to build an answer to the user about a particular characteristic of one particular vaccine, it checks that information by looking at the line which corresponds to that vaccine and the column that corresponds to the required information. Using natural language templates, as shown in Fig. 1, after querying the necessary information, the chatbot can build and provide an answer to the user in natural language.

---

[7] The complete list of sources of information used in this study is available in [13], including a variety of tables with the information used to answer each one of the questions from Table 1.

[8] Some columns are: number of doses, efficacy by age, efficacy against variants, dose interval, collateral effects, tecnology used to produce the vacines, etc.

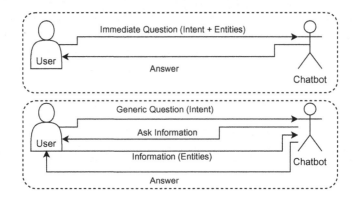

Fig. 2. Generic dialogue strategy.

For example, when users ask about the efficacy of the Pfizer vaccine after the second dose, using the following question: ''What is the efficacy of Pfizer after the second dose?'', the chatbot recognise the user intention as ''immediate_efficacy'' (user intents to know about the efficacy of a provided vaccine), and the entities [Pfizer](vaccine) and [second dose](dose). At this moment, the chatbot can look for that information in the database, looking for the line corresponding to the vaccine's name Pfizer, column efficacy after the second dose. With that information, the chatbot can use natural language templates to build and present the answer to the user. In our example, the chatbot uses the following natural language template:

```
The vaccine <VACCINE> has an efficacy of <EFFICACY>% after
                     the second dose.
```

Instantiating <VACCINE> and <EFFICACY> according to our example, the chatbot will provide the following answer to the user:

```
The vaccine PFizer has an efficacy of 95% after the second dose.
```

With the proposed (modular) architecture, the database can be easily updated with new vaccines and new information about the vaccines. When new vaccines (lines) are added to the database, the chatbot automatically can answer questions about them. When new information (column) about vaccines is added to the database, it is necessary to train the NLU to recognise those new intents of the users (and entities when necessary), as well as to provide natural language templates the chatbot will use to build and provide that particular information.

4.1   Dialogue Strategies

There are different ways users can ask for information about the COVID-19 vaccines, varying in their forms (the natural language expression used to make the question) and in their granularity (providing its intent, and entities, directly or during the dialogue), as we have identified in our survey.

Different forms of expressions are treated by the chatbot NLU, as we described in Sect. 3.2. However, the granularity, i.e., how the user gives the information necessary to the chatbot to understand the answer the user is looking for, is treated by a dialogue strategy (and chatbot polices) we will describe in this section.

The decision-making (intelligence) behind how a chatbot responds to users is modelled towards the so-called *stories*. Stories provide examples of conversation scenarios between the chatbot and the user. Training the chatbot with a variety of stories, allows it to learn how to interact with users in a variety of situations [1]. They also implement dialogue strategies proposed by the designer of the chatbot, and different stories can be combined by sophisticated dialogue polices, for example, polices based on machine learning such as the Transformer Embedding Dialogue (TED) policy[9] [20].

Figure 2 shows two dialogue strategies used to implement the proposed chatbot. In Fig. 2, we can observe that there will be dialogues in which users directly provide, during their first interaction, all information necessary for the chatbot to answer the user (their intents and entities). In contrast, there will be dialogues in which the user will ask generic questions, which allows the chatbot to understand the user's intent, but it requires more information to answer the user properly. Thus, the chatbot will ask those missing information (entities) until it has all information necessary to answer the user.

Below, we show how this dialogue strategy has been implemented, in which we developed three different stories used to train the chatbot to answer the question D2 from Table 1.

```
- story: Know collateral effect - specific vaccine immediate
  steps:
  - intent: know_collateral_effects_immediate
  - action: vaccinename_form
  - active_loop: vaccinename_form
  - slot_was_set:
    - requested_slot: vaccinename
  - activate_loop: null
  - action: action_inform_collateral_effects
```

The first story simulates situations in which the user asks the question, and, at the same time, provides the name of the vaccine they would like to know the collateral effects. For example, asking ``Which are the collateral effects for the Pfizer vaccine?'' allows the chatbot to understand the user intent (i.e., user wants to know about collateral effects of vaccines) and extract the entity necessary to provide that information (i.e., the name of the vaccine).

```
- story: Know collateral effect - all vaccines
  steps:
  - intent: know_collateral_effects
```

---

[9] https://rasa.com/docs/rasa/policies/.

```
- action: utter_one_or_all
- intent: know_all
- action: action_inform_collateral_effects_all

- story: Know collateral effect - specific vaccine
  steps:
  - intent: know_collateral_effects
  - action: utter_one_or_all
  - intent: know_specific
  - action: vaccinename_form
  - active_loop: vaccinename_form
  - slot_was_set:
    - requested_slot: vaccinename
  - activate_loop: null
  - action: action_inform_collateral_effects
```

The second and third stories simulate situations in which the user asks a generic question, the chatbot can understand the user's intent (i.e., know about collateral effects), but it has not been provided either the name of the vaccine or the information that the user wants to know that information about all vaccines. For example, telling ''I am wondering about the collateral effects of the vaccines''. Then the chatbot asks that missing information and: (i) when the user informs a specific vaccine, according to the second story, the chatbot answers with the information about that particular vaccine; or (ii) when the user selects all vaccines, according to the third story, the chatbot answers with that information for all vaccines available in its database.

5   Empirical Evaluation

We empirically evaluated the proposed chatbot using the test tools available with Rasa Framework[10]. Rasa testing tools provide means to test both the Natural Language Understanding (NLU) model and the stories (dialogue strategies).

5.1   Evaluating the NLU

The natural language understanding model can be tested separately, using standard machine learning methodology. Once the chatbot is deployed in the real world, it will be processing messages that it has not seen in the training data. To evaluate its behaviour, we can set aside some part of the data set of natural language expressions for testing.

To evaluate the NLU, we split the data set into train and test sets using 80% of the data set to train the natural language unit model, and 20% to test it (i.e., the standard configuration from the Rasa test tools).

---

[10] https://rasa.com/docs/rasa/testing-your-assistant/.

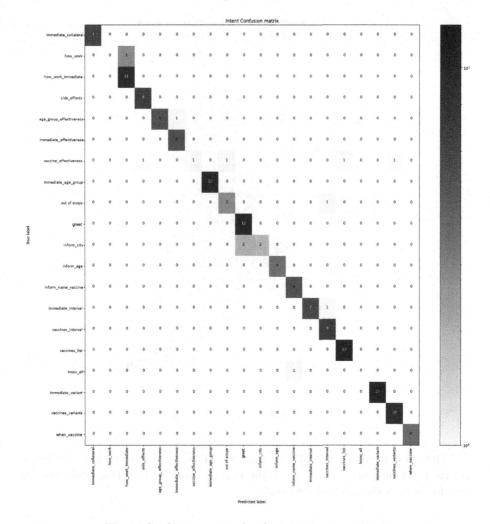

Fig. 3. Confusion matrix for the intention recognition.

Figure 3 shows the intent confusion matrix resulting from tests over the NLU, in which lines represent the expected classification of intents, and columns represent the predicted intent during the tests. In Fig. 3, the main diagonal contains the number of intents correctly predicted by the NLU model.

Further, Rasa test tools provide the intent prediction histogram shown in Fig. 4, which allows the visualisation of the confidence for all predictions, with the correct and incorrect predictions being displayed by blue and red bars respectively. Our results show that there were only incorrect predictions with low prediction confidence, that is, the chatbot did not predict an intent correctly when the algorithm provided confidence in classifying that intent below 0.44 (for one intent). Also, most of the incorrect predictions had confidence below 0.31.

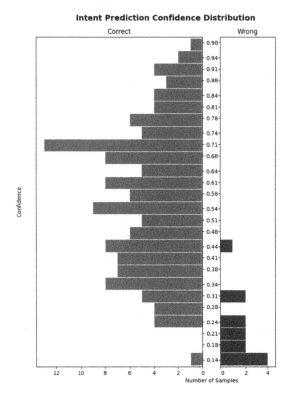

Fig. 4. Intent prediction histogram.

With these results[11], we validated the NLU model. Also, those low confidence incorrect predictions did not affect the overall result of resulting dialogues, as we will show in the next subsection.

5.2 Evaluating the Dialogue Strategies

To evaluate the dialogue strategies implemented through the stories and verify how the proposed chatbot will act in different conversation scenarios, we wrote different test stories (12 different stories), which are stories used to test the chatbot. Tests stories are written in a modified story format, and they allow us to provide entire conversations to evaluate the chatbot. Test stories provide certain user input, after they verify if the chatbot behaves as expected, providing new user input, and so on. Test stories are especially important when introducing more complicated stories from user conversations, such as those implemented in this work.

---

[11] Results also include (i) intent evaluation with accuracy of 0.8, f1-score of 0.8, and precision of 0.82, and (ii) entities evaluation with accuracy of 0.97, f1-score of 0.77, and precision of 0.74.

Test stories are very similar to stories used to train the chatbot (e.g., those presented in Subsect. 4.1), but they include the user's messages as well, as shown in the test story below, about the efficacy of vaccines, i.e., D1 from Table 1.

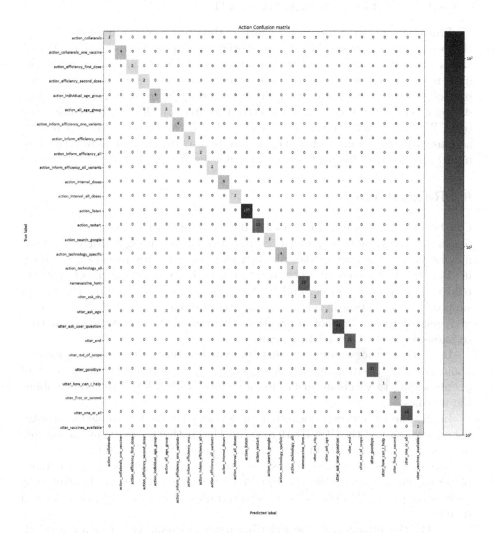

Fig. 5. Action confusion matrix.

```
- story: efficacy of all vaccines
  steps:
  - user: |
      What is the efficacy of COVID-19 vaccines?
  - intent: know_efficacy
  - action: utter_one_or_all
```

```
- user: |
    I would like to know all!
- intent: know_all
- action: action_inform_efficacy_all
```

Providing the test stories to the chatbot, Rasa test tools evaluate whether the chatbot acts or not as specified by the stories used to train the chatbot. Figure 5 shows the action confusion matrix resulting from this evaluation, in which the lines represent the actions the chatbot should execute, and the columns the predicted actions (executed by the chatbot) during the tests. In Fig. 5, the main diagonal represents the actions correctly executed by the chatbot during the tests. We can observe satisfactory results. Also, it can be observed that default actions such as `action_listen` (i.e., action used by the chatbot to receive users' input) are the most frequent actions executed by the chatbot.

6 Related Work

Chatbot technologies have been used in different application domains. For example, in [19], the authors proposed a chatbot aimed at providing information about movies; in [12], the authors describe how chatbots have been used for teaching; in [16], the authors describe how users engage in the development of chatbots related to customer service; and in [8,9], the authors proposed a chatbot to support bed allocation in hospital, using the interface built in [6,7]. In particular, related to our chatbot application, there are few works applying chatbot technologies in the COVID-19 domain.

In [2] the authors propose a chatbot aimed at answering general questions about the COVID-19 domain. Consequently, it has resulted in the increase of people willing to vaccinate in France, a country with a lot of reluctance about the vaccines.

In [21], the authors describe a chatbot developed in the USA aimed at answering questions about the COVID-19 to users. The chatbot proposed by the authors focuses on telling users about the symptoms caused by the COVID-19 virus. The authors described that they have satisfactory results in increasing the number of people intending to take the COVID19 vaccines. Also, they describe that their study proves that chatbots are a powerful tool for answering questions and doubts.

In [11], the authors describe a chatbot named Corona-Kun, which was developed to face the problem of a large number of people hesitating on taking the COVID19 vaccines in Japan. According to the authors, the chatbot had over 59k people using it. This chatbot aim at answering questions like: "when the person could get vaccinated?" or "what are the symptoms of COVID-19 virus?". The authors describe to be massive the scope of the chatbot, and that it proved to be helpful in providing that information to people. Similar to other chatbots mentioned, those used in France and USA, Corona-Kun increased the acceptance of the vaccines and decreased the number of non-vaccinated people.

In [14], the authors proposed a chatbot called Clara. The proposed chatbot aimed at checking the users' symptoms, helping them to identify a possible positive COVID-19 infection. The authors describe that the chatbot proved itself to be useful for that task.

Our work differs from [2,11,14,21] in its general goal and architecture. In our approach, we focus on answering questions about COVID-19 vaccines, also focusing on an easy-to-update database present in the modular chatbot extended architecture. All those other chatbots were aimed at answering general questions about the COVID-19 domain. For example, questions about the COVID-19 symptoms, vaccination calendar, medication, etc. The chatbot proposed in this paper aim at answering questions about COVID-19 vaccines, in particular, describing how they work, their side effects, their efficiency ratio, the vaccines available in Brazil, etc.

7   Conclusion

In this work, we proposed a chatbot aimed at answering questions about COVID-19 vaccines. To develop the proposed chatbot, we survey the most common question people have about COVID-19 vaccines, also collecting a data set of natural language expressions used to train the natural language understanding model of the chatbot. After understanding the most common questions people have about the vaccines, we built an easy-to-update database with information about COVID-19 vaccines extracted from trustworthy sources of information, such as the webpage of the vaccine manufacturing companies, recognised public health organisations, and scientific studies published by recognised institutions (research institutions and universities). Further, we propose a modular chatbot extended architecture composed of the built database and natural language templates, in which the chatbot can build and provide answers about the COVID-19 vaccines to users using natural language. The modular approach allows us to update the chatbot application database (e.g., with new vaccines) without the need for retraining the chatbot. Furthermore, we propose a dialogue strategy in which the chatbot can directly answer questions to users when the user provides all information necessary to the chatbot doing so, or the chatbot asks the missing information, receives that information, and then provides the answer to the user.

Finally, we empirically evaluated the implementation of the proposed chatbot, demonstrating its efficiency in predicting user intents and acting according to expected in a large diversity of dialogue situations, covering all questions identified through the execution of the survey, described in Table 1.

References

1. Adamopoulou, E., Moussiades, L.: Chatbots: history, technology, and applications. Mach. Learn. Appl. **2**, 100006 (2020)
2. Altay, S., Hacquin, A.S., Chevallier, C., Mercier, H.: Information delivered by a chatbot has a positive impact on COVID-19 vaccines attitudes and intentions. J. Exp. Psychol.: Appl. (2021)
3. Bansal, H., Khan, R.: A review paper on human computer interaction. Int. J. Adv. Res. Comput. Sci. Softw. Eng. **8**, 53–56 (2018)

4. Brandtzaeg, P.B., Følstad, A.: Why people use chatbots. In: Kompatsiaris, I., et al. (eds.) INSCI 2017. LNCS, vol. 10673, pp. 377–392. Springer, Cham (2017). https://doi.org/10.1007/978-3-319-70284-1_30

5. Cook, J., Ecker, U., Lewandowsky, S.: Misinformation and how to correct it. Emerging Trends in the Social and Behavioral Sciences: An Interdisciplinary, Searchable, and Linkable Resource, pp. 1–17 (2015)

6. Engelmann, D., et al.: Dial4JaCa – a demonstration. In: Dignum, F., Corchado, J.M., De La Prieta, F. (eds.) PAAMS 2021. LNCS (LNAI), vol. 12946, pp. 346–350. Springer, Cham (2021). https://doi.org/10.1007/978-3-030-85739-4_29

7. Engelmann, D., et al.: Dial4JaCa – a communication interface between multi-agent systems and chatbots. In: Dignum, F., Corchado, J.M., De La Prieta, F. (eds.) PAAMS 2021. LNCS (LNAI), vol. 12946, pp. 77–88. Springer, Cham (2021). https://doi.org/10.1007/978-3-030-85739-4_7

8. Engelmann, D.C., Cezar, L.D., Panisson, A.R., Bordini, R.H.: A conversational agent to support hospital bed allocation. In: Britto, A., Valdivia Delgado, K. (eds.) BRACIS 2021. LNCS (LNAI), vol. 13073, pp. 3–17. Springer, Cham (2021). https://doi.org/10.1007/978-3-030-91702-9_1

9. Ferreira, C.E.A., Panisson, A.R., Engelmann, D.C., Vieira, R., Mascardi, V., Bordini, R.H.: Explaining semantic reasoning using argumentation. In: Dignum, F., Mathieu, P., Corchado, J.M., De La Prieta, F. (eds.) PAAMS 2022. LNCS, pp. 153–165. Springer, Cham (2021). https://doi.org/10.1007/978-3-031-18192-4_13

10. Khanna, A., Pandey, B., Vashishta, K., Kalia, K., Pradeepkumar, B., Das, T.: A study of today's AI through chatbots and rediscovery of machine intelligence. Int. J. u-and e-Serv. Sci. Technol. 8(7), 277–284 (2015)

11. Kobayashi, T., et al.: 439. Corowa-kun: impact of a COVID-19 vaccine information chatbot on vaccine hesitancy, Japan 2021. In: Open Forum Infectious Diseases, vol. 8, pp. 321–322 (2021)

12. Kuyven, N.L., Antunes, C.A., de Barros Vanzin, V.J., da Silva, J.L.T., Krassmann, A.L., Tarouco, L.M.R.: Chatbots na educação: uma revisão sistemática da literatura. RENOTE 16(1) (2018)

13. Letzov Peloso, M., et al.: Utilizando tecnologias chatbot para responder dúvidas sobre as vacinas do covid-19 (2022)

14. Miner, A.S., Laranjo, L., Kocaballi, A.B.: Chatbots in the fight against the Covid-19 pandemic. NPJ Digit. Med. 3(1), 1–4 (2020)

15. Ranoliya, B.R., Raghuwanshi, N., Singh, S.: Chatbot for university related FAQs. In: 2017 International Conference on Advances in Computing, Communications and Informatics (ICACCI), pp. 1525–1530. IEEE (2017)

16. Rossato, D.M., et al.: Engajamento de trabalhadores na implementação de chatbot para atendimento aos usuários de uma universidade (2020)

17. Sharma, R.K., Joshi, M.: An analytical study and review of open source chatbot framework, RASA. Int. J. Eng. Res. 9(06) (2020)

18. Shawar, B.A., Atwell, E.: Chatbots: are they really useful? In: LDV Forum, vol. 22, pp. 29–49 (2007)

19. de Souza, G.O., Ribeiro, V.M., Breternitz, V.J.: Um estudo e implementação de chatbots utilizando uma arquitetura orientada a serviços (2019)

20. Vlasov, V., Mosig, J.E., Nichol, A.: Dialogue transformers. arXiv preprint arXiv:1910.00486 (2019)

21. VolppKevin, G., et al.: Asked and answered: building a chatbot to address Covid-19-related concerns. NEJM Catalyst Innovations in Care Delivery (2020)

Analysis of Neutrality of AutoML Search Spaces with Local Optima Networks

Matheus Cândido Teixeira$^{(\boxtimes)}$ and Gisele Lobo Pappa

Universidade Federal de Minas Gerais, Belo Horizonte, Brazil
{matheus.candido,glpappa}@dcc.ufmg.br

Abstract. AutoML tackles the problem of automatically configuring machine learning pipelines to specific data analysis problems. Optimization methods are used to explore this space of hyperparameters but little is known about this search space. Understanding the characteristics that make a space difficult to search can help developing better search methods besides better understaing the parameter and their interactions. This work uses a technique, named Local Optima Network (LON), which builds a graph from the fitness landscape of the problem. The fitness landscape is a way to represent the quality of all possible solutions in the search space. In particular, we use derivatives of the original LON, Monotonic LON (MLON) and Compressed MLON (CMLON). These variants have the advantage of dealing with the neutrality present in search spaces, i.e., regions of the space with solutions of the same quality. This paper analyzes the use of MLON and CMLON built from the fitness landscape of AutoML problems to better understand the search space of AutoML problems. The results indicate the presence of neutrality in many datasets and that it may have links to the fitness variance in the search space.

Keywords: AutoML · Local optima networks · Analysis of neutrality

1 Introduction

The use of methods to automate the process of hyperparameter optimization has been lately an almost mandatory process in machine learning tasks. These methods are studied under the area of Automated Machine Leaning (AutoML) [6], which in the past years has turned its attention specially to the area of Neural Architecture Search (NAS) [9,19].

State-of-the art methods for hyperparameters tuning are based on Bayesian Optimization, Evolutionary Computation, and Reinforcement Learning techniques [6]. Although many different methods of AutoML following these approaches have been proposed, we still understand very little about the search spaces generated by these machine learning hyperparameters and explored by these algorithms [12]. One way to analyze this space is using techniques of fitness landscape analysis [16].

© The Author(s), under exclusive license to Springer Nature Switzerland AG 2022
J. C. Xavier-Junior and R. A. Rios (Eds.): BRACIS 2022, LNAI 13653, pp. 473–487, 2022.
https://doi.org/10.1007/978-3-031-21686-2_33

The concept of a fitness landscape was first introduced for characterizing the distribution of fitness values over the space of genotypes in (natural) evolution, and was later mapped onto a generic framework for optimization problems [8]. In optimization problems, the fitness landscape is defined by a tuple $(\mathcal{S}, f, N)$, where $\mathcal{S}$ is the set of all possible solutions (*i.e.* the search space), $f : \mathcal{S} \to \mathbb{R}$ is a function that attributes a real valued performance estimation for each solution in $\mathcal{S}$, and $N(x)$ is a notion of neighborhood between solutions, usually defined as a distance metric $N(x) = \{y \in \mathcal{S} | d(x,y) \leq \epsilon\}$ for a sufficiently small ϵ.

Knowing the fitness landscape of a problem, one can understand its characteristics, including roughness (the frequency of peaks and valleys), modality (the number of peaks), and neutrality (adjacent regions in the configuration space where there is little or no fitness variation). Fitness landscape analysis (FLA) encompasses a set of metrics and methods to extract these features from the fitness landscape, and it has showed to be very useful for characterizing and analyzing search spaces.

There are a few studies in the literature that have looked at the landscape of AutoML problems to generate machine learning pipelines, where a pipeline is defined by a sequence of tasks including methods for data preprocessing, classification (or clustering, regression, etc.), and postprocessing. Garciarena et al. [5] performed an analysis of a subset of the search space explored by an AutoML system named TPOT, and found many regions of very high fitness but prone to overfitting. Pimenta et al. [12] used FLA techniques to look at AutoML search spaces. They measured fitness distance correlation (FDC) [7] and neutrality in a search space composed of machine learning pipelines for classification, and found FDC to be a poor metric for analyzing this kind of space. In parallel, other studies have looked at the loss landscape of neural networks (NN) as architecture and hyperparameters are changed [9,14].

All the works that perform some type of FLA have common limitations. First, they focus on extracting mainly local measures from the fitness landscapes, which might be restrictive and miss the global vision of the space [12,13]. Second, most metrics are computationally expensive and, per definition, do not focus their analysis on the most relevant regions of the search space, such as the ones close to the local optima [9].

In order to deal with fitness landscapes that have a number of solutions prohibitive to enumerate and to provide a more global view of the search space, the authors in [11] proposed a new way to analyze and visualize search spaces, named Local Optima Networks (LONs). LONs represent the search space by taking into account the basic concepts of network analysis, and allow us to extract a great number of metrics that can help understand the search space.

A LON is a graph where nodes represent local optima and edges represent relationships between edges, such as their probability of transition [11,21]. Different ways to assign weights to edges have been previously proposed in the literature according to the space to be analyzed. For example, basin transition and escape edges, for example, are used in combinatorial problems while perturbation edges are recommended for continuous problems [1,21].

A recent paper proposed to use LONs to analyze the search space of AutoML problems [18]. The authors use the LON proposed in [11]. This original version of LON has the drawback of not supporting search spaces with neutrality (i.e., spaces with regions of points with very similar values of fitness, which generate plateaus). As the literature has previously discussed that AutoML search spaces can have high neutrality [12,13], in this paper we propose to use two other variations of LONs - namely Monotonic Local Optima Network (MLON) and Compressed Monotonic Local Optima Network (CMLON), capable of dealing with the aforementioned problems, to analyze the fitness landscape of AutoML problems. Compared to the original model, MLONs discard edges that connect solutions where the quality of the final node is smaller than the quality in the initial node. CMLONs, in turn, compress the search space generated by MLONs by collapsing nodes that of same quality (value of fitness) into a single local optima. MLON defines an intermediate representation needed to build CMLON. Thus, the space characteristics are extracted from the final representation, i.e.,the CMLON.

The remainder of this paper is organized as follows. Section 2 presents related work into fitness landscape analysis and AutoML. Section 3 formally defines the problem and the fitness landscape of the space of solutions we are going to analyze. Section 4 introduces traditional LON and two of its variants, MLON and CMLON, and how they can be used to better understand AutoML spaces. Finally, Sect. 5 presents the experimental results and Sect. 6 draws conclusions and presents directions of future work.

2   Related Work

One of the few works that analyzes fitness landscapes based on LONs and is somehow related to ours is [20], where the authors adapt LONs to analyze the global structure of parameter configuration spaces. They looked at the metrics extracted from LONs and FDC, and observed large differences when tuning the same algorithm for different problem instances. For complex scenarios, they found a large number of sub-optimal funnels, while simpler problems had a single global funnel. With this same objective, the authors in [3] looked at parameter spaces for Particle Swarm Intelligence (PSO), and found that PSO's parameter landscapes are relatively simple at the macro level but a lot more complex at the micro level, making parameter tuning more difficult than they initially assumed.

The authors of [19] also used LONs as a tool to perform FLA on neural architecture research (NAS). They also proposed a set of characteristics to describe the search space. Some of the features are the overall fitness, ruggedness, cardinal of optima, among others. The authors of [17] studied the effect of the pivoting rule on the sampling configuration process of the search space. They identified that maximum expansion remains the most efficient technique to reach good-quality solutions.

The authors of [9] used FLA to analyze the NAS, where the solutions are Graph Convolutional Networks (GCN). They measure the neutrality of space using the method proposed by [15] and conclude that the analysis of the neutrality ratio indicates that the space is not neutral and suggest the need to use more elaborate techniques. The authors of [13] analyzed the AutoML loss fitness landscape to verify if the scenario is generally unimodal or convex and if most of the hyperparameters are independent of each other. They also empirically demonstrate that FDC has limitations in characterizing certain spaces, which highlights the need for new methods with LON.

3   Problem Definition and Fitness Landscape

The first step to apply FLA to a problem is to define the solution configuration space (the search space) and its fitness landscape. As previously defined, the fitness landscape is composed of three main components: the set of possible solutions (or configuration space)[1], the definition of a fitness function and the definition of a neighborhood, as explained next.

The set of solutions generated aim to solve a generalization of the Combined Algorithm Selection and Hyperparameter optimization (CASH) problem [4]. In its original definition, given a set $\mathcal{A} = \{A^{(1)}, A^{(2)}, \ldots, A^{(k)}\}$ of learning algorithms, where each algorithm $A^{(j)}$ has a hyperparameter space $\Lambda^{(j)} = \{\lambda^{(1)}, \ldots, \lambda^{(S)}\}$, defined from the full set of algorithm's hyper-parameters Ω, the CASH problem is defined as in Eq. 1[2]:

$$A^*_{\lambda^*} = \operatorname*{argmax}_{A^{(j)} \in \mathcal{A}, \lambda \in \Lambda^{(i)}} \frac{1}{k} \sum_{i=1}^{k} \mathcal{F}\left(A^{(j)}_{\lambda}, \mathcal{D}^{(i)}_{train}, \mathcal{D}^{(i)}_{valid}\right) \tag{1}$$

where $\mathcal{F}(A^{(j)}_{\lambda}, \mathcal{D}^{(i)}_{train}, \mathcal{D}^{(i)}_{valid})$ is the gain achieved when a learning algorithm A, with hyperparameters Λ, is trained and validated on disjoint training and validation sets $\mathcal{D}^{(i)}_{train}$ and $\mathcal{D}^{(i)}_{valid}$, respectively, on each partition $1 \leq i \leq k$ of a k-fold cross-validation procedure.

A generalization can be made if we replace $\mathcal{A}$ by a set of pipelines $\mathcal{P} = \{P^{(1)}, \ldots, P^{(V)}\}$, which includes a subset of algorithms from $\mathcal{A}$ and their respective set of hyperparameters $\Gamma^{(i)} = \{\Lambda^{(1)}, \ldots, \Lambda^{(S)}\}$, represented by the full set Ψ, as defined in Eq. 2

$$\mathbf{P}^*_{\Gamma^*} = \operatorname*{argmax}_{\mathcal{P}^{(i)} \subseteq \mathbf{P}, \Gamma^{(i)} \subseteq \Psi} \frac{1}{K} \cdot \sum_{j=1}^{K} \mathcal{F}(\mathbf{P}^{(i)}_{\Gamma^{(i)}}, D^{(j)}_{train}, D^{(j)}_{valid}) \tag{2}$$

According to this definition, we need to choose the set of algorithms and their respective hyperparameters that can be present into a pipeline. Depending on the number of algorithms and hyperparameters defined, the number of solutions to be generated may be prohibitive to enumerate. We chose to work with a space

---

[1] Solution and configuration are synonyms in fitness landscape terminology.
[2] The original definition casts the problem as a minimization one. Here we replace the loss function by a gain function.

large enough to be optimized but small enough to be fully enumerated. The main rationale behind that is to understand how well the LON and metrics extracted from it characterize the space in terms of local and global optima.

Search Space Definition: The space is defined by a grammar to avoid the generation of invalid solutions, and allow for preprocessing methods and classification algorithms and their hyperparameters. Each solution is represented by a derivation tree extracted from the grammar. The grammar has 38 production rules[3]. In terms of preprocessing, it includes algorithms that deal with feature scaling and dimensionality reduction, such as PCA and Select K-Best. It is also possible for a pipeline to use no preprocessing algorithms. In terms of classification methods, there are five possible options: Logistic Regression, Multilayer Perceptron, K-Nearest Neighbors (KNN), Random Forest, and Ada Boost. The number of hyperparameters varies from one classification algorithm to another, going from two (Ada Boost) to 7 (Random Forest). It can generate up to 69,960 solutions.

In this paper, the definition of neighborhood depends on the concept of mutation and the method adopted was the same used by Pimenta et. al [12], however, after some experiments, it was possible to observe that the cost of mutating a pipeline is a bottleneck in the experiments, so the strategy adopted to deal with this problem was to calculate the PMF of the mutation of pipeline u to generate v. Mutation occurs through the random selection of a node in the tree, where the probability of a node being selected is inversely proportional to its distance from the root of the tree, and generating a subtree from that node. Thus, the probability is proportional to the product of selecting a given node by the number of subtrees that can be built from it. This process can be repeated until reaching the root of the tree, where the probability of generating the pipeline v is the probability of generating any tree, that is, $p = 1/\#tree$.

The process is illustrated in the Algorithm 1. It calculates the probability of mutation between any two pipelines. Initially the algorithm looks for nodes with different values in the trees passed as an argument and returns them, for example, in Fig. 1 the pipelines differ in the hyperparameters (leaf nodes) represented with dashed borders. Then, the algorithm identifies the path from the node to the root common to all nodes identified by the previous function. In the figure, these nodes are represented by the shaded background color. Then, for each of these nodes present in the path, the function accumulates the probability considering the probability of the node being selected (GET_PROB_SEL) multiplied by the probability of a subtree of v rooted at the node common to both pipelines is generated (GET_NUM_COMB).

Neighborhood: The neighborhood operator $\mathcal{N}$ is defined as all pipelines generated from the mutation operator. A consequence of the neighborhood definition is that neighboring pipelines are more likely to have the same algorithms, since

---

[3] The complete grammar in BNF is available at https://bit.ly/38Fo3U.

Algorithm 1. Mutation probability
```
 1: function MUTATION_PROB(p₁, p₂)
 2:     diffNodes ← FIND_DIFF_NODES(p₁, p₂)
 3:     mutationPath ← COMMON_ANCESTOR(diffNodes)
 4:     prob ← 0
 5:     for node in mutationPath do
 6:         probSel ← GET_PROB_SEL(node)
 7:         numComb ← GET_NUM_COMB(node)
 8:         prob += probSel · 1/numComb
 9:     end for
10:     return prob
11: end function
```

the nodes most likely to be selected are the leaf nodes, which represent the hyper-parameters, although it is still possible that pipelines with completely different algorithms be neighbors. For example, Fig. 1 shows the process of calculating the probability of mutation between two pipelines. Observe that the point selected for mutation is the <ada_boost> and, from it, an entire subtree is generated according to the grammar. Note, however, that if any other parent node, direct or not, of <ada_boost> were selected (the nodes highlighted by the shaded background color), it would be possible to get the same neighbor if the same subtree was generated, but the probability of selecting nodes closer to the root is smaller than that of selecting nodes further away from the root.

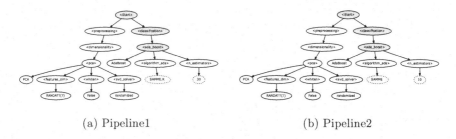

(a) Pipeline1   (b) Pipeline2

Fig. 1. Example of solutions. Both are using PCA as preprocessing algorithm and AdaBoost as classification algorithm. The hyperparameter can be seen in the leaf nodes.

Fitness: The fitness function adopted was the micro F1. As many datasets have more than two classes, the One-vs-All method was adopted to deal with these cases.

4   Local Optimal Networks

Having the fitness landscape, the next step to perform FLA, in our case, is to build the LON. One way to see a fitness landscape if using a graph $G = (V, E)$, where each configuration v of a solution space represents a node in the graph, i.e., $v \in V$, and there is a directed edge (u, v) if $v \in \mathcal{N}(u)$ to $u, v \in V$. The weight w_{uv} of an edge can represent, for example, a distance between the two nodes, and depends on how the operator $\mathcal{N}$ is defined.

In the local optima networks, given the graph G, we need to filter the nodes so that only the local optima remain. Hence, we can define the LON as LON = (V', E') where $V' = \{v \mid v \in V \text{ and } \mathcal{F}(v) \geq \mathcal{F}(\mathcal{N}(u))$ and $\mathcal{F} : S \to \mathbb{R}$ is a function that maps each solution in the search space to a quality metric, in our case, the fitness.

As we are working with an enumerable space, we evaluate all solutions and identify the LO in the LON instead of sampling the space – which is the standard for larger spaces. Having the LO, we need to understand the concept of basins of attraction. A basin of attraction has all solutions from the space that are "attracted" into its direction, i.e., all the solutions that, after a local search (LS), end up in LO_u. Formally:

$$\text{Basin of attraction LO}_u := \{v \in V \mid LS(v) = LO_u\} \tag{3}$$

where $LS : S \to S$ performs a local search: given a solution, it returns its LO. In practice, LS is usually defined as the Hill-Climb (HC) [2,10,11]. The HC can be implemented in different ways, including the use of First Improvement, Best Improvement, Worst Improvement, Approximate Worst Improvement and Max-Expansion. The results in [17] show Max-Expansion as the most effective to find high quality solutions. Here we use the Best Improvement due to its simplicity and the use of other strategies is left for future work.

Given the definitions above, observe there is a direct relationship between the number of LO and the roughness of the search space. According to the literature, roughness is a characteristic that can make search difficult for different search methods, specially those that follow a local search approach. As the LON is a graph, we can then borrow a lot of metrics from the network analysis literature to characterize the search space of the generalized CASH problem. For example, the results reported in [18] show that neighborhood size affects the roughness of the space, as more neighbors generate bigger basins of attraction. However, one of the main goals of AutoML is to minimize the number of fitness evaluations (which can be computationally expensive in many problems), which can be achieved by minimizing the number of neighbors. However, LONs do not provide ways to analyze neutrality. For this reason, CMLONs – which are networks derived from LONs – were proposed.

4.1   Basin Transition and Escape Edges

Another characteristic of the search space that is interesting to be studied is the relationship between basins of attraction, i.e., the ability or probability of a solutions going from one basin of attraction to another. The relationship between basins of attraction is represented by weighted directed edges between LO. Here these relationships are captured using two different types of edges: (i) basin-transition and (ii) escape edges.

Given an edge (LO_u, LO_v) that connects two local optima u and v, basin-transition edges receive weights given by the sum of the probabilities of a solution in the basin of attraction of u end up in the basin of attraction of v after a perturbation (mutation) is applied to the solution. That is,

$$p(LO_u, LO_v) = \frac{1}{|LO_u|} \sum_{u \in LO_u} \sum_{v \in LO_v} p(u \to v)$$

where $p(u \to v)$ is the probability of mutation, LO_u and LO_v are basin of attractions.

Escape Edges, in turn, define the weight of (LO_u, LO_v) proportional to the number of solutions s that are within a distance D of LO_u, $d(s, LO_u) \leq D$, but that are also within the basin of attraction of LO_v.

4.2   Monotonic LON (MLON) and Compressed MLON (CMLON)

Although LON reduces the configuration space looking only at the LOs, in many problems even the space composed of only LOs still has a high number of nodes and edges, making it difficult to visualize the problem and not providing means for analyzing the neutrality of the space. Because of that, the Monotonic LON (MLON) was proposed. The main difference of the MLON compared to the original LON is that is does not include edges where the value of fitness from one LO to the next decreases, i.e., a LO only connects to another with improved fitness values. Hence, MLONs are more compact than the original LON.

In many cases, even the reduction of the graph from LON to MLON was still not enough to improve space visualization, which motivated the introduction of Compressed MLONs. cMLONs compress adjacent nodes that have the same value of fitness. This type of compression is very effective in networks with high neutrality, i.e., many solutions with neighbors with the same fitness. Neutral regions of the network are identified using an adaption of a Breadth First Search (BFS), which receives a node u and returns all the nodes with the same fitness and that can be reached from u.

The regions where LOs are compressed for having the same fitness are known as plateaus [2,10], and the sets of LOs that are in the basins of attraction of other LOs after edges with decreasing fitness are removed are named funnels [10].

4.3   Space Metrics

Here we use the same metrics used in the work of [10] to analyze the structure of a LON, MLON and CMLON. These metrics help to measure the neutrality of space. The first two metrics are independent from the type of edge used to build the network. They are: (1) the number of LO (**noptima**) and (2) the space modality (**nglobal**).

Additionally, for each type of edge being considered in the standard LON representation, the following metrics are analyzed:

- **edgesi**: measures the sum of the weights of the edges that leave a solution u to a solution v, where $\mathcal{F}(u) < \mathcal{F}(v)$;
- **edgesn**: measures the total weight of neural edges, i.e., edges between nodes with same fitness; and
- **edgesw**: measures the total weight of edges that connect solutions in nodes u and v, where $\mathcal{F}(u) > \mathcal{F}(v)$.

Apart from the metrics related to traditional LONs, five other can be extracted from CMLONs:

- **ncoptima**: is the number of compressed LO, i.e., the number of LO that are neighbors of other LO with the same value of fitness and were merged in the CMLON;
- **ncglobal**: is a subset of *ncoptima* that accounts only for the number of global optima compressed;
- **ncedges**: calculates the weight between the LONs that were compressed when building the CMLON from the MLON.
- **neutrality**: is given by the ratio between the number of compressed LO and the number of LO in the space;
- **lplateau**: measures the size of the plateaus – compressed regions of the CMLON – and identifies the regions with the highest number of LO.

5   Experimental Results

Experiments were performed in 7 classification datasets from the UCI repository[4], listed in Table 1. As the majority of the datasets are multi-class, the fitness used was the micro-f1 together with the one versus all approach. From the data in the table, observe that most datasets have more than one global optimum, making the search spaces multi-modal.

All LONs were generated 30 times to guarantee statistical validity of the results, and hence all results reported are an average over 30 runs. Note that although we have the full space enumerated, we simulate the process followed by the LON technique, which is based on a sample of the space. In this case, three sizes of neighborhood were tested: 15, 20, and 25.

The first experiment was performed to verify if the size of the basins of attraction from different executions come from the same distribution. It is important to check whether small changes in the input graph cause large changes in the LON. For that, a two-sample Kolmogorov-Smirnov test was applied to each LON built from a different neighborhood size. The results show that, with 95% confidence, we cannot reject H_0 for more than 92.87% of cases. If we increase the confidence to 99%, we cannot reject H_0 in 97.47% of the cases. This is an important

---

[4] https://archive.ics.uci.edu/ml/datasets.php.

Table 1. Characteristics of the datasets.

Dataset	Instances	Features	Classes	Optimum	#Optimum
Diabetes (DB)	768	8	2	0.8011	1
Ml-prove (MP)	6,118	51	6	0.4478	21
Statlog-segment (ST)	2,310	19	7	0.9696	40
Texture (TX)	5,500	40	11	0.9980	2
Vehicle (VH)	846	18	4	0.7993	1
Wilt (WL)	4,839	5	2	0.9890	1
Wine-quality-red (WN)	1,599	11	6	0.6446	16

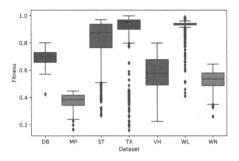

Fig. 2. Boxplot of fitness obtained in each of the datasets.

result and indicates that even changing the sampling performed in the configuration space, the structure of the network remains the same, demonstrating the robustness of the LON over small changes in the structure of the graph.

Turning to the analysis of the LON obtained from the 7 datasets studied in this paper, Fig. 3a shows the number of LO per dataset. Observe that by changing the neighborhood size we reduce the number of LOs in all datasets. However, this reduction is more significant in a few datasets. For example, for dataset DB, the reduction in the number of LO is of 43.77%, while for dataset WL we get a reduction of 65.90% when changing the neighborhood size from 15 to 25. This shows that increasing the neighborhood size can bring significant gains for some datasets. Also note that the drop in the number of LOs when neighborhood size increases from 15 to 20 is higher than when it goes from 20 to 25, suggesting there might be an ideal trade-off between neighborhood size and space roughness. Figure 3b shows the size of the basins of attraction, and is complementary to Fig. 3a. Since the number of configurations is constant, as the number of LO decreases the concentration of solutions in each basin of attraction increases. Raising neighborhood size results in a few LO being attached to others, creating "super" basins of attraction.

Table 2 presents the metrics extracted from the CMLON built using the basin-transition method. In the table, it is possible to observe that the dataset with the largest number of compressed local optima (ncoptima) is WL followed

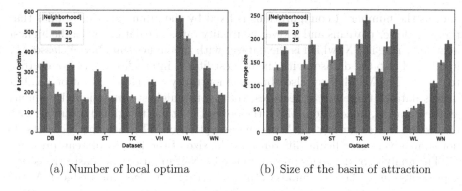

(a) Number of local optima   (b) Size of the basin of attraction

Fig. 3. Characteristics of the LON generated from the fully-enumerated space.

by DB and WN. It is interesting to note that, in Fig. 2, the WL dataset has the lowest variance ($\sigma^2 = 0.00007$), meaning that in this dataset there are many solutions with the same fitness (99.99% of the solutions have their fitness value duplicated), including the local optima and, consequently, many end up being compressed in the construction of the CMLON.

Regarding the number of compressed global optima (ncglobal), the metric is zero in the DB, WH and WL datasets, as these datasets have only one global optimum. As for the others, it is possible to observe that the dataset with the largest number of compressed global optima is ST followed by MP and WN. This result agrees with those presented in Table 1, as these datasets are the ones with the highest number of global optima. It is interesting to observe that the global optima in these datasets are formed by pipelines with the same classifier – only with variations in their hyperparameters. Regarding the preprocessing algorithms, in the ST dataset there are different global optima with and without preprocessing, showing that the fitness remains the same even when removing an entire task. This indicates that under the presence of a certain sets of hyperparameters, others do not affect the fitness of the solution.

Also note, from the table, that the total weight of the compressed edges (ncedges) is proportional to the number of compressed local optima (ncoptima). The WL dataset has the highest total compressed weight due to the number of compressed local optima. Now, when the DB dataset is compared to WN, it is possible to observe that DB has more compressed solutions. However, it is the WN dataset that has the greatest weight. This indicates that for WN the solutions compressed have a higher probability of transition between the basins of attraction than in the DB dataset, as the weight of the edges is greater.

The neutrality rate, represented by the neutrality column in the table, is the ratio of the number of plateaus to the number of local optima. Although the WL dataset has the highest number of compressed local optima, this dataset has the lowest neutrality rate because the number of plateaus is small when compared to the number of local optima present in the space. Furthermore, in this dataset the size of the largest plateau is significantly larger than that of the other datasets and, as the number of configurations is fixed by the grammar, this implies that there are fewer plateaus and a lower neutrality rate. The dataset with the highest neutrality rate is TX, which is the dataset with the highest number of classes (11 in all) and also the dataset with the highest fitness interval, ranging from 0.1603 to 0.9980. The MP dataset also obtains a neutrality rate close to that of TX, but in this dataset there are 51 features, that is, the highest dimensionality among all the datasets. In addition, MP is the dataset that has the global optimum with the lowest fitness among all. With the exception of WL, the other datasets do not show a very significant difference in the size of the largest plateau present.

The analysis of the edges present in the LON graph using the basin-transition is presented in the last 3 columns of the table (iedges, nedges and wedges). As the absolute value does not provide much information, in this work the proportion of each type of edge in the graph is presented, so the sum of the three always results in 1. In all cases, it is possible to observe that the weight of the iedges is greater than that of the others. The WL dataset has the highest proportion of iedges in relation to the other datasets and is also the dataset with the lowest proportion of wedges. This shows that there is a greater probability of transition from a basin of attraction to another with higher fitness.

Table 2. Metrics for CMLONs with basin transition.

Dataset	$|\mathcal{N}|$	ncoptima	ncglobal	ncedges	neutrality	lplateau	iedges	nedges	wedges
DB	15	322.37	0.00	27.49	0.12	33.37	0.61	0.23	0.16
	20	236.03	0.00	21.27	0.13	29.33	0.60	0.23	0.17
	25	188.13	0.00	17.82	0.14	24.33	0.61	0.23	0.17
MP	15	250.40	7.00	22.70	0.16	38.13	0.69	0.20	0.11
	20	175.57	6.90	19.45	0.17	36.53	0.68	0.22	0.10
	25	145.93	7.20	17.62	0.18	34.13	0.67	0.23	0.10
ST	15	286.50	14.33	25.26	0.12	34.80	0.62	0.21	0.17
	20	211.53	14.40	19.48	0.12	29.00	0.63	0.21	0.17
	25	171.30	13.57	16.24	0.13	26.73	0.64	0.20	0.16
TX	15	244.63	0.93	26.03	0.17	29.93	0.61	0.24	0.15
	20	168.53	1.07	24.52	0.18	30.40	0.58	0.27	0.15
	25	138.53	1.40	24.33	0.18	30.73	0.55	0.30	0.15
VH	15	233.87	0.00	21.58	0.14	22.63	0.57	0.26	0.17
	20	172.70	0.00	16.66	0.16	19.20	0.56	0.25	0.18
	25	145.37	0.00	13.61	0.16	18.33	0.57	0.24	0.19
WL	15	539.00	0.00	74.64	0.07	215.80	0.71	0.23	0.07
	20	448.27	0.00	55.54	0.07	187.60	0.74	0.20	0.06
	25	359.30	0.00	41.16	0.08	157.67	0.77	0.18	0.05
WN	15	292.33	4.67	30.52	0.12	33.73	0.60	0.26	0.14
	20	219.87	5.60	26.44	0.13	27.43	0.58	0.27	0.15
	25	180.77	6.03	24.59	0.14	25.77	0.56	0.27	0.16

Table 3. Metrics for CMLONs with escape edges (D = 2).

Dataset	$|\mathcal{N}|$	ncoptima	ncglobal	ncedges	neutrality	lplateau	iedges	nedges	wedges
DB	15	290.07	0.00	30.23	0.21	27.33	0.58	0.34	0.08
	20	224.70	0.00	24.38	0.20	26.40	0.54	0.39	0.07
	25	181.33	0.00	19.57	0.20	21.70	0.52	0.42	0.06
MP	15	207.33	7.50	27.44	0.18	28.57	0.64	0.29	0.07
	20	156.80	6.90	23.15	0.20	29.20	0.58	0.36	0.06
	25	136.10	7.40	19.99	0.20	30.43	0.54	0.41	0.05
ST	15	262.87	14.83	28.03	0.19	31.60	0.60	0.33	0.07
	20	202.70	14.40	21.60	0.18	28.23	0.57	0.36	0.07
	25	166.03	13.87	18.02	0.19	26.30	0.56	0.38	0.06
TX	15	218.60	0.93	25.97	0.22	29.73	0.59	0.33	0.08
	20	160.73	1.07	24.39	0.23	30.40	0.52	0.41	0.06
	25	136.47	1.40	24.36	0.21	30.73	0.47	0.47	0.06
VH	15	220.73	0.00	21.25	0.20	21.50	0.55	0.37	0.08
	20	167.73	0.00	16.93	0.21	18.63	0.51	0.41	0.07
	25	142.83	0.00	13.75	0.20	17.93	0.50	0.44	0.07
WL	15	449.10	0.00	71.55	0.13	154.10	0.71	0.26	0.03
	20	394.53	0.00	52.83	0.13	145.80	0.73	0.25	0.03
	25	318.30	0.00	40.30	0.13	117.23	0.75	0.23	0.02
WN	15	273.07	4.93	31.99	0.20	27.70	0.55	0.37	0.08
	20	214.73	5.67	26.66	0.20	26.20	0.50	0.43	0.08
	25	178.40	6.10	25.07	0.19	25.77	0.45	0.47	0.08

The results obtained from escape-edges are shown in Table 3. It is possible to verify that, as in the basin-transition, the WL dataset is the one with the largest number of compressed local optima, followed by DB and WN. It is possible to observe a reduction in the magnitude of the values and this is due to solutions that are at a distance $D > 2$ from the local optimum, not satisfying the escape-edge condition and, therefore, not causing an increase in the weight of the edge.

The number of compressed global optima is also similar to the one obtained using the previous method. The same scenario is repeated when analyzing the number of compressed edges, as the WL dataset is still the one with the highest total weight followed by the WN and DB datasets. The neutrality rate indicates that the TX dataset is the largest followed by DB and VH. The WL dataset, in turn, has the lowest neutrality rate in relation to the others, but in the escape-edges the difference between this dataset and the others is less significant. As for the analysis of the largest plateau of the graph, WL still has the largest region of neutrality among the datasets, but at a lower intensity than in the basin-transition.

Regarding the edges assigned with the escape-edges, the columns iedges, nedges, and wedges represent the proportion of edges of each type present in the graph. It is possible to observe that, in comparison with the basin-transition, the number of neutral edges increased while the number of wedges decreased, i.e., the cut in $D \leq 2$ reduced the relation of global optima with others with lower fitness. The datasets that have the highest proportion of neutral edges are TX and WN with a neighborhood size equals to 25. Analyzing the proportion of edge weights, observe that there is a tendency of transition between basins of attraction with the same fitness (neutral), although there is also a significant percentage of transitions to basins of attraction with higher fitness.

6   Conclusions and Future Work

This work presents the analysis of the fitness landscape of AutoML problems. The structure of MLON and CMLON were used to analyze the neutrality of AutoML search spaces. CMLON proved to be a valid tool for the analysis of the neutrality of spaces, as it allows the extraction of several metrics that provide more robust information than those provided by other techniques used in previous works. However, the neutrality rate can be misleading, as it considers the number of neutral LOs but does not account for the size of the plateaus. This can result in considering a space with few plateaus of significant size to be indicated as a space with little neutrality. One solution would be to consider the number and magnitude of plateaus when calculating the neutrality rate.

In short, results show that the space has high neutrality, with the datasets with the largest plateaus being those with low fitness variance, such as WL. The analyses with the CMLON provided several metrics to estimate the neutrality of the spaces studied. However, new metrics can still be derived to improve the analyses made using the CMLON.

References

1. Adair, J., Ochoa, G., Malan, K.M.: Local optima networks for continuous fitness landscapes. In: Proceedings of the Genetic and Evolutionary Computation Conference Companion, pp. 1407–1414. ACM, New York (2019). https://doi.org/10.1145/3319619.3326852
2. Chicano, F., Ochoa, G., Tomassini, M.: Real-like MAX-SAT instances and the landscape structure across the phase transition. In: Proceedings of GECCO, pp. 207–215 (2021)
3. Cleghorn, C.W., Ochoa, G.: Understanding parameter spaces using local optima networks. In: Proceedings of GECCO Companion, New York, NY, USA, pp. 1657–1664 (2021)
4. Feurer, M., Klein, A., Eggensperger, K., Springenberg, J., Blum, M., Hutter, F.: Efficient and robust automated machine learning. Adv. Neural Inf. Process. Syst. **28**, 2962–2970 (2015)
5. Garciarena, U., Santana, R., Mendiburu, A.: Analysis of the complexity of the automatic pipeline generation problem. In: 2018 IEEE Congress on Evolutionary Computation (CEC), pp. 1–8. IEEE (2018)
6. Hutter, F., Kotthoff, L., Vanschoren, J. (eds.): Automated Machine Learning: Methods, Systems, Challenges. Springer, Heidelberg (2018). in press, https://doi.org/10.1007/978-3-030-05318-5. http://automl.org/book

7. Jones, T., Forrest, S., et al.: Fitness distance correlation as a measure of problem difficulty for genetic algorithms. In: ICGA, vol. 95, pp. 184–192 (1995)
8. Malan, K., Engelbrecht, A.P.: A survey of techniques for characterising fitness landscapes and some possible ways forward. Inf. Sci. **241**, 148–163 (2013)
9. Nunes, M., Fraga, P.M., Pappa, G.L.: Fitness landscape analysis of graph neural network architecture search spaces. In: Proceedings of GECCO, New York, NY, USA, pp. 876–884 (2021)
10. Ochoa, G., Chicano, F.: Local optima network analysis for MAX-SAT. In: Proceedings of the GECCO Companion, pp. 1430–1437 (2019)
11. Ochoa, G., Verel, S., Daolio, F., Tomassini, M.: Local optima networks: a new model of combinatorial fitness landscapes, pp. 233–262 (2014)
12. Pimenta, C.G., de Sá, A.G.C., Ochoa, G., Pappa, G.L.: Fitness landscape analysis of automated machine learning search spaces, pp. 114–130 (2020)
13. Pushak, Y., Hoos, H.H.: AutoML loss landscapes (2022)
14. Rakitianskaia, A., Bekker, E., Malan, K.M., Engelbrecht, A.: Analysis of error landscapes in multi-layered neural networks for classification. In: 2016 IEEE Congress on Evolutionary Computation (CEC), pp. 5270–5277. IEEE (2016)
15. Reidys, C.M., Stadler, P.F.: Neutrality in fitness landscapes. Appl. Math. Comput. **117**(2–3), 321–350 (2001)
16. Richter, H.: Fitness landscapes: from evolutionary biology to evolutionary computation. In: Richter, H., Engelbrecht, A. (eds.) Recent Advances in the Theory and Application of Fitness Landscapes. ECC, vol. 6, pp. 3–31. Springer, Heidelberg (2014). https://doi.org/10.1007/978-3-642-41888-4_1
17. Tari, S., Ochoa, G.: Local search pivoting rules and the landscape global structure. In: Proceedings of the Genetic and Evolutionary Computation Conference, pp. 278–286 (2021)
18. Teixeira, M.C., Pappa, G.L.: Understanding AutoML search spaces with local optima networks. In: Genetic and Evolutionary Computation Conference (2022)
19. Traoré, K.R., Camero, A., Zhu, X.X.: Fitness landscape footprint: a framework to compare neural architecture search problems (2021). http://arxiv.org/abs/2111.01584
20. Treimun-Costa, G., Montero, E., Ochoa, G., Rojas-Morales, N.: Modelling parameter configuration spaces with local optima networks. In: Proceedings of GECCO, pp. 751–759 (2020)
21. Vérel, S., Daolio, F., Ochoa, G., Tomassini, M.: Local optima networks with escape edges, pp. 49–60 (2012)

Dealing with Inconsistencies in $ASPIC^+$

Rafael Silva$^{(\boxtimes)}$ and João Alcântara

Department of Computer Science, Federal University of Ceará, Fortaleza, Brazil
{rafaels,jnando}@lia.ufc.br

Abstract. Inconsistencies in argumentation theory have been a recurrent topic in the literature. The $ASPIC^+$ being one of the most well know argumentation formalisms is frequently used to deal with inconsistencies. However, the existing approaches consider a limited version of $ASPIC^+$. In our work, we managed to deal with inconsistencies in a very general scenario. To do this, we impose in $ASPIC^+$ some reasonable conditions to the relations between arguments to adjust how arguments interact with each other. As consequence, we avoid inconsistent arguments interfere with consistent arguments by neutralizing the inconsistent argument. Then we show that under simple conditions, $ASPIC^+$ preserves current results on the satisfaction of fundamental properties of consistency and logical closure.

Keywords: Argumentation · $ASPIC^+$ · Inconsistencies

1 Introduction

As noticed in [1], inconsistencies can be considered under the mantle of many points of view: as a consequence of the only correct description of a contradictory world, as a temporary state of our knowledge, as the outcome of a particular language which we have chosen to describe the world, as the result of conflicting observational criteria, as the superposition of world-views, or as the result from the best theories available at a given moment.

Argumentation is a form of reasoning which is usually referred to as a natural approach to handling inconsistencies and uncertain information. In argumentation, a debate between people with opposing opinions can be represented by a directed graph, in which each argument is a node and the attacks between arguments are the edges. Such a graph can then be analyzed to determine which arguments are acceptable according to some general criteria. This approach to model arguments and their interactions was proposed by Dung in [2]. These arguments are said to be abstract as they have no internal structure.

Part of what makes Dung's work interesting is its level of abstraction, since the argumentation framework can be instantiated by a wide range of logical formalisms. Despite the popularity of Dung's frameworks, their abstract nature is something that does not gives us an idea of what kind of instantiation satisfies

This research was partly financed by FUNCAP.

J. C. Xavier-Junior and R. A. Rios (Eds.): BRACIS 2022, LNAI 13653, pp. 488–503, 2022.
https://doi.org/10.1007/978-3-031-21686-2_34

intuitive rational properties. Then, as an alternative, the structured argumentation framework $ASPIC$ [3] was developed, which provides an intermediate level of abstraction. In [4], $ASPIC^+$ framework was presented, which generalizes $ASPIC$ framework to accommodate a broader range of instantiations.

As observed in [5], in this framework is not a system but a framework for specifying systems. In $ASPIC^+$, arguments represent inference trees formed by chaining applications of inference rules to premises. It also permits to compare arguments through preference relations. In $ASPIC^+$, preference relations play a major role in the interaction between arguments.

Although $ASPIC^+$ is a prominent approach in structured argumentation, in [6,7] it is noticed that when mixing certain and uncertain information as for example strict and defeasible rules, $ASPIC^+$ may lead to unintuitive results such as inconsistent arguments interfering with the acceptability of consistent arguments. To deal with such anomalies, in [3], some rationality postulates were defined. These postulates are intended to guarantee some basic suitability of the outputs of any argumentation formalism. According to [3], any argumentation formalism should satisfy these postulates. Another challenge is how to handle inconsistencies in $ASPIC^+$. Over the years, some proposals have been presented to handle them [6–8]. However, these proposals only considered restricted versions of $ASPIC^+$.

In our work, for the first time it is presented an argumentation formalism based on $ASPIC^+$ robust enough to work with inconsistent arguments, to allow for the instantiation of preference relations between arguments, and to satisfy the fundamental rationality postulates of [3]. We achieved this by imposing some reasonable conditions to neutralize the effect of inconsistent arguments over consistent arguments. As we are working with a very general version of $ASPIC^+$, our approach is ready to accommodate any logical language with these reasonable properties.

The rest of the paper is organised as follows: in Sect. 2, $ASPIC^+$ framework is presented. Section 3 is focused on proving the satisfaction of the rationality postulates described in [3]. In the next section, we discuss related works. Finally, we summarize our contributions and themes for future developments.

2   The $ASPIC^+$ Framework

$ASPIC^+$ is a general framework for specifying systems. It defines the notion of an abstract argumentation system as a structure consisting of a logical language $\mathcal{L}$ with a binary contrary relation, a set of strict and defeasible rules, two types of premises: firm and plausible, and a partial function. In $ASPIC^+$ no assumptions are made about how these elements are defined.

Definition 1 (Argumentation System). [5] *An argumentation system is a tuple $AS = (\mathcal{L}, ^-, \mathcal{R}, n)$, in which*

- *$\mathcal{L}$ is a logical language with a unary negation symbol $\neg$.*
- *$^-$ is a function from $\mathcal{L}$ to $2^{\mathcal{L}}$, such that*

- φ is a contrary of ψ if $\varphi \in \overline{\psi}$, $\psi \notin \overline{\varphi}$;
- φ is a contradictory of ψ if $\varphi \in \overline{\psi}$, $\psi \in \overline{\varphi}$;

- $\mathcal{R} = \mathcal{R}_s \cup \mathcal{R}_d$ is a set of strict ($\mathcal{R}_s$) and defeasible ($\mathcal{R}_d$) inference rules of the form $\phi_1, \ldots, \phi_n \to \phi$ and $\phi_1, \ldots, \phi_n \Rightarrow \phi$ respectively (in which $\phi_1, \ldots, \phi_n, \phi$ are meta-variables ranging over wff in $\mathcal{L}$), and $\mathcal{R}_s \cap \mathcal{R}_d = \emptyset$.
- n is a partial function such that $n : \mathcal{R}_d \longrightarrow \mathcal{L}$.

For any formula $\phi \in \mathcal{L}$, we say $\phi \in \overline{\psi}$ if either $\phi = \neg\psi$ or $\neg\phi = \psi$. By $\mathbb{C}(\phi) = \{\psi \mid \psi \in \overline{\phi} \text{ and } \phi \in \overline{\psi}\}$ we mean the set of all contradictory formulas of ϕ. We will refer to a contradictory of ϕ as $-\phi$ ($-\phi \in \mathbb{C}(\phi)$). Note for any $\phi \in \mathcal{L}$, ϕ is a contradictory of $\neg\phi$.

It is also required a knowledge base to provide premises for the arguments.

Definition 2 (Knowledge Base). [5] *A knowledge base in an argumentation system $AS = (\mathcal{L}, ^-, \mathcal{R}, n)$ is a set $\mathcal{K} \subseteq \mathcal{L}$ consisting of two disjoint subsets $\mathcal{K}_n$ (the axioms) and $\mathcal{K}_p$ (the ordinary premises).*

Ordinary premises are uncertain knowledge which can be attacked, and axioms are certain and cannot be attacked. Now we define an argumentation theory:

Definition 3 (Argumentation Theory). [5] *An argumentation theory is a tuple $AT = (AS, \mathcal{K})$ where AS is an argumentation system and $\mathcal{K}$ is a knowledge base in AS.*

In Definition 4, arguments are defined together with some functions. Informally, functions `Prem`, `Conc`, `Sub`, `Rules`, `TopRule` and `DefR` return respectively the set of premises, sub-arguments, defeasible rules, conclusion and top-rule of an argument. In $ASPIC^+$, arguments are constructed recursively from an argumentation theory by the successive application of construction rules:

Definition 4 (Argument). [5] *An argument A on the basis of an argumentation theory $(AS, \mathcal{K})$ and an argumentation system $(\mathcal{L}, ^-, \mathcal{R}, n)$ is*

1. *ϕ if $\phi \in \mathcal{K}$ with $\text{Prem}(A) = \{\phi\}$, $\text{Conc}(A) = \phi$, $\text{Sub}(A) = \{\phi\}$, $\text{DefR}(A) = \emptyset$, $\text{Rules}(A) = \emptyset$, $\text{TopRule}(A) = undefined$.*
2. *$A_1, \ldots, A_n \to \psi$ if $n \geq 1$ and $A_1, \ldots, A_n$ are arguments s.t. there is a strict rule $\text{Conc}(A_1), \ldots, \text{Conc}(A_n) \to \psi \in \mathcal{R}_s$; $\text{Prem}(A) = \text{Prem}(A_1) \cup \cdots \cup \text{Prem}(A_n)$; $\text{Conc}(A) = \psi$; $\text{Sub}(A) = \text{Sub}(A_1) \cup \cdots \cup \text{Sub}(A_n) \cup \{A\}$; $\text{Rules}(A) = \text{Rules}(A_1) \cup \cdots \cup \text{Rules}(A_n) \cup \{\text{Conc}(A_1), \ldots, \text{Conc}(A_n) \to \psi\}$; $\text{TopRule}(A) = \text{Conc}(A_1), \ldots, \text{Conc}(A_n) \to \psi$.*
3. *$A_1, \ldots, A_n \Rightarrow \psi$ if $n \geq 1$ and $A_1, \ldots, A_n$ are arguments such that there exists a defeasible rule $\text{Conc}(A_1), \ldots, \text{Conc}(A_n) \Rightarrow \psi \in \mathcal{R}_d$; $\text{Prem}(A) = \text{Prem}(A_1) \cup \cdots \cup \text{Prem}(A_n)$; $\text{Conc}(A) = \psi$; $\text{Sub}(A) = \text{Sub}(A_1) \cup \cdots \cup \text{Sub}(A_n) \cup \{A\}$; $\text{Rules}(A) = \text{Rules}(A_1) \cup \cdots \cup \text{Rules}(A_n) \cup \{\text{Conc}(A_1), \ldots, \text{Conc}(A_n) \Rightarrow \psi\}$; $\text{TopRule}(A) = \text{Conc}(A_1), \ldots, \text{Conc}(A_n) \Rightarrow \psi$.*

For any argument A we define $\text{Prem}_n(A) = \text{Prem}(A) \cap \mathcal{K}_n$; $\text{Prem}_p(A) = \text{Prem}(A) \cap \mathcal{K}_p$; $\text{DefR}(A) = \{r \in \mathcal{R}_d \mid r \in \text{Rules}(A)\}$ and $\text{StR}(A) = \{r \in \mathcal{R}_s \mid r \in \text{Rules}(A)\}$.

Example 1. Consider the argumentation system $AS = (\mathcal{L}, ^-, \mathcal{R}, n)$, in which

- $\mathcal{L} = \{a, b, f, w, \neg a, \neg b, \neg f, \neg w, \sim a, \sim b, \sim f, \sim w, \sim \neg a, \sim \neg b, \sim \neg f, \sim \neg w\}$. The symbols $\neg$ and $\sim$ respectively denote strong and weak negation.
- $\forall \phi \in \mathcal{L}$ and $\forall \psi \in \mathcal{L}$, $\phi \in \overline{\psi}$ iff (a) $\psi = \neg\phi$ or $\phi = \neg\psi$; or (b) $\psi = \sim \phi$.
- $\mathcal{R}_s = \{\neg f \rightarrow \neg w; b \rightarrow a\}$ and $\mathcal{R}_d = \{a \Rightarrow \neg f; b, \sim \neg w \Rightarrow w; \neg f \Rightarrow \neg w\}$.

Let $\mathcal{K}$ be the knowledge base such that $\mathcal{K}_n = \emptyset$ and $\mathcal{K}_p = \{b, \sim \neg w\}$. The arguments defined on the basis of $\mathcal{K}$ and AS are $A_1 = [b]$, $A_2 = [\sim \neg w]$, $A_3 = [A_1 \rightarrow a]$, $A_4 = [A_3 \Rightarrow \neg f]$, $A_5 = [A_1, A_2 \Rightarrow w]$ and $A_6 = [A_4 \Rightarrow \neg w]$.

An argument A is for ϕ if $\text{Conc}(A) = \phi$; it is *strict* if $\text{DefR}(A) = \emptyset$; *defeasible* if $\text{DefR}(A) \neq \emptyset$; *firm* if $\text{Prem}(A) \subseteq \mathcal{K}_n$; *plausible* if $\text{Prem}(A) \cap \mathcal{K}_p \neq \emptyset$. An argument is *fallible* if it is defeasible or plausible and *infallible* otherwise. We write $\mathfrak{S} \vdash \phi$ if there is a strict argument for ϕ with all premises taken from $\mathfrak{S}$, and $\mathfrak{S} \mathrel{\vert\!\sim} \phi$ if there is a defeasible argument for ϕ with all premises taken from $\mathfrak{S}$.

Definition 5 (Inconsistent set of formulas). [8] *Let $\mathcal{L}$ be a language and $\mathfrak{S} \subseteq \mathcal{L}$. We say that $\mathfrak{S}$ is inconsistent if for some $\phi, \psi \in \mathcal{L}$, $\mathfrak{S} \vdash \phi$ and $\mathfrak{S} \vdash \psi$ such that $\phi \in \overline{\psi}$. Otherwise, it is consistent. We say $\mathfrak{S} \subseteq \mathcal{L}$ is minimally inconsistent iff $\mathfrak{S}$ is inconsistent and $\forall \mathfrak{S}' \subset \mathfrak{S}$, $\mathfrak{S}'$ is consistent.*

With Definition 5, we can define an inconsistent set of arguments.

Definition 6 (Inconsistent set of arguments). [8] *An argument A is consistent iff $\{\text{Conc}(A') \mid A' \in \text{Sub}(A)\}$ is consistent. Otherwise A is inconsistent. For a set S of arguments, we define $\text{Concs}(S) = \{\text{Conc}(A) \mid A \in S\}$. $S = \{A_1, \ldots, A_n\}$ is consistent if $\text{Concs}(\text{Sub}(A_1)) \cup \ldots \cup \text{Concs}(\text{Sub}(A_n))$ is consistent, otherwise S is inconsistent.*

Definition 7 (c-consistent arguments). [5] *Let A be an argument on the basis of an argumentation theory $(AS, \mathcal{K})$ and an AS $(\mathcal{L}, ^-, \mathcal{R}, n)$. The argument A is c-consistent iff $\text{Prem}(A)$ is consistent.*

Note that if an argument A is consistent, then A is also c-consistent. However, if an argument A is c-consistent, it does not mean that A is also consistent.

Example 2. Let A and B be arguments on the basis of an argumentation theory $(AS, \mathcal{K})$ and an argumentation system $(\mathcal{L}, ^-, \mathcal{R}, n)$, in which

- $\mathcal{L} = \{a, b, c, f, w, \neg w\}$,
- $\forall \phi \in \mathcal{L}$ and $\forall \psi \in \mathcal{L}$, (1) $\phi \in \overline{\psi}$ iff (a) $\psi = \neg\phi$ or $\phi = \neg\psi$; or (b) $\psi = \sim \phi$,
- $\mathcal{R}_s = \{a \rightarrow f; b, c \rightarrow w; w, \neg w \rightarrow \neg f\}$, and $\mathcal{R}_d = \{f \Rightarrow \neg w\}$.

Let $\mathcal{K}$ be the knowledge such that $\mathcal{K}_n = \emptyset$ and $\mathcal{K}_p = \{a, b, c\}$. The arguments defined on the basis of $\mathcal{K}$ and AS are $A_1 = [a]$, $A_2 = [b]$, $A_3 = [c]$, $A_4 = [A_1 \rightarrow f]$, $A_5 = [A_2, A_3 \rightarrow w]$, $A_6 = [A_4 \Rightarrow \neg w]$, and $A_7 = [A_5, A_6 \rightarrow \neg f]$.

Observe that A_7 is c-consistent, however $A_5, A_6 \in \text{Sub}(A_7)$. As $\{\text{Conc}(A_5), \text{Conc}(A_6)\}$ is inconsistent, A_7 is an inconsistent argument.

Definition 8. [5] *Consider the argumention system $(\mathcal{L}, ^-, \mathcal{R}, n)$. For any $\mathfrak{S} \subseteq \mathcal{L}$, let the closure of $\mathfrak{S}$ under strict rules, denoted $Cl_{R_s}(\mathfrak{S})$, be the smallest set containing $\mathfrak{S}$ and the consequent of any strict rule in $\mathcal{R}_s$ whose antecedents are in $Cl_{R_s}(\mathfrak{S})$. Then, 1) a set $\mathfrak{S} \subseteq \mathcal{L}$ is* directly consistent *iff $\not\exists \psi, \varphi \in \mathfrak{S}$ such that $\psi \in \overline{\varphi}$; 2)* indirectly consistent *iff $Cl_{R_s}(\mathfrak{S})$ is directly consistent.*

In Example 2, we assume $\mathfrak{S} = \{\text{Conc}(A_1), \text{Conc}(A_2), \text{Conc}(A_3), \text{Conc}(A_4), \text{Conc}(A_6)\}$. Note $\mathfrak{S}$ is directly consistent as $\forall \phi \in \mathfrak{S}$, $\not\exists \psi \in \mathfrak{S}$ s.t. $\psi \in \overline{\phi}$. However $\mathfrak{S}$ is not indirectly consistent as $\text{Conc}(A_5) \in Cl_{\mathcal{R}_s}(\mathfrak{S})$ and $\text{Conc}(A_5) \in \overline{\text{Conc}(A_6)}$.

2.1 Attacks and Defeats

In $ASPIC^+$ arguments are related to each other by attacks:

Definition 9 (Attacks). [5] *Consider the arguments A and B. We say A attacks B iff A undercuts, undermines and rebuts B, in which*

- *A undercuts B (on B') iff $\text{Conc}(A) \in \overline{n(r)}$ for some $B' \in \text{Sub}(B)$ such that B''s top rule r is defeasible.*
- *A undermines B (on ϕ) iff $\text{Conc}(A) \in \overline{\phi}$ and $\phi \in \text{Prem}_p(B)$. In such a case, A contrary-undermines B iff $\text{Conc}(A)$ is a contrary of ϕ.*
- *A rebuts B (on B') iff $\text{Conc}(A) \in \overline{\phi}$ for some $B' \in \text{Sub}(B)$ of the form $B_1'', \dots, B_n'' \Rightarrow \phi$. In such a case, A contrary-rebuts B iff $\text{Conc}(A)$ is a contrary of ϕ.*

Example 3. Recalling Example 1, we have A_5 rebuts A_6 and A_6 rebuts A_5. Besides, A_6 contrary-undermines A_2 and A_5 on $\sim \neg w$. If in addition, one had the argument $A_7 = [A_4 \rightarrow \neg w]$, then A_7 (like A_6) would rebut A_5 on A_5; however, A_7 (unlike A_6) would not be rebutted by A_5.

Definition 10. [5] *A (c-)structured argumentation framework ((c-)SAF) defined by an argumentation theory $AT = (AS, \mathcal{K})$ is a tuple $(\mathcal{A}, \mathcal{C}, \preceq)$, in which*

- *In a SAF (resp. c-SAF), $\mathcal{A}$ is the set of all arguments (resp. c-consistent arguments) constructed from $\mathcal{K}$ in AS satisfying Definition 4;*
- *$(X, Y) \in \mathcal{C}$ iff X attacks Y;*
- *$\preceq$ is a preference ordering on $\mathcal{A}$.*

It is clear a c-SAF is a SAF in which all arguments are required to have a consistent set of premises. Next, we define the corresponding defeat relation, which is inspired in the defeat relation already defined in [5]:

Definition 11 (Defeat). *Let $A, B \in \mathcal{A}$ and A attacks B. If A is consistent and A undercut, contrary-rebut or contrary-undermine attacks B on $B' \in \text{Sub}(B)$ then A is said to preference-independent attack B on B'; otherwise A is said to preference-dependent attack B on B'. A defeats B iff for some $B' \in \text{Sub}(B)$ either A preference-independent attacks B on B' or A preference-dependent attacks B on B' and $A \not\prec B'$.*

Having defined the defeat relation we proceed to the definition of argumentation framework associated to a $(c\text{-})SAF$.

Definition 12 (Argumentation frameworks). *An abstract argumentation framework (AF) corresponding to a $(c\text{-})SAF = (\mathcal{A}, \mathcal{C}, \preceq)$ is a tuple $(\mathcal{A}, \mathcal{D})$ such that $\mathcal{D} = \{(X, Y) \in \mathcal{C} \mid X \text{ defeats } Y\}$.*

Example 4 (Example 3 continued). Let $\preceq = \{(A_6, A_2)\}$ (i.e., $A_6 \prec A_2$) be a preference ordering on $\mathcal{A} = \{A_1, A_2, A_3, A_4, A_5, A_6\}$. In the $c\text{-}SAF$ $(\mathcal{A}, \mathcal{C}, \preceq)$ defined by AT, we have $\mathcal{C} = \{(A_6, A_2), (A_5, A_6), (A_6, A_5)\}$. As (A_6, A_2) is a preference independent attack, we obtain $\mathcal{D} = \mathcal{C}$.

Traditional approaches to argumentation semantics ensure conflicts are not tolerated in the same set, which is said to be conflict-free.

Definition 13 (Conflict-free sets). *Let $\triangle = (\mathcal{A}, \mathcal{C}, \preceq)$ be a $(c\text{-})SAF$, $(\mathcal{A}, \mathcal{D})$ be the AF corresponding to $\triangle$, $A \in \mathcal{A}$, and $\mathcal{S} \subseteq \mathcal{A}$. We define $\mathcal{C}^+(A) = \{B \in \mathcal{A} \mid (A, B) \in \mathcal{C}\}$ and $\mathcal{C}^+(\mathcal{S}) = \{B \in \mathcal{A} \mid (A, B) \in \mathcal{C} \text{ for some } A \in \mathcal{S}\}$. We say that $\mathcal{S}$ is att-conflict-free (in $\triangle$) iff $\mathcal{C}^+(\mathcal{S}) \cap \mathcal{S} = \emptyset$. We also define $\mathcal{D}^+(A) = \{B \in \mathcal{A} \mid (A, B) \in \mathcal{D}\}$ and $\mathcal{D}^+(\mathcal{S}) = \{B \in \mathcal{A} \mid (A, B) \in \mathcal{D} \text{ for some } A \in \mathcal{S}\}$. We say that $\mathcal{S}$ is def-conflict-free (in AF) iff $\mathcal{D}^+(\mathcal{S}) \cap \mathcal{S} = \emptyset$.*

Arguments are evaluated on the basis of the extensions of a Dung framework:

Definition 14 (Semantics). *Let $\triangle = (\mathcal{A}, \mathcal{C}, \preceq)$ be a $(c\text{-})SAF$ and $(\mathcal{A}, \mathcal{D})$ be the AF corresponding to $\triangle$. We say $X \in \mathcal{A}$ is acceptable w.r.t. $\mathcal{E} \subseteq \mathcal{A}$ iff $\forall Y \in \mathcal{A}$ such that $(Y, X) \in \mathcal{D} : \exists Z \in \mathcal{E}$ such that $(Z, Y) \in \mathcal{D}$.*

We define $f_{AF}(\mathcal{E}) = \{A \in \mathcal{A} \mid A \text{ is acceptable w.r.t. } \mathcal{E}\}$. Let $x \in \{att, def\}$. For a x-conflict-free set $\mathcal{E}$ in AF, we say 1) $\mathcal{E}$ is an x-admissible extension of AF iff $\mathcal{E} \subseteq f_{AF}(\mathcal{E})$; 2) $\mathcal{E}$ is a x-complete extension of AF iff $f_{AF}(\mathcal{E}) = \mathcal{E}$; 3) $\mathcal{E}$ is a x-preferred extension of AF iff it is a set inclusion maximal x-complete extension of AF; 4) $\mathcal{E}$ is the x-grounded extension of AF iff it is the set inclusion minimal x-complete extension of AF; 5) $\mathcal{E}$ is a x-semi-stable extension iff it is a x-complete extension of AF such that there is no x-complete extension $\mathcal{E}_1$ of AF in which $\mathcal{E} \cup \mathcal{X}^+(\mathcal{E}) \subset \mathcal{E}_1 \cup \mathcal{X}^+(\mathcal{E}_1)$ with $\mathcal{X}^+ = \mathcal{C}^+$ if $x = att$ and $\mathcal{X}^+ = \mathcal{D}^+$ if $x = def$; 6) $\mathcal{E}$ is an att-stable (resp. def-stable) extension iff $\mathcal{E}$ is an att-complete (resp. a def-complete) extension of AF and $\forall Y \notin \mathcal{E}, \exists X \in \mathcal{E} \text{ s.t. } (X, Y) \in \mathcal{C}$ (resp. $(X, Y) \in \mathcal{D}$).

Notice the basic distinction between an att-extension and a def-extension is that an att-extension is att-conflict-free and def-extension is def-conflict-free.

Example 5 (Example 4 continued)
    Regarding the *AF* constructed in Example 4, we obtain the following results:

- def-complete extensions: $\{A_1, A_3, A_4\}$, $\{A_1, A_3, A_4, A_5\}$, $\{A_1, A_3, A_4, A_6\}$;
- def-grounded extension: $\{A_1, A_3, A_4\}$;
- def-preferred extensions: $\{A_1, A_3, A_4, A_5\}$, $\{A_1, A_3, A_4, A_6\}$;
- def-stable and def-semi-stable extension: $\{A_1, A_3, A_4, A_6\}$.

As it will be clear latter, def-extensions will coincide with att-extensions.

3   Rationality Postulates

Caminada and Amgoud [3] proposed four postulates to constraint on any extension of an argumentation framework corresponding to an argumentation theory. It is shown in [4,5] under which conditions these postulates hold in $ASPIC^+$.
    We proceed by constraining our attention to well defined $(c\text{-})SAF$s:

Definition 15 (Well defined $(c\text{-})SAF$). *Let $AT = (AS, \mathcal{K})$ be an argumentation theory, where $AS = (\mathcal{L}, {}^-, \mathcal{R}, n)$. We say that AT is*

- *closed under contraposition iff for all $\mathcal{S} \subseteq \mathcal{L}$, $s \in \mathcal{S}$ and $\phi \in \mathcal{L}$, if $\mathcal{S} \vdash \phi$, then for each $-\phi \in \mathbb{C}(\phi)$ and each $-s \in \mathbb{C}(s)$, it holds $\mathcal{S}\backslash\{s\} \cup \{-\phi\} \vdash -s$.*
- *closed under transposition iff if $\phi_1, \ldots, \phi_n \to \psi \in \mathcal{R}_s$, then for $i = 1 \ldots n$, for any $-\psi \in \mathbb{C}(\phi)$ and any contradictory $-\phi \in \mathbb{C}(\phi)$, it holds $\phi_1, \ldots, \phi_{i-1}, -\psi, \phi_{i+1}, \ldots, \phi_n \to -\phi_i \in \mathcal{R}_s$;*
- *axiom consistent iff $Cl_{R_s}(\mathcal{K}_n)$ is consistent.*
- *c-classical iff for any minimal inconsistent $\mathcal{S} \subseteq \mathcal{L}$ and for any $\varphi \in \mathcal{S}$, for each $-\varphi \in \mathbb{C}(\varphi)$, it holds that $\mathcal{S}\backslash\{\varphi\} \vdash -\varphi$*
- *well-formed if for $\varphi, \psi \in \mathcal{L}$, whenever φ is a contrary of ψ then $\psi \notin \mathcal{K}_n$ and ψ is not the consequent of a strict rule.*

    A $(c\text{-})SAF$ is well defined if it is defined by an AT that is c-classical, axiom consistent, well-formed and closed under contraposition or transposition.

To prove our results, we resort to the maximal fallible sub-arguments.

Definition 16 (Maximal fallible sub-arguments). [9] *For any argument A, the set $M(A)$ of maximal fallible sub-arguments of A is inductively defined as:*

1. *If $A \in \mathcal{K}_n$, then $M(A) = \emptyset$;*
2. *If $A \in \mathcal{K}_p$ or A has a defeasible top rule, then $M(A) = \{A\}$;*
3. *otherwise, i.e., if $A = A_1, \ldots, A_n \to \varphi$, then $M(A) = M(A_1) \cup \ldots \cup M(A_n)$.*

Next we define what is a strict continuation of a set $\mathcal{S}$ of arguments:

Definition 17 (Strict continuations). [9] *The set of strict continuations of a set of arguments is the smallest set satisfying the following conditions:*

1. *Any argument A is a strict continuation of $\{A\}$.*
2. *If $A_1, \ldots, A_n$ and $S_1, \ldots, S_n$ are such that for each $i \in \{1, \ldots, n\}$, A_i is a strict continuation of S_i and $\{B_{n+1}, \ldots, B_m\}$ is a (possibly empty) set of strict-and-firm arguments, and $\texttt{Conc}(A_1), \ldots, \texttt{Conc}(A_n), \texttt{Conc}(B_{n+1}), \ldots,$ $\texttt{Conc}(B_m)) \to \varphi$ is a strict rule in $\mathcal{R}_s$, then $A_1, \ldots, A_n, B_{n+1}, \ldots, B_m \to \varphi$ is a strict continuation of $S_1 \cup \ldots \cup S_n$.*

In addition, $\preceq$ should satisfy properties that one might expect to hold of orderings over arguments composed from fallible, infallible, consistent and inconsistent elements. The next definition is an adaptation of [5] to ensure any inconsistent argument is strictly less preferred than any consistent argument.

Definition 18 (Reasonable Argument Ordering). *An argument ordering $\preceq$ is reasonable iff*

1. (a) *$\forall A, B$, if A is consistent and B is inconsistent, then $B \prec A$.*
   (b) *$\forall A, B$, if B is strict, firm and consistent then $B \nprec A$;*
   (c) *$\forall A, A', B$ such that A' is a strict continuation of $\{A\}$, if $A \nprec B$ and A' is consistent, then $A' \nprec B$ and if $B \nprec A$ then $B \nprec A'$;*
2. *Let $\{C_1, \ldots, C_n\}$ be a finite subset of $\mathcal{A}$, and for $i = 1 \ldots n$, let $C^{+\backslash i}$ be some strict continuation of $\{C_1, \ldots, C_{i-1}, C_{i+1}, \ldots, C_n\}$. Then it is not the case that: $\forall i$ $(1 \leq i \leq n)$ such that $C^{+\backslash i}$ is consistent, $C^{+\backslash i} \prec C_i$.*

From now on we will assume $\preceq$ is reasonable. The following lemmas provide us with some results required to show that our approach satisfies the rationality postulates of [3]. Due to space restrictions, the simplest proofs of some results have been omitted. In addition, we highlight that despite Definition 18 is different from the original (Definition 18 from [5]), the results from [5] are preserved.

Lemma 1. *[5] Let $\triangle = (\mathcal{A}, \mathcal{C}, \preceq)$ be an well defined (c-)SAF, $(\mathcal{A}, \mathcal{D})$ the AF corresponding to $\triangle$, and $A \in \mathcal{A}$.*

1. *If A is acceptable w.r.t. $\mathcal{S} \subseteq \mathcal{A}$, A is acceptable w.r.t. any superset of $\mathcal{S}$.*
2. *If $(A, B) \in \mathcal{D}$, $(A, B') \in \mathcal{D}$ for some $B' \in \texttt{Sub}(B)$ and if $(A, B') \in \mathcal{D}$, for some $B' \in \texttt{Sub}(B)$, $(A, B) \in \mathcal{D}$.*
3. *If A is acceptable w.r.t. $\mathcal{S} \subseteq \mathcal{A}$ and $A' \in \texttt{Sub}(A)$, A' is acceptable w.r.t. $\mathcal{S}$.*

Lemma 2. *[5] Let $\triangle = (\mathcal{A}, \mathcal{C}, \preceq)$ be a well defined (c-)SAF and $(\mathcal{A}, \mathcal{D})$ be the corresponding AF. For $A, B \in \mathcal{A}$, suppose B is consistent, B attacks A on some $A' \in \texttt{Sub}(A)$, and if A and B are defined as in Definition 7, then $\texttt{Prem}(A) \cup \texttt{Prem}(B)$ is consistent. If $(B, A) \notin \mathcal{D}$ then, either:*

1. *$(A', B) \in \mathcal{D}$, or;*
2. *For some $B' \in M(B)$, there is a strict continuation $A'^+_{B'}$ of $(M(B) \backslash \{B'\}) \cup M(A')$ s.t. $(A'^+_{B'}, B) \in \mathcal{D}$.*

Lemma 3. *Let $\triangle = (\mathcal{A}, \mathcal{C}, \preceq)$ be a well defined (c-)SAF defined by the argumentation theory $(AS, \mathcal{K})$, in which $AS = (\mathcal{L}, {}^-, \mathcal{R}, n)$ and $\mathcal{K} = \mathcal{K}_p \cup \mathcal{K}_n$. Let $\{\phi_1, \ldots, \phi_n, \phi_{n+1}\} \subseteq \mathcal{L}$ be inconsistent, but $\{\phi_1, \ldots, \phi_n\}$ is consistent. Then $\{\phi_1, \ldots, \phi_n\} \vdash -\phi_{n+1}$.*

Lemma 4 shows that the set of conclusions of a consistent set of formulas is also consistent. This result is employed to prove Proposition 3.

Lemma 4. *Let* $\triangle = (\mathcal{A}, \mathcal{C}, \preceq)$ *be a well defined* $(c\text{-})SAF$ *defined by the argumentation theory* $(AS, \mathcal{K})$, *in which* $AS = (\mathcal{L}, ^-, \mathcal{R}, n)$ *and* $\mathcal{K} = \mathcal{K}_p \cup \mathcal{K}_n$. *Let* $\{\phi_1, \ldots, \phi_n, \phi_{n+1}\} \subseteq \mathcal{L}$ *be inconsistent, but* $\{\phi_1, \ldots, \phi_n\}$ *is consistent. Then* $\{\phi_1, \ldots, \phi_n, -\phi_{n+1}\}$ *is also consistent.*

Lemma 5 is employed to prove Proposition 1.

Lemma 5. *Let* $\triangle = (\mathcal{A}, \mathcal{C}, \preceq)$ *be a well defined* $(c\text{-})SAF$ *defined by the argumentation theory* $(AS, \mathcal{K})$, *in which* $AS = (\mathcal{L}, ^-, \mathcal{R}, n)$ *and* $\mathcal{K} = \mathcal{K}_p \cup \mathcal{K}_n$. *Let* $\mathcal{B} = \{B_1, \ldots, B_m, B_{m+1}\} \subseteq \mathcal{A}$ *be an inconsistent set of arguments such that for each* $B_i \in \mathcal{B}$, *it holds* $\text{Sub}(B_i) \subseteq \mathcal{B}$, $\{B_1, \ldots, B_m\}$ *is consistent and for each* $B_j \in \mathcal{B} - \{B_{m+1}\}$, *it is the case* $B_{m+1} \notin \text{Sub}(B_j)$. *Then* $\text{TopRule}(B_{m+1}) \notin \mathcal{R}_s$.

Proof. By absurd, assume $\text{TopRule}(B_{m+1}) \in \mathcal{R}_s$. Then, B_{m+1} is an argument of the form $B'_1, \ldots, B'_n \to \text{Conc}(B_{m+1})$ with strict top rule $\text{Conc}(B'_1), \ldots, \text{Conc}(B'_n)$ $\to \text{Conc}(B_{m+1})$. Let $\varGamma = \{\text{Conc}(B_1), \ldots, \text{Conc}(B_m), \text{Conc}(B_{m+1})\}$ and $\varGamma' = \{\text{Conc}(B_1), \ldots, \text{Conc}(B_m)\}$. First we show $\varGamma$ is inconsistent and $\varGamma'$ is consistent.

As $\mathcal{B}$ is inconsistent, by Definition 6, we have $\mathfrak{S} = \text{Concs}(\text{Sub}(B_1)) \cup \ldots \cup \text{Concs}(\text{Sub}(B_m)) \cup \text{Concs}(\text{Sub}(B_{m+1}))$ is inconsistent. Given $\forall B_i \in \mathcal{B}$, $\text{Sub}(B_i) \subseteq \mathcal{B}$, we obtain $\forall B_i \in \mathcal{B}$, $\text{Concs}(\text{Sub}(B_i)) \subseteq \varGamma$. Thus, $\mathfrak{S} \subseteq \varGamma$, and so $\varGamma$ inconsistent.

Given $\{B_1, \ldots, B_m\}$ is consistent, by Definition 6, $\mathfrak{S}' = \text{Concs}(\text{Sub}(B_1)) \cup \ldots \cup \text{Concs}(\text{Sub}(B_m))$ is consistent. Note that $\forall B_i \in \{B_1, \ldots, B_m\}$, $\text{Conc}(B_i) \in \text{Concs}(\text{Sub}(B_i))$. It implies $\varGamma' \subseteq \mathfrak{S}'$. As $\mathfrak{S}'$ is consistent, $\varGamma'$ is also consistent.

Then by Lemma 3, $\varGamma' \vdash -\text{Conc}(B_{m+1})$. As $\{\text{Conc}(B'_1), \ldots, \text{Conc}(B'_n)\} \subset \varGamma'$ and $\text{Conc}(B'_1), \ldots, \text{Conc}(B'_n) \vdash \text{Conc}(B_{m+1})$, $\varGamma' \vdash \text{Conc}(B_{m+1})$. But then by Definition 5, $\varGamma'$ is inconsistent. It is an absurd as $\varGamma'$ is consistent.   □

The next result is fundamental prove Propositions 3 and 4.

Proposition 1. *Let* $\triangle = (\mathcal{A}, \mathcal{C}, \preceq)$ *be a well defined* $(c\text{-})SAF$ *and* $(\mathcal{A}, \mathcal{D})$ *be the corresponding* AF, $B \in \mathcal{A}$ *is an inconsistent argument, and* $\preceq$ *is reasonable. There exists a consistent argument* B' *s.t.* $(B', B) \in \mathcal{D}$ *and* $M(B') \subseteq \text{Sub}(B)$.

Proof. Let $\mathcal{S}_1 = \{A \in \text{Sub}(B) \mid A \text{ is firm and strict}\}$. Note $\varGamma_A = \text{Concs}(\mathcal{S}_1) \subseteq Cl_{\mathcal{R}_s}(\mathcal{K}_n)$. As $\triangle$ is well defined (Definition 15), $Cl_{\mathcal{R}_s}(\mathcal{K}_n)$ is consistent. Thus, $\varGamma_A$ is consistent and by Definition 6, $\mathcal{S}_1$ is consistent. By construction, there exists $\mathcal{S}_2 \subseteq \text{Sub}(B)$ and $B'' \in \text{Sub}(B)$ such that $\mathcal{S}_2$ is consistent, $\mathcal{S}_1 \subseteq \mathcal{S}_2$, $\mathcal{S}_2 \cup \{B''\}$ is inconsistent, $\forall B_i \in \mathcal{S}_2 \cup \{B''\}$, $\text{Sub}(B_i) \subseteq \mathcal{S}_2 \cup \{B''\}$, and $\forall B_i \in \mathcal{S}_2$, $B'' \notin \text{Sub}(B_i)$. Then, by Lemma 5, $\text{TopRule}(B'') \notin \mathcal{R}_s$. Note that $B'' \notin \mathcal{K}_n$, otherwise $\mathcal{S}_2$ would be inconsistent. Thus, $\text{TopRule}(B'') \in \mathcal{R}_d$ or $B'' \in \mathcal{K}_p$.

Let $\varGamma = \text{Concs}(\mathcal{S}_2) \cup \text{Conc}(B'')$ and $\varGamma' = \text{Concs}(\mathcal{S}_2)$. Given $\varGamma$ is inconsistent, $\varGamma'$ is consistent, by Lemma 3, $\varGamma' \vdash -\text{Conc}(B'')$. By Lemma 4, $\varGamma' \cup \{-\text{Conc}(B'')\}$ is consistent. Then, there exists an argument B' of the form $B'_1, \ldots, B'_n \to -\text{Conc}(B'')$ s.t. $\{B'_1, \ldots, B'_n\} \subseteq \mathcal{S}_2 \subset \text{Sub}(B)$. As $\text{Concs}(\text{Sub}(B')) \subseteq \varGamma' \cup \{-\text{Conc}(B'')\}$, B' is consistent. There are two possibilities:

- $(B', B'') \in \mathcal{D}$. As $B'' \in \text{Sub}(B)$, $(B', B) \in \mathcal{D}$. Given $\text{TopRule}(B') \in \mathcal{R}_s$, $M(B') \subseteq \text{Sub}(B_1') \cup \ldots \cup \text{Sub}(B_n')$. Also, as $\{B_1', \ldots, B_n'\} \subseteq \text{Sub}(B)$, we obtain $M(B') \subseteq \text{Sub}(B)$.
- $(B', B'') \notin \mathcal{D}$. Then, by Lemma 2, for some $C \in \text{Sub}(B'')$, either
  - $(C, B') \in \mathcal{D}$. It follows $\exists B^* \in \text{Sub}(B')$ s.t. $\text{TopRule}(B^*) \in \mathcal{R}_d$ or $B^* \in \mathcal{K}_p$, C attacks B' on B^* and $(C, B^*) \in \mathcal{D}$. As $\text{TopRule}(B') \in \mathcal{R}_s$, it must be that $B^* \in \text{Sub}(B_1') \cup \ldots \cup \text{Sub}(B_n') \subset \text{Sub}(B)$. Then, $(C, B) \in \mathcal{D}$. Also, given $C \in \text{Sub}(B'')$ and $B'' \in \text{Sub}(B)$, $\text{Sub}(C) \subseteq \text{Sub}(B)$. It implies $M(C) \subseteq \text{Sub}(B)$. In addition, given that B' is consistent, $(C, B') \in \mathcal{D}$ and $\preceq$ is reasonable, we have that C is consistent.
  - $\exists D \in M(B')$ s.t. C_D^+ is a strict continuation of $(M(B') \backslash \{D\}) \cup M(C)$ s.t. $(C_D^+, B') \in \mathcal{D}$. It implies $\exists B^* \in \text{Sub}(B')$ s.t. $\text{TopRule}(B^*) \in \mathcal{R}_d$ or $B^* \in \mathcal{K}_p$, C_D^+ attacks B' on B^* and $(C_D^+, B^*) \in \mathcal{D}$. As $\text{TopRule}(B') \in \mathcal{R}_s$, it must be that $B^* \in \text{Sub}(B_1') \cup \ldots \cup \text{Sub}(B_n') \subset \text{Sub}(B)$. Then, $(C_D^+, B) \in \mathcal{D}$. Note $M(C_D^+) = M(B') \backslash \{D\} \cup M(C)$. Given $C \in \text{Sub}(B'')$ and $B'' \in \text{Sub}(B)$, $M(C) \subseteq \text{Sub}(B)$. Again, as $\text{TopRule}(B') \in \mathcal{R}_s$, $M(B') \subseteq \text{Sub}(B_1') \cup \ldots \cup \text{Sub}(B_n') \subset \text{Sub}(B)$, $M(B') \subseteq \text{Sub}(B)$. Thus, $M(C_D^+) \subset \text{Sub}(B)$. In addition, given that B' is consistent, $(C_D^+, B') \in \mathcal{D}$ and $\preceq$ is reasonable, we have that C_D^+ is consistent. $\square$

Lemma 6 comes from the fact that if B defeats a strict continuation A of $\{A_1, \ldots, A_n\}$, then B defeats A on some $A_i \in \{A_1, \ldots, A_n\}$.

Lemma 6. [5] Let $\triangle = (\mathcal{A}, \mathcal{C}, \preceq)$ be a well defined (c-)SAF. Let $A \in \mathcal{A}$ be a strict continuation of $\{A_1, \ldots, A_n\} \subseteq \mathcal{A}$, and $\forall A_i$ $(1 \leq i \leq n)$, A_i is acceptable w.r.t. $\mathcal{E} \subseteq \mathcal{A}$. Then A is acceptable w.r.t. $\mathcal{E}$.

The next lemma is employed to prove Proposition 3.

Lemma 7. [5] Let $(\mathcal{A}, \mathcal{C}, \preceq)$ be a well defined (c-)SAF and $A \in \mathcal{A}$ be acceptable w.r.t. an att-admissible extension $\mathcal{E} \subseteq \mathcal{A}$ of the corresponding AF $(\mathcal{A}, \mathcal{D})$. Then $\forall B \in \mathcal{E} \cup \{A\}$, neither $(A, B) \in \mathcal{D}$ nor $(B, A) \in \mathcal{D}$.

For the following proposition, recall that by assumption, any c-SAF is well defined and so satisfies c-classicality (Definition 15).

Proposition 2. [5] Let $\triangle = (\mathcal{A}, \mathcal{C}, \preceq)$ be a well defined c-SAF. If $\exists \mathcal{E} \subseteq \mathcal{A}$ s.t. $A_1, \ldots, A_n \in \mathcal{A}$ are acceptable w.r.t. $\mathcal{E}$, then $\bigcup_{i=1}^{n} \text{Prem}(A_i)$ is consistent.

Proposition 3 shows that if an argument A is acceptable w.r.t. to an att-admissible extension $\mathcal{E}$, then $\mathcal{E} \cup \{A\}$ is att-conflict-free.

Proposition 3. Let $A \in \mathcal{A}$ be acceptable w.r.t. an att-admissible extension $\mathcal{E}$ of an AF $(\mathcal{A}, \mathcal{D})$ corresponding to a well defined (c-)SAF $\triangle = (\mathcal{A}, \mathcal{C}, \preceq)$. Then, $\mathcal{E}' = \mathcal{E} \cup \{A\}$ is att-conflict-free.

Proof. Firstly, since for any $B \in \mathcal{E}$, B is acceptable w.r.t. $\mathcal{E}$, in case $\triangle$ is a *c-SAF*, by Proposition 2, $\mathtt{Prem}(A) \cup \mathtt{Prem}(B)$ is consistent. By absurd, suppose $\mathcal{E}'$ is not att-conflict-free. It is clear $\mathcal{E}$ is att-conflict-free as it is an att-admissible extension. Given $A \not\kern-0.3em\curlyvee A$, A can not attack itself since we would have $(A, A) \in \mathcal{D}$, contradicting Lemma 7. Hence, we have two possibilities:

- $\exists B \in \mathcal{E}$, $(B, A) \in \mathcal{C}$. By Lemma 7, $(B, A) \notin \mathcal{D}$. Observe that $\mathtt{Prem}(A) \cup \mathtt{Prem}(B)$ is consistent when $\triangle$ is well defined.
  - If B is consistent, by Lemma 2, for some $A' \in \mathtt{Sub}(A)$, either
    1. $(A', B) \in \mathcal{D}$. Given B is acceptable w.r.t. $\mathcal{E}$, $\exists C \in \mathcal{E}$ s.t. $(C, A') \in \mathcal{D}$. Thus, by Lemma 1–2, $(C, A) \in \mathcal{D}$, contradicting Lemma 7.
    2. $\exists B' \in M(B)$ s.t. $A'^+_{B'}$ is a strict continuation of $(M(B) \setminus \{B'\}) \cup M(A')$ s.t. $(A'^+_{B'}, B) \in \mathcal{D}$. As B is acceptable w.r.t. $\mathcal{E}$, $\exists C \in \mathcal{E}$ s.t. $(C, A'^+_{B'}) \in \mathcal{D}$. By Lemma 1–2, for some $Z \in \mathtt{Sub}(A'^+_{B'})$, $(C, Z) \in \mathcal{D}$. It follows $(C, Z) \in \mathcal{C}$. As $A'^+_{B'}$ is a strict continuation of $(M(B) \setminus \{B'\}) \cup M(A')$, by Definition 17, $Z \in \mathtt{Sub}(A) \cup \mathtt{Sub}(B)$. Hence, by Lemma 1–2, either $(C, B) \in \mathcal{D}$, from which follows $(C, B) \in \mathcal{C}$, contradicting $\mathcal{E}$ is att-conflict-free, or $(C, A) \in \mathcal{D}$, contradicting Lemma 7.
  - If B is inconsistent, by Proposition 1, there exists B'' s.t. $(B'', B) \in \mathcal{D}$ and $M(B'') \subseteq \mathtt{Sub}(B)$. As B is acceptable w.r.t. $\mathcal{E}$, $\exists C \in \mathcal{E}$ s.t. $(C, B'') \in \mathcal{D}$. This implies $\exists Z \in \mathtt{Sub}(B'')$ s.t. $\mathtt{TopRule}(Z) \in \mathcal{R}_d$ or $Z \in \mathcal{K}_p$, C attacks B'' on Z and $(C, Z) \in \mathcal{D}$. It must be that for some $B^* \in M(B'')$, $Z \in \mathtt{Sub}(B^*)$. Given $M(B'') \subseteq \mathtt{Sub}(B)$, $Z \in \mathtt{Sub}(B)$. By Lemma 1–2, $(C, B) \in \mathcal{D}$, hence $(C, B) \in \mathcal{C}$, contradicting $\mathcal{E}$ is att-conflict free.
- $\exists B \in \mathcal{E}$, $(A, B) \in \mathcal{C}$. The proof is similar to the previous case.   □

Theorems 1, 2, 3, and 4 show $ASPIC^+$ satisfies the postulates of [3]. Theorem states that for an att-complete extension $\mathcal{E}$, $\forall A \in \mathcal{E}$, $\forall A' \in \mathtt{Sub}(A)$, $A' \in \mathcal{E}$.

Theorem 1 (Sub-argument closure). [5] *Let $\triangle = (\mathcal{A}, \mathcal{C}, \preceq)$ be a well defined (c-)SAF and $\mathcal{E}$ an att-complete extension of $\triangle$. Then $\forall A \in \mathcal{E}$: if $A' \in \mathtt{Sub}(A)$ then $A' \in \mathcal{E}$.*

Proof. According to Lemma 1–3, A' is acceptable w.r.t. $\mathcal{E}$, and $\mathcal{E} \cup \{A'\}$ is att-conflict-free (Proposition 3). As $\mathcal{E}$ is an att-complete extension, $A' \in \mathcal{E}$.   □

Theorem 2 states that the conclusions of arguments in an att-complete extension are closed under strict rules.

Theorem 2 (Closure under strict rules). [5] *Let $\triangle = (\mathcal{A}, \mathcal{C}, \preceq)$ be a well defined (c-)SAF and $\mathcal{E}$ an att-complete extension of $\triangle$. Then $\{\mathtt{Conc}(A) \mid A \in \mathcal{E}\} = Cl_{\mathcal{R}_s}(\{\mathtt{Conc}(A) \mid A \in \mathcal{E}\})$.*

Proof. We will show for any strict continuation X of $\mathcal{E}$, $X \in \mathcal{E}$. Any such X is acceptable with relation to $\mathcal{E}$ (Lemma 6), and $\mathcal{E} \cup \{X\}$ is att-conflict-free (Proposition 3). As $\mathcal{E}$ is an att-complete extension, $X \in \mathcal{E}$.   □

The following lemma is employed to prove Theorem 3.

Lemma 8. *Let $\triangle = (\mathcal{A}, \mathcal{C}, \preceq)$ be a well defined $(c\text{-})SAF$, $\mathcal{E}$ an att-admissible extension of $\triangle$, and $\exists Y \in \mathcal{E}$ s.t. Y is defeasible or plausible, and $\text{TopRule}(Y) \in \mathcal{R}_s$. Then $\forall X \in \mathcal{E}$, it holds $\text{Conc}(X) \notin \overline{\text{Conc}(Y)}$.*

Theorem 3 below, states that the conclusions of all arguments in an att-admissible extension are mutually consistent.

Theorem 3 (Direct consistency). *For a well defined $(c\text{-})SAF$ $\triangle = (\mathcal{A}, \mathcal{C}, \preceq)$ and $\mathcal{E}$ an att-admissible extension of $\triangle$, $\{\text{Conc}(A) \mid A \in \mathcal{E}\}$ is directly consistent.*

Proof. By absurd, suppose $A, B \in \mathcal{E}$, and $\text{Conc}(A) \in \overline{\text{Conc}(B)}$.

- If A is firm and strict, and 1) B is strict and firm. It is an absurd as it contradicts axiom consistency (Definition 15). 2) B is plausible or defeasible. By Lemma 8, $\text{TopRule}(B) \notin \mathcal{R}_s$ i.e., $B \in \mathcal{K}_p$ or $\text{TopRule}(B) \in \mathcal{R}_d$. Thus, $(A, B) \in \mathcal{C}$, which is an absurd as $\mathcal{E}$ is att-conflict-free.
- A is plausible or defeasible, and
  - B is strict and firm. From Definition 15, $\text{Conc}(A)$ and $\text{Conc}(B)$ are a contradictory of each other. By Lemma 8, $\text{TopRule}(A) \notin \mathcal{R}_s$ i.e., $A \in \mathcal{K}_p$ or $\text{TopRule}(A) \in \mathcal{R}_d$. Then $(B, A) \in \mathcal{C}$, an absurd as $\mathcal{E}$ is att-conflict-free.
  - B is plausible or defeasible. By Lemma 8, $\text{TopRule}(B) \notin \mathcal{R}_s$ i.e., $B \in \mathcal{K}_p$ or $\text{TopRule}(B) \in \mathcal{R}_d$. We have $(A, B) \in \mathcal{C}$, contradicting $\mathcal{E}$ is att-conflict-free. Again by Lemma 8, $\text{TopRule}(B) \notin \mathcal{R}_s$. □

Next, we employ Theorems 2 and 3 to show the closure under strict rules of conclusions of arguments in an att-complete extension is directly consistent.

Theorem 4 (Indirect consistency). *Let $\triangle = (\mathcal{A}, \mathcal{C}, \preceq)$ be a well defined $(c\text{-})SAF$ and $\mathcal{E}$ an att-complete extension of $\triangle$. Then $Cl_{\mathcal{R}_s}(\{\text{Conc}(A) \mid A \in \mathcal{E}\})$ is directly consistent.*

Proof. It follows from Theorems 2 and 3. □

3.1 Relation Between Att-conflict-free and Def-conflict-free

Now we will show the rationality postulates of [3] for $(c\text{-})SAF$s when the notion of conflict-free is based on the defeat relation. In Proposition 4 we show for a well defined $(c\text{-})SAF$ $(\mathcal{A}, \mathcal{C}, \preceq)$, $\mathcal{E} \subseteq \mathcal{A}$ is an att extension iff $\mathcal{E}$ is a def extension.

Proposition 4. *Let $\triangle = (\mathcal{A}, \mathcal{C}, \preceq)$ be a well defined $(c\text{-})SAF$ and $(\mathcal{A}, \mathcal{D})$ the corresponding AF. For $T \in \{admissible, complete, grounded, preferred, stable, semi-stable\}$, $\mathcal{E}$ is an att-T extension of $\triangle$ iff $\mathcal{E}$ is a def-T extension of $\triangle$.*

Proof. We first show that $\mathcal{E}$ is att-conflict-free iff $\mathcal{E}$ is def-conflict-free.

- If $\mathcal{E}$ is att-conflict-free, then $\mathcal{E}$ is def-conflict-free.
- If $\mathcal{E}$ is def-conflict-free, then $\mathcal{E}$ is att-conflict-free: By absurd, suppose $B, A \in \mathcal{E}$ such that $(B, A) \in \mathcal{C}$ and $(B, A) \notin \mathcal{D}$. Since $A, B \in f_{AF}(\mathcal{E})$, if $\triangle$ is a $c\text{-}SAF$ in which A and B are defined as in Definition 7, $\text{Prem}(A) \cup \text{Prem}(B)$ is consistent (Proposition 2). With relation to B, there are two possibilities:

- If B is consistent, by Lemma 2, for some $A' \in \mathrm{Sub}(A)$, either
  1. $(A', B) \in \mathcal{D}$: As B is acceptable w.t.t. $\mathcal{E}$, $\exists C \in \mathcal{E}$ s.t. $(C, A') \in \mathcal{D}$. By Lemma 1–2, $(C, A) \in \mathcal{D}$, contradicting $\mathcal{E}$ is def-conflict-free.
  2. $\exists B' \in M(B)$ s.t. $A'^{+}_{B'}$ is a strict continuation of $(M(B) \backslash \{B'\}) \cup M(A')$ s.t. $(A'^{+}_{B'}, B) \in \mathcal{D}$. Given B is acceptable w.r.t. $\mathcal{E}$, $\exists C \in \mathcal{E}$ s.t. $(C, A'^{+}_{B'}) \in \mathcal{D}$. By Lemma 1–2, $\exists Z \in \mathrm{Sub}(A'^{+}_{B'})$, $(C, Z) \in \mathcal{D}$. It follows $(C, Z) \in \mathcal{C}$. As $A'^{+}_{B'}$ is a strict continuation of $(M(B) \backslash \{B'\}) \cup M(A')$, by Definition 17, $Z \in \mathrm{Sub}(A') \cup \mathrm{Sub}(B)$. Hence, by Lemma 1–2, either $(C, B) \in \mathcal{D}$ or $(C, A) \in \mathcal{D}$, contradicting $\mathcal{E}$ is def-conflict-free.
- If B is inconsistent, by Proposition 1, there exists B'' such that $(B'', B) \subseteq \mathcal{D}$ and $M(B'') \in \mathrm{Sub}(B)$. As B is acceptable w.r.t. $\mathcal{E}$, $\exists C \in \mathcal{E}$ s.t. $(C, A'') \in \mathcal{D}$. It implies $\exists Z \in \mathrm{Sub}(B'')$ s.t. $\mathrm{TopRule}(Z) \in \mathcal{R}_d$ or $Z \in \mathcal{K}_p$, C attacks B'' on Z and $(C, Z) \in \mathcal{D}$. It must be that for some $B^* \in M(B'')$, $Z \in \mathrm{Sub}(B^*)$. Given $M(B'') \subseteq \mathrm{Sub}(B)$, $Z \in \mathrm{Sub}(B)$. By Lemma 1–2, $(C, B) \in \mathcal{D}$, contradicting $\mathcal{E}$ is def-conflict-free.

Let $\mathcal{E}$ be att-conflict-free. From Definition 14 and the result above, $\mathcal{E}$ is att-admissible iff $\mathcal{E} \subseteq f_{AF}(\mathcal{E})$ iff $\mathcal{E}$ is def-admissible. As any att-complete extension is att-admissible, $\mathcal{E}$ is att-complete iff $\mathcal{E}$ is def-complete. As any preferred/grounded/stable/semi-stable extension is also complete, the proposition holds for these extensions. □

By Proposition 4 we obtain that the postulates of [3] also hold for $(c\text{-})SAF$s when conflicts are defined under the defeat relation.

Corollary 1. *Let $\triangle$ be a well defined $(c\text{-})SAF$. Then Theorems 1, 2, 3, and 4 hold for the def-admissible and def-complete semantics of $\triangle$.*

Proof. It follows straightforwardly from Proposition 4. □

4 Related Work and Discussion

Over the years, diverse proposals have been proposed to handle inconsistencies in argumentation [6–8,10–16]. In [10], a simplified version of $ASPIC^{+}$ named $ASPIC^{-}$ allowing arguments with a strict top rule to be rebutted was introduced. This suffices to avoid inconsistent extensions, but as observed in [11], in $ASPIC^{-}$ inconsistencies may interfere with the acceptability of consistent arguments. In [11], the $ASPIC^{-}$ is improved to remedy such a problem. The resulting approach satisfies the rationality postulates of [3] and also makes it possible to compare arguments using a preference relation. As downside, it is limited to total preorderings, while in our work we can employ any reasonable preordering.

Another approach taken by Arieli in [14–16] considers a sequent-based argumentation framework to accommodate different types of languages, including paraconsistent logics. However, unlike our work, this approach does not make it possible to compare arguments with a preference relation. A different rout taken in [6,7] employs the paraconsistent logic W presented in [17] to deal with inconsistencies. Although this approach allows for the instantiation of preferences relations, it fails to satisfy any of the postulates in [3]. In contrast, our work is not limited to a single logic and also satisfies the rationality postulates of [3].

In [8] the authors requires that for each argument, the set of conclusions of all its sub-arguments are classically consistent. They show this solution satisfies the rationality postulates of [3] for a restricted version of $ASPIC^+$ without preferences, but give counterexamples to the consistency postulates for the case with preferences. In what follows, we compare this approach with our work when considering preferences.

Example 6. [8] Let $\mathcal{R}_d = \{p \Rightarrow q\}$ and $\mathcal{K}_p = \{p; \neg p \vee \neg q\}$. The corresponding AF includes the arguments in Table 1 with their respective last defeasible links and preference, which is given by the last link principle [4]. We assume that p has priority 1, $\neg p \vee \neg q$ has priority 2 and $p \Rightarrow q$ has priority 3.

Table 1. Arguments and their preferences

Argument	Last Defeasible Link	Preference
$A_1 = [p]$	p	(1)
$A_2 = [A_1 \Rightarrow q]$	$p \Rightarrow q$	(3)
$A_3 = [\neg p \vee \neg q]$	$\neg p \vee \neg q$	(2)
$A_4 = [A_1, A_2 \rightarrow \neg(\neg p \vee \neg q)]$	$p; p \Rightarrow q$	(1)
$A_5 = [A_1, A_3 \rightarrow \neg q]$	$p; \neg p \vee \neg q$	(1)
$A_6 = [A_2, A_3 \rightarrow \neg p]$	$p \Rightarrow q; \neg p \vee \neg q$	(2)

As A_6 is inconsistent, the solution proposed in [8] is to discard A_6 from the framework. The resulting framework is

whose the only complete extension is $\mathcal{E} = \{A_1, A_2, A_3, A_4, A_5\}$. Observe that $\mathcal{E}$ is inconsistent, since $A_3, A_4 \in \mathcal{E}$ and $\mathrm{Conc}(A_4) \in \overline{\mathrm{Conc}}(A_3)$. In addition, $\mathcal{E}$ is not closed under strict rules as $A_2, A_3 \in \mathcal{E}$ and $A_6 \notin \mathcal{E}$.

In our approach, unlike in [8], the argument A_6 will not be eliminated, but neutralized by attributing to it a preference value lower than any preference assigned to consistent arguments. Note that A_4, A_5 and A_6 are strict continuations of $\{A_1, A_2\}$, $\{A_1, A_3\}$ and $\{A_2, A_3\}$ respectively. Also, A_4 and A_5 are the only consistent arguments among A_4, A_5 and A_6. According to Definition 18 (item 2), it should be that $A_4 \not\prec A_3$ or $A_5 \not\prec A_2$. However, as it can be seen in Table 1, $A_4 \prec A_3$ and $A_5 \prec A_2$. Thus, we obtain the last link principle is not a reasonable preference ordering. If we replace in Table 1 the last link principle by a reasonable preference ordering, for example, by imposing $A_4 \not\prec A_3$,

$(A_4, A_3) \in \mathcal{D}$ in the resulting framework, whose only complete extension will be $\mathcal{E}' = \{A_1, A_2, A_4, A_5\}$. As expected, $\mathcal{E}'$ satisfies the postulates of Sect. 3 (see Corollary 1). In particular, $\mathcal{E}'$ is consistent and closed under strict rules.

5   Conclusion and Future Works

Inconsistency may occur for many reasons such as evolving information and merging information from multiple sources. In the literature, argumentation is frequently referred to as a natural approach to dealing with inconsistency. In this work, we impose some conditions on the relations between arguments in $ASPIC^+$ to prevent inconsistent arguments from interfering with consistent arguments.

By guaranteeing that any inconsistent argument is less preferred than any consistent argument, we have neutralized undesirable consequences of the inconsistent arguments. We also showed our proposal satisfies all the rationality postulates described in [3]. Consequently, our proposal is the first one to present an argumentation formalism based on $ASPIC^+$ robust enough to work with inconsistent arguments, to allow for the instantiation of preference relations between arguments, and to satisfy the fundamental rationality postulates of [3].

Future developments encompass identifying a preference relation that is reasonable according to our definition of reasonable ordering. Another venture is to verify under which conditions our approach satisfies the important principles of Non-interference and Crash-Resistance of [18].

References

1. Carnielli, W., Marcos, J.: A taxonomy of C-systems. In: Paraconsistency, pp. 24–117. CRC Press (2002)
2. Dung, P.: On the acceptability of arguments and its fundamental role in non-monotonic reasoning, logic programming and N-person games. Artif. Intell. **77**(2), 321–357 (1995)
3. Caminada, M., Amgoud, L.: On the evaluation of argumentation formalisms. Artif. Intell. **171**(5–6), 286–310 (2007)
4. Prakken, H.: An abstract framework for argumentation with structured arguments. Argument Comput. **1**(2), 93–124 (2010)
5. Modgil, S., Prakken, H.: A general account of argumentation with preferences. Artif. Intell. **195**, 361–397 (2013)
6. Grooters, D., Prakken, H.: Combining paraconsistent logic with argumentation. In: COMMA, pp. 301–312 (2014)
7. Grooters, D., Prakken, H.: Two aspects of relevance in structured argumentation: minimality and paraconsistency. J. Artif. Intell. Res. **56**, 197–245 (2016)
8. Wu, Y., Podlaszewski, M.: Implementing crash-resistance and non-interference in logic-based argumentation. J. Logic Comput. **25**(2), 303–333 (2015)
9. Prakken, H.: Rethinking the rationality postulates for argumentation-based inference. In: COMMA, pp. 419–430 (2016)
10. Caminada, M., Modgil, S., Oren, N.: Preferences and unrestricted rebut. Computational Models of Argument (2014)

11. Heyninck, J., Straßer, C.: Revisiting unrestricted rebut and preferences in structured argumentation. In: IJCAI, pp. 1088–1092 (2017)
12. Arieli, O.: Conflict-tolerant semantics for argumentation frameworks. In: del Cerro, L.F., Herzig, A., Mengin, J. (eds.) JELIA 2012. LNCS (LNAI), vol. 7519, pp. 28–40. Springer, Heidelberg (2012). https://doi.org/10.1007/978-3-642-33353-8_3
13. Arieli, O.: Conflict-free and conflict-tolerant semantics for constrained argumentation frameworks. J. Appl. Log. **13**(4), 582–604 (2015)
14. Arieli, O., Straßer, C.: Sequent-based logical argumentation. Argument Comput. **6**(1), 73–99 (2015)
15. Arieli, O., Straßer, C.: Logical argumentation by dynamic proof systems. Theor. Comput. Sci. **781**, 63–91 (2019)
16. Borg, A., Straßer, C., Arieli, O.: A generalized proof-theoretic approach to logical argumentation based on hypersequents. Stud. Logica **109**(1), 167–238 (2021)
17. Rescher, N., Manor, R.: On inference from inconsistent premises. Theor. Decis. **1**(2), 179–217 (1970)
18. Caminada, M., Carnielli, W., Dunne, P.: Semi-stable semantics. J. Log. Comput. **22**(5), 1207–1254 (2012)

A Grammar-based Genetic Programming Hyper-Heuristic for Corridor Allocation Problem

Rafael F. R. Correa[✉][ID], Heder S. Bernardino[ID], João M. de Freitas[ID], Stênio S. R. F. Soares[ID], Luciana B. Gonçalves[ID], and Lorenza L. O. Moreno[ID]

Universidade Federal de Juiz de Fora (UFJF), Juiz de Fora, Brazil
{rafaelfreesz,heder,joao,ssoares,lbrugiolo,lorenza}@ice.ufjf.br

Abstract. Layout problems are the physical arrangement of facilities along a given area commonly used in practice. The Corridor Allocation Problem (CAP) is a class of layout problems in which no overlapping of rooms is allowed, no empty spaces are allowed between the rooms, and the two first facilities (one on each side) are placed on zero abscissa. This combinatorial problem is usually solved using heuristics, but designing and selecting the appropriate parameters is a complex task. Hyper-Heuristic can be used to alleviate this task by generating heuristics automatically. Thus, we propose a Grammar-based Genetic Programming Hyper-Heuristic (GGPHH) to generate heuristics for CAP. We investigate (i) the generation of heuristics using a subset of the instances of the problem and (ii) using a single instance. The results show that the proposed approach generates competitive heuristics, mainly when a subset of instances are used. Also, we found a single instance that can be used to generate heuristics that generalize to other cases.

Keywords: Corridor Allocation Problem · Genetic Programming · Combinatorial problem

1 Introduction

Layout problems are the physical arrangement of facilities along a given area, forming a layout [10]. Many applications of these problems can be found in the literature [3,10], such as the arrangement of rooms in buildings, schools and hospitals, positioning and ordering of machines in production lines, arrangement of books on shelves, and positioning of semiconductors in circuits printed. In a practical context, a good layout configuration represents cost reductions, improvement of work efficiency, and reduction of necessary spaces. Thus, one may increase the competitiveness of the enterprises that optimizes their layouts.

Considering the layout problems, the Double-Row Facility Layout Problem (DRFLP) deals with the arrangement of facilities along two rows [11], as shown in Fig. 1. The cost of a layout is defined as the cost of all arranged facility pairs, in which the cost of each pair of facilities is given by the product of their distance by

© The Author(s), under exclusive license to Springer Nature Switzerland AG 2022
J. C. Xavier-Junior and R. A. Rios (Eds.): BRACIS 2022, LNAI 13653, pp. 504–519, 2022.
https://doi.org/10.1007/978-3-031-21686-2_35

a coefficient. The understanding of this coefficient depends on the context of the problem being solved. For instance, it may represent the volume of demand, the average daily traffic, and the priority in functional dependencies. This problem contains a single constraint: no overlapping of rooms is allowed.

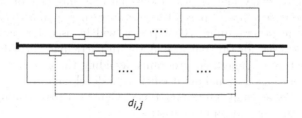

Fig. 1. Illustrative example of an DRFLP problem.

The Corridor Allocation Problem (CAP) is a DRFLP variant and is handled here. Two constraints are added to this problem [3]: (i) no empty spaces are allowed between the rooms, and (ii) the two first facilities, one in each side of the landscape, are placed on zero abscissa. This problem is illustrated in Fig. 2.

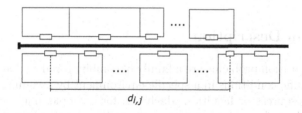

Fig. 2. Illustrative example of an CAP problem.

The layout problems are NP-Hard [16] and, in this way, the search for the best solution through the application of exact methods can result in high execution times, making them hard to be applied in real-world situations [3]. Usually, this type of problem is solved by obtaining good solutions in a reduced time. For this purpose, heuristics are commonly used, in which candidate solutions are obtained by exploring neighborhoods and no guarantee of optimality is provided. On the other hand, a good solution is found with a small processing time [22].

One can find related work in the literature applying heuristics for the development of efficient solutions for CAP. For instance, the use of Genetic Algorithms, Local Searches, Scatter Search and Path Relinking is reported in [14]. Also, a Simulated Annealing is proposed to solve CAP problems in [17]. Finally, in [16] a Variable Neighborhood Search (VNS) is presented.

The challenge in using heuristics is both in a large number of user-defined parameters, and the definition of the proper heuristic, as its components may vary according to the problem being solved. A solution to this problem lies in automating the development of heuristics through hyper-heuristics [8].

Hyper-heuristics is a search methodology for automating the process of selecting or combining low level heuristics to solve hard optimization problems [9]. This type of technique searches a space of heuristics rather than a space of solutions directly. Genetic programming (GP) [18] is widely used for program generation and can be used for this purpose [8,15]. In particular, formal grammar can limit the structure of generated programs, avoiding the generation of syntactically invalid solutions and making it efficient in developing solutions to complex problems [13]. As a result, Grammar-Based Genetic Programming (GGP) [23] is suitable for the generation of heuristics.

Although CAP has already been explored using meta-heuristics, to the best of our knowledge the design of heuristics through GP and GGP was not investigated. The low-level heuristics presented in previous work can be explored by HHs to improve the solutions of CAP. Thus, we propose here a Grammar-based Genetic Programming Hyper-heuristic (GGPHH) for designing heuristics to solve CAP problems. In addition, we define a subset of representative problems to be used in the generation of the heuristics.

Finally, computational experiments are performed with 89 instances from the literature [1–3,5,6,21].

2   Problem Description

Layout configuration problems are a family of combinatorial problems. They aim to improve traffic conditions in a specific environment by assigning geographical locations to resources or facilities. Machines, for example, may be relocated to optimize the flow of products on factory floor. The location of departments in a building or examination rooms in a hospital may have an impact on the overall traffic of people, products, or patients. The alignment of bays on semiconductors production is also an interesting application of these problems [24].

Facilities are organized side-by-side along several rows on several variants of layout configuration problems, such as the Space-Free Multi-Row Facility Layout Problem (SF-MRFLP) [16], the Single Row Facility Layout Problem (SRFLP) [19], Double Row Facility Layout Problem (DRFLP) [11] and the Corridor Allocation Problem (CAP) [3]. They differ on the number and alignment of the rows and on the possibility of including free spaces among the facilities.

The Corridor Allocation Problem (CAP) was proposed by Amaral [3]. It distributes a set of n non-overlapping facilities (F) alongside the two sides of a corridor. Both sides start at the same point (i.e., rows are left-aligned on zero abscissa) and no spaces are allowed between consecutive facilities. The length of the corridor facing wall of each facility ($l_i, \forall i \in F$) and the predicted flow of people for each pair of facilities ($f_{ij}, \forall i, j \in F, i \neq j$) are known.

In a feasible solution, the abscissa of a facility $(x_i, \forall i \in F)$ is the distance between its middle point and the corridor starting point. The distance between two facilities does not consider the corridor width and is the difference of their respective abscissas $(d_{ij} = |x_i - x_j|, \forall i, j \in F, i \neq j)$. The CAP aims to minimize the overall traffic z based on flow and distance between each pair of facilities, as

$$\min \sum_{i \in F} \sum_{j \in F} f_{ij} \times d_{ij}. \tag{1}$$

A solution to the problem consists of a permutation of facilities distribution alongside the corridor and it can be represented by a single permutation sequence and an integer t, that indicates where the second side of the corridor starts. Therefore, facilities on indexes up to $t - 1$ are ordered from left to right on one side of the corridor and the remaining facilities (from index t up to the end of the sequence) are ordered from left to right on the other side.

3   Methods

The approaches used here are described in the following sections. First, the genetic programming technique used to generate programs is described. In the sequence, the components adopted in this work to compose the heuristics are defined. We considered perturbation and local search strategies.

3.1   Grammar-Based Genetic Programming

Genetic Programming (GP) [18] is a nature-inspired meta-heuristic for the generation of programs. Each individual in the population represents a solution to the problem, usually represented by trees. The tree encodes the solution, where the internal nodes are operations (such as a program, a control, and mathematical operators) and the leaves are operands (as variables and numeric constants).

The fitness of an individual indicates its performance, based on an objective function. Its calculation usually evaluates the model over a dataset or executes a controller for performing a given task.

GP evolves programs similar to other populational meta-heuristics. Initially, a population is created and evaluated according to an objective function. Individuals are selected from the population to become parents. The parents are recombined (crossover) and the offspring is mutated. The new individuals are evaluated and the current population is updated using the offspring. The selection, crossover, mutation, evaluation, and replacement steps are repeated while a stop criterion is not met.

In GP, usually, two individuals are selected, and produce two offspring. With the tree representation, the crossover is based on selecting a node in both parents, and then swapping them, creating two new individuals based on their code. The mutation operates over one individual. In GP, commonly it is an individual recently created by crossover. One sub-tree is randomly selected, and a new derivation is created from that node.

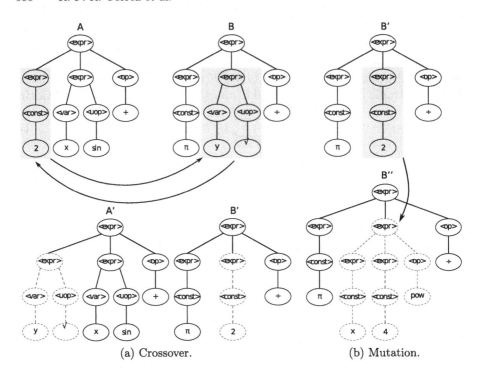

Fig. 3. Examples of the crossover (a) and mutation (b) operators of GGP.

Grammar-based Genetic Programming [23] (GGP) is a GP variant that uses a grammar to create individuals. Using a formal grammar makes it possible to constrain the search space and introduce previous knowledge.

A grammar is formed by a set of rules that arrange terminal and nonterminal symbols. The structure of a GGP individual is a derivation tree, created according to the grammar rules. In this case, the internal nodes are nonterminals, and the leaves are terminals.

The crossover operator of GGP is similar to that of standard GP. However, the crossover of GGP is guided by a formal grammar. An internal node is randomly selected in the first parent and a node of the same type (representing the same nonterminal) is randomly chosen in the second parent. The sub-trees of those nodes are swapped, generating two new individuals. The mutation operator is also guided by the grammar. A random internal node is selected and the sub-tree is replaced by another one, derived from the rule of the chosen nonterminal. Figure 3 presents illustrative examples of crossover and mutation of GGP. In (a), the sub-trees swapped between the two parents are those not shaded. In (b), the shaded area is the sub-tree replaced by a new randomly generated one.

3.2   Perturbation Strategies

Perturbation methods are atomic or block movements that seek to change the layout, regardless of the cost obtained after its application. This type of movement avoids the solutions becoming trapped in local optima, making it possible to search in other neighborhoods. The approaches used here are:

- **Swap:** a pair of facilities of the layout are swapped. It represents a unique movement with predefined indexes;
- **Shake:** n pairs of facilities are randomly swapped;
- **Shift:** based on the representation of the solution on a single vector and an index t that indicates the beginning of the second side of the corridor, this movement receives a parameter r, which shifts the cut-off point of the array to the index $t + r$. For $r > 0$, the offset moves the first r facilities located at the beginning of the second side of the corridor to the end of the first side. For $r < 0$, the last r facilities of the first side of the corridor are moved to the beginning of the second side; and
- **Reconstruction:** r facilities are randomly selected and removed from the layout, and these selected facilities are relocated one at a time, each one on the side with the smallest number of facilities.

3.3   Local Search Approaches

Local search techniques are movements in a layout to find better solutions in the search space. For each move that improves the total cost, the process is restarted using the solution obtained.

This set of operations is performed repeatedly until a local optimum is reached. The approaches used here are:

- **Neighbour Swap:** this movement swaps every pair of neighbour facilities on the same side of the layout, restarting the procedure when an improvement is obtained and finalizing the search when no improvement is reached for all pairs of neighbour facilities.
- **Non-Neighbour Swap:** the facilities in positions i and j, with $j > (i + 1)$, are swapped, and the procedure is restarted when an improvement occurs and concluded when no improvement is found (as in Neighbour Swap);
- **Swap Between Opposite Sides:** every pair of facilities placed on different sides of the layout are swapped, and its restarting and stop criterion is the same as the previous swap operators;
- **Insertion:** every facility on one side of the layout is removed and inserted in every position on the other side, with the same restarting and stop criterion of the previous approaches;
- **Random Variable Neighborhood Descent (RVND):** Neighbour Swap, Non-neighbour Swap, Swap Between Opposite Sides, and Insertion are randomly applied, and RVND is restarted when an improvement is observed when applying any of these four movement operators; similarly to the other cases, the search stops when no improvement is observed by using the four movement operators;

Algorithm 1. Pseudo-code for Genetic Programming.

1: **procedure** GP
2:     Create initial population *pop* of heuristics of size *PS*
3:     evaluate(*pop*)
4:     **while** stop criterion is not met **do**
5:         *parents* = selection(*pop*)
6:         *new_individuals* = crossover(*parents*)
7:         *new_individuals* = mutation(*new_individuals*)
8:         evaluate(*new_individuals*)
9:         replacement(*pop*, *new_individuals*)
10:     **end while**
11: **end procedure**

– **Full Permutation:** pairs of positions of the layout are randomly selected and swapped, the procedure is restarted when an improvement is observed and is concluded when no improvement is reached for all pairs of neighbour facilities.

4   The Proposed GGPHH

We propose here a Grammar-based Genetic Programming Hyper-Heuristic (GGPHH) for generating heuristics for CAP. GGPHH evolves heuristics according to the GGP search procedure. Thus, heuristics are randomly generated to compose the initial population, these candidate heuristics are evaluated and, while a stop criterion is not met, the following steps are performed: candidate solutions are selected to be recombined, the solutions generated are mutated, the new individuals are evaluated and the current population is updated using these new heuristics. Algorithm 1 presents a pseudo-code for the proposed GGPHH.

An important component of GP is the representation of its candidate solutions. Here, we adopted a GGP with a context-free grammar to represent the candidate heuristics, that are composed by the movements presented in Sects. 3.2 and 3.3. The nonterminal <do> starts the grammar proposed in this work for generating heuristics for CAP and its production rules are as

$$<do> ::= <pert> | <ref> | <do> <do>$$

$$<pert> ::= shake(<cons>) | swap(<cons>, <cons>)$$
$$| shift(<signOp> <cons>) | reconstruction(<cons>)$$

$$<ref> ::= nbrSwap() | nonNbrSwap() | oppSidesSwap()$$
$$| insertion() | rvnd() | fullPermut();$$

$$<signOp> ::= + | -$$

$$<cons> ::= 0.00 | 0.05 | ... | 0.95 | 1.00$$

The start symbol $<do>$ generates a combination of one or more perturbation or local search strategies. Each nonterminal $<pert>$ is replaced by a call to a single perturbation strategy (*shake, swap, shift* or *reconstruction*, described in Sect. 3.2). Similarly, each nonterminal $<ref>$ produces a call to a single local search (*nbrSwap, nonNbrSwap, oppSidesSwap, insertion, rvnd* or *fullPermut*, detailed in Sect. 3.3). The nonterminal $<signOp>$ allows the production of negative numbers by introducing positive or negative signs on numerical constants. The nonterminal $<cons>$ is replaced by one of the real values listed, that represents a percentage of the size of the problem.

For each individual, the grammar is used to derive a heuristic to the problem, which is interpreted and executed during the GP algorithm. The fitness of each individual, based on heuristic testing, allows for the selection of the best individuals.

Two distinct strategies were used, based on the computation of the fitness of each individual. The first one focuses on a single problem instance during the execution of the grammar-derived heuristic. In this case, the fitness f is defined as the gap between the literature best-known result $z_{p_i}^*$ and the solution cost z_{p_i} is the overall traffic z (Eq. 1) of the individual heuristic when solving instance p_i as

$$f_1(p_i) = (z_{p_i} - z_{p_i}^*)/z_{p_i}^*. \tag{2}$$

The second strategy uses a set of instances $(P = \{p_1, ..., p_{|P|}\})$ to evaluate the performance of each individual. The heuristic is executed once per instance and the fitness is the average of the computed instances gaps, as

$$f_2(P) = \sum_{i=1}^{|P|} f_1(p_i) \; / \; |P|. \tag{3}$$

The development of two different strategies aims to compare the performance of the GP algorithm depending on the use of a single instance or a set of instances. Since each instance must be executed for each individual of each generation, the first approach tends to run much faster. Nevertheless, the quality of the algorithm based on a single instance may vary depending on the chosen instance and has to be compared to the second approach.

5   Computational Experiments

Computational experiments were performed to generate heuristics for CAP and to analyze the results obtained by the proposed GGPHH technique considering its variants. We used the following set of 89 instances from the literature: $S\alpha$, $\alpha \in \{9, 9H, 10, 11\}$ [21]; $Am\beta$, $\beta \in \{12a, 12b, 13a, 13b, 15\}$ [3]; $N\gamma.\delta$, $\gamma \in \{25, 30, 40\}$, $1 \leq \delta \leq 5$ [5]; $Sko.\epsilon.\zeta$, $\epsilon \in \{42, 49, 56\}$ and $1 \leq \zeta \leq 5$ and $Sko.64.5$ [1]; $P\theta$, $1 \leq \theta \leq 6$ [2]; $ste36.0\eta$, $1 \leq \eta \leq 6$, [6]; $CAP.\iota.\kappa.\lambda$, $\iota = \{30, 60, 90\}$, $\kappa = \{30, 40, 50, 60\}$ and $1 \leq \lambda \leq 3$ [1]; and finally, $AKV.\mu.05$, $\mu \in \{60, 70\}$ [4]; with $n \in \{9, \ldots, 70\}$ the number of facilities. From these problems, a subset of representative instances was selected to generate the GGPHH variants. The following ones were chosen: $S9$, $S10$, $S11$, $Am15$, $N25.01$, $Sko42.1$, $Sko42.2$, $Sko42.3$, $Sko42.4$, and $Sko49.1$. Finally, the *General* variant evolves the heuristics using all these 10 instances.

The experiments were performed on a PC with an AMD Ryzen 5 2600 3.40GHz CPU, 16.0GB DDR4-1066GHz, and Ubuntu 22.04 LTS. The proposals were implemented in C++11. The source code of the proposals and the results are public available[1]. A supplementary material is also provided.

GP evolved 100 individuals during 100 generations. We defined the other parameters, as follows: maximum tree depth equals 10, the crossover probability is 90%, the mutation probability is 10%, and 5% of the best individuals remains in the population (elitism). It is important to highlight that any generation of new candidate solutions (in the initial population, or by the crossover and mutation operators) is performed without violating the maximum tree depth.

5.1 Analysis of the Results

The best and average values of the results obtained by the variants of the proposed GGPHH are presented in Tables 1, 2, 3, and 4. The columns indicate the instance used to evaluate the candidate solutions. Column *General* represents the GGPHH variant that uses a set of 10 instances to evaluate each individual. The best results are highlighted in boldface and the non-parametric Kruskal-Wallis was performed to evaluate if the results with the best average results are statistically better than the other ones. The results statistically similar to the best ones are marked with a star (∗). In addition, the number of instances/problems each variant presented the best values are presented in Table 5.

According to these results, the General variant found the best results in most of the instances (30 of the 89 instances), the Sko42.03 variant in 28 instances and the Sko49.01 variant in 26 instances. Also, the General variant reached the best average results in most problems (30 of the 89 instances) and the Sko42.01 variant obtained the best average results in 19 instances. Thus, the General variant presented the best values in most cases. It is interesting to notice that the Sko42.01 variant uses a single instance to generate the heuristics but it is also capable to reach competitive results when compared to the General variant.

Performance Profiles (PPs) [12] are used to analyze the relative performance of the search techniques when the set of problems is large. With a set S of solvers s_i, $i \in \{1, \ldots, n_s\}$, and a set P of problems/instances p_j, $j \in \{1, \ldots, n_p\}$, $t_{p,s}$ can be defined as a performance measurement for the method $s \in S$ when solving the problem $p \in P$. The performance ratio $r_{p,s} = \frac{t_{p,s}}{\max\{t_{p,s}:s \in S\}}$ is the relative performance of the method $s \in S$. The probability that the performance ratio $r_{p,s}$ of method s is within a factor $\tau > 0$ of the best ratio observed for every technique in S can be defined as $\rho_s(\tau) = \frac{1}{n_p} |\{p \in P : r_{p,s} \leq \tau\}|$, where $\rho_s(\tau)$ denotes the PPs curves. From PPs [7], are identified: (i) the approach that obtained the best results for most problems (largest $\rho(1)$), (ii) the most reliable

---

[1] https://github.com/rafaelfreesz/capPG.git.

Table 1. Best solutions found for the instances with size $n \leq 56$.

Instance	GGPHH Variant										
	S9	S10	S11	Am15	N25.01	Sko42.01	Sko42.02	Sko42.03	Sko42.04	Sko49.01	General
S9	1181.5	1181.5	1181.5	1181.5	1181.5	1181.5	1181.5	1181.5	1181.5	1181.5	1181.5
S9H	2294.5	2294.5	2294.5	2329.5	2294.5	2294.5	2294.5	2294.5	2294.5	2294.5	2294.5
S10	1374.5	1374.5	1374.5	1374.5	1374.5	1374.5	1374.5	1374.5	1374.5	1374.5	1374.5
S11	3466.5	3439.5	3439.5	3479.5	3439.5	3439.5	3439.5	3439.5	3439.5	3439.5	3439.5
Am12a	1529.0	1529.0	1529.0	1529.0	1529.0	1529.0	1529.0	1529.0	1529.0	1529.0	1529.0
Am12b	1609.5	1609.5	1609.5	1609.5	1609.5	1609.5	1609.5	1609.5	1609.5	1609.5	1609.5
Am13a	2472.5	2467.5	2467.5	2467.5	2467.5	2467.5	2467.5	2467.5	2467.5	2467.5	2467.5
Am13b	2871.0	2870.0	2870.0	2870.0	2870.0	2870.0	2870.0	2870.0	2870.0	2870.0	2870.0
Am15	3207.0	3195.0	3195.0	3195.0	3195.0	3195.0	3195.0	3195.0	3195.0	3195.0	3195.0
N25.01	2306.0	2302.0	2302.0	2302.0	2302.0	2302.0	2302.0	2302.0	2302.0	2302.0	2302.0
N25.02	18662.5	18600.5	18600.5	18626.5	18595.5	18601.5	18595.5	18601.5	18601.5	18595.5	18595.5
N25.03	12195.0	12138.0	12136.0	12138.0	12137.0	12133.0	12128.0	12124.0	12139.0	12138.0	12116.0
N25.04	24344.5	24192.5	24214.5	24290.5	24196.5	24216.5	24218.5	24216.5	24218.5	24234.5	24214.5
N25.05	7852.0	7819.0	7819.0	7825.0	7819.0	7819.0	7827.0	7819.0	7819.0	7819.0	7819.0
N30.01	4127.0	4115.0	4115.0	4115.0	4115.0	4115.0	4115.0	4115.0	4115.0	4115.0	4115.0
N30.02	10829.5	10790.5	10789.5	10784.5	10784.5	10791.5	10789.5	10793.5	10785.5	10785.5	10781.5
N30.03	22725.0	22706.0	22706.0	22711.0	22706.0	22706.0	22713.0	22706.0	22706.0	22708.0	22702.0
N30.04	28558.5	28401.5	28417.5	28404.5	28401.5	28401.5	28404.5	28401.5	28401.5	28404.5	28411.5
N30.05	57752.0	57480.0	57471.0	57456.0	57453.0	57436.0	57481.0	57424.0	57447.0	57434.0	57437.0
P1	30422.5	30316.5	30290.5	30318.5	30310.5	30303.5	30309.5	30290.5	30295.5	30316.5	30312.5
P2	34135.0	34012.0	33991.0	34015.0	33997.0	33980.0	34018.0	33976.0	33986.0	33979.0	33985.0
P3	35131.5	35099.5	35081.5	35093.5	35080.5	35073.5	35098.5	35062.5	35074.5	35081.5	35068.5
P4	34812.5	34667.5	34677.5	34701.5	34670.5	34675.5	34675.5	34668.5	34679.5	34680.5	34675.5
P5	30880.0	30812.0	30791.0	30826.0	30814.0	30806.0	30801.0	30830.0	30803.0	30786.0	30787.0
P6	34506.5	34437.5	34437.5	34447.5	34424.5	34441.5	34447.5	34441.5	34437.5	34424.5	34438.5
ste36.01	5182.0	4966.0	4966.0	5486.0	5003.0	5014.0	5003.0	4966.0	4966.0	5003.0	4966.0
ste36.02	90682.0	88382.0	88431.0	89270.0	88382.0	88977.0	89041.0	88819.0	89152.0	88939.0	88466.0
ste36.03	51777.5	50824.5	50727.5	51362.5	50686.5	50529.5	50426.5	50688.5	50616.5	50703.5	50713.5
ste36.04	49093.5	46884.5	47306.5	48059.5	47224.5	47156.5	47007.5	47193.5	46940.5	47102.5	46992.5
ste36.05	46537.5	45183.5	45030.5	45398.5	44710.5	44964.5	44804.5	44688.5	44619.5	44565.5	44974.5
N40.01	53943.5	53746.5	53769.5	53858.5	53755.5	53748.5	53743.5	53745.5	53753.5	53764.5	53740.5
N40.02	49082.0	48932.0	48942.0	48973.0	48924.0	48954.0	48942.0	48924.0	48908.0	48923.0	48925.0
N40.03	39357.5	39282.5	39260.5	39270.5	39280.5	39264.5	39275.5	39266.5	39261.5	39265.5	39263.5
N40.04	38485.0	38381.0	38380.0	38381.0	38361.0	38355.0	38380.0	38379.0	38384.0	38377.0	38383.0
N40.05	51688.0	51529.0	51511.0	51544.0	51505.0	51532.0	51532.0	51505.0	51515.0	51518.0	51544.0
Sko42.01	12760.0	12731.0	12731.0	12741.0	12731.0	12731.0	12731.0	12731.0	12731.0	12731.0	12731.0
Sko42.02	108359.5	108055.5	108050.5	108083.5	108082.5	108059.5	108049.5	108053.5	108069.5	108043.5	108071.5
Sko42.03	86921.5	86667.5	86677.5	86744.5	86676.5	86683.5	86700.5	86672.5	86678.5	86704.5	86684.5
Sko42.04	69111.0	68785.0	68842.0	68808.0	68780.0	68771.0	68767.0	68753.0	68795.0	68811.0	68771.0
Sko42.05	124313.5	124111.5	124064.5	124105.5	124088.5	124118.5	124055.5	124095.5	124098.5	124123.5	124075.5
Sko49.01	20506.0	20478.0	20482.0	20478.0	20478.0	20478.0	20482.0	20478.0	20482.0	20470.0	20470.0
Sko49.02	208910.0	208159.0	208241.0	208450.0	208217.0	208176.0	208179.0	208242.0	208192.0	208157.0	208160.0
Sko49.03	162517.0	162430.0	162332.0	162631.0	162299.0	162349.0	162238.0	162389.0	162405.0	162273.0	162344.0
Sko49.04	118556.5	118345.5	118315.5	118370.5	118324.5	118376.5	118298.5	118337.5	118319.5	118306.5	118307.5
Sko49.05	333418.0	332879.0	332941.0	332946.0	333038.0	332943.0	333016.0	332953.0	332936.0	333020.0	333033.0
Sko56.01	32031.0	31983.0	31983.0	31984.0	31983.0	31977.0	31986.0	31985.0	31982.0	31984.0	31983.0
Sko56.02	248791.0	248334.0	248311.0	248628.0	248329.0	248310.0	248309.0	248286.0	248301.0	248285.0	248279.0
Sko56.03	85472.0	85215.0	85223.0	85318.0	85212.0	85242.0	85227.0	85221.0	85214.0	85210.0	85218.0
Sko56.04	157052.0	156770.0	156701.0	156804.0	156730.0	156707.0	156746.0	156812.0	156782.0	156684.0	156758.0
Sko56.05	296678.5	296283.5	296347.5	296428.5	296304.5	296350.5	296280.5	296298.5	296267.5	296316.5	296270.5

approach (smallest τ such that $\rho(\tau) = 1$), and (iii) the best overall performance (largest area under the PPs curves).

PPs are presented in Fig. 4, where one can observe that the General variant obtained the best average results in most of the problems (30 of the 89 instances) and reached the largest area under the PPs curves. Also, this variant is the second most reliable one, with $\rho(1.006787) = 1$. On the other hand, the variant that uses Sko42.01 obtained the second largest area under the PPs curves and is the second one that found the best average results in most of the problems (19 instances). However, this variant reached $\rho(\tau) = 1$ only when $\tau = 1.010377$. Sko42.03 is the second one that found the best results in most of the problems (28 instances) and obtained the third largest area under the PPs curves. It is also the most reliable one, reaching $\rho(1.005836) = 1$ with the smallest τ.

Table 2. Best solutions obtained for the instances with size $n \geq 60$.

Instance	GGPHH Variant										
	S9	S10	S11	Am15	N25.01	Sko42.01	Sko42.02	Sko42.03	Sko42.04	Sko49.01	General
AKV.60.05	161497.0	159805.0	159806.0	160093.0	159777.0	159771.0	159823.0	159789.0	159766.0	**159761.0**	159763.0
CAP.30.30.1	205190.0	204308.0	204277.0	205459.0	204346.0	204293.0	204308.0	204306.0	**204245.0**	204451.0	204287.0
CAP.30.30.2	194606.5	193619.5	193469.5	193457.5	193466.5	193543.5	193460.5	193492.5	193430.5	193433.5	**193407.5**
CAP.30.30.3	162627.5	161735.5	161702.5	161744.5	161724.5	161705.5	**161600.5**	161779.5	161717.5	161634.5	161793.5
CAP.30.40.1	135977.5	135409.5	**135221.5**	135319.5	135330.5	135338.5	135291.5	135345.5	135312.5	135390.5	135294.5
CAP.30.40.2	160655.0	**159261.0**	159359.0	159388.0	159299.0	159368.0	159338.0	159331.0	159331.0	159357.0	159324.0
CAP.30.40.3	160339.5	159331.5	159188.5	159402.5	159149.5	159172.5	159116.5	**159093.5**	159406.5	159143.5	159103.5
CAP.30.50.1	111618.5	110816.5	110715.5	110983.5	**110617.5**	110898.5	110768.5	110840.5	110758.5	110732.5	110736.5
CAP.30.50.2	115951.0	115533.0	115503.0	115775.0	115507.0	**115439.0**	115486.0	115448.0	115489.0	115480.0	115513.0
CAP.30.50.3	115412.0	114455.0	114340.0	114749.0	114375.0	114401.0	114424.0	114365.0	114401.0	114406.0	**114337.0**
CAP.30.60.1	108788.0	108153.0	108203.0	**108123.0**	108241.0	108192.0	108214.0	108194.0	108168.0	108221.0	108184.0
CAP.30.60.2	110739.5	110124.5	110079.5	110512.5	110115.5	**110036.5**	110088.5	110094.5	110099.5	110091.5	110049.5
CAP.30.60.3	92510.0	92009.0	91920.0	92259.0	91967.0	91978.0	91891.0	**91801.0**	91996.0	91845.0	91882.0
CAP.60.30.1	446198.5	445655.5	445448.5	446109.5	445580.5	**445440.5**	445493.5	445687.5	445548.5	445609.5	445473.5
CAP.60.30.2	409318.5	408056.5	408005.5	408426.5	408020.5	408011.5	**407953.5**	408073.5	408034.5	408018.5	408051.5
CAP.60.30.3	417897.5	417045.5	417027.5	417067.5	417117.5	417054.5	417070.5	417059.5	417037.5	417068.5	**417003.5**
CAP.60.40.1	313945.0	313416.0	313381.0	313435.0	313472.0	313399.0	313432.0	313392.0	313384.0	313403.0	**313371.0**
CAP.60.40.2	321370.5	320875.5	320827.5	**320785.5**	320953.5	320815.5	320818.5	320832.5	320811.5	320829.5	320807.5
CAP.60.40.3	363827.5	363216.5	363173.5	363218.5	**363088.5**	363169.5	363219.5	363255.5	363288.5	363335.5	363137.5
CAP.60.50.1	274006.0	273587.0	273542.0	273576.0	273566.0	273548.0	273532.0	273575.0	273527.0	**273518.0**	273527.0
CAP.60.50.2	270218.5	269908.5	269838.5	269834.5	269886.5	269817.5	**269782.5**	269860.5	269821.5	269842.5	269917.5
CAP.60.50.3	295968.0	295569.0	295567.0	295568.0	295569.0	295536.0	295507.0	**295501.0**	295516.0	295567.0	295552.0
CAP.60.60.1	228230.0	227981.0	227979.0	228133.0	227973.0	227953.0	227978.0	228012.0	227980.0	227992.0	**227945.0**
CAP.60.60.2	247057.0	246712.0	**246625.0**	246978.0	246629.0	246659.0	246651.0	246707.0	246646.0	246655.0	246628.0
CAP.60.60.3	206959.5	206609.5	206603.5	206923.5	206593.5	**206576.5**	206604.5	206591.5	206624.5	206589.5	206635.5
CAP.90.30.1	629500.0	628957.0	628965.0	629079.0	629026.0	629059.0	628994.0	629075.0	628983.0	628958.0	**628934.0**
CAP.90.30.2	561801.5	561306.5	561269.5	561483.5	561275.5	561287.5	561245.5	561335.5	561306.5	561217.5	**561212.5**
CAP.90.30.3	588096.5	587861.5	587908.5	587908.5	587869.5	587841.5	587854.5	587849.5	587878.5	**587838.5**	587852.5
CAP.90.40.1	474373.0	474130.0	474096.0	474135.0	474138.0	474127.0	**474066.0**	474153.0	474098.0	474083.0	474088.0
CAP.90.40.2	480865.0	480011.0	**480006.0**	480469.0	480135.0	480034.0	480015.0	480243.0	480051.0	480044.0	480022.0
CAP.90.40.3	513048.0	512492.0	512483.0	512555.0	512523.0	512480.0	512529.0	**512475.0**	512493.0	512510.0	512509.0
CAP.90.50.1	480630.0	479820.0	**479705.0**	479835.0	479787.0	479780.0	479778.0	479780.0	479749.0	479742.0	479768.0
CAP.90.50.2	445505.0	445201.0	445151.0	445304.0	445128.0	445163.0	445099.0	445183.0	445140.0	**445094.0**	445105.0
CAP.90.50.3	495349.5	495153.5	495084.5	495245.5	495109.5	495097.5	495094.5	495138.5	495078.5	495122.5	**495054.5**
CAP.90.60.1	385720.5	385449.5	385450.5	385469.5	385467.5	385445.5	385455.5	**385441.5**	385445.5	385454.5	385456.5
CAP.90.60.2	344880.0	344788.0	344804.0	344817.0	344840.0	344837.0	344802.0	344811.0	344812.0	344808.0	**344784.0**
CAP.90.60.3	411582.0	411322.0	411268.0	411462.0	411323.0	**411259.0**	411273.0	411283.0	411280.0	411267.0	411284.0
Sko64.05	251144.5	250950.5	250923.5	250913.5	251014.5	250913.5	250942.5	**250904.5**	250912.5	250905.5	250924.5
AKV.70.05	2120309.5	2110546.5	2111485.5	2117174.5	2110405.5	2111787.5	2111356.5	**2110367.5**	2110894.5	2111087.5	2110528.5

5.2   Analysis of the Best Heuristic Created

One of the main advantages of GP techniques is its symbolic solutions. Here, we present in Fig. 5 the best heuristic generated by GGPHH, which was obtained using the General variant. This solution contains a pattern in the four blocks of instructions clustered in lines 4–7. Each of these lines starts with a perturbation and follows with one or more local searches. These steps can be observed in well-known search techniques, such as Iterated Local Search (ILS) [22]. On the other hand, there is a predominance of using permutations by reconstructions of the solutions (lines 4–6) and these instructions are common in Iterated Greedy (IG), as IG performs deconstruction and reconstruction operations on some elements of the solution until a stopping criterion is reached [20].

Table 3. Average solutions obtained for the instances of size $n \le 56$.

Instance	GGPHH Variant										
	S9	S10	S11	Am15	N25.01	Sko42.01	Sko42.02	Sko42.03	Sko42.04	Sko49.01	General
S9	1190.13	1181.50*	1181.50*	1181.50*	1181.50*	1181.50*	1181.50*	1181.50*	1181.50*	1181.50*	1181.50*
S9H	2298.70	2295.00*	2295.00*	2329.50	2295.10	2294.60*	2295.20	2294.80*	2295.00*	2294.60*	2294.60*
S10	1386.47	1374.50*	1374.77*	1378.07	1374.63*	1375.53	1374.50*	1374.63*	1374.50*	1374.50*	1374.90*
S11	3524.67	3451.87*	3448.30*	3489.03	3453.70*	3445.33*	3449.27*	3447.10*	3450.73*	3448.87*	3445.27*
Am12a	1531.57*	1531.90*	1533.20*	1532.90*	1531.73*	1534.00	1533.50*	1533.57*	1533.03*	1533.03*	1531.30*
Am12b	1661.87	1612.10*	1619.97	1617.53*	1612.27*	1620.03	1622.13	1616.47*	1618.77*	1627.13	1614.87*
Am13a	2485.83	2467.70*	2467.80*	2469.50	2467.70*	2467.93*	2467.90*	2467.67*	2467.70*	2467.70*	2468.00*
Am13b	2904.93	2870.50*	2870.83*	2878.77	2871.87*	2871.17*	2872.00*	2870.60*	2871.50*	2872.53*	2870.47*
Am15	3243.53	3201.83*	3205.57	3205.13*	3206.70	3204.00*	3205.50*	3204.67	3207.70	3208.33	3199.90*
N25.01	2317.60	2308.20	2305.97	2310.47	2306.63	2302.83*	2306.67	2305.87	2306.13	2307.37	2303.60*
N25.02	18801.27	18647.67*	18651.27*	18659.97	18635.77*	18643.53*	18652.47*	18652.47	18650.87*	18651.33*	18639.47*
N25.03	12300.70	12169.10	12167.47	12187.67	12167.80*	12164.77*	12167.37	12166.23	12165.47	12163.50*	12157.57*
N25.04	24471.57	24273.43*	24306.13	24319.33	24276.60*	24294.50	24300.60	24306.73	24299.83	24308.23	24272.43*
N25.05	7909.60	7834.87*	7844.70	7847.67	7836.73*	7843.87	7845.87	7842.03	7838.23*	7839.33*	7835.67*
N30.01	4162.20	4115.87*	4116.83*	4117.73	4116.70	4116.03*	4116.07*	4116.33*	4116.03*	4117.27	4115.80*
N30.02	10901.33	10805.13*	10813.03	10800.33*	10803.97	10809.03	10812.63	10812.27	10808.60	10809.63	10805.07*
N30.03	22883.53	22723.70*	22728.53*	22745.40	22725.93*	22723.20*	22733.50	22729.40*	22722.90*	22725.03*	22723.73*
N30.04	28814.80	28459.87	28456.27	28503.73	28456.57*	28439.90*	28454.43*	28460.57	28442.90*	28454.03	28447.10*
N30.05	58078.23	57627.63*	57652.83	57605.00*	57619.00*	57591.70*	57623.73*	57659.13	57588.90*	57635.30*	57596.13*
P1	30519.27	30383.37	30360.17*	30406.27	30374.67	30368.07*	30356.07*	30367.67*	30363.10*	30368.33*	30359.87*
P2	34282.67	34073.17	34061.27*	34137.43	34069.27	34042.73*	34064.97*	34057.47*	34043.60*	34054.50*	34048.27*
P3	35415.90	35161.70	35146.23*	35344.20	35149.57*	35139.50*	35165.17*	35157.50	35175.87*	35142.40*	35143.53*
P4	34980.63	34734.17	34738.00	34791.20	34745.93	34719.37*	34752.67	34757.73	34754.97	34736.83*	34725.67*
P5	30976.63	30873.00	30866.87*	30883.47	30874.07	30857.23*	30864.03*	30860.00*	30862.00*	30863.73*	30856.60*
P6	34620.77	34478.93*	34486.27	34504.47	34468.60*	34471.43*	34485.47	34474.97*	34477.73*	34473.23*	34470.27*
ste36.01	5324.90	5133.00	5137.33	5541.00	5106.33*	5140.83	5160.03	5088.03*	5151.23	5156.90	5122.57*
ste36.02	96169.23	91018.17	90937.17	93826.20	90430.40*	90878.30	90557.47*	90201.60*	90697.57	90759.97	90303.13*
ste36.03	52561.57	51531.33*	51525.97	53319.37	51470.70*	51576.93	51640.00	51301.17*	51519.07	51578.70	51392.97*
ste36.04	50855.73	48409.63*	48561.90*	49995.83	48237.17*	48414.50*	48912.67	48488.63*	48253.97*	48537.57*	48207.30*
ste36.05	48249.23	46096.47	45943.40	47473.53	46090.00	45888.20	45847.37	45501.90*	45749.93*	45883.80	45694.90*
N40.01	54296.17	53845.97*	53851.80*	54633.50	53838.60*	53820.07*	53861.43	53864.00	53862.80	53831.77*	53828.20*
N40.02	49377.07	49002.27	49006.07	49049.17	48986.70*	48980.43*	48994.43	49008.40	48984.83*	48984.40*	48990.40*
N40.03	39534.33	39304.60	39304.00	39293.77*	39319.93	39307.77	39302.53*	39298.00*	39306.17	39301.80*	39301.50*
N40.04	38698.77	38429.20*	38420.10*	38478.90	38413.77*	38427.57*	38442.00	38424.57*	38419.93*	38419.93*	38419.50*
N40.05	51948.33	51595.80	51589.50*	51969.03	51578.93*	51569.97*	51584.83*	51589.93*	51574.07*	51576.40*	51576.50*
Sko42.01	12859.70	12744.13	12747.87	12760.17	12744.30	12734.63*	12747.20	12745.00	12742.77	12746.17	12742.40
Sko42.02	108600.90	108168.43*	108158.10*	108221.67	108189.77	108156.20*	108175.20*	108180.50	108141.43*	108160.07*	108158.73*
Sko42.03	87285.23	86870.80	86836.77*	86938.43	86848.17	86836.50*	86847.43	86820.17*	86835.23*	86840.40*	86794.50*
Sko42.04	69315.93	68900.20	68926.73	69046.47	68887.17*	68883.43	68888.47*	68919.23	68905.10	68890.60*	68876.23*
Sko42.05	125046.60	124250.27*	124235.20*	124584.53	124227.60*	124236.40*	124198.43*	124225.63*	124232.80*	124215.40*	124225.87*
Sko49.01	20550.17	20495.20*	20498.37	20506.90	20496.07*	20489.60*	20498.20	20492.67*	20499.50	20495.77	20489.97*
Sko49.02	209659.57	208544.57*	208472.70*	209527.33	208504.77*	208439.67*	208506.77*	208584.77*	208461.73*	208468.00*	208486.17*
Sko49.03	163539.87	162603.40	162549.63	163403.00	162480.70*	162489.23*	162586.87*	162604.47	162622.30	162505.20*	162528.53*
Sko49.04	118882.17	118503.17	118475.20*	118649.03	118458.60*	118521.20	118478.20*	118468.63*	118466.17*	118465.57*	118441.57*
Sko49.05	334706.50	333302.10*	333389.50	333324.07	333316.27	333156.23*	333312.23	333272.33	333241.77*	333344.33	333263.97*
Sko56.01	32163.27	32010.47	32024.03	32010.40	32022.33	31998.90*	32021.00	32010.50	32022.27	32012.57	32008.10
Sko56.02	249867.20	248603.10	248506.20*	250451.13	248562.97*	248474.87*	248481.67*	248605.73	248495.03*	248537.57	248448.23*
Sko56.03	85670.10	85380.07	85368.07	85574.73	85309.50*	85355.93	85355.10	85418.57	85339.60*	85366.60	85321.63*
Sko56.04	157604.40	156911.97	156878.77*	157053.57	156908.43	156858.03*	156915.37	156960.87	156897.97	156851.63*	156861.17*
Sko56.05	298502.87	296615.67	296527.70	297572.67	296431.53*	296473.30*	296454.97*	296657.07	296449.63*	296518.37*	296425.97*

Table 4. Average solutions found for the instances of size $n \geq 60$.

Instance	GGPHH Variant										General
	S9	S10	S11	Am15	N25.01	Sko42.01	Sko42.02	Sko42.03	Sko42.04	Sko49.01	
AKV.60.05	162939.60	159999.80	160179.10	160744.27	**159921.13***	160074.70	160130.03	160135.83	160140.57	160029.47*	160091.30
CAP.30.30.1	207162.30	205051.17	204863.30	206542.60	204999.87	**204576.83***	204929.57	205068.97	204911.30	204986.37	204772.20*
CAP.30.30.2	195460.10	194030.90	193900.43	194614.80	194079.47	193823.63*	193859.70*	193881.43*	193886.37*	193933.73	**193768.87***
CAP.30.30.3	164022.00	162506.00	162422.77*	162512.93	162544.47	162372.73*	**162210.20***	162469.47*	162332.20*	162309.70*	162277.60*
CAP.30.40.1	136956.50	135785.37	135588.13*	135722.00	135732.60	**135527.10***	135666.40	135720.37	135738.73	135658.30	135650.23
CAP.30.40.2	161618.30	160002.03	159843.53	160161.80	159844.93	159807.57	159739.83	159893.33	159622.87*	159589.13*	**159567.23***
CAP.30.40.3	161196.80	159935.70*	159793.93*	160322.70	159898.40	**159765.03***	159939.07*	159904.97*	159940.80*	159971.83*	159799.60*
CAP.30.50.1	112343.10	111329.53	111118.73*	111800.37	111182.33*	111228.87	111134.13*	111366.83	111080.13*	111092.13*	**111079.37***
CAP.30.50.2	117288.60	115826.73	115683.77*	116616.03	115860.43	**115623.83***	115682.00*	115761.67	115673.97*	115683.40*	115643.17*
CAP.30.50.3	116124.80	114814.17	114676.70*	115722.90	114727.87*	114680.40*	114689.40*	114756.63*	114675.53*	114682.27*	**114672.93***
CAP.30.60.1	109472.33	108382.90	108351.07*	108435.33	108377.83	108412.33	108354.10*	108405.63	108315.27*	108372.67	**108309.13***
CAP.30.60.2	111516.47	110608.17	110444.03	111374.73	110485.23	110307.13*	110399.27	110550.90	110407.03	110425.87	**110273.40***
CAP.30.60.3	93209.17	92382.60	92328.00	92753.77	92402.47	92228.00*	92343.53	92306.43*	92311.37*	92269.00*	**92199.77***
CAP.60.30.1	447474.50	446211.87	445945.07*	447057.63	446015.43*	446034.50	**445871.57***	446342.50	445994.97*	445975.77*	445940.83*
CAP.60.30.2	410759.47	409007.93	408500.33*	409964.13	408577.53*	408628.83*	**408370.97***	408921.50	408527.10*	408506.60*	408382.00*
CAP.60.30.3	419055.23	417273.73	417223.90*	418315.77	417328.53	417215.10*	417234.13*	417344.80	417222.70*	417203.77*	**417200.13***
CAP.60.40.1	314557.63	313692.57	**313570.93***	313627.20*	313823.30	313678.73*	313628.70*	313605.03*	313633.20	313602.87*	313670.33*
CAP.60.40.2	321860.13	321150.20	320995.47*	321221.27	321154.00	321051.93	321007.07*	321221.20	321036.00*	**320981.47***	321031.00*
CAP.60.40.3	365073.17	363750.37	**363517.97***	364387.33	363707.83	363655.43*	363555.27*	363941.17	363518.80*	363554.57*	363554.57*
CAP.60.50.1	274682.83	273846.50	**273701.50***	274255.80	274008.53	273718.43*	273738.23*	273853.00	273713.07*	273774.33*	273713.20*
CAP.60.50.2	271264.37	270321.37	270143.80*	270296.60	270344.13	270150.67*	**270091.00***	270331.30	270131.23*	270202.87*	270145.33*
CAP.60.50.3	296749.93	295740.97*	295701.00*	295927.17	295765.20	295697.17*	295699.43*	295751.20*	295710.10*	295713.87*	**295680.60***
CAP.60.60.1	228867.57	228139.63	228123.20	228277.07	228095.60*	228103.77*	228121.30	228207.07	228130.87	228123.67	**228078.37***
CAP.60.60.2	248104.40	247097.93	246938.20*	247630.23	246971.50	247027.80	246912.50*	247280.03	246937.77	246908.63*	**246820.03***
CAP.60.60.3	207583.23	206804.07*	**206768.13***	207379.73	206826.23*	206783.97*	206759.10*	206833.33*	206890.50	206830.93*	206825.80*
CAP.90.30.1	630429.30	629220.70	629185.90*	629529.83	629251.43	629247.03	629195.00*	629286.20	629177.57*	629177.90*	**629142.10***
CAP.90.30.2	563100.30	561650.43	**561466.03***	562536.37	561665.93	561586.67*	561530.80*	561810.50	561514.87*	561524.87*	561498.57*
CAP.90.30.3	589359.93	587966.33	587073.87	588880.40	588055.00	588138.83	**587953.70***	588138.67	587958.67*	587979.33*	587996.00
CAP.90.40.1	475573.90	474496.43	474320.97*	474865.40	474556.80	474293.53*	**474289.77***	474716.03	474321.40*	474373.03*	474304.17*
CAP.90.40.2	481478.90	480485.27*	480518.70*	481098.93	480627.67	480502.00*	480462.67*	480717.33	480502.43*	**480453.37***	480465.67*
CAP.90.40.3	514154.07	512995.20*	**512756.47***	513583.37	512970.87	513009.33*	512811.40*	513156.07	512853.23*	512798.73*	512767.67*
CAP.90.50.1	481386.33	480068.30	479966.07*	480797.57	480012.37	479994.97*	479991.73*	480114.27	480036.60*	**479954.70***	480016.97*
CAP.90.50.2	446432.77	445620.50	445491.23	446184.73	445475.40*	445510.17	445466.13*	445652.47	**445383.63***	445480.70*	445463.17*
CAP.90.50.3	496209.43	495570.03	**495402.27***	495964.43	495464.40*	495516.73*	495458.13*	495556.37*	495510.97*	495521.30*	495442.97*
CAP.90.60.1	386222.77	385601.97	385551.83*	386057.43	385597.70	385537.30*	385552.27*	385609.93*	385550.27*	385562.90*	**385537.07***
CAP.90.60.2	345175.43	**344852.13***	344868.20*	344857.67*	344935.03	344872.20	344868.93*	344853.77*	344874.80	344874.80	344875.77
CAP.90.60.3	412100.47	411474.10*	411438.80*	411667.37	411469.53	411470.83	411424.70*	411504.10	411420.43*	**411414.43***	411434.80*
Sko64.05	251745.90	251088.13	251095.43	251080.70	251291.40	251021.83*	251053.60	251145.13	251036.50*	251037.53*	**250997.17***
AKV.70.05	2124248.43	2112045.67	2114540.07	2119937.77	**2111324.63***	2117322.53	2114944.67	2112827.83	2114852.27	2115584.83	2115388.40

For comparison purposes, the results obtained by the General variant are compared to those of [16], which presents the best costs in the literature. Tables containing these results are available in the supplementary material, where one can observe gaps smaller than 1.0% for all the instances tested here, except for ste36.02, ste36.03, and ste36.05. The average gap for all instances is 0.09% and, thus, the results found by the proposal are similar to the best ones from the literature.

Table 5. Number of instances each variant reached the best values, where #Best represents the counting for the best case, #Avg. is the counting for the average values, and #Stat.Test. is the number of instances each technique found the best average values or its solutions are statistically similar to the best ones.

Score	GGPHH Variant										General
	S9	S10	S11	Am15	N25.01	Sko42.01	Sko42.02	Sko42.03	Sko42.04	Sko49.01	
#Best	5	22	20	11	20	21	21	28	18	26	**30**
#Avg	0	5	8	3	8	19	10	6	6	8	**30**
#Stat.Test	1	34	54	9	42	66	57	39	61	63	**82**

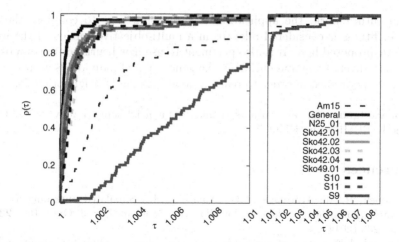

Fig. 4. PPs using the average results as performance measurement. The normalized areas under the PPs curves are 1.0000, 0.9951, 0.9949, 0.9949, 0.9946, 0.9930, 0.9927, 0.9919, 0.9915, 0.9405, and 0.8931, respectively, for General, Sko42.01, Sko42.03, Sko42.04, N25.01, S11, Sko49.01, Sko42.02, S10, Am15, and S9.

```
1    void execute(I){
2        S = randomlyGenerate(I); Sb=S;
3        for(i={1,...,50}){
4            S.reconstruction(0.04); S.rvnd(); S.fullPermut(); S.insertion();
5            S.reconstrucion(0.65); S.rvnd(); S.insertion();
6            S.reconstruction(0.5); S.rvnd();
7            S.swap(0.40,0.25); S.rvnd();
8            if (F(S)<F(Sb)){Sb=S;}
9        }
10   }
```

Fig. 5. Best heuristic generated by GGPHH.

6 Concluding Remarks and Future Works

The proposed GGPHH was able to produce efficient heuristics to CAP. Two strategies were tested: one using a single instance during the GP evaluation and another one training each individual on a set of instances.

We observed that the General GGPHH variant, that evolved using information of ten instances, performed better than the other ones. The generated heuristic reached the highest amount of cost averages when used with all instances of the problem. Based on this information, we can conclude that the exploration of the problem under the perspective of its different characteristics allows the development of more robust heuristics.

On the other hand, one of the variants with a single instance also obtained good results. In particular, we observed that using instances Sko42.01 and Sko42.03 during the evolution allows for the creation of good heuristics.

For future work, the application of processing time to compose the individual's fitness is considered relevant in a multi-objective version of the hyper-heuristic proposed here. It is also pertinent to use new heuristics to compose the grammar, diversifying and enriching the generation of individuals with a variety of larger operations, in order to reach a greater range of neighborhoods.

Acknowledgements. The authors thank the financial support provided by CNPq, Capes, FAPEMIG, and UFJF.

References

1. Ahonen, H., de Alvarenga, A.G., Amaral, A.R.: Simulated annealing and tabu search approaches for the corridor allocation problem. Eur. J. Oper. Res. **232**(1), 221–233 (2014)
2. Amaral, A.R.: An exact approach to the one-dimensional facility layout problem. Oper. Res. **56**(4), 1026–1033 (2008)
3. Amaral, A.R.: The corridor allocation problem. Comput. Oper. Res. **39**(12), 3325–3330 (2012)
4. Anjos, M.F., Kennings, A., Vannelli, A.: A semidefinite optimization approach for the single-row layout problem with unequal dimensions. Discret. Optim. **2**(2), 113–122 (2005)
5. Anjos, M.F., Vannelli, A.: Computing globally optimal solutions for single-row layout problems using semidefinite programming and cutting planes. INFORMS J. Comput. **20**(4), 611–617 (2008)
6. Anjos, M.F., Yen, G.: Provably near-optimal solutions for very large single-row facility layout problems. Optim. Meth. Softw. **24**(4–5), 805–817 (2009)
7. Barbosa, H.J.C., Bernardino, H.S., Barreto, A.M.S.: Using performance profiles to analyze the results of the 2006 CEC constrained optimization competition. In: Evolutionary Computation (CEC), 2010 IEEE Congress on, pp. 1–8. IEEE (2010)
8. Burke, E.K., et al.: Hyper-heuristics: a survey of the state of the art. J. Oper. Res. Soc. **64**(12), 1695–1724 (2013)
9. Burke, E.K., Hyde, M.R., Kendall, G., Ochoa, G., Ozcan, E., Woodward, J.R.: Exploring hyper-heuristic methodologies with genetic programming. In: Mumford, C.L., Jain, L.C. (eds.) Computational Intelligence. Intelligent Systems Reference Library, vol. 1, pp. 177–201. Springer, Berlin Heidelberg (2009). https://doi.org/10.1007/978-3-642-01799-5_6
10. Cellin, B.D.P.: Métodos para Resolução eficiente de Problemas de Layout. Master's thesis, Universidade Federal do Espírito Santo, Vitória (2017)
11. Chung, J., Tanchoco, J.: The double row layout problem. Int. J. Prod. Res. **48**(3), 709–727 (2010)
12. Dolan, E.D., Moré, J.J.: Benchmarking optimization software with performance profiles. Math. Program. **91**(2), 201–213 (2002). https://doi.org/10.1007/s101070100263
13. de Freitas, J.M., de Souza, F.R., Bernardino, H.S.: Evolving controllers for mario AI using grammar-based genetic programming. In: 2018 IEEE Congress on Evolutionary Computation (CEC), pp. 1–8. IEEE (2018)
14. Ghosh, D., Kothari, R.: Population heuristics for the corridor allocation problem. Technical report, Indian Institute of Management at Ahmedabad, Research and Publication Department (2012)

15. Harris, S., Bueter, T., Tauritz, D.R.: A comparison of genetic programming variants for hyper-heuristics. In: Proceedings of the Companion Publication of the Conference on Genetic and Evolutionary Computation, pp. 1043–1050 (2015)
16. Herrán, A., Colmenar, J.M., Duarte, A.: An efficient variable neighborhood search for the space-free multi-row facility layout problem. Eur. J. Oper. Res. **295**(3), 893–907 (2021)
17. Kothari, R., Ghosh, D.: A scatter search algorithm for the single row facility layout problem. J. Heuristics **20**(2), 125–142 (2014)
18. Koza, J.R.: Genetic Programming: On the Programming of Computers by Means of Natural Selection. The MIT Press, Cambridge (1992)
19. Kusiak, A., Heragu, S.S.: The facility layout problem. Eur. J. Oper. Res. **29**(3), 229–251 (1987)
20. Lozano, M., Molina, D., García-Martínez, C.: Iterated greedy for the maximum diversity problem. Eur. J. Oper. Res. **214**(1), 31–38 (2011)
21. Simmons, D.M.: One-dimensional space allocation: an ordering algorithm. Oper. Res. **17**(5), 812–826 (1969)
22. Talbi, E.G.: Metaheuristics: From Design to Implementation. John Wiley, Hoboken (2009)
23. Whigham, P.A.: Grammatically-based genetic programming. In: Rosca, J.P. (ed.) Proceedings of the Workshop on Genetic Programming: From Theory to Real-World Applications, pp. 33–41. University of Rochester, Tahoe City, USA (1995)
24. Yang, T., Peters, B.A.: A spine layout design method for semiconductor fabrication facilities containing automated material-handling systems. Int. J. Oper. Prod. Manage. **17**(5), 490–501 (1997)

Generalising Semantics to Weighted Bipolar Argumentation Frameworks

Renan Cordeiro$^{(\boxtimes)}$ and João Alcântara

Department of Computer Science, Federal University of Ceará, Fortaleza, Brazil
`renandcsc@alu.ufc.br`, `jnando@lia.ufc.br`

Abstract. In Bipolar Argumentation Frameworks (*BAF*s), arguments cannot only attack, but support other arguments. In this paper, we generalise the fundamental semantics for *BAF*s to deal with Weighted Bipolar Argumentation Frameworks (*WBAF*s), in which real numbers are assigned as weights to arguments. A distinguishing aspect of our approach is that our semantics are determined in terms of two measures: one is employed to assess the degree of acceptability of an argument, and the other, its degree of rejection. Then, as opposed to previous proposals, we have more expressive semantics that naturally generalise their corresponding semantics for *BAF*s and are defined for each *WBAF*.

Keywords: Argumentation · Semantics · Bipolarity

1 Introduction

Computational argumentation provides reasoning models by which arguments are constructed, compared and evaluated. Over the last decades, argumentation has become a prominent Artificial Intelligence research area. It has been employed in a wide range of problems and applications such as reasoning with inconsistent information [1], reasoning with defeasible information [2], decision making [3], classification [4] computational persuasion [5], recommender systems [6] and review aggregation [7].

Most of these computational models of arguments are settled on the seminal Dung's work [8] on Abstract Argumentation Frameworks (*AAF*s), which ignore the content of the arguments and focuses solely on their relationships. Indeed, Abstract Argumentation Frameworks can be understood as a directed graph whose nodes are arguments and edges are the attack relation between arguments. Despite their success, *AAF*s are not immune to criticisms. A contentious issue refers to their alleged limited expressivity as they lack features which are common in almost every form of argumentation found in practice [9]. Indeed, in *AAF*s, the only interaction between atomic arguments is given by the attack relation.

As consequence, several extended versions of Dung's framework have been proposed [10–19]. One of them is the Bipolar Argumentation Framework [11, 12, 19], in which besides attacking, an argument can support another argument. However, introducing the notion of support between arguments within abstract frameworks has been controversial and counter-intuitive results have been obtained. Recently, in [19], the

J. C. Xavier-Junior and R. A. Rios (Eds.): BRACIS 2022, LNAI 13653, pp. 520–534, 2022.
https://doi.org/10.1007/978-3-031-21686-2_36

authors proposed some semantics for *BAF*s that treat attacks and supports symmetrically and directly lead to natural generalisations of the corresponding semantics for *AAF*s.

In [14–18], *BAF*s have been generalised to Weighted (or Gradual) Bipolar Argumentation Frameworks (*WBAF*s) that assign real numbers as weights to arguments. As noticed in [17], these initial weights may be interpreted as representing the acceptability of an argument on its own, that is without considering the effects of supports or attacks by other arguments.

In *AAF*s, deciding which arguments to accept based only on the attack relation has been extensively studied in [8], where some semantics have been proposed. In contradistinction, in the gradual setting [17], the so-called acceptability semantics define two separate functions for specifying the acceptability of an argument: an aggregation and an influence function. The first aggregates the influence from other arguments in the framework based on the relations held between them, and the second combines the aggregated result with the weight of an argument (representing its initial strength from pre-conceptions) to determine the influence of its neighbours on its acceptability. The challenge is to find acceptability semantics defined for most or even all argumentation frameworks.

It is clear there are many possible acceptability semantics (see [17] for an extensive list), but not all of them are defined for all argumentation frameworks. Besides, among the semantics studied, none of them generalises the semantics proposed in [19] for *BAF*s. In fact, none of them generalises most of the semantics proposed for *AAF*s. This is a very disappointing result as *WBAF*s are generalisations of *BAF*s and *AAF*s; then one would expect the acceptability semantics for *WBAF*s would coincide with the semantics for *BAF*s (resp. *AAF*s) when restricted to the *BAF* (resp. *AAF*) setting.

In order to solve this problem, we propose a couple of new acceptability semantics for *WBAF*s. Besides being defined for every *WBAF*, they have been tailored to generalise the semantics proposed in [19] for *BAF*s as well as the semantics for *AAF*s. A distinguishing aspect of our approach is that we resort not only to the degree of acceptability of an argument, but also to its degree of rejection and employ both measures to characterise the acceptability semantics for *WBAF*s. As result, we have more expressive semantics that naturally generalise their corresponding semantics for *BAF*s and *AAF*s and are more flexible to be adapted to different scenarios.

The outline of the paper is as follows. We introduce in Sect. 2 the basic notions related to *BAF*s and *AAF* as well as their main semantics. In Sect. 3, we recall the notion of Weighted Argumentation Frameworks and present the main contribution of this work: a family of new semantics that generalise the semantics for *BAF*s found in [19] as well as those semantics for *AAF*s. Next, we focus on proving properties of the proposed semantics; in particular, we prove they are generalisations of semantics for *BAF*s and *AAF*s. Related works are discussed in Sect. 5. Finally, Sect. 6 closes our work with a conclusion and an account on future works.

2 Preliminaries

We proceed by presenting the fundamental notions of Abstract Argumentation Frameworks and Bipolar Argumentation Frameworks.

2.1 Abstract Argumentation Frameworks (*AAF*s)

Abstract Argumentation Frameworks were introduced in [8] and can be understood simply as graph where the nodes are arguments and the edges are attacks. To simplify things, we will restrict ourselves to finite argumentation frameworks.

Definition 1 (*AAF*). *[8] An* Abstract Argumentation Framework *(AAF) is a pair* $(\mathscr{A}, Att)$, *in which* $\mathscr{A}$ *is a finite set of arguments and* $Att \subseteq \mathscr{A} \times \mathscr{A}$.

Given an *AAF*, we are mainly interested in computing its semantics, which can be given in terms of labellings:

Definition 2 (**Labellings**). *[20] A labelling is a function* $\mathscr{L} : \mathscr{A} \to \{\text{in}, \text{out}, \text{undec}\}$ *that assigns to each argument in* $\mathscr{A}$ *a label. We say an argument A is accepted if* $\mathscr{L}(A) = \text{in}$, *A is rejected if* $\mathscr{L}(A) = \text{out}$ *and A is undecided if* $\mathscr{L}(A) = \text{undec}$. *We also define* $\text{in}(\mathscr{L}) = \{A \in \mathscr{A} \mid \mathscr{L}(A) = \text{in}\}$, $\text{out}(\mathscr{L}) = \{A \in \mathscr{A} \mid \mathscr{L}(A) = \text{out}\}$ *and* $\text{undec}(\mathscr{L}) = \{A \in \mathscr{A} \mid \mathscr{L}(A) = \text{undec}\}$.

Now we can define the main semantics for *AAF*s:

Definition 3 (**Semantics for *AAF*s**). *[20] Let* $\mathscr{A} = (\mathscr{A}, Att)$ *be an AAF. A labelling* $\mathscr{L} : \mathscr{A} \to \{\text{in}, \text{out}, \text{undec}\}$ *is a* complete labelling *of* $\mathscr{A}$ *iff for each* $A \in \mathscr{A}$ *it holds*

- $\mathscr{L}(A) = \text{in}$ *iff* $\mathscr{L}(B) = \text{out}$ *for every* $B \in Att(A)$.
- $\mathscr{L}(A) = \text{out}$ *iff* $\mathscr{L}(B) = \text{in}$ *for some* $B \in Att(A)$.

We say a complete labelling $\mathscr{L}$ *of* $\mathscr{A}$ *is*

Grounded *if* $\text{in}(\mathscr{L})$ *is minimal (w. r. t.* $\subseteq$*) among the complete labellings of* $\mathscr{A}$.
Preferred *if* $\text{in}(\mathscr{L})$ *is maximal (w. r. t.* $\subseteq$*) among the complete labellings of* $\mathscr{A}$.
Semi-stable *if* $\text{undec}(\mathscr{L})$ *is minimal (w. r. t.* $\subseteq$*) among the complete labellings of* $\mathscr{A}$.
Stable *if* $\text{undec}(\mathscr{L}) = \emptyset$.

It is clear complete labellings have a pivotal role in defining semantics for *AAF*s. As it will be shown in the sequel, complete labellings are also fundamental for Bipolar Argumentation Frameworks.

2.2 Bipolar Argumentation Frameworks (*BAF*s)

Many works (see [12, 19, 21, 22]) motivated the convenience of considering not only attack relations, but also support relations in argumentation frameworks.

Definition 4 (*BAF*). *[12] A* Bipolar Argumentation Framework *(BAF) is a tuple* $(\mathscr{A}, Att, Sup)$, *in which* $\mathscr{A}$ *is a finite set of arguments,* $Att \subseteq \mathscr{A} \times \mathscr{A}$ *is the attack relation and* $Sup \subseteq \mathscr{A} \times \mathscr{A}$ *is the support relation. For an argument* $A \in \mathscr{A}$, *we define* $Sup(A) = \{B \in \mathscr{A} \mid (B, A) \in Sup\}$.

In this paper, we will take into account the semantics introduced in [19] for *BAF*s; they are settled on the notions of attackers' domination and supporters' domination:

Definition 5 (Attackers' Domination). *[19] Let $\mathcal{B} = (\mathcal{A}, Att, Sup)$ be a BAF and $\mathcal{L}$ a labelling. We say the attackers of $A \in \mathcal{A}$ dominate its supporters according to $\mathcal{L}$ if*

$$|\{B \in Att(A) \mid \mathcal{L}(B) = \text{in}\}| > |\{B \in Sup(A) \mid \mathcal{L}(B) \neq \text{out}\}|$$

Definition 6 (Supporters' Domination). *[19] Let $\mathcal{B} = (\mathcal{A}, Att, Sup)$ be a BAF and $\mathcal{L}$ a labelling. We say the supporters of $A \in \mathcal{A}$ dominate its attackers according to $\mathcal{L}$ if*

$$|\{B \in Sup(A) \mid \mathcal{L}(B) = \text{in}\}| > |\{B \in Att(A) \mid \mathcal{L}(B) \neq \text{out}\}|$$

With these definitions in mind, a generalisation of those semantics for *AAF*s to *BAF*s can be presented as follows:

Definition 7 (Semantics for BAFs). *[19] Let $\mathcal{B} = (\mathcal{A}, Att, Sup)$ be a BAF. A labelling $\mathcal{L} : \mathcal{A} \rightarrow \{\text{in}, \text{out}, \text{undec}\}$ is a bi-complete labelling of $\mathcal{B}$ if for any $A \in \mathcal{A}$,*

- *$\mathcal{L}(A) = \text{in}$ if and only if $\mathcal{L}(B) = \text{out}$ for all $B \in Att(A)$ or A's supporters dominate its attackers.*
- *$\mathcal{L}(A) = \text{out}$ if and only if A's attackers dominate its supporters.*

We say a bi-complete labelling $\mathcal{L}$ of $\mathcal{B}$ is

Bi-Grounded *if $\text{in}(\mathcal{L})$ is minimal (w. r. t. $\subseteq$) among the bi-complete labellings of $\mathcal{B}$.*
Bi-Preferred *if $\text{in}(\mathcal{L})$ is maximal (w. r. t. $\subseteq$) among the bi-complete labellings of $\mathcal{B}$.*
Bi-Semi-stable *if $\text{undec}(\mathcal{L})$ is minimal (w. r. t. $\subseteq$) among the bi-complete labellings of $\mathcal{B}$.*
Bi-Stable *if $\text{undec}(\mathcal{L}) = \emptyset$.*

Besides being generalisations, these semantics for *BAF*s are strictly more expressive than their corresponding semantics for *AAF*s [19]. Now we can move to the next section, where we will give a step further and define semantics for Weighted Bipolar Argumentation Frameworks.

3   Semantics for Weighted Bipolar Argumentation Frameworks

Next, we present one of the main contributions of this paper: a generalisation of those semantics for *BAF*s to Weighted Bipolar Argumentation Frameworks (*WBAF*s). The general idea is to extend the notions of attackers' domination and supporters' domination to deal with weighted arguments. We start by recalling *WBAF*s:

Definition 8 (WBAF). *[17] A Weighted Bipolar Argumentation Framework (WBAF) (over $\mathbb{D}$) is a triple $\langle \mathcal{A}, G, W \rangle$, in which*

- *$\mathcal{A}$ is an n-dimensional ($n \in \mathbb{N}^{+}$) vector of arguments such that all components of $\mathcal{A}$ are pairwise distinct.*
- *$G = \{G_{ij}\}$ is a square matrix of order n with $G_{ij} \in \{-1, 0, 1\}$, in which $G_{ij} = -1$ means argument A_j attacks A_i; $G_{ij} = 1$ means argument A_j supports A_i; and $G_{ij} = 0$ if neither of these;*

– $W \in \mathbb{D}^n$ *is a vector of the initial weights of the arguments in $\mathscr{A}$.*
– $\mathbb{D}$ *is a set with a minimum (denoted by $\perp$) and a maximum (denoted by $\top$).*

In this work, we restrict our attention to arguments whose weights are in $\mathbb{D} = [0, 1]$.

Example 1. Let $\langle \mathscr{A}, G, W \rangle$ be a *WBAF* defined over $[0, 1]$, in which

$$\mathscr{A} = \begin{bmatrix} A_1 \\ A_2 \\ A_3 \end{bmatrix} \qquad G = \begin{bmatrix} 0 & 0 & 0 \\ 0 & 0 & 0 \\ -1 & 1 & 0 \end{bmatrix} \qquad W = \begin{bmatrix} 0.8 \\ 0.4 \\ 1.0 \end{bmatrix}$$

It can be represented graphically as bellow, where the continuous edge is an attack, the dashed edge is a support and the weights are next to the corresponding argument:

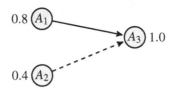

In the sequel, we disclose our definition of semantics (in general) for *WBAFs*. A distinguishing aspect of our proposal is that we resort not only to the degree of acceptability of an argument, but also to its degree of rejection and employ both measures to characterise the acceptability semantics for *WBAFs*.

Definition 9 (*WBAF Semantics*). *An acceptability semantics is a function* S *transforming any WBAF $Q = \langle \mathscr{A}, G, W \rangle$ over $\mathbb{D}$ into a set (possibly empty) $\mathrm{Deg}_Q^{\mathrm{S}}$ of vectors in $(\mathbb{D} \times \mathbb{D})^n$, in which n is the number of arguments in $\mathscr{A}$, i.e., the output of S is*

$$\mathrm{Deg}_Q^{\mathrm{S}} = \{[(\sigma_1, \pi_1), \ldots, (\sigma_n, \pi_n)] \mid [(\sigma_1, \pi_1), \ldots, (\sigma_n, \pi_n)] \in \mathrm{S}(Q)\}.$$

Besides, for any argument $a_i \in \mathscr{A}$,

$$(\mathrm{Deg}_Q^{\mathrm{S}})_i = \{(\sigma_i, \pi_i) \mid [(\sigma_1, \pi_1), \ldots, (\sigma_n, \pi_n)] \in \mathrm{S}(Q)\}$$

is the set of pairs (σ_i, π_i) assigned to a_i by the semantics S. Intuitively, σ_i is intended to assess the degree of acceptability of a_i (how much it is in) and π its degree of rejection (how much it is out).

We will propose a function as a semantics to generalise the bi-complete semantics to *WBAFs*. Before, however, present the notion of valuation:

Definition 10 (*WBAF Valuations*). *Let $Q = \langle \mathscr{A}, G, W \rangle$ be a WBAF over $\mathbb{D} = [0, 1]$ with $\mathscr{A} = [a_1, \ldots, a_n]$. A valuation $d = [(\sigma_1, \pi_1), \ldots, (\sigma_n, \pi_n)]$ of Q is a vector in $(\mathbb{D} \times \mathbb{D})^n$ such that for each $i \in \{1, \ldots, n\}$, $\sigma_i + \pi_i \leq 1$.*

Given a *WBAF* $\langle \mathscr{A}, G, W \rangle$, we will refer to the *i*-th row of G as $G_i = [G_{i1}, \ldots, G_{in}]$ and we will refer to the *i*-th component of the valuation *d* as $d_i = (\sigma_i, \pi_i)$. The next two definitions are natural generalisations of the notions of attackers' domination (Definition 5) and supporters' domination (Definition 6) to *WBAF*.

Definition 11 (Degree of Attackers' Domination). *Let* $Q = \langle \mathscr{A}, G, W \rangle$ *be a WBAF over* $[0,1]$ *and* $d = [(\sigma_1, \pi_1), \ldots, (\sigma_n, \pi_n))$ *a valuation of Q. Let also* $G_i = [G_{i1}, \ldots, G_{in}]$. *We define*

$$D_{Att}(G_i, d) = \min\left\{\max\left\{0, \sum_{G_{ij}=-1} \sigma_j - \sum_{G_{ij}=1}(1 - \pi_j)\right\}, 1\right\}$$

to denote the degree of domination of the attackers of an argument a_i *in comparison with its supporters according to d.*

Definition 12 (Degree of Supporters' Domination). *Let* $Q = \langle \mathscr{A}, G, W \rangle$ *be a WBAF over* $[0,1]$ *and* $d = [(\sigma_1, \pi_1), \ldots, (\sigma_n, \pi_n))$ *a valuation of Q. Let also* $G_i = [G_{i1}, \ldots, G_{in}]$. *We define*

$$D_{Sup}(G_i, d) = \min\left\{\max\left\{0, \sum_{G_{ij}=1} \sigma_j - \sum_{G_{ij}=-1}(1 - \pi_j)\right\}, 1\right\}$$

to denote the degree of domination of the supporters of an argument a_i *in comparison with its attackers according to d.*

In order to define the Weighted Bi-complete Semantics, we resort to two separate functions: an aggregation and an influence function. The first aggregates the influence from other arguments in the framework based on the relations held between them, and the second combines the aggregated result with the initial weight of an argument to determine the influence of its neighbours on its acceptability/rejection.

Definition 13 (Aggregation Function). *Let* $\mathbb{D} = [0,1]$. *We define the aggregation function* $\alpha : \{-1, 0, 1\}^n \times (\mathbb{D} \times \mathbb{D})^n \to [-1, 1]$ *as follows:*

$$\alpha(g, d) = D_{Sup}(g, d) + \gamma(g, d) - 1,$$

in which $d = [(\sigma_1, \pi_1), (\sigma_2, \pi_2), \ldots, (\sigma_n, \pi_n)]$ *and* $\gamma(g, d) = \min\{\pi_j \mid g_j = -1\}$. *We assume* $\min\{\} = 1$.

Definition 14 (Influence Function). *Let* $\mathbb{D} = [0,1]$. *We define the influence function* $\iota : [-1, 1] \times \mathbb{D} \to \mathbb{D}$ *as*

$$\iota(x, w) = \begin{cases} \dfrac{w + x}{1 + x} & \text{if } x \geq 0 \\[2mm] \dfrac{w(1 + x)}{1 - x} & \text{if } x \leq 0 \end{cases}$$

The Weighted Bi-complete semantics can be defined as follows:

Definition 15 (Weighted Bi-complete). *Let* $Q = \langle \mathscr{A}, G, W \rangle$ *be a WBAF over* $[0,1]$, *in which* $W = [W_1, \ldots, W_n]$. *We say* $d = [(\sigma_1, \pi_1), (\sigma_2, \pi_2), \ldots, (\sigma_n, \pi_n)]$ *is a weighted bi-complete valuation of Q if for each row* G_i *in G,*

$$\sigma_i = \iota(\alpha(G_i, d), W_i) \text{ and } \pi_i = D_{Att}(G_i, d)$$

We define

$$\mathrm{Deg}_Q^{\mathrm{com}} = \{d \mid d \text{ is a weighted bi-complete valuation of } Q\}.$$

as the set of weighted bi-complete valuations of Q.

Now we have to define $\sqsubseteq$ and $\sqsubseteq_u$ orderings to characterise the remaining semantics:

Definition 16 ($\sqsubseteq$ and $\sqsubseteq_u$). *Let $Q = \langle \mathscr{A}, G, W \rangle$ a WBAF over $[0,1]$ and $d' = [(\sigma_1', \pi_1'),$ $\dots, (\sigma_n', \pi_n')]$ and $d'' = [(\sigma_1'', \pi_1''), \dots, (\sigma_n'', \pi_n'')]$ be valuations of Q. We define*

- *$d' \sqsubseteq d''$ iff for each $i \in \{1, \dots, n\}$, it holds $\sigma_i' \le \sigma_i''$ and $\pi_i' \le \pi_i''$.*
- *$d' \sqsubseteq_u d''$ iff for each $i \in \{1, \dots, n\}$, it holds $\sigma_i' + \pi_i' \le \sigma_i'' + \pi_i''$.*

Then the weighted bi-grounded, weighted bi-preferred, weighted bi-semi-stable and weighted bi-stable semantics can be defined as natural generalisations of the corresponding semantics for *BAF*s:

Definition 17 (Refinements of Weighted Bi-Complete Valuations). *Let $Q = \langle \mathscr{A}, G, W \rangle$ be a WBAF over $[0,1]$ and $d = [(\sigma_1, \pi_1), (\sigma_2, \pi_2), \dots, (\sigma_n, \pi_n)]$ be a weighted bi-complete valuation of Q. We say d is a*

- *Weighted Bi-Grounded valuation of Q iff d is $\sqsubseteq$-minimal among the weighted bi-complete valuations of Q.*
- *Weighted Bi-Preferred valuation of Q iff d is $\sqsubseteq$-maximal among the weighted bi-complete valuations of Q.*
- *Weighted Bi-Semi-Stable valuation of Q iff d is $\sqsubseteq_u$-maximal among the weighted bi-complete valuations of Q.*
- *Weighted Bi-Stable valuation of Q iff d is a weighted bi-complete valuation of Q such that for each $i \in \{1, \dots, n\}$, $\sigma_i + \pi_i = 1$.*

Consider the following example:

Example 2. Let Q be the *WBAF* over $[0,1]$ depicted below:

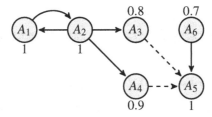

As the reader can check, the weighted bi-complete valuations of Q are

- $V_1 = [(1,0), (0,1), (0.8,0), (0.9,0), (0.54,0), (0.7,0)]$
- $V_2 = [(0,1), (1,0), (0,1), (0,1), (0,0.7), (0.7,0)]$
- $V_3 = [(0,0), (0,0), (0,0), (0,0), (0,0), (0.7,0)]$

and the remaining semantics are

- Weighted bi-grounded: $\{V_3\}$
- Weighted bi-preferred: $\{V_1, V_2\}$
- Weighted bi-semi-stable: $\{V_2\}$
- Weighted bi-stable: $\emptyset$

4 Results

In this section, we will prove some properties of our proposal. In particular, we will show the semantics we defined for *WBAFs* are generalisations of the corresponding semantics for *BAF*. We will also show they are defined for every *WBAF* (the unique exception is the weighted bi-stable). Next, we guarantee that it is not the case that both $D_{Sup}(g,d) \neq 0$ and $D_{Att}(g,d) \neq 0$:

Theorem 1. *Let* $\langle \mathscr{A}, G, W \rangle$ *a WBAF over* $[0,1]$ *and* d *a valuation of* Q. *If* $D_{Sup}(g,d) \neq 0$, *then* $D_{Att}(g,d) = 0$.

Proof. Assume $D_{Sup}(g,d) > 0$. Note that

$$\forall i(\sigma_i \leq 1 - \pi_i) \Longrightarrow \sum_{g_j=-1} \sigma_j \leq \sum_{g_j=-1} (1-\pi_j) \Longrightarrow -\sum_{g_j=-1}(1-\pi_j) \leq -\sum_{g_j=-1}\sigma_j$$

Thus, from $0 < D_{Sup}(g,d)$ it follows that

$$0 < \sum_{g_j=1}\sigma_j - \sum_{g_j=-1}(1-\pi_j) \leq \sum_{g_j=1}(1-\pi_j) - \sum_{g_j=-1}\sigma_j = -\left(\sum_{g_j=-1}\sigma_j - \sum_{g_j=1}(1-\pi_j)\right)$$

Then,

$$\sum_{g_j=-1}\sigma_j - \sum_{g_j=1}(1-\pi_j) < 0 \Longrightarrow D_{Att}(g,d) = 0$$

$\square$

Now we will show the bi-complete semantics for *BAF* coincides with the weighted bi-complete semantics for the corresponding *WBAF*:

Definition 18 (Corresponding WBAF). *Let* $B = (\mathscr{A}, Att, Sup)$ *be a BAF. The corresponding WBAF is* $Q_B = \langle \mathscr{A}_B, G, W \rangle$, *in which*

- $\mathscr{A}_B = [A_1, \ldots, A_n]$ *is a vector with all the arguments in* $\mathscr{A} = \{A_1, \ldots, A_n\}$.
- $G_{ij} = \begin{cases} -1 & \text{if } (A_j, A_i) \in Att \\ 1 & \text{if } (A_j, A_i) \in Sup \\ 0 & \text{otherwise} \end{cases}$
- $W = \underbrace{[1, \ldots, 1]}_{n}$

Theorem 2. *Let* $B = (\mathscr{A}, Att, Sup)$ *be a BAF and* $Q_B = \langle \mathscr{A}_B, G, W \rangle$ *be the corresponding WBAF. For a labelling* $\mathscr{L}$ *of* B, *let* $d_{\mathscr{L}} = [(\sigma_1, \pi_1), \ldots, (\sigma_n, \pi_n)]$ *be a valuation of* Q_B *such that for each argument* $A_i \in \mathscr{A}$,

- $\mathscr{L}(A_i) = \text{in iff } \sigma_i = 1 \text{ and } \pi_i = 0$;
- $\mathscr{L}(A_i) = \text{out iff } \sigma_i = 0 \text{ and } \pi_i = 1$;
- $\mathscr{L}(A_i) = \text{undec iff } \sigma_i = 0 \text{ and } \pi_i = 0$.

It holds

- $\mathscr{L}$ is a bi-complete labelling of B iff $d_{\mathscr{L}}$ is a weighted bi-complete valuation of Q_B.
- $\mathscr{L}$ is a bi-grounded labelling of B iff $d_{\mathscr{L}}$ is a weighted bi-grounded valuation of Q_B.
- $\mathscr{L}$ is a bi-preferred labelling of B iff $d_{\mathscr{L}}$ is a weighted bi-preferred valuation of Q_B.
- $\mathscr{L}$ is a bi-semi-stable labelling of B iff $d_{\mathscr{L}}$ is a weighted bi-semi-stable valuation of Q_B.
- $\mathscr{L}$ is a bi-stable labelling of B iff $d_{\mathscr{L}}$ is a weighted bi-stable valuation of Q_B.

Proof. Note that $\sigma_i, \pi_i \in \{0,1\}$ and $W_i = 1$ for every $i \in \{1, \ldots, n\}$.

($\Longrightarrow$) Assume $\mathscr{L}$ is a bi-complete labelling of Q.

$$\sigma_i = 1 \wedge \pi_i = 0 \Longleftrightarrow \mathscr{L}(A_i) = \mathtt{in}$$
$$\Longleftrightarrow |\{B \in Sup(A_i) \mid \mathscr{L}(B) = \mathtt{in}\}| > |\{B \in Att(A_i) \mid \mathscr{L}(B) \neq \mathtt{out}\}|$$
$$\vee \forall B \in Att(A_i)(\mathscr{L}(B) = \mathtt{out})$$
$$\Longleftrightarrow D_{Sup}(G_i, d_{\mathscr{L}}) + \gamma(G_i, d_{\mathscr{L}}) \geq 1$$
$$\Longleftrightarrow \iota(D_{Sup}(G_i, d_{\mathscr{L}}) + \gamma(G_i, d_{\mathscr{L}}) - 1, W_i) = 1 \wedge D_{Att}(G_i, d_{\mathscr{L}}) = 0$$

$$\sigma_i = 0 \wedge \pi_i = 1 \Longleftrightarrow \mathscr{L}(A_i) = \mathtt{out}$$
$$\Longleftrightarrow |\{B \in Att(A_i) \mid \mathscr{L}(B) = \mathtt{in}\}| > |\{B \in Sup(A_i) \mid \mathscr{L}(B) \neq \mathtt{out}\}|$$
$$\Longleftrightarrow D_{Sup}(G_i, d_{\mathscr{L}}) = \gamma(G_i, d_{\mathscr{L}}) = 0 \wedge D_{Att}(G_i, d_{\mathscr{L}}) = 1$$
$$\Longleftrightarrow \iota(D_{Sup}(G_i, d_{\mathscr{L}}) + \gamma(G_i, d_{\mathscr{L}}) - 1, W_i) = 0 \wedge D_{Att}(G_i, d_{\mathscr{L}}) = 1$$

$$\sigma_i = 0 \wedge \pi_i = 0 \Longleftrightarrow \mathscr{L}(A_i) = \mathtt{undec}$$
$$\Longleftrightarrow \mathscr{L}(A_i) \neq in \wedge \mathscr{L}(A_i) \neq out$$
$$\Longleftrightarrow \iota(D_{Sup}(G_i, d_{\mathscr{L}}) + \gamma(G_i, d_{\mathscr{L}}) - 1, W_i) = 0 \wedge D_{Att}(G_i, d_{\mathscr{L}}) = 0$$

Since $(d_{\mathscr{L}})_i = (\iota(\alpha(G_i, d_{\mathscr{L}}), W_i), D_{Att}(G_i, d_{\mathscr{L}}))$ for every i, $d_{\mathscr{L}}$ is a weighted bi-complete labelling.

($\Longleftarrow$) Assume $d_{\mathscr{L}}$ is a weighted bi-complete valuation of Q, i.e., for every $i \in \{1, \ldots, n\}$, $(d_{\mathscr{L}})_i = (\iota(\alpha(G_i, d_{\mathscr{L}}), W_i), D_{Att}(G_i, d_{\mathscr{L}}))$.

$$\mathscr{L}(A_i) = \mathtt{in} \Longleftrightarrow \sigma_i = 1 \wedge \pi_i = 0$$
$$\Longleftrightarrow \iota(D_{Sup}(G_i, d_{\mathscr{L}}) + \gamma(G_i, d_{\mathscr{L}}) - 1, W_i) = 1 \wedge D_{Att}(G_i, d_{\mathscr{L}}) = 0$$
$$\Longleftrightarrow D_{Sup}(G_i, d_{\mathscr{L}}) + \gamma(G_i, d_{\mathscr{L}}) \geq 1$$
$$\Longleftrightarrow |\{B \in Sup(A_i) \mid \mathscr{L}(B) = \mathtt{in}\}| > |\{B \in Att(A_i) \mid \mathscr{L}(B) \neq \mathtt{out}\}|$$
$$\vee \forall B \in Att(A_i)(\mathscr{L}(B) = \mathtt{out})$$

$$\mathscr{L}(A_i) = \mathtt{out} \Longleftrightarrow \sigma_i = 0 \wedge \pi_i = 1$$
$$\Longleftrightarrow \iota(D_{Sup}(G_i, d_{\mathscr{L}}) + \gamma(G_i, d_{\mathscr{L}}) - 1, W_i) = 0 \wedge D_{Att}(G_i, d_{\mathscr{L}}) = 1$$
$$\Longleftrightarrow D_{Sup}(G_i, d_{\mathscr{L}}) = \gamma(G_i, d_{\mathscr{L}}) = 0 \wedge D_{Att}(G_i, d_{\mathscr{L}}) = 1$$
$$\Longleftrightarrow |\{B \in Att(A_i) \mid \mathscr{L}(B) = \mathtt{in}\}| > |\{B \in Sup(A_i) \mid \mathscr{L}(B) \neq \mathtt{out}\}|$$

Thus, $\mathscr{L}$ is a bi-complete labelling of Q. As the other semantics are defined in terms of (weighted) bi-complete, the remaining results follow straightforwardly. $\qquad\square$

As in the *BAF* setting, the valuations in the bi-complete semantics can be defined on *WBAF*s in terms of the fixpoints of the Γ operator as defined below.

Definition 19 (Γ Operator). *Let* $Q = \langle \mathscr{A}, G, W \rangle$ *be a WBAF over* $[0,1]$ *and* $d = [(\sigma_1, \pi_1), \ldots, (\sigma_n, \pi_n)]$ *a valuation of Q. We define*

$$\Gamma_Q(g,d) = [\Omega_Q(g_1,d), \ldots, \Omega_Q(g_n,d)],$$

in which

$$\Omega_Q(g_i,d) = (\iota(\alpha(g_i,d)), D_{Att}(g_i,d)).$$

We will prove the monotonicity of Γ operator and employ this result to guarantee the weighted bi-grounded valuation is uniquely defined for every *WBAF*:

Theorem 3 (Monotonicity of Γ). *Let* $Q = \langle \mathscr{A}, G, W \rangle$ *a WBAF over* $[0,1]$ *and* $d' = [(\sigma'_1, \pi'_1), \ldots, (\sigma'_n, \pi'_n)]$ *and* $d'' = [(\sigma''_1, \pi''_1), \ldots, (\sigma''_n, \pi''_n)]$ *be valuations of Q. If* $d' \sqsubseteq d''$, *then* $\Gamma_Q(g,d') \sqsubseteq \Gamma_Q(g,d'')$.

Proof. Let $d' = (\sigma', \pi')$ and $d'' = (\sigma'', \pi'')$. Assume $d' \sqsubseteq d''$.

$$D_{Sup}(G_i,d') = \sum_{G_{ij}=1} \sigma'_j - \sum_{G_{ij}=-1} (1-\pi'_j) \leq \sum_{G_{ij}=1} \sigma''_j - \sum_{G_{ij}=-1} (1-\pi''_j) = D_{Sup}(G_i,d'')$$

$$D_{Att}(G_i,d') = \sum_{G_{ij}=-1} \sigma'_j - \sum_{G_{ij}=1} (1-\pi'_j) \leq \sum_{G_{ij}=-1} \sigma''_j - \sum_{G_{ij}=1} (1-\pi''_j) = D_{Att}(G_i,d'')$$

$$\gamma(G_i,d') = \min\{\pi'_j \mid G_{ij} = -1\} \leq \min\{\pi''_j \mid G_{ij} = -1\} = \gamma(G_i,d'')$$

Thus

$$\alpha(G_i,d') = D_{Sup}(G_i,d') + \gamma(G_i,d') - 1 \leq D_{Sup}(G_i,d'') + \gamma(G_i,d'') - 1 = \alpha(G_i,d'')$$

As $s \leq s' \rightarrow \iota(s,w) \leq \iota(s',w)$, it holds $\iota(\alpha(G_i,d'),W_i) \leq \iota(\alpha(G_i,d''),W_i)$. Hence, for all $i \in \{1, \cdots, |\mathscr{A}|\}$,

$$\iota(\alpha(G_i,d'),W_i) \leq \iota(\alpha(G_i,d''),W_i) \text{ and } D_{Att}(G_i,d') \leq D_{Att}(G_i,d''),$$

i.e., $\Gamma_Q(G,d') \sqsubseteq \Gamma_Q(G,d'')$. □

We can resort to the well known Knaster-Tarski theorem [23] to show the least fixpoint of Γ is guaranteed to exist, corresponding to the weighted bi-grounded valuation:

Theorem 4. *Every WBAF* $Q = \langle \mathscr{A}, G, W \rangle$ *has a unique weighted bi-grounded valuation.*

Proof. From Theorem 3, we know Γ_Q is monotonic w.r.t. $\sqsubseteq$. Then, according to the Knaster-Tarski Theorem [23], a least (w.r.t. $\sqsubseteq$) fixpoint of Γ_Q operator is guaranteed to exist. By definition, it corresponds to the unique weighted bi-grounded valuation of Q.

As the bi-complete labelling is also uniquely defined for each *BAF*, Theorem 4 assures this unicity is preserved when moving to *WBAF*. Besides, as every *WBAF* has a unique weighted bi-grounded valuation and weighted bi-grounded valuations are weighted bi-complete valuations, we obtain the following result:

Corollary 1. *Each WBAF has at least one weighted bi-complete valuation, weighted bi-preferred valuation and weighted bi-semi-stable valuation.*

Proof. It follows from Theorem 4 and from Definition 17.

Nonetheless, as we can notice in Example 2, we cannot say the same for weighted bi-stable valuations, i.e., they are not defined for every *WBAF*. This is an expected result as bi-stable labellings are not defined for every *BAF* (see [19]).

5   Related Work

We have introduced several semantics that are defined for every *WBAF* and are not only natural extensions of the corresponding semantics for *BAF*, but also preserve some of their properties. Now we will compare our approach with other semantics for *WBAFs*.

The continuous approach to the acceptability of arguments in the bipolar setting has been extensively studied in [24]. Many semantics in the literature for *WBAFs* are not defined for all graphs. This is the case with the semantics Euler-based [25], direct aggregation, positive direct aggregation, sigmoid direct aggregation and quadratic energy [18]. Other semantics are defined for all graphs, such as those in Tables 1 and 2: max Euler-based, damped max-based and sigmoid damped max-based, but they neither generalise the bi-complete semantics nor the complete semantics.

Table 1. Overview of convergent semantics' aggregation function.

Semantics	α's range	α
Max Euler-based	$\{-1,0,1\}^n \times \mathbb{R}^n \to \mathbb{R}$	$\alpha(g,d) = top(g,d)d$
Damped max-based	$\{-1,0,1\}^n \times \mathbb{R}^n \to \mathbb{R}$	$\alpha(g,d) = top(g,d)d$
Sigmoid damped max-based	$\{-1,0,1\}^n \times (-1,1)^n \to \mathbb{R}$	$\alpha(g,d) = top(g,\sigma^{-1}(d))\sigma^{-1}(d)$

where

$$top(g,d)_j = \begin{cases} g_j & \text{if } d_k < d_j \text{ for } 1 \leq k < j, \, sgn(g_k) = sgn(g_j) \\ & \text{and } d_k \leq d_j \text{ for } j < k \leq n, \, sgn(g_k) = sgn(g_j) \\ & \text{and } d_j \geq 0 \\ 0 & \text{otherwise} \end{cases}$$

and $sgn(g_j)$ is the sign function and outputs the sign of g_j. Furthermore, $\sigma : \mathbb{R} \to (-1,1)$ is continuous and strictly increasing (for concreteness, let $\sigma(x) = \tanh(x)$).

The next result shows these three main semantics for *WBAFs* are not generalisations of any semantics for *BAF* and *AAF* which are not uniquely defined:

Table 2. Overview of convergent semantics' influence function.

Semantics	ι's range	ι
Max Euler-based	$\mathbb{R} \times [0,1) \to [w^2, 1]$	$\iota(s,w) = 1 - \dfrac{1 - w^2}{1 + w \cdot e^s}$
Damped max-based	$\mathbb{R} \times \mathbb{R} \to \mathbb{R}$	$\iota(s,w) = \dfrac{s}{\delta} + w$ with $\delta > 2$
Sigmoid damped max-based	$\mathbb{R} \times (-1,1) \to (-1,1)$	$\iota(s,w) = \sigma\left(\dfrac{s}{\delta} + \sigma^{-1}(w)\right)$ with $\delta > 2$

Proposition 1. *The damped max-based semantics, sigmoid damped max-based semantics and Euler max-based semantics do not generalise the complete semantics, bicomplete semantics, preferred, bi-preferred, semi-stable, bi-semi-stable, stable and bistable semantics.*

Proof. The result is immediate as the damped max-based semantics, sigmoid damped max-based semantics and Euler max-based semantics are uniquely defined for every *WBAF*s, but the other semantics are not uniquely defined for every *AAF* (or *BAF*). □

One would ask if Euler max-based, damped max-based and sigmoid damped max-based semantics generalise a uniquely defined semantics as the bi-grounded. In the sequel, we show it is not the case for any top-based semantics as the Euler max-based and Damped max-based semantics:

Proposition 2. *Any top-based semantics (with $\alpha(g,d) = top(g,d)d$) and $\iota(0,w) = w$ does not generalise the bi-grounded semantics.*

Proof. Let $S = (\alpha, \iota)$ be a semantics with $\alpha(g,d) = top(g,d)d$ and $\iota(0,w) = w$. Let
$$G = \begin{bmatrix} 0 & 0 & 0 \\ 0 & 0 & 0 \\ -1 & 1 & 0 \end{bmatrix}$$
be an argumentation matrix of a *WBAF* Q. Consider all weights to be $w \geq 0$.

Let $d = S(Q)$ be the valuation of Q according to S. Then, $d_1 = d_2 = \iota(0,w) = w$, since arguments A_1 and A_2 are neither attacked nor supported.

Note that $\alpha(G_3, d) = top(G_3, d)d = -w + w = 0$, because the strongest attacker and strongest supporter of A_3 have the same acceptability w. Thus, $d_3 = \iota(0,w) = w$ and $d_1 = d_2 = d_3$. However, in the unique bi-grounded labelling of Q, A_1 and A_2 are in, but A_3 is undec. Therefore they do not have the same acceptability, and as consequence, S does not generalise the bi-grounded semantics. □

After showing that the Euler max-based and the Damped max-based semantics do not generalise the bi-grounded semantics, it is time to prove a similar result for any σ-top-based semantics as it is the case of the sigmoid damped max-based semantics:

Proposition 3. *Any σ-top-based semantics (with $\alpha(g,d) = top(g, \sigma^{-1}(d))\sigma^{-1}(d)$), $\iota(0,w) = w$ and $\sigma(0) = 0$, does not generalise the bi-grounded semantics.*

Proof. Let $S = (\alpha, \iota)$ be a semantics with $\alpha(g,d) = top(g, \sigma^{-1}(d))\sigma^{-1}(d)$ and $\iota(0,w) = w$. Let $G = \begin{bmatrix} 0 & 0 & 0 \\ 0 & 0 & 0 \\ -1 & 1 & 0 \end{bmatrix}$ be an argumentation matrix of a *WBAF* Q. Consider all weights to be $w \geq 0$.

Let $d = S(Q)$ be the valuation of Q according to S. Then, $d_1 = d_2 = \iota(\sigma^{-1}(0), w) = \iota(0,w) = w$, since arguments A_1 and A_2 are neither attacked nor supported.

Note that $\alpha(G_3, d) = top(G_3, \sigma^{-1}(d))\sigma^{-1}(d) = -\sigma^{-1}(w) + \sigma^{-1}(w) = 0$, because the strongest attacker and strongest supporter of A_3 have the same acceptability $\sigma^{-1}(w)$. Thus, $d_3 = \iota(0,w) = w$ and $d_1 = d_2 = d_3$. However, in the unique bi-grounded labelling of Q, A_1 and A_2 are in, but A_3 is undec. Therefore they do not have the same acceptability, and as consequence, S does not generalise the bi-grounded semantics. □

Hence, unlike our approach (see Theorem 2), Euler max-based, damped max-based and sigmoid damped max-based semantics do not generalise any of the semantics proposed to *BAFs* in [19]. They also do not generalise the semantics complete, preferred, semi-stable and stable for *AAFs*. A question naturally arises: do they have some relation with the semantics for *AAFs*? The next proposition provides a positive answer for damped max-based semantics by relating it with one of the complete labellings:

Proposition 4. *Let d be a valuation in the damped max-based semantics and $\mathscr{L}$ a labelling such that for every argument A_i in an attack-only WBAF when all arguments' weights are $w > 0$ and $\delta > 1$, it holds 1) $\mathscr{L}(A_i) = $ in iff $d_i > w\dfrac{\delta}{\delta+1}$; 2) $\mathscr{L}(A_i) = $ undec iff $d_i = w\dfrac{\delta}{\delta+1}$; 3) $\mathscr{L}(A_i) = $ out iff $d_i < w\dfrac{\delta}{\delta+1}$. Then $\mathscr{L}$ is a complete labelling.*

Proof.

$$\forall j(G_{ij} = -1 \to \mathscr{L}(A_j) = \text{out}) \iff \forall j(G_{ij} = -1 \to d_j < w\frac{\delta}{\delta+1})$$

$$\iff \alpha(G_i, d) = top(G_i, d)d > -w\frac{\delta}{\delta+1}$$

$$\iff d_i = \iota(\alpha(G_i, d), w) > \iota\left(-w\frac{\delta}{\delta+1}, w\right) = w\frac{\delta}{\delta+1}$$

$$\iff \mathscr{L}(A_i) = \text{in}$$

$$\exists j(G_{ij} = -1 \wedge \mathscr{L}(A_j) = \text{in}) \iff \exists j(G_{ij} = -1 \wedge d_j > w\frac{\delta}{\delta+1})$$

$$\iff \alpha(G_i, d) = top(G_i, d)d < -w\frac{\delta}{\delta+1}$$

$$\iff d_i = \iota(\alpha(G_i, d), w) < \iota\left(-w\frac{\delta}{\delta+1}, w\right) = w\frac{\delta}{\delta+1}$$

$$\iff \mathscr{L}(A_i) = \text{out}$$

$\mathscr{L}(A_i) = \text{undec} \iff d_i = w\dfrac{\delta}{\delta+1}$ follows from the last two statements.

6   Conclusion and Future Works

Many works [14–18] have argued about the significance of extending Bipolar Argumentation Frameworks (*BAFs*) [19] by weighting arguments. The resulting frameworks are called Weighted Bipolar Argumentation Frameworks (*WBAFs*). Notwithstanding, although *WBAFs* are extensions of *BAFs*, the semantics proposed to *WBAFs* are not generalisations of the semantics proposed to *BAFs*. In this work, we have solved this problem by defining the weighted bi-complete, weighted bi-preferred, weighted bi-grounded, weighted bi-semi-stable and weighted bi-stable semantics for *WBAFs*, which are respectively generalisations of the bi-complete, bi-preferred, bi-grounded, bi-semi-stable and bi-stable semantics for *BAFs*. As these semantics for *BAFs* are generalisations of the corresponding semantics for Abstract Argumentation Frameworks (*AAFs*) [8], our semantics also encompass them. Many semantics in the literature for *WBAFs* are not defined for all graphs. Another improvement our proposal provides to these computational models of arguments is that except for the weighted bi-stable semantics, these new semantics are defined for every *WBAF*.

Future developments include investigating computational complexity issues on the proposed semantics and analysing which properties are satisfied by them and which are impossible to be satisfied among those pointed in [17] as structured, essential and optional properties for *WBAFs* semantics. Another important line of research is to examine the consequence of introducing weights not only to the arguments, but also to the attack and support relations.

References

1. Simari, G.R., Loui, R.P.: A mathematical treatment of defeasible reasoning and its implementation. Artif. Intell. **53**(2–3), 125–157 (1992)
2. García, A.J., Simari, G.R.: Defeasible logic programming: an argumentative approach. Theory Pract. Log. Programm. **4**(1–2), 95–138 (2004)
3. Amgoud, L., Prade, H.: Using arguments for making and explaining decisions. Artif. Intell. **173**(3–4), 413–436 (2009)
4. Spieler, J., Potyka, N., Staab, S.: Learning gradual argumentation frameworks using genetic algorithms. arXiv preprint arXiv:2106.13585 (2021)
5. Hadoux, E., Hunter, A.: Learning and updating user models for subpopulations in persuasive argumentation using beta distribution. In: Proceedings of the 17th International Conference on Autonomous Agents and Multiagent Systems, vol. 17, pp. 1141–1149. Association for Computing Machinery (ACM) (2018)
6. Rago, A., Cocarascu, O., Toni, F.: Argumentation-based recommendations: fantastic explanations and how to find them. In Proceedings of the Twenty-Seventh International Joint Conference on Artificial Intelligence, pp. 1949–1955 (2018)
7. Cocarascu, O., Rago, A., Toni, F.: Extracting dialogical explanations for review aggregations with argumentative dialogical agents. In: Proceedings of the 18th International Conference on Autonomous Agents and MultiAgent Systems, pp. 1261–1269. Association for Computing Machinery (2019)
8. Phan Minh Dung: On the acceptability of arguments and its fundamental role in nonmonotonic reasoning, logic programming and n-person games. Artif. Intell. **77**(2), 321–357 (1995)

9. Brewka, G., Woltran, S.: Abstract dialectical frameworks. In: Twelfth International Conference on the Principles of Knowledge Representation and Reasoning, pp. 102–111. AAAI Press (2010)

10. Bench-Capon, T.J.M.: Persuasion in practical argument using value-based argumentation frameworks. J. Log. Comput. **13**(3), 429–448 (2003)

11. Cayrol, C., Lagasquie-Schiex, M.C.: On the acceptability of arguments in bipolar argumentation frameworks. In: Godo, L. (ed.) ECSQARU 2005. LNCS (LNAI), vol. 3571, pp. 378–389. Springer, Heidelberg (2005). https://doi.org/10.1007/11518655_33

12. Amgoud, L., Cayrol, C., Lagasquie-Schiex, M.-C., Livet, P.: On bipolarity in argumentation frameworks. Int. J. Intell. Syst. **23**(10), 1062–1093 (2008)

13. Kaci, S., van der Torre, L.: Preference-based argumentation: arguments supporting multiple values. Int. J. Approx. Reason. **48**(3), 730–751 (2008)

14. Baroni, P., Romano, M., Toni, F., Aurisicchio, M., Bertanza, G.: Automatic evaluation of design alternatives with quantitative argumentation. Argum. Comput. **6**(1), 24–49 (2015)

15. Rago, A., Toni, F., Aurisicchio, M., Baroni, P.: Discontinuity-free decision support with quantitative argumentation debates. In: Fifteenth International Conference on the Principles of Knowledge Representation and Reasoning (2016)

16. Amgoud, L., Ben-Naim, J., Doder, D., Vesic, S.: Acceptability semantics for weighted argumentation frameworks. In: Twenty-Sixth International Joint Conference on Artificial Intelligence (2017)

17. Mossakowski, T., Neuhaus, F.: Modular semantics and characteristics for bipolar weighted argumentation graphs. arXiv preprint arXiv:1807.06685 (2018)

18. Potyka, N.: Continuous dynamical systems for weighted bipolar argumentation. In: Sixteenth International Conference on Principles of Knowledge Representation and Reasoning (2018)

19. Potyka, N.: Generalizing complete semantics to bipolar argumentation frameworks. In: Vejnarová, J., Wilson, N. (eds.) ECSQARU 2021. LNCS (LNAI), vol. 12897, pp. 130–143. Springer, Cham (2021). https://doi.org/10.1007/978-3-030-86772-0_10

20. Caminada, M.W.A., Gabbay, D.M.: A logical account of formal argumentation. Studia Logica. **93**(2–3), 109 (2009)

21. Boella, G., Gabbay, D.M., van der Torre, L., Villata, S.: Support in abstract argumentation. In: Proceedings of the Third International Conference on Computational Models of Argument (COMMA 2010), pp. 40–51. Frontiers in Artificial Intelligence and Applications, IOS Press (2010)

22. Cayrol, C., Lagasquie-Schiex, M.-C.: Bipolarity in argumentation graphs: towards a better understanding. Int. J. Approx. Reason. **54**(7), 876–899 (2013)

23. Tarski, A.: A lattice-theoretical fixpoint theorem and its applications. Pac. J. Math. **5**(2), 285–309 (1955)

24. Amgoud, L., Ben-Naim, J.: Ranking-based semantics for argumentation frameworks. In: Liu, W., Subrahmanian, V.S., Wijsen, J. (eds.) SUM 2013. LNCS (LNAI), vol. 8078, pp. 134–147. Springer, Heidelberg (2013). https://doi.org/10.1007/978-3-642-40381-1_11

25. Amgoud, L., Ben-Naim, J.: Evaluation of arguments in weighted bipolar graphs, pp. 25–35 (2017)

The Use of Multiple Criteria Decision Aiding Methods in Recommender Systems: A Literature Review

Renata Pelissari[1]([✉]), Paulo S. Alencar[2], Sarah Ben Amor[3],
and Leonardo Tomazeli Duarte[1]

[1] School of Applied Sciences, University of Campinas, Limeira, Brazil
`renatapelissari@gmail.com`, `leonardo.duarte@fca.unicamp.br`
[2] Computer Science Department, University of Waterloo, Waterloo, Canada
`palencar@uwaterloo.ca`
[3] Telfer School of Management, University of Ottawa, Ottawa, Canada
`BenAmor@telfer.uottawa.ca`

Abstract. Multiple Criteria Decision Making (MCDA) methods have been increasingly applied to improve recommendations when multiple criteria are considered in Recommender Systems (RSs). This study presents the preliminary results of a systematic literature review, following Kitchenham's guidelines, regarding the application of MCDA methods in RSs over the last two decades. Based on our findings, MCDA methods can be applied in two RS phases: the preference elicitation and the recommendation phases. In the former, RSs usually have a strong interaction with the user, which results in more personalized recommendations, ensuring higher user satisfaction and contributing to address the cold-start challenge in RSs. Regarding the recommendation phase, while most RSs are based on ranking approaches, there is a trend to apply sorting methods in order to avoid an additional step involving a filtering application that selects a subset of alternatives. Future research could focus on applying preference learning combined with MCDA methods for exploring improvements in prediction and recommendation phases, and also in quality and processing time.

Keywords: Multiple criteria decision making · Literature review · Preference learning

1 Introduction

While the majority of existing Recommender Systems (RSs) depend only on one single criterion numerical rating as input information, there has been an increasing interest over the last two decades in taking into consideration a rating based on multiple criteria, since the user's preferences might cover more than one perspective [2,3]. Thus, the recommendation process can be approached

© The Author(s), under exclusive license to Springer Nature Switzerland AG 2022
J. C. Xavier-Junior and R. A. Rios (Eds.): BRACIS 2022, LNAI 13653, pp. 535–549, 2022.
https://doi.org/10.1007/978-3-031-21686-2_37

as a Multi-criteria Decision Aiding (MCDA) problem [35,53], in which MCDA methods are used as part of the recommendation algorithm.

In MCDA, alternatives are evaluated by the decision-maker (DM) according to several criteria, usually conflicting to each other, with the goal of either ranking the alternatives (ranking problematic) or sorting them into predefined and ordered categories (sorting problematic) [63]. RSs based on multiple criteria have been pointed out as one of the promising research areas in RSs [3]. Since then, many studies have considered merging MCDA and RSs. Towards this direction, Manouselis and Costopoulou [53] proposed a review in order to analyze and classify MCDA-based RSs. As this study was published 2007, there is an opportunity for an up-to-date review.

Given this context, the overall objective of this paper is to identify the MCDA methods that have been used in RSs and the advantages and disadvantages of employing these methods. To that end, we have performed a systematic literature review.

The rest of the paper is structured as it follows. In Sect. 2 and 3, we give an overview of RSs and MCDA methods. In Sect. 4, we present the adopted methodology for conducting the literature review. The results are presented in Sect. 5, and the limitations of the study and future improvements are presented in Sect. 5. We conclude our study in Sect. 6.

2   Recommender Systems

RSs are expert systems able to suggest items that a user may be interested in. More specifically, the basic idea of RSs is to utilize user feedback about or the act of a user buying or browsing an item, watching a movie, listing to a song, to infer customer interests [5].

RSs can be divided into four main types: collaborative, content-based, utility-based, and hybrid filtering RSs [4]. Collaborative filtering recommender systems (CFRSs) aim to identify users' preferences, considering the ratings given by them to the items they have already interacted with. Then, the closest users are identified based on similar preferences, and predictions regarding new items are made based on these estimated proximities. In content-based RSs the descriptive attributes of items are used to make recommendations. Utility-based RSs provide recommendations based on the computation of the utility of each item for the user. Finally, hybrid systems combine the strengths of various types of RSs in order to create more robust techniques [14].

While the majority of existing RSs depend only on one single criterion rating as input information, there has been an increasing interest in taking into consideration a rating based on multiple criteria, since the users' preferences usually cover more than one point of view [2,3]. For instance, in the context of restaurant recommendation, instead of giving a single rating representing their opinion about the restaurant, the user rates different characteristics of the restaurant such as price, place and quality of the food. Under this perspective, the recommendation process can be addressed as MCDA problem [2,35,53].

3   Multiple Criteria Decision Aiding

Multiple Criteria Decision Aiding (MCDA) is a research field in Operational Research that encompasses the development and application of methods with the aim of supporting decision processes. MCDA methods are characterized by considering multiple criteria, usually conflicting to each other, and taking into consideration the preferences of the involved persons [63].

MCDA methods can be classified–among other possible classifications–into: scoring, distance-based, pairwise comparison-based, utility-based and outranking methods. Scoring methods, such as Simple Additive Weighting (SAW) [28] and Complex Proportional Assessment (COPRAS) [79], find a score for each alternative by applying basic arithmetic; then, this score can be used to rank or sort the alternatives. Distance-based methods, like Technique for Order of Preference by Similarity to Ideal Solution (TOPSIS) and Multi-criteria Optimization and Compromise Solution (VIKOR), rank the alternatives based on the distance to both the optimal and the worst existing solutions [57].

Pairwise comparison-based methods compare each pair of alternatives to provide a final score. Examples of these methods are AHP (Analytic Hierarchy Process) [65] and ANP (Analytic Network Process) [66]. Outranking methods, such as PROMETHEE (The Preference Ranking Organization METHod for Enrichment of Evaluations) [12,13] and ELECTRE (ÉLimination Et Choix Traduisant La RÉalité) [33,62], are also based on pairwise comparisons, but they are mainly characterized by establishing a degree of dominance between alternatives and consider these degrees to find the final ranking. Utility-based methods use the degree of satisfaction expected by each alternative to form the scores as in UTA (UTilité Additive) [21], Choquet integral [18], SMART (Simple Multi-attribute Rating Technique) and its variation SMARTER (SMART Exploiting Ranks) [25]. A detailed description of these and other MCDA methods can be found in [19].

The methods descried above are suitable for ranking problems, but they also have their variations for sorting: AHPSort [44], TOPSIS Sort [67], ELECTRE-TRI [64,78], FlowSort [55] and UTADIS (UTilités Additives DIScriminantes) [21]. A complete list of MCDA sorting methods and a detailed review of their characteristics are presented in [8].

4   Methodology

Our study followed the principles of a systematic literature review proposed in [46], which is a well-defined approach to identify, evaluate and interpret all relevant studies regarding a particular research question, topic area or phenomenon of interest. Following Kitchenham's guidelines, a research protocol must be defined and must contain the generic steps listed as follows:

1. **Step 1:** Definition of research questions that the review is intended to answer, an that will guide the study.

2. **Step 2:** Definition of the search strategy, including the databases and the search terms used to identify and select papers.
3. **Step 3:** Definition of the study selection criteria, determining criteria for excluding a study from the review.
4. **Step 4:** Definition of how to categorize the studies, which information shall be extracted from the studies, and how this information will be synthesized and analyzed.

The protocol definition can be seen as the first phase of the literature review, which is followed by two more phases, the review conduction and the review report.

4.1   Phase 1: Protocol Definition

The first step of the protocol definition starts with the selection of research questions that will guide the study. This review examines the following seve sets of research questions:

RQ1. How frequently have MCDA methods been implemented in the use or research of RSs over the last 20 years? What are the trends in terms of publication venues?

RQ2. What are the MCDA methods employed in RSs? And which are employed most?

RQ3. What purpose do the MCDA methods serve in RSs?

RQ4. What are the types and main characteristics of RSs where MCDA methods are employed?

RQ5. What are the evaluation metrics employed in the selected studies?

RQ6. What are the contributions of MCDA methods to RSs regarding some of their challenges such as cold-start, data sparsity, scalability, and privacy?

RQ7. What are the trends and gaps in the use or research of MCDA methods in RSs?

We focused our search on Scopus, since it is the largest curated, peer-reviewed abstract and indexing database available to academia. The list of keywords used in our search is: ("recommender system" OR "recommendation system" OR "recommender systems" OR "recommendation systems") AND ("multicriteria decision making" OR "multiple criteria decision making" OR "multi-criteria decision making" OR "multicriteria decision analysis" OR "multiple criteria decision analysis" OR "multi-criteria decision analysis" OR "multicriteria decision aiding" OR "multiple criteria decision aiding" OR "multi-criteria decision aiding" OR promethee OR electre OR ahp OR vikor OR topsis OR anp OR uta OR utadis OR maut OR dematel OR "multi-attribute utility" OR "multi-attribute utility").

We were only interested in the publications of the last twenty years describing either an application or a theoretical development, once an MCDA method has been used as a core or at least as an important part of the system. Based on that, we defined the following exclusion criteria:

EC1. Papers published later than 2002.

EC2. Papers written in languages other than English.

EC3. Non-primary studies., e.g., surveys of literature.

EC4. Papers whose document type is other than "Article", e.g., conference papers, book chapters, conference reviews, etc.

EC5. Papers whose abstract does not provide enough information in order to verify whether the paper is related to the review goal.

EC6. Papers that do not describe or consider an RS approach.

EC7. Papers that do not describe or consider an MCDA method.

EC8. Papers that do not apply MCDA methods as a core or at least as an important part of the system, e.g., application of MCDA methods for tool selection.

As the last step of the protocol definition, we defined that the papers shall be categorized by the year of publication, source of the paper, the MCDA method applied and the type of RS considered. Moreover, from each paper, the following information shall be extracted: main contribution of applying MCDA to RSs, whether and which evaluation metrics were employed in the system evaluation, and whether the paper considered matters related to cold-start, data sparsity, scalability, and privacy.

4.2 Phase 2: Review Conduction

On March 17, 2022, we queried the digital library Scopus using the search terms presented in Sect. 4.1, and our search returned 313 papers.

The exclusion criteria from EC1 to EC4 were automatically applied using search resources from the Scopus database itself, reducing the number of studies from 313 to 135. Abstracts of the remaining papers were read, and the exclusion criteria from EC5 to EC8 were manually applied. In the end, 96 studies were retained. The 96 selected papers were then read in full. Throughout the reading process, papers that met at least one of the exclusion criteria EC6 to EC8, and that were not identified in the previous step, were excluded. From these, 49 studies were finally selected.

5   Results

In this section, we present the main results of our literature review based on the defined research questions.

RQ1. How frequently have MCDA methods been implemented in the use or research of RSs over the last 20 years? What are the trends in terms of publication venues?

Figure 1 shows the number of studies applying MCDA methods to RSs over the last two decades. We can see continuing growth over time in the number of published papers, with a clear increase from 2018 on. About 60% (29 papers) of the total number of selected papers have been published in the last 4 years.

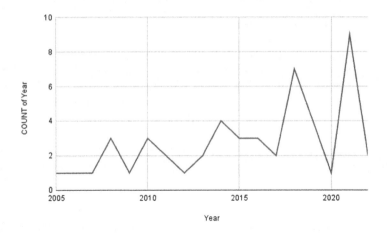

Fig. 1. Number of publications of studies regarding MCDA in RSs over the years.

Regarding the journal of publication, the papers have been published in 47 different journals, indicating that there are not specific journals concentrating on the topic discussed here. The journal with the most published papers is IEEE Access, with 4 papers, followed by Expert Systems with Applications, Applied Artificial Intelligence, Electronic Commerce Research and Applications, with 2 papers each.

RQ2. What are the MCDA methods employed in RSs? And which are employed most?

Based on the MCDA types of methods defined in Sect. 2, we verified that all types of MCDA methods have been employed in the RS context. Table 1 shows the number of papers that applied each method and their references. As a paper may have applied more than one method, the total number of papers presented in Table 1 is greater than the number of selected papers. The main observation is that AHP and TOPSIS are by far the most popular applied MCDA methods with 34% and 22% of the studies, respectively.

RQ3. What purpose do the MCDA methods serve in RSs?

Generally speaking, MCDA methods have been applied to RSs in order to consider multiple criteria and to take the preferences of the user into consideration, thus providing more personalized and accurate recommendations. However, depending on the applied method, other reasons were identified, as presented below.

The AHP method is applied mainly to estimate the users' preferences regarding the criteria weights. AHP allows the preferences to be elicited through pairwise comparisons, which requires less cognitive effort from the user, but, at the same time, results in a longer user-interaction time with the system. In [36], the AHP method was applied for preference elicitation in a group decision making problem–a type of decision problem when multiple DMs/users are involved. AHP was shown to effectively reduce the complexity of the group decision problem and make the DM roles clearer.

Table 1. MCDA methods employed in the RS context.

Type of MCDA method	MCMA method	Number of papers	References
Pairwise comparison	AHP	17	[7, 16, 17, 17, 24, 36, 41, 42, 50, 56, 58, 58, 61, 73, 74, 76, 77]
Distance-based	TOPSIS	11	[6, 10, 11, 22, 27, 31, 36, 49, 60, 69, 75]
Outranking	PROMETHEE	5	[9, 15, 68, 68, 70]
Utility-based	MAUT/UTA	5	[1, 23, 47, 52, 54]
Outranking	ELECTRE	4	[37, 38, 45, 72]
Utility-based	Choquet integral	3	[29, 38, 39]
Scoring	WSM	3	[41, 43, 71]
Utility-based	SMARTER	1	[40]
Scoring	Dominance intensity	1	[51]
Other	Consensus method	1	[48]
Distance-based	VIKOR	1	[34]

A few studies also applied AHP to the recommendation phase, to rank the items [7,42]. A sorting version of AHP, the AHPSort method, was applied as a first step of the recommendation phase, filtering out items which were not appropriate to the current user characteristics [77]; then, the recommendation was made based on the remaining items by applying an optimization model.

Another reason for applying AHP is for the purpose of organizing the criteria hierarchically. When the decision problem is based on a large set of features, it is difficult to solve the problem using a single step where all criteria are taken into account simultaneously. Therefore, organizing the criteria hierarchically is a better way to model the problem.

While most of the papers which applied AHP are for preference elicitation, most of the papers applied TOPSIS for making the recommendation. In [6], the authors apply TOPSIS to substitute prediction methods used in conventional content-based RSs, and the results pointed to a more accurate system than baseline content-based methods. Two variations of the TOPSIS, GDSS-TOPSIS and Dynamic Weight TOPSIS, have also been applied to RSs. The former is a variation of TOPSIS for group decision-making [22]. The Dynamic Weight TOPSIS method is a variation for decisions in which the evaluations of criteria change over time. Therefore, it is a suitable tool to be applied to tourism RSs given the important seasonal component of tourism [10].

Combinations of the already cited methods were also employed in RSs, for instance, AHP and TOPSIS [36], and AHP and TOPSIS integrated to the Fuzzy Sets theory [26, 27, 31, 36, 59, 60].

Outranking methods have been also used in RSs. ELECTRE was applied for the purpose of supporting user classification in CFRSs [37], using the indifference relation of ELECTRE to identify discrimination or similarity between any two users. In [72] (see also [20]), ELECTRE-TRI-B-H, a sorting method based on ELECTRE, was applied in order to directly assign items to a set of defined categories, avoiding the need of an additional phase when a ranking method is applied. These categories indicate the degree of fulfillment of the user's expectations and directly lead to the best alternatives to recommend. The other well-known MCDA outranking method, PROMETHEE, was applied to RSs mainly to rank and make the recommendation [9,15,68,70]). In [68], the Fuzzy Sets theory was combined with PROMETHEE to conduct sentiment orientation analysis.

Utility-based methods have also been highly used in RSs. The framework proposed in [47] aims at modeling user's preferences together with the CF technique in order to identify the most preferred unknown items for every user and consists of four phases: data acquisition, multi-criteria user modeling, clustering and recommendation. In the data acquisition phase, users are asked to evaluate items on the different criteria and to rank all items in order of preference. In the second phase, the ratings based on multiple criteria are processed in order to obtain a weight vector for each user, which represents their preferences regarding the different criteria. Criteria weight vectors are then used, in the third phase, to identify clusters of users with similar preferences by applying the k-means cluster algorithm. In the last phase, items are recommended by implementing the CF philosophy inside each user cluster, defining the user's neighborhood. One can note that the second phase requires the application of a multi-criteria method in order to learn preferences; in this case, the UTA method was applied. Similar ideas using utility-based methods are proposed in [1,23,52,54].

The study proposed in [43] discusses the properties of CFRSs and shows that the weighted sum model (WSM) meets them and, therefore, is suitable for applications in these systems. WSM is also applied in [41,71]. Another scoring method, the Choquet integral, is applied in RSs with the aim of modeling interacting criteria [29,38,39].

RQ4. What are the types and main characteristics of RSs where MCDA methods are employed?

Most RSs using MCDA methods are customer-oriented and characterized for a highly direct interaction with the user. Regarding the type of RSs, results point to a significant number of CF RSs (12 papers), hybrid RSs (9 papers) and content-based RSs (2 papers). These numbers are based on the classification given by the authors of the selected papers. Many papers, however, did not clearly define the type of RS adopted. However, most of these cases can be considered hybrid since more than one method usually used in RSs were combined.

RQ5. What are the evaluation metrics employed in the selected studies?

Evaluation metrics typically employed in RSs are used in about 60% of the selected studies (30 papers). In Table 2, we present the papers and the combination of metrics adopted by them.

Table 2. Performance metrics used in the selected papers. RMSE - Root-mean-square deviation; nDCG - normalized Discounted Cumulative Gain; MRR - Mean Reciprocal Rank.

Combination of the used metrics	References
Precision (or precision-in-top-N), recall, F1-measure, Accuracy	[6, 10, 23, 24, 39, 47, 48, 50, 52, 73]
Accuracy	[27, 30, 38, 45, 59, 68, 69, 75]
Rank consistency test and accuracy	[22]
Precision, novelty and diversity	[15]
Precision-in-top-N	[37]
nDCG, MRR, MAP	[34]
Effectiveness	[36]
Content and system satisfaction, realiability (Cronbach's alpha), convergent and discriminant validity	[17]
Accuracy (MAE), coverage and response-time	[54]
Accuracy (Spearman-rank-correlation), scalability and usability (survey - Cronbach's alpha coefficient)	[17]
Accuracy, interesting and satisfaction	[71]
Accuracy, useful and usability (effectiveness, efficiency, satisfaction)	[58]
Accuracy, response-time and user perceptions (satisfaction, usefulness, and trustworthiness)	[40]
Accuracy (MEA, Spearman-rank-correlation), response-time and precision	[9]

RQ6. What are the contributions of MCDA methods to RSs regarding some of their challenges such as cold-start, data sparsity, scalability, and privacy?

Scalability and privacy have not been addressed in any of the papers. More than that, scalability proved to be a problem in the analyzed RSs since many of them require high interaction with the user, which may lead to slowness.

On the other hand, MCDA can contribute to the cold-start and sparsity problems. Indeed, these problems can be addressed by applying utility-based methods, since preferences (utilities) of new users are learned from previous users [40, 47]. Frameworks in which the users are invited to input their preferences are also able to deal with the cold-start problem [22, 42].

RQ7. What are the trends and gaps in the use or research of MCDA methods in RSs?

The vast majority of MCDA methods employed in RSs are for ranking. However, producing a ranking of alternatives can be seen as a disadvantage since, in most cases, a second stage of filtering is required to finally select a subset of recommended items. In order to avoid the need of this additional phase, it seems to be more appropriate to use sorting methods that directly assign alternatives to a set of defined categories. This approach is also indicated as more scalable for large sets of alternatives and/or users. Another trend in the use of MCDA methods in RSs regards the modeling of hierarchical criteria. Despite that, few methods have been explored for this purpose so far. Exploring how MCDA methods can be used to address some of the RS challenges, such as cold-start, data sparsity, scalability, and privacy, is also an open question, which naturally open up possibilities for future studies.

As already discussed, although applying MCDA methods to RSs may improve personalizing the system when taking user preferences into consideration, these methods also leads to problems regarding system response time. Moreover, those systems usually ask for the user their preferences and do not take advantage of information already available in historical data. In order to improve those aspects, the application of preference learning combining to MCDA in RSs comes up as an interesting possibility of future research. Indeed, the research field called "preference learning", which can be considered a sub-field of the machine learning research area, concerns with the acquisition of preference models from data–it involves learning from observations that reveal information about the preferences of an individual or a class of individuals, and building models that generalize beyond such training data [32]. A RSs based on preference learning and MCDA would allow at the same time to learn preferences from historical data and to use preferences established by the user, offering a large and promising scope to be explored. Bringing together contributions involving these two areas to RSs presents itself as a good solution for recommendations in high personalized learning environments. It is important to note that, although the topic preference learning has already being explored in RSs, throughout this literature review we could see that the integration of preference learning and MCDA applied to RSs is still an unexplored topic in the literature.

6   Conclusions

In this paper, we have conducted a systematic literature review of the application of MCDA methods to RSs. Typically, a small sample size affects the generalizability of the research results. In order to overcome this problem, we intend to include more studies identified through a snowballing approach and take into account all the main conferences on software engineering. We also intend to extend this study in order to better discuss open issues and opportunities for future work. To do so, we want to build a novel taxonomy on multi-criteria RSs, including RSs based on MCDA methods, analyze strengths and weakness associated to each category of the taxonomy, and point to immediate future works to be developed, which could later help the research community to obtain importance advances in this research area.

Acknowledgment. This work was supported by the São Paulo Research Foundation (FAPESP), grants #2018/23447 and #2020/01089-9, and the Brazilian National Council for Scientific and Technological Development (CNPq). This project is also part of the Brazilian Institute of Data Science, grant #2020/09838-0, São Paulo Research Foundation (FAPESP).

References

1. Abbas, A., Bilal, K., Zhang, L., Khan, S.: A cloud based health insurance plan recommendation system: a user centered approach. Futur. Gener. Comput. Syst. **43–44**, 99–109 (2015)
2. Adomavicius, G., Kwon, Y.: New recommendation techniques for multicriteria rating systems. IEEE Intell. Syst. **22**(3), 48–55 (2007)
3. Adomavicius, G., Manouselis, N., Kwon, Y.O.: Multi-criteria recommender systems. In: Ricci, F., Rokach, L., Shapira, B., Kantor, P.B. (eds.) Recommender Systems Handbook, pp. 769–803. Springer, Boston, MA (2011). https://doi.org/10.1007/978-0-387-85820-3_24
4. Adomavicius, G., Tuzhilin, A.: Toward the next generation of recommender systems: a survey of the state-of-the-art and possible extensions. IEEE Trans. Knowl. Data Eng. **17**(6), 734–749 (2005). https://doi.org/10.1109/TKDE.2005.99
5. Aggarwal, C.C.: Recommender Systems. Springer, Cham (2016). https://doi.org/10.1007/978-3-319-29659-3
6. Al-Bashiri, H., Abdulgabber, M., Romli, A., Kahtan, H.: An improved memory-based collaborative filtering method based on the TOPSIS technique. PLoS ONE **13**(10), e0204434 (2018)
7. Angskun, T., Angskun, J.: A qualitative attraction ranking model for personalized recommendations. J. Hosp. Tour. Technol. **9**, 2648352 (2018)
8. Anselmo Alvarez, P., Ishizaka, A., Martínez, L.: Multiple-criteria decision-making sorting methods: a survey. Expert Syst. Appl. **183**, 115368 (2021). https://doi.org/10.1016/j.eswa.2021.115368
9. Arentze, T., Kemperman, A., Aksenov, P.: Estimating a latent-class user model for travel recommender systems. Inf. Technol. Tour. **19**(1–4), 61–82 (2018)
10. Arif, Y., Harini, S., Nugroho, S., Hariadi, M.: An automatic scenario control in serious game to visualize tourism destinations recommendation. IEEE Access **9**, 89941–89957 (2021)
11. Baczkiewicz, A., Kizielewicz, B., Shekhovtsov, A., Watróbski, J., Sałabun, W.: Methodical aspects of MCDM based E-commerce recommender system. J. Theor. Appl. Electron. Commer. Res. **16**(6), 2192–2229 (2021)
12. Behzadian, M., Kazemzadeh, R., Albadvi, A., Aghdasi, M.: PROMETHEE: a comprehensive literature review on methodologies and applications. Eur. J. Oper. Res. **200**(1), 198–215 (2010)
13. Brans, J., Vincke, P., Mareschal, B.: How to select and how to rank projects: the PROMETHEE method. Eur. J. Oper. Res. **24**(2), 228–238 (1986)
14. Burke, R.: Hybrid recommender systems: survey and experiments. User Model. User-Adapt. Interact. **12**(4), 331–370 (2002). https://doi.org/10.1023/A:1021240730564
15. Chai, Z., Li, Y., Zhu, S.: P-MOIA-RS: a multi-objective optimization and decision-making algorithm for recommendation systems. J. Ambient. Intell. Humaniz. Comput. **12**(1), 443–454 (2021). https://doi.org/10.1007/s12652-020-01997-x

16. Chaimae Lamaakchaoui, A.A., Jarroudi, M.E.: The AHP method for the evaluation and selection of complementary products. Int. J. Serv. Sci. Manag. Eng. Technol. **9**(3), 96695–96711 (2018)
17. Chen, D.N., Hu, P.H., Kuo, Y.R., Liang, T.P.: A web-based personalized recommendation system for mobile phone selection: design, implementation, and evaluation. Expert Syst. Appl. **37**(12), 8201–8210 (2010)
18. Choquet, G.: Theory of capacities. Ann. Inst. Fourier **5**, 131–295 (1954)
19. Cinelli, M., Kadziński, M., Miebs, G., Gonzalez, M., Słowiński, R.: Recommending multiple criteria decision analysis methods with a new taxonomy-based decision support system. Eur. J. Oper. Res. **302**(2), 633–651 (2022)
20. Del Vasto-Terrientes, L., Valls, A., Zielniewicz, P., Borràs, J.: Erratum to: A hierarchical multi-criteria sorting approach for recommender systems. J. Intell. Inf. Syst. **46**(2), 347–348 (2016)
21. Devaud, J., Groussaud, G., Jacquet-Lagrèze, E.: UTADIS: Une méthode de construction de fonctions d'utilité additives rendant compte de jugements globaux. European Working Group on Multicriteria Decision Aid (1980)
22. Dewi, R., Ananta, M., Fanani, L., Brata, K., Priandani, N.: The development of mobile culinary recommendation system based on group decision support system. Int. J. Interact. Mob. Technol. **12**(3), 209–216 (2018)
23. Dixit, V.S., Mehta, H., Bedi, P.: A proposed framework for group-based multi-criteria recommendations. Appl. Artif. Intell. **28**(10), 917–956 (2014)
24. Ebrahimi, F., Asemi, A., Nezarat, A., Ko, A.: Developing a mathematical model of the co-author recommender system using graph mining techniques and big data applications. J. Big Data **8**(1), 1–15 (2021)
25. Edwards, W., Barron, F.: Smarts and smarter: improved simple methods for multiattribute utility measurement. Organ. Behav. Hum. Decis. Process. **60**(3), 306–325 (1994)
26. Effendy, F., Kartono, K., Herawatie, D.: Mobile apps for boarding house recommendation. Int. J. Interact. Mob. Technol. **14**(11), 32–47 (2020)
27. Effendy, F., Nuqoba, B.: Taufik: culinary recommendation application based on user preferences using fuzzy topsis. IIUM Eng. J. **20**(2), 163–175 (2019)
28. Fishburn, P.C.: Additive utilities with incomplete product sets: application to priorities and assignments. Oper. Res. **15**(3), 537–542 (1967)
29. Fomba, S., Zarate, P., Kilgour, M., Camilleri, G., Konate, J., Tangara, F.: A recommender system based on multi-criteria aggregation. Int. J. Decis. Support Syst. Technol. **9**(4), 1–15 (2017)
30. Forouzandeh, S., Berahmand, K., Nasiri, E., Rostami, M.: A hotel recommender system for tourists using the artificial bee colony algorithm and fuzzy TOPSIS model: a case study of tripadvisor. Int. J. Inf. Technol. Decis. Mak. **20**(1), 399–429 (2021)
31. Forouzandeh, S., Rostami, M., Berahmand, K.: A hybrid method for recommendation systems based on tourism with an evolutionary algorithm and TOPSIS model. Fuzzy Inf. Eng. **14**(1), 26–50 (2022)
32. Fürnkranz, J., Hüllermeier, E.: Preference learning (2011)
33. Govindan, K., Jepsen, M.B.: ELECTRE: a comprehensive literature review on methodologies and applications. Eur. J. Oper. Res. **250**(1), 1–29 (2016)
34. Guo, Z., Tang, C., Tang, H., Fu, Y., Niu, W.: A novel group recommendation mechanism from the perspective of preference distribution. IEEE Access **6**, 5865–5878 (2018)
35. Gupta, S., Kant, V.: Credibility score based multi-criteria recommender system. Knowl.-Based Syst. **196**, 105756 (2020)

36. Hong, Y., Zeng, X., Bruniaux, P., Chen, Y., Zhang, X.: Development of a new knowledge-based fabric recommendation system by integrating the collaborative design process and multi-criteria decision support. Text. Res. J. **88**(23), 2682–2698 (2018)
37. Hu, Y.C.: A multicriteria collaborative filtering approach using the indifference relation and its application to initiator recommendation for group-buying. Appl. Artif. Intell. **28**(10), 992–1008 (2014)
38. Hu, Y.C.: Nonadditive similarity-based single-layer perceptron for multi-criteria collaborative filtering. Neurocomputing **129**, 306–314 (2014)
39. Hu, Y.C.: A novel nonadditive collaborative-filtering approach using multicriteria ratings. Math. Probl. Eng. **2013** (2013)
40. Huang, S.L.: Designing utility-based recommender systems for e-commerce: evaluation of preference-elicitation methods. Electron. Commer. Res. Appl. **10**(4), 398–407 (2011)
41. Huang, Y., Wang, N.N., Zhang, H., Wang, J.: A novel product recommendation model consolidating price, trust and online reviews. Kybernetes **48**(6), 1355–1372 (2019)
42. Huang, Y., Bian, L.: A Bayesian network and analytic hierarchy process based personalized recommendations for tourist attractions over the internet. Expert Syst. Appl. **36**(1), 933–943 (2009)
43. Iijima, J., Ho, S.: Common structure and properties of filtering systems. Electron. Commer. Res. Appl. **6**(2), 139–145 (2007)
44. Ishizaka, A., Nemery, P., Pearman, C.: AHPSort: an AHP based method for sorting problems. Int. J. Prod. Res. **50**(17), 4767–4784 (2012)
45. Ke, C.K., Chang, C.M.: Optimizing target selection complexity of a recommendation system by skyline query and multi-criteria decision analysis. J. Supercomput. **76**(8), 6453–6474 (2020)
46. Kitchenham, B., Charters, S.: Guidelines for performing systematic literature reviews in software engineering (2007)
47. Lakiotaki, K., Matsatsinis, N.F., Tsoukiàs, A.: Multicriteria user modeling in recommender systems. IEEE Intell. Syst. **26**(2), 64–76 (2011)
48. Lee, S.K., Cho, Y.H., Kim, S.H.: Collaborative filtering with ordinal scale-based implicit ratings for mobile music recommendations. Inf. Sci. **180**(11), 2142–2155 (2010)
49. Li, S., Pham, T., Chuang, H., Wang, Z.W.: Does reliable information matter? Towards a trustworthy co-created recommendation model by mining unboxing reviews. Inf. Syst. e-Bus. Manag. **14**(1), 71–99 (2016). https://doi.org/10.1007/s10257-015-0275-6
50. Liu, D.R., Shih, Y.Y.: Integrating AHP and data mining for product recommendation based on customer lifetime value. Inf. Manag. **42**(3), 387–400 (2005)
51. Mahajan, P., Kaur, P.D.: Three-tier IoT-edge-cloud (3T-IEC) architectural paradigm for real-time event recommendation in event-based social networks. J. Ambient. Intell. Humaniz. Comput. **12**(1), 1363–1386 (2020). https://doi.org/10.1007/s12652-020-02202-9
52. Manouselis, N.: Deploying and evaluating multiattribute product recommendation in e-markets. Int. J. Manag. Decis. Mak. **9**(1), 43–61 (2008)
53. Manouselis, N., Costopoulou, C.: Analysis and classification of multi-criteria recommender systems. World Wide Web **10**(4), 415–441 (2007)
54. Manouselis, N., Costopoulou, C.: marService: multiattribute utility recommendation for e-markets. Int. J. Comput. Appl. Technol. **33**(2–3), 176–189 (2008)

55. Nemery, P., Lamboray, C.: Flow sort: a flow-based sorting method with limiting or central profiles. TOP **16**(1), 90–113 (2008). https://doi.org/10.1007/s11750-007-0036-x

56. Olugbara, O.O., Ojo, S.O., Mphahlele, M.I.: Exploiting image content in location-based shopping recommender systems for mobile users. Int. J. Inf. Technol. Decis. Mak. **09**(05), 759–778 (2010)

57. Opricovic, S., Tzeng, G.H.: Compromise solution by MCDM methods: a comparative analysis of VIKOR and TOPSIS. Eur. J. Oper. Res. **156**(2), 445–455 (2004)

58. Park, H.S., Park, M.H., Cho, S.B.: Mobile information recommendation using multi-criteria decision making with Bayesian network. Int. J. Inf. Technol. Decis. Mak. **14**(2), 317–338 (2015)

59. Pinandito, A., Ananta, M., Brata, K., Fanani, L.: Alternatives weighting in analytic hierarchy process of mobile culinary recommendation system using fuzzy. ARPN J. Eng. Appl. Sci. **10**(19), 8791–8798 (2015)

60. Qin, Y., Wang, X., Xu, Z.: Ranking tourist attractions through online reviews: a novel method with intuitionistic and hesitant fuzzy information based on sentiment analysis. Int. J. Fuzzy Syst. **24**(2), 755–777 (2022)

61. Rizvi, S., Zehra, S., Olariu, S.: ASPIRE: an agent-oriented smart parking recommendation system for smart cities. IEEE Intell. Transp. Syst. Mag. **11**(4), 48–61 (2019)

62. Roy, B.: Electre iii: Un algorithme de classements fondé sur une représentation floue des préférences en présence de critères multiples. Cahiers du Centre d'Etudes de Recherche Opérationnelle **20**(1), 3–24 (1978)

63. Roy, B.: Multicriteria Methodology Goes Decision Aiding, 1st edn. Kluwer Academic Publishers, The Netherlands (1996)

64. Roy, B., Bouyssou, D.: Aide multicritère à la décision: méthodes et cas, 1st edn. Econômica, Paris (1993)

65. Saaty, R.: The analytic hierarchy process—what it is and how it is used. Math. Model. **9**(3), 161–176 (1987)

66. Saaty, T.L.: Decision Making with Dependence and Feedback: The Analytic Network Process, vol. 4922. RWS publications, Pittsburgh (1996)

67. Sabokbar, H., Hosseini, A., Banaitis, A., Banaitiene, N.: A novel sorting method TOPSIS-SORT: an application for Tehran environmental quality evaluation. E a M: Econ. Manag. **19**(2), 87–104 (2016)

68. Serrano-Guerrero, J., Bani-Doumi, M., Romero, F., Olivas, J.: A fuzzy aspect-based approach for recommending hospitals. Int. J. Intell. Syst. **37**(4), 2885–2910 (2022)

69. Showafah, M., Sihwi, S.: Winarno: Ontology-based daily menu recommendation system for complementary food according to nutritional needs using naïve bayes and topsis. Int. J. Adv. Comput. Sci. Appl. **12**(11), 638–645 (2021)

70. Tian, Y., Wang, W., Gong, X., Que, X., Ma, J.: An enhanced personal photo recommendation system by fusing contextual and textual features on mobile device. IEEE Trans. Consum. Electron. **59**(1), 220–228 (2013)

71. Troussas, C., Krouska, A., Sgouropoulou, C.: Enhancing human-computer interaction in digital repositories through a MCDA-based recommender system. Adv. Hum.-Comput. Interact. **2021** (2021)

72. Vasto-Terrientes, L., Valls, A., Zielniewicz, P., Borràs, J.: A hierarchical multi-criteria sorting approach for recommender systems. J. Intell. Inf. Syst. **46**(2), 313–346 (2016)

73. Verma, P., Sood, S., Kalra, S.: Student career path recommendation in engineering stream based on three-dimensional model. Comput. Appl. Eng. Educ. **25**(4), 578–593 (2017)
74. Wang, N.: Ideological and political education recommendation system based on AHP and improved collaborative filtering algorithm. Sci. Program. **2021**, 2648352 (2021)
75. Wang, L., Zhang, R., Ruan, H.: A personalized recommendation model in E commerce based on TOPSIS algorithm. J. Electron. Commer. Organ. **12**(2), 89–100 (2014)
76. Yang, L., Yeung, K., Ndzi, D.: A proactive personalised mobile recommendation system using analytic hierarchy process and Bayesian network. J. Internet Serv. Appl. **3**(2), 195–214 (2012)
77. Yera Toledo, R., Alzahrani, A.A., Martínez, L.: A food recommender system considering nutritional information and user preferences. IEEE Access **7**, 96695–96711 (2019)
78. Yu, W.: Aide multicritére à la décision dans le cadre de la problématique du tri: Concepts, méthodes et applications. PhD dissertation, Université Paris-Dauphine (1992)
79. Zavadskas, E.K., Kaklauskas, A., Sarka, V.: The new method of multicriteria complex proportional assessment of projects. Technol. Econ. Dev. Econ. **1**, 131–139 (1994)

Explaining Learning Performance with Local Performance Regions and Maximally Relevant Meta-Rules

Ricardo B. C. Prudêncio[1(⊠)] and Telmo M. Silva Filho[2]

[1] Centro de Informática, Universidade Federal de Pernambuco, Recife, Brazil
`rbcp@cin.ufpe.br`
[2] Department of Engineering Mathematics, University of Bristol, Bristol, UK
`telmo.silvafilho@bristol.ac.uk`

Abstract. Identifying instances in a learning task that are difficult to predict is important to avoid critical errors at deployment time. Additionally, providing explanations for good or bad predictions of a model can be useful to understand its behavior and to plan how to improve it (e.g., by data augmentation in specific areas of instances). In this paper, we propose a method to provide explanations for a model's predictive performance based on the induction of meta-rules. Each meta-rule identifies a local region in the instance space, called Local Performance Region (LPR). The meta-rules are induced using a reduced number of attributes, in such a way that each LPR can be inspected by, e.g., plotting a pairwise attribute plot. Additionally, given a group of instances to explain (or eventually an individual instance), we propose a greedy-search algorithm that finds the subset of non-redundant LPRs that maximally covers the instances. By explaining the (in)correctness of model predictions, LPRs constitute a novel use of meta-learning and a novel application in explainable AI. Experiments show the usefulness of LPRs while explaining inaccurate class predictions of Random Forest in a benchmark dataset, demonstrating a special case of LPRs, called Local Hard Regions (LHRs).

Keywords: Meta-learning · Explainability · Rule learning

1 Introduction

Predicting the performance of Machine Learning (ML) algorithms is an important task to support algorithm selection and to understand the limits of each algorithm of interest. This task has been treated by meta-learning [2] as another supervised learning task. In this approach, a set of training meta-examples is produced from experiments performed to evaluate a set of algorithms on a set of learning problems of interest. A meta-learner is then built to predict algorithm performance for new problems. In this paper, we are focused on the instance-level meta-learning approach [3,6,14], which is specific to predict algorithm performance for instances in a single learning problem of interest. So, before using a

© The Author(s), under exclusive license to Springer Nature Switzerland AG 2022
J. C. Xavier-Junior and R. A. Rios (Eds.): BRACIS 2022, LNAI 13653, pp. 550–564, 2022.
https://doi.org/10.1007/978-3-031-21686-2_38

candidate model to predict an input instance, the meta-learner could be used to foresee if the model is actually adequate for that instance. Additionally, such an approach can be used to perform dynamic algorithm selection [5].

Previous work on meta-learning has focused on optimizing the predictive performance of meta-models, while neglecting explainability, which should be an important issue in meta-learning. In fact, a primary objective of meta-learning is actually to *understand* algorithm performance. However previous works are limited for example to learn meta-models that are directly interpretable, e.g., decision trees. The same challenges that motivated the topic of explainable ML [12] are also applied to meta-learning, and thus, further investigations are necessary to propose procedures for explaining meta-models. Hence, important practical questions such as when an algorithm fails could be answered more properly.

This paper proposes a new approach to explain learning performance by extracting meta-rules for a learned model in a dataset. Each meta-rule defines a Local Performance Region (LPR) in terms of subsets of instances and features for which a learned model has remarkable poor predictive performance (Local Hard Region – LHR) or strong predictive performance (Local Easy Region – LER). By producing meta-rules with one or two attributes, LPRs can be inspected in visual plots. As a second contribution, we proposed a greedy-search procedure to find the subset of LPRs which covers the maximal number of instances given by the user. Depending on the dataset complexity, the number of LPRs can be high but it is possible that different users aims at each time to explain specific groups of instances or even a single one. Thus, filtering non-redundant LPRs can be beneficial for satisfying each user's demand.

Current state-of-the-art explainers, such as LIME [16], act at the instance level and focus on which features were important for the class prediction given by the model. However this explanation is produced without taking into account the accuracy/error of the model's prediction. This is an important distinction between these explainers and our approach, which actually aims to explain performance, instead of individual predictions. This means that LPRs and LIME (or similar explainers) are complementary and can be used together for a full picture of the explained model's predictions.

Experiments were performed in the Statlog (Heart) classification dataset, in which 14 LHRs were identified for the Random Forest (RF) algorithm. Each LHR indicates a specific area in the instance space where RF produced relatively poor class probabilities. Each LHR could be explained by one or two attributes in the dataset, which could be inspected individually in the experiments. By adopting the maximal coverage search procedure, we identified 3 non-redundant LHRs which explained more than 80% of the RF's errors. In the experiments, we discussed how the LHRs can be used to explain the RF's performance for groups of hard instances or even for individual instances.

This paper is organized as follows. Section 2 presents the related work on meta-learning, followed by Sect. 3 in which we describe the proposed solutions. Section 4 in turn presents the performed experiments. Finally Sect. 5 concludes the paper with final considerations and future work.

2   Meta-Learning

Meta-learning predicts algorithm performance in a supervised way, usually applied to select algorithms at the dataset level [2]. In this case, a meta-learner is built from a set of meta-examples, in which each meta-example is related to a dataset and stores: (1) the characteristics describing the dataset, called meta-features (e.g., the number of attributes and examples, correlation between attributes, class entropy,...); (2) the candidate algorithm with best empirical performance for that dataset, usually estimated by cross-validation. A meta-learner model learned from a set of such meta-examples is a classifier that predicts the best algorithm for new datasets based on their meta-features. Alternatively, the meta-target stored in the meta-example can be the empirical performance estimation of a single algorithm. In this case, the meta-learner is a regression model which predicts the algorithm performance for new datasets.

In the instance-learning approach, meta-models are adopted to perform algorithm selection for each instance in a task of interest [3]. Thus, each meta-example is related to a single instance in a dataset. The meta-features in turn can be the original features of the instance or other features related to the models, like model confidence degree. The meta-target usually indicates the best algorithm to predict that instance. Alternatively, the meta-target can be a loss measure specific for an algorithm (e.g., $1|0$ loss, absolute error,...). In this case, the meta-learner predicts the loss measure for each instance given as input.

Instance-level meta-learning has been adopted to perform dynamic selection of models in literature [5], where a meta-learner is used to select a candidate algorithm for each instance to classify [6,9,11,14,14]. Although related, our objectives are different (and complementary). Instead of identifying areas of competence for a query instance to classify, we aim to find all the local areas in instance space where a given model presented poor performance. Obviously such information can be used in dynamic classifier selection, for instance by discarding a model if the instance belongs to a known hard area for that model.

Finally, the current work is related to previous work dedicated to measure instance hardness [17]. Such previous work ranges from measuring the performance of a single model or a pool of models for each instance [15,17], instance hardness based on item response theory [4,10,13] and the use of data complexity measures [1]. The main focus of the previous work is to detect individual instances or groups of instances that are difficult to predict. Our work shares the objective of detecting hard areas in the instance space. However our proposed methods provide mechanisms to explain the instance hardness for a given model, which is not deeply investigated in literature.

3   Finding Local Performance Regions in Instance Space

In this paper, we propose an original method to explain algorithm performance for instance-level meta-learners. The main objective is to identify regions in the instance space that explain model performance. Hence the general question

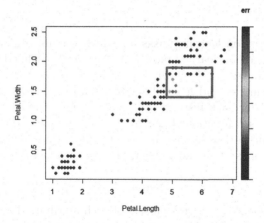

Fig. 1. Local Hard Region (LHR) in the Iris dataset.

addressed in the paper is to know when a learned model fails or succeeds in its prediction. For this, we propose a method to detect Local Performance Regions (LPRs) and their special cases Local Hard Regions (LHRs) and Local Easy Regions (LERs). These local regions are subsets of instances in a dataset, defined by rules involving a reduced number of attributes (e.g., one or two), in such a way that model performance can be inspected in one or two dimensional scatter plots with the chosen attributes.

As a motivating example, consider the Iris dataset, which has four attributes and three classes. We train a Random Forest (RF) model and we aim to explain when it fails in its predictions. We start by assuming a bad prediction when the correct class probability predicted by RF is lower than 0.9 (or error greater than 0.1). This is obviously a strict criterion for defining that a model had a bad performance but notice that the Iris dataset is a very simple classification task. By using such criterion, we observed 10 hard examples, where 4 instances belong to the Virginica class and 6 belong to the Versicolor class. Figure 1 presents a possible LHR (the red frame) for inspecting the hard examples, using the Petal Length and the Petal Width attributes. This LHR covers 14% of the Iris dataset, with average RF error of 0.12. Additionally it covers 9 out of 10 hard examples, i.e., a single 2-D plot could explain 90% of the poor predictions of the RF model.

In this paper, we propose a method to find LPRs by learning discriminating meta-rules to explain the model performance (see Algorithm 1). Additionally, as several LPRs can be found for a single dataset and model of interest, we proposed a greedy-search method to select a reduced subset of LPRs to explain a group of instances given by the user. This is done by searching for the subset of LPRs that maximizes the total relevance for the given examples (see Algorithm 2). This algorithm can be eventually used to explain the (in)correctness for a single instance, by selecting a single LPR for that instance.

3.1 Definitions

Our work is focused on learning tasks for which we aim to derive explanations for the performance of a learned model. A task can be defined in terms of a set of predictor attributes $\{a_i\}_{i=1}^m$, where each attribute a_i has a domain A_i. Let $\mathcal{M}$ be a learned model and let D be a test dataset to collect the predictions and evaluate $\mathcal{M}$. For each instance $d \in D$, we collect the model's performance according to a chosen function: $L(\mathcal{M}, d)$.

An LPR in our work is defined by a meta-rule r that considers in its head a pair of attributes a_i and a_j and predicts the performance of model $\mathcal{M}$:

$$r : per = e \text{ when } a_i \in A_i^r \ \& \ a_j \in A_j^r,$$

where $A_i^r \subseteq A_i$ and $A_j^r \subseteq A_j$ are subsets of each attribute domain. We adopted two attributes in the LPRs for two reasons: (1) to provide simple explanations; and (2) to inspect the meta-rule in an LPR plot of two dimensions. Notice that a meta-rule with one antecedent is a special case of LPR in which $A_i^r = A_i$.

The model performance e returned by the meta-rule can be estimated as average $L(\mathcal{M}, d)$ for the instances where the meta-rule's conditions apply:

$$e(\mathcal{M}, r, D) = \frac{1}{n_r} \sum_{d \in D_r} L(\mathcal{M}, d), \tag{1}$$

where $D_r \in D$ is the subset of instances covered by r and $n_r = |D_r|$. The meta-rule coverage is the proportion of examples covered by the meta-rule:

$$Cov(r, D) = \frac{n_r}{n}, \tag{2}$$

where $n = |D|$.

3.2 Finding LPRs

Algorithm 1 shows the steps to find LPRs. The performance function $L(\mathcal{M}, d)$ can be chosen to measure model *incorrectness*, i.e., lower values are better, in which case Algorithm 1 outputs LHRs. Alternatively, if the user wishes to obtain LERs, $L(\mathcal{M}, d)$ can be chosen to measure model *correctness*, such that the higher its values, the better. As an example, the LHR presented in Fig. 1 for the Iris problem is defined by the meta-rule:

$$r : per = 0.12 \text{ when Petal Length} \in [4.5; 6.3] \ \& \ \text{Petal Width} \in [1.4; 1.8].$$

A number of $n_r = 21$ instances are covered by this meta-rule and hence coverage $Cov(r, D_{iris}) = 21/270$. The average incorrectness returned by training the RF model for these 21 instances is 0.12 (i.e. $e(\mathcal{M}, r, D_{iris}) = 0.12$), with the chosen performance function being $L(\mathcal{M}, d) = 1 - p(y|d)$, where $p(y|d)$ is the probability given by $\mathcal{M}$ to the correct class y of instance d.

As mentioned above each LPR is related to a pair of attributes in the dataset. Given a pair of attributes a_i and a_j, a meta-dataset D_{ij} is produced where those attributes are used as predictors and the performance $L(\mathcal{M}, d)$ of model $\mathcal{M}$ for every instance in the dataset is used as the target attribute. A rule learning algorithm is applied on D_{ij} to derive a set of candidate meta-rules to predict the model performance based on a_i and a_j.

For LHRs, each candidate meta-rule is returned when the corresponding model *incorrectness* is greater than a pre-defined threshold ($thrE$), that is if $e(\mathcal{M}, r_k, D) > thrE$. In this case, the meta-rule actually distinguishes a bad performance region. We highlight that several LHRs can be found for a single attribute pair, depending on the value of $thrE$. The lower the value of this parameter, the higher the number of LHRs found by the algorithm. Obviously, $thrE$ has to be set according to the application domain, based on the model error that is tolerated. Naturally, the opposite happens for LERs, where a candidate meta-rule is returned when the corresponding model *correctness* is greater than the pre-defined threshold. Thus, as mentioned above, one must choose a suitable performance function in each case.

For LHRs, for example, the performance function $L(\mathcal{M}, d)$ can be chosen as any instance-wise error metric, such as a binary error function (1 if prediction was wrong, 0 otherwise), proper scoring rules (Brier score, log-loss, . . .), or the complement of the probability predicted for the actual class of the instance, which we employ in our experiments in Sect. 4. Whichever metric is chosen, $L(\mathcal{M}, d)$ needs to be calculated on a dataset with available ground truth. However, our approach also allows for a trained meta-learner, which can be used to estimate $\hat{L}(\mathcal{M}, d)$ for new instances, meaning LPRs can be found to explain when the meta-learner thinks that $\mathcal{M}$ will predict poorly or not.

3.3   Finding Maximally Relevant LPRs

In this section we describe the method proposed to select a reduced subset $R \subseteq RuleSet$ to explain the set of examples D_u provided by the user. For example, D_u could be the set of all instances in D for which the model $\mathcal{M}$ returned errors greater than the threshold $thrE$, i.e. $L(\mathcal{M}, d) > thrE$. Alternatively the user could inform specific groups of instances to explain (e.g., all instances that are female and the model returned bad predictions). The proposed method is relevant when there are correlated predictor attributes in a dataset, in such a way that meta-rules extracted by the method proposed in previous section can be redundant.

The proposed method is closely related to [7], which proposed a greedy-search method to find 2-D plots of features to visualize outliers in a dataset. Our work adapted this idea to find subsets of LPRs that cover the instances provided by the user. The objective function $TotalRel(R, D_u)$ is based on a function that measures the relevance of R for each instance $d \in D_u$. The function considers the most relevant LPR that covers the example d:

$$Rel(\mathcal{M}, d, R) = \max_{r \in R} e(\mathcal{M}, r_k, D) \times Cov(r, d), \qquad (3)$$

Algorithm 1: Find Local Performance Regions.

Input: Model $\mathcal{M}$; $n \times m$ test dataset D, in which n is the number of instances and m is the number of attributes; threshold $thrE$ for predicted model performance; and loss function $L(\mathcal{M}, d)$.

```
/* Test model M on D to collect the performances per (a n × 1
   vector).                                                      */
```
1   $\textbf{per} \leftarrow (\textbf{L}(\mathcal{M}, \textbf{d}) \mid \textbf{d} \in \textbf{D})$
```
/* Initialize the set of meta-rules with empty set.             */
```
2   $RuleSet \leftarrow \{\}$
3   **for** *each attribute pair* (a_i, a_j) **do**
```
    /* Produce a meta-dataset from (aᵢ, aⱼ) and per.            */
```
4      $D_{ij} \leftarrow$ Dataset built using the attributes a_i and a_j as predictors and model performance *per* as the target attribute
```
    /* Extract meta-rules.                                      */
```
5      $RuleSet_{ij} \leftarrow$ Rules extracted by applying a rule learning algorithm on D_{ij}.
```
    /* Select the meta-rules based on threshold thrE.           */
```
6   **for** *rule* r_k *in* $RuleSet_{ij}$ **do**
7   **if** $e(\mathcal{M}, r_k, D) > thrE$ **then**
8              $RuleSet \leftarrow RuleSet \bigcup r_k$

Output: $RuleSet$

---

where $Cov(r, d) = 1$ if the rule r covers the example d and it is 0, otherwise. The relevance function is useful to distinguish among several LPRs that compete to cover the example d. For example, the most relevant LHR for a hard instance is the one that has the highest predicted error. Finally the objective function for a subset R is defined as the total relevance:

$$TotalRel(R, D_{hard}) = \sum_{d \in D_u} Rel(\mathcal{M}, d, R) \tag{4}$$

Algorithm 2 is a greedy-search algorithm that starts with an empty set of selected meta-rules and iteratively includes a candidate meta-rule that obtains the marginal total relevance for the user examples (see lines 6 and 7). A meta-rule is added at each iteration until a number b of meta-rules is selected.

3.4   Explaining LPRs Using Meta-Features

Algorithm 1 finds meta-rules defined using the features available in the dataset. As a complementary analysis, one could add an additional explanation layer to describe each LPR in terms of meta-features extracted from the data in the LPR. For example, instances in the minority class can be more frequent in a given LPR. If the model is biased to respond well for the majority class, the predictive model performance will be low for that region. Also, a classification task can be linear for certain regions of instances while more complex for other instances. In such analysis, the instances' features cause (high or low) data complexity, which in turn can explain the model performance in a different perspective.

Algorithm 2: Find the Meta-Rule Subset to Maximize the Total Relevance for User Examples

---

Input: *RuleSet*, set of meta-rules extracted by Algorithm 1; budget b, number
     of meta-rules to select; dataset D_u, set of user examples to cover.

   /* Initialization   */

1 $candR \leftarrow RuleSet$ /* Subset of candidate meta-rules to select   */

2 $selectR \leftarrow \{\}$ /* Subset of selected meta-rules   */

3 $totalRel = 0$ /* Total relevance for the user examples   */

   /* Select b meta-rules, by maximizing marginal relevance for the
     user examples at each iteration   */

4 **while** $size(selectR) < b$ **do**

      /* Compute the marginal relevance score of each candidate
        meta-rule   */

5   **for** *each* $r_k \in candR$ **do**

6         $auxR \leftarrow selectR \bigcup r_k$

7         $margRel[r_k] \leftarrow TotalRel(auxR, D_u) - totalRel$

      /* Find the meta-rule in candidate set with maximal marginal
        relevance score   */

8     $r^* \leftarrow argmax_{r_k \in candR} margRel[r_k]$

      /* Update the selected and candidate meta-rule sets   */

9     $selectR \leftarrow selectR \bigcup r^*$

10    $candR \leftarrow candR \setminus r^*$

      /* Update the total relevance of the selected meta-rule set   */

11    $totalRel = totalRel + margRel[r^*]$

Output: selectR

Different candidate meta-features are available in literature to analyze data complexity at the instance level, including metrics of feature relevance, linearity-based measures, class balancing measures, neighborhood-based measures, among others [1,17]. In the end-user perspective, meta-rules defined using the original features can be easier to interpret. In the data analyst perspective, identifying relevant meta-features in a LPR can be useful to understand model performance also considering the eventual biases the model has and the consequences to deal with the instances belonging to specific regions in the instance space.

4 Experiments

In this section we present an experiment to produce explanations for RF applied to the Statlog(Heart) dataset. This dataset is a binary classification task, containing 270 instances and 13 predictor attributes. In the experiments, we performed 10-fold cross-validation to collect the prediction errors returned by RF. For each instance d we computed the error obtained by RF:

$$L(\mathcal{M}, d) = 1 - p(y|d), \tag{5}$$

where y is the true class label of instance d and $p(y|d)$ is the predicted class probability returned by RF. In the experiment, we adopted the randomForest package in R, with default parameters.

The errors obtained by RF in the Heart dataset had an average value of 0.27. The distribution is skewed and the RF errors tend to be low. About 60% of errors were lower than average. More extreme errors (e.g., greater than 0.8) were observed for 2.9% of the instances.

4.1   Finding LHRs: Algorithm 1

For finding the LHRs (Algorithm 1), we adopted the rpart package in R for rule induction. Initially a decision tree is induced using D_{ij}. Then the rpart.rules function is used to extract the meta-rules from the induced decision tree. All parameters of rpart were defined as the default values, apart from the *minbucket* parameter, which controls the coverage of instances in each terminal node. In our experiments, minbucket was set to 5% of the instances, in such a way that each returned LHR has a minimum level of representativeness in the dataset (i.e., $Cov(r, D) > 0.05$). This is to avoid very small LHRs that pinpoint single very hard instances. When few instances are present in a LHR, the rules can have a lower reliability. In future work, we aim to address this challenge by adopting data augmentation procedures to fill in the instance space with more instances, and potentially extracting more reliable LHRs.

Finally, in order to select the LHRs we adopted $thrE = 0.3$, which is about 10% greater than the average error. Table 1 shows the meta-rules returned by Algorithm 1. A total number of 14 meta-rules were induced. Notice that some extracted rules have only one variable in their condition statements as less relevant attributes were discarded by the rule learning algorithm. The predicted errors associated to these meta-rules ranged from 0.31 to 0.38, while the coverage values ranged from 14% to 39%. Rules are ordered by predicted error. Figure 2(a) shows the hardest LHR (Meta-rule #1, $err = 0.38$). Figure 2(b) presents the distribution of errors for the examples associated to this meta-rule, which, as expected, tend to be higher.

4.2   Filtering Meta-Rules: Algorithm 2

In this section, we discuss the results obtained by Algorithm 2 to maximize the coverage of hard examples. In this experiment, we defined a budget of 7 rules to select. The set of examples D_u was composed by all instances for which RF returned an error greater than 0.3 (i.e., the same threshold $thrE$ adopted in Algorithm 1). A total number of 98 instances were considered as hard examples in this set.

The 7 selected meta-rules were sufficient to explain most hard examples in D_u (coverage higher than 0.90). Table 2 presents the top three meta-rules filtered by Algorithm 2, which covered more than 80% of the hard examples. Meta-rule #12 (Fig. 3) covers alone 16% of the given examples. The rule's conditions bring

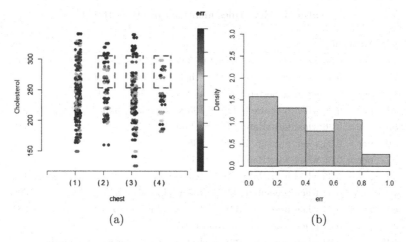

Fig. 2. LHR for Meta-rule #1 and histogram of errors. Color represents the RF error, ranging from 0 (dark blue) to 0.6 (dark red). (Color figure online)

contradictory signals about the risk of heart disease, which make the LHR challenging for RF to provide good predictions. In order to understand this conflict, we first discuss the roles of each attribute to predict the disease class. First, in the Heart dataset the patient's pressure is alone an indicator that suggests a higher risk of heart disease. In turn, the attribute electrocardiographic has three distinct values (left-ventr-hyperthophy, normal and wave-abnormal), but only the value left-ventr-hyperthophy is clearly related to a high risk of heart disease. Hence, the pressure condition in Meta-rule #12 indicates a higher chance of disease, while the condition on electrocardiographic indicates a lower risk of disease. These conflicting signals cause high errors for the RF algorithm in both classes. The average RF error in this LHR was 0.32 for instances in the absence class and 0.33 for instances in the presence class.

Meta-rule #11 defines an LHR with medium-to-high values of cholesterol, with a prevalence of the presence class (56%) over the absence class (44%). The class entropy in this LHR causes a difficulty for the RF algorithm, with higher errors in this case for the absence class. Finally, Meta-rule #4 is similar to Rule #12 in the sense that it also brings conflicting conditions. The value 'upsloping' of the slope attribute is also an indicator of lower risk of disease, while high pressure is an indicator of higher risk. The class imbalance inside LHR #4 is higher than the other LHRs, with a prevalence of the absence class (75%). In this case, average RF error was higher for the presence class (0.38).

4.3   Explaining Single Instances

Algorithm 2 can be adopted to choose an LPR to explain a single instance. This is done by setting D_u as the instance of interest. In this case, Algorithm 2 will return the most relevant LPR that covers that instance. In order to illustrate, we choose an instance for which RF returned a very high error. Specifically, we

Table 1. Meta-rules identified by Algorithm 1

N.	Meta-rule	Cov.
#1	per = 0.38 when cholesterol is 253 to 305 & chest is not asymptomatic	14%
#2	per = 0.38 when heartRate ≥134 & cholesterol is 253 to 301	20%
#3	per = 0.37 when cholesterol is 253 to 305 & oldpeak <0.95	18%
#4	per = 0.36 when pressure is 116 to 135 & slope is upsloping	21%
#5	per = 0.36 when heartRate ≥162 & oldpeak <0.05	18%
#6	per = 0.36 when pressure is 130 to 135	15%
#7	per = 0.35 when heartRate is 134 to 150	17%
#8	per = 0.35 when heartRate ≥162 & chest is asymptom. or atyp-angina	18%
#9	per = 0.34 when pressure is 116 to 135 & oldpeak <1.45	30%
#10	per = 0.34 when pressure ≥116 & cholestoral is 253 to 305	28%
#11	per = 0.34 when cholesterol is 253 to 305	32%
#12	per = 0.33 when pressure ≥116 & electrocardio is normal or waveabn	39%
#13	per = 0.31 when heartRate ≥162	34%
#14	per = 0.31 when pressure ≥116 & cholesterol <225	24%

Table 2. Meta-rules filtered by Algorithm 2

Selected meta-rule	Marginal incrimination	Marginal coverage	Absence (%)	Presence (%)	Avg. error of absence class	Avg. error of presence class
#12	0.16	0.49	61%	39%	0.32	0.33
#11	0.09	0.25	44%	56%	0.35	0.32
#4	0.03	0.07	75%	25%	0.34	0.38

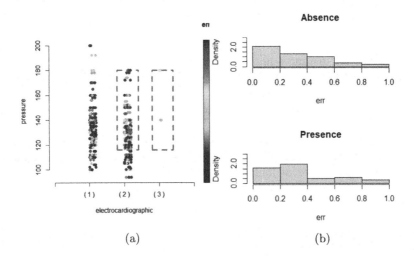

Fig. 3. LHR for the Rule #12 and histograms of errors per class.

inspect the RF performance for an old man (77-years old), who has high pressure (125) and high cholesterol (304). This patient has heart disease (presence class),

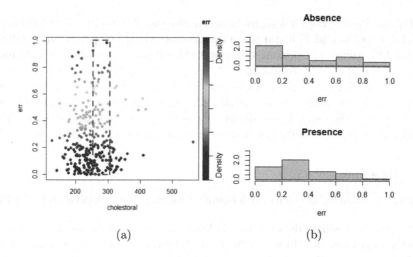

Fig. 4. LHR for the Rule #11 and histograms of errors per class.

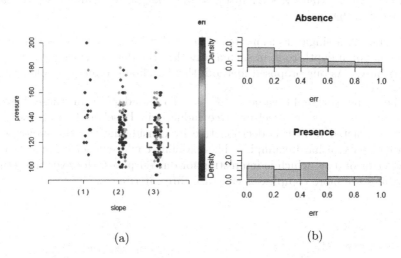

Fig. 5. LHR for the Rule #4 and histograms of errors per class.

which could be expected by considering these indicators in isolation. The error returned for RF for this patient was 0.73, which is actually a very high error. Algorithm 2 returned for this patient the Meta-rule #3 in Table 1. Again, this meta-rule has conflicting conditions associated to the risk of heart disease. The oldpeak attribute measures the heart stress during exercise, and low values indicate low risk of disease. In fact, in the Heart dataset, when oldpeak < 0.95, 106 instances belong to the absence class and 39 belong to the presence class, which corresponds to a low probability of 0.27 for the disease. On the other hand, when oldpeak ≥0.95, 44 instances belong to the absence class, while 81 belong to the presence class, i.e., probability of disease is 0.64.

By considering the cholesterol between 253 and 305, there are 38 instances in the absence class and 47 instances in the presence class, i.e., probability of disease is 0.55. Otherwise, 112 instances belong to the absence class and 73 instances in the presence class, i.e., probability of disease is 0.64. High cholesterol suggests high risk of disease. Although the patient of interest has characteristics that suggest a high risk of disease, he also has an exam result (oldpeak attribute) that suggests the opposite (oldpeak is equal to zero for this patient). These conflicting indicators make this patient a hard instance to be predicted by RF.

4.4   Baselines

Finally we evaluate two natural baselines that could be adopted to find LPRs:

- Baseline 1: a single decision tree is learned using all attributes and the rules which matched the threshold $thrE$ are returned. This baseline tends to produce more complex rules as no limit is defined in the number of attributes in each rule. Some rules for instance may not produce an LPR that can be plotted in 2D space;

- Baseline 2: a single decision tree is learned using all attributes but limiting the maximal depth to 2, in such a way that the derived rules have at most 2 attributes. As our proposed method, this baseline produces simple rules.

The same threshold value $thrE = 0.3$ is adopted in these baselines. Figure 6(a) presents the decision tree induced in Baseline 1. Two meta-rules can be extracted from this decision tree by considering the thresholds, covering about 65% of hard examples. This baseline could extract a hard area with model error of 0.38, which is interesting for our purposes. However the returned meta-rules are more complex as they use three attributes.

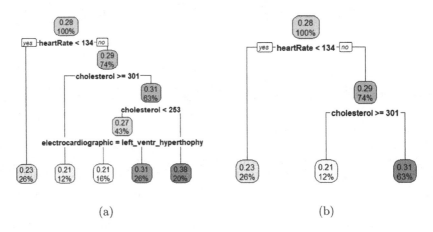

(a)	(b)

Fig. 6. Decision tree induced by (a) Baseline 1 and (b) Baseline 2. Each node contains the model error prediction and the coverage of examples in percentage.

Figure 6(b) presents the decision tree induced in Baseline 2, which is similar to Baseline 1, but limiting the depth. A single meta-rule is returned in this decision tree, covering 75% of the hard examples. The proposed method to find LHRs provided alternative explanations that covered the remaining hard examples. Additionally the proposed method could find harder hard areas (e.g., Meta-Rule #1 in Table 1, which had a model error of 0.38). The proposed method is more flexible than the baselines since it finds LPRs in a pairwise way over the attributes. As a consequence, it can find more diverse explanations for the instances provided by the user.

5   Conclusion

In this paper we proposed an original method to find local areas in the instance space that explain model performance. The proposed method is based on learning meta-rules to predict model performance. By using a reduced number of attributes, graphical explanations for the model errors are provided in the form of LPR plots. Experiments were performed in a benchmark dataset to explain the bad predictions of random forest. The proposed method could return a variety of meta-rules to explain the bad RF predictions. Additionally, the proposed greedy-search algorithm could select a reduced number of non-redundant meta-rules to cover most of the hard instances.

As future work, we intend to investigate other rule learning algorithms or subgroup discovery methods, to identify LPRs. Additionally experiments will be performed in more benchmarks and algorithms, extracting LHRs, as well as LERs. The proposed methods can be investigated as a component of a dynamic algorithm selection procedure. Also, it can be combined with data augmentation procedures in order to improve robustness of the model to the hardest instances identified in a problem of interest. Finally, the proposed methods can be applied to explain meta-models that predict instance hardness measures. In this case, the proposed methods could not only identify LHRs specific to a single model, but instance regions in a learning task that are intrinsically hard to predict.

Although we presented our proof of concept by finding LHRs for RF in a classification task, Algorithm 1 can easily be used to find LHRs and LERs for regression tasks, as long as a suitable performance function is chosen. Additionally, LERs can be used to explain feature importance, by checking which attributes appear more frequently in the meta-rules. Finally, it is possible to use our method with meta-rules defined by more than two features and still visualize the resulting higher-dimensional LPRs, using graphical tools such as parallel coordinates [8]. We intend to investigate all of these possibilities as future works.

References

1. Arruda, J.L.M., Prudêncio, R.B.C., Lorena, A.C.: Measuring instance hardness using data complexity measures. In: Cerri, R., Prati, R.C. (eds.) BRACIS 2020. LNCS (LNAI), vol. 12320, pp. 483–497. Springer, Cham (2020). https://doi.org/10.1007/978-3-030-61380-8_33

2. Brazdil, P., van Rijn, J.N., Soares, C., Vanschoren, J.: Metalearning: applications to automated machine learning and data mining. Springer, Cham (2022). https://doi.org/10.1007/978-3-030-67024-5

3. Brazdil, P., van Rijn, J.N., Soares, C., Vanschoren, J.: Metalearning in ensemble methods. In: Metalearning. Cognitive Technologies. Springer, Cham (2022). https://doi.org/10.1007/978-3-030-67024-5_10

4. Chen, Y., Silva Filho, T., Prudêncio, R.B., Diethe, T., Flach, P.: β^3-irt: a new item response model and its applications. In: The 22nd International Conference on Artificial Intelligence and Statistics, vol. 89, pp. 1013–1021 (2019)

5. Cruz, R.M., Sabourin, R., Cavalcanti, G.D.: Dynamic classifier selection: recent advances and perspectives. Inf. Fusion **41**, 195–216 (2018)

6. Cruz, R.M., Sabourin, R., Cavalcanti, G.D., Ing Ren, T.: META-DES: a dynamic ensemble selection framework using meta-learning. Pattern Recogn. **48**(5), 1925–1935 (2015)

7. Gupta, N., Eswaran, D., Shah, N., Akoglu, L., Faloutsos, C.: Beyond outlier detection: LookOut for pictorial explanation. In: Berlingerio, M., Bonchi, F., Gärtner, T., Hurley, N., Ifrim, G. (eds.) ECML PKDD 2018. LNCS (LNAI), vol. 11051, pp. 122–138. Springer, Cham (2019). https://doi.org/10.1007/978-3-030-10925-7_8

8. Inselberg, A.: The plane with parallel coordinates. Vis. Comput. **1**(2), 69–91 (1985)

9. Khiari, J., Moreira-Matias, L., Shaker, A., Ženko, B., Džeroski, S.: MetaBags: bagged meta-decision trees for regression. In: Berlingerio, M., Bonchi, F., Gärtner, T., Hurley, N., Ifrim, G. (eds.) ECML PKDD 2018. LNCS (LNAI), vol. 11051, pp. 637–652. Springer, Cham (2019). https://doi.org/10.1007/978-3-030-10925-7_39

10. Martínez-Plumed, F., Prudêncio, R.B.C., Martínez-Usó, A., Hernández-Orallo, J.: Item response theory in AI: analysing machine learning classifiers at the instance level. Artif. Intell. **271**, 18–42 (2019)

11. Merz, C.J.: Dynamical learning bias selection. In: Proceedings of Machine Learning Research Intelligence and Statistics. vol. R0, pp. 386–395 (1995)

12. Molnar, C.: Interpretable Machine Learning, 2 edn (2022). https://christophm.github.io/interpretable-ml-book

13. Moraes, J.V., Reinaldo, J.T., Ferreira-Junior, M., Filho, T.S., Prudêncio, R.B.: Evaluating regression algorithms at the instance level using item response theory. Knowl.-Based Syst. **240**, 108076 (2022)

14. Pinto, F., Soares, C., Mendes-Moreira, J.: CHADE: metalearning with classifier chains for dynamic combination of classifiers. In: Frasconi, P., Landwehr, N., Manco, G., Vreeken, J. (eds.) ECML PKDD 2016. LNCS (LNAI), vol. 9851, pp. 410–425. Springer, Cham (2016). https://doi.org/10.1007/978-3-319-46128-1_26

15. Prudêncio, R.B.C.: Cost sensitive evaluation of instance hardness in machine learning. In: Brefeld, U., Fromont, E., Hotho, A., Knobbe, A., Maathuis, M., Robardet, C. (eds.) ECML PKDD 2019. LNCS (LNAI), vol. 11907, pp. 86–102. Springer, Cham (2020). https://doi.org/10.1007/978-3-030-46147-8_6

16. Ribeiro, M.T., Singh, S., Guestrin, C.: "Why should i trust you?": explaining the predictions of any classifier. In: Proceedings of the 22nd ACM SIGKDD International Conference on Knowledge Discovery and Data Mining, pp. 1135–1144 (2016)

17. Smith, M.R., Martinez, T., Giraud-Carrier, C.: An instance level analysis of data complexity. Mach. Learn. **95**(2), 225–256 (2014)

Artificial Intelligence, Algorithmic Transparency and Public Policies: The Case of Facial Recognition Technologies in the Public Transportation System of Large Brazilian Municipalities

Rodrigo Brandão[1,2]([✉]) [iD], Cristina Oliveira[1,2] [iD], Sarajane Marques Peres[1,2] [iD],
Leôncio da Silva Junior[1] [iD], Marco Papp[1] [iD], João Paulo Cândia Veiga[1,2] [iD],
Rubens Beçak[1,2] [iD], and Laura Camargo[1,2] [iD]

[1] Universidade de São Paulo, São Paulo, Brazil
brandao-cs@usp.br
[2] Center for Artificial Intelligence (C4AI-USP), São Paulo, Brazil

Abstract. Reports of errors committed in public contexts by facial recognition systems based on machine learning techniques have multiplied. Still, these systems have been increasingly used by the Brazilian public administration. Consequently, the following key problem is established: how can errors committed by facial recognition systems be prevented or mitigated when these systems are used for the elaboration and implementation of public policies? Guided by the understanding that algorithmic transparency is key to preventing and mitigating these errors, we empirically analysed whether, or not, the Brazilian General Data Protection Law (Lei Geral de Proteção de Dados Pessoais – LGPD, in the Portuguese acronym) has been used to promote this kind of transparency in situations in which facial recognition systems are employed. We circumscribed our study to the public transportation sector of 30 large Brazilian municipalities. To gather information, we sent a questionnaire to the municipal public agencies responsible for the public transportation system with questions about how the LGPD works in this public policy area. We used the Access to Information Law to do that. Upon legal analyses, we built an algorithmic transparency scale and found that, in the sector studied, the level of transparency is "Very Low" in most municipalities. This research finding indicates that the risk of lack of control over errors made by facial recognition systems is high. It suggests that the Brazilian public administration does not know how to use the systems in question ethically, and that this lack of knowledge may apply to other Artificial Intelligence systems.

Keywords: Ethics · Algorithmic transparency · Public policy

1 Introduction

Artificial Intelligence (AI) can be defined as "a machine-based system that can, for a given set of human-defined objectives, make predictions, recommendations, or decisions

J. C. Xavier-Junior and R. A. Rios (Eds.): BRACIS 2022, LNAI 13653, pp. 565–579, 2022.
https://doi.org/10.1007/978-3-031-21686-2_39

influencing real or virtual environments. AI systems are designed to operate with varying levels of autonomy" (OECD 2019)[1]. This technology has different applications. One of them is facial recognition (FR) systems that use pattern recognition algorithms, often implemented with machine learning techniques. This subset of AI has attracted the attention of different stakeholders, especially because of the errors it makes when used in public contexts. In his mapping of such errors, Silva (2020) an HP computer in an electronics store. The MediaSmart face motion tracking feature could identify the white woman's face, but not the Black man's face". The author also reminds us that, six years later, in 2015, the automatic tagging feature of the Google Photos app tagged Black people as "gorillas". Three years later, cases like these were observed in Face++ and Microsoft's technologies, which associated negative emotions with Black people, reinforcing the social stigma that they are angry by nature; and in Google Vision, which mistook black hair for wigs.

Despite being old and resilient, errors such as those described above have not prevented FR systems from being increasingly used by the Brazilian public sector. According to the think-and-do tank Igarapé Institute (2019), the number of cases in which FR technologies are used to operationalize public policies jumped from one, in 2011, to 47, in 2019. The institution also noted that these uses are widespread among 30 municipalities and 16 states, and are concentrated in four public policy areas: transportation, security, border control, and education. A use so expressive as this demands attention as the mistakes made by FR technologies during the elaboration and implementation of State actions can make it a reproducer of social wounds, such as racism and misogyny. After all, stereotypes unfavourable to Blacks, for example, are reinforced every time a Black student is mistakenly taken as a possible fraudster of a free student pass to which he/she is legally entitled to, or when a young Black man, equally mistakenly, is confused by technology, leading the police to approach him (Brandão et al. 2022). The use of AI systems with such consequences might be considered unethical or, in Floridi et al.'s (2020) terms, contrary to the idea of "Artificial Intelligence for Social Good", according to which the design, development, and employment of AI systems should "(i) prevent, mitigate, or solve problems adversely affecting human life and/or the wellbeing of the natural world, and/or (ii) enable socially preferable and/or environmentally sustainable development". The work of Shearer et al. (2020), in turn, allows us to say that, if the use of FR systems by governments increases social problems, it is irresponsible and thus unethical.

The prevention and mitigation of errors such as those mapped by Silva (2020) and the ethical risks linked to them depend, at least in part, on the existence of information that makes it clear to citizens that algorithm-based technologies are in use and how they inform public decision-making so that citizens can demand answers and justifications about the use and operation of these technologies (Ada Lovelace Institute 2021).

---

[1] There are different definitions for the term "Artificial Intelligence", which were mapped by Russell and Norvig (2016) and Sweeney (2003). As we will see throughout this section, few works have investigated the use of AI and its applications by the Brazilian public sector. In this article, we cannot analyse each of them. We only mention that, in all of them, the AI systems analysed seem to adhere to the OECD (Organization for Economic Co-operation and Development) definition. For this reason, we also adopted this definition.

Following the UK's Central Digital and Data Office, we use the term "algorithmic transparency" to refer to this body of information. Guided by this concept, we ask: what is the level of algorithmic transparency in the use of FR systems by the Brazilian public administration? To build initial answers to this question, we analysed, specifically, the use of FR technologies in the municipal public transportation sector to combat fraud in discounts and gratuities assured by law to specific audiences, such as students and the elderly.

Two reasons justify the specificity of our analytical focus. The first is that the use of FR systems for the aforementioned purpose is numerous. Among the 47 cases identified by the Igarapé Institute (2019), 21 "are focused on fighting fraud in public transportation gratuities, especially in intercity transportation". The information gathered by Brandão and Oliveira (2021) reinforces this finding[2]. The authors focused only on the 17 Brazilian municipalities with at least one million inhabitants and searched occurrences of the term "facial recognition" in editions of electronic Official Diaries published between January 2010 and December 2020. In ten localities, they identified official uses of FR systems that were no longer happening, in progress or under discussion. In nine of them, the use of this technology was associated with the fight against fraud in discounts and gratuities in public transportation; in four, with objectives related to public safety; in two, with the operationalization of health and education policies; and in one, with actions in the area of social assistance.

Despite being significant, the use of FR technologies in municipal public transportation is not analysed enough, and this is precisely the second reason why we have chosen this public policy area for study. The literature on the use of AI applications, in general, and FR systems, in particular, by the Brazilian public sector is insipient. Besides that, it has given greater attention to public safety (Nunes 2019; Francisco et al. 2020; Nunes et al. 2022). Even studies that analyse different public policy areas – such as Coelho and Burg (2020) and Reis et al. (2021) – end up devoting greater attention to it.

Certainly, analyses of the public safety area are crucial to understanding how algorithmic transparency can be improved in the public sector. However, the lessons learned in this public policy area are not easily transferable to other sectors, for two reasons. Firstly, the type of FR systems used varies from one public policy sector to another. In the case of public transport, the technology used is the type "one-to-one". It means it seeks to authenticate the identity of specific persons – in general, in closed spaces. In public safety, in turn, the use of systems that try to identify individuals in large groups of people is not unusual. In the second place, the functioning of the numerous public policy areas and their stakeholders are different. For these reasons, there is an urgent need for the elaboration of sectoral case studies on the challenges of algorithmic transparency in the use of FR systems.

To partially fill this gap, we analysed the role of the Brazilian General Data Protection Law (Lei Geral de Proteção de Dados Pessoais – LGPD, in the Portuguese acronym) in promoting algorithmic transparency in the public transportation system. We circumscribed our study to the 30 Brazilian municipalities that meet at least one of two criteria:

---

[2] The Igarapé Institute (2019) does not make explicit in which states and municipalities FR systems have been used in the public transportation system. The survey carried out by Brandão and Oliveira (2021) partially fills this gap.

having at least one million inhabitants and/or being a state capital. We conducted this study to test two hypotheses:

> H1: The level of algorithmic transparency in the municipal public transportation system is low.
>
> H2: Two of the most important mechanisms within the LGPD to promote algorithmic transparency – obtaining free and informed consent and preparing personal data protection impact reports – are frequently neglected in the municipal public transportation system.

The remainder of this article proceeds as follows. In Sect. 2, we discuss the reasons why the LGPD can be understood as a mechanism for promoting algorithmic transparency and the reasons that underpin our hypotheses. In Sect. 3, we present the methodology used to gather information regarding the LGPD in the sector studied. In Sect. 4, we present the pieces of information gathered by us and evaluate whether, or not, each of them promotes algorithmic transparency. As we will see, this evaluation strategy allowed us to create a scale of algorithmic transparency. In the Sect. 5, we discuss the reasons why the pieces of information gathered partially confirm our hypotheses. In the Sect. 6, we conclude the article, pointing out research paths revealed by our study.

2   The LGPD as an Algorithmic Transparency Instrument

Like most countries, Brazil does not have a specific legal regulation that disciplines the use of AI applications. However, the country has the LGPD. Instituted in 2018, Law No. 13,709 "provides for the processing of personal data, including in digital media, by a natural person or legal entity of public or private law, to protect the fundamental rights of freedom and privacy and the free development of the personality of the natural person".

The expression "processing of personal data" refers to any operation performed with personal data, such as the collection, storage, and diffusion of this kind of information. The term "personal data", in turn, refers to pieces of information related to natural persons, that is, to individuals. Some types of personal data are considered sensitive. This group includes information "on racial or ethnic origin, religious belief, political opinion, membership in a union or religious, philosophical or political organization, data concerning health or sex life, genetic or biometric data, when linked to a natural person" (Article 5, Item II).

As we can see, the LGPD establishes guidelines to be observed by public and private entities interested in collecting and using individuals' data and information[3], which makes it strategic to most of the discussions about AI. The centrality of this law can be illustrated with two hypothetical examples. If a team of developers is interested in creating image recognition technologies that can identify or authenticate images of cats, they do not need to pay attention to the LGPD. However, if they are interested in developing systems that can identify or authenticate human faces, they should do

---

[3] Despite its name, the purpose of the LGPD is to protect the personal data subject, and not the personal data itself.

so, keeping in mind that images of this nature represent biometric information and are therefore sensitive data.

In general, the LGPD states that personal data can only be processed after the data subject has agreed to this. The consent must be given in a free and informed way and must be related to a specific purpose. In some cases, as in the use of personal data for the operationalization of public policies, the processing of personal data may take place without the prior specific consent of the data subject. However, even in these cases, some premises must be complied with, such as that found in Paragraph 2 of Article 11, which imposes on public agencies and entities the obligation to publicize the waiver of consent. In addition, according to Item I of Article 23, the public sector must make it clear, preferably using its official websites, whenever it processes personal data.

While obtaining consent is not mandatory for the public sector, it can be useful in promoting algorithmic transparency in the context of public policies for two reasons. Firstly, public agencies are not prohibited from collecting citizens' consent for specific purposes if they want to do so. In the second place, even when it decides to exercise its legal right not to collect individual consent, the public sector must publicize this decision, which indirectly communicates that it is using technologies whose operation depends on the use of sensitive personal data. For these reasons, the mechanism in question has the potential to increase the visibility of the use of AI systems, thus increasing the chances of citizens demanding answers and justifications about the use and operation of these technologies. Since the collection of consent can inform "users about the data processing and the working of the system upfront" and it can describe "how the AI system reaches decisions in general", it can be called a "prospective transparency" mechanism, in Felzmann et al.'s (2019) terms.

Another mechanism of the LGPD critical to the promotion of algorithmic transparency is the legal authorisation it gives to the National Data Protection Authority (Agência Nacional de Proteção de Dados – ANPD, in the Portuguese acronym) to request controllers[4] of personal (sensitive) data to prepare personal data protection impact reports (Relatórios de Impacto à Proteção de Dados Pessoais – RIPDPs, in the Portuguese acronym). These reports must comprise documents that contain "the description of the personal data processing processes that may generate risks to civil liberties and fundamental rights, as well as measures, safeguards and risk mitigation mechanisms" (Article 5, Item XVII). The ANPD may require public and private entities to prepare these reports at any time and whenever it deems them relevant. That is, when using AI technologies that rely on sensitive personal data, the public sector is not obliged to produce RIPDPs, even though – in our assessment – it should do so. After all, in its RIPDPs, the public sector can make it clear how it intends to prevent or mitigate errors commonly made by FR technologies, like the ones presented in the introductory section of this article, whenever it contracts or develops this technology. As can be seen, the elaboration of RIPDPs can also be considered a "prospective transparency" mechanism, but it can be understood as a "retrospective transparency" tool as well, as it can be useful for providing post hoc explanations and rationales of AI systems' outputs (Felzmann et al. 2019).

---

[4] A controller of personal data is any "natural or legal person [...] who is responsible for decisions concerning the processing of personal data" (Article 5, Item VI).

The fact that consent collection and the elaboration of RIPDPs are desirable but not mandatory makes us believe that the Brazilian public administration, especially at the municipal level, has not mobilized these mechanisms to increase algorithmic transparency in the context of public policies. We highlight the potential difficulty of municipalities because, in most cases, they have less budget and personnel structure to deal with Information and Communications Technology (ICT) agendas, when compared to the Union and the states (NIC.BR 2020).

3   Methods

Between August and October 2021, we requested information from 30 municipalities about the use of FR systems in the public transportation system to prevent frauds in discounts and gratuities guaranteed by law to specific audiences. The information requests were addressed to the public transportation offices of all state capitals and all municipalities with at least one million inhabitants and were anchored in the Access to Information Law (Lei de Acesso à Informação – LAI, in the Portuguese acronym). In general terms, this legislation gives any citizen the right to request information from the State about its activities. For the purposes of this paper, we just highlight that this request can be made through different channels, such as in person, by telephone, and through electronic platforms. In our research, we used the latter channel – both to request and to receive the requested information.

Our questionnaire comprised around 40 questions and was based on guides for the responsible use of AI in the public sector, such as the works of Reisman et al. (2018) and Leslie (2019). The questions were divided into six blocks: (i) general information about the use of FR systems, such as the starting date of use; (ii) general characteristics of the FR system used; (iii) measures adopted prior to the employment of the system, such as offering training to public agents to use it; (iv) measures adopted to make the use of FR aligned with the purposes of LGPD; (v) how the information generated by the FR system is supervised by humans; (vi) number of frauds identified by the system and how the holder of public transportation benefits is communicated when she/he has allegedly committed fraud. In this paper, we only address the questions and answers pertaining to block IV[5]. In the next section, we present them in conjunction with the answers provided by the municipalities and analyse whether, or not, the different pieces of information we could gather are neutral, favourable, or contrary to the promotion of algorithmic transparency. As we will see, our evaluation culminated in the creation of a scale of algorithmic transparency divided into five intervals: "Very Low", "Low", "Medium", "High", and "Very High".

Before moving on, a clarification is needed. Some public policy areas are operated directly by the public administration. This is the case with public safety. In other areas, such as municipal public transport, the Public Power is the Granting Power as it grants companies the legal right to provide public services. For this reason, we could have

---

[5] The survey described in this section gave rise to two other papers: Brandão (2022) and Brandão et al. (2022). In three of them, the description of the data collection procedure is similar. However, this is the only work in which LGPD data are presented in a systematic way and analysed in depth.

addressed our request for information to the companies that provide transportation services. We preferred to approach the municipal bodies responsible for the transportation sector as we understand these bodies integrate the Granting Authority and therefore must be aware of the operations of the concessionary companies, including the technical aspects of their operations. Our understanding is based on Article 30, Item V, of the Federal Constitution, and on Laws No. 8,987/1995 and 11,079/2004.

4   Results

By December 2021, four municipalities completely ignored our request for information (Belém, Florianópolis, Macapá, and Natal). In six municipalities, technical problems linked to the electronic platforms through which LAI is operationalized made it impossible to obtain answers (Cuiabá, São Gonçalo, Fortaleza, Recife, São Luís, and Teresina).

The other 20 municipalities responded fully or partially to the questionnaire. Out of them, four stated they do not use the technology in question (Aracaju, Belo Horizonte, Boa Vista, Vitória), one indicated that it was implementing it (Curitiba), and another one did not make it clear whether, or not, it uses FR systems in the public transportation system (Goiânia). The other 14 municipalities use FR tools to avoid fraud in discounts and gratuities: Brasília, Campinas, Campo Grande, Guarulhos, João Pessoa, Maceió, Manaus, Palmas, Porto Alegre, Porto Velho, Rio Branco, Rio de Janeiro, Salvador, and São Paulo. Among these 14 municipalities[6], eight did not submit any answer about the LGPD or submitted very incomplete answers to questions regarding this law.

The Municipal Superintendence of Transport and Traffic of Maceió offered a single answer to all the questions presented, which did not include any reference to the protection of personal data. The Municipal Secretariat of Traffic, Mobility, and Transport of Porto Velho, on the other hand, answered most of the questions. It did not provide any information, however, on the working of the LGPD. The Secretariat of Transparency and Internal Control of Palmas pointed out that questions regarding the protection of personal data could be submitted to the Agency for Regulation, Control, and Inspection of Public Services of the municipality. Finally, the city hall of Campo Grande, the Executive Superintendence of Urban Mobility of João Pessoa, the Institute of Urban Mobility of Manaus, the Municipal Superintendence of Transportation and Traffic of Rio Branco, and the Municipal Secretariat of Mobility of Salvador signalled that answers regarding questions of this nature would be up to the companies responsible for operating the municipal public transportation system. In some of these cases, we filed an appeal claiming that, as the Granting Authority, the municipality should be aware of this type of information. The measure had no effect.

Among the other six cases, the Secretary of State for Transport and Mobility of the Federal District, the Secretary of Transport and Urban Mobility of Guarulhos, and the Public Transport and Circulation Company of Porto Alegre directly answered our

---

[6] In Campinas, João Pessoa, and Rio Branco, the use of FR systems in public transportation was suspended at times during the Covid-19 pandemic, as the use of masks negatively interfered with the functioning of the technology.

questions. In the cases of Rio de Janeiro and São Paulo, the local secretariats for mobility contacted the companies involved in the operation of the transportation system and demanded them to provide us with answers. In the former, Riopar Participações S/A (successor by incorporation of Riocard Administradora de Cartões e Benefícios S/A) was responsible for doing that, whereas in the latter it was the São Paulo Transportes S/A (SPTrans). Finally, in Campinas, the Municipal Development Company of Campinas (EMDEC) alleged, initially, that answers regarding the protection of personal data should be provided by the Association of Urban Transportation Companies of Campinas (TRANSURC). After we contested their position, the agency provided the requested information.

Below we present the pieces of information collected from these six municipalities. In addition, we evaluate the answers received in order to create an algorithmic transparency score, shown in Table 1. To this end, we assigned points to each of them: we assigned (+1) to answers that refer to elements and conducts that promote algorithmic transparency; (−1) to answers that contain elements and conducts that hinder algorithmic transparency; and (0) to neutral measures and conducts.

As will see in the next section ("Discussion"), our score must be read as a preliminary indicator of the preparation of the investigated municipalities for the use of FR systems in a transparent way, and not as a definitive legal analysis of the level of transparency of each one of them, since we did not analyse a series of information relevant to the promotion of algorithmic transparency, such as the contracts between the municipality and the companies that operate public transportation services and the communication policies of these companies and of the transportation departments.

QUESTION 1: Have measures been taken to bring the transportation service into compliance with the LGPD? – All six municipalities answered "yes". However, only some of them detailed the measures taken. To these, we assigned the score (+1), even in cases where the description of the measures was generic. We adopted this conduct because we understand that every action to adapt to the LGPD favours algorithmic transparency, even if some actions are more effective than others. To the municipalities that did not detail their answers, we assigned (0), and would have scored negatively municipalities that had not reported any measure at all.

QUESTION 2: Were RIPDPs produced? – Only Guarulhos and Rio de Janeiro answered "yes". We attributed (+1) to them. Campinas and São Paulo, on the other hand, answered "no". Therefore, they scored (−1). Brasilia and Porto Alegre pointed out that they are developing this kind of measure – an answer that we considered neutral.

QUESTION 3: Who is in charge of data processing? – While Brasília, Rio de Janeiro, and São Paulo presented a specific name, Porto Alegre pointed to the name of an interim person in charge. We have assigned (+1) to the answers of these four municipalities. Campinas answered "To be defined", so it does not have a person in charge. For this reason, it scored (−1). Finally, Guarulhos answered only "An employee of Guarupass". Guarupass is the company responsible for issuing all public transportation tickets in the municipality. In our evaluation, the municipality's answer indicates that there is someone that is in charge of data processing, but his/her identity is not easy to be accessed. Therefore, we assigned (0) for Guarulhos' response.

QUESTION 4: What are the security measures adopted for privacy and data protection? – Some of the respondents specified the measures adopted:

Brasilia – Restriction, control, and registration of access to the systems of the Undersecretariat for Information Technology.

Campinas – Implementation of Kaspersky and Firewall antiviruses and employee training and qualification.

Guarulhos – Human verification performed by Guarupass' employees in cases of blocking due to improper use of the discount/gratuity cards.

Porto Alegre – Suggested direct contact with the Association of Passenger Transport Carriers, responsible for the FR system.

Rio de Janeiro – The municipality stated that "access to biometric data collected from users is restricted to a specific group of people. In addition, the images captured by the biometric facial identification system are purged from the system after 30 days, keeping for compliance with legal determinations [...] only those that configured the improper use of the gratuity benefits".

São Paulo – SPTrans said it does not share personal data with other companies and that it has monitoring and surveillance procedures over its systems and databases.

We understand that security measures for personal data protection are different from information security measures. For this reason, we assigned (+1) only to Rio de Janeiro, and (0) to Brasilia, Campinas, and São Paulo, where information security measures were highlighted. We also assigned (0) to Guarulhos, because we understand that human intervention is not, by essence, favourable or contrary to data or information security. Finally, we assigned (−1) only for Porto Alegre, which did not answer the question.

QUESTION 5: Is there a collection of consent from service users for the use of their biometric data in facial recognition systems, in the terms of the LGPD? If "yes", by what technical and legal means does the collection of consent occur? - Most municipalities provided detailed answers:

Brasilia – Did not offer any response.

Campinas – Answered affirmatively, mentioning Article 11, Item II, Letters "a", "b" and "d", in addition to the municipality's Decree No. 19,316/2016.

Guarulhos – Does not perform consent collection. It pointed out that consent is waived under the terms of Article 11, Letter "g", of LGPD, but it informs the data subject about the need for the collection and its purpose (prevention of fraud linked to gratuity cards).

Porto Alegre – Pointed out that the waiver of consent is supported by Article 11, item II, Letter "b". However, in some situations, the user must fill out a consent form by electronic means. The municipality did not specify what these situations are.

Rio de Janeiro – Does not collect consent. Mentioning Article 11, Item II, Letter "g", of the LGPD, the municipality answered that the prevailing legal understanding does not oblige the performance of this operation.

São Paulo – Indicated that only the images of beneficiaries – such as students and the elderly – are collected and that, in the benefit request term, they are informed about the need for biometric data collection.

Considering that the collection of consent is not mandatory, Rio de Janeiro scored (0). Only in Campinas this operation is actively and deliberately carried out, which made us assign (+1) to the municipality. We also assigned this score to Porto Alegre and São Paulo because on some occasions consent is obtained in these two cities, and also to Guarulhos, where the data subject is informed about the necessity and purpose of the collection of his/her biometric data. As for Brasilia, we granted (−1) due to the lack of response.

QUESTION 6: What are the available mechanisms for the user to exercise the rights of: i) access to the personal data used, ii) revocation of consent? – While it can be argued that consent is not mandatory for the use of biometric data in the public transportation system, the right of access is a common condition regardless of the legal basis used by the municipalities. For this reason, we assigned (−1) only to Porto Alegre, which informed us that the mechanisms in question are under construction. The other municipalities scored (+1) for having presented mentions – albeit generic, in some cases – of channels such as e-mail, telephone, and ombudsmen.

QUESTION 7: Is user data shared with third parties? If "yes", which ones? – We received the following answers:

Brasilia – The answer presented was contradictory and confusing. The Secretariat of Transportation and Mobility sent four different documents, each one with a different date: 08/23/2021, 08/25/2021, 08/26/2021, and 08/27/2021. In the documents dated 08/23/2021 and 08/27/2021, the public entity gave the following answer: "No, according to Article 3, of Ordinance No 15, of 04/30/18, the biometric records […] will be used by operators exclusively for the implementation of the Facial Biometric Compatibility Verification, and the data cession to third parties is forbidden, as well as its commercialization, in any way, without the consent of the Granting Authority". In the document dated 08/26/2021, the answer was: "Yes, they are shared". And finally, in the document of 08/25/2021, the agency presented the following note: "Bus operators and BRB mobility (which deals with the operation of the ticketing system). Police and control bodies have access to the data".

Campinas – The sharing is done with the Granting Public Power, by virtue of a public contract.

Guarulhos – The data is shared with the Public Power. Besides, it is shared with companies that use transport vouchers. Without giving more details, the municipality mentioned that there is also sharing when there is a request to meet a specific and justified purpose, observing the anonymization of the data when possible.

Porto Alegre – The FR system is under the responsibility of the Association of Passenger Transport Companies, and the data collected are shared with the Public Company for Transportation and Circulation, responsible for regulating and supervising the activities related to traffic and transport in the city.

Rio de Janeiro – The granting of gratuities involves secretariats from different areas, such as the Secretariats of Education and Health. The data collected by Riopar Participações S/A is shared with employees designated by these secretariats.

São Paulo – Answered that it does not share, pointing out that "only in the case of a judicial request may there be sharing with the requesting authority".

We assigned (−1) only to Brasilia, because the confusion in its answers made them inconclusive. As for São Paulo, we assigned (+1), because, in general, there is no sharing of personal data – except in cases where there is a judicial request. To the other four municipalities, we assigned (0) because we understand that data sharing, in itself, has no impact on transparency, as long as the competencies and responsibilities are correctly attributed in the data processing chains. Since we did not have access to more information about these chains, we chose to classify as neutral the information provided by Campinas, Guarulhos, Porto Alegre, and Rio de Janeiro.

Table 1. Score of algorithmic transparency.

	Brasília	Campinas	Guarulhos	Porto Alegre	Rio de Janeiro	São Paulo
Question 1	1	0	1	1	0	1
Question 2	0	−1	1	0	1	−1
Question 3	1	−1	0	1	1	1
Question 4	0	0	0	−1	1	0
Question 5	−1	1	1	1	0	1
Question 6	1	1	1	−1	1	1
Question 7	−1	0	0	0	0	1
Total	1	0	4	1	4	4

Due to the way we built our score, it may vary from (−7) to (+7). Based on this variation, we built Table 2, which comprises an algorithmic transparency scale. We distributed the 30 municipalities studied along it. As we have seen, in most cases we reserved (−1) for the lack of response. By this logic, a score (−7) should be assigned to the 18 municipalities in which we did not get any information on the functioning of the LGPD. They are: Belém, Florianópolis, Macapá, and Natal (they completely ignored our request for information); Cuiabá, São Gonçalo, Fortaleza, Recife, São Luís, and Teresina (their electronic platforms devoted to the LAI presented technical problems that prevented us from obtaining information); and Campo Grande, João Pessoa, Maceió, Manaus, Palmas, Porto Velho, Rio Branco, and Salvador (they did not provide any information about the LGPD or provided too incomplete information about this law). For the reasons registered at the beginning of this section, our score could not be applied to six municipalities: Aracajú, Belo Horizonte, Boa Vista, Vitória, Curitiba, and Goiânia.

Table 2. Algorithmic transparency scale – Quantitative.

Scores	(−7) (−6) (−5)	(−4) (−3) (−2)	(−1) (0) (1)	(2) (3) (4)	(5) (6) (7)	Does not apply
Level	Very low	Low	Medium	High	Very high	
Municipalities	18	0	3	3	0	6

5 Discussion

Among the 30 municipalities investigated, the level of algorithmic transparency is "Very Low" in 18 of them (60% of the total). This information seems to confirm our first hypothesis ("The level of algorithmic transparency in the municipal public transportation system is low"), thus revealing a scenario in which the risk of lack of control over the prevention and mitigation of errors commonly made by FR systems is high.

In our second hypothesis, obtaining informed consent and elaborating RIPDPs are presented as independent variables capable of explaining, at least partially, the level of algorithmic transparency (dependent variable). However, in the 18 cases above, we did not get any information about these variables, and in the six cases in which FR systems are not used or are being implemented, this analysis could not be applied. In other words, we had only six cases (Brasilia, Campinas, Guarulhos, Porto Alegre, Rio de Janeiro, and São Paulo) in which we could statistically test the relationship between these variables. Because it is a low N, we would not obtain statistically relevant results. For this reason, it does not seem possible to confirm or refute our second hypothesis. Even so, the answers of the municipalities allow qualitative evaluations about the obtaining of free and informed consent and the elaboration of RIPDPs.

Regarding the first of these two mechanisms, it is worth mentioning that the answers received seem to indicate that the fact that obtaining consent is not mandatory discourages the adoption of this mechanism. Only Campinas resorts to it voluntarily, while three municipalities (Guarulhos, Porto Alegre, and Rio de Janeiro) presented interpretations of the LGPD to justify the reasons why they do not collect or do not need to collect citizens' consent to process their biometric data. We did not investigate whether, or not, there is publicization of the consent waiver in these three municipalities. If not, algorithmic transparency is threatened, especially in Rio de Janeiro, given that in Porto Alegre consent is obtained in some cases, and that in Guarulhos data subjects are informed about the purpose of the collection of their biometric data.

Concerning the elaboration of RIPDPs, it should be noted that this mechanism is used in two municipalities (Guarulhos and Rio de Janeiro) and is in the process of being adopted in other two (Brasília and Porto Alegre). This is not an entirely discouraging scenario, especially when we consider that there are considerable uncertainties associated with this instrument. Firstly, its potential to reduce risks, such as those caused by FR system errors, is not yet known. For the time being, the bet is on its ability to do so, but there is as yet no empirical evidence to provide any certainty about this. In addition, there is no consensus – either in Brazil or in other countries – about which risks should be considered high and/or unacceptable, and what is the most appropriate methodology to define them. Finally, it is not yet known how strict the ANPD will be regarding the request of RIPDPs.

Finally, our survey has identified eight cases (Campinas, Campo Grande, Guarulhos, João Pessoa, Manaus, Porto Alegre, Rio Branco, and Salvador) in which public agencies are not able to provide any information on how concessionary companies deal with the LGPD or seem to face some level of difficulty to do so. It suggests that coordination problems among essential actors for the construction and operation of algorithmic transparency may have been ongoing in the municipal public transportation system, which indicates that future studies must investigate whether, or not, concession contracts can be a hindrance to the promotion of algorithmic transparency.

As can be seen, our research findings are useful for enriching the discussions on general legal requirements related to algorithmic transparency. However, even in this realm, our contribution is limited, as important legal elements were not included in our analysis. We have not studied, for instance, the communication policies of the concessionary companies and of the transportation departments on the use of FR systems. Besides that, our approach to "transparency" presents it as the mere disclosure of information, which is not sufficient for making individuals understand how AI systems work and how their personal data are used in the functioning of this technology. In other words, disclosing information is not enough to make citizens able to challenge algorithmic technologies (Ananny and Crawford 2018). Nevertheless, this is a necessary step towards that direction – and that is why it is important to analyse the legal requirements related to algorithmic transparency, as we did above.

6 Conclusion

Based on information about the operation of LGPD in municipal public transportation, we constructed a preliminary algorithmic transparency score and scale regarding the use of FR systems. Despite the limitations of our initiative, it was able to point out that, in general, the level of algorithmic transparency is low in the sector studied, which increases the chances that mistakes made by FR technologies are not challenged by citizens or other stakeholders. In addition, our analysis identified how municipalities position themselves in relation to each other regarding algorithmic transparency, which gives rise to the following question: how have Guarulhos, São Paulo, and Rio de Janeiro managed to prepare themselves to use the LGPD in a reasonably successful way in promoting algorithmic transparency? Among different explanatory variables that can be analysed, we will devote attention in future studies to one in particular: the institutional capacity of municipalities, measured upon different information, such as the municipal budget dedicated to areas related to ICTs and the existence, or not, of planning and management tools in departments of these areas. Research of this kind is still rare. One of the few exceptions is the index of government readiness for the responsible use of AI prepared by Shearer et al. (2020). Even so, the authors work with indicators that could hardly be used to study the Brazilian municipal reality.

If the variable in question proves to be relevant, municipalities with higher institutional capabilities should present higher levels of algorithmic transparency, which leads us to an initial and speculative questioning: if São Paulo and Rio de Janeiro – which are among the richest municipalities in Brazil – were not able to achieve higher scores in our algorithmic transparency score, what will be the reality of the poorer municipalities? In addition, we need to investigate whether the low levels of algorithmic transparency observed by us in the municipal public transportation sector are present in other public policy areas and when other AI applications are used. Answers to such questions are essential to understanding how the Brazilian public administration can and should prepare itself to use AI systems – such as FR systems – to successfully prevent, mitigate and solve social problems.

Acknowledgements. The authors are grateful to Professor Glauco Arbix for his invaluable comments on previous versions of this work.

Funding. The authors of this work would like to thank the C4AI-USP and the support from the São Paulo Research Foundation (FAPESP grant #2019/07665-4) and from the IBM Corporation.

References

Ada Lovelace Institute: Algorithmic accountability for the public sector – learning from the first wave of policy implementation. Technical report. Ada Lovelace Institute, AI Now and Open Government Partnership. London, New York, and Washington, DC (2021)

Ananny, M., Crawford, K.: Seeing without knowing: limitations of the transparency ideal and its application to algorithmic accountability. New Media Soc. **20**(3), 973–989 (2018)

Brandão, R., Oliveira, J.L.: Reconhecimento facial e viés algorítmico em grandes municípios brasileiros. In: Anais do II Workshop sobre as Implicações da Computação na Sociedade, pp. 122–127 (2021)

Brandão, R.: Artificial intelligence, human oversight, and public policies: facial recognition systems in Brazilian cities. In: Proceedings of the 35th Canadian Conference on Artificial Intelligence (2022)

Brandão, R., et al.: Reconhecimento facial, viés algorítmico e intervenção humana no transporte municipal (2022). Article under peer-review

Central Digital and Data Office's homepage. https://www.gov.uk/government/collections/algorithmic-transparency-standard. Accessed 25 May 2022

Coelho, J., Burg, T.: Uso de inteligência artificial pelo poder público. Technical report. Transparência Brasil, Rio de Janeiro (2020)

Felzmann, H., et al.: Transparency you can trust: transparency requirements for artificial intelligence between legal norms and contextual concerns. Big Data Soc. **6**(1), 1–14 (2019)

Floridi, L., Cowls, J., King, T.C., Taddeo, M.: How to Design AI for Social Good: Seven Essential Factors. Sci. Eng. Ethics **26**(3), 1771–1796 (2020). https://doi.org/10.1007/s11948-020-00213-5

Francisco, P., et al.: Regulação do reconhecimento facial no setor público: avaliação de experiências internacionais. Instituto Igarapé and DataPrivacy BR Research. Technical report, Rio de Janeiro and São Paulo (2020)

Leslie, D.: Understanding artificial intelligence ethics and safety: a guide for the responsible design and implementation of AI systems in the public sector. Technical Report. The Alan Turing Institute, London (2019)

LGPD's official link. http://www.planalto.gov.br/ccivil_03/_ato2015-2018/2018/lei/l13709.htm. Accessed 25 May 2022

Igarapé Institute's 2019 homepage on facial recognition in Brazil. https://igarape.org.br/infografico-reconhecimento-facial-no-brasil/. Accessed 25 May 2022

Nunes, P.: Novas ferramentas, velhas práticas: reconhecimento facial e policiamento no Brasil. Retratos da Violência: cinco meses de monitoramento, análises e descobertas – Junho a Outubro 2019. Technical report. CeSEC – Centro de Estudos de Segurança e Cidadania, Rio de Janeiro (2019)

NIC.BR.: ICT Electronic Government – Survey on the Use of Information and Communication Technologies in Brazilian Public Sector. Technical Report, São Paulo (2020)

Nunes, P., et al.: Um Rio de olhos seletivos – uso de reconhecimento facial pela polícia fluminense. Digital book. CeSEC – Centro de Estudos de Segurança e Cidadania, Rio de Janeiro (2022)

OECD's homepage on legal instruments. https://legalinstruments.oecd.org/en/instruments/oecd-legal-0449

Reis, C., et al.: Relatório sobre o uso de tecnologias de reconhecimento facial e câmeras de vigilância pela administração pública no Brasil. Technical report. LAPIN – Laboratório de Políticas Públicas e Internet, Brasília (2021)

Reisman, D., et al.: Algorithmic Impact Assessments: A Practical Framework for Public Agency Accountability. Technical Report. AI Now, New York (2018)

Russell, S., Norvig, P.: Artificial Intelligence: A Modern Approach. Pearson Education Limited, Malaysia (2016)

Shearer, E., et al.: Government AI Readiness Index 2020. Technical Report. IDRC & Oxford Insights, Canada & London (2020)

Silva, T.: Visão computacional e racismo algorítmico: branquitude e opacidade no aprendizado de máquina. Revista da ABPN **12**(31), 428–448 (2020)

Sweeney, L.: That's AI?: a history and critique of the field. Technical Report. Carnegie Mellon University, School of Computer Science, CMU-CS-03-106, Pittsburgh (2003)

Resource Allocation Optimization in Business Processes Supported by Reinforcement Learning and Process Mining

Thais Rodrigues Neubauer$^{(\boxtimes)}$, Valdinei Freire da Silva,
Marcelo Fantinato, and Sarajane Marques Peres

School of Arts, Sciences and Humanities, University of Sao Paulo,
Arlindo Béttio, 1000, Sao Paulo, SP 03828-000, Brazil
{thais.neubauer,valdinei.freire,m.fantinato,sarajane}@usp.br

Abstract. Resource allocation to execute business processes is increasingly crucial for organizations. As the cost of executing process tasks relies on several dynamic factors, optimizing resource allocation can be addressed as a sequential decision process. Process mining can aid this optimization with the use of data from the event log, which records historical data related to the corresponding business process executions. Probabilistic approaches are relevant to solve process mining issues, especially when applied to the usually unstructured and noisy real-world business processes. We present an approach in which the problem of resource allocation in a business process is modeled as a Markovian decision process and batch reinforcement learning algorithm is applied to get a resource allocation policy that minimizes the cycle time. With batch reinforcement learning algorithms, the knowledge underlying the event log data is used both during policy learning procedures and to model the environment. Resource allocation is performed considering the task to be executed and the resources' current workload. The results with both Fitted Q-Iteration and Neural Fitted Q-Iteration batch reinforcement learning algorithms demonstrate that this approach enables a resource allocation more adherent to the business interests. Per the evaluation we performed on data of a real-world business process, if our approach had been used, up to 37.2% of the time spent to execute all the tasks could have been avoided compared to what is represented in the historical data at the event log.

Keywords: Reinforcement learning · Process mining · Resource allocation · Business processes

1 Introduction

Allocating resources to execute business processes is an essential task which has proven benefits if done properly [14], especially when it comes to human resources

This study was partially supported by CAPES (Finance Code 001) and FAPESP (Process Number 2020/05248-4).

J. C. Xavier-Junior and R. A. Rios (Eds.): BRACIS 2022, LNAI 13653, pp. 580–595, 2022.
https://doi.org/10.1007/978-3-031-21686-2_40

[26]. Determining the proper decisions to supply effective resource allocation, considering the main goal of the business process, is not a trivial task. There are several aspects to consider when designing the problem to be solved and when evaluating its results, such as resource costs, workload, capacity, idleness, relations, and expertise. In addition, we need to understand how these aspects relate to business main interest and how each decision made affects the others [10]. Accordingly, the resource allocation optimization in this context should be addressed as a sequential decision process, as the execution cost of a given task is not only related to a single decision point [10].

Resource allocation can be enabled by the knowledge revealed from the event logs' analysis provided by process mining [7]. Process mining has enabled satisfactory results for low noise, structured business processes [21], rare characteristics in real-world business processes. In this regard, probabilistic models hold an inspiring promise for business processes context and process mining [21] due to the high level of flexibility offered. Probabilistic models better designate issues related to complexity resulting from noise and business process structuredness and allow to consider sequences and temporal order of events [11].

Markovian decision processes are applied to model sequential decision problems [10]. There are a number of algorithms to find optimal policies for this kind of decision process, e.g., dynamic programming. However, defining whether transition probabilities are sufficiently reliable to the resource allocation context is not trivial due to the interactivity involved in a specific work item being assigned to a specific resource at a given time [24]. Reinforcement learning helps to overcome issues related to these algorithms by supplying a straightforward learning framework based on interactions in the environment [10].

Huang et al. [10], Firouzian, Zahedi and Hassanpour [7], and Liu et al. [16] addressed resource allocation optimization in business process with Q-learning algorithm and presented results showing this technique leads to effective resource allocation. This paper addresses the optimization of resource allocation for a real-world business process supported by reinforcement learning based on process mining concepts. The business process is modeled as a Markovian decision process and two batch reinforcement learning algorithms are used – *Fitted Q-Iteration* (FQI) [6] and *Neural Fitted Q-Iteration* (NFQ) [19] – to get a resource allocation policy capable of minimizing the business process cycle time, which refers to the mean duration of its cases. By applying batch reinforcement learning algorithms, we are using the event log as a source of historical data for learning the resource allocation policy. This highlights the contribution of the approach discussed herein. As to the best of our knowledge, previous works used event logs only for problem modeling, not to provide inputs for policy learning.

This paper is structured as follows: Sect. 2 introduces basic concepts of process mining, Markovian decision processes, and batch reinforcement learning. Section 3 discusses related work. Section 4 details the real-world business process explored, particularly how resource allocation for this process was modeled as a Markovian decision process to apply the batch reinforcement learning. Section 5 reports experiments results and analyses. Finally, Sect. 6 concludes the paper.

2   Preliminaries

This section presents the basic concepts used in this paper.

2.1   Business Process Management and Process Mining

According to Weske [23], a *business process* corresponds to a set of tasks (or activities), which are performed in coordination in an organizational and technical environment, and together achieve a business goal. A business process can be subject to analysis, improvement, and enactment through the lens of the business process management and its lifecycle.

An effective management of business process guarantees consistent results and allows value aggregation to the organization and its clients [5]. Traditionally, business processes are enacted manually, guided by the knowledge of company staff, and aided by organizational regulations and procedures. However, more benefits can be obtained if automation is adopted to coordinate the business processes' tasks and if process mining is employed to support lifecycle tasks with factual and historical data about process executions, registered in an event log. Event logs are sequential files in which events related to business process tasks executions are registered.

The goal of the business process mining is to enhance the business process comprehension through knowledge extraction from event logs [1]. According to van der Aalst [1], an event $e \in \mathcal{E}$, where $\mathcal{E}$ is the event universe, corresponds to the execution of a task $\tau \in \mathcal{T}$, where $\mathcal{T}$ is the set of tasks that occur in a given business process. Events are commonly associated with non-mandatory attributes (e.g., task, timestamp, resources, cost, and transaction type). Transaction type refers to the task's life-cycle associated with the event. Possible values for such an attribute are "start", "complete", "abort" and "suspend". The values "start" and "complete" are crucial in the modeling discussed in this paper.

A case $c \in \mathcal{C}$, where $\mathcal{C}$ is the case universe, comprises a sequence of events e^c. A case refers to a process instance which is an actual execution of the process. All cases must have a mandatory attribute called trace, denoted by σ. A trace is a finite sequence of events $\sigma \in \mathcal{E}^*$ and $1 \leq i < j \leq |\sigma| : \sigma(i) \neq \sigma(j)$. An event log L is a set of cases $L \subset \mathcal{C}$ and for any $c_1, c_2 \in L$, if $c_1 \neq c_2$ then $\delta(e^{c_1}) \cap \delta(e^{c_2}) = \emptyset$, where δ is an operator for converting a sequence to a multiset. A classifier is a function mapping the attributes of an event onto a label or value. In this paper, we adopt $e.[attribute\_notation]$ to show a classifier function. Thus, $e.\tau$ shows the task name associated with the event e.

Playing a central and strategic role for organizations, process mining uses the event logs data to: (a) discover process models (*discovery*), (b) replay and analyze data from the process models or related to them (*conformance*), and (c) find information that enables process improvements (*enhancement*) [1,9,17]. In this paper, we are especially interested in enabling process improvements in resource allocation optimization.

2.2   Markovian Decision Process and Batch Reinforcement Learning

Markovian decision process is applied in different contexts to model sequential decision-making problems [18]. A Markovian decision process is defined by a tuple $<S, A, P, C>$, where:

- S is the set of possible states of the environment;
- A is the set of possible actions;
- $P{:}S{\times}A{\times}S{\rightarrow}[0,1]$ is a state transition function that defines a probability distribution over the states to each pair (s, a) – $P(s'|s, a)$ is the probability of ending in state s' given that the system starts in state s and takes action a; and
- $C{:}S{\times}A{\rightarrow}\mathbb{R}^+$ is the real value immediate cost function and $C(s, a)$ is the expected immediate cost for taking action a in state s.

To use the data registered in the event log as examples of the process behavior under analysis, the *Fitted Q-Iteration* batch reinforcement learning algorithm was applied [6]. This algorithm is part the off-policy with approximation class of reinforcement algorithms [15,22]. The set F built in the algorithm is based on the tasks execution history H registered on the event log data L, as described in Sect. 4. In this algorithm, the $\hat{Q}$ function estimation is refined at each learning iteration using supervised learning and the inputs F and $\hat{Q}$, both calculated in the earlier iteration. Experiments were also conducted using the algorithm *Neural Fitted Q-Iteration* [19], which uses artificial neural networks – more specifically the Resilient Propagation (RPROP) algorithm [20] – as the regression strategy for the $\hat{Q}$ estimation.

3   Related Work

Resource allocation has been widely studied in job-shop scheduling, a classic problem in operation management area [3]. A job-shop model is a decision-making process which allocates limited resources over time to perform a set of jobs to be processed on a set of machines [25]. Reinforcement learning has been applied in job-shop scheduling [2,25]. Zhang and Dietterich [25] applied temporal difference algorithm TD(λ) and conducted an evaluation on problems from NASA space shuttle payload processing task. The results show efficiency in constructing high-performance scheduling. Aydin and Ozteme [2] applied Q-learning algorithm to deal with dynamic scheduling - when the scheduler does not have detailed information about the jobs. The results show better performance than other approaches based on dispatching rules.

In business process contexts, the dynamic nature of a business process execution need to be considered when allocating resources: (i) the execution path is determined at the runtime and different processes may have different paths within particular execution scenarios; (ii) resource behaviors affect the resource allocation, since resources' preferences, capability, and availability, are also dynamic [10]. Huang et al. [10] present a mechanism to model resource allocation as a Markovian decision process and proposed to solve it with a Reinforcement Learning Based Resource Allocation Mechanism (RLRAM). RLRAM aims to minimize long-term cost, measured by cycle time, i.e. time to execute all tasks of a process case. The experiments were conducted for a synthetic and for a real-world business process related to a radiology examination. The results show the proposed approach may indeed improve the current state of business process management and outperform well-known heuristics and hand-coded strategies.

Liu et al. [16] presented an approach applying process mining by the use of an event log information to model the business process as Markovian decision process. The effect of social relation among the resources was also considered. The analyzed business process was the one described in the event log of the BPI Challenge 2012, the same selected for this paper. The Q-Learning algorithm was applied and the results show improvement on the minimization of the business process' cycle time. Yaghoubi and Zahedi [24], and Firouzian, Zahedi and Hassanpour [7] presented approaches similar to this with the addition of considering task similarities and workload balance. Experimental results on the same business context show the proposed method leads to the reduction in cycle time, compared to some other well-known algorithms.

Koschmider et al. [13] proposed an organizational model that considers resources' competences, skills and knowledge to assign roles to process activities based on this model and hidden Markov model. The event log is pointed to be used to obtain the hidden Markov model transition matrix by counting the frequency of direct role transition during execution. A toy example of application of the proposed approach show its effectiveness.

To the best of our knowledge, there is no approach using real-world event log data to directly support the reinforcement learning algorithm in policy learning steps, or directly build an actual real-world based evaluation environment. Therefore, we propose to use this data both during policy learning procedures and to model the environment within batch reinforcement learning. In addition, we describe a realistic environment for testing the approach in which we are able to compare the policy-based decisions to the historical decisions.

4   Problem Definition

This section presents the real-world business process under analysis, the respective event log and preprocessing procedures adopted, the Markovian decision process modeling that defines the resource allocation optimization problem, the details about the reinforcement learning application, and how the optimization results were evaluated through a simulation procedure.

4.1   Real-World Business Process Context

To evaluate how the application of reinforcement learning algorithms can help in the resource allocation optimization problem in a real-world business process, the event log presented in the BPI Challenge 2012[1] [4] was used. This event log has 262,200 events in 13,087 cases and corresponds to a process of a Dutch finance institution on loan requisitions opened by its clients. An excerpt of the event log and the business process model discovered by applying Apromore[2] process mining tool in the event log is presented in Fig. 1.

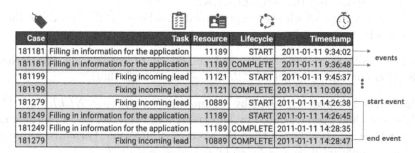

Case	Task	Resource	Lifecycle	Timestamp
181181	Filling in information for the application	11189	START	2011-01-11 9:34:02
181181	Filling in information for the application	11189	COMPLETE	2011-01-11 9:36:48
181199	Fixing incoming lead	11121	START	2011-01-11 9:45:37
181199	Fixing incoming lead	11121	COMPLETE	2011-01-11 10:06:00
181279	Fixing incoming lead	10889	START	2011-01-11 14:26:38
181249	Filling in information for the application	11189	START	2011-01-11 14:26:45
181249	Filling in information for the application	11189	COMPLETE	2011-01-11 14:28:35
181279	Fixing incoming lead	10889	COMPLETE	2011-01-11 14:28:47

(a) Excerpt of the event log

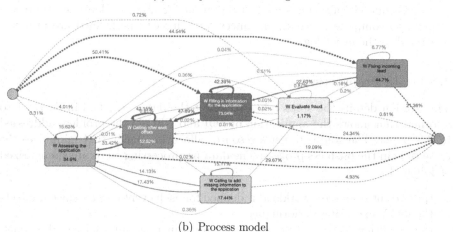

(b) Process model

Fig. 1. (a) Excerpt of the event log presented in the BPI Challenge 2012; (b) Process model discovered by applying Apromore tool on the event log provided in BPI Challenge 2012. Rectangles represent tasks, arcs represent transitions between two sequential tasks in a case, colors and textures represent frequency of each task and transition in the event log, and circles represent the beginning and the end of the process instances. (Color figure online)

Available Information. Each line in L corresponds to an event e described by the following attributes:

---

[1] www.win.tue.nl/bpi/2012/challenge.
[2] https://apromore.org/.

- $e.\tau$: task $\tau \in \mathcal{T}_L$ executed, where $\mathcal{T}_L$ is the set of tasks that occur in the event log L;
- $e.\psi$: human resource $\psi \in \Psi_L$ that executed the task $e.\tau$, where Ψ_L is the set of resources that exists in the event log L;
- $e.lc$: the lifecycle stage in $\{START, COMPLETE\}$, which identifies if the event is registering, respectively, the start or the end of a task execution;
- $e.ts$: the timestamp the event occurred;
- $e.c$: the case c to which the event is related, where $c \in \mathcal{C}_L$ and $\mathcal{C}_L$ is the set of cases in L;

To use only pertinent, stable, and reliable data considering the designed application context, a subset $L' \subset L$ was selected, such that:

- the events had been executed by a human resource: $e.\psi \neq NULL \; \forall \, e \in L'$;
- the events had been executed by an *experienced* resource, i.e., a resource that presents occupation larger than the average occupation. Formally, a resource ψ_i is experienced if $|\{e|e \in \mathcal{E}_L \text{ and } e.\psi = \psi_i\}| > |\mathcal{E}_L|/|\Psi_L|$.

We assume all cases registered in the event log are complete, i.e., they correspond to complete processes instances and all tasks performed are properly registered in the event log.

4.2   Data Preprocessing

This section details the steps taken to adapt the data described in Sect. 4.1 to a format that allows the application of FQI and NFQ in the problem addressed in this paper. The goal is to get a resource allocation policy that minimizes the cycle time in the business process. The data preprocessing steps are summarized as followed:

- the current resources' workload is defined for each event e in L', since it affects the task's execution performance;
- the execution history H is built by disregarding the information about lifecycle, thus considering only one event to represent an activity execution;
- the history H is split in: (i) one segment to be used by the learning procedures, conducted through the application of FQI and NFQ (cf. Sect. 4.4); and (ii) the other segment to be used for evaluation conducted through a simulation procedure (cf. Sect. 4.5).

Because of factors such as preference, competence and cooperation among resources, the task's execution performance may be affected by the workload of the resource executing it. The raised hypothesis in this scenario is the greater the workload of the resource ψ during the execution of the task τ, the longer will take for ψ to conclude the execution [10]. Thus, the resources' workload set was included as an event attribute $e.wl = (\mathrm{WL}_f(\psi, e.ts) \ \forall \ \psi \in \Psi)$, where $\mathrm{WL}_f(\psi, e.ts)$ is a function that determines the ψ workload at moment $e.ts$:

$$WL_f(\psi, e.ts) = \begin{cases} FREE, |\psi.wl| = 0 \\ LOW, |\psi.wl| \neq 0, |\psi.wl| \leq avg\_wl \\ HIGH, |\psi.wl| > avg\_wl \end{cases} \tag{1}$$

where $|\psi.wl|$ is the number of tasks being executed by ψ at $e.ts$ and $avg\_wl = \dfrac{\sum_{\psi \in \Psi} |\psi.wl|}{|\Psi|}$. $e.wl(\psi)$ denotes the workload of ψ.

The execution history H was defined from the identification of the pairs of start ($e1$) and end ($e2$) events of each task execution in L:

- $e2.lc = COMPLETE$, $e1.lc = START$;
- $e2.\tau = e1.\tau$, $e2.c = e1.c$;
- $\nexists e \in L$, such that $e.ts < e2.ts$, $e.c = e1.c$, $e.\tau = e1.\tau$ and $e.lc = COMPLETE$.

To each task execution $h \in H$ and its corresponding identified pair of start ($e1$) and end ($e2$) events in L', the duration $h.d$ was defined as: $h.d = e2.ts - e1.ts$. Besides $h.d$, h is described by:

- $h.\tau = e1.\tau = e2.\tau$, composing the set of tasks T_H in H;
- $h.\psi = e1.\psi = e2.\psi$, composing the set of resources Ψ_H in H;
- $h.c = e1.c = e2.c$, composing the set of cases C_H in H;
- $h.wl = e1.wl$, composing the set of resources workload lists WL_H in H;
- $h.ts = e1.ts$.

The final quantity of $h \in H$ elements in H is $12,100$. As for business process cases $i \in I$, we have $3,480$. 30 resources $\psi \in \Psi$ were selected and 6 different kinds of tasks $\tau \in T$ are registered in the H. Regarding the lists of workloads $wl \in \mathrm{WL}$, $4,783$ different lists were obtained.

In order to apply H for learning (i.e. application of the reinforcement learning algorithms FQI and NFQ) and also for evaluating the resulting policy, H was split in a training set $H_{tr} \subseteq H$ and a test set $H_{te} \subseteq H$. This split is based on the timestamp attribute and the two subsets are defined as: $h.ts < T \ \forall \ h \in H_{tr}$ and $h.ts \geq T \ \forall \ h \in H_{te}$, where T is the timestamp corresponding to the upper quartile (Q3) of the distribution of timestamps of task executions ($h.ts$) in H.

Resulting from this split, the quantity of task executions in each subset is: $|H_{tr}| = 9,086$ and $|H_{te}| = 3,014$. The split was made based only on the timestamp, thus some cases $c \in C$ have part of their executions in the training subset H_{tr} and another part in the test subset H_{te}.

4.3   Definition of the Markovian Decision Process

Regarding mapping the decision process of resource allocation into a Markovian decision process, the set of states S, the set of actions A and the cost function C were defined:

- Let S be the set of states, $S \subseteq T_H \times \mathrm{WL}_H$. For instance, a particular state $s = (\tau, wl)$, where τ is a task and wl is the list of resources' workloads at the moment that execution of the task τ started. Thus, an execution $h \in H$ can be mapped to a specific state at a certain time point, such that $\tau = h.\tau, wl = h.w$.
- Let A be the set of actions, $A \subseteq \Psi_H$, such that $a = \psi \, \forall \, a \in A, \, \psi \in \Psi_H$.
- Let C be the cost function, $C(s, a) = d$, where $s = (\tau, wl)$, $a = \psi$ and d is the execution duration of τ by ψ. Thus, an execution $h \in H$ can be mapped to a specific state $s = (\tau = h.\tau, wl = h.wl)$, a specific $a = \psi = h.\psi$ and a specific cost $C = h.d$.

4.4   Application of Reinforcement Learning Algorithms

The cost function for $\mathcal{C}_{H_{tr}}$ minimizes total cycle time of items in:

$$H_{tr}: \min \sum_{c \in \mathcal{C}_{H_{tr}}} \sum_{h \in H_{tr}|h.c=c} h.d.$$

FQI and NFQ algorithms were applied using H_{tr} as input to estimate $\hat{Q}$. To initialize the values of $\hat{Q}$, the greater value of the distribution of the costs $h.d$ in H_{tr} was used ($243.19\,\mathrm{s}$) and the value adopted for the discount factor γ was 0.9. The stop condition determined for the reinforcement learning algorithms was to reach ten hours of execution or $\Delta = \sum_{s \in S, a \in A} ||\hat{Q}_N(s, a) - \hat{Q}_{N-1}(s, a)|| < \theta$, where θ is the first quantile of $h.d$ distribution in H_{tr}, $\theta = 63$.

To apply a regression strategy, the pairs of state-action (s, a) must be converted to a vector. This representation was built in two steps: applying the concept of one hot encoding to the task τ of state S composition to get a binary matrix in which the features (columns) are the different values in T and the cells are marked with 1 if the corresponding column represents the task τ or 0, otherwise; applying a similar idea to the workload wl of the state S composition, resulting in a matrix which features identify the resource ψ and each cell is valued with the resources' workload in the event converted to the scale $0, 1, 2$ in which $0 = \mathrm{FREE}$, $1 = \mathrm{LOW}$ and $2 = \mathrm{HIGH}$.

Figure 2 presents an illustration of the vectorization for two states, $s1$ and $s2$. For $s1$, the task is "Fixing incoming lead" and the workload for the resources "10910" and "11003" are LOW and HIGH, respectively, and FREE for the other resource present at the illustration. For the second state $s2$, the task is "Filling in information for the application" and the workload for the resource "11003" is LOW and FREE for the other resources visible in the illustration.

State	Tasks				Resources				
	Calling to add missing information to the application	Fixing incoming lead	Filling in information for the application	...	10910	10929	11003	11009	...
s1	0	1	0	...	1	0	2	0	...
s2	0	0	1	...	0	0	1	0	...

Fig. 2. Illustration of the vector representation used as input for regression, which estimate $\hat{Q}$

All experiments were conducted in *Python*, using *Google Colaboratory*[3] plat-form[4]. To apply FQI, the regression strategies adopted were Linear Regression and Random Forest and for applying NFQ, the neural network strategies *Multilayer Perceptron* (MLP) and *RPROP*. For Linear Regression, Random Forest and MLP, we used the implementations available on *ScikitLearn* library. For RPROP, we used the implementation of Neupy library[5]. The parameters values applied that were different from those indicated by the used libraries' implementations of the regression strategies are listed in column "Parameters" of Table 1.

4.5   Simulation of Policy Application

For each estimation of $\hat{Q}$ got from the iterations of the reinforcement learning algorithm, using a certain regression strategy and a certain combination of its parameters, a resource allocation policy π was defined. To evaluate the obtained policies, we developed a simulation. The simulation consists of applying the policy over H_{te} whenever the state s corresponding to h can be found in the policy, i.e., the policy is applied when possible; the original resource assignment and task duration are considered otherwise. For this simulation, we considered:

- The resources' workload when simulation starts is the same as in H_{tr} end.
- The workload $h.wl$ associated with each task execution h during the simulation may be different of the original workloads in H_{te}, as it depends on the assignments to resources; the simulation aim is to assign task executions to resources according to the policy.
- The sequence of executions is not changed by the task's execution performance.
- When the state s can be found in the policy, we have a policy indication a (resource ψ) to the task execution h; applying this action may affect the duration $h.d$, estimated by: $h.d = \dfrac{\sum_{i=1}^{|H'|} h'_i.d}{|H'|}$, where $H' = \{h'|h' \in H_{tr}, h'_i.\tau = h.\tau, h'_i.\psi = h.\psi, h'_i.\mathrm{wl}(\psi) = h.wl(\psi)\}$.
- The change in $h.d$ may affect only the start moment of the upcoming executions in $h.c$.

5   Results and Discussions

This section is dedicated to present and discuss the results of all the 13 combinations of batch reinforcement learning algorithms, regression strategies to estimate the $\hat{Q}$ function and regression strategies' parameters tested, as listed

---

[3] Colaboratory virtual machine with CPU Intel Xeon processor, 2.30 GHz of frequency, two cores, RAM 12 GB and 25 GB of HD.

[4] Developed code available at: https://github.com/pm-usp/RL-resource-allocation.

[5] More information about the implementations of Linear Regression, Random Forest, MLP and RPROP used are available at: https://bit.ly/3imZQRx, https://bit.ly/3hJqMMi, https://bit.ly/3xNfGvh and https://bit.ly/36Gr6VR.

in Table 1. The graph in Fig. 3 shows that the reinforcement learning algorithms achieved training stability and convergence for all combinations since Δ converged towards the defined threshold θ (cf. Sect. 4.4). The results presented in Table 1 give us confidence to analyze the policy behavior on H_{te}.

Table 1. Algorithms of reinforcement learning, parameters of the regression strategies and results got from the simulation over H_{te}. **Parameters**: values applied to regression strategy parameters when they differ from the default of the implementation used; **N**: number of iterations of the algorithm; **Runtime**: algorithm runtime; π: rate of events in H_{te} for which the simulator found a corresponding state in the policy; **Opt.**: rate of task executions in H_{te} for which the application of the policy π decreases cycle time comparing to the original; **Time reduction**: estimation in minutes of the optimization achieved per case. LR = Linear Regression and RF = Random Forest.

Algorithm and model	Parameters	N	Runtime	π	Opt.	Time reduction
FQI-LR	-	30	07 m 29 s	0.501	0.143	14.658
FQI-RF	trees = 10	11	03 m 02 s	0.403	0.138	4.556
FQI-RF	trees = 20	15	03 m 56 s	0.422	0.143	2.376
FQI-RF	trees = 100	14	04 m 57 s	0.421	0.145	8.161
NFQ-MLP	layers = (10,10), func. = relu	22	07 m 50 s	0.508	0.164	12.947
NFQ-MLP	layers = (50,50), func. = relu	26	14 m 12 s	0.455	0.145	12.597
NFQ-MLP	layers = (100), func. = relu	24	31 m 26 s	0.445	0.151	9.829
NFQ-MLP	layers = (50,50), func. = sigmoid	5	03 m 12 s	0.317	0.119	−10.342
NFQ-MLP	layers = (100), func. = sigmoid	7	04 m 13 s	0.321	0.127	−4.016
NFQ-RPROP	layers = (50,50), func. = relu	13	04 m 54 s	0.478	0.144	9.147
NFQ-RPROP	layers = (100), func. = sigmoid	12	03 m 46 s	0.520	0.152	14.598
NFQ-RPROP	layers=(500,500), func. = sigmoid	12	04 m 05 s	0.524	0.153	15.562

The last three columns of Table 1 concern the results obtained from the application of the simulator over H_{te}. The values in column π **applied** show a lower performance for MLP with sigmoid activation function. The difficulty faced by the resulting policy in these cases reflects negatively on the final optimization, as seen in **cycle time reduction**. Regarding the percentages of tasks optimized by the application of the policy (cf. column **Optimized**), there is regularity for all combinations of algorithms and regression strategies.

In general, 11% to 16% of tasks were executed more efficiently when the resource is allocated by the policy as opposed to the original allocation presented in H_{te}. In addition, as showed in column **Cycle time reduction**, some combinations of reinforcement learning algorithm, regression strategy and parametrization stand out for achieving higher cycle time reduction, from 14 to 15 minutes: FQI with Linear Regression and NFQ with RPROP (with sigmoid activation function). Finally, we state the cycle time reduction obtained is certainly associated with the success in the policy's application, since the greatest optimization

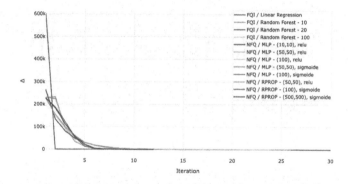

Fig. 3. Δ values over reinforcement learning iterations

gains occur from policies which have high π **applied** rate, i.e. high correspondence between states available in the policy and states in the simulation.

An analysis based on the total time of executions' duration also reveals the usefulness of applying the policy for effective allocation of resources in the business process registered in L. For the executions in H_{te}, the original time taken to perform all tasks is 24.46 days. With application of the policy, a reduction of up to 9.1 (37%) days was registered. Considering all the performed experiments, the mean reduction was 4.25 days, with a standard deviation of 4.55.

For further analysis related to resources' workload, one of the resulting policies was selected. By the Occam's Razor principle, the policy resulting from the application of FQI with Linear Regression was selected since it is the most simple combination in terms of algorithms' complexity while it is among the top tree results considering the cycle time reduction presented in Table 1. Regarding the workload impact in decision-making, the experiments showed the policy application increased the workload associated with some resources.

Figures 4(a), 4(b) and 4(c) show the workload distribution among the resources considering H_{tr}, H_{te} and the result of the simulation over H_{te}, respectively. The distributions reveal seven and 13 people did not perform more than one task at a time on H_{tr} and on H_{te}, respectively. As for the simulation application, only three people did not perform more than one task at a time. On the other hand, in both H_{tr} and H_{te}, only one person was overloaded with concurrent tasks, surpassing a 0.5 rate. After applying the policy, two people who had a high workload started to work with maximum workload, implying they remained in this condition for the entire working period covered in H_{te}.

Although the policy application increased the workload of the resources, the reduction in tasks' duration noticed by the cycle time reduction reflected in the total resources' working time. Originally, the total working time in H_{te} ranged from 15 m 32 s to 2 d 13 h. After policy application, the range is 12 m 12 s to 19 h 19 s.

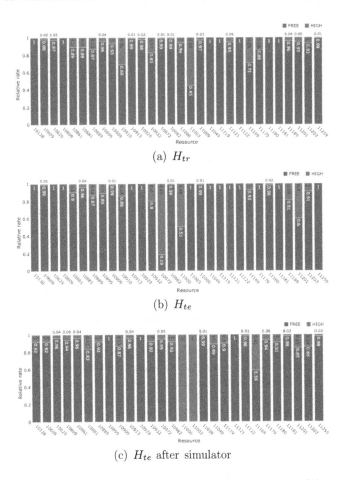

(a) H_{tr}

(b) H_{te}

(c) H_{te} after simulator

Fig. 4. Workload distribution among resources in (a) training and (b) test subsets of H and in (c) the test subset after applying the simulator.

As for the relation between workload level and task executions' duration, Fig. 5 shows the workloads *HIGH* in H_{tr} diverge from our premise "the greater the resource's workload, the greater the time needed for the resource to complete the task". Meanwhile, the cycle time resulting from the simulation minimizes the hypothetical harmful effects of a work backlog.

The results confirm the hypothesis of Justin and Wickens [12] and posted by Liu et al. [16]: the workload of the human will affect the human's ability, namely Yerkes-Dodson Law of Arousal, which shows that a worker will take less time to execute a task if he/she is under some work pressure. However, if the pressure is too high, the worker's performance may degrade." Given that, we state the policy is effective to allocate the right resource for the right task.

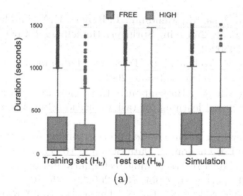

Fig. 5. Duration by workload: the resources performance by workload in each subset of H and in the simulation

6 Conclusions

In this paper, we presented an approach for resource allocation optimization in business process context. The contribution brought by our approach is the application of batch reinforcement learning algorithms and process mining principles to allow information about past process executions to be used in learning a policy capable of indicating how to effectively use resources. Experiments with real-world business process were conducted, and the results attested the suitability and efficiency of the approach. The downside of the policy is that it does not evenly distribute the workload among the resources and does not consider restrictions related to resource limitations, particularities of tasks or organization's rules and regulations. However, as recently pointed out by Folino and Pontieri [8], reinforcement learning offers a solid basis for exploiting feedbacks coming from business environments, to continuously learning in a context-awareness condition. Given this and the results discussed in this paper, future works should consider extend resource allocation optimization with reinforcement learning strategies to embrace context, legal and ethical requirements to make the policy fairer and more realistic.

References

1. van der Aalst, W.M.P.: Process Mining: Data Science in Action, 2nd edn. Springer, Heidelberg (2016). https://doi.org/10.1007/978-3-662-49851-4
2. Aydin, M., Öztemel, E.: Dynamic job-shop scheduling using reinforcement learning agents. Robot. Auton. Syst. **33**(2), 169–178 (2000)
3. Baker, K.R.: Introduction to Sequencing and Scheduling, 1st edn. Wiley, Hoboken (1974)
4. van Dongen, B.: BPI challenge 2012. 4TU.ResearchData.Dataset (2012). https://doi.org/10.4121/uuid:3926db30-f712-4394-aebc-75976070e91f

5. Dumas, M., de La Rosa, M., Mendling, J., Reijers, H.A.: Fundamentals of Business Process Management, 2nd edn. Springer, Heidelberg (2018). https://doi.org/10.1007/978-3-662-56509-4

6. Ernst, D., Geurts, P., Wehenkel, L.: Tree-based batch mode reinforcement learning. J. Mach. Learn. Res. **6**(18), 503–556 (2005)

7. Firouzian, I., Zahedi, M., Hassanpour, H.: Cycle time optimization of processes using an entropy-based learning for task allocation. Int. J. Eng. **32**(8), 1090–1100 (2019)

8. Folino, F., Pontieri, L.: Ai-empowered process mining for complex application scenarios: survey and discussion. J. Data Semant. **10**, 77–106 (2021)

9. Garcia, C.d.S., et al.: Process mining techniques and applications - a systematic mapping study. Expert Syst. Appl. **133**, 260–295 (2019)

10. Huang, Z., van der Aalst, W., Lu, X., Duan, H.: Reinforcement learning based resource allocation in business process management. Data Knowl. Eng. **70**(1), 127–145 (2011)

11. Jaramillo, J., Arias, J.: Automatic classification of event logs sequences for failure detection in WfM/BPM systems. In: Proceedings of the IEEE Colombian Conference on Applications of Computational Intelligence, pp. 1–6. IEEE (2019)

12. Justin, G.H., Wickens, C.D.: Engineering Psychology and Human Performance. Pearson, Tokyo (1999)

13. Koschmider, A., Yingbo, L., Schuster, T.: Role assignment in business process models. In: Daniel, F., Barkaoui, K., Dustdar, S. (eds.) BPM 2011. LNBIP, vol. 99, pp. 37–49. Springer, Heidelberg (2012). https://doi.org/10.1007/978-3-642-28108-2_4

14. Kumar, A., van der Aalst, W., Verbeek, H.: Dynamic work distribution in workflow management systems: how to balance quality and performance. J. Manag. Inf. Syst. **18**(3), 157–194 (2002)

15. Levine, S., Kumar, A., Tucker, G., Fu, J.: Offline reinforcement learning: tutorial, review, and perspectives on open problems. CoRR abs/2005.01643 (2020)

16. Liu, X., Chen, J., Ji, Yu., Yu, Y.: Q-learning algorithm for task allocation based on social relation. In: Cao, J., Wen, L., Liu, X. (eds.) PAS 2014. CCIS, vol. 495, pp. 49–58. Springer, Heidelberg (2015). https://doi.org/10.1007/978-3-662-46170-9_5

17. Maita, A.R.C., Martins, L.C., Paz, C.R.L., Rafferty, L., Hung, P.C.K., Peres, S.M.: A systematic mapping study of process mining. Enterp. Inf. Syst. **12**, 1–45 (2017)

18. Puterman, M.L.: Markov Decision Processes. Wiley, Hoboken (1994)

19. Riedmiller, M.: Neural fitted Q iteration – first experiences with a data efficient neural reinforcement learning method. In: Gama, J., Camacho, R., Brazdil, P.B., Jorge, A.M., Torgo, L. (eds.) ECML 2005. LNCS (LNAI), vol. 3720, pp. 317–328. Springer, Heidelberg (2005). https://doi.org/10.1007/11564096_32

20. Riedmiller, M., Braun, H.: A direct adaptive method for faster backpropagation learning: the RPROP algorithm. In: Proceedings of the IEEE International Conference on Neural Networks, vol. 1, pp. 586–591 (1993)

21. da Silva, G.A., Ferreira, D.R.: Applying hidden Markov models to process mining. In: Proceedings of the 4th Iberian Conference on Information Systems and Technologies, pp. 207–210. AISTI (2009)

22. Sutton, R.S., Barto, A.G.: Reinforcement Learning: An Introduction. MIT Press, Cambridge (2018)

23. Weske, M.: Business Process Management: Concepts, Languages, Architectures, 2nd edn. Springer, Heidelberg (2007). https://doi.org/10.1007/978-3-642-28616-2

24. Yaghoubi, M., Zahedi, M.: Resource allocation using task similarity distance in business process management systems. In: Proceedings of the 2nd International Conference of Signal Processing and Intelligent Systems, pp. 1–5 (2016)
25. Zhang, W., Dietterich, T.G.: A reinforcement learning approach to job-shop scheduling. In: Proceedings of the 14th International Joint Conference on Artificial Intelligence, vol. 2, pp. 1114–1120. Morgan Kaufmann, Burlington (1995)
26. Zhao, W., Pu, S., Jiang, D.: A human resource allocation method for business processes using team faultlines. Appl. Intell. **50**(9), 2887–2900 (2020). https://doi.org/10.1007/s10489-020-01686-4

Exploitability Assessment with Genetically Tuned Interconnected Neural Networks

Thiago Figueiredo Costa$^{(\boxtimes)}$ and Mateus Tymburibá

Departamento de Computação,
Centro Federal de Educação Tecnológica de Minas Gerais,
Belo Horizonte MG 30510-000, Brazil
thiagofigcosta@gmail.com, mateustymbu@cefetmg.br

Abstract. With the ever-increasing digitalization of data, processes, and services, people rely more than ever on computerized systems. These systems are practical and efficient, so people and institutions have come to depend on them substantially. However, computerized systems are not perfect, they can have cybersecurity vulnerabilities that allow hackers to attack these systems. Since those systems have so many responsibilities, they can also cause serious consequences when invaded by malicious people. Considering the huge volume of newly discovered vulnerabilities, it is impossible to create and apply security patches for all of them in a timely manner. Hence we need to prioritize security patches. The state of the art for vulnerability prioritization is the CVSS, an incomplete system, which has no scientific evidence and considers only the severity of a vulnerability, without taking into consideration its probability of exploitation, greatly hampering its efficiency. In this paper we propose V-REx, an open-source software that uses multiple neural networks tuned by a modified genetic algorithm to predict the exploitation probability of a vulnerability and therefore helping to prioritize them. Our experiments unveil that V-REx was more efficient than CVSS in prioritizing the vulnerabilities published in year of 2019.

Keywords: Vulnerability prioritization · Exploitability assessment · Neural networks · Neural architecture search · Genetic algorithm

1 Introduction

Cybersecurity vulnerabilities keep being discovered every day, in fact, hundreds of them are discovered monthly [10]. By exploiting vulnerabilities, hackers can get access to private data, shutdown systems (Denial of Service), hijack and erase data, discredit a person or a institution, cause financial loss, apply scams, among other consequences. Nowadays, multiple essential services, companies,

Thanks to the Universidade Federal de Minas Gerais for providing hardware.

people and governments rely on digital services, increasing the harm that people with bad intentions can cause. Creating and applying patches that can fix a vulnerability, and therefore prevent attacks, is very expensive. When creating a patch, several specialized programmers must identify and fix the vulnerability. Applying an update patch for a software or library can also be disruptive, specially on big networks. Applying a patch may cause systems to go down or get unstable, workstations may become unavailable, and incompatibilities may occur, making it necessary for other software or hardware to be updated, causing files to be incompatible or making it impossible to update in the short term.

The identification date of a vulnerability is when someone discovers that a flaw exists and then reports the discovery to the developers or to the affected security company. The details of the vulnerability are kept secret, being known only to a few security professionals until the vulnerability is published. Exploits are published on average 23 days after Common Vulnerabilities and Exposures (CVE) publication. This time is even shorter if the vulnerability can be exploited remotely [20], since hackers strive harder due to higher returns. Because of that, it is crucial to precisely assign a repair priority to vulnerabilities, helping to apply a fix timely. Due to the risks of attacks, the costs of applying patches, the urgency of developing a fix and the huge amount of new vulnerabilities, an effective way of prioritizing which vulnerabilities should receive a patch first is needed. There are not enough resources to patch every single vulnerability [8].

To classify vulnerabilities, we depend on the Common Vulnerability Scoring System (CVSS) [3], which is the state of the art. CVSS classifies vulnerability according to its severity, without considering the odds of exploitation [9], the vulnerability lifecycle [13], the software popularity and without taking into consideration inter-dependant vulnerabilities. CVSS lacks of transparent documentation, it is computed using a biased and subjective method that does not have scientific evidence [2,26], relying on incomplete data and subjective opinions of experts [9]. A CVSS based strategy (CVSS+) consists of patching every vulnerability above a certain severity threshold. Nevertheless, mitigating a vulnerability above a severity score is not better than mitigating them randomly [8].

In 2021, on Brazil, hackers managed to take down the ConecteSUS system, a health platform for Brazilian citizens and health professionals [1]. This system had a significant role on the combat of the COVID-19 pandemics for providing vaccination certificates, which were a mandatory requirement for traveling, working on office, accessing places and other activities. Unfortunately, due to the attack this system went down for more than 12 days. Another example is the case of WannaCry, a virus designed for the EternalBlue vulnerability. WannaCry affected the entire world, even hospitals, causing an estimated damage of 4 billion dollars across the world [14]. The EternalBlue vulnerability was already known two years before the attack [24], and therefore could be avoided if we had an efficient vulnerability prioritization system.

This work presents and discusses the Vulnerabilities' Risk of Exploitation System (V-Rex), an open-source software whose the goal is to prioritize software vulnerabilities patching by predicting the odds of exploitation of a vulnerability, i.e. Exploitability Assessment, using only public domain data from multiple sources ranging from 1999 to 2020. Our approach uses genetically tuned inter-connected neural networks, and was able to overcome CVSS based strategies. V-REx and its raw database are available under the MIT license through https://github.com/thiagofigcosta/V-REx-v2.

2   Related Work

On [2] the authors use a union of public and private vulnerability datasets, aim-ing to predict the presence of an exploit. They extracted textual features from the datasets to feed a linguistic processing neural network (fastText). The out-put of this network alongside with other non-textual features went through a LightGBM Classifier to compute the final result. The approach of processing different types of features on multiple machine learning models, is similar to our interconnected neural networks. A grid search with 10-fold cross-validation was done to determine hyperparameters of their model. To determine our hyperpa-rameters, we opted for a Genetic Algorithm, which has a better optimization performance, specially when the search space increases [17].

The Exploit Prediction Scoring System (EPSS)[1] was created by [9] to predict the odds of a vulnerability to have an exploit. They achieve this goal by using a logistic regression on data from private cybersecurity companies. They executed a manual feature selection and an Elastic Net regularization on the data. Our proposal uses Neural Networks to predict the presence of an exploit instead of the logistic regression, which is limited to linear problems. Public data besides allowing the reproducibility of the experiments, also avoid the bias that comes within private data [2], that, in addition, is very sensitive and hard to obtain. Their cross-validation method was the rolling forecasting origin technique [7], which consists of a rolling windows on the train dataset with a validation fold on the end of the window. This technique preserves the chronological aspect of the data, thus avoiding the drift phenomenon [2].

The authors of [8] used private data from more than 100,000 corporate net-works alongside with public domain data to predict the existence of an exploit using machine learning. They used tags extracted from the textual description of vulnerabilities together with other non-textual features to classify the vul-nerabilities through gradient boosted trees, generated with Extreme Gradient Boosting. Due to the sparseness of some features, they performed 5-fold stratified cross-validation, repeated 5 times, to limit possible over-fitting in the training, in contrast with our approach which is the rolling forecasting origin technique. V-REx uses 319 multi-word expressions as inputs for the machine learning model in comparison with the 191 tags used by [8], this higher dimensionality makes the model harder to train, but it may encode more information on the problem.

Instead of predicting the likelihood of an exploit, the authors of [6] catego-rized vulnerabilities using Deep Neural Networks. Their goal was slightly differ-ent from the present work, however, their means to achieve it are quite similar. They used the Term Frequency - Inverse Document Frequency (TF-IDF) [22]

---

[1] See: https://www.kennaresearch.com/tools/epss-calculator/.

to extract meaningful words from vulnerabilities' descriptions an then use this data for the classification. Our strength in comparison to their work is to use multi-word expressions instead of single words. Using the Rapid Automatic Keyword Extraction (RAKE) algorithm [23] with the Smart Stop World List [4] it is possible to extract meaningful expressions from the vulnerability description, and then compute the TF-IDF for those multi-word expressions.

3   Data

We crawled data ranging from 1999 to 2020, inclusively, from seven public data sources of cybersecurity vulnerabilities: (i) Common Vulnerabilities and Exposures (CVE) database from MITRE[2] (ii) Common Weakness Enumerations (CWE) database from MITRE[3] (iii) National Vulnerability Database from United States (NVD)[4] (iv) Exploit database[5] (v) Common Attack Pattern Enumeration and Classification (CAPEC) database from MITRE[6] (vi) Open Vulnerability and Assessment Language (OVAL) database[7] and (vii) CVE Details database[8] Using data from all those multiple sources is important, since there is no complete and homogeneous vulnerabilities and exploits dataset [19]. By combining multiple sources, we can achieve a more robust dataset. The vulnerability prioritization problem is very hard due to lack of ground truth and more accurate data [12], since we can only trust on the positive labels, because we might not know of the existence of exploits for negative labels yet.

As mentioned before, we extracted multi-word expressions from the textual features. Then we performed a lemmatization of the text and filtered manually meaningless expressions. The use of multi-word expressions is very important for expressions such as "Buffer Overflow", since they do not have the same meaning and importance when analyzed word-by-word. A number ranging from 0 to 1, was assigned to each of those expressions. This number was obtained from the TF-IDF formula $\left(w_{i,j} = tf_{i,j} * log\left(\frac{N}{df_i}\right)\right)$, where for an expression i in a document j we have tf as the number of occurrences of an expression, df as the number of documents containing the expression and N as the total amount of documents [6]. We also extracted numeric features, date intervals, vendors' data, CVSS scores and the amount of CVE references. All the features were normalized using the MinMax normalization. Features with null values, low variance and sparse values were removed from the dataset.

The dataset ended up with 878 features and a label, which represents the presence of an exploit, which can be extrapolated to the likelihood of the existence of an exploit for that particular vulnerability. The features were grouped

---

[2] See: https://cve.mitre.org/.
[3] See: https://cwe.mitre.org/index.html.
[4] See: https://nvd.nist.gov/.
[5] See: https://www.exploit-db.com/.
[6] See: http://capec.mitre.org/.
[7] See: https://oval.cisecurity.org/repository/search.
[8] See: https://www.cvedetails.com/.

into five groups: (i) CVSS/ENUM features; (ii) Description (text) features; (iii) Reference features; (iv) Vendor features; and, (v) Other features. The amount of features in each group is respectively 43, 319, 88, 393 and 35.

4   Enhanced Genetic Algorithm

The hyperparameters have enormous influence on the results of a Neural Network. Its search is not a trivial problem [25]. Therefore, the algorithm used to optimize them has a direct influence on the results. We proposed changes to the Standard Genetic Algorithm (SGA) to improve the search. This modified algorithm was named as Enhanced Genetic Algorithm (EGA). One of the differences from the SGA is the use of mitochondrial DNA (mtDNA). The advantage of using the mtDNA is to increase the diversity, preventing the algorithm converge to a local minima. This idea was inspired by [25]. On the first generation, each individual receives a random mtDNA, which is an UUID. Then, before making the crossover between two individuals, their mtDNA is compared. If their mtDNA are the same, it means that they are related to each other, and therefore the reproduction cannot occur. Else, as in nature, the children inherit the mtDNA only from the genetic material of the mother. If the mtDNA is not reset at some point, all individuals would be related to each other and no more crossover would occur. To avoid this, the mtDNA is reset when the odds of two randomly selected individuals not be relatives is below the mtDNA Reset Threshold.

To avoid genetic drift as in [15], we also adopted an age structure in the genetic algorithm, maintaining diversity. A given individual of some generation will only leave the population when they die, instead of giving place to their descendants after the crossover. During the crossover process a random number r between zero and one is generated for each gene. Through this number the gene of the child is defined as a weighted average between the values of the parents' genes, where the weights are r and $1 - r$. Another advantage of this technique happens when tuning neural networks, since an individual can have its weights trained and reutilized over several generations. This can help finding optimum points in cases where the network's hyperparameters are good, but were not trained sufficiently yet. Differently from the work of [15] we include in the DNA a maximum age gene, that as all the other genes goes through the process of mutation and crossover, causing the death age of individuals to be different among them and somewhat related to their parents. At each generation, the age of the individual increases, and when it becomes equal to the individual's maximum age, it dies and is removed from the population.

We also added a gene to control the number of children that a couple of individuals can have. The amount of children that a couple will have depends on the quantity coded in their DNA. The value extracted from the parents' DNA depends on a random portioning that decides how much of the gene to take from each of the parents. The final amount of progeny is then defined by the combination of the parents' DNA multiplied by a population limit variable which reduces the amount of progeny to be generated as the population size approaches a limit, following a Logistic Population Growth curve. When the limit is still exceeded the worst individuals are removed from the population. Another technique that has been used to increase diversity and escape local minima is to make the random number generator have a small chance of generating larger numbers than usual, causing the mutation process to rarely make an individual's genes to go very far from its current location on the search space.

In [11] the authors developed a technique to optimize hyperparameters, which is similar to genetic algorithms, the population-based training. On this technique when a neural network has a low performance it does a process of exploit, in which it copies parameters and weights of another network, and then an explore process that is similar to mutation, where it differs from the copied individual randomly. Based on these concepts the EGA adapted these methods so that the worst individuals at the end of each generation have a random chance, called recycling rate, to copy the best elite individuals and then mutate them. A great advantage of this method is that the exploitation process happens before the death of the individual, avoiding it to exist in the population for a long time if it has a low performance.

After so many modifications to increase diversity, two methods of reducing diversity were also implemented to maintain a balance. The first method is to keep alive, even after the maximum age has been reached, a tiny part of the elite of the population, so that there will always be good individuals to be exploited by the worse ones. The second feature, was the creation of the Hall Of Fame, a structure that eternalizes a specific amount of the best individuals among all generations. At the end of each generation the Hall of Fame is updated with a copy of the best individuals, so that it is always possible to recover the best ones. It is important to point out that the individuals in the Hall of Fame do not suffer any kind of modification, their only purpose is to be a record of a configuration that had good results.

4.1 Optimizing the Enhanced Genetic Algorithm

The proposed EGA has more parameters to tune than the SGA. Since the EGA is a new algorithm, there is no reference values to use. Therefore we built a search space Π for the EGA parameters based on the well-known SGA parameters. Then, we used the SGA to find a good parameter set $\mathbf{p}$, by creating a population of 100 individuals for 50 generations, with a mutation rate of 0.2 and a crossover rate of 0.6. Each of the SGA individuals represents two EGA experiments, one optimizing the EggHolder function (EHf), given by Eq. 1, and the other optimizing the Easom function (EAf), given by Eq. 2. Both functions are good benchmarks for optimization since they are hard to find the global minima and they have several local minima. They also have a fast computation time which is important since this population of populations can rapidly increases the execution time. The goal is to tune Eq. 3 with a fitness weighted utility function $u : \Pi \rightarrow \mathbb{R}$ given by Eq. 4, where RPD (Result Percentage Distance) is given by the ratio between the optimized value and the known global minima, and SPD (Speed Percentage Distance) is given by the inverse of the ratio between maximum amount of generations and the difference of the maximum amount of generations and the generation that found the best value.

$$EHf = -(x_2 + 47)\sin\left(\sqrt{\left|x_2 + \frac{x_1}{2} + 47\right|}\right) - x_1\sin\left(\sqrt{|x_1 - (x_2 + 47)|}\right) \quad (1)$$
$$x_i \in [-512, 512] \; \forall i = 1, 2$$

$$EAf = -\cos x_1 \cos x_2 \exp\left(-(x_1 - \pi)^2 - (x_2 - \pi)^2\right) \quad (2)$$
$$x_i \in [-100, 100] \; \forall i = 1, 2$$

$$\max u(\mathbf{p}) \quad s.t. : \; \mathbf{p} \in \Pi \quad (3)$$

$$u(\mathbf{p}) = -\frac{5 \cdot RPD_{EHf}(\mathbf{p}) + SPD_{EHf}(\mathbf{p}) + 5 \cdot RPD_{EAf}(\mathbf{p}) + SPD_{EAf}(\mathbf{p})}{12} \quad (4)$$

The EGA population start size was set to 200 and the maximum generations to 80, the search space for other the parameters can be found on Table 1 as well as the their regions, i.e. search boundaries, and the optimized values. The fitness of the optimized parameter set was 99 out of 100, indicating that the EGA managed to get close to the global minima in most cases. The next experiments using the EGA will be carried on using the optimized parameters.

Table 1. Search space and optimized results for EGA parameters

Parameter	Region	Optimized parameter
mtDNA reset threshold	[1, 50]	35.897
Max age	[2, 10]	5
Max children	[2, 6]	4
Mutation rate	[0.05, 0.35]	0.200
Recycle rate	[0.05, 0.35]	0.332
Crossover rate	[0.5, 0.9]	0.829
Elite individuals ratio	[0.01, 0.2]	0.111
Individuals to recycle ratio	[0.01, 0.2]	0.190

5   Neural Network Base Models

We proposed three base Neural Network models to classify the vulnerabilities: (i) Standard Neural Networks (SNN); (ii) Enhanced Neural Networks (ENN); and (iii) Interconnected Enhanced Neural Networks (IENN). The SNN is basically a regular Neural Network with dense layers, between each dense layer there is also a dropout layer. There are always 878 input neurons on the SNN, one for each feature, and a single output neuron for the label. Since this problem was modulated as a binary classification, the loss function for every neural network is the Binary Cross-entropy, and the activation function of the last layer is the

Sigmoid function. When the activation of the output neuron of a network surpasses a threshold of 0.5, a vulnerability is considered to have an exploit. The activation value can be interpreted as the probability of existence of an exploit for the given vulnerability. This odds can be sorted descending to become a patch priority list. Stop and Checkpoint callbacks were implemented in order to stop the training after a given amount of Patience Epochs without improvements on the validation loss, after stopping the training, the network weights of the best model so far are restored. Besides the loss value, the network can also provide training F1-Score, Recall, Precision or Accuracy measurements. There is no shuffle on the dataset, in order to preserve the chronological aspects of the data. To avoid overfitting we clip the gradient values to 1, and we also implemented the leaky ReLU activation function instead of the traditional ReLU. All the other aspects of the Neural network are given by the hyperparameters, such as the batch size, the learning rate, the optimizer, the amount of training epochs, the patience epochs, the layer sizes, the use of bias, the dropout values, and the activation functions of the non-output layers.

On the dataset, only 13.59% of the entries have an exploit available. Due to this reason the dataset is highly unbalanced. To overcome this, we proposed the ENNs, which are a fork of the SNNs, but with one difference on the loss function. When a vulnerability does not have an exploit, only 30% of the loss value is applied during the backpropagation. When the label is positive, i.e. there is an exploit, 100% of the loss is used. This approach makes the model to learn more when the dataset entry have an exploit, in comparison when it does not. With this change we expect to increase the recall, since the priority is to detect vulnerabilities that have an exploit, but it is acceptable to classify some vulnerabilities that do not have an exploit as having one. In turn, IENNs are a fork of ENNs, but instead of having a single neural network, they have six. There is one neural network for each of the five groups of features, the group definitions can be found on Sect. 3. The outputs of those networks are the inputs of the sixth network, the Concatenation network, which has a single output neuron to classify the vulnerability as having or not an exploit. The IENNs have the following advantages: (i) allow to increase the amount of layers without increasing the model complexity excessively, since neurons from different networks are not connected; (ii) each network can grow independently, since each network input might have different needs; and, (iii) the dimensionality of each input group can be reduced or increased independently before joining the processed features into the Concatenation network. The Appendix Fig. 1 highlights an example of IENN model architecture. There, the inputs and outputs represents the shape that each layer receives or produces, and the question mark represents "any amount" as the size of fed entries is not fixed.

5.1   Neural Architecture Search

The Neural Architecture Search (NAS) is the process of using some algorithm to search for a neural network architecture, we opted to do it using Genetic Algorithms since they can improve the results of a Neural Network, such as on

[16]. During a NAS, neural networks and genetic algorithms are combined by considering each genetic individual as a neural network, and its output function as some metric of the neural model. To do this, a certain number of neural network training epochs must occur in each generation of the genetic algorithm. Each network/individual has its own weights that are persisted across generations, that is, even if the network configuration is different, it starts out already pre-trained [11]. During the crossover process the weights from the mother are passed to her children. To make possible the reuse of the weights in networks with different numbers of neurons and layers, the weights are always stored with the highest dimension for the given search space, but only the relevant weights are applied. The hyperparameters are encoded in the DNA. The DNA is a float array that represent each hyperparameter. The hyperparameters are then inserted in a known order in the DNA. As there are unique hyperparameters per layer, such as the size of a layer, they are repeated for the maximum possible amount of layers, where each one is associated with a layer in an ordered way. After every iteration the values are truncated according to their minimums and maximums. Besides its DNA, an individual is also composed of its neural network weights.

6   Testing the Enhanced Genetic Algorithm

The EGA was proposed and its parameters were tuned. Now, we need compare its performance with the SGA, to decide which will be used on further experiments. The Sect. 6.1 is dedicated to compare both algorithms using the benchmark functions aforementioned, whereas the Sect. 6.2 focus on the experiments that compare both performances on the actual NAS problem.

6.1   On Optimization Benchmark Functions

Four types of optimizations were tested: (i) for EAf Eq. 2 with a mutation rate of 0.1; (ii) for EAf Eq. 2 with a mutation rate of 0.2; (iii) for EHf Eq. 1 with a mutation rate of 0.1; and (iv) for EHf Eq. 1 with a mutation rate of 0.2. The other EGA parameters can be found on Table 1, the population start size was 100. Since the EGA population size varies during the execution, the SGA population size, which is fixed, was set to be the average of the population sizes of each EGA generation. The SGA crossover rate was set to be the same as EGA, 0.829. Each of the optimizations were done 300 times, and the average of the results can be found on Table 2. The results show that the EGA is worse when considering runtime, which is obvious considering that it has more computations to do during its generations. In general, the EGA had better results for the EHf, both in the result and on the amount of generations to converge. On the other hand, the SGA had better results for the EAf. This results could indicate that both have similar performances, however, by analyzing the percentage distance from the actual global minima (-1 for EAf and -959.6407 for EHf) we can see that EGA got closer to the global minima of EAf, than the SGA got from the global minima of EHf. This indicates that the EGA had a slightly better performance, with the trade off of having a higher runtime.

Table 2. EGA versus SGA on benchmark functions

Measurement	SGA-0.1	SGA-0.2	EGA-0.1	EGA-0.2
EAf minima	-0.999999997	-0.9999999881	-0.9999212342	-0.9995891618
EHf minima	-948.5131826	-950.4962868	-955.8765385	-957.4322888
EAf best gen.	20.23	22.68	48.63	24.43
EHf best gen.	42.40	37.85	34.49	27.1
Gen. runtime (ms)	432.36	438.88	652.02	654.39

6.2   On Neural Architecture Search

The EGA has proven to be a bit better than the SGA on the benchmark functions, now we need to test its performance on a real application, NAS. Since the runtime of training a new network for each individual and for each generation is much higher than the cost of the benchmark function, this experiment was ran only once. The NAS was done on a small search space that contains the hyperparameters for a SNN, they can be found on Table 3. The Neural Network was trained using the Rolling Forecasting Origin Technique, with the F1-Score as the maximization goal, using an unbalanced vulnerabilities train data from 2018 and 2019. The SGA was configured to run for 50 generations, with a population of 100 individuals, a mutation rate of 0.2 and a crossover rate of 0.7. The EGA parameters can be found on Table 1. The results revealed a cross-validation F1-Score of 0.9403 and 0.9271 for EGA and SGA respectively. This result reinforces a minor superiority of EGA. Thus, the EGA will be used on all following experiments. On further work, the optimization of the EGA parameters using a more complex function instead of the fast selected benchmark functions could improve the results.

Table 3. Search space for SNN hyperparameters

Parameter	Region
Batch size	[0, 128]
Learning rate	[0.0001, 0.1]
Optimizer	{SGD, Adam, RMSProp}
Epochs	[30, 70]
Patience epochs	[20, 30]
Amount layers	[1, 5]
Per layer hyperparameters	
Layer size	[1, 200]
Bias	{False, True}
Activation	{ReLU, Softmax, Sigmoid, Tanh}
Dropout	[0, 0.3]

7   Experiments and Results

With an optimization algorithm defined and properly optimized, we could run two experiments to assess the proposed Neural Network models. The strategy based on our model consists of applying a patch for every CVE with a predicted exploit. One of the experiments use data from 2016 and 2017 and the other was done with data from 2018 and 2019. The search spaces can be found on Table 4. The measurement used on this experiment was the validation Recall. The results shown on Table 5 reveal that, as expected, the recall increases when using the ENN instead of the SNN, the recall also increases when using the IENN model. The "physical" separation of feature groups on IENN make easier for the neural network to select relevant features and to train hidden neurons with meaningful abstractions on the input networks. The hyperparameter set of the IENN model can be found on Table 6.

Table 4. Neural networks base model search space

Parameter	SNN & ENN region	IENN region
Batch size	[0, 128]	
Learning rate	[0.0001, 0.1]	
Optimizer	{SGD, Adam, RMSProp}	
Epochs	[100, 600]	
Patience epochs	[20, 60]	
Amount layers	[1, 10]	[1,5], [1,5], [2,13], [2,10], [1,5], [1,5]
Loss function[a]	-	{Mean Squared Error, Categorical Cross Entropy}
Per layer hyperparameters		
Layer size	[1, 1000]	[1,50], [1,60], [1,400], [1,120], [1,420], [1,100]
Bias	{False, True}	
Activation	{ReLU, Softmax, Sigmoid, Tanh}	
Dropout	[0, 0.3]	

[a] Only for the five input neural networks.

Table 5. Neural network base models results

Measurement	SNN	ENN	IENN
	2017 & 2018		
Recall	0.88731	0.90289	0.92365
Best gen.	13	50	26
	2018 & 2019		
Recall	0.92543	0.93615	0.94750
Best gen.	4	11	43

Table 6. IENN hyperparameter set

Parameter	Region
Batch size	128
Learning rate	0.0607, 0.0614, 0.0080, 0.0822, 0.0537, 0.0598
Optimizer	Adam on the concatenation, SGD on the others
Epochs	103
Patience epochs	60
Amount layers	1, 2, 2, 2, 1, 1
Loss function	Categorical Cross Entropy on all input networks
Per layer hyperparameters	
Layer size	43, (25, 1), (1, 181), (96, 55), 262, 1
Bias	False, (True, False), (False, False), (True, True), False, True
Activation	ReLU on all layers of the input networks
Dropout	0, (0.026, 0.136), (0, 0.275), (0.214, 0.269), 0, 0

Even with its flaws, CVSS is still the state of the art on vulnerability classification. We compared the V-REx IENN performance with CVSS based remediation strategies. For this experiment we used data ranging from 2015 to 2018 to train the models and data from 2019 to test. We trained three IENN models: (i) IENN - balanced, with the training dataset balanced, i.e. the same amount of positive and negative labels; (ii) IENN - unbalanced, without balancing the dataset; and (iii) IENN - unbalanced - 2018 only, training only with 2018 unbalanced data. The results available on Table 7 shows that both V-REx implementations with an unbalanced dataset outperformed every CVSS based strategy. It is also noticeable that the F1-Score dropped significantly when comparing with the results of Sect. 6.2, this is due to the unbalanced characteristic of the problem, there are few positive samples on the validation fold of the rolling window, causing the score to be higher when compared with a test dataset.

Table 7. Comparison between IENN and CVSS+ strategies

Model	Precision	Recall	F1-score
IENN - balanced	0.058	0.779	0.108
IENN - unbalanced	0.386	0.383	0.384
IENN - unbalanced - 2018 only	0.391	0.519	0.446
CVSS 10+	0.095	0.149	0.116
CVSS 9+	0.085	0.25	0.127
CVSS 8+	0.082	0.489	0.14
CVSS 7+	0.062	0.627	0.113
CVSS 6+	0.054	0.641	0.1
CVSS 5+	0.049	0.79	0.092
CVSS 4+	0.043	0.967	0.082

8   Conclusion and Further Work

In this work, we presented V-REx, an open-source classification system that uses public data to predict the odds of a known vulnerability to be exploited. Three base Neural Network models were developed and tested. The IENN model achieved the best result among the others, in every case scenario. We also proposed a EGA to genetically tune the Neural Network models. The EGA proved to have a slightly better performance than the SGA, both on benchmark functions and on a NAS. We lack of a complete risk analysis that considers not only the severity, as CVSS does, but also considers the probability of exploitation, besides having a clear methodology and being reliable [26]. V-REx contributes to this goal by predicting the exploitation probability, and being able to perform better than CVSS based strategies on prioritizing vulnerabilities. Which is a very relevant task, since poor system administration, i.e. patching risky vulnerabilities in a timely manner, leads to the occurrence of attacks [13].

Even with promising results, there is a lot of space for improvements. For instance, natural language processing could be done better by using Convolutional Neural Networks (CNN) instead of using multi-word expressions to encode textual features. CNNs can process the descriptions directly as in [5,18]. It is also possible to use CNN at the character-level as in [21] which, according to the authors, performs better than CNN in word-level. Our NAS is done by optimizing a single measurement, an improvement for that is to use multi-objective functions, for example, instead of optimizing the recall only, we could use a combination of recall, accuracy, loss and runtime. The classification threshold could be variable as in [8], instead of a fixed value. We could also use two classes to encode the presence of an exploit as in [2], turning the problem into a multi-class classification instead of binary classification.

References

1. Abril, V.: Conectesus segue fora do ar há 12 dias após ataque hacker. Online (2021). https://veja.abril.com.br/coluna/maquiavel/conecte-sus-segue-fora-do-ar-ha-12-dias-apos-ataque-hacker/
2. Fang, Y., Liu, Y., Huang, C., Liu, L.: Fastembed: predicting vulnerability exploitation possibility based on ensemble machine learning algorithm. PLoS ONE **15**(2), e0228439 (2020)
3. FIRST: Common vulnerability scoring system sig. FIRST's Web Page (2022). https://www.first.org/cvss/
4. Fox, C.: A stop list for general text. In: ACM SIGIR forum, vol. 24, pp. 19–21. ACM, New York (1989)
5. Han, Z., Li, X., Xing, Z., Liu, H., Feng, Z.: Learning to predict severity of software vulnerability using only vulnerability description. In: 2017 IEEE International Conference on Software Maintenance and Evolution (ICSME), pp. 125–136 (2017)
6. Huang, G., Li, Y., Wang, Q., Ren, J., Cheng, Y., Zhao, X.: Automatic classification method for software vulnerability based on deep neural network. IEEE Access **7**, 28291–28298 (2019)

7. Hyndman, R.J., Athanasopoulos, G.: Forecasting: Principles and Practice. OTexts (2018)
8. Jacobs, J., Romanosky, S., Adjerid, I., Baker, W.: Improving vulnerability remediation through better exploit prediction. J. Cybersecur. **6**(1) (2020)
9. Jacobs, J., Romanosky, S., Edwards, B., Roytman, M., Adjerid, I.: Exploit prediction scoring system (epss). arXiv preprint arXiv:1908.04856 (2019)
10. Jacobs, J., Roytman, M.: The etiology of vulnerability exploitation. In: RSA Conference 2019 (2019). https://published-prd.lanyonevents.com/published/rsaus19/sessionsFiles/14122/HT-F03_The_Etiology_of_Vulnerability_Exploitation.pdf
11. Jaderberg, M., et al.: Population based training of neural networks. arXiv preprint arXiv:1711.09846 (2017)
12. Johnson, P., Lagerström, R., Ekstedt, M., Franke, U.: Can the common vulnerability scoring system be trusted? A Bayesian analysis. IEEE Trans. Depend. Secure Comput. **15**(6), 1002–1015 (2016)
13. Jumratjaroenvanit, A., Teng-Amnuay, Y.: Probability of attack based on system vulnerability life cycle. In: 2008 International Symposium on Electronic Commerce and Security, pp. 531–535. IEEE (2008)
14. Kaspersky: What is wannacry ransomware? Kaspersky's Web Page (2020). https://www.kaspersky.com/resource-center/threats/ransomware-wannacry
15. Kubota, N., Fukuda, T.: Genetic algorithms with age structure. Soft. Comput. **1**(4), 155–161 (1997)
16. Li, J., Liang, C., Zhang, B., Wang, Z., Xiang, F., Chu, X.: Neural architecture search on acoustic scene classification. arXiv preprint arXiv:1912.12825 (2019)
17. Liashchynskyi, P., Liashchynskyi, P.: Grid search, random search, genetic algorithm: a big comparison for NAS. arXiv preprint arXiv:1912.06059 (2019)
18. Liu, K., Zhou, Y., Wang, Q., Zhu, X.: Vulnerability severity prediction with deep neural network. In: 2019 5th International Conference on Big Data and Information Analytics (BigDIA), pp. 114–119. IEEE (2019)
19. Marconato, G.V., Kaâniche, M., Nicomette, V.: A vulnerability life cycle-based security modeling and evaluation approach. Comput. J. **56**(4) (2013)
20. Marconato, G.V., Nicomette, V., Kaâniche, M.: Security-related vulnerability life cycle analysis. In: 2012 7th International Conference on Risks and Security of Internet and Systems (CRiSIS), pp. 1–8. IEEE (2012)
21. Nakagawa, S., et al.: Character-level convolutional neural network for predicting severity of software vulnerability from vulnerability description. Trans. Inf. Syst. **102**(9), 1679–1682 (2019)
22. Ramos, J., et al.: Using TF-IDF to determine word relevance in document queries. In: Proceedings of the First Instructional Conference on Machine Learning, New Jersey, USA, vol. 242, pp. 29–48 (2003)
23. Rose, S., Engel, D., Cramer, N., Cowley, W.: Automatic keyword extraction from individual documents. Text Min. Appl. Theory **1**, 1–20 (2010)
24. Sawadogo, A.D., et al.: Learning to catch security patches. arXiv preprint arXiv:2001.09148 (2020)
25. Shrestha, A., Mahmood, A.: Optimizing deep neural network architecture with enhanced genetic algorithm. In: 2019 18th IEEE International Conference On Machine Learning And Applications (ICMLA), pp. 1365–1370. IEEE (2019)
26. Spring, J., Hatleback, A., Manion, A., Shic, D.: Towards improving CVSs. Technical report, Software Engineering Institute, Carnegie Mellon University (2018)

Appendix

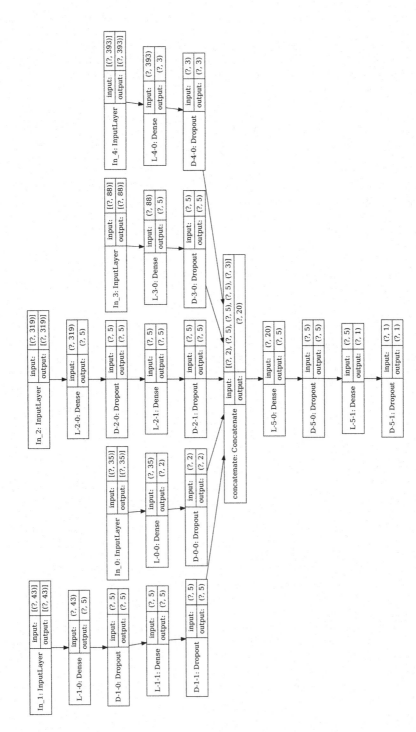

Fig. 1. Example model of an IENN architecture

Predicting Compatibility of Cultivars in Grafting Processes Using Kernel Methods and Collaborative Filtering

Thiago B. R. Silva[1]($\boxtimes$), Nina I. Verslype[2], André C. A. Nascimento[3],
and Ricardo B. C. Prudêncio[1]

[1] Center of Informatics, UFPE, Recife, PE, Brazil
tbrs@cin.ufpe.br
[2] UFRPE, Recife, PE, Brazil
[3] Department of Informatics, UFRPE, Recife, PE, Brazil

Abstract. Viticulture is the cultivation and harvesting of grapes for use in the production of juices, wines and other derivatives, with great socioeconomic importance. Grafting techniques have been applied to increase productivity and quality in the sector, but the process of finding compatible cultivars is slow and costly. Although Machine Learning (ML) methods have already been applied in several applications in agriculture, their use to support grafting processes is still very scarce. This work investigates ML-based recommender systems to address the problem of scion and rootstock compatibility in grafting processes in viticulture. In the experiments, collaborative filtering algorithms and kernel-based methods were evaluated on a dataset of 251 rated interactions, reaching a F1-score of approximately 96% for the best model. The results indicated advantages of kernel-based models over standard collaborative filtering models, as well as demonstrated the feasibility of a decision support tool to guide the choice of the best cultivars for grafting.

Keywords: Scion and rootstock compatibility · Grafting · Agroinformatics · Recommendation system · Decision support tool · Supervised learning · Machine learning · Intelligent systems · Artificial intelligence

1 Introduction

The vine (*Vitis* spp.) occupies a prominent place among the most important agricultural species in the world, as it is a versatile perennial fruit tree, with high added value to its products and great socioeconomic importance [1–3]. The grape is considered to be the oldest known domesticated fruit [1]. Despite being cultivated and shaped by man since antiquity, the usage of grafting techniques is considered relatively recent in the viticulture, considering its extensive cultivation history [2,3].

Grafting is a simple technology that adds several advantages, through the association of the scion and the rootstock cultivars [2,3,5]. Among these advantages, one can mention: the capacity of the rootstock to transmit its vigor to the

© The Author(s), under exclusive license to Springer Nature Switzerland AG 2022
J. C. Xavier-Junior and R. A. Rios (Eds.): BRACIS 2022, LNAI 13653, pp. 611–625, 2022.
https://doi.org/10.1007/978-3-031-21686-2_42

grafted scion cultivars; resistance to pests, diseases and adverse abiotic conditions; the influence on early production and plant size; increased productivity; improved grape quality; and adaptation to different types of soils [5,6].

The choice of vine rootstocks for grafting purposes should be done carefully, being a slow and costly process. This choice requires a specific research for a given region, due to the importance of environmental aspects on the combination responses between the scion and the rootstock. Besides, establishing a vineyard is a long-term investment [5,7], thus, the response of such combinations can take many years to be fully understood. Hence, the development of computational methods to accelerate such process is of high relevance.

Based on the above motivation, this work proposes to model the scion-rootstock compatibility problem as a recommendation task, which can be dealt with Recommender Systems (RS). In literature, RS has already been applied for different purposes to increase efficiency in agriculture, e.g., in decision making to choose cultivars, disease detection, crop forecast, fertilizer recommendation, among other applications [8,9]. However, the investigation of RS or any other Machine Learning (ML) technique in the context of grafting is still new.

In the proposed solution, we investigated and compared the results of Collaborative Filtering (CF) algorithms and kernel methods in the task of identifying compatible pairs of rootstocks and scions. In the CF context, scions were considered as users of a classic RS model, while rootstocks were treated as items for recommendation. The scions-rootstocks compatibility matrix, used as input for both CF and kernel methods, stores the degrees of compatibility between scions and rootstocks, based on interactions already known and extracted from the literature. This 2-dimensional matrix was constructed from the interactions between 40 species of vines (scions) and 31 species of other plants used as rootstocks. In all, degrees of compatibility were collected from the literature for 251 pairs of scions and rootstocks, thus forming a sparse matrix with only about 20% of the possible interactions between the identified cultivars.

In the experiments carried out, 6 models based on standard CF algorithms (k-NN, SVD and SlopeOne) and 3 kernel-based methods (Linear, Gaussian and Polynomial) were trained and tested. The metrics applied to evaluate the results of RS model were root mean squared error (RMSE), precision, recall, F1 score and area under the ROC curve (AUC). The analysis of these results suggests that kernel-based methods are potentially better than standard CF methods as they can take advantage on available scions/rootstocks features. It also demonstrates the feasibility of a decision support tool, such as a RS, to help grafting specialists make better choices regarding which rootstock to use for a given crop.

The main article contributions can thus be summarized as follows:

1. First, we dealt with the problem of choosing suitable rootstocks for vine scions as a recommendation task;
2. Second, the performance of CF algorithms and kernel methods were evaluated for pairs of rootstocks and scions with known compatibility;
3. Finally, we demonstrated the viability of using RS in the grafting process.

The rest of the article is organized as follows: Sect. 2 presents the most common uses of RS in agriculture; Sect. 3 describes the problem itself and how it was approached in the present work; Sect. 4 explains the data, the methods and the experimental methodology; Sects. 5 and 6 present the results, evaluates its applicability and proposes future works.

2 Recommender Systems in Agriculture

The literature on the use of RS covers applications for a wide variety of problems. In this section, we will introduce the most common areas where RS has been used in agriculture.

RS has been widely used by researchers, notably in India, to support the choice of the best crop for a region, based on its geographic and/or climatic conditions [10,11]. For example, in [12], the authors propose a RS that uses Pearson's correlation to identify the top-n regions most similar to those of a given user. Then, the crops that should best develop for the user specific conditions are recommended, based on the highest yielding crops in the top-n similar regions in a specific season. In [13], an ensemble-based recommendation system, which uses the majority vote technique, was proposed. This RS compares the results of the decision tree, CHAID, k-nearest neighbors (k-NN) and Naive Bayes algorithms to recommend the best crop to be explored in a region, according to the soil characteristics of rural properties.

Another common application is to use RS to choose the most suitable fertilizers, pesticides and other chemical agents to intervene in a specific case. In [14], the authors built an ontology that models the interactions between crops, pests and pesticides, and developed a recommendation system in order to facilitate the identification of pests and the choice of the most appropriate treatment. In [9], the authors developed a system capable of generating location-specific fertilizer recommendations for selected crops by analyzing the national soil database from Bangladesh. This system requires farmer field location, soil and land type, crop type and variety information to generate instant crop-specific fertilizer recommendations.

Other applications include the use of RS to support the choice of the most appropriate time to start planting or harvesting crops in general. In [15], a system that uses support vector regression was proposed to provide useful recommendations for farmers. This RS uses data about weather conditions and how it affects the crops to predict the best cultivation and harvesting times. In [16], the authors measured the effectiveness of a system that used a ML algorithm to predict the harvesting times of the rice crop. This study indicates that the system's recommendations are close enough to those ones made by experienced farmer and experts.

There are many other relevant works that apply RS in completely different contexts. For example, in [17], a RS based on collaborative filtering was used to recommend the best government programs available to farmers in India. This system requires farmer profile information and program admission characteristics to compute the best matches.

Even though there are many works applying RS in agriculture, to the best of our knowledge, there are no published studies demonstrating the application of this ML technique to the specific problem of scion and rootstock compatibility in grafting processes.

3 Proposed Solution

As mentioned in the previous sections, the cultivation of vines for the production of juices, wines and other derivatives (i.e., viticulture) has great relevance for culture, industry and the economy in almost every continent. Grafting is a very common practice in this type of culture, however, selecting the cultivars that will be used in the process is not an easy task. In general, this is done empirically and the result of experimentation is slow and costly. Therefore, the development of tools capable of guiding the choice of cultivars with a greater chance of compatibility can be a valuable resource for researchers and entrepreneurs in the area.

This work presents a RS approach based on CF and kernel methods to specifically address the problem of choosing compatible pairs of scions and rootstocks for grafting processes in viticulture. Figure 1 presents an overview of the work developed and each step is detailed below:

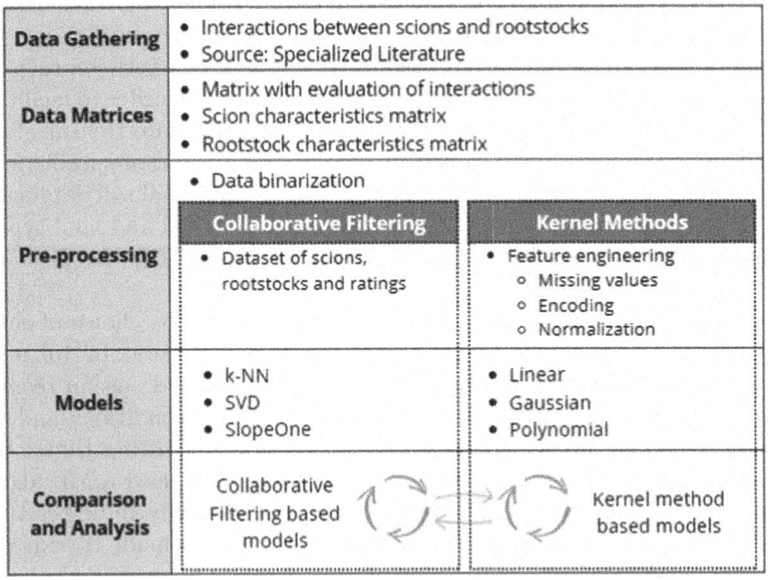

Data Gathering	• Interactions between scions and rootstocks • Source: Specialized Literature	
Data Matrices	• Matrix with evaluation of interactions • Scion characteristics matrix • Rootstock characteristics matrix	
Pre-processing	• Data binarization	
	Collaborative Filtering	**Kernel Methods**

	• Dataset of scions, rootstocks and ratings	• Feature engineering
		  ○ Missing values
		  ○ Encoding
		  ○ Normalization

Models	• k-NN • SVD • SlopeOne	• Linear • Gaussian • Polynomial
Comparison and Analysis	Collaborative Filtering based models	Kernel method based models

Fig. 1. Overview of the proposed approach.

1. **Data gathering:** Collect data from books, scientific articles and specialized websites about the degree of affinity between cultivars in grafting processes in viticulture;
2. **Interactions and features matrices:** Split the data into 3 different matrices: one to store the interactions between scion and rootstock cultivars and their degrees of compatibility; and another two to store the characteristics of the scions and rootstock cultivars;
3. **Pre-processing:** Perform the transformations and feature engineering necessary to adjust the structure of the matrices and the format of the data in order to use them to train and test the models.
4. **Models training and testing:** Train different classifier models based on collaborative filtering and kernel methods, using different hyperparameter configurations and resampling techniques.
5. **Comparison and analysis:** Based on the metrics calculated in the tests, compare the results obtained by each model and algorithm class as a whole.

The solution proposed in this article is an application of CF and kernel methods in a RS approach on an original problem. To the best of our knowledge, there are no previous works in the literature that used RS or other ML technique to address the problem of scion-rootstock compatibility, whether in viticulture or any other crop that uses grafting.

4 Experiments

In this section, we present the experiments performed to evaluate different RS techniques in a case study of viticulture grafting.

4.1 Dataset

The data used in this experiment were collected from the literature and represent degrees of compatibility between scion-rootstock pairs [29]. This information was split in 3 matrices to better represent the collected data: interaction matrix, scion features matrix and rootstock features matrix.

The interaction matrix $\mathcal{I}$ is a sparse matrix $|S| \times |R|$, where S represents the scions set and R represents the rootstocks set. Each element a_{ij} of matrix $\mathcal{I}$, being $\{a_{ij} \in \mathbb{N} | 1 \leq a_{ij} \leq 5\}$, indicates the degree of compatibility of the i-th scion with the j-th rootstock and can receive values from 1 to 5, in ascending order of compatibility. The degree of compatibility may also be unknown for many pairs of scions and rootstocks. These missing interactions will be candidates for recommendation. Table 1 gives a sample of the original interaction matrix, showing some degrees of compatibility and some missing values (in gray) between scions and rootstocks. In the interaction matrix, degrees of compatibility greater than 4 indicates good or excellent compatibility, values equals to 3 indicates weak compatibility, values equals to 1 and 2 indicates absolute incompatibility.

Table 1. Sample of the original matrix of collected data. Grayed cells represent unknown or unevaluated interactions.

		Rootstocks				
		1103P	110R	420A	5BB	SO4
	BORDO	5		4		
	ISABEL	5				
Scions	MERLOT	5	5	5		5
	SYRAH	5	4	3		4
	TANNAT	5		5		5

The scion features matrix S has dimension $|S| \times |F|$, where S represents the scions set and F represents the features set of the scion cultivars. The rootstock features matrix $\mathcal{R}$ has dimension $|R| \times |F|$, where R represents the rootstocks set and F represents the features set, which can be used to describe both scion and rootstocks. In these two matrices, the element e_{ij} recorded at the intersection between $S \times F$ or $R \times F$, represent the value of the j-th feature of the i-th scion or rootstock and it can be ordinal, numerical or categorical, depending on the type of the corresponding feature.

The features set is composed of 6 plant characteristics, common to scions and rootstocks, and 7 specific characteristics, according to the purpose of the cultivar in the grafting process. These features relate to morphological, genetic, immunological and life cycle aspects of the cultivars, such as: species, vigor, nematode resistance, acidity, etc. The complete list of features is available in a public electronic appendix[1].

Table 2 presents some statistics calculated from the interaction matrix $\mathcal{I}$. In all, 251° of compatibility were collected from the literature for 40 scions and 31 rootstocks. That is, only about 20% of the 1.240 possible interactions are known. The hypothesis that emerges is that, among the 80% of unexplored interactions, there could be others scion-rootstock pairs with high degree of compatibility.

The degree of compatibility histogram (see Fig. 2) indicates a high level of imbalance among the known interactions. In fact, 220 out of 251 interactions (87,5%) assume the values 4 and 5, being 144 interactions equal to 5 and 76 equals to 4. In turn, 31 interactions are equals to 1, 2 or 3. The mean degree of compatibility, $\approx 4,34$, directly reflects the imbalance of degree distribution. Also, the standard deviation of ≈ 1 allows us to conclude that most of the data lies close to the higher degrees. As in this dataset the degree of compatibility only vary between 1 and 5, there are no outliers to worry about.

The interactions between cultivars can also be represented as an undirected bipartite graph $\mathcal{G}$ of type $\mathcal{G} = (S, R, E)$, where S and R are sets of vertices, each $s_i \in S$ represents a scion and each $r_j \in R$ represents a rootstock. E is a set of edges indicating a know interaction between the elements of S and R, thus $E \rightarrow \{(s_i, r_j) | a_{ij} \; is \; not \; missing\}$. This graph representation allows us to calculate the connectivity degrees shown in Table 2, indicating the number of

---

[1] https://github.com/thiagobrs/grafting-recommender.

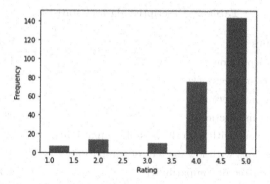

Fig. 2. Degrees of compatibility (ratings) distribution histogram.

known interactions that each node has. The average degree is ≈ 7 interactions per vertex, with a minimum of 1 and a maximum of 23 connections (see Fig. 3). An edge density of 0.101 confirms the fact that a low number of interactions (edges) are known and the average clustering coefficient of zero shows that this graph has no clusters.

A graph visualization is presented in Fig. 4, filtered to show only the edges related to good or excellent interactions. The figure suggests the existence of elements with greater predisposition for interactions. Both scions and rootstocks with higher connectivity degree seems to easily form positive connections not only between them, but also with the low degree vertices from the other group. The number of edges between the low degree vertices of both groups is visibly smaller (edges between the bottom and the left side of Fig. 4).

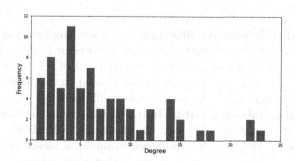

Fig. 3. Node degree distribution for the rootstock-scions interaction graph.

4.2   Methods

In this section, we present the RS methods adopted in our work for recommending scion-rootstock interactions. CF is considered to be the most popular and

Table 2. Descriptive dataset statistics.

Statistics	Values
Number of scions	40
Number of rootstocks	31
Known interactions	251
Missing interactions	989
Interactions with high degrees of compatibility (≥ 4)	220
Interactions with low degrees of compatibility (< 4)	31
Mean degree of compatibility	4.338645
Standard deviation of degrees of compatibility	0.988363
Average connectivity degree of graph nodes	7.07
Scions max. connectivity degree	22
Scions avg. connectivity degree	6.27
Rootstocks max. connectivity degree	23
Rootstocks avg. connectivity degree	8.1
Edge density of the graph	0.101
Average clustering coefficient of the graph	0.0

widely implemented technique in RS. Basically, it consists in predicting the preferences of a user based on the preferences of other users or similarity between items [18]. Drawing a parallel, in our context the rootstocks would be items that the system should recommend to the scion "users" and the degrees of compatibility would be the ratings of each interaction. Two main kinds of CF algorithms can be adopted:

1. **User-based collaborative filtering:** makes recommendations based on the preferences of other users with similar preferences or profiles. For the scion-rootstock example, it means to learn the pattern of rootstock preference of each scion, and recommend new rootstocks based on the preferences of other scions that demonstrated similar rootstock affinities;
2. **Item-based collaborative filtering:** makes recommendations based on the similarity between items the user has demonstrated affinity for and other unknown items. For the scion-rootstock example, it means to learn the similarity between the rootstocks and, given that a scion has good interactions with rootstock X, the RS recommends other rootstocks that are similar to X.

In this work, we adopted the implementations of 3 CF algorithms available in the *SurpriseLib*[2] library for Python. These algorithms hyperparameters were optimized through the execution of a *cross-validated grid-search* method, using RMSE and Fraction of Concordant Pairs as evaluation criteria to select the best

---

[2] http://surpriselib.com/.

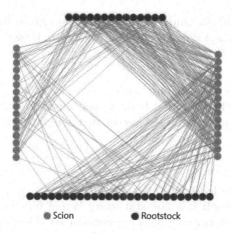

Fig. 4. Bipartite graph representation: high-low connectivity subsets. Scions vertices with high degree (>6) are positioned on the right. Rootstocks vertices with high connectivity degree are positioned on the top.

combination of hyperparameters. Below, we present the collaborative filtering algorithms used and their selected hyperparameters:

- **k-NN:** basic implementation of the k-nearest neighbors algorithm. List of hyperparameters:
  - Maximum number of neighbors: 9;
  - Minimum number of neighbors: 1;
  - Similarity function: Cosine similarity [20];
  - User based similarity;
  - Minimum number of common items: 1.
- **SVD:** an algorithm based on matrix factorization. List of hyperparameters:
  - Number of iterations: 40;
  - Learning rate: 0.01;
  - Regularization term for all parameters: 0.03;
  - Whether to use baselines (or biases): True.
- **SlopeOne:** direct implementation of the proposed algorithm in [19], there are no hyperparameters to configure.

Kernel methods are a class of algorithms that uses kernel functions to map the input data into a high-dimensional space, in which data is linearly separable. This mapping allows simpler models to learn from the new feature space, occasionally increasing the performance of the models [21,22]. In this work, we used the implementation of 3 kernel methods available in the RLScore[3] package for Python to calculate kernel matrices from the interactions data and from the scion and rootstock features data. The following are the kernel methods used and a brief description of each, based on the work published in [28]:

---

[3] http://staff.cs.utu.fi/~aatapa/software/RLScore/index.html.

- **Linear kernel:** is the simplest kernel function, given by the inner product of the entries (x_1, x_2) plus an optional constant c (bias).

$$k(x_1, x_2) = x_1^T x_2 + c \qquad (1)$$

Value used for the hyperparameter: $c = 1$.
- **Gaussian kernel:** is a radial basis kernel function, classified as stationary, as it depends exclusively on the vector that separates the two examples.

$$k(x_1, x_2) = exp\left(-\frac{\|x_1 - x_2\|^2}{\theta}\right) \qquad (2)$$

Value used for the hyperparameter: $\theta = 1$.
- **Polynomial kernel:** is a non-stationary kernel function, more suitable for problems where all training data are normalized. The adjustable parameters are the slope α, the constant term c, and the polynomial degree d.

$$k(x_1, x_2) = (\alpha x_1^T x_2 + c)^d \qquad (3)$$

Value used for the hyperparameters: $\alpha = 1$, $c = 0$ e $d = 2$.

4.3   Experimental Methodology

In our work, we treated the recommendation problem as a binary classification problem, i.e., we distinguished between compatible and incompatible scion-rootstock pairs. Hence, we binarized the degrees of compatibility stored in the interaction matrix: value 0 was assigned for bad interactions (i.e., ratings 1, 2 and 3) and value 1 was assigned for good interactions (i.e., ratings 4 and 5).

Two resampling methods were used to train and test the CF models: k-fold cross validation (CV) and leave-one-out (LOO). The k-fold cross-validation consists in randomly partition the original sample into k equal sized subsamples, also called folds. The CV process is repeated k times, each time a different fold is used as validation data and the others are used as training data. Then, the average of all k results for any chosen metric are used as an estimation [23]. LOO is a particular case of CV where k is equal to the size of the original sample. In this paper, the k value for the cross validation method was set to 5 while the k value for the leave one out, by definition, had to be 251 (number of labeled examples, see Table 2).

For models based on kernel methods, it was necessary to carry out a feature engineering step on the scion and rootstock feature matrices before models training and testing. First, we adopted the following strategy to deal with the missing values problem: for categorical data, the mode value of the corresponding feature was used; and for numerical data, the mean value was used. Then, we encoded the categorical data using the one-hot-encoding strategy. Lastly, we normalized the numerical data using L2 normalization.

The training and testing steps for kernel-based models were performed through a specific function, also available in the RLScore package, called CGKro-nRLS. This function is an iterative Kronecker RLS training algorithm specially designed to work with pairwise incomplete data, like ours, and is based on the work published in [24].

To evaluate the models, the following metrics were averaged over 10 independent executions of CV, LOO and CGKronRLS routines:

- **RMSE:** although the main task is to give correct classifications of good and bad interactions between scions and rootstocks, the RMSE metric was used to assess the accuracy of the models around values 0 and 1;
- **Precision and recall:** precision and recall are performance metrics for evaluating classifiers. Precision shows how good the classifier is at not labeling a sample that is negative as positive, while recall shows how good the classifier is at finding all the positive samples [25];
- **F1 score:** also known as balanced F-score or F-measure, is the harmonic mean between precision and recall, the closest to 1 indicates a better balance between precision and recall [25];
- **Area under the ROC curve (AUC-ROC) score:** computes the relation between the classifier's specificity and sensitivity at different decision thresholds. The closer to 1 the AUC-ROC score is, the better the performance of the model at distinguishing between the positive and negative classes [26].

5 Results

Table 3 presents the results of each model for the metrics mentioned in Sect. 4.3. The best overall results for each metric are highlighted in bold and yellow. Regarding RMSE, the model with the best result was the one trained with Polynomial Kernel ($RMSE = 0.315074$). Regarding the classification performance,

Table 3. Metric results for each model in the test set.

| | CF based models | | | | | |
| | KNN | | SVD | | SlopeOne | |
	CV	LOO	CV	LOO	CV	LOO
RMSE	0.342427	0.324082	0.327919	0.322533	0.346852	0.329052
Precision	0.914859	0.945553	0.931515	0.931670	**0.954545**	0.954198
Recall	0.786363	0.710454	0.636818	0.780909	0.544090	0.568181
F1-score	0.845758	0.811315	0.756479	0.849653	0.693109	0.712250
AUC-ROC score	0.629300	0.698027	0.701468	0.726308	0.716509	0.727639

| | Kernel based models | | |
	Linear Kernel	Gaussian Kernel	Polynomial Kernel
RMSE	0.329891	0.318331	**0.315074**
Precision	0.933333	0.938462	0.935484
Recall	0.903226	**0.983871**	0.935484
F1-score	0.918033	**0.960630**	0.935484
AUC-ROC score	0.729391	**0.769713**	0.745520

SlopeOne-LOO model had the best result for precision, but very low recall. The kernel method trained with Gaussian kernel had the best results for recall, f1-score and AUC-ROC score and a competitive value of precision. Additionally, RMSE score was approximately the same as the one obtained by the Polynomial Kernel model $(0, 318331)$.

As, on average, the results of models belonging to the same category (CF or kernel-based) were very similar, we selected the best model of each category and performed a deeper investigation of performance differences. The Gaussian Kernel model was chosen among the kernel-based methods, due to the results discussed above. In turn, the SVD-LOO model was considered the best model within the class of CF algorithms, by considering the overall results available in Table 3.

Figure 5 presents the scatter diagrams for SVD-LOO and Gaussian Kernel models classifications over the test set. The classifiers output is given in terms of approximations or probabilities in the continuous space in relation to target classes 0 (bad interactions) and 1 (good interactions). Therefore, a threshold was calculated using ROC curve and Youden's J statistic, as described in [27], to establish the boundary between the predicted classes. Thus, predictions situated below the threshold line belong to class 0 and predictions above the threshold line belong to class 1.

It is possible to visualize through Fig. 5 how different the Gaussian Kernel and the SVD-LOO classifiers performed in the task of identifying examples of the positive class. The same pattern was observed for the other kernel and CF-based models. The recall metric expresses this difference in numbers (see Table 3), as kernel-based models had higher recall scores. Looking at the issue from another perspective, CF-based models committed false negatives (FN) at a rate between 40–50%, while kernel-based models between 3–10%.

Figure 6 shows the ROC curve of the SVD-LOO and Gaussian Kernel models, through which it is possible to visualize the relationship between the true positive rate (TPR) and the false positive rate (FPR) obtained by these classifiers. Once again, the pattern was similar for the other CF and kernel-based models.

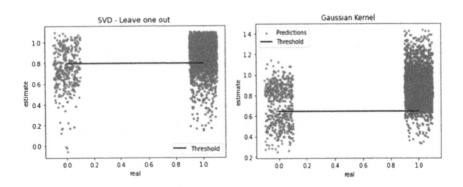

Fig. 5. Scatter diagrams between predicted and actual values.

Contrary to what the isolated analysis of the AUC-ROC score metric suggests (see Table 3), which showed a relatively low difference for the AUC-ROC scores, the graphical analysis of the ROC curve indicates that the tradeoff between TPR and FPR is more costly for the CF-based models, since a small increase in TPR implies a large increase in FPR.

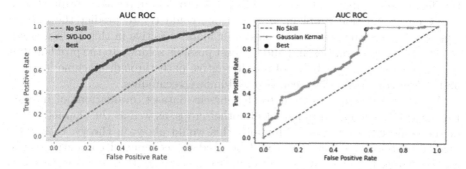

Fig. 6. ROC curves.

Therefore, the results obtained indicate an advantage for the models that used the SVD algorithm, among the CF-based models, in a context where the features of the inputs were not used. Additionally, among the models based on kernel methods, which took advantage of the features of the inputs, the one that used the Gaussian Kernel offered the best response. One hypothesis that the results may suggest is that models based on kernel methods with input features were better than those based on CF. However, as the resampling methods used were different for the two classes of algorithms, it was not possible to state this with complete certainty.

It was not part of the scope of this work to evaluate the effect of data imbalance on the results of the models, mainly because imbalance is an expected characteristic in datasets for this problem domain. Therefore, as a rule, it would be desirable for the predictive engine of a rootstock RS to be able to perform well even if the data imbalance was large and there were few labeled examples available.

6   Conclusion

In the present work, we used a RS approach to analyze the performance of CF algorithms and kernel-based methods to address the problem of identifying compatible pairs of scions and rootstocks in viticulture. This approach has been little explored until then, probably unprecedented, in this domain.

The results obtained in the experiments conducted, considering exclusively the available data and the chosen methods, suggest that, in a context where input features are unavailable, CF-based classifiers using SVD algorithms are a better choice. However, if input resources are available, Gaussian kernel-based classifiers are preferable. Furthermore, the F1-score of 96% for the Gaussian Kernel model and 85% for the SVD-LOO model are good indicators that they could be tested as a decision support tool for researchers in the field of viticulture.

Obviously, other approaches could have been applied, for example, modelling the problem as a regression problem, in which the objective would be to predict the degrees of compatibility, or to address the problem in terms of graph theory, i.e., a link prediction problem, in which one could try to predict the missing edges and their strength. Nonetheless, our main objective in this work was not to exhaust the possibilities and find the best possible solution to the problem, but to demonstrate the feasibility of an alternative path through machine learning and offer a comparison between two well-known techniques.

Future work could also add to the current dataset, evaluations of interactions between scions and rootstocks in other crops, as well as the characteristics of the respective cultivars. Ultimately, this would allow, in theory, to build a general-purpose classifier that could be used to recommend compatible scion and rootstock pairs in any type of crop. Even more so if, in addition to learning from the characteristics of the cultivars, the classifier models also learned from the context, for example, through environmental information capable of interfering with the success of grafting, such as climate, soil and relief.

RS methods are powerful and versatile techniques. We tried through this work to explore it in a different context, still lacking references in the literature, and we chose a problem, of undeniable economic and social relevance, to offer a method based on machine learning for planning agricultural grafting processes.

References

1. Leão, P.C.S., Borges, R.M.E.: Melhoramento genético da videira. Embrapa Semiárido-Documentos (2009)
2. Dougherty, P.H.: The Geography of Wine Regions, Terroir and Techniques. Springer, Dordrecht (2012). https://doi.org/10.1007/978-94-007-0464-0
3. Cantu, D., Walker, M.A.: The Grape Genome. Springer, Cham (2019). https://doi.org/10.1007/978-3-030-18601-2
4. FAO - Food and Agriculture Organization of the United Nations. http://www.Fao.org/faostat/en/#data/QC/visualize. Accessed 16 June 2021
5. Soares, J.M., Leão, P.C.S.: A vitivinicultura no Semiárido Brasileiro. Embrapa Semiárido, Petrolina (2009)
6. Hernandes, J.L., Martins, F.P., Pedro Júnior, M.J.: Uso de porta-enxertos: Tecnologia simples e fundamental na cultura da videira. Instituto Agronômico de Campinas, Jundiaí (2011)
7. Serra, I., Strever, A., Myburgh, P.A., Deloire, A. : The interaction between rootstocks and cultivars (Vitis vinifera L.) to enhance drought tolerance in grapevine. Australian J. Grape Wine Res. (2014)
8. Mohanty, S., Chatterjee, J., Jain, S., Elngar, A., Gupta, P.: Recommender System with Machine Learning and Artificial Intelligence. Wiley-Scrivener, Hoboken (2020)

9. Bondre, D. A., Mahagaonkar, S.: Prediction of crop yield and fertilizer recommendation using machine learning algorithms. Int. J. Eng. Appl. Sci. Technol. (2019)
10. Vivek, M.V.R., Harsha, D.V.V.S.S.S., Maran, P.S.: A survey on crop recommendation using machine learning. Int. J. Recent Technol. Eng. 120–125 (2019)
11. Patel, K., Patel, H.B.: A state-of-the-art survey on recommendation system and prospective extensions. Comput. Electron. Agric. (2020)
12. Mokarrama, M.J., Arefin, M.S.: RSF: a recommendation system for farmers. In: 5th IEEE Region 10 Humanitarian Technology Conference, pp. 843–850 (2018)
13. Pudumalar, S., Ramanujam, E., Rajashree, R.H., Kavya, C., Kiruthika, T., Nisha, J.: Crop recommendation system for precision agriculture. In: 8th International Conference on Advanced Computing, pp. 32–36 (2017)
14. Lacasta, J., Lopez-Pellicer, F.J., Espejo-García, B., Nogueras-Iso, J., Zarazaga-Soria, F.J.: Agricultural recommendation system for crop protection. Comput. Electron. Agric. 82–89 (2018)
15. Osman, H., El-Bendary, N., Fakharany, E.E., Emam, M.E. Ontology based recommendation system for predicting cultivation and harvesting timings using support vector regression. Softw. Eng. Perspect. Intell. Syst. (2020). Advances in Intelligent Systems and Computing, vol. 1295. Springer
16. Saravanakumar R., et. al.: Estimating the efficiency of machine learning in forecasting harvesting time of rice. Int. J. Mod. Agric. 10(2), 1930–1937 (2021)
17. Jaiswal, S., Kharade, T., Kotambe, N., Shinde, S.: Collaborative recommendation system for agriculture sector. In: ITM Web of Conferences (2020)
18. Ricci, F., Rokach, L., Shapira, B.: Introduction to recommender systems handbook. In: Ricci, F., Rokach, L., Shapira, B., Kantor, P. (eds.) Recommender Systems Handbook, pp. 1–35. Springer, Boston (2011). https://doi.org/10.1007/978-0-387-85820-3_1
19. Lemire, D., Maclachlan, A.: Slope One Predictors for Online Rating-Based Collaborative Filtering. In: Proceedings of the SIAM Data Mining Conference (2005)
20. Cosine Similarity. https://en.wikipedia.org/wiki/Cosine_similarity#Definition. Accessed 21 May 2022
21. Genton, M.G.: Classes of kernels for machine learning: a statistics perspective. Journal of Mach. Learn. Res. 2 (2001)
22. Hofmann, T., Scholkopf, B., Smola, A.J.: Kernel methods in machine learning. Inst. Math. Stat. Ann. Stat. 36(3), 1171–1220 (2008)
23. Devijver, P.A., Kittler, J.: Pattern Recognition: A Statistical Approach. Prentice-Hall, London (1982). ISBN 0-13-654236-0
24. Airola, A., Pahikkala, T.: Fast Kronecker product kernel methods via generalized vec trick. IEEE Trans. Neural Netw. Learn. Syst. 29(8), 3374–3387 (2018)
25. Powers, D.M.W.: Evaluation: from precision, recall and F-measure to ROC, informedness, markedness & correlation. J. Mach. Learn. Technol. 2(1), 37–63 (2011)
26. Fawcett, T.: An introduction to ROC analysis. Pattern Recogn. Lett. 27(8), 861–874 (2006)
27. Brownlee, J.: Imbalanced classification with python: better metrics, balance skewed classes, cost-sensitive learning. Mach. Learn. Mastery, 248–258 (2020)
28. Genton, M.G.: Classes of kernels for machine learning: a statistics perspective. J. Mach. Learn. Res. 2, 299–312 (2001)
29. Verslype, N.I.: Avaliação e seleção de porta-enxertos de videira (Vitis spp.) tolerantes ao déficit hídrico através de aprendizagem de máquina. 2021, p. 140, UFRPE (2021)

Cross-validation Strategies for Balanced and Imbalanced Datasets

Thomas Fontanari[1,2], Tiago Comassetto Fróes[1],
and Mariana Recamonde-Mendoza[1,2(✉)]

[1] Institute of Informatics, Universidade Federal do Rio Grande do Sul,
Porto Alegre, RS, Brazil
{tvfontanari,tiago.froes,mrmendoza}@inf.ufrgs.br
[2] Bioinformatics Core, Hospital de Clínicas de Porto Alegre, Porto Alegre, RS, Brazil

Abstract. Cross-validation (CV) is a widely used technique in machine learning pipelines. However, some of its drawbacks have been recognized in the last decades. In particular, CV may generate folds unrepresentative of the whole dataset, which led some works to propose methods that attempt to produce more distribution-balanced folds. In this work, we propose an adaption of a cluster-based technique for cross-validation based on mini-batch k-means that is more computationally efficient. Furthermore, we compare our adaptation with other splitting strategies previously not compared and also analyze whether class imbalance may influence the quality of the estimators. Our results indicate that the more elaborate CV strategies show potential gains when a small number of folds is used, but stratified cross-validation is preferable for 10-fold CV or in imbalanced scenarios. Finally, our adaptation of the cluster-based splitter reduces its computational cost while retaining similar performance.

Keywords: Model evaluation · Cross-validation · Class imbalance

1 Introduction

Splitting a dataset into various subsets for training and validation is a fundamental part of machine learning and is present in multiple tasks, such as model evaluation, model comparison, and hyperparameter tuning. Some traditional methods to split datasets into training and test sets are holdout, bootstrap, and cross-validation (CV).

Although cross-validation is arguably the most popular partitioning method, it has some relevant drawbacks that have been studied in the last decades. Given

This study was financed in part by the Coordenação de Aperfeiçoamento de Pessoal de Nível Superior - Brasil (CAPES) - Finance Code 001, and by grants from the Fundação de Amparo à Pesquisa do Estado do Rio Grande do Sul - FAPERGS [21/2551-0002052-0] and Conselho Nacional de Desenvolvimento Científico e Tecnológico (CNPq) [308075/2021-8].

© The Author(s), under exclusive license to Springer Nature Switzerland AG 2022
J. C. Xavier-Junior and R. A. Rios (Eds.): BRACIS 2022, LNAI 13653, pp. 626–640, 2022.
https://doi.org/10.1007/978-3-031-21686-2_43

its stochastic nature, CV may lead to poor estimates because of a partition-induced dataset shift [14], that is, some of the generated folds are not representative of the data. One generally handles this by using repeated cross-validation, However, since applying cross-validation is already computationally expensive, repeating it multiple times may be prohibitive.

The aforementioned issues are related to the randomized steps of CV. Therefore, a few methods have been proposed that attempt to improve the estimates of cross-validation by introducing a more deterministic process of generating folds. In one of the first works on the topic, Diamantidis *et al.* [9] introduced a clustering-based technique that relied on k-means (here referred to as CBD-SCV). However, k-means can be expensive when used as a splitting strategy for CV. At a similar time, Zeng *et al.* [24] proposed distribution-balanced stratified cross-validation (DBSCV), which was later adapted by Moreno-Torres *et al.* [14], introducing the distribution optimally balanced stratified cross-validation (DOB-SCV). Although DBSCV and DOBSCV have been compared before, there has been no direct comparison between them and the cluster-based methods.

In our work, we propose the use of mini-batch k-means as a way of reducing the computational cost of CBDSCV. Furthermore, we provide a comparison between CBDSCV, DOBSCV, and DBSCV, besides the traditional cross-validation techniques, on 20 datasets of various sizes, class imbalance levels, number of features, and number of classes. Our experiments aim to assess whether any cross-validation splitting strategy tends to outperform the others in terms of bias, variance, or computational cost.

The rest of the paper is structured as follows. In Sect. 2 we present the theoretical background of our work, followed by a description of our experiments in Sect. 3. Next, in Sect. 4, we present and discuss our results for balanced and imbalanced datasets. Section 5 revises other papers that presented efforts towards proposing cross-validation splitting strategies and were not directly compared in our experiments. Finally, Sect. 6 presents our conclusions and directions for future works.

2 k-fold Cross-validation Partitioning Methods

The traditional k-fold cross-validation (CV) [10] consists in dividing the given dataset into k folds. Each fold is then used once as the validation set, while the remaining $k - 1$ folds are used for training. Finally, the average performance obtained for each fold is the performance estimate of the k-fold CV. In general, k is set as 5 or 10, which makes it much more computationally tractable than leave-one-out cross-validation (LOOCV), besides showing less variance than LOOCV estimates. Furthermore, it is less biased than the holdout method, since it is able to use more instances for training than the holdout. K-fold cross-validation can also be used in a stratified fashion (k-fold SCV) to guarantee that the proportion of instances of each class is the same for all folds.

However, the instances assigned to each fold by traditional k-fold CV and SCV are selected randomly, which can cause some folds not to be good representatives of the whole dataset [14]. For instance, it is not guaranteed that all the regions in the input space will be appropriately represented over all folds. This phenomenon may impact performance estimates and thus has been considered in various works (see Sect. 5 for references), leading to new splitting strategies based on the features of the data and not only on their class labels. The methods we have considered in this work are reviewed in the following sections, and we also describe the adaptation we developed.

2.1   Distribution-Balanced Stratified Cross-validation

Following Moreno-Torres et al. [14], the distribution-balanced stratified cross-validation (DBSCV) [24] attempts to generate folds representative of the full dataset by assigning neighboring instances to different folds. Specifically, DBSCV randomly selects an instance and assigns it to a fold; it then jumps to the nearest instance of the same class and assigns it to the next fold. These steps are repeated until all instances of that class have been assigned to a fold. The same process is applied to the other classes so that the folds have approximately the same number of instances per class. Assuming a balanced distribution of the instances, building pairwise distance matrices for each class has complexity $\mathcal{O}(C(\frac{N}{C})^2) = \mathcal{O}(\frac{N^2}{C})$, where N and C are the numbers of instances and classes. The search-and-hop step has complexity $\mathcal{O}(C\frac{N}{C}\frac{N}{C})$, so that the final complexity of the algorithm is $\mathcal{O}(\frac{N^2}{C})$.

2.2   Distribution Optimally Balanced Stratified Cross-validation

The distribution optimally balanced stratified cross-validation (DOBSCV) is a modification of DBSCV. It also starts on a random instance of the datasets, but instead of hopping to the closest one of the same class, DOBSCV finds the (k-1) nearest neighbors of the current instance belonging to the same class and assigns each of them to a different fold. This process is repeated independently for each class, similarly to DBSCV, until all instances have been assigned to a fold. Our implementation of DOBSCV also uses a pairwise distance matrix for each class. Assuming balanced classes, building the matrices has complexity $\mathcal{O}(\frac{N^2}{C})$ and searching the k-NN for the selected instances in each class can be done in $\mathcal{O}(C\frac{N}{kC}\frac{kN}{C})$, resulting in an overall asymptotic complexity of $\mathcal{O}(\frac{N^2}{C})$.

2.3   Clustering-Based Approaches

Diamantidis et al. [9] introduced unsupervised stratification for cross-validation, based on dataset clustering. Although they have also explored hierarchical clustering, their main proposed algorithm is using k-means to cluster the dataset into M clusters. The instances inside each cluster are then sorted by their distances to the cluster center in ascending order. Finally, they assign adjacent instances to different folds, i.e., they make a pass over the sorted list of instances assigning each to a different fold. Note that the number of folds K, clusters M, and

classes C need not be equal. We refer to this method as cluster-based stratified cross-validation (CBDSCV). The unsupervised stratification process, however, does not guarantee that the classes are stratified in the usual sense, *i.e.*,
the method does not necessarily generates folds with the same proportion of
instances per class as the original dataset. K-Means has an average complexity of $\mathcal{O}(MNT)$, where T is the number of iterations, and sorting each cluster
can be done in $\mathcal{O}(MN \log N)$. Therefore, CBDSCV has a complexity given by
$\mathcal{O}(MN(T + \log N))$.

Mini-Batch CBDSCV. The running time of the CBDSCV algorithm is generally dominated by k-means. Therefore, we propose the use of mini-batch k-means
[21] as a way of reducing the cost of performing CBDSCV. Mini-batch k-means
is an adaptation of k-means with two major differences. First, at each iteration,
it selects only a batch of samples instead of the whole dataset. These samples
are then assigned to the nearest centroid. Then, instead of computing the new
centroid as the mean of all instances assigned to a cluster, it iterates over the
instances of the cluster, updating the centroid at each instance using a learning rate η inversely proportional to the number of times this centroid has been
updated previously. Mini-batch k-means converges faster than k-means while
producing results that are only slightly worse [2, 21]. In the following sections,
we will refer to our adaptation of CBDSCV as CBDSCV_Mini.

3   Experiments

The experiments performed here were designed to evaluate whether there is
a cross-validation splitting strategy which generally outperforms the others in
terms of bias, variance, or computational cost. Moreover, we also wish to study
whether the imbalance of the datasets may influence the quality of the splitters
estimations. Since the estimations they produce may depend on the dataset,
classifier, and also on the metric being estimated, we experimented with 20
different datasets (from PMLB [16]) and 4 different classifiers. The datasets
were selected so that two groups would be apparent, one with balanced and the
other with imbalanced datasets. The complete list is shown in Table 1.

   Note that we use the same class imbalance measure $I \in [0, 1]$ as in [16],
defined by $I = \frac{K}{K-1} \sum_{i=1}^{K} \left(\frac{n_i}{N} - \frac{1}{K} \right)^2$, where K is the number of classes, n_i is
the number of instances in class i, and N is the dataset size. Imbalance is 0 when
the classes are equally distributed and approaches 1 when almost all instances
belong to the same class. When analyzing balanced datasets, we evaluated the
splitters in terms of their accuracy estimations, as this is the most common and
traditional metric. However, when handling imbalanced datasets, we used the
F1 score since accuracy is inappropriate in these cases. We used the average
between the F1 scores computed for each class, *i.e.*, the macro average.

   We chose learning algorithms that presented different biases and variance
levels. Specifically, we experimented using Logistic Regression (LR), Decision
Trees (DT), Support Vector Machines (SVC), and Random Forests (RF). We

Table 1. List of datasets used in the experiments. Datasets with Imbalance higher than 0.20 were considered imbalanced.

Dataset	Examples	Attributes	Classes	Imbalance	Clusters
allrep	3772	29	4	0.91	7
analcatdata_cyyoung8092	97	10	2	0.26	3
analcatdata_dmft	797	4	6	0.00	5
analcatdata_germangss	400	5	4	0.00	4
analcatdata_happiness	60	3	3	0.00	4
analcatdata_japansolvent	52	9	2	0.00	3
appendicitis	106	7	2	0.36	6
backache	180	32	2	0.52	5
car	1728	6	4	0.39	6
chess	3196	36	2	0.00	4
colic	368	22	2	0.07	5
dna	3186	180	3	0.08	1
flare	1066	10	2	0.43	5
hepatitis	155	19	2	0.34	5
movement_libras	360	90	15	0.00	5
new_thyroid	215	5	3	0.30	6
page_blocks	5473	10	5	0.76	4
postoperative_patient_data	88	8	2	0.21	4
vote	435	16	2	0.05	2
vowel	990	13	11	0.00	4

have used only the RBF kernel with SVC since linear decision functions could be represented by Logistic Regression. To avoid overfitting when handling class-imbalanced datasets, weights associated with the instances of each class during training were set to be inversely proportional to the class frequencies in the training set. Prior to the experiments, we tuned each classifier to each dataset using the entire data and grid search. The performance of each hyperparameter set was evaluated using 5-fold cross-validation and the hyperparameters which showed the highest F1 score were chosen. These selected hyperparameters for a classifier-dataset pair were fixed for the experiments so that the classifiers were always trained with the same hyperparameters independently of the splitting strategy being analyzed. With this approach, we aim to capture performance differences caused by variation in the splitting strategy rather than by variation in the hyperparameters values.

Finally, we compared three splitting strategies, CBDSCV, DBSCV, and DOBSCV, against traditional k-fold cross-validation and stratified k-fold cross-validation. We also included our adaptation of CBDSCV, which uses mini-batch k-means for faster computation of the clusters, using batches of size 100. Therefore, six different cross-validation splitting strategies were compared in terms of their bias and variance, as well as computational cost. Our implementations of the splitting strategies, the selected hyperparameters, and the code used in experiments are available online[1].

3.1 Estimating the Bias and the Variance

The cross-validation methods considered here attempt to estimate the test performance of the learning algorithms fitted to the datasets. The bias of a cross-validation method is defined as the difference between the expected estimate and the (true) test performance [11]. Since we are working with real datasets, it is unfeasible to obtain the test performance. However, we can compute estimations for it using repeated holdout a large number of times, similarly to [3,11]. Specifically, we estimated the true performance for each dataset and classifier by repeating a stratified holdout 100 times, using 90% of the dataset for training, and getting the mean value. We chose a small test set in order to reduce the bias caused by using smaller training sets, while we expect that the high number of repetitions will attenuate the variance of the holdout.

The expected estimate of each cross-validation technique was computed for each dataset by resampling 90% of the dataset without repetition 20 times and applying the cross-validation technique to obtain the estimates of the true performance. The average value of the 20 estimates was used as the expected estimate of the cross-validation method. That is, let CV_i be the performance estimate of running k-fold cross-validation on a given dataset and learning algorithm, with a chosen splitting strategy, then we approximated the expected cross-validation estimate as

$$\overline{CV} = \frac{1}{20} \sum_{i=1}^{20} CV_i. \tag{1}$$

Finally, we computed the bias using $b_{CV} = \overline{CV} - \hat{P}$, where $\hat{P}$ is the estimation of the true performance that was computed using 100-times repeated stratified holdout, as described above.

The other important quantity that determines the quality of an estimator is its variance. We computed the variance of the cross-validation estimates using

$$s_{CV}^2 = \frac{1}{20-1} \sum_{i=1}^{20} (CV_i - \overline{CV})^2. \tag{2}$$

In this paper, however, we will work with the standard deviation (std) s, since we believe it is more easily readable. Note that an estimator with high variance may give poor results even if it has a low bias since one may not have the luck to obtain one of the estimates closer to the true value.

We evaluated the bias and variance of the six different dataset partitioning strategies over 20 different datasets and four classifiers. For each k-fold cross-validation strategy, we experimented with 2, 5, and 10 folds. Finally, we used accuracy and F1 as the performance metrics.

---

[1] https://github.com/froestiago/K-Fold-Partitioning-Methods/tree/bracis22.

3.2   Defining the Number of Clusters

The cluster-based method requires a number of clusters to be given as input. Ideally, we would compute the number of clusters right before each splitting is performed. However, this would be too computationally expensive, since the number of experiments performed is already large. Therefore, we have chosen to estimate the number of clusters for each dataset prior to the main experiments, and use this number (rounded to the nearest integer) for all cluster-based splitter methods. We have followed the same strategy as Diamantidis *et al.* [9] to estimate the number of clusters, which was based on repeatedly applying hierarchical clustering to small samples of the datasets and using a threshold on the similarity between clusters being merged to determine the number of clusters. The resulting number of clusters for each dataset is shown in Table 1.

4   Results and Discussion

The experiments performed as described in the previous section result in 80 different samples of bias and variance for each k-fold splitter, where $k = 2$, 5, and 10. Each of the 80 samples corresponds to a dataset-classifier pair. In the next sections, we describe the results grouped by balanced and imbalanced datasets in terms of class labels, resulting in 40 dataset-classifier pairs for each group. We focus mainly on the results with 2 and 10 folds, but the Figures for the 5-folds case are available in our git page (See footnote 1).

4.1   Balanced Datasets

The bias and standard deviations of each 10-fold cross-validation splitting strategy for all datasets and classifiers are summarized in Fig. 1. All the methods showed a general tendency to very low bias and similar standard deviations, indicating that there is no solution that consistently performs better than all others.

Note, however, that this does not imply that the accuracy (or F1) estimates produced by each partitioning strategy is not different. The p-values for the Friedman tests [7] comparing the estimates of the splitters are shown in Table 2. In particular, the p-value for the estimates considered here is 0.0279, suggesting that the bias estimates differ depending on the splitting strategy. However, there is no significant difference in terms of the standard deviations. Table 3 shows the number of times each method performed the best. For 10 folds, accuracy, and balanced datasets, stratified 10-fold CV had the most wins for both bias and std.

Reducing the number of folds increases the bias (in absolute terms) and the standard deviations, as shown in Fig. 2. However, the methods still have similar performance overall. We note, however, that DOBSCV and CBDSCV had an increase in the number of times they had the best results, while stratified CV showed worse results compared with its performance in the 10-folds scenario,

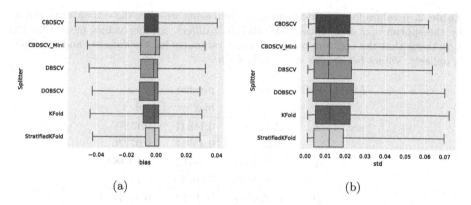

Fig. 1. (a) Bias and (b) standard deviation of each splitter method across all balanced datasets and classifiers. Each splitter runs 10-folds.

particularly with respect to the standard deviation of the estimates. This is an indication that the DOBSCV and CBDSCV can be useful when a reduced number of folds is desired so that the computational cost resulting from training various models can be reduced.

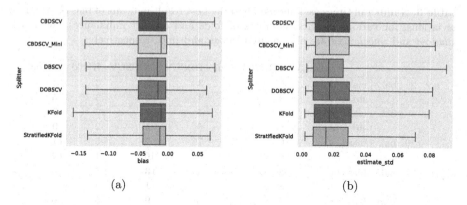

Fig. 2. (a) Bias and (b) standard deviation of each splitter method across all balanced datasets and classifiers. Each splitter runs 2-folds.

4.2   Imbalanced Datasets

The bias and standard deviations for 10-folds and imbalanced datasets are shown in Fig. 3. Since accuracy is not appropriate for studying imbalanced datasets, the bias and std were calculated for the f1-score estimations. Stratified 10-folds showed the best results for both bias and standard deviation. Tables 2 shows that the difference between the splitters is significant in the imbalanced cases, and

Table 2. p-values for the Friedman tests comparing whether the estimates produced by the splitters for each dataset-classifier pair differs. Smaller values mean that the hypothesis that the splitters produce similar estimates for the datasets and classifiers is unlikely. Values below 0.05 are in bold form.

Metric	Splits	Balance	p-value	
			bias	std
acc.	2	Balanced	0.11841	0.13278
		Imbalanced	0.45917	0.07770
	5	Balanced	0.58700	**0.01481**
		Imbalanced	0.09520	0.10409
	10	Balanced	**0.02790**	0.45271
		Imbalanced	0.72980	0.07762
f1	2	Balanced	**0.01858**	0.24038
		Imbalanced	0.10906	0.20719
	5	Balanced	**0.00462**	**0.00055**
		Imbalanced	**<0.00001**	**0.00158**
	10	Balanced	**<0.00001**	**0.03294**
		Imbalanced	**<0.00001**	**0.00069**

Table 3. Number of times each method had the best result in terms of bias or standard deviations, for various metrics, numbers of folds and dataset imbalance. The words *balanced* and *imbalanced* are abbreviated to *bal.* and *imb.*, respectively.

				CBDSCV	CBDSCV_Mini	DBSCV	DOBSCV	KFold	SKFold
acc	2	bal.	bias	2	9	3	**12**	3	11
			std	7	9	4	**12**	3	5
	5	bal.	bias	6	5	7	7	7	**8**
			std	8	8	8	5	2	**9**
	10	bal.	bias	4	6	7	7	5	**11**
			std	5	6	6	10	1	**12**
f1	2	bal.	bias	0	8	4	**13**	5	10
			std	8	8	5	**11**	3	5
		imb.	bias	6	5	**9**	**9**	5	6
			std	3	7	9	8	2	**11**
	5	bal.	bias	6	5	2	5	7	**15**
			std	11	8	5	5	1	10
		imb.	bias	6	6	5	1	7	**15**
			std	6	6	8	3	3	**14**
	10	bal.	bias	2	4	4	5	4	**21**
			std	5	8	5	9	2	**11**
		imb.	bias	4	4	2	1	5	**24**
			std	8	7	1	5	3	**16**

Table 3 shows that indeed stratified 10-fold presents the less biased and most consistent estimates for most datasets and classifiers. It is interesting to note that DBSCV and DOBSCV deal with class stratification by performing their splitting strategies per class. The fact that their performance was worse than stratified cross-validation suggests that there may be more appropriate ways to develop stratified versions of DBSCV and DOBSCV. The CBDSCV techniques, however, do not handle class imbalance directly.

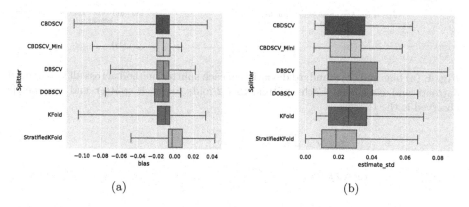

Fig. 3. (a) bias and (b) standard deviation of each splitter method across all imbalanced datasets and classifiers. Each splitter runs 10-folds for each splitter and the metric observed is the f1 score.

When one reduces the number of folds to 2, both the bias (absolute value) and the std of the estimates increase. More interestingly, the advantage that the stratified cross-validation had almost disappeared. This pattern is similar to the one observed for the balanced datasets. Table 3 shows that the number of times the SCV performs best indeed reduces when compared to the 10-folds case (Fig. 4).

4.3 Running Times

We also compared the running times of each splitting strategy. The running time of a k-fold splitting strategy for a dataset was obtained by averaging the running times of the splitting process over all the 20 runs and the 4 classifiers. We noticed that the running times obtained for 2, 5, and 10 folds were very similar, and therefore we consider only the 10-folds case in the following discussions. Figure 5 shows the running times of each splitter method for the 20 datasets considered. One can see that the classical strategies, KFold and StratifiedK-Fold, have negligible running times when compared to the others. Furthermore, we note that DBSCV and DOBSCV have higher variability depending on the

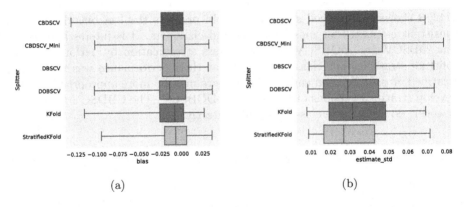

Fig. 4. (a) bias and (b) standard deviation of each splitter method across all imbalanced datasets and classifiers. Each splitter runs 2-folds for each splitter and the metric observed is the f1 score.

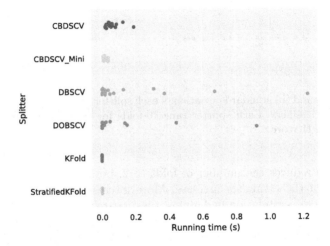

Fig. 5. Average running times in seconds across all the 20 datasets. The running times correspond to the use of 10 folds.

dataset, reaching the highest running times of all methods. In comparison, CBD-SCV and CBDSCV_Mini have closer run times for all datasets.

Ignoring KFold and StratifiedKFold, however, DOBSCV was actually the fastest method in 14 out of 20 datasets, while CBDSCV_Mini was the quickest in the other six. Specifically, it was fastest in 'analcatdata_germangss', 'movement_libras', 'analcatdata_dmft', 'appendicitis', 'page_blocks', and 'postoperative_patient_data', which are the datasets with more instances. This is expected if one considers the algorithmic complexity of each method: DOBSCV scales quadratically with the number of samples.

4.4   Cluster-Based Splitters

When comparing only the cluster-based approaches, CBDSCV and CBD-SCV_Mini, we observed that the running times on the minibatch version were smaller for all datasets, as expected. Specifically, it ran on average 2.4 times faster than CBDSCV. Furthermore, we have detected no significant change between estimations given by CBDSCV and CBDSCV_Mini. Table 4 shows the number of times each method performed better than the other, for all datasets and classifiers, and the p-value computed using the Wilcoxon test.

Table 4. Number of times each cluster-based method had the best result in terms of bias and variance. Only the cluster-based methods are considered here. The last columns shows the p-value for the two-sided Wilcoxon signed-rank test.

			CBDSCV	CBDSCV_Mini	p-value
Accuracy	2	bias	32	48	0.07208
		std	39	41	0.94265
	5	bias	43	37	0.88180
		std	40	40	0.91216
	10	bias	44	36	0.18399
		std	35	45	0.36078
f1	2	bias	35	45	0.04396
		std	38	42	0.64348
	5	bias	43	37	0.79371
		std	43	37	0.91596
	10	bias	33	47	0.12133
		std	38	42	0.61814

5   Related Work

Besides more recent theoretical works on traditional cross-validation estimates [4,20,22], various works have proposed cross-validation splitting strategies for different scenarios which are not directly related to our cases. Motl et al. [15] developed a technique based on linear programming for performing label-based stratification when the instances have more than one label at the same time, i.e., multi-label datasets. Specific methods have also been developed for data drift situations, where some instances become obsolete over time [13], or for the specific case of dataset shift in credit card validation [19]. Cross-validation over graphs [8,12] has also seen some recent works, as well as methods for reducing cross-validation computational cost in deep learning [6]. Cross-validation adaptations have been developed in order to handle duplicate data in medical records [1], for calibration models in chemistry, [23] and for infrared and mass-spectroscopy images [17,18]. All the methods cited above, however, are developed for different specific scenarios, whereas the methods explored here aim at tabular data and single-label datasets. Closer to the methods explored in this work are the proposals by Budka *et al.* [3] and Cervellera *et al.* [5]. In both works, the methods attempt to partition a dataset by generating samples whose distributions are as similar as possible to the distribution of the original data. We were not able to include them in this work due to lack of compatibility with the framework we had developed for our cross-validation methods comparison. Nevertheless, we will be including them in future works we are developing in this area of research. We note also that these two approaches have not been compared with each other yet.

6   Conclusion and Future Work

In this work, we proposed an adaptation of a cluster-based technique for splitting a dataset for cross-validation. We also compared various CV strategies using different classifiers for balanced and imbalanced datasets. We found that no method consistently outperforms all others in terms of bias or standard deviation when estimating accuracy using 10 folds and balanced datasets. In these cases, traditional stratified cross-validation remains a good choice. When the number of folds is reduced to 2, however, stratified cross-validation may produce accuracy estimates with higher variance than DOBSCV and the cluster-based techniques.

When considering F1 score estimates, traditional stratified cross-validation produced the best results in terms of bias and variance for most datasets and classifiers when used with 5 and 10 folds, for both balanced and imbalanced datasets. When the number of folds is reduced, however, F1 scores in balanced datasets may be better estimated by other methods such as DOBSCV and the cluster-based splitter. For imbalanced datasets, SCV remained the most frequent winner. In particular, traditional SCV was most significantly better when F1 score and imbalanced datasets were present. This suggests that better class-based stratification adaptations can be developed for DBSCV and DOBSCV. The development of a supervised version of CBDSCV is also an interesting topic for further work. Finally, we found no significant change in the quality of the estimations produced by CBDSCV and CBDSCV_Mini, whereas the mini-batch version is significantly less expensive in terms of computational cost.

We have not studied dataset characteristics such as the presence of subconcepts in the input space, as this kind of information is not easily extracted from a dataset. Those characteristics, however, may be relevant in determining which performance estimator should be used and may provide deeper insight into the use cases of each method. Similarly, we haven't analyzed deeper whether some splitting strategies may work better for each classifier or for datasets with different sample sizes. Finally, in future works, we intend to expand our experimental comparison and explore other approaches proposed in the literature [3,5].

References

1. Bey, R., Goussault, R., Grolleau, F., Benchoufi, M., Porcher, R.: Fold-stratified cross-validation for unbiased and privacy-preserving federated learning. J. Am. Med. Inform. Assoc. **27**(8), 1244–1251 (2020). https://doi.org/10.1093/jamia/ocaa096

2. Bottou, L., Bengio, Y.: Convergence properties of the k-means algorithms. In: Tesauro, G., Touretzky, D., Leen, T. (eds.) Advances in Neural Information Processing Systems, vol. 7. MIT Press (1994)

3. Budka, M., Gabrys, B.: Density-preserving sampling: robust and efficient alternative to cross-validation for error estimation. IEEE Trans. Neural Netw. Learn. Syst. **24**(1), 22–34 (2013). https://doi.org/10.1109/TNNLS.2012.2222925

4. Celisse, A., Mary-Huard, T.: Theoretical analysis of cross-validation for estimating the risk of the k-nearest neighbor classifier. J. Mach. Learn. Res. **19**(1), 2373–2426 (2018). JMLR. org

5. Cervellera, C., Maccio, D.: Distribution-preserving stratified sampling for learning problems. IEEE Trans. Neural Netw. Learn. Syst. 1–10 (2017). https://doi.org/10.1109/TNNLS.2017.2706964

6. Cheng, J., et al.: dwt-cv: dense weight transfer-based cross validation strategy for model selection in biomedical data analysis. Futur. Gener. Comput. Syst. **135**, 20–29 (2022). https://doi.org/10.1016/j.future.2022.04.025

7. Corder, G.W., Foreman, D.I.: Nonparametric statistics for non-statisticians (2011)

8. Dabbs, B., Junker, B.: Comparison of cross-validation methods for stochastic block models. Technical report, arXiv:1605.03000, arXiv (May 2016), arXiv:1605.03000 [stat] type: article

9. Diamantidis, N., Karlis, D., Giakoumakis, E.: Unsupervised stratification of cross-validation for accuracy estimation. Artif. Intell. **116**(1–2), 1–16 (2000). https://doi.org/10.1016/S0004-3702(99)00094-6

10. James, G., Witten, D., Hastie, T., Tibshirani, R.: An Introduction to Statistical Learning, vol. 112, chap. 5. Springer, New York (2013). https://doi.org/10.1007/978-1-4614-7138-7

11. Kohavi, R., others: A study of cross-validation and bootstrap for accuracy estimation and model selection. In: IJCAI, vol. 14, no. 2, pp. 1137–1145. Montreal, Canada (1995)

12. Li, T., Levina, E., Zhu, J.: Network cross-validation by edge sampling. Biometrika **107**(2), 257–276 (2020). https://doi.org/10.1093/biomet/asaa006

13. Maldonado, S., López, J., Iturriaga, A.: Out-of-time cross-validation strategies for classification in the presence of dataset shift. Appl. Intell. **52**(5), 5770–5783 (2021). https://doi.org/10.1007/s10489-021-02735-2

14. Moreno-Torres, J.G., Saez, J.A., Herrera, F.: Study on the impact of partition-induced dataset shift on k-fold cross-validation. IEEE Trans. Neural Netw. Learn. Syst. **23**(8), 1304–1312 (2012). https://doi.org/10.1109/TNNLS.2012.2199516

15. Motl, J., Kordík, P.: Stratified cross-validation on multiple columns. In: 2021 IEEE 33rd International Conference on Tools with Artificial Intelligence (ICTAI), pp. 26–31, November 2021. https://doi.org/10.1109/ICTAI52525.2021.00012

16. Olson, R.S., La Cava, W., Orzechowski, P., Urbanowicz, R.J., Moore, J.H.: PMLB: a large benchmark suite for machine learning evaluation and comparison. BioData mining **10**(1), 1–13 (2017)

17. Pérez-Guaita, D., Kuligowski, J., Lendl, B., Wood, B.R., Quintás, G.: Assessment of discriminant models in infrared imaging using constrained repeated random sampling-cross validation. Analytica Chimica Acta **1033**, 156–164 (2018). Elsevier

18. Pérez-Guaita, D., Quintás, G., Kuligowski, J.: Discriminant analysis and feature selection in mass spectrometry imaging using constrained repeated random sampling - Cross validation (CORRS-CV). Anal. Chim. Acta **1097**, 30–36 (2020). https://doi.org/10.1016/j.aca.2019.10.039
19. Qian, H., Wang, B., Ma, P., Peng, L., Gao, S., Song, Y.: Managing dataset shift by adversarial validation for credit scoring (2021). https://doi.org/10.48550/ARXIV.2112.10078
20. Santos, M.S., Soares, J.P., Abreu, P.H., Araujo, H., Santos, J.: Cross-Validation for Imbalanced Datasets: avoiding overoptimistic and overfitting approaches [research frontier]. IEEE Comput. Intell. Mag. **13**(4), 59–76 (2018). https://doi.org/10.1109/MCI.2018.2866730
21. Sculley, D.: Web-scale k-means clustering. In: Proceedings of the 19th International Conference on World Wide Web, pp. 1177–1178 (2010)
22. Wong, T.T., Yeh, P.Y.: Reliable accuracy estimates from k-fold cross validation. IEEE Trans. Knowl. Data Eng. **32**(8), 1586–1594 (2020). https://doi.org/10.1109/TKDE.2019.2912815
23. Xu, Q.S., Liang, Y.Z.: Monte Carlo cross validation. Chemom. Intell. Lab. Syst. **56**(1), 1–11 (2001). https://doi.org/10.1016/S0169-7439(00)00122-2
24. Zeng, X., Martinez, T.R.: Distribution-balanced stratified cross-validation for accuracy estimation. J. Exp. Theor. Artif. Intell. **12**(1), 1–12 (2000). https://doi.org/10.1080/095281300146272

Geographic Context-Based Stacking Learning for Election Prediction from Socio-economic Data

Tiago Pinho da Silva[1]([⊠]) [iD], Antonio R. S. Parmezan[1] [iD],
and Gustavo E. A. P. A. Batista[2] [iD]

[1] University of São Paulo, São Carlos, Brazil
tpinho@usp.br
[2] University of New South Wales, Sydney, Australia

Abstract. Voting behavior analysis involves understanding factors influencing an election to identify possible trends, new features, and extrapolations. A growing body of research has joined efforts to automate this process from high-dimensional spatial data. Although some studies have investigated machine learning methods, the capability of this artificial intelligence subarea has not been fully explored due to the challenges posed by the spatial autocorrelation structure prevalent in the data. This paper advances the current literature by proposing a geographic context-based stacking learning approach for predicting election outcomes from census data. Our proposal models data in spatial contexts of different dimensions and operates on them at two levels. First, it captures local patterns extracted from spatial contexts. Then, at the meta-level, it globally captures information from the K contexts nearest to a region we want to predict. We introduce a spatial cross-validation-driven experimental setup to assess and compare the stacking approach with state-of-the-art methods fairly. This validation mechanism aims to diminish spatial dependence's influence and avoid overoptimistic results. We estimated a considerable multi-criteria performance of our proposal concerning baseline and reference models taking data from the second round of the 2018 Brazilian presidential elections into account. The stacking approach presented the best overall performance, being able to generalize better than the compared ones. It also provided intelligible and coherent predictions in challenging regions, emphasizing its interpretability. These results evidence the potential use of our proposal to support social research.

Keywords: Ensemble learning · Metalearning · Preferential voting · Spatial dependence · Voting behavior

1 Introduction

Elections are non-trivial processes essential to any representative democracy, which can provide the best expression of public opinion and party involvement. A

This study was financed in part by the Coordenação de Aperfeiçoamento de Pessoal de Nível Superior – Brasil (CAPES) – Finance Code 001.

J. C. Xavier-Junior and R. A. Rios (Eds.): BRACIS 2022, LNAI 13653, pp. 641–656, 2022.
https://doi.org/10.1007/978-3-031-21686-2_44

post-election data analysis allows us to describe voting behavior and the aspects that guide it [4]. Understanding voting behavior is vital to identifying trends and factors influencing election results [6,10].

Researchers consider that the electoral processes are associated with the population characteristics regarding the locations where they occur. Thus, an electoral process comprises aspects that indicate local patterns related to spatial autocorrelation and local relationships across space [13]. In this perspective, people from the same region tend to present similar voting behavior, while those from distinct areas may have different vote distributions.

Considering how people are geographically contextualized and the data's spatial characteristics can enrich our understanding of electoral processes. We have witnessed an increasing number of interdisciplinary studies aimed at predictive modeling election features from thousands of explanatory spatial features [7,12]. However, the high dimensionality and spatial autocorrelation structure inherent in such data limit the ability of conventional learning models to capture the relationships between spatial features completely.

Many econometric and machine learning methods, which can deal with the curse of dimensionality, totally ignore the geography present in electoral data, such as spatial boundaries, clustering effects, and distance measures [1,3]. Consequently, they treat data separated into regions as independent and identically distributed. In the opposite direction, recent studies have suggested using spectral and spatial filtering Graph Convolutional Neural Network (GCNN) methodologies to enrich election data modeling [7]. Such methods seem to adequately fit the problem at hand, given the intrinsic graph structure of electoral data.

This work advances the literature on voting behavior analysis by proposing a geographic context-based stacking learning approach to describe election outcomes from thousands of census features. Our proposal models data in spatial contexts of different dimensions and operates on them at two levels: (i) at the base level, it captures local patterns extracted from spatial contexts; (ii) at the meta-level, it globally captures information from the K contexts nearest to a region we want to predict. Furthermore, we introduce a spatial cross-validation-driven experimental setup to assess and compare the stacking approach with state-of-the-art methods fairly. This validation mechanism can generate robust assessments by diminishing the spatial dependence's influence and consequently avoiding overoptimistic results [11,12].

We estimated a considerable multi-criteria performance of our proposal concerning two baselines and the state-of-the-art Hierarchical GCNN method taking data from the second round of the 2018 Brazilian presidential elections into account. The stacking approach exhibited the best overall performance, being able to generalize better than the compared ones. It also led to intelligible and coherent predictions in challenging regions, highlighting its interpretability. These results demonstrate the potential use of our proposal to support social research.

The rest of this paper is organized as follows: Sect. 2 introduces the background and related work. Section 3 describes our geographic context-based stacking learning approach. Section 4 reports the case study involving data from the second round of the 2018 Brazilian presidential election. Finally, Sect. 5 concludes the study and highlights future work.

2   Background and Current Trends

This section defines the mathematical notation that models election voting behavior considering the spatial characteristics of the data. It also discusses related work covering the most recent advances in the literature.

2.1   Problem Definition and Research Challenges

We can formulate the problem in question as follows. First, let us specify a set of lattice-type spatial objects O, where each object o_i is a polygon that delimits a region in the spatial domain (*e.g.*, neighborhoods, districts and cities). Note that the spatial intersection between any distinct objects o_i and $o_j \in O$ is the empty set ($\emptyset$). Now, let us assume a spatial dataset $D = \{X, Y\}$ that characterizes each of the n objects in O. The target feature, $Y \in \mathbb{R}$, reflects the vote shares (vote percentage) for each spatial object in O from a given candidate or party. The explanatory features, $X \in \mathbb{R}^m$, where $m > 0$ is the number of characteristics, describes the spatial objects from O in another election-related domain. Let us also consider a set of spatial contexts C with boundaries that segment D in the geographic space, where C can be defined based on preexisting boundaries (*e.g.*, states and macro-regions). The objective is to generate a model $F(D, C)$ that learns local relationship patterns from the spatial contexts present in D.

Modeling local relationships between explanatory features and the target feature (vote shares) is not a trivial task. These relationships may vary across spatial contexts, meaning that a relevant characteristic that can describe the vote shares from one context may not be useful to another [13]. Furthermore, in a conventional machine learning approach, local relationships are disregarded in favor of those that describe the vote shares globally [12,13].

Another challenge in modeling voting behavior relates to using Spatial Cross-Validation (SCV) as a sampling technique. While it is the most suitable procedure for assessing machine learning models built from spatial data, it generates unseen correlated distributions in the test set. This scenario happens because spatial boundaries determine the folds, and a removing buffer region is defined as a strategy to diminish the spatial dependence between the test and training sets [11,12]. Consequently, the test set distribution is not observed in the training set, and there are only correlated distributions.

Studies on applying machine learning methods for analyzing voting behavior are maturing through scientific debate. Section 2.2 briefly summarizes some related work in this field, emphasizing the challenges they brought.

2.2   Related Work

The vast literature on voting behavior varies from standard econometric techniques [3] to regression analysis [13] and machine learning models [1,7]. Econometrics and regression analysis studies usually focus on national-level estimators using surveys and economic features to understand election results. Although these methods are well established [3], applying them to thousands of features in several locations is challenging. Conversely, machine learning models can deal with the curse of dimensionality more naturally. However, most works employs social media data and sentiment analysis to understand voting behavior. Their results are commonly explored on a national scale, and spatial aspects are not considered [1].

Recently, researchers recommended using a hierarchical GCNN-based approach that can be considered state-of-the-art in voting behavior analysis via machine learning [7]. The authors combined the inherited hierarchical characteristic of the census and election data with the GCNN capability to learn local patterns and generate a model capable of predicting the vote shares from the 2016 Australia congress election with low error rates.

In contrast to existing analytical models, here we design a descriptive approach that considers thousands of socio-economic explanatory features and the involved spatial characteristic to analyze locally and comprehensively election outcomes across multiple locations.

3   Proposed Approach

We have identified two main challenges linked to the problem formalized in Sect. 2.1: (i) capturing local patterns that are occluded when globally modeling the data; and (ii) building a model that can generalize over different spatial contexts. This paper addresses these challenges by proposing a geographic context-based stacking approach to model local relationships at the ensemble level and globally capture information from contexts employing a meta-regressor.

When applied to regression tasks, the conventional stacking approach builds an ensemble using the entire training set to fit each base regressor. We typically choose regression algorithms from different paradigms to introduce diversity, generating a heterogeneous ensemble [2]. The predictions of each base regressor on a validation set give rise to an attribute-value table, which is employed to train a meta-regressor. The meta-regressor, in turn, learns how to ponder the base regressors' predictions to issue final predictions.

Our approach differs from traditional ones in the following aspects. First, we define the ensemble by the K nearest spatial contexts to the test set region. Such a strategy is based on the first law of geography, which states that "everything is related to everything else, but near things are more related than distant things" [14]. Second, we use spatial context to build the base regression models so that each model can capture local patterns related to the contexts. Lastly, the ensemble is homogeneous, meaning we adopt the same base regressor method. However, the diversity comes from the spatial contexts that present different dimensions (#instances × #features), following the idea that a different set of features may describe each context.

Figure 1 outlines the steps of our approach. In Step 1, we group the training data in agreement with a pre-defined set of spatial contexts and select the K

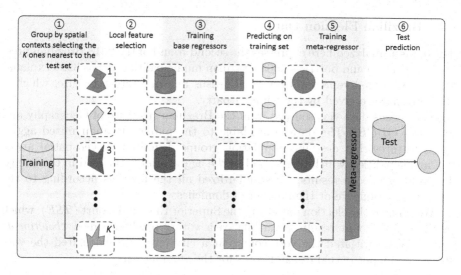

Fig. 1. Proposed approach.

ones nearest to the test set, considering geographic proximity; the number of instances in each spatial context can vary. In Step 2, we run a feature selection method for each spatial context data, generating K spatial context with different dimensions. In Step 3, we build an ensemble where each base regressor is fitted to a spatial context. In Step 4, we use the base regressors that make up the ensemble to predict the entire training set, creating an attribute-value table. In Step 5, we employ the data table to train a meta-regressor that will seek to ponder the local knowledge learned by the base regressors to maximize the generalization potential. Finally, in Step 6, the meta-regressor uses the predictions from the base regressors on the test set to provide final predictions.

As can be seen, our approach operates on two levels. The first level learns local patterns from geographically contextualized data samples, *i.e.*, regions containing mutually exclusive instances described by an optimal feature subset. In a complementary way, the second level extracts global knowledge of the local patterns to predict a region of interest.

4   Case Study

In 2018, Brazilians went to the polls to vote for their president. The final result was 55.13% for the Social Liberal Party (Jair Bolsonaro) and 44.87% for the Worker's Party (Fernando Haddad). This election, however, was marked by a highly polarized environment and flooded by distrust in the voting system [4]. Understanding outcomes in this context is essential to discuss the external factors influencing voting decisions and identify the geographic and socio-economic extensions of the processes that undermine democratic foundations.

4.1   Brazilian Election Data

The dataset analyzed here expresses the second round of the 2018 Brazilian presidential election and portrays 5565 Brazilian municipalities. It has 3999 explanatory features that represents the 2010 census and one target feature, which is the vote share received by the winning party.

The census data was sourced from the Brazilian Institute of Geography and Statistics (*IBGE*). The data is available to the public via anonymized aggregated features that describe population groups delimited by geospatial areas, *e.g.* municipalities, which correspond to the aggregation level used in this study. To avoid erroneous results, we standardized all the features according to the city's population size or the number of domiciles.

We sourced the election data from the Superior Electoral Court (*TSE*), which provides vote count results regarding each voting machine called *"boletim da urna"*. We aggregated the vote counts at a city level and calculated the vote shares as the percentual of valid votes for the winning party.

4.2   Machine Learning Approaches and Algorithms

Standard econometric methods are not comparable with our approach. They often focus on regression analysis and employ data at higher aggregation to provide national-level predictions [3]. The HIERARCHICAL GCNN model [7], in turn, can be used as a representative of state-of-the-art applied machine learning research. We adopted this method parameterized according to the best results in [7], named variation 2. Furthermore, we considered city-level data as the prediction layer and state-level data to create the second aggregation layer.

We also defined two baselines, GLOBAL and LOCAL MEAN, to be compared with our proposal, addressed from now on as LOCAL META. GLOBAL is the conventional approach that selects features and fits models favoring global relationships. LOCAL MEAN is an ensemble of contextual models that employs the average as a fusion function to compose the final predictions. These baselines can help us understand in which situations LOCAL META performs best and explain how the stacking strategy increases performance and improves generalization.

We investigated two configurations involving the number of spatial contexts (K) for LOCAL MEAN and LOCAL META. The first uses all the contexts in the training set ($K = C$), while the second employs the seven closest contexts ($K = 7$) to the prediction area. Note that 7 is the mean of each context's neighbors. This configuration, in particular, aims to answer whether filtering contexts based on the prediction area proximity can enhance results.

Concerning the base regressors, we considered nine belonging to different machine learning paradigms. Table 1 lists these algorithms and their parameters. As the meta-regressor for LOCAL META, we chose Ordinary Least Squares (OLS) since it is a simple and parameterless model. We adopted the Correlation-based Feature Selection (CFS) method to reduce the attribute space. CFS aims to find a minimal optimal subset of features that are highly correlated with the target and not very redundant with each other.

Table 1. Base regressors and their parameters. The acronyms not yet defined are: k-Nearest Neighbors (kNN), Least Absolute Shrinkage and Selection Operator (LASSO), Decision Tree (DT), Gradient Boosting DT (GBDT), Multiyear Perceptron (MLP), and Support Vector Regression (SVR).

Base regressor	Parameter	Value
kNN	Number of nearest neighbors (k)	3
OLS	—	—
LASSO	Regularization strength (α)	1
Ridge	Regularization strength (α)	1
ElasticNet	Constant that multiplies the penalty terms (α)	1
	Mixing parameter ($l1\_ratio$)	0.5
DT	Split criterion	GINI
GBDT	Number of boosted trees to fit ($n\_estimators$)	100
	Learning rate (ε)	0.1
MLP	Hidden layers (h)	1
	Hidden layer size (n)	$M/2$
	Learning rate (ε)	0.001
SVR	Kernel	RBF
	Gaussian's width of the radial basis kernel function (σ)	$1/(M*X.var())$
	Regularization parameter ($\mathbb{C}$)	1

Finally, we defined the spatial contexts for the ensemble approaches as the 26 Brazilian states save the Federal District, which has only one city. This decision comprises the understanding that, at a higher level, stakeholders such as political scientists and journalists are more interested in analyzing the election results considering known spatial boundaries like states.

4.3   Evaluation Measures

We assessed the approaches described in Sect. 4.2 using four individual performance measures: Mean Squared Error (MSE), Mean Error Standard Deviation (MESD), SPearman correlation (SP), and SPearman correlation Standard Deviation (SPSD). MSE expresses the approaches' performance in predicting the correct vote-share scale, while MESD reflects their stability regarding MSE. MSE does not indicate whether the order of the achieved predictions matches those in the ground truth. That is, if city A received more votes than city B, MSE does not tell us whether the approaches were able to capture this order. Thus, we employed SP to assess the order of the predictions yielded by the approaches. Finally, SPSD provides information on how SP is distributed over the folds.

We also applied a Multi-Criteria Performance Measure (MCPM) [9] to combine the four metrics mentioned above and thus guide the choice of adequate approaches. MCPM reflects the sum of the total area of an irregular polygon whose vertices comprise individual performance indexes. In this work, lower total area values indicate better predictive performances. Unlike the MSE and Standard Deviation measures, in which resulting values must be minimized, SP (ρ) must be maximized. Hence, we applied the SP complement: $1 - \rho$.

To understand the predictive power of our proposal, we employed the SHapley Additive exPlanation (SHAP) Values technique to analyze the results in the best and worst fold scenarios [8]. SHAP Values is a unified measure of feature importance widely used to comprehend predictions made by models. We believe that examining it within the scope of our application is indispensable, as it can reveal biased models and avoid misinterpretations. Especially for our approach, SHAP Values can be employed in the meta and base regressors to explain the most important spatial contexts to predict a given fold and the most relevant features of that context.

4.4   Experimental Setup

Figure 2 illustrates our experimental setup, which considers the space's role in evaluating models designed to predict election outcomes. As depicted in this figure, we used the dataset prepared in agreement with Sect. 4.1 (Step 1) to assess the approaches parameterized according to Sect. 4.2 (Step 2). We applied an SCV technique, which da Silva *et al.* [12] explicitly proposed for the application in question, to estimate the performance of the models (Step 3). We reported these results via the evaluation metrics described in Sect. 4.3 (Step 4). This experimental protocol assesses the investigated approaches more rigorously, as it avoids overoptimistic results by reducing the spatial dependence between test and training sets. While our multi-criteria analysis compares these models taking into account two important characteristics – the scale and the order of the vote shares –, our interpretability analysis is necessary to understand the patterns found and uncover biased models.

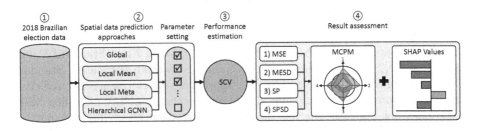

Fig. 2. Experimental setup.

The main difference between the SCV adopted in this work and the standard cross-validation lies in the fold definition so that in the former, the folds are determined based on preexisting geographic boundaries. Here, each spatial fold follows the geographic boundaries of the 26 Brazilian states, also employed as spatial contexts to build contextual base models in the ensemble approaches. Unlike the traditional cross-validation, in the SCV, the folds may have different distributions and sizes, creating a more challenging scenario for the approaches.

This study did not use folds 12 (Acre) and 13 (Rondonia) to calculate MSE and MSDE values because they present ambiguous distributions [5]; *i.e.*, they describe a population with similar socio-economic characteristics to the northeastern but with vote shares similar to the southern states. This fact requires the approaches to learn the opposite of what they observed in the training set. Scenarios like these are challenging and exhibit incredibly high error rates, impacting empirical assessments, specifically the choice of the best regressor to compose GLOBAL. We decided to keep the analysis of SP and SPSD on such folds considering that the raised issue is linked to scale and not to order.

Finally, we implemented the experimental setup of Fig. 2 employing the Python programming language combined with the following libraries: Pandas, GeoPandas, SciPy, PySAL, and Scikit-learn. Our code and supplementary material are available on the GitHub platform[1].

4.5 Results and Discussion

As we can see from the averaged values of the individual performance metrics (Table 2), LOCAL META $K = 7$ achieved the best MSE and MESD results, LOCAL MEAN $K = C$ presented the highest SP values, and LOCAL MEAN $K = 7$ stood out in terms of SPSD. We obtained all these results using MLP as a base regressor. However, there was no consensus regarding the best model configuration – approach and base regressor combination – concerning all the metrics. To identify the most promising model, we evaluated the configurations under three perspectives: (i) the MCPM to determine the best overall model; (ii) the performance per fold to identify the best context-level configuration; (iii) the model interpretability to understand the best configuration results.

Table 2. Overall results of the approaches considering each base regressor and the following metrics: MSE, MESD, SP, and SPSD. Green cells symbolize the best results.

Base regressors	GLOBAL				LOCAL MEAN K = ALL				LOCAL MEAN K = 7				LOCAL META K = ALL				LOCAL MEAN K = 7			
	MSE	MESD	SP	SPSD	MSE	MESD	SP	SPSD	MSE	MESD	SP	SPSD	MSE	MESD	SP	SPSD	MSE	MESD	SP	SPSD
kNN	239.68	267.93	0.46	0.16	272.04	187.92	0.58	0.15	303.78	216.51	0.50	0.17	179.50	157.93	0.56	0.17	173.60	162.65	0.51	0.17
OLS	121.81	134.17	0.59	0.15	271.08	197.79	0.60	0.19	219.96	170.12	0.59	0.13	224.67	195.01	0.55	0.20	125.81	141.26	0.57	0.13
LASSO	965.26	465.45	0.02	0.50	949.42	457.43	0.38	0.24	949.24	456.68	0.26	0.24	876.07	405.93	0.35	0.25	837.60	427.85	0.27	0.25
Ridge	304.48	194.94	0.58	0.16	355.77	214.79	0.64	0.12	428.35	247.88	0.62	0.13	124.69	129.78	0.57	0.16	132.77	128.39	0.61	0.14
ElasticNet	964.61	465.43	0.23	0.29	961.42	463.94	0.48	0.27	961.99	464.01	0.36	0.35	299.24	216.62	0.50	0.21	528.13	317.23	0.43	0.34
DT	269.36	355.57	0.32	0.22	223.96	178.53	0.52	0.17	287.03	232.21	0.41	0.19	213.99	189.83	0.48	0.19	230.87	201.68	0.43	0.19
GBDT	171.56	159.91	0.51	0.17	196.32	153.97	0.61	0.16	249.46	186.00	0.56	0.17	159.09	149.73	0.54	0.18	142.39	140.13	0.56	0.18
MLP	133.87	138.18	0.61	0.14	240.22	170.30	0.64	0.13	311.43	207.35	0.62	0.12	127.01	132.60	0.57	0.17	111.22	127.16	0.59	0.14
SVR	243.97	206.58	0.55	0.18	398.58	238.68	0.62	0.14	447.63	261.98	0.59	0.15	168.97	149.09	0.52	0.23	164.13	146.32	0.55	0.18

---

[1] https://github.com/tpinhoda/Spatial_Context_Stacking_Approach.

Multi-criteria Performance. Figure 3 shows, for each approach configuration, the MCPM values ranked in descending order of importance. LOCAL META $K = 7$ presented a more consistent behavior occupying the first and second positions for most configurations, with its lowest position being the third employing DT. On the other hand, the GLOBAL and LOCAL META $K = C$ approaches exhibited high variance across the multi-criteria ranks, indicating a sensibility to the choice of the base regressor. Furthermore, the ensemble approaches that adopted the average-based voting strategy (LOCAL MEAN $K = C$ and LOCAL MEAN $K = 7$) yielded the poorest MCPM values, occupying the fourth and fifth positions for most configurations.

Ordinal ranking	kNN	OLS	LASSO	Ridge	ElasticNet	DT	GBDT	MLP	SVR	Approach
①	0.0907	0.0628	0.5127	0.0617	0.1585	0.1133	0.0823	0.0573	0.0893	■ Global
②	0.0975	0.0647	0.5436	0.0702	0.3325	0.1246	0.0832	0.0627	0.1094	■ Local Mean $K = C$
③	0.1112	0.0887	0.5576	0.1183	0.5274	0.1389	0.0883	0.0728	0.1196	■ Local Mean $K = 7$
④	0.1388	0.1183	0.6148	0.1229	0.6262	0.1674	0.0953	0.0834	0.1432	■ Local Meta $K = C$
⑤	0.1487	0.1205	0.9194	0.1512	0.6712	0.2207	0.1137	0.1097	0.1745	■ Local Meta $K = 7$

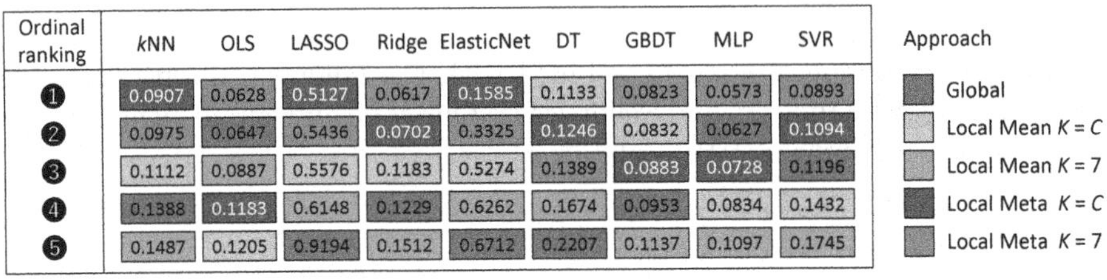

Fig. 3. MCPM values ranked in descending order of importance for each approach regarding different base regressors.

To compare our proposal and the baseline approaches with the state-of-the-art model, we selected their best configurations in terms of base regressors and arranged their results in Table 3. LOCAL META $K = 7$ achieved the best performances in four out of five metrics, including MCPM, and presented the second-best SP result. HIERARCHICAL GCNN, in turn, presented the worst performance across all the metrics. We must emphasize that the method had parameter values following the best results reported in [7], which considered the 2019 Australian election and the traditional cross-validation. Therefore, the present work did not apply a fine-tuning step for HIERARCHICAL GCNN since it is not a step performed in our experimental protocol.

Table 3. Most promising approaches based on overall configuration performances. Green cells denote the best results.

Approach	Base regressor	MSE	MESD	SP	SPSD	MCPM
GLOBAL	MLP	133.87114	138.17511	0.5907	0.1531	0.0681
LOCAL MEAN $K = C$	GBDT	196.31742	153.97060	0.6090	0.1560	0.0832
LOCAL MEAN $K = 7$	OLS	219.95581	170.12371	0.5850	0.1298	0.0887
LOCAL META $K = C$	Ridge	124.69428	129.78054	0.5701	0.1646	0.0702
LOCAL META $K = 7$	MLP	111.22174	127.19046	0.5911	0.1355	0.0573
HIERARCHICAL GCNN	GCNN	229.11080	209.11220	0.4917	0.1822	0.1279

In summary, our approach proved to be more stable against base regressors from different paradigms than the baselines. Additionally, LOCAL META $K = 7$ configured with MLP culminated in the best overall results compared to the best configurations of the other approaches.

Performance per Fold. Figure 4 displays the fold-level results of the best-instantiated approaches indicated in Table 3. The performances are reported according to MSE, MESD, and SP. We disregard the SPSD metric here since we cannot produce its values per fold.

Concerning MSE (Fig. 4(a)) and MESD (Fig. 4(b)), LOCAL META $K = 7$ behaved stably over the folds, followed by GLOBAL. LOCAL META $K = 7$ also achieved the best results on most folds, performing exceptionally well in the northeastern states (samples 21 to 29), where it exhibited the best or second best MSE and MESD. Folds 17, 31 and 50, for which LOCAL META $K = 7$ was ranked lower regarding MSE, demonstrated close values. Thus, there was no discrepant difference between the investigated approaches. We observed the same in folds 16, 17, and 30 concerning MESD.

In terms of SP (Fig. 4(c)), the GLOBAL approach obtained better results than the two variations of our proposal in most folds. However, it was closely followed by LOCAL META $K = 7$, specifically in the northeastern states (samples 21 to 29). This fact is reinforced by both approaches' relatively close average performances (GLOBAL: 0.61; LOCAL META $K = 7$: 0.59).

As we can see, the per-fold analysis of the three individual performance measures corroborates the one guided by MCPM, indicating that the two variations of our approach perform better than the other baseline models. Our proposal exhibited better MSE and MESD results than the other approaches. However, despite presenting lower SP values when compared with GLOBAL, the two variations of our proposal showed close results in most folds and performed better in some folds from the Southeast and Northeast.

Model Interpretability. Aiming to understand the predictions assigned by the best approach configuration (LOCAL META $K = 7$ with MLP), we considered the SHAP Values technique to produce an in-depth interpretability analysis. We sought to understand the most important context and the most important relevant features from this context in the best (sample 23 or Maranhão) and worst (sample 16 or Amapá) folds regarding MSE. Figures 5 and 6 comprise four plots. The first is the actual vote share distribution, while the second is the predicted distribution. The third concerns the feature importance given by the meta-regressor for each spatial context when predicting the fold. The fourth and last plot presents the top-five most relevant features according to the base regressor fitted to the most important context. Besides, Tables 4 and 5 list the top-five most important features of Figs. 5 and 6, respectively.

Concerning the best-case scenario (Fig. 5), our approach predicted vote shares slightly above the actual values, observed by the number of dark regions in the prediction map in relation to the ground-truth map. The meta-regressor chose

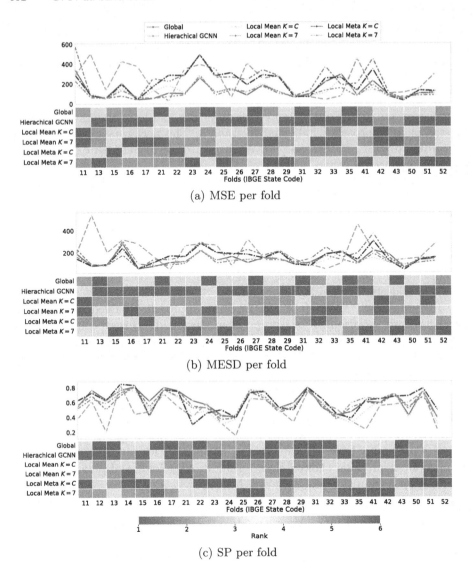

Fig. 4. Metrics per fold coming from the approaches whose configurations were considered the best by MCPM.

the geographic context 15 (Amazonas) as the most important, and the most relevant feature from context 15 was `PessoaRenda V045`. The selection of Amazonas as the most important context to predict the vote shares in Maranhão is coherent since both states present similar vote shares and related socio-economic characteristics. We should note that the first and third most relevant features from context 15 describe the women with income per capita less than half of the minimum wage (Table 4). This result is in line with research that points to the

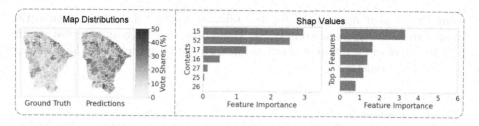

Fig. 5. Vote share distribution and SHAP Values for the best-case scenario (fold 23) in terms of LOCAL META $K = 7$ with MLP. The plots cover, from left to right, the following information: (a.1) actual vote share distribution, (a.2) predicted vote share distribution, (b.1) feature importance given by the meta-regressor for each spatial context, and (b.2) top-five most relevant features according to the base regressor fitted to the most important context (sample 15). The feature names are presented in the same order as in Table 4.

relationship between low-income women and lower votes for the winning party in the 2018 Brazilian presidential election [6, 10]. The remaining characteristics still need to be carefully analyzed to verify if they are proxies for other known related features such as poverty (Domicilio02 V057 and Domicilio02 V052) or a local relationship (Entorno05 V977).

Table 4. Top-five features from the most important context in the best-case scenario (fold 23).

Feature	Description
PessoaRenda V045	Women over ten years with a nominal monthly income of up to half minimum wage
Domicilio02 V057	Men living in permanent private households with water supply from a well or spring on the property
ResplRenda V055	Total nominal monthly income of responsible women with a nominal monthly income of up to 1/2 minimum wage
Domicilio02 V052	Men residing in rented permanent private homes
Entorno05 V977	Asian residents in permanent private homes with street lighting

In the worst-case scenario (Fig. 6), our approach predicted much higher vote shares than the ground truth, especially in the northern region. The meta-regressor chose context 24 (Rio Grande do Norte) as the most important, and the most relevant feature from context 24 was Entorno04 V490 followed close by Domicilio02 V040. The selection of Rio Grande do Norte as the most important context to predict the vote shares in Amapá was not a good decision, given the high error rates. Its features were insufficient to provide a good performance and cannot deliver insights into Amapá's voting results. We can see from Table 5 that most of the top-five relevant features describe particularities related to rural regions of context 24. These proprieties may not be observed in fold 16 or present a different relationship with the target, causing the approach performance to deteriorate.

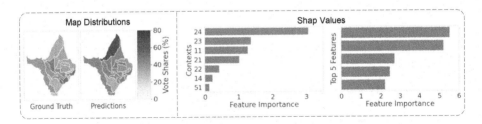

Fig. 6. Vote share distribution and SHAP Values for the worst-case scenario (fold 16) in terms of LOCAL META $K = 7$ with MLP. The plots cover, from left to right, the following information: (a.1) actual vote share distribution, (a.2) predicted vote share distribution, (b.1) feature importance given by the meta-regressor for each spatial context, and (b.2) top-five most relevant features according to the base regressor fitted to the most important context (sample 24). The feature names are presented in the same order as in Table 5.

Despite the challenge in modeling local relationships from socio-economic and election data, the in-depth assessment of SHAP Values indicated that our proposal is intelligible and, at best, predictions are based on coherent features that can aid in understanding electoral outcomes.

Table 5. Top-five features from the most important context in the worst-case scenario (fold 16).

Feature	Description
Entorno03 V490	Number of residents in private households without permanent public lighting with a well or spring on the property
Domicilio02 V040	Residents in permanent private households with electricity from other sources
Domicilio01 V026	Permanent private homes with two bathrooms for the exclusive use of residents
Domicilio01 V162	Permanent private dwellings, such as village houses or condominiums, without a bathroom for the exclusive use of residents
Entorno04 V730	Residents without nominal monthly household income per capita in permanent private households without sidewalks

5   Conclusion

This work proposed a geographic context-based stacking learning approach to predict election outcomes using socio-economic features. Our model is built in levels and dynamically selects contexts according to a data sample we want to predict. This modeling allows the generation of more realistic descriptive models whose relationships enable a more accurate understanding of voting behavior. We also introduced a spatial cross-validation-driven experimental setup to fairly assess and compare geographically contextualized approaches. Despite the challenging nature of the problem, by considering the second round of the 2018 Brazilian presidential election, our proposal experimentally showed promising results, including intelligible and coherent predictions in the best-case scenario and stable performance over the remaining folds compared with the reference models.

However, there is still room for further improvement. Our approach does not deal with ambiguous distributions, an aspect that often appears in modeling voting behavior. Furthermore, this paper was restricted to analyzing a single dataset, and studies with other election databases may be beneficial. Sophisticated machine learning methods such as Graph Neural Networks should also be better evaluated as they have shown satisfactory results for spatial data.

References

1. Chauhan, P., Sharma, N., Sikka, G.: The emergence of social media data and sentiment analysis in election prediction. J. Ambient. Intell. Humaniz. Comput. **12**(2), 2601–2627 (2021)
2. Dieterich, T.G.: Ensemble methods in machine learning. In: Kittler, J., Roli, F. (eds.) MCS 2000. LNCS, vol. 1857, pp. 1–15. Springer, Heidelberg (2000). https://doi.org/10.1007/3-540-45014-9_1
3. Graefe, A., Green, K.C., Armstrong, J.S.: Accuracy gains from conservative forecasting: Tests using variations of 19 econometric models to predict 154 elections in 10 countries. PLoS ONE **14**(1), e0209850 (2019)
4. Jacintho, L.H.M., Silva, T.P., Parmezan, A.R.S., Batista, G.E.A.P.A.: Analysing spatio-temporal voting patterns in Brazilian elections through a simple data science pipeline. J. Inf. Data Manag. 1–16 (2021)
5. Jiang, Z., Sainju, A.M., Li, Y., Shekhar, S., Knight, J.: Spatial ensemble learning for heterogeneous geographic data with class ambiguity. ACM Trans. Intell. Syst. Technol. **10**(4), 1–25 (2019)
6. Layton, M.L., Smith, A.E., Moseley, M.W., Cohen, M.J.: Demographic polarization and the rise of the far right: Brazil's 2018 presidential election. Res. Politics **8**(1), 2053168021990204 (2021)
7. Li, M., Perrier, E., Xu, C.: Deep hierarchical graph convolution for election prediction from geospatial census data. In: AAAI Conference on Artificial Intelligence, vol. 33, pp. 647–654 (2019)
8. Lundberg, S.M., Lee, S.I.: A unified approach to interpreting model predictions. In: Advance Neural Information Processing System, vol. 30 (2017)

9. Parmezan, A.R.S., Lee, H.D., Wu, F.C.: Metalearning for choosing feature selection algorithms in data mining: Proposal of a new framework. Expert Syst. Appl. **75**, 1–24 (2017)
10. Pinheiro-Machado, R., Scalco, L.M.: From hope to hate: the rise of conservative subjectivity in Brazil. HAU: J. Ethnogr. Theory **10**(1), 21–31 (2020)
11. Ploton, P., Mortier, F., et al.: Spatial validation reveals poor predictive performance of large-scale ecological mapping models. Nat. Commun. **11**(1), 1–11 (2020)
12. da Silva, T.P., Parmezan, A.R.S., Batista, G.E.A.P.A.: A graph-based spatial cross-validation approach for assessing models learned with selected features to understand election results. In: International Conference on Machine Learning and Applications, pp. 909–915. IEEE (2021)
13. Stewart Fotheringham, A., Li, Z., Wolf, L.J.: Scale, context, and heterogeneity: a spatial analytical perspective on the 2016 us presidential election. Ann. Am. Assoc. Geogr. **111**(6), 1602–1621 (2021)
14. Tobler, W.R.: A computer movie simulating urban growth in the Detroit region. Econ. Geogr. **46**(sup1), 234–240 (1970)

Author Index

Printed in the United States
by Baker & Taylor Publisher Services